Geometric Data Analysis

Geometric Data Analysis

From Correspondence Analysis to Structured Data Analysis

Brigitte Le Roux

MAPS 5 (CNRS) Department of Mathematics and Computer Science,
Université René Descartes, Paris, France

and

Henry Rouanet

CRIP 5 Department of Mathematics and Computer Science,
Université René Descartes, Paris, France

KLUWER ACADEMIC PUBLISHERS

DORDRECHT / BOSTON / LONDON

A C.I.P. Catalogue record for this book is available from the Library of Congress.

ISBN 978-90-481-6619-0 (PB)
ISBN 978-1-4020-2236-4 (e-book)

Published by Kluwer Academic Publishers,
P.O. Box 17, 3300 AA Dordrecht, The Netherlands.

Sold and distributed in North, Central and South America
by Kluwer Academic Publishers,
101 Philip Drive, Norwell, MA 02061, U.S.A.

In all other countries, sold and distributed
by Kluwer Academic Publishers,
P.O. Box 322, 3300 AH Dordrecht, The Netherlands.

Printed on acid-free paper

Printed in the Netherlands.

Contents

Foreword

Geometric Data Analysis (GDA) is the name I have proposed to designate the approach to Multivariate Statistics initiated by Benzécri as Correspondence Analysis, an approach that has become more used and appreciated over the years.

After numerous working sessions with Brigitte Le Roux and Henry Rouanet, both in Paris and in Stanford, it was evident that they were highly qualified to write a reference book about GDA that should meet the following two requirements: first, present in full the formalization of GDA in terms of the structures of linear algebra, which is an essential part of the mathematical foundations; and second, show how conventional statistical methods applicable to structured data analysis, i.e., analysis of variance and statistical inference, can be used in conjunction with GDA.

The richness of the actual content of the book they have written far exceeds these requirements. For example, Chapter 9, Research Case Studies, is nearly a book in itself. It presents the methodology in action with three well chosen extensive applications, one from medicine, one from political science, and one from education. The authors have taken time and effort to make this book accessible to a wide audience of practicing scientists. The mathematical framework is carefully explained. It is an important and much needed contribution to the statistical use of geometric ideas in the description and analysis of scientific data.

PATRICK SUPPES

Stanford, California
February, 2004

Preface

In our computer age, all research areas are replete with massive and complex data sets. Statistical packages offer myriads of methods for processing "multivariate data". The problem has now become: Which statistical method to choose, to make sense of data in the most meaningful way?

To say that "reality is multidimensional" is a truism. Yet, statistical thinking remains permeated with an ideology for which — alleging that "everything that exists, exists in some amount" — doing scientific work means quantifying phenomena. The achievements of this approach often fall short of promises. Indeed, the "reduction to unidimensionality" is sometimes so futile that it leads some good minds to the wholesale rejection of any statistical analysis, as reflected in sentences like this: "Intelligence is multidimensional, therefore it cannot be measured."

Beyond the opposition "quality" vs "quantity", there is *geometry*, whose objects (points, lines, planes, geometric figures) may be described by numbers, but are not reducible to numbers. Geometric thinking in statistics, with the idea that for transmitting information, a good picture may be more efficient than lots of numbers, is probably as old as statistics itself, and is historically traceable with the advent of scatter diagrams, charts and pictorial representations of statistical results. In the computer age, to meet the multidimensionality challenge, a more elegant way than a sterile retreat to a "qualitative approach" is offered by "l'Analyse des Données": the approach of multivariate statistics that Jean-Paul Benzécri, the geometer–statistician, initiated in the 1960s, and that we call *Geometric Data Analysis* (GDA)[1].

To cope with multivariate data, GDA consists in modeling data sets as *clouds of points* in multidimensional Euclidean spaces, and in basing the interpretation of data on the clouds of points. Clouds of points are

[1]The name *"Geometric Data Analysis"*, which marks the unique thrust of the approach, was suggested to us by Patrick Suppes.

not ready–made geometric objects, they are constructed from data tables, and the construction is based on the mathematical structures of abstract linear algebra. The formalization of these structures is an integral part of the approach; properly speaking, GDA is the *formal–geometric approach* of multivariate data analysis. At the same time, *clouds of points are not mere graphical displays*, like temperature charts (where coordinate scales may be changed arbitrarily); they have a well–defined distance scale, like geographic maps.

Why a new book? Since the 1970s, Geometric Data Analysis has enjoyed a sustainable success in France, where "Analyse des Données" is taught both in statistics and in applied research departments, from biometry to economics and social sciences. In the international scientific community, Correspondence Analysis (CA) (the "leading case" of GDA), has been appreciated more and more widely over the years. The phrase *Correspondence Analysis* is now well–rooted, and CA is renowned as a powerful method for visualizing data. Yet GDA, as a comprehensive set of methods for multivariate statistics, remains largely to be discovered, both from the theoretical and practical viewpoints. Accordingly, the following topics have been emphasized in this book.

• *Formalization*, which is the most valuable guide at the crucial stages of the construction of clouds and of the determination of principal axes.

• *Aids to interpretation*, which are indispensable constituents of GDA.

• *Multiple Correspondence Analysis*, which is so efficient for analyzing large questionnaires.

• *Structured Data Analysis*, a synthesis of GDA and analysis of variance.

• *Integration of statistical inference* into GDA.

• *Full size research studies* (the largest chapter of the book), detailing the strategy of data analysis.

This book should thus provide a reference text to all those who use or/and teach Multivariate Statistics, as well as to mathematics students interested in applications, and applied science students specialized in statistical analysis.

The *mathematical prerequisites* are essentially some acquaintance with linear algebra; the specific background gathered in the Mathematical Bases chapter should render the book self–contained in this respect. There are *no statistical inference prerequisites!* Inference procedures are only used in the Research Case Studies chapter, and their principles recalled in the preceding chapter.

About the Authors and Acknowledgements

Brigitte Le Roux is Maître de Conférences at the Laboratoire de Mathématiques Appliquées de Paris 5 (MAP5), Université René Descartes and CNRS, Paris. E-mail: `lerb@math-info.univ-paris5.fr`.
Henry Rouanet is guest researcher at the Centre de Recherches en Informatique de Paris 5 (CRIP5), Université René Descartes, Paris. E-mail: `rouanet@math-info.univ-paris5.fr`.

The authors gratefully acknowledge the support of the laboratories (MAP5 and CRIP5) to which they belong.

This book has profited from our teaching experience over the years at the Department of 'Mathématiques & Informatique' of our University; we especially thank Dominique Seret, the Head of Department, who supported the book project by arranging a six-month sabbatical leave for B. Le Roux.

We are also grateful to our mathematician colleagues, especially Pierre Cazes (University Paris-Dauphine), who scrutinized the entire manuscript, and Bernard Bru (University René Descartes), for many discussions.

Above all, our gratitude goes to the experts thanks to whom we made so much progress in the theory and practice of Geometric Data Analysis: the late Pierre Bourdieu (Collège de France), and Frédéric Lebaron (Centre de Sociologie Européenne); Jean Chiche and Pascal Perrineau (Sciences Politiques, CEVIPOF); Werner Ackermann (Centre de Sociologie des Organisations); Jeanine Louis-Sylvestre and Michèle Chabert (Laboratoire de Physiologie du comportement alimentaire, EPHE).

We also wish to thank, for their remarks, Philippe Bonnet, Jean-Luc Durand and Pierre Vrignaud. Special thanks are due to Geneviève Vincent (Language Center of our University), for assistance in English.

Our utmost gratitude goes to Patrick Suppes (Stanford University), whose energetic encouragements have been the efficient cause of this book; the EPGY case study in the book is a reflection of the collaborative work undertaken with him.

We thank Jean-Paul Benzécri who, during our friendly meetings in the Loire Valley, expressed his blessings for our enterprise.

Last but not least, we thank Paul Roos, James Finlay, Inge Hardon and Anneke Pot of Kluwer, for their helpful attention and smiling patience.

BRIGITTE LE ROUX & HENRY ROUANET

Paris
December 28, 2003

[1]The book has been composed in LaTeX; our thanks go to AsTeXassociation and especially to Michel Lavaud (University of Orléans).

Chapter 1

Overview of Geometric Data Analysis

('*OVERVIEW*')

The more abstract the truth you want to teach,
the more you will have to win over the senses in its favor.
Nietzsche

Introduction

Multivariate Statistics is the branch of statistics involving *several variables* (in a broad sense, covering categorized as well as numerical variables). The basic data sets in Multivariate Statistics are Individuals×Variables tables; derived data sets are contingency tables, scatter diagrams, etc. Classical chapters of Multivariate Statistics are Regression, Principal Component Analysis, etc. Besides, there is a link between Multivariate Statistics and observational data (questionnaires, etc.), as opposed to experimental data; the thrust of the present book will be on observational data.

By *Geometric Data Analysis* (GDA), we designate the approach of Multivariate Statistics that represents multivariate data sets as *clouds of points* and bases the interpretation of data on these clouds. The methods developed by Benzécri and his colleagues around the "leading case" of *Correspondence Analysis* (CA) and Euclidean classification — known in France as "Analyse des Données"[1] — are for us the core of GDA. Beside CA, there

[1]In English, "Data Analysis" — the literal translation of "Analyse des Données" — does not sound specific enough. As mentioned in the preface of this book, the name "Geometric Data Analysis" has been suggested to us by Patrick Suppes.

are related methods that fit appropriately in the framework of GDA. In fact, most of Multivariate Statistics can be considered from the viewpoint of GDA: see Rouanet, Le Roux (1993) and Rouanet & al (2002).

The objective of this book is to present the theoretical principles of GDA, together with comprehensive application to research studies. In this introductory chapter, we briefly illustrate the application of Correspondence Analysis to a historical data set (§1.1). We present the three key ideas of Geometric Data Analysis (§1.2), and the three paradigms of GDA (§1.3). We go on with a historical sketch (§1.4). We discuss the methodological strong points of GDA (§1.5). From Descriptive Data Analysis, we move on to a discussion of Inductive Data Analysis (§1.6). Lastly, we present the organization of the book (§1.7).

1.1 CA of a Historical Data Set

1.1.1 The "Hair & Eye Color" Data Table

To immediately give the flavor of GDA, we now present the CA of Tocher's historical contingency table "Hair and Eye Color", analyzed by Fisher (1940) and Maung (1941), with 5387 children cross–classified according to the variables "Hair Color" (5 modalities: fair, red, medium, dark, black) and "Eye Color" (4 modalities: blue, light, medium, dark) (Table 1.1).

		Eye colors				
		Blue	*Light*	*Medium*	*Dark*	
Hair colors	(Fair)	326	688	343	98	1455
	(Red)	38	116	84	48	286
	(Medium)	241	584	909	403	2137
	(Dark)	110	188	412	681	1391
	(Black)	3	4	26	85	118
		718	1580	1774	1315	5387

Table 1.1. Hair and Eye Color: joint and marginal absolute frequencies

We briefly comment on the table, then present the main CA results.

1.1.2 From Table to Clouds

Examination of the table

• *Looking at cells.* A casual examination of the table shows that there are strong associations between some Hair and Eye modalities. For instance, *Fair hair* goes with *Light eyes*, and *Medium hair* with *Medium eyes*. The modalities of such pairs, with comparatively high joint frequencies, will be said to be in "attraction" with each other. Some other joint frequencies are comparatively low, such as (Dark hair, Blue eyes), expressing "repulsions" between the modalities in the pair.

As a descriptive index of the overall association between the two variables "Hair Color" and "Eye Color" we will take the classical *Mean square contingency coefficient*[2], found here to be equal to $\Phi^2 = 0.230$.

• *Looking at columns or rows.* In the two columns *Blue* and *Light* eyes, high frequencies for *Blue* correspond to high ones for *Light*, and low frequencies for *Blue* to low ones for *Light*; more specifically, the absolute frequencies in column *Light eyes* are roughly twice those of *Blue eyes* (688 vs 326, etc.). The *Blue* and *Light* eye colors will be said to have "close profiles".

If we now look at the two rows *Fair* and *Black* hair, we see on the contrary that high frequencies for *Fair* tend to go with low ones for *Black*, and vice versa. These two hair colors will be said to have "distant profiles"[3].

Main results of the correspondence analysis of the table

Eigenvalues. The first outcome of the CA procedure is the set of *eigenvalues*: $\lambda_1 = .199$, $\lambda_2 = .030$, $\lambda_3 = .001$. The number L of eigenvalues (here 3) is obtained from the minimum of row and column numbers (here 4) and subtracting 1; hence $L = 4 - 1 = 3$.

The sum of the eigenvalues is equal to the Φ^2–coefficient of the table.

Clouds of points. The CA procedure generates *two clouds of points* (corresponding to the rows and the columns of the table respectively), here a cloud of 5 hair–color points and a cloud of 4 eye–color points. Both spaces have the same dimensionality, equal to the number L of eigenvalues (here 3). In the geometric construction, coefficient Φ^2 becomes the *overall variance* of each cloud, and the eigenvalues are the variances of each cloud along its *principal directions*.

Here the first eigenvalue amounts to 86.6% of Φ^2, and the sum of the first two amounts to 99.6% of Φ^2. These results imply that each cloud can be represented roughly on a *line* (one–dimensional space), and quite accurately in a *plane* (two–dimensional space). See Figure 1.1.

---

[2]The coefficient Φ^2 is related to the usual χ^2 test–statistic by the formula $\Phi^2 = \chi^2/n$ (where n denotes the sample size); we have here $\chi^2 = 1240$, $n = 5387$, and $\Phi^2 = 1240/5387 = 0.230$.

[3]The terms introduced here informally, such as attractions, repulsions, profiles, will be defined precisely as technical terms in chapter 2 (*CORRESPONDENCE*), as well as eigenvalues, Euclidean cloud, χ^2–metric, principal axes, etc.

Principal coordinates. The CA procedure gives *principal coordinates*, that is, the coordinates of the points of each cloud, with respect to *principal axes* (Table 1.2)

Hair colors	Freq.	Axis 1	Axis 2
Fair	.270	+0.544	+0.174
Red	.053	+0.233	+0.048
Medium	.397	+0.042	−0.208
Dark	.258	−0.589	+0.104
Black	.022	−1.094	+0.286

Eye colors	Freq.	Axis 1	Axis 2
Blue	.133	+0.400	+0.165
Light	.293	+0.441	+0.088
Medium	.329	−0.034	−0.245
Dark	.244	−0.703	+0.134

Table 1.2. Coordinates of hair colors and eyes colors on axes 1 and 2.

On the first axis, the principal coordinates of hair colors define the first *principal variable* for "Hair Color"; those of eye colors define the first principal variable for "Eye Color". Both principal variables are centered (their means are zero), and their variances are both equal to λ_1 (first eigenvalue)[4]. The second principal variables, for "Hair Color" and "Eye Color", are defined similarly, with variances equal to λ_2.

Simultaneous representation. Figure 1.1 represents the projections of the two clouds in principal plane 1-2.

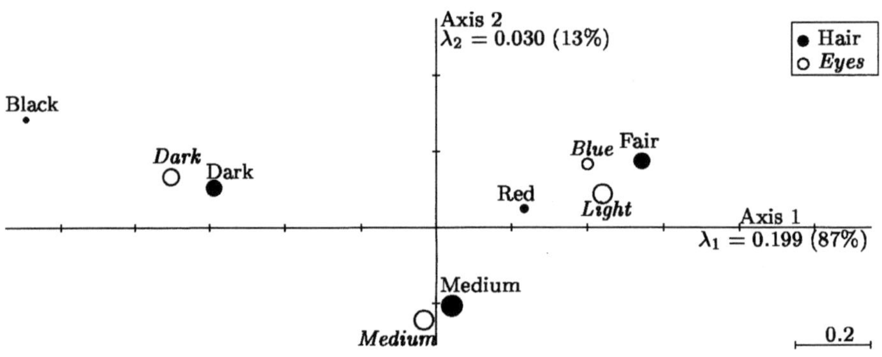

Figure 1.1. Simultaneous representation

The shapes of the clouds suggest an ordering among modalities corresponding to increasing degrees of coloration, for hair as well as for eyes. Close profiles (rows or columns) in the table are represented by points that are near each other, such as *Blue* and *Light* eye points. Distant profiles in the table are represented by points that are remote from each other, such as the *Fair* and *Dark* hair points.

[4]Means and variances are calculated taking the marginal frequencies as weights.

Interpretation of Axes. The first principal axis (horizontal axis in Figure 1.1) places *Dark* Hair and Eyes in opposition to *Light* ones. On Figure 1.1, dark colors have been plotted on the left and light ones on the right; but the converse could have been done as well, i.e. the orientations of axes are arbitrary[5].

The second principal axis (vertical axis on figure), much less important, is specific to *Medium* hair and *Medium* eye colors, in the sense that it places these two modalities in opposition to all the others and expresses their association.

As a general rule, when two modalities of the same variable have similar profiles, the corresponding points are close in the geometric representation; whereas when they have dissimilar profiles, the points are remote[6]. And as for points of different clouds, when they are far enough from the center of the cloud (origin of axes), their proximities in the simultaneous representation reflect attractions between the corresponding modalities[7].

1.1.3 Fisher as a Precursor of CA

Indeed, Fisher's and Maung's papers contain the basic equations of CA, from which the principal coordinates can be deduced. Yet there is no geometrical display in these papers. The reason is that they are concerned with another (though closely related) problem, namely, the *optimal scaling* of the two categorized variables answering the question (in Fisher's words): "For what eye colour scores are the five hair colour classes most distinct?". For this problem, the solutions of the equations do not have to be defined on a specific scale and there is no reason to construct a space of points with a Euclidean structure.

As a conclusion, Fisher is a precursor of CA as a method for "quantifying categorical data", not as a method for modeling data geometrically.

1.2 The Three Key Ideas of GDA

The main stream of Multivariate Statistics, namely *Multivariate Analysis*, is *quantitative*, uses the *matrix approach*, and is *sampling-oriented*.

[5] As a consequence, a cloud produced by CA is always defined "up to its mirror image".

[6] The latter property applies to the representation in the full geometric space; it will need qualification when projected clouds are considered, as will be seen below (contracting distance property).

[7] Further comments upon the "Hair & Eye Color" data will be given in chapter *CORRESPONDENCE*, §2.3.5, p.51.

1. *Quantitative.* Numbers are the basic ingredients and the end products of procedures[8].

2. *Matrix approach.* Procedures are defined and proofs are conducted exclusively by matrix operations.

3. *Sampling–oriented.* Any data set is reputed to be a sample and treated as such: modeling, fitting and testing (this is literally the "sample–minded approach" to statistics...)

By contrast, the GDA approach to multivariate statistics is *geometric, formal*, and *description–oriented.*

1. *Geometric.* Data are represented as clouds of points in geometric spaces.

2. *Formal.* Procedures and proofs are guided by mathematical structures. Structures govern procedures!

3. *Description–oriented.* The descriptive analysis of data is the first and indispensable phase of data analysis.

   Description comes first, Statistics is not Probability!

We will comment on the three key ideas of GDA right now, before proceeding to a historical sketch (§1.4).

1.2.1 Geometric Modeling of Data

The geometric modeling of data, that is to say, the representation of data as *clouds of points* in a multidimensional Euclidean space, is the most distinctive characteristic of GDA with respect to Multivariate analysis.

Starting from a two–way table, with two sets indexing the entries of the table (rows and columns), the first phase of GDA is *metric specification*, that is, the determination of Euclidean distances between the elements of each of these two sets, the χ^2–metric in CA being the most famous.

In Euclidean representation, these elements become points in a multidimensional space, defining clouds of points.

To sum up the construction:

[8]The so–called "geometric representations" of sums of squares, F–ratios, etc., occasionally found in statistical textbooks, do not reflect a genuine geometric approach to statistics, if only because they do not represent individual points; at best, by being confined to variables they only tell one side of the geometric story.

Data Table ⟶ Clouds of points

Table

326	688	343	98
38	116	84	48
241	584	909	403
110	188	412	681
3	4	26	85

⟶

Clouds (• & ∘)

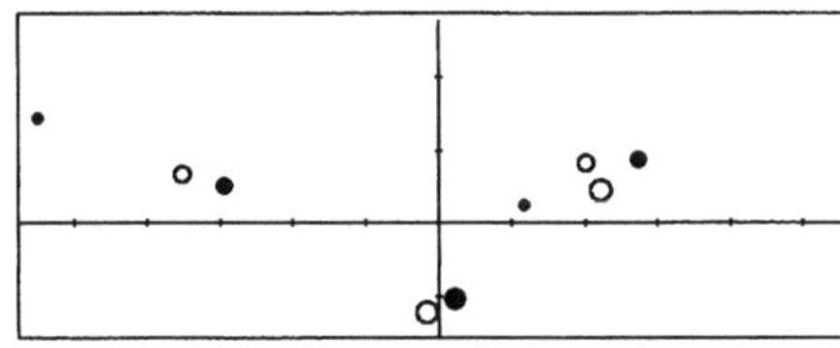

Once constructed, clouds of points become the fundamental objects for interpreting and exploring data. A basic characteristic of a cloud of points is its *dimensionality* (number of dimensions). The simplest case is that of a *one-dimensional* cloud, whose points lie on a *line*. Then comes the case of a *two-dimensional* cloud, whose points lie in a *plane*, as exemplified (approximately) by the "Hair & Eye Color" data.

In many research studies, GDA yields high-dimensional clouds. In order to study them, one performs orthogonal projections of the clouds onto well chosen lower-dimensional subspaces, mostly principal axes or planes. One then interprets proximities of points on axes or in planes, keeping in mind the *contracting distance* property of orthogonal projection: Two points close in projection may not be close in the whole space (whereas two points remote in projection are necessarily remote in space)[9].

Priority is not exclusivity! The geometric key idea specifies priority; this does not mean that quantitative procedures are excluded.

To begin with, there are the *elementary procedures* that should always be performed at the early phase of any statistical data analysis.

Then, as we have seen in the "Hair & Eye Color" example, there are the *basic descriptive statistics* (Φ^2, eigenvalues, principal coordinates) of the geometric construction.

Lastly, in addition to the visual examination of clouds, GDA involves calculating *contributions*, as an essential *aid to interpretation*. The point is that all numerical procedures make sense in terms of the geometric model; their role is to sustain (possibly qualify) the conclusions.

Sometimes the explicit objective of a research study is to construct numerical indices; for this purpose, the principal variables of a GDA, with their optimal scaling properties, offer responses that may be more effective than indices concocted a priori (Benzécri, 1973).

[9]A geometric representation in two or more dimensions is not a simple "graphical display" like a temperature chart (whose coordinate scales may be changed arbitrarily); it is more akin to a geographic map with a distance well-defined in space.

1.2.2 Formal Approach and Abstract Linear Algebra

The formal approach to statistics consists in founding statistical procedures on a mathematical theory, as is done when elementary procedures are founded on *scale structures*: ordinal scale for the median, interval scale for the mean, etc. In multivariate statistics, the relevant mathematical theory is *abstract linear algebra* — a well-constituted mathematical theory if ever there was one — with the concepts of vector space, linear mapping, scalar product, quadratic form, etc., making clear distinctions between *vectors* (elements of a vector space), *points* (elements of a geometric space), and *sets of numbers*, e.g. coordinates (See chapter *MATH.BASES*).

The intimate alliance between the two key ideas (geometric and formal) is the quintessence of GDA, which is properly speaking the *formal–geometric approach* to multivariate statistics. This approach is crucial at the following two phases: (i) *Construction of clouds and metric specification*[10]. (ii) *Spectral decomposition* (the "mathematical heart" of GDA), i.e. principal axes and coordinates; the statistical procedures readily follow from the mathematical theory, which carries the burden of proof. This means a fantastic saving in intellectual investment.

Measure vs variable duality. Measure vs variable duality (based on the duality of vector spaces) is the cornerstone of the formal approach in GDA. In physics, when two quantities of liquid are mixed, their two volumes *add up*, whereas their two temperatures *average up*. In statistics, the same distinction occurs in connection with the operation of *grouping units*. There are some quantities, such as weights, frequencies, etc., for which grouping units entails summing (adding up) values, and which we call *measures* (as in measure theory); there are other quantities, such as scores, rates, etc., for which grouping units entails averaging values, and which we call *variables*[11]. *Duality notation*, or *transition notation*, with subscripts denoting measures and superscripts variables, highlights this fundamental distinction.

Methodologically, the measure vs variable duality provides operational guidelines. For nonmathematically–oriented users, the formal approach is not as "abstract" as might be thought. Looking for a classification in harmony with a geometric representation, or figuring out whether in a study about lifestyles in Europe the populations of countries should enter the analysis either as weights (measures) or as scores (variables): Such

[10] For CA, the χ^2–metric follows directly from the structures of linear algebra, without any additional assumption, as will be seen in chapter *CORRESPONDENCE*.

[11] The distinction is akin to that of extensive vs intensive measurement in Mathematical Psychology: see Suppes & al (1989). In this context, "variables" are numerical variables.

tasks, in our experience, offer more gratifying challenges than debugging row and column intricacies in a matrix formula.

The matrix approach to statistics. Multivariate statistics, as a whole, ignores abstract linear algebra and relies exclusively on matrix manipulations to characterize statistical procedures and perform proofs. In the matrix approach, vectors are confused with numerical n–tuples, and points are confused with vectors. Tukey warned in vain: "An arrow is not a set of coordinates"; to which we may add "A point is not an arrow." Clearly the formal approach has not penetrated statistical research, not to mention textbooks[12]. This is a severe obstacle to a correct appreciation of GDA in multivariate statistics. Geometric figures, when disconnected from their structural rationale, are little more than pictorial illustrations (cf. note p.6 about "geometric representations"). Blurred structures hinder understanding of procedures.

For instance, in CA, marginal frequencies serve firstly to provide reference measures, secondly to go from measures to variables and vice versa, and thirdly to define the χ^2–metric on two clouds. Translating these different properties into matrix formulas is an easy task, and provides a compact format for communicating the algorithm of CA[13] — but no more than the algorithm. The converse "translation", i.e. from matrix formulas, deciphering the rationale of the procedure, is more of a headache. Again, in spectral decomposition, expressing the geometric projection onto a principal subspace in matrix terms readily follows from abstract linear algebra, whereas the converse "translation" is virtually hopeless. Indeed, the matrix approach is not powerful enough to cope with geometric structures[14]. It is a pity that, in order to comply with the dominant statistical style, most texts about CA in English — certainly not Benzécri's (1992)! — have confined themselves to this "matrix presentation"; the trouble is that in the translation a major feature of CA has evaporated.

1.2.3 Data Analysis Philosophy

In the description–oriented approach to statistics, the descriptive analysis of data should be conducted first, as a fully–fledged analysis. Statistics is not

---

[12]Dempster's treatise (1969) is an outstanding exception. There are also the *coordinate–free* essays by Kruskal (1968), Eaton (1983), or Stone (1983). Yet, the thrust of the formal approach in GDA is not so much the coordinate–free idea, but the "summing vs averaging" duality, with intrinsic bases, though with nonelementary scalar products.

[13]This is what we will do in the present book, using matrix formulas as shorthand.

[14]Cf. Benzécri & al (1973 p.58): "Critique du Calcul matriciel".

probability! This assertion marks a definite departure from the sampling–oriented mainstream approach to statistics[15].

In multivariate statistics, where data often seemingly go (literally) "in all directions", the geometric model provides an efficient frame for descriptive analysis. Determining principal axes immediately enables us to master multidimensionality, and — in Benzécri's (1969) words — yields "a tool to make patterns emerge from data", especially when used in conjunction with a classification. The descriptive key idea is in harmony with the *Data Analysis Philosophy*, with its message: Let the data speak for themselves! Statistical analysis should remain "impervious to the expectations and prejudices of researchers" (Benzécri, 1973) and encourage an *exploratory attitude* toward data.

Like the geometric key idea, the descriptive one means priority, not exclusivity. Geometric modeling precedes probabilistic modeling; it does not preclude statistical inference[16]. We will mention these topics further in the present chapter (§1.6), and discuss them in detail in chapter *INDUCTIVE*.

1.3 Three Paradigms of GDA

Once the analysis of two–way frequency tables by CA was established, the geometric approach was extended to various kinds of tables, in particular to Individuals×Variables tables, the basic tables of multivariate statistics.

On the one hand, for *numerical variables*, the method of *Principal Component Analysis* (PCA), already developed in Multivariate Analysis, was recast in a geometric framework, which led to representing variables in a "circle of correlations", and also — in principle at least — to constructing a cloud of individuals.

On the other hand, for *categorized variables*, the method of CA was applied to the associated table of indicator variables (disjunctive coding) and/or to the associated Modality×Modality frequency table ("Burt table"). This method, known as "Analyse des Correspondances Multiples" (Lebart, 1975), that is, *Multiple Correspondence Analysis* (MCA), became another major GDA paradigm, as a counterpart of PCA for categorized variables, and was widely used in the analysis of questionnaires.

[15] This antagonism is perennial; the history of statistics (see Stigler, 1986) amounts to the story of the recurrent stormy affair between random thinking and data analysis.

[16] For example, Principal Component Analysis, formalized in Multivariate Analysis as a probabilistic model, becomes in GDA a primarily descriptive model, which does not preclude inferential developments.

Eventually, three GDA paradigms associated with three types of data tables emerged:

— CA (with χ^2–metric): *two–way frequency, contingency tables*;

— PCA (various metrics): *Individuals×Numerical Variables* tables;

— MCA: *Individuals× Categorized Variables* tables.

Moving on from CA on frequency tables to MCA on Individuals×Variables tables means a shift in methodological emphasis. In MCA as well as in PCA, the individuals carry all the information, and it is on the cloud of individuals that interpretation and exploration deserve to be concentrated.

1.4 Historical Sketch

Predecessors and contemporaries. At this point, a historical sketch of GDA is in order. In a well thought–out text, entitled "Histoire et Préhistoire de l'*Analyse des Données*", Benzécri (1982a) recalls the great historical figures of statistics, especially K. Pearson (1901) — "Should we need an Anglo–Saxon patronage for *Analyse des Données*, we would be pleased to turn to the great Karl Pearson." — and Fisher; he situates his approach in the *psychometric tradition*, with *Factor Analysis* from Spearman to Thurstone and Burt (1950), and *scaling methods* with Hirschfeld (1935), Eckart & Young (1936) and especially Guttman (1941). He also makes due reference to related contemporary works, such as the *quantification method* developed around Hayashi (1952) in Japan, and proximity analysis — also known as *MultiDimensional Scaling* (MDS) — developed by Torgerson (1958), Shepard (1962), and others[17]. Admittedly, the list of references could be enlarged to Guttman (1959) for a synthesis of early literature, to Dempster (1969) for the formal key idea, or to Tukey (1960) for the Data Analysis Philosophy. But the conclusion that clearly emerges from this review is that, beyond its similarities with several anterior or contemporary undertakings, the geometric construction around CA was most original and brought an in–depth renewal of multivariate statistics.

As far as the *history* (strictly speaking) of GDA is concerned, it can be divided into three periods (landmark years are indicative).

First period: Emergence (1963–1973). The hard core of CA was achieved in the mid–sixties, with the six lectures given by Benzécri at the

[17] MDS (see Shepard, 1966, 1980) — with its variants like Small Space Analysis (SSA) — is a case in point; it unquestionably belongs to GDA, without being an outgrowth of CA.

Collège de France, B. Cordier–Escofier's dissertation (1964, 1969), and a host of mimeographed reports that were widely circulating. A brief account in English of the first developments can be found in Benzécri (1969). Toward 1973, the emergence of "Analyse des Données" around CA, combined with classification methods, was completed with the publication of the monumental treatise by Benzécri & Coll. (1973) in two volumes (Taxinomy and Correspondence Analysis). Meanwhile, the first statistical textbook incorporating CA had appeared: Lebart & Fénelon (1971), followed by many others: Berthier & Bouroche (1975), Cailliez & Pagès (1976), Lebart, Morineau & Tabard (1977), etc. (all of them in French).

Second period: Splendid isolation (1973–1980). The movement of "Analyse des Données" enjoyed a golden age in France. Benzécri's laboratory at the "Institut de Statistique de l'Université de Paris" (ISUP) was for many years a place of creative dialogue between statisticians and researchers. An innovative statistical tradition developed, around a body of expert knowledge: contributions, supplementary elements, Guttman effect, disjunctive coding, Burt table, etc. Procedures were implemented in software (whose sources were free). Statistical work was published mostly in *Cahiers d'Analyse des Données* and *Revue de Statistique Appliquée.* "Analyse des Données" began to be taught in graduate statistics curricula. In applications, CA (especially MCA) became a major tool for analyzing multivariate data such as questionnaires (cf. §1.5).

Throughout the first two periods, in the relevant literature in English, there were very few published reactions to the work done in France, even though some "joint display spirit" came to be floating in the air, as reflected in Gower (1966), Good (1969), Gabriel (1971). In the seventies, CA is ignored in MDS publications, such as Shepard, Romney, Nerlove (1972), Kruskal & Wish (1978), Shepard (1980); also ignored in the encyclopedic treatise by Kendall & Stuart (1976). The silence about CA was conspicuously broken by Hill (1974), who — encouraged by Gower? — launched the English phrase "Correspondence Analysis" (perhaps its first appearance in print) and emphatically announced that CA was a "neglected method".

Third period: Bounded international recognition (since 1981). International recognition eventually came in the eighties. Books in English were published that directly stemmed from the work done in France: Greenacre (1984) was specifically devoted to CA, whereas Lebart & al (1984) dealt more generally with "multivariate descriptive analysis"; then came Jambu (1991); then Benzécri (1992): the translation of the introductory book of 1984. In the meantime, Malinvaud (1980) and Deville & Malinvaud (1983) had discussed CA in official statistics and referred to "Data Analysis" as "Econometrics without stochastic models". In psychometry, there was the valiant paper by Tenenhaus & Young (1985), etc.

Recognition also came from other lines of work that incorporated CA into their own systems, especially *Dual Scaling*, developed by Nishisato (1980), who compiled a giant bibliography on "quantification of categorical data" (Nishisato, 1986), and *Homogeneity Analysis*, developed by the Leyden group led by de Leeuw and reflected in Gifi (1981/1990).

Most important, CA penetrated such areas as marketing research (Hoffman & Franke, 1986), where *MultiDimensional Scaling* (MDS) techniques had long been dominating. In the late eighties, the MDS group came to adopt CA as an authentic (even though "less conventional") MDS method, as reflected in Carroll & Green (1988)[18], Weller & Romney (1990), etc. In the 1984 edition of Kendall & Stuart, a casual reference to Lebart & Fénelon (1971) can be spotted in Volume 3 (p.418).

This sort of recognition continued in the nineties, with e.g. Gower & Hand (1996) presenting CA and MCA as "biplot" methods among others.

Where do we stand now? In 2003, the situation calls for a mixed assessment.

On the positive side, the phrases "Correspondence Analysis" and even "Multiple Correspondence analysis" are well rooted in English. The basic procedure of CA can be found in international statistical software. CA is definitely renowned for the visual exploration of data. It is now commonplace to discuss topics explicitly related to CA, such as stability, choice of metrics (Rao, 1995), canonical analysis (Goodman, 1991), etc. International conferences specifically oriented to "CA and related methods" are organized outside France; see e.g. Blasius and Greenacre (1998), and the recent conference organized by Greenacre (2003).

On the other side, CA still remains isolated in the field of Multivariate Statistics. In spite of increasing demand from users, popular books and international software all too often offer imperfect versions of the method. Frankly speaking, for MCA the situation is really defective. This method, which is so powerful for analyzing large scale questionnaires, is still hardly ever discussed and therefore remains underutilized, as does most GDA expert knowledge.

---

[18]The concluding sentence of this paper is that INDSCAL (an MDS variant) should be used "if only as an adjunct to the more conventional MCA"; a sentence that speaks volumes for the increasing popularity of CA and MCA in marketing research.

To sum up, in the international scientific community, CA is now recognized and used, but GDA largely remains to be discovered.

1.5 Methodological Strong Points

We will now consider GDA from the *user's viewpoint*, and discuss GDA as a geometric frame model (including Euclidean classification).

In social sciences, GDA has generated a statistical practice in sharp contrast with the conventional one. In the latter, *numerical indicators* (regression coefficients, etc.) occupy center–stage, together with significance levels (the *star system*: * significant at .05 level, ** significant at .01, etc.). In GDA, *clouds of points* are central. In social sciences, this contrast reflects two distinct conceptions of the role of statistics, namely sustaining a "sociology of variables" versus *constructing a social space.*

1.5.1 GDA as a Frame Model

Any empirical study of a research field involves some theoretical framework, which, by summarizing relevant knowledge, guides the collection of data and the interpretation of results. When this framework is formalized in mathematical terms, we call it a *mathematical frame model.* For example, regression or ANOVA models are frame models that summarize the relevant knowledge in terms of variables; whereas GDA models summarize this knowledge in geometric terms. The frame model concept serves as a reminder that GDA methods, like all statistical ones, can only be fruitful if they deal with *relevant data*; performing a geometric analysis does not mean gathering disparate data and see 'what comes out' from the computer.
Homogeneity and Exhaustiveness. According to Benzécri (1973, p.21), two principles should be fulfilled. i) *Homogeneity*: All variables in the table should be of the same nature. ii) *Exhaustiveness*: The data should constitute an exhaustive or at least a representative inventory of a real research field. The exhaustiveness principle is seen to be quite at odds with the *parsimony* principle often advanced in the conventional statistical methodology, especially in connection with regression procedures.

Large–size tables. GDA is eminently apt at revealing the structural complexities of large–size tables[19]; all authors stress this fact[20]. Small tables

[19] "Large–size tables" should of course not be confused with "large frequency tables"; a 5 × 4 contingency table involving millions of individuals is still a small–size table!

[20] See Benzécri (passim). Lebart & al. (1984): "Large data sets often contain so many

(such as the 5 × 4 "Hair & Eye" table), that are convenient for teaching procedures, are simply not appropriate for showing the power of geometric frame models in large-scale research studies. Lebart & al (1998) put this nicely: "It is extremely difficult to demonstrate the effectiveness of a fish net in a living–room aquarium!"

The construction of social spaces. Since the very beginnings, GDA has been applied to a large range of domains, such as medicine[21], lexicology[22], marketing research, econometry and social sciences. In this latter domain, the work of Bourdieu is exemplary of the "elective affinities" between the spatial conception of social space and GDA representations: See Rouanet, Ackermann & Le Roux (2000). These affinities have led Bourdieu and his school to use CA (especially MCA) consistently since 1976[23]. For instance, the sociological frame model of *La Distinction* (Bourdieu, 1979, 1984) specifies three types of capital (economic, cultural and social) as the basic dimensions of the social space of lifestyles, and guides the choice of data to be collected. Then, at the stage of statistical analyses, Individuals×Modalities tables are analyzed, taking a set of modalities sufficiently ample to cover the overall social space under study, and the interpretations of analyses refer to the basic dimensions of the frame model[24]. Examples of recent research work using GDA with this philosophy are Chiche & al (2000), Lebaron (2000, 2001), Rosenlund (2000), Hjellbrekke & Korsnes (2003).

About GDA in social sciences, three points can be made.

i) *Rehabilitation of individuals.* The frame model is always formulated at the *individual level*, as opposed to models framed at an aggregated level, where individuals mainly serve as a residual term.

ii) *Specific analyses.* Sociological variables induce subclouds of individuals that can be studied by specific analyses (as subfields within the overall social space).

iii) *Explanatory schemes.* Once a social space (such as a space of lifestyles) has been constructed, the geometric model of data can give rise to new investigations, especially on the following two kinds of problems.

First kind: How can the individual positions in the social space be

interrelationships that it is almost impossible to interpret them at first sight." See also Deville & Malinvaud (1983).

[21]See e.g. Leclerc (1975), Benzécri & Coll. (1992), Kaminska & al. (1999).

[22]See e.g. Benzécri & Coll. (1981), Lebart & Salem (1988), Lebart & al. (1998).

[23]From Bourdieu & Saint–Martin (1976) to Bourdieu (1999) via Bourdieu (1979), etc. See Bourdieu (2001).

[24]For the upper class, the first three principal axes are interpreted as the "global volume" of capital, its "composition", and the "ancientness in the class", respectively.

explained by external factors, such as education, age, etc.?

Second kind: How can these individual positions explain in turn the position–takings of individuals about various (political, etc.) issues (one of Bourdieu's famous claims)?

1.5.2 Euclidean Classification

The concern for classification — putting together objects (e.g. individuals of a survey) that "go together" — is a most natural one. GDA soon came to be used in relation with this concern. Guided by spatial intuition, researchers spontaneously proceeded to ***visual clustering***, drawing contours on diagrams around high density zones to group close objects and separate remote ones. Nowadays, efficient methods of classification exist, such as Ascending Hierarchical Classification (AHC) methods, which yield systems of nested partitions (as in natural sciences). When a classification procedure is used in conjunction with a geometric representation, the continuous geometric substratum allows the researcher to qualify the sharpness of the cuttings into classes, and to exert the sense of approximation that should prevail in all applications of mathematics.

AHC methods differ according to the distance index between objects and the criterion chosen for aggregating classes. Among AHC methods with Euclidean distance, the ***variance criterion*** (also known as ***Ward criterion***) is most in harmony with geometric structures. The joint interpretation of axes and classes is in this case especially enlightening. Accordingly, we call it ***Euclidean classification***; for Benzécri (1992), it is the "companion method" of CA.

1.5.3 GDA and the Computer

Applying GDA to real research studies involves higher–rank matrices to diagonalize, hence resorting to high–speed computers. In the sixties, the advent of the computer rendered the "Analyse des Données" project feasible. As theoretical developments of CA went on, software was developed, bringing statistics and computer science closer. Toward the end of the seventies, the state of the art was advanced enough so that, for dealing with the large–size sociological questionnaires, "everything was possible" (in Bourdieu's words). And nowadays, some formerly thankless tasks — such as modality coding (crucial for successful analyses) — can be carried out very comfortably, and highly readable graphical results can be edited. More than ever, GDA methods are accessible and user–friendly.

1.6 From Descriptive to Inductive Data Analysis

The outstanding capacity of CA for revealing unknown (or only partly known) structures of multivariate data is well recognized today. On the other hand, the priority granted to descriptive analysis, and even the successes of CA in data exploration, have given rise to the myth that CA would be no more than a "descriptive and exploratory method", cut off from conventional statistical methods, and precluding (so to say "by essence") both statistical inference and explanatory interpretation. This myth is unfounded, although elucidating comments are in order. To close this overview of GDA, we will tackle moot topics in connection with the following three questions frequently raised about CA.

1. *Can* GDA *deal with structured data?*
2. *Is* GDA *amenable to statistical inference?*
3. *Can* GDA *be used for explanatory purposes?*

The answers to each of the three questions are "*Yes*, absolutely", for the reasons that we outline hereafter; more specifically, we will contrast "descriptive" to "inferential", and "exploratory" to "explanatory".

1.6.1 GDA for Structured Data

In a table analyzed by a GDA method (cf. the three paradigms, p.10), there are often characteristics of the basic two sets of the table that are not used to construct the clouds. For instance, in an attitude questionnaire analyzed by MCA, identification characteristics of individuals (age, gender, etc.) do not serve to define the metric; their role is similar to the factors of an experimental design. We call such characteristics ***structuring factors***, and data tables with structuring factors ***structured data***. Evidently, structured data are the rule rather than the exception.

For structured data, the issue of investigating the *effects* of structuring factors on "dependent variables" naturally arises. Now there exist traditional methods (analysis of variance and regression) that have been devised for this purpose. From then on, confronted with structured data, is it worth using geometric methods[25]? Yes, it is worth using geometric methods for structured data. We hold that a geometric model should in any case be constructed as the fundamental representation of data. Then, the use of ***supplementary variables*** should be a first step toward Structured Data Analysis. As a further step, ANOVA procedures, making full use of the

[25]Equivocal declarations can be found in literature that tend to cast doubt on the efficiency of geometric methods when applied to structured data.

relations beween factors (nesting, crossing, etc.), while accounting for the specificities of observational data (e.g. nonorthogonality), can be embedded in the geometric representations to investigate the effects of structuring factors geometrically. Structured Data Analysis will be presented in the *STRUCTURED* chapter, and applied in the *CASE STUDIES* chapter, to enrich the geometric model of data, in accordance with the scheme:

Data Table → Clouds of points → Structured Data Analysis

1.6.2 Statistical Inference in GDA

At this point, the issue of statistical inference in GDA must be faced. Statistical inference, especially significance testing, so widespread in empirical research, is conspicuously absent from much research using CA or MCA. This is an awkward situation[26].

No doubt, the basic statistics of GDA methods are ***descriptive statistics***, in the technical sense that they do not depend on the sample size[27]. For instance, if in a contingency table all entries are doubled, the eigenvalues and principal coordinates of CA remain the same. Similar properties, however, hold for the basic statistics of all statistical methods. For example, in regression, the β–coefficients and the multiple R^2 are descriptive statistics, even though the method of regression is not confined to descriptive statistics. The same can be said of CA or more generally GDA. There is no such thing as a "descriptive only" statistical method.

The real question about statistical inference is not whether it can be used or not in GDA — it certainly can, as reflected e.g. in PCA — but when and how to use it in a fruitful way; and our point is that the mechanical application of the "fit–&–test" technique so widely spread today is not especially fruitful. In our opinion, statistical inference should be conducted as an ***Inductive Data analysis***, that is, in the line of the Data Analysis Philosophy, as a natural extension of Descriptive Data Analysis (cf. Rouanet, Bernard, Le Roux, 1990). In fact, there is a tradition of inference in research, in the line of Fisherian thinking, that is still very alive in laboratory practice — even though, it must be said, it is rather belittled in statistical

[26]The norms prevailing in scientific publications virtually impose inference procedures as the accepted formal way to substantiate inductive conclusions, that is, conclusions that go beyond the data set under study; they tend to underrate any research that — rightly or wrongly — does not comply with this requirement as mere exploratory work, good enough to "clear the ground" before the serious work begins.

[27]The distinction ***descriptive vs inductive*** statistics and procedures will be taken up in chapter *INDUCTIVE*.

teaching (in spite of Tukey's valiant efforts). In this researcher's tradition, significance testing is the natural tool for extending the descriptive conclusion of *existence of effects*[28]. The first thing to do, we submit, should be to enforce this tradition of inference in research. Then, to improve its efficiency, an advantageous complement should be found in *Bayesian methods*, especially for assessing the *importance of effects* (Rouanet & al, 1997, 1998). Inductive Data Analysis will be outlined in the INDUCTIVE chapter, then applied in the CASE STUDIES chapter, to assess the effects brought to light by structured data analysis. As a result of Inductive Data Analysis, the geometric model of data, already enriched by Structured Data Analysis, will be enriched further still, in accordance with the scheme:

Data Table ⟶ Clouds of points ⟶ Structured ⟶ Inductive

"The model should follow the data, not the reverse!"[29].

1.6.3 Exploratory versus Confirmatory and Explanatory

Lastly, let us allude to the alleged "exploratory only" character of geometric methods, as opposed to the "confirmatory" and "explanatory" atrributes that would be the prerogative of other methods.

Exploratory vs Confirmatory. The distinction *Exploratory vs Confirmatory* — to which the name of Tukey is associated — basically refers to *research attitudes* in the process of the statistical analysis of data, namely uncovering unknown features of data vs investigating hypotheses. The distinction between these two attitudes may be useful, even though its practical implementation is often elusive, as pointed out by Cox (1977, p.56), Guttman (1977), Lebart & al (1995). When applied to statistical methods, however, the distinction is misguided. No doubt, GDA is often conducted in an exploratory way, but the same can be said of any other method. It is simply untenable to assign by decree each of the great statistical methods to one of the two varieties, e.g. to oppose "exploratory CA" to "confirmatory regression". There is no such thing as an "exploratory only" statistical method.

[28] In this tradition, the role of significance testing is different from its use in the fit–&–test technique, where descriptive conclusions are simply ignored.

[29] With increasingly large data bases, the Data Analysis Philosophy is undoubtedly gaining ground in statistical thinking. If this trend continues, the perverse effects of the "fit–&–test" technique, — for which the famous K. Pearson's (1900) paper on the χ^2 criterion is certainly not responsible — might well in the future be remembered as the *physicalist utopia* (if not the opprobrium) of twentieth–century statistics!

Exploratory vs Explanatory. The explanatory language is clear enough in reference to a substantive (physical, biological, etc.) theory: Temperature explains the elongation of a metallic bar, not the reverse. When applied to statistical methods, it is often confusing. The "explanatory" phraseology is the plague of regression. Elongation can be regressed on temperature, but temperature can be regressed on elongation as well[30]. It seems a wise methodological position, especially for observational data, to hold that *explanatory schemes* can be investigated by various statistical methods (geometric methods, regression, discrimination, etc.), but do not belong to statistical methods *per se.* Any statistical method can be used for an explanatory purpose, as the case may be, but there is no such thing as an "explanatory statistical method."[31] On this point we certainly subscribe to the sentence by Lebart & al (1995, p.209): "Statistics does not explain anything — but provides potential elements for explanation."

1.7 Organization of the Book

What is Geometric Data Analysis? We are now in a position to provide an elaborate answer to the question: "What is Geometric Data Analysis?" — a question that echoes the title of the book by Fénelon (1981): "Qu'est-ce que l'Analyse des Données?". Our answer is not far from that of Fénelon. The core of GDA is the Euclidean cloud, with the determination of principal axes ("axes factoriels") and Euclidean classifications; clouds are constructed from multivariate data tables of the type Individuals×Variables, or from derived tables such as contingency tables. Our own contribution consists in highlighting the unified conceptual framework of GDA on the one hand, and in integrating not only statistical tools for structured data but also inference methods on the other hand.

The aim of the book is not to review all work relevant to GDA, but to explain what GDA is about, from theoretical bases to methodological potentialities. Having finished the book, readers should be in a position to proceed to creative codings of data sets, to construct and interpret meaningful geometric models, with a judicious utilization of simultaneous representation. Last but not least, they should be able to adopt a constructive critical viewpoint to assess research papers making use of geometric analyses.

[30]When Galton regressed parent heights on child heights, he surely did not mean the latter variable to explain the former.

[31]Needless to say, this statement does not concern the formalization of causality (Granger, 1969, Suppes, 1993), or explicit "causal models" (e.g. Aris, 2001).

The chapters. Below is a summary of the nine chapters that follow the present introductory chapter[32].

Chapter 2 (*CORRESPONDENCE*) deals with *Correspondence Analysis*, making full use of the measure/variable duality; the domain of applicability of CA is discussed, and the probabilistic modeling of CA is evoked.

Chapter 3 (*CLOUD*) is devoted to the central object of GDA, namely the *Euclidean Cloud*; after the descriptive basic statistics and properties (mean point, variance, inter–intra breakdown), the theory of principal axes and variables is developed, and Euclidean classification presented.

Chapter 4 (*PRINCIPAL*) and **Chapter 5** (*MULTIPLE*) are devoted to *Principal Component Analysis* and *Multiple Correspondence Analysis* respectively; starting from an Individuals×Variables table, the construction of the Euclidean cloud is developed, then the properties of methods are presented in detail.

Chapter 6 (*STRUCTURED*) presents methods to analyze *structured data*.

Chapter 7 (*STABILITY*) tackles the topic of the *stability* of the methods for determining principal axes and variables, and investigates the sensitivity to various types of modifications such as deletion/adjunction or grouping of points.

Chapter 8 (*INDUCTIVE*) deals with *Inductive Data Analysis*, and shows how statistical inference conducted in the line of the Data Analysis Philosophy can be performed in the context of GDA, from significance testing to Bayesian methods.

Chapter 9 (*CASE STUDIES*) illustrates "GDA in practice" in full–scale *Research Case Studies*.

Chapter 10 (*MATH.BASES*) gathers in a compact way the *Mathematical bases*, that is, the mathematical background necessary for reading this book.

We have tried to allow for a "non–linear reading" of the book, at the expense of some redundancy. At this point, the reader will perhaps wish to have a look at Chapter 9 (*CASE STUDIES*), before turning to the more technical chapters that follow the present one.

Prerequisites. There are mathematical and statistical prerequisites.

The *mathematical background* mainly consists in *abstract linear algebra*. The *MATH.BASES* chapter should make the book self–contained.

The *statistical background* consists in elementary multivariate descriptive statistics: means, variances, covariances, correlation and regression

[32]See Table of Contents, p.V.

coefficients. There are *no statistical inference prerequisites*! Inference procedures are solely used in Chapter 9 (*CASE STUDIES*), and their principles are recalled in Chapter 8 (*INDUCTIVE*), making the book self-contained in this respect too.

Exercises. Exercises are proposed at the end of chapters 2 to 5, most of them with solutions and comments. Some exercises are *direct applications* of the main text. Others point to *theoretical deepenings.* Others still are *data analysis* exercises on small but real data sets, including "historical data".

About Software. Applying GDA methods to real data requires computer use. Which software is recommended?

Many international packages now incorporate the basic algorithm of CA; however, to efficiently perform the procedures specific to GDA, especially MCA, it seems only fair to prefer software that is historically rooted in the development of GDA methods and provides the indispensable expert knowledge — at least contributions and supplementary elements — that goes with serious geometric analysis. As far as we are concerned, for the case studies in this book we used:

• The SPSS software for the elementary procedures, as well as for the codings of variables;

• the ADDAD software for CA, MCA, PCA and AHC[33].

• the EyeLID software, specifically developed for structured data analysis and visual exploration: see Bernard, Baldy, Rouanet (1988)[34].

Let us add that the current version of the MacSAIF software due to J-P. Benzécri should be soon available from the website (now under construction) devoted to Benzécri's work.

---

[33] ADDAD and SPAD software perform geometric analyses, with aids to interpretation. ADDAD is a very comprehensive "tool box" for all the methods developed around Benzécri.

[34] Free versions of ADDAD and EyeLID are available (see `ftp` address, p.335).

Chapter 2

Correspondence Analysis

(*'CORRESPONDENCE'*)

Introduction

This chapter presents Correspondence Analysis (CA), the leading case of Geometric Data Analysis (GDA). Since 1964, Correspondence Analysis has been developed by Benzécri and his research group, and this method is extensively presented in the two–volume treatise by Benzécri & al. (1973). However, most of the early CA developments were written in French, one notable exception being Benzécri (1969). Furthermore, they were expressed in abstract mathematical language, leaving aside matrix formulations. This "Bourbakist language" erected a second language barrier which, for a long time, impeded access to CA in the international statistical community. Since the eighties, there have been CA presentations that have targeted this community by painstakingly "translating" the abstract formulations into matrix notation, as is the case in the often quoted book by Greenacre (1984). Thanks to these presentations, CA is nowadays recognized as a powerful visualization method, which is no mean achievement, but these presentations only deliver part of the message. The Benzécri handbook (1992) is just the translation of the 1980 introductory textbook. The "fundamental writings", such as the 1973 treatise, still await translation.

As is clear from the fundamental writings, the abstract formulations used for developing CA refer to an essential part of the approach, not an accidental matter of presentation. Indeed, the force of CA lies in the strong alliance between the mathematical underlying structures and their geometric representations. In our presentation, we would like to communicate the enlightenment and enrichment provided by the formal approach.

First of all, we characterize the statistical objects taken as primitives, namely measures and variables, by the summing vs averaging property, which leads to constructing two mutually dual vector spaces, and we discuss the geometry of measures (§2.1). Following that, formalization is applied to a measure over a Cartesian product with strictly positive margins — a paradigm being the contingency table — and we construct the two associated Euclidean clouds of points (§2.2). Then we proceed to Correspondence Analysis, with the joint search for the principal axes of the two clouds; geometric properties are emphasized, then statistical characterizations, with the detailed analysis of the *Hair & Eye Color* example (§2.3). The final section is devoted to extensions of CA and concluding comments (§2.4).

The formal approach of CA requires the standard undergraduate curriculum knowledge, not advanced Mathematics. The necessary material is recalled in chapter 10 (*MATH.BASES*): *abstract linear algebra, multidimensional geometry* and *spectral theorem*. Another topic with which some familiarity should be helpful is *mathematical measure theory* (see e.g. Cramér, 1946), even though it is only used in the finite case. The additive property of measures and the concept of density of a measure with respect to a strictly positive measure provide the key to the *measure vs variable duality* that is the cornerstone of the formal approach of CA.

2.1 Measure vs Variable Duality

In linear formalization, well–chosen statistical objects, taken as primitives, are recast in the framework of abstract linear algebra and thus become isomorphic to mathematical objects; then the statistical theory is formally constituted as a mathematical one. In this section, we present the linear formalization of measures and variables.

2.1.1 Summing and Averaging

If two quantities of a liquid are pooled together their volumes add up, whereas their temperatures average up. In physics, it is classical to distinguish magnitudes that are proportional to the quantities of matter involved (volume, mass, energy, etc.), and magnitudes that are independent of those quantities (temperature, density, etc.).

Similarly in statistics, if J denotes a nonempty finite set of statistical units, numerical functions $J \to \mathbb{R}$ can be distinguished with respect to the operation of *grouping* units, i.e. pooling together elements of J. As an example, consider the individuals in a firm who are university graduates across the various personnel categories of the firm. With each category $j \in J$, we may associate the *number* (absolute frequency) of graduates on the one hand, and their *percentage* in the category on the other hand. If some categories are grouped (pooled) into a class c, denoted $J{<}c{>}$[1], the number of graduates in class c is the *sum* of the corresponding numbers in categories $j \in J{<}c{>}$, whereas their percentage in class c is the *average* — with respect to the frequencies of the various categories — of the corresponding percentages in categories $j \in J{<}c{>}$.

As a general rule, among the functions $J \to \mathbb{R}$ we distinguish those for which grouping units entails *summing* (adding up) values, and those for which grouping units entails *averaging* values. The former are called *measures* over J and the latter *variables* over J. The whole construction of CA is founded on this dual characterization of *measures vs variables*.

2.1.2 Measures over J

The property of grouping by summing is nothing else than the well–known additivity property of a signed measure in measure theory, here a measure over the measurable space $(J, \mathcal{P}(J))$ where $\mathcal{P}(J)$ is the algebra of all subsets of the finite set J. Therefore, in the formalization, a numerical function on J for which grouping is done by summing is a *measure over J*.

The *mass of j* is denoted u_j (with subscript j), and the measure is denoted u_J as the indexed family of masses u_j: $u_J = (u_j)_{j \in J}$. With any $J{<}c{>} \subseteq J$ there is associated $\sum\limits_{j \in J<c>} u_j$, that is, the mass of u_J on $J{<}c{>}$. The *total mass* of measure u_J is denoted u (dropping index)[2]: $\sum\limits_{j \in J} u_j = u$.

Particular cases of measures are the *null measure* $0_J = (0)_{j \in J}$, and the *Dirac measure at j*, denoted $\delta^j_J = (\delta^j_{j'})_{j' \in J}$, with $\delta^j_j = 1$ and $\delta^j_{j'} = 0$ for $j' \neq j$ (Kronecker symbol); a measure u_J is a *contrast over J* if its total mass is 0: $\sum\limits_{j \in J} u_j = 0$; a *normalized* measure over J if its total mass is 1: $\sum\limits_{j \in J} u_j - 1$. The difference between two normalized measures u_J and v_J is the *contrast* $c_J = u_J - v_J$, and conversely the sum of a normalized

[1] *Nesting notation.* The subset of elements of J which belong to class c is denoted $J{<}c{>}$ (read "J in c").

[2] *Summing notation (subscripts).* We simply write u (rather than $u_.$); the omission of any subscript indicates summing over this subscript. Thus, starting with n_{jk} we write: $n_j = \sum\limits_{k \in K} n_{jk}$, $n = \sum\limits_{j \in J} \sum\limits_{k \in K} n_{jk}$, etc. Contrary to the dot notation ($n_{j.}$, $n_{..}$, etc.), subscripts are position independent.

measure u_J and a contrast c_J is the normalized measure $v_J = u_J + c_J$. A non–null measure is a *positive measure* if $\forall j \in J : u_j \geq 0$; a *strictly positive* measure if $\forall j \in J : u_j > 0$. In statistics, a family $(n_j)_{j\in J}$ of *absolute frequencies*, where n_j is the number of observations in category j, defines a positive measure n_J, with total mass $n = \sum_{j\in J} n_j$. The *relative frequencies* $(f_j)_{j\in J}$, with $f_j = n_j/n$, define a positive normalized measure f_J over J, since $\sum_{j\in J} f_j = 1$.

2.1.3 Densities and Variables over J

From now on we assume that there is a special strictly positive measure over J, denoted $\varpi_J = (\varpi_j)_{j\in J}$ with $\forall j \in J : \varpi_j > 0$, and called the *fundamental weighting* on J. The numerical function on J defined by $(u_j/\varpi_j)_{j\in J}$ — adopting measure–theoretic language — is the *density* of measure u_J with respect to ϖ_J. We denote this density $u^J = (u^j)_{j\in J}$, now putting j (and J) as upper indices (superscripts), using the *duality notation*[3]. In particular, the ϖ_J–density of ϖ_J is given by $\forall j : \varpi_j/\varpi_j = 1$, that is, the density of the fundamental weighting is the *constant function on J* everywhere equal to 1, denoted $1^J = (1)_{j\in J}$. More generally, with any subset $J<c>$ of J, we associate the ratio $\sum_{j\in J<c>} u_j / \sum_{j\in J<c>} \varpi_j$, that is, $\sum_{j\in J<c>} \varpi_j u^j / \sum_{j\in J<c>} \varpi_j$, which is the average of density u^J over $J<c>$ and denoted u^c. For densities, grouping is done by *averaging* with respect to fundamental weighting. The formalized notion of a numerical *variable over J* is the extension of that of density. A numerical function on J for which grouping is done by averaging with respect to some strictly positive measure is called a *variable over J*, and denoted with upper indices. More precisely, assuming a fundamental weighting ϖ_J, we say that $x^J = (x^j)_{j\in J}$ is a variable over J if with any subset we associate $\sum_{j\in J<c>} \varpi_j x^j / \sum_{j\in J<c>} \varpi_j$, denoted x^c. The numbers x^j are called the *values* of the variable. Contrary to summing, the averaging procedure involves the fundamental weighting up to a multiplicative factor. If x^J is a variable over J, the function x_J on J defined by $\forall j \in J : x_j = \varpi_j x^j$ (term by term product of ϖ_J by x^J) is a measure over J.

As particular cases of variables, we define the *indicator variable of j*,

[3] The *duality notation*, or *transition notation* (subscripts–superscripts), was introduced in CA by Benzécri (1966). The use of superscripts to denote conditioning is also found in probability texts such as Neveu (1964). Duality notation avoids writing meaningless expressions, which would appear as syntactic mistakes.

denoted $\delta_j^J = (\delta_j^{j'})_{j' \in J}$, with $\delta_j^j = 1$ and $\delta_j^{j'} = 0$ for $j' \neq j$ (Notice the interchange of upper and lower indices with Dirac measures). A variable x^J is ϖ_J*–centered* if its ϖ_J–mean over J is null, i.e. if $\sum\limits_{j \in J} \varpi_j x^j = 0$. If the variable x^J is ϖ_J–centered, the measure $x_J = (\varpi_j\, x^j)_{j \in J}$ is a contrast, whose ϖ_J–density is the variable x^J.

Last, with each variable x^J and each measure u_J, we associate the numerical product $\sum\limits_{j \in J} u_j\, x^j$, which in measure–theoretic terms is the integral — here a finite sum — of the fonction x^J with respect to measure u_J. This product of variable x^J by measure u_J — or of measure u_J by variable x^J — is denoted $[x^J | u_J]$ or $[u_J | x^J]$.

2.1.4 Vector Spaces of Measures and of Variables over J

The set of mappings $J \to \mathbb{R}$ is naturally structured as a vector space on $\mathbb{R}$ of dimension[4] J. The vector space of the mappings $J \to \mathbb{R}$ which are measures over J is denoted $\mathbb{R}_J$. The null measure 0_J is the null vector of $\mathbb{R}_J$, and the family of Dirac measures $(\delta_J^j)_{j \in J}$ is its canonical basis, since $\forall u_J \in \mathbb{R}_J$: $u_J = \sum\limits_{j \in J} u_j \delta_J^j$, and coefficients $(u_j)_{j \in J}$ are unique. We denote $\mathbb{R}_J^0$ the *subspace of contrasts* and $\mathbb{R}_J^1$ the *set of normalized measures.*

The *dual vector space* of $\mathbb{R}_J$ consists of all *linear forms* on $\mathbb{R}_J$ (linear mappings from $\mathbb{R}_J$ to $\mathbb{R}$, see Definition 10.5, p.424). Now let x^J be a variable over J; x^J defines the linear form, also denoted x^J, such that:

$$
\begin{array}{rcccl}
x^J & : & \mathbb{R}_J & \longrightarrow & \mathbb{R} \\
x^J & : & u_J & \longmapsto & \sum\limits_{j \in J} u_j x^j
\end{array}
$$

The variables over J can be identified with the linear forms on $\mathbb{R}_J$. The *vector space of variables* over J, that is, the vector space of linear forms on $\mathbb{R}_J$, is denoted $\mathbb{R}^J$. The canonical basis of $\mathbb{R}^J$, that is, the dual basis of the Dirac basis $(\delta_J^j)_{j \in J}$ of $\mathbb{R}_J$, is the family of indicator variables $(\delta_j^J)_{j \in J}$. As a linear form, δ_j^J associates which each u_J the mass u_j ($\delta_j^J : u_J \mapsto u_j$). The numerical product $[u_J | x^J] = \sum\limits_{j \in J} u_j x^j$ is the value of the canonical bilinear form $[\cdot|\cdot]$ between the mutually dual spaces $\mathbb{R}^J$ and $\mathbb{R}_J$.

[4]As a general convention, we denote the cardinalities of finite sets like the sets themselves: The cardinality of J, often denoted $/J/$, will be denoted J.

Now let us introduce the *fundamental weighting measure* ϖ_J ($\forall j \in J$: $\varpi_j \neq 0$). To begin with, ϖ_J induces the following two reciprocal isomorphisms $\varpi_J^{-1} : \mathbb{R}_J \rightarrow \mathbb{R}^J$ and $\varpi_J : \mathbb{R}^J \rightarrow \mathbb{R}_J$, such that:

$$\begin{array}{llll} \varpi_J^{-1}: & u_J & \longmapsto & u^J = (u^j)_{j\in J} \quad \text{with } u^j = u_j/\varpi_j \\ \varpi_J: & x^J & \longmapsto & x_J = (x_j)_{j\in J} \quad \text{with } x_j = \varpi_j x^j \end{array}$$

Making use of those isomorphisms, and of the canonical bilinear form, the fundamental weighting ϖ_J induces *scalar products* on both $\mathbb{R}^J$ and $\mathbb{R}_J$ — at this point, the strict positivity of ϖ_J is required — that we both denote $\langle\cdot|\cdot\rangle$ and define as follows. If x^J and y^J are two variables over J, letting $x_J = (\varpi_j x^j)_{j\in J}$ and $y_J = (\varpi_j y^j)_{j\in J}$, their ϖ_J–scalar product is defined by:

$$\forall x^J \in \mathbb{R}^J \ \ \forall y^J \in \mathbb{R}^J: \ \langle x^J|y^J\rangle = [x^J|y_J] = [x_J|y^J] = \sum_{j\in J} \varpi_j x^j y^j \qquad (2.1)$$

The ϖ_J–scalar product between two measures u_J and v_J is defined by:

$$\langle u_J|v_J\rangle = [u_J|v^J] = [u^J|v_J] = \sum_{j\in J} u_j v_j/\varpi_J \qquad (2.2)$$

Definitions of ϖ_J–Euclidean norms on $\mathbb{R}^J$ and $\mathbb{R}_J$ readily follow, letting $\|x^J\|^2 = \sum_{j\in J} \varpi_j (x^j)^2$ and $\|u_J\|^2 = \sum_{j\in J} (u_j)^2/\varpi_j$.

— *Particular cases.*

The Dirac measures $(\delta_J^j)_{j\in J}$ are ϖ_J–orthogonal, since $\langle\delta_J^j|\delta_J^{j'}\rangle = 0$ for $j' \neq j$; but they are not (in general) orthonormal, since $\|\delta_J^j\|^2 = 1/\varpi_j$. Similarly the indicator variables $(\delta_j^J)_{j\in J}$ are ϖ_J–orthogonal but not orthonormal; we have $\langle\delta_j^J|\delta_{j'}^J\rangle = 0$ for $j \neq j'$, and $\|\delta_j^J\|^2 = \varpi_j$.

2.1.5 The Geometry of Measures over J

The geometric representation of measures over J is obtained by constructing the affine space underlying the vector space $\mathbb{R}_J$.

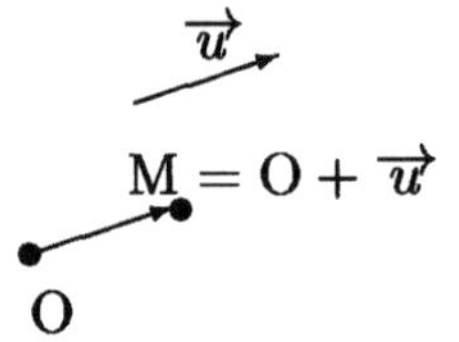

Letting O be the point representing the null measure, measure u_J is then represented by a geometric vector $\overrightarrow{u}$ ("arrowed letter"), and/or by the point M defined by $\mathrm{M} = \mathrm{O} + \overrightarrow{u}$.

The geometric representations of normalized measures and of contrasts will be of prime interest. We call *ideal points* the J points defined by $\mathrm{I}^j = \mathrm{O} + \delta_J^j$. The ideal points $(\mathrm{I}^j)_{j\in J}$ span a hyperplane $\mathcal{H}$, such that every point $\mathrm{M} \in \mathcal{H}$ is the barycenter of ideal points with (positive or negative) weights (cf. MATH.BASES, §10.15, p.436).

With a normalized measure $u_J = \sum_{j\in J} u_j \delta_J^j$ ($\sum_{j\in J} u_j = 1$), there is associated the barycenter M of ideal points $(\mathrm{I}^j)_{j\in J}$ with weights $(u_j)_{j\in J}$, that is $\mathrm{M} = \sum_{j\in J} u_j \mathrm{I}^j \in \mathcal{H}$. With the set $\mathbb{R}_J^{+1}$ of *positive normalized measures* over J, there is associated the *simplex* whose vertices are the ideal points.

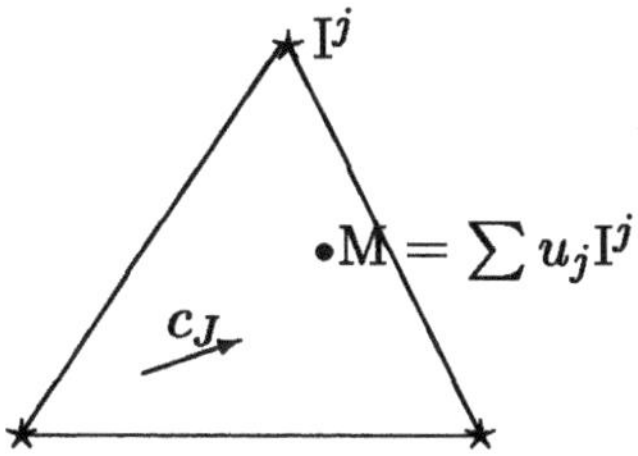

Figure 2.1. Hyperplane $\mathcal{H}$ and simplex $\mathcal{S}$ for $J = 3$.

The underlying vector subspace of the hyperplane $\mathcal{H}$ is the subspace $\mathbb{R}_J^0$ of contrasts over J; the dimension of this space is equal to $J - 1$.

> *To sum up the geometric representation of measures*: Normalized measures over J are represented by points of the hyperplane spanned by ideal points, and contrasts over J are represented by vectors of the underlying vector space.

Φ^2–metric over normalized measures. As a fundamental weighting, let us take a strictly positive normalized measure $p_J = (p_j)_{j\in J}$. The *distance of center p_J* between two normalized measures u_J and v_J is defined as the p_J–norm of the difference:

$$\|u_J - v_J\|^2 = \sum_{j\in J} \frac{(u_j - v_j)^2}{p_j} \tag{2.3}$$

which also defines the distance MN between the two points $\mathrm{M} = \sum_{j\in J} u_j \mathrm{I}^j$ and $\mathrm{N} = \sum_{j\in J} v_j \mathrm{I}^j$. If f_J is a distribution of observed relative frequencies and p_J a reference distribution, we have:

$$\|f_J - p_J\|^2 = \sum_{j\in J} \frac{(f_j - p_j)^2}{p_j}$$

which is the descriptive goodness–of–fit statistic $\Phi^2 = \chi^2/n$ for comparing an observed distribution to a reference distribution. This is why this metric is named the Φ^2–metric — also called the χ^2 metric — with center p_J.

2.1.6 Linear Formalization of Grouping

A grouping on J is defined by a surjective mapping $h : J \to C$ (C nonempty finite set). The summing operation of measures consists in associating with measure u_J over J the measure u_C over C defined by $\forall c \in C$: $u_c = \sum_{j \in J<c>} u_j$, with $J<c> = h^{-1}(c)$; in measure–theoretic terms, u_C is the *image measure* of u_J by mapping h. In linear formalization, the mapping $u_J \mapsto u_C$ is a homomorphism from $\mathbb{R}_J$ (space of measures over J) to $\mathbb{R}_C$, that is denoted δ_C^J; $\mathbb{R}_C$ is the vector space of measures over C with its canonical basis $(\delta_C^c)_{c \in C}$ equipped with the scalar product:

$$\forall u_C \in \mathbb{R}_C, \forall v_c \in \mathbb{R}_C : \langle u_C | v_C \rangle = \sum_{c \in C} u_c v_c / \varpi_c \quad \text{with } \varpi_c = \sum_{j \in J<c>} \varpi_j$$

This homomorphism $\delta_C^J : \mathbb{R}_J \to \mathbb{R}_C$ can be represented by the endomorphism p, ϖ_J–orthogonal projection, on $\mathbb{R}_J$, along the following theorem.

Theorem 2.1. *Let* $\forall c \in C : \delta_J^c = \sum_{j \in J<c>} \varpi_j \delta_J^j / \varpi_c$, *and* $\mathbb{R}_J^C$ *the subspace of* $\mathbb{R}_J$ *spanned by* $(\delta_J^c)_{c \in C}$. *One has* $\delta_C^J = i \circ p$, *where* p *is the* ϖ_J*-orthogonal projection onto* $\mathbb{R}_J^C$ *and* i *is the isomorphism on* $\mathbb{R}_J^C$ *onto* $\mathbb{R}_C$ *defined by:* $\forall c \in C : \delta_J^c \mapsto \delta_C^c$.

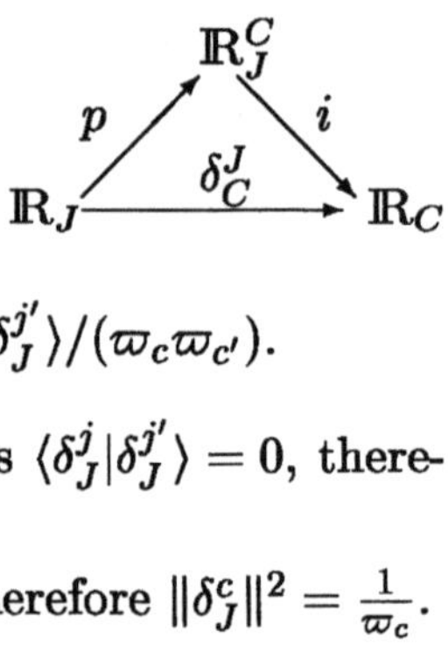

Proof (i) One has $\langle \delta_J^c | \delta_J^{c'} \rangle = \sum_{j \in J<c>} \sum_{j' \in J<c'>} \varpi_j \varpi_{j'} \langle \delta_J^j | \delta_J^{j'} \rangle / (\varpi_c \varpi_{c'})$.

Now $\forall j \in J<c>$, $\forall j' \in J<c'>$, with $c \neq c'$, one has $\langle \delta_J^j | \delta_J^{j'} \rangle = 0$, therefore $\langle \delta_J^c | \delta_J^{c'} \rangle = 0$: vectors δ_J^c and $\delta_J^{c'}$ are orthogonal.
$\|\delta_J^c\|^2 = \frac{1}{\varpi_c^2} \sum_{j \in J<c>} (\varpi_j)^2 \|\delta_J^j\|^2 = \frac{1}{(\varpi_c)^2} \sum_{j \in J<c>} \frac{(\varpi_j)^2}{\varpi_j}$. Therefore $\|\delta_J^c\|^2 = \frac{1}{\varpi_c}$.

(ii) The projection of u_J onto δ_J^c is, by definition: $\langle u_J | \delta_J^c \rangle / \|\delta_J^c\|^2 \, \delta_J^c$, with
$\frac{\langle u_J | \delta_J^c \rangle}{\|\delta_J^c\|^2} = \langle \sum_{j \in J} u_j \delta_J^j | \sum_{j' \in J<c>} \frac{\varpi_{j'}}{\varpi_c} \delta_J^{j'} \rangle / (1/\varpi_c) = \sum_{j \in J} \sum_{j' \in J<c>} u_j \varpi_{j'} \langle \delta_J^j | \delta_J^{j'} \rangle$
$= \sum_{j \in J<c>} u_j \varpi_j \|\delta_J^j\|^2 = \sum_{j \in J<c>} u_j \varpi_j \frac{1}{\varpi_j} = u_c$
therefore $p(u_J) = \sum_{c \in C} u_c \delta_J^c$.

(iii) One has $\langle p(u_J) | p(v_J) \rangle = \sum_{c \in C} \sum_{c' \in C} u_c v_{c'} \langle \delta_J^c | \delta_J^{c'} \rangle = \sum_{c \in C} \frac{u_c v_c}{\varpi_c} = \langle u_C | v_C \rangle$, ϖ_C–scalar product of measures u_C and v_C, hence i is an isomorphism. ◁

Corollary 2.1. $u_J = p(u_J) \iff \forall c \in C, \forall j \in J<c>$: $u^j = u_j / \varpi_j$ *only depends on* c.

Proof. $u_J = \sum_{j\in J} u_j \delta^j_J$ $\qquad p(u_J) = \sum_{c\in C} u_c \delta^c_J = \sum_{c\in C} u_c \sum_{j\in J<c>} \varpi_j \delta^j_J / \varpi_c$

$= \sum_{c\in C} \sum_{j\in J<c>} u_j \delta^j_J$ $\qquad = \sum_{c\in C} \sum_{j\in J<c>} (u_c/\varpi_c) \times \varpi_j \delta^j_J$

hence $u_J = p(u_J) \iff \forall c \in C; \forall j \in J<c>: u_j = (u_c/\varpi_c) \times \varpi_j$ (since $(\delta^j_J)_{j\in J}$ are linearly independent). ◁

2.2 Measure over a Cartesian Product

2.2.1 Basic Definitions

Measure over a Cartesian product. We now apply the foregoing formalization to a *measure over a Cartesian product* with strictly positive margins, which is the central object of CA. A stronger positivity assumption will be made only when necessary (cf. §2.3.6, p.57). Let J (of elements $j \in J$) and K (of elements $k \in K$) be two finite sets, and $J \times K$ be their Cartesian product (of elements $jk \in J \times K$). Let $n_{JK} = (n_{jk})_{j\in J, k\in K}$ be a measure over $J \times K$ of masses n_{jk}. Denoting $n_j = \sum_{k\in K} n_{jk}$ and $n_k = \sum_{j\in J} n_{jk}$, the families $n_J = (n_j)_{j\in J}$ and $n_K = (n_k)_{k\in K}$ define the *marginal measures* over J and over K, with $\forall j \in J : n_j > 0$ and $\forall k \in K : n_k > 0$.

The paradigm of contingency tables. The statistical paradigm of measure over a Cartesian product is the *contingency table*, where j and k are observable modalities and the mass n_{jk} is the number of observations corresponding to the joint occurrence of modalities j and k. This defines a *measure over* $J \times K$, since when cells are grouped, the numbers of observations add up. This is why CA is traditionally presented in terms of a contingency table.

Now for a contingency table the measure over $J \times K$ is a *positive measure*, that is, $\forall jk \in J \times K : n_{jk} \geq 0$. Even though we will not assume this positivity property in our general presentation, it will be convenient to adopt the terminology of contingency tables. Thus n_{jk} is called the *absolute joint frequency* (j, k), and n_j and n_k the absolute *marginal frequencies* of j and k. We denote n the total mass of the measure n_{JK} (total number of observations), with $n = \sum_j \sum_k n_{jk} = \sum_j n_j = \sum_k n_k$. Dividing absolute frequencies by n yields the *relative frequencies*, which are similarly denoted, replacing n by f:

$$f_{jk} = n_{jk}/n,\ f_j = n_j/n,\ f_k = n_k/n \text{ with } \sum_j \sum_k f_{jk} = \sum_j f_j = \sum_k f_k = 1$$

hence the measures f_{JK} (normalized measure over $J \times K$), f_J and f_K (strictly positive normalized measures over J and over K).

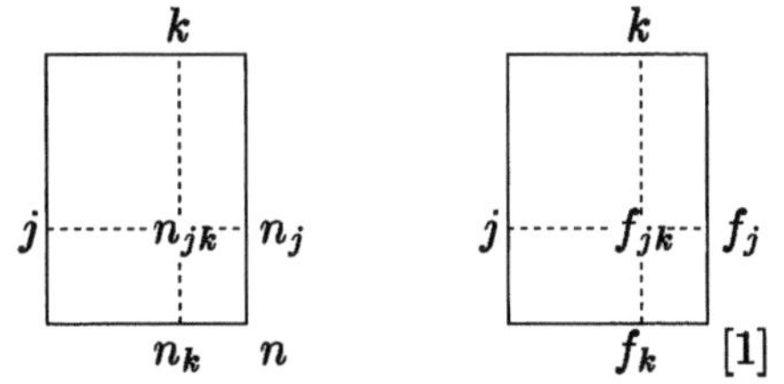

$$f_{jk} = \frac{n_{jk}}{n}$$

$$f_j = \frac{n_j}{n} \quad f_k = \frac{n_k}{n}$$

Figure 2.2. Absolute and relative frequencies (joint and marginal).

Product–measure and association rates. The product of f_J by f_K (marginal measures) defines the measure over $J \times K$ denoted $f_J \otimes f_K$. The association between modalities j and k is null if $f_{jk} = f_j\, f_k$; positive (attraction) if $f_{jk} > f_j\, f_k$; negative (repulsion) if $f_{jk} < f_j\, f_k$. In a sense, CA is the multidimensional display of the deviation of f_{JK} from the product–measure $f_J \otimes f_K$. The family of ratios $(f_{jk}/f_j\, f_k)_{jk \in J \times K}$ is the density of the measure f_{JK} with respect to the product–measure $f_J \otimes f_K$. The *association rate* between j and k is defined by

$$t^{jk} = (f_{jk} - f_j\, f_k)/f_j\, f_k \tag{2.4}$$

hence the variable $t^{JK} = (t^{jk})_{jk \in J \times K}$. The variance of the variable t^{JK} with respect to the product–measure is seen to be $\Phi^2 = \chi^2/n$, the classical mean square contingency coefficient.

Conditional frequencies, profiles and transitions. The *conditional frequency* of k given j, defined as $n_{jk}/n_j = f_{jk}/f_j$, is denoted f^j_k — with superscript j indicating conditioning — and read "f of k if j"; thus we have: $f^j_k = n_{jk}/n_j = f_{jk}/f_j$. The $J \times K$ family $f^J_K = (f^j_k)_{j \in J, k \in K}$ defines the *transition* from J to K. Given j (row of transition table), the family of conditional frequencies $f^j_K = (f^j_k)_{k \in K}$ defines the *profile of* j (over K); f^j_K is a normalized *measure over* K, since if k–modalities are pooled, the f^j_k add up. On the other hand, given k (column of transition table), the family $f^J_k = (f^j_k)_{j \in J}$ is a *variable over* J, since if j–modalities are pooled, the f^j_k average up with respect to the weighting f_J (the variable f^J_k is the density of measure n_{Jk} with respect to n_J). Consider now the family of marginal frequencies $f_K = (f_k)_{k \in K}$; one has $\forall k \in K\, :\, f_k = \sum_j f_j\, f^j_k$, hence

$f_K = \sum_j f_j f_K^j$, that is, f_K is the f_J–weighted mean of the profiles $(f_K^j)_{j \in J}$. The family f_K is therefore the *mean profile* over K. A similar construction applies to the conditional frequencies f_j^k (read "f of j if k"), which define the transition from K to J, leading to define f_J^k (profile of k over J), f_J (mean profile over J), etc. Table 2.1 summarizes these notions.

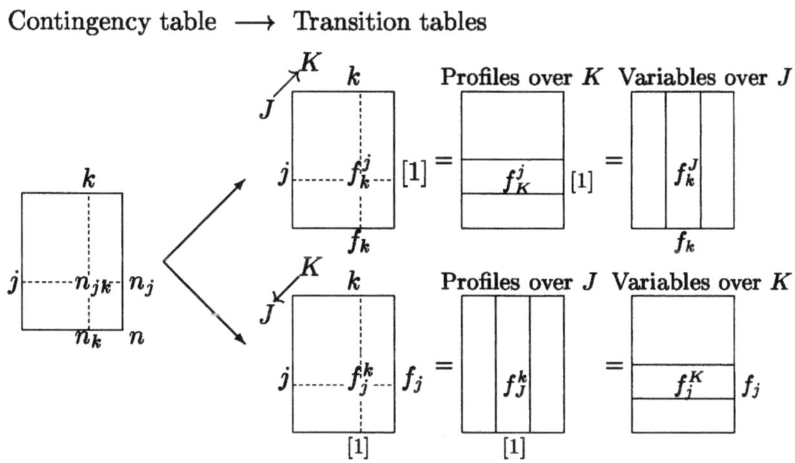

Table 2.1. Summary of profiles and transitions associated with contingency table.

2.2.2 The Construction of the Two Clouds

In the affine space canonically associated with the vector space $\mathbb{R}_K$, the profile f_K^j is represented by a point denoted M^j weighted by f_j, which we call *point* j; the Dirac measure δ_K^k is represented by the *ideal point* I^k, which we call *ideal point* k. The family of the J points $(\mathrm{M}^j)_{j \in J}$ defines the $\mathcal{J}$*–cloud* (cf. Figure 2.3).

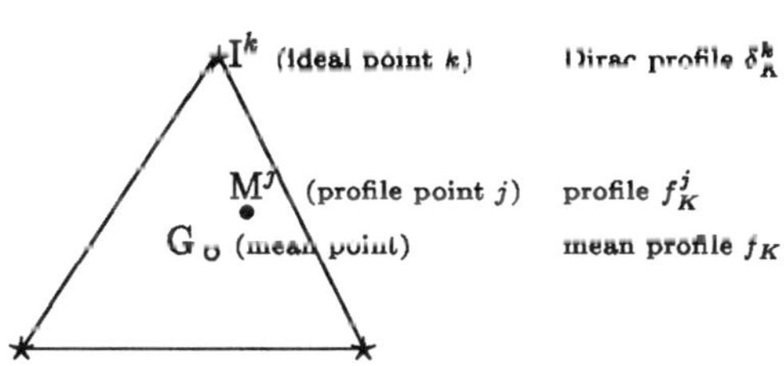

Figure 2.3. $\mathcal{J}$–cloud for $K = 3$.

Each point M^j is the barycenter of the ideal points $(\mathrm{I}^k)_{k \in K}$ weighted by the conditional frequencies $(f_k^j)_{k \in K}$, that is, $\mathrm{M}^j = \sum_{k \in K} f_k^j \mathrm{I}^k$ (barycentric relation). The mean profile f_K is represented by a point denoted G which is the barycenter of the ideal points $(\mathrm{I}^k)_{k\in K}$ weighted by the marginal frequencies $(f_k)_{k\in K}$, that is, $\mathrm{G} = \sum_{k\in K} f_k \mathrm{I}^k$. Since $f_K = \sum_{j\in J} f_j f_K^j$, point G is also the f_J–barycenter of points $(\mathrm{M}^j)_{j\in J}$, that is, the mean point of the cloud. The whole $\mathcal{J}$–cloud lies within the simplex if and only if the positivity property $(\forall j \in J\, \forall k \in K : f_k^j \geq 0)$ holds.

The *distance* $\mathrm{M}^j\mathrm{M}^{j'}$ between points M^j and $\mathrm{M}^{j'}$ is obtained from the Φ^2–distance between profiles f_K^j and $f_K^{j'}$ (measures over K); its square $(\mathrm{M}^j\mathrm{M}^{j'})^2$ is given by

$$||f_K^j - f_K^{j'}||^2 = \sum_{k\in K} \frac{(f_k^j - f_k^{j'})^2}{f_k} \tag{2.5}$$

Similarly, the distance GM^j between the point M^j and the mean point G is given by its square $||f_K^j - f_K||^2 = \sum_{k\in K} \frac{(f_k^j - f_k)^2}{f_k}$.

The *variance of the cloud*, that is, the weighted mean of squares of distances $\mathrm{GM}^j = ||f_K^j - f_K||$, is equal to $\sum_{j\in J}\sum_{k\in K} \frac{(f_{jk} - f_j f_k)^2}{f_j f_k} = \Phi^2 = \chi^2/n$ (mean square contingency coefficient).

The foregoing Euclidean distance is referred to as the Φ^2–distance (also called χ^2–distance) associated with the contingency table.

Exchanging j and k, a similar construction applies to the K profiles over J, so we end up with *two clouds*, namely the $\mathcal{J}$–cloud of the J profiles over K (in $\mathbb{R}_K^1$), and the $\mathcal{K}$–cloud of the K profiles over J (in $\mathbb{R}_J^1$).

2.2.3 Relations between the two Clouds

Theorem 2.2. *Grouping two columns k and k' into a single column k_0 with $\forall j \in J : n_{jk0} = n_{jk} + n_{jk'}$ amounts to the following:*

(i) In $\mathbb{R}_J$, replacing the two profiles f_J^k and $f_J^{k'}$ by their barycenter $f_J^{k0} = (f_k f_J^k + f_{k'} f_J^{k'})/(f_k + f_{k'})$, weighted by $f_{k0} = f_k + f_{k'}$.

(ii) In $\mathbb{R}_K$, orthogonally projecting all profiles f_K^j onto the subspace $\mathcal{E}$ spanned by $(\delta_K^{k''})_{k''\in K-\{k,k'\}}$ and $\delta_K^{k0} = (f_k \delta_K^k + f_{k'}\delta_K^{k'})/(f_k + f_{k'})$.

(i) is immediate; *(ii)* results from Theorem 2.1 (p.30).

Corollary 2.2 (Distributional Equivalence). *The $\mathcal{J}$–cloud is invariant by the grouping of k and k' if and only if $\forall j \in J : f_j^k = f_j^{k'}$, that is, if n_{Jk} and $n_{Jk'}$ are proportional.*

Proof. From Corollary 2.1 (p.30), the $\mathcal{J}$–cloud is invariant if and only if $\forall j \in J$: $f_k^j/f_k = f_{k'}^j/f_{k'}$, or equivalently $f_{jk}/(f_j f_k) = f_{jk'}/(f_j f_{k'})$, that is, if the profiles f_J^k and $f_J^{k'}$ are equal, or equivalently, if the measures n_{Jk} and $n_{Jk'}$ are proportional. ◁

Distributional equivalence plays a central role in CA (Benzécri, 1992, p.51); quasi–distributional equivalence has been investigated by Escofier & Le Roux (1972), see also chapter *STABILITY*, §7.1 (p.270) with an application to the "Hair and Eye Color" example (p.276).

Definition 2.1 (Transitions). *Let us consider the homomorphism from $\mathbb{R}_J$ to $\mathbb{R}_K$ that associates with the Dirac measure δ_J^j the profile f_K^j; we call this homomorphism transition from J to K and denote it f_K^J. Similarly, let us consider f_J^K, homomorphism (transition) from K to J, such that $\delta_K^k \mapsto f_J^k$.*

$$f_K^J : \mathbb{R}_J \longrightarrow \mathbb{R}_K \qquad\qquad f_J^K : \mathbb{R}_K \longrightarrow \mathbb{R}_J$$
$$\delta_J^j \longmapsto f_K^j \qquad\qquad\qquad \delta_K^k \longmapsto f_J^k$$

Since $f_J = \sum_{j \in J} f_j \delta_J^j$, we have $f_K^J(f_J) = \sum_{j \in J} f_j f_K^J(\delta_J^j) = \sum_{j \in J} f_j f_K^j = f_K$, that is, we have $f_K^J(f_J) = f_K$, and similarly $f_J^K(f_K) = f_J$. In geometric terms, through the homomorphismsf_K^J and f_J^K, the ideal points of one cloud are mapped into the profile points of the other cloud, and vice versa; and the two mean points correspond to each other (Figure 2.4).

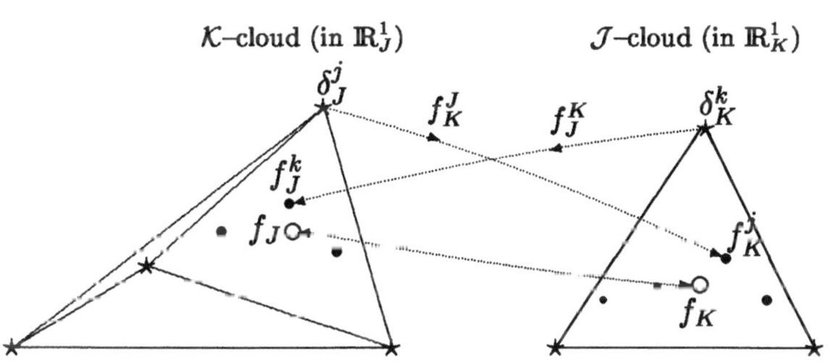

Figure 2.4. Adjoint transitions with $J = 4$ and $K = 3$.

Theorem 2.3 (Adjoint Transitions). *The transitions f_K^J and f_J^K are mutually adjoint homomorphisms.*

Proof. (see MATH.BASES, Theorem 10.3, p.432). $\forall u_K \in \mathbb{R}_K$, $\forall u_J \in \mathbb{R}_J$:
$\langle f_K^J(u_J)|u_K\rangle = \sum\limits_{j\in J}\sum\limits_{k\in K} \frac{f_{jk}}{f_j f_k} u_j u_k = \langle u_J|f_J^K(u_K)\rangle$ ◁

Theorem 2.4 (Variance of projected Cloud). *Let z_K be a contrast of unit-norm; then the variance of the $\mathcal{J}$-cloud along the direction of z_K is $\|f_J^K(z_K)\|^2$. Similarly, if z_J is a contrast of norm 1, the variance of the $\mathcal{K}$-cloud along the direction of z_J is $\|f_K^J(z_J)\|^2$.*

Proof. Let y^j denote the coordinate of the orthogonal projection of point M^j on the axis directed by z_K and with origin point G (mean point of $\mathcal{J}$-cloud); then $y^j = \langle f_K^j - f_K|z_K\rangle = \langle f_K^j|z_K\rangle = \sum\limits_{k\in K} f_k^j z_k/f_k = \sum\limits_{k\in K} f_j^k z_k/f_j$.

The image of z_K by the transition f_J^K is the contrast $y_J = \sum\limits_{k\in K} z_k f_J^k$, and the variable $y^J = (y^j)_{j\in J}$ is therefore the density of this contrast.
By definition, the variance of the projected cloud is equal to $\mathrm{Var}\, y^J = \|y^J\|^2$, hence by duality to $\|y_J\|^2 = \|f_J^K(z_K)\|^2$.

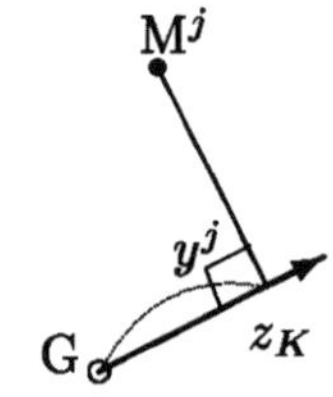

Similarly, for the $\mathcal{K}$-cloud (exchanging the roles of j and k). ◁

2.3 Correspondence Analysis

We now present CA as a GDA method, dealing successively with the joint search of principal axes and variables of the two clouds (§2.3.1), matrix formulations (§2.3.2), geometric representations (§2.3.3), interpretation of CA (§2.3.4), and the detailed analysis of the *Hair & Eye Color* example (§2.3.5). Last, we recall the statistical characterizations of CA (§2.3.6).

2.3.1 Principal Axes and Variables

Theorem 2.5. *The principal axes of the two clouds are spanned by the eigenvectors of the adjoint homomorphisms f_J^K and f_K^J, that is the solutions of the following system of equations:*

$$\begin{cases} f_K^J(z_J) &= \xi z_K \\ f_J^K(z_K) &= \xi z_J \end{cases} \quad (\text{with } \|z_K\| = \|z_J\|) \tag{2.6}$$

Proof. Since the points of the $\mathcal{J}$–cloud lie in the hyperplane $\mathcal{H}$ spanned by the ideal points $(\mathrm{I}^k)_{k\in K}$, the principal axes are generated by *contrasts* over K (and similarly for the $\mathcal{K}$–cloud). We therefore first discard the improper solution $(f_K, f_J, 1)$, which expresses the property that the mean points of the two clouds correspond to each other. Unit–norm contrasts may be sought without loss of generality. By definition, the first principal axis of a cloud is the one along which the projected cloud is of maximal variance. The principal axes of order 1 of the two clouds are therefore defined by contrasts z_K and z_J such that the variances $\|f_J^K(z_K)\|^2$ and $\|f_K^J(z_J)\|^2$ are maximum. These variances are maximal for contrasts z_K and z_J that are eigenvectors of homomorphisms f_J^K and f_K^J associated with the greatest singular value (see MATH.BASES, Property 10.32, p.446), hence the system of principal equations. ◁

The principal axes of order 2, orthogonal to the axes of order 1, are associated with ξ_2, second singular value, etc.

— *Remark.* These equations define L orthogonal contrasts, called *principal contrasts*, L denotes the rank of the transition homomorphisms (dimension of clouds).

Theorem 2.6. *The variance of each cloud along the ℓ–th principal direction is equal to eigenvalue λ_ℓ (square of the singular value ξ_ℓ).*

This immediately follows from property 10.32 (see MATH.BASES, p.446).

Principal variables. Let (z_J, z_K, ξ) with $\|z_J\| = \|z_K\| = 1$ be a solution of the system of Equations (2.6). By duality, we define the variables z^J and z^K such that $z^j = z_j/f_j$, $z^k = z_k/f_k$. Since z_J is a contrast, by duality the variable z^J is f_J–centered. Since $\|z_J\|^2 = 1$, we also have $\|z^J\|^2 = 1$, hence $\sum\limits_{j\in J} f_j(z^j)^2 = 1$, that is, variable z^J is centered and of variance 1 (similarly for variable z^K). The variables z^J and z^K are the *standard principal variables* associated with eigenvalue $\lambda = \xi^2$.

Transition formulas (Reciprocal averaging). In terms of variables, the principal equations read (denoting $\circ$ the mapping composition):

$$\begin{cases} z^K \circ f_K^J &= \xi z^J \\ z^J \circ f_J^K &= \xi z^K \end{cases} \tag{2.7}$$

Let $y^J = \xi z^J$ and $y^K = \xi z^K$. The variables y^J and y^K are centered and of variance $\lambda = \xi^2$. They are the *scaled principal variables* associated with eigenvalue $\lambda = \xi^2$. We may write the principal equations as follows:

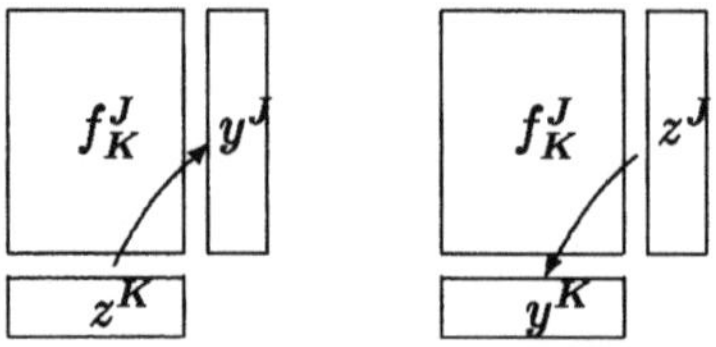

$$\begin{cases} \sum\limits_{k\in K} z^k f_k^J & = y^J \\ \sum\limits_{j\in J} z^j f_j^K & = y^K \end{cases} \tag{2.8}$$

In CA literature, these equations are known as *transition formulas.* They express the characterization of CA as *reciprocal averaging.*

Duality scheme. It summarizes the foregoing properties.

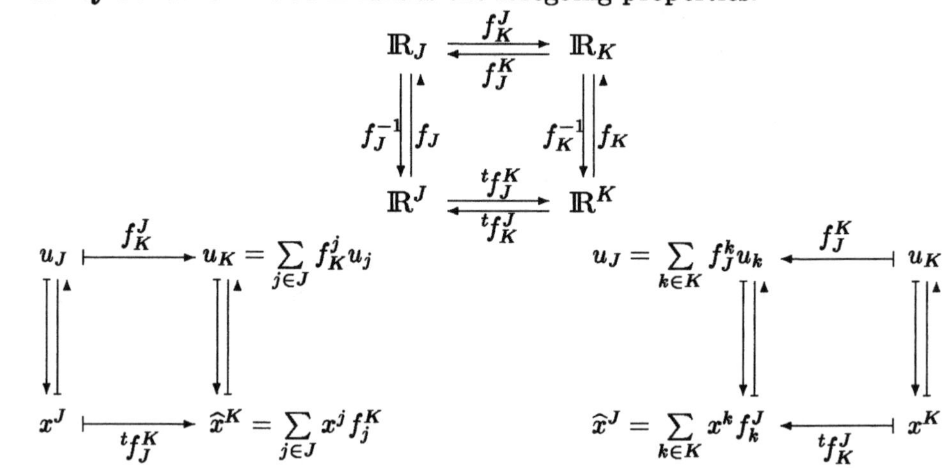

Reconstitution formulas. If $(z_{K\ell})_{\ell=1,\dots L}$ is an orthonormal basis of principal contrasts of the support of the cloud, we have: $\langle f_K^j - f_K | z_{K\ell}\rangle = \sum\limits_{k\in K}(f_k^j - f_k)z_{k\ell}/f_k = \sum\limits_{k\in K} f_k^j z_\ell^k = y_\ell^j$ (see the above transition formula (2.8)). Then we deduce the expression of profile f_K^j in terms of its L principal coordinates $f_K^j = f_K + \sum\limits_{\ell=1}^{L} y_\ell^j z_{K\ell}$. Now $z_{k\ell} = (1/\xi_\ell) f_k y_\ell^k$, hence $f_k^j = f_k(1+\sum\limits_{\ell=1}^{L}\frac{1}{\xi_\ell}y_\ell^j y_\ell^k)$, $f_j^k = f_j(1+\sum\limits_{\ell=1}^{L}\frac{1}{\xi_\ell}y_\ell^j y_\ell^k)$ and the reconstitution formula for the contingency table:

$$f_{jk} = f_j f_k (1 + \sum_{\ell=1}^{L} \frac{1}{\xi_\ell} y_\ell^j y_\ell^k) \tag{2.9}$$

From such reconstitution formulas, one also gets the reconstitution formula of the associated Φ^2, namely $\Phi^2 = \sum\limits_{\ell=1}^{L} \lambda_\ell$.

2.3.2 Matrix Formulations

So far the whole theory has been developed in terms of abstract linear algebra. To obtain matrix formulations, it suffices to refer the homomorphisms involved in the theory to appropriate bases. The canonical bases of $\mathbb{R}_J$ and $\mathbb{R}_K$ are orthogonal but not orthonormal. Taking the associated orthonormal bases, the mutually adjoint homomorphisms f_K^J and f_J^K are expressed by mutually transposed matrices. Then matrix formulations of CA readily follow from the abstract theory.

Matrix notation. Standard notation for matrices will be used, namely bold uppercase letters in the general case, bold lowercase letters for column–matrices, and *prime symbol* ($'$) for transposition (*MATH.BASES*, §10.1, p.420).

• $\mathbf{N}$ denotes the $J \times K$ matrix with general term n_{jk}; $\mathbf{R}$ ("Row" margin) the J–diagonal matrix with diagonal terms $(n_j)_{j\in J}$; $\mathbf{C}$ ("Column" margin) the K–diagonal matrix with diagonal terms $(n_k)_{k\in K}$;

• $\mathbf{F} = \frac{1}{n}\mathbf{N}$ denotes the $J \times K$ matrix with general term $f_{jk} = n_{jk}/n$; $\mathbf{F}_J$ the J–diagonal matrix, $\mathbf{f}_J$ the J–column with terms $(f_j = n_j/n)_{j\in J}$; $\mathbf{F}_K$ the K–diagonal matrix, $\mathbf{f}_K$ the K–column with terms $(f_k = n_k/n)_{k\in K}$.

• $\mathbf{e}_K$ denotes the K–column of 1, and $\mathbf{e}_J$ the J–column of 1.

• $\mathbf{F}_J^{-1}$ denotes the matrix of the isomorphism $\mathbb{R}_J \to \mathbb{R}^J$ and $\mathbf{F}_K^{-1}$ the matrix of the isomorphism $\mathbb{R}_K \to \mathbb{R}^K$.

Retracing the main steps previously developed in this chapter, we now derive the matrix formulations.

First step: From canonical bases to orthonormal bases. In the canonical bases $(\delta_J^j)_{j\in J}$ of $\mathbb{R}_J$ and $(\delta_K^k)_{k\in K}$ of $\mathbb{R}_K$, the $J \times K$ matrix of the transition f_J^K is $\mathbf{T} = \mathbf{N}\mathbf{C}^{-1} = \mathbf{F}\mathbf{F}_K^{-1}$, and the $K \times J$ matrix of the transition f_K^J is $\mathbf{T}^\star = \mathbf{N}'\mathbf{R}^{-1} = \mathbf{F}'\mathbf{F}_J^{-1}$ (Definition 2.1, p.35).

From the canonical bases, we get the orthonormal bases $(\sqrt{f_j}\,\delta_J^j)_{j\in J}$ and $(\sqrt{f_k}\,\delta_K^k)_{k\in K}$. One goes from coordinates in the canonical basis to coordinates in the orthonormal basis by means of diagonal matrices $\mathbf{F}_J^{-1/2}$ and $\mathbf{F}_K^{-1/2}$. With these orthonormal bases, the $J \times K$ matrix of the transition f_J^K is denoted $\mathbf{M}$, with: $\mathbf{M} = \mathbf{F}_J^{-1/2}\mathbf{T}\,(\mathbf{F}_K^{-1/2})^{-1} = \mathbf{F}_J^{-1/2}\mathbf{F}\,\mathbf{F}_K^{-1/2} = \mathbf{R}^{-1/2}\mathbf{N}\,\mathbf{C}^{-1/2}$. The $K \times J$ matrix of transition f_K^J is equal to $\mathbf{M}'$, since: $\mathbf{F}_K^{-1/2}\mathbf{T}^\star(\mathbf{F}_J^{-1/2})^{-1} = \mathbf{F}_K^{-1/2}\mathbf{F}'\,\mathbf{F}_J^{-1/2} = \mathbf{C}^{-1/2}\mathbf{N}'\mathbf{R}^{-1/2}$.

Second step: From orthonormal bases to principal bases. In the orthonormal bases, let the J–column $\mathbf{a}$ represent any principal contrasts of norm 1 on J and the K–column $\mathbf{b}$ represent any principal contrasts of norm 1 on K; then, the ***system of principal equations*** (2.6) (p.36) reads:

$$\begin{cases} \mathbf{M}\,\mathbf{b} & = \xi\,\mathbf{a} \\ \mathbf{M}'\mathbf{a} & = \xi\,\mathbf{b} \end{cases} \quad \text{with } \mathbf{a}'\mathbf{a} = 1 = \mathbf{b}'\mathbf{b} \tag{2.10}$$

A trivial solution of the system is $(\mathbf{F}_J^{1/2}\mathbf{e}_J, \mathbf{F}_K^{1/2}\mathbf{e}_K, 1)$. As a solution of principal equations, the singular value decomposition (SVD) of matrix $\mathbf{M}$ yields — beside the trivial solution — the three matrices $\mathbf{\Xi}$ (L–diagonal of singular values $(\xi_\ell)_{\ell=1,\dots L}$), $\mathbf{A}$ $(J \times L)$ and $\mathbf{B}$ $(K \times L)$ such that:

$$\begin{cases} \mathbf{M} & = \mathbf{F}_J^{1/2}\mathbf{e}_J(\mathbf{F}_K^{1/2}\mathbf{e}_K)' + \sum_{\ell=1}^{L} \xi_\ell \mathbf{a}_\ell \mathbf{b}'_\ell \\ & = \mathbf{F}_J^{1/2}\mathbf{e}_J\mathbf{e}'_K\mathbf{F}_K^{1/2} + \mathbf{A}\mathbf{\Xi}\mathbf{B}' \end{cases} \quad \text{with } \mathbf{A}'\mathbf{A} = \mathbf{I}_L = \mathbf{B}'\mathbf{B}. \tag{2.11}$$

Notice that, in terms of symmetric matrices $\mathbf{S} = \mathbf{M}'\mathbf{M}$ with general term $s_{kk'} = (1/\sqrt{n_k n_{k'}}) \sum_{j \in J} n_{jk}n_{jk'}/n_j$ and $\mathbf{W} = \mathbf{M}\mathbf{M}'$ with general term $w_{jj'} = (1/\sqrt{n_j n_{j'}}) \sum_{k \in K} n_{jk}n_{j'k}/n_k$, we have:

$$\begin{cases} \mathbf{S}\mathbf{b} = \mathbf{M}'\mathbf{M}\,\mathbf{b} & = \xi^2\,\mathbf{b} \\ \mathbf{W}\mathbf{a} = \mathbf{M}\mathbf{M}'\,\mathbf{a} & = \xi^2\,\mathbf{a} \end{cases} \quad \text{with } \mathbf{a}'\mathbf{a} = 1 = \mathbf{b}'\mathbf{b} \tag{2.12}$$

Third Step: Principal contrasts and principal variables. In the canonical bases, let z_J and z_K be any principal contrasts of norm 1, represented by columns $\mathbf{z}_J$ and $\mathbf{z}_K$; then we have:

$$\mathbf{z}_J = \mathbf{F}_J^{1/2}\mathbf{a} \quad \text{and} \quad \mathbf{z}_K = \mathbf{F}_K^{1/2}\mathbf{b} \tag{2.13}$$

with $\mathbf{e}'_K\mathbf{z}_K = 0 = \mathbf{e}'_J\mathbf{z}_J$ and $\mathbf{z}'_K\mathbf{F}_K^{-1}\mathbf{z}_K = 1 = \mathbf{z}'_J\mathbf{F}_J^{-1}\mathbf{z}_J$

In canonical bases, the matrix expression of Equations (2.6) for principal contrasts (Theorem 2.5, p.36) is the following:

$$\begin{cases} \mathbf{T}\,\mathbf{z}_K & = \xi\,\mathbf{z}_J \\ \mathbf{T}^\star\mathbf{z}_J & = \xi\,\mathbf{z}_K \end{cases} \quad \text{with } \mathbf{z}'_K\mathbf{F}_K^{-1}\mathbf{z}_K = 1 = \mathbf{z}'_J\mathbf{F}_J^{-1}\mathbf{z}_J \tag{2.14}$$

By duality, we get any standard principal variables z^J denoted by the J–column $\mathbf{u}$ and z^Kdenoted by the K–column $\mathbf{v}$, that is, the *principal coordinates of ideal points* (see below p.42), given by: $\mathbf{u} = \mathbf{F}_J^{-1}\mathbf{z}_J = \mathbf{F}_J^{-1/2}\mathbf{a}$ and $\mathbf{v} = \mathbf{F}_K^{-1}\mathbf{z}_K = \mathbf{F}_K^{n-1/2}\mathbf{b}$ with $\mathbf{f}'_J\mathbf{u} = 0 = \mathbf{f}'_K\mathbf{v}$ and $\mathbf{u}'\mathbf{F}_J\mathbf{u} = 1 = \mathbf{v}'\mathbf{F}_K\mathbf{v}$.

Hence the matrices $\mathbf{U} = \mathbf{F}_J^{-1/2}\mathbf{A}$ and $\mathbf{V} = \mathbf{F}_K^{-1/2}\mathbf{B}$ of the principal coordinates of ideal points. The J–column $\overline{\mathbf{u}} = \xi\mathbf{u} = \xi\mathbf{F}_J^{-1/2}\mathbf{a}$ gives the principal coordinates of profile points M^j; the K–column $\overline{\mathbf{v}} = \xi\mathbf{v} = \xi\mathbf{F}_K^{-1/2}\mathbf{b}$ the principal coordinates of profile points k (p.37). Hence the matrices $\overline{\mathbf{U}} = \mathbf{F}_J^{-1/2}\mathbf{A}\Xi$ and $\overline{\mathbf{V}} = \mathbf{F}_K^{-1/2}\mathbf{B}\Xi$ of the *scaled principal coordinates of profile points*, that is, the principal coordinates of points (see below p.42).

Transition formulas. (cf. Formulas (2.8), p.38)

$$\begin{cases} \mathbf{T}^{*\prime}\mathbf{V} & = \xi\,\mathbf{U} \\ \mathbf{T}'\mathbf{U} & = \xi\,\mathbf{V} \end{cases} \quad \text{with } \mathbf{V}'\mathbf{F}_K\mathbf{V} = \mathbf{I} = \mathbf{U}'\mathbf{F}_J\mathbf{U} \tag{2.15}$$

Reconstitution formulas. (cf. p.38)
$\mathbf{F} = \mathbf{f}_J\mathbf{f}'_K + \mathbf{F}_J\,\overline{\mathbf{U}}\Xi^{-1}\overline{\mathbf{V}}'\,\mathbf{F}_K = \mathbf{f}_J\mathbf{f}'_K + \mathbf{Z}_J\Xi\mathbf{Z}'_K$ etc.

Summary of computation procedure.

- **Step 1.** Compute $\mathbf{M} = \mathbf{R}^{-1/2}\mathbf{N}\,\mathbf{C}^{-1/2}$.
- **Step 2.** Proceed to the SVD of $\mathbf{M}$, or of $\mathbf{M}_0 = \mathbf{M} - \frac{1}{n}\mathbf{R}^{1/2}\mathbf{e}_J\,\mathbf{e}'_K\mathbf{C}^{1/2}$, that is, $\mathbf{M}_0 = \mathbf{A}\Xi\mathbf{B}'$ with Ξ (diagonal of singular values), $\mathbf{A}$ and $\mathbf{B}$ (matrices of eigenvectors) such that $\mathbf{A}'\mathbf{A} = \mathbf{B}'\mathbf{B} = \mathbf{I}$.
- **Step 3.** The scaled principal coordinates are given by the columns of

$$\overline{\mathbf{U}} = \sqrt{n}\,\mathbf{R}^{-1/2}\mathbf{A}\Xi \quad \text{and} \quad \overline{\mathbf{V}} = \sqrt{n}\,\mathbf{C}^{-1/2}\mathbf{B}\Xi$$

2.3.3 Geometric Representations

At this point we assume $\forall jk \in J \times K : f_{jk} \geq 0$ (positivity assumption).

Theorem 2.7. *If $\forall jk \in J \times K : f_{jk} \geq 0$, then all singular values are less than or equal to 1.*

Proof. From transition formulas (2.8): $\forall j \in J\ :\ \xi z^j = \sum_{k\in K} f_k^j z^k \leq \max_{k\in K} z^k$ (property of average), hence: $\xi \max z^j \leq \max z^k$. Similarly, one has: $\xi \max z^k \leq \max z^j$. From these two relations, it follows that $\xi^2 \max z^j \leq \xi \max z^k \leq \max z^j$, which implies $\xi^2 \leq 1$. ◁

The *converse* of this theorem is false, as the following *counterexample* shows. Consider a 2×2 table, with $J = \{j, j'\}$ and $K = \{k, k'\}$, letting j and k be such that $f_j \leq 1/2$ and $f_k \leq 1/2$. If f_{jk} is such that $f_j f_k - \sqrt{f_j(1-f_j)f_k(1-f_k)} < f_{jk} < 0$, then $\xi < 1$. It is only when $f_{jk} < f_j f_k -$

$\sqrt{f_j(1-f_j)f_k(1-f_k)}$ that $\xi > 1$. For further discussion, see the positivity issue in §2.4. The positivity property entails the statistical interpretation of the singular values in terms of correlation coefficients (§2.3.6).

Principal coordinates of points. The principal geometric representation of the $\mathcal{J}$–cloud consists in referring the points $(\mathrm{M}^j = \sum_{k\in K} f_k^j \mathrm{I}^k)_{j\in J}$ to a principal orthonormal Cartesian frame, that is, having mean point G as origin–point and principal normal contrasts $(z_{K\ell})_{\ell=1,\dots L}$ as a system of orthonormal vectors.

The *principal coordinate of the profile point* M^j on the principal axis generated by $z_{K\ell}$ is the value y_ℓ^j of the scaled principal variable y_ℓ^J:

$$\langle f_K^j - f_K | z_{K\ell}\rangle = \sum_{k\in K} (f_k^j - f_k) z_{k\ell}/f_k = \sum_{k\in K} f_k^j z_\ell^k = y_\ell^j \quad \text{(transition formula)}$$

The *principal coordinate of the ideal point* I^k on this principal axis is z_ℓ^k:

$$\langle \delta_K^k | z_{K\ell}\rangle = z_{k\ell}/f_k = z_\ell^k$$

Similarly the ℓ–th coordinate of the profile point M^k is equal to y_ℓ^k, and that of the ideal point I^j is equal to z_ℓ^j.

Barycentric diagrams. In practice, only the first L' principal axes (with $L' < L$) are retained, and the principal geometric representation of each cloud is restricted to the corresponding subspace. When the projected cloud of the points $(\mathrm{M}^j)_{j\in J}$ onto the L'- -principal subspace is represented together with the projected cloud of the ideal points $(\mathrm{I}^k)_{k\in K}$, we get a *barycentric diagram* (Benzécri, 1973, Vol 2, p.41; Rouanet, Le Roux, 1993, p.226). And similarly for the cloud $(\mathrm{M}^k)_{k\in K}$ (cf. Figure 2.5).

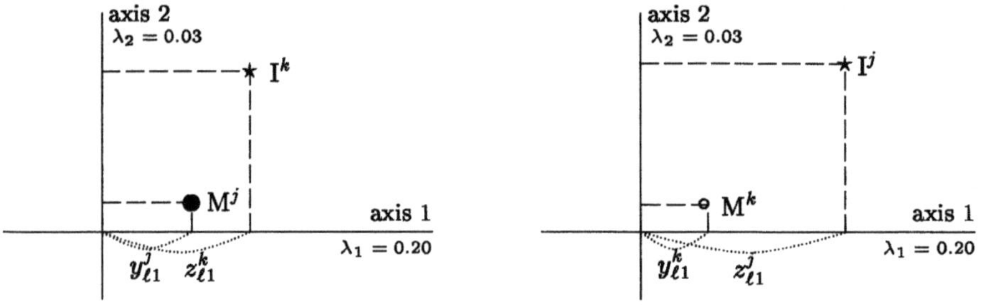

Figure 2.5. Ideal and profile points (up to projections) in plane 1-2.

The two barycentric representations are *dissymmetrical representations* involving both J and K. They are genuine geometric representations in the sense that the distances between the point M^j and the ideal point I^k, and between the point M^k and the ideal point I^j can be read on the diagram up to the distorsions due to projection, especially for ideal points.

Joint representations. Building *joint representations* where J and K play *symmetric* roles was the main motivation that led to developing CA. The fundamental joint representation will be obtained by embedding $\mathbb{R}_J$ and $\mathbb{R}_K$ in the space $\mathbb{R}_{JK}$ of the measures over $J \times K$ with its associated basis $(\delta_{JK}^{jk})_{jk \in J \times K}$.

With the projection $J \times K \to J$ there is associated the summation over K that with each measure $u_{JK} \in \mathbb{R}_{JK}$ associates the measure $u_J \in \mathbb{R}_J$, with $u_j = \sum_{k \in K} u_{jk}$. This homomorphism is denoted δ_J^{JK}.

$$\begin{array}{rccl} \delta_J^{JK}: & \mathbb{R}_{JK} & \longrightarrow & \mathbb{R}_J \\ & u_{JK} & \longmapsto & u_J = (\sum_{k \in K} u_{jk})_{j \in J} \end{array}$$

The homomorphism $i_{JK}^J : \mathbb{R}_J \to \mathbb{R}_{JK}$ is defined by:

$$\begin{array}{rccl} i_{JK}^J: & \delta_J^j & \longmapsto & \delta_{JK}^j = \sum_{k \in K} f_k^j \delta_{JK}^{jk} \\ & u_J & \longmapsto & u_{JK} = (u_j f_k^j)_{jk \in J \times K} \end{array}$$

Its subspace image, denoted $\mathbb{R}_{JK}^J$, admits the basis $(\delta_{JK}^j = \sum_{k \in K} f_k^j \delta_{JK}^{jk})_{j \in J}$. A similar construction applies for the Cartesian projection $J \times K \to K$.

Let $S = \{(j,k) : j \in J, k \in K, f_{jk} > 0\}$. The subspace $\mathbb{R}_S \subseteq \mathbb{R}_{JK}$ is then equipped with the scalar product defined by

$$\forall u_{JK} \in \mathbb{R}_S \ \forall v_{JK} \in \mathbb{R}_S : \ \langle u_{JK} | v_{JK} \rangle = \sum_{jk \in S} u_{jk} v_{jk} / f_{jk}$$

with the orthogonal basis $(\delta_S^{jk})_{jk \in S}$ ($\|\delta_S^{jk}\| = 1/f_{jk}$). The subspace $\mathbb{R}_S^J = i_S^J(\mathbb{R}_J)$ is spanned by $(\delta_S^j - \sum_{k \in K_j} f_k^j \delta_S^{jk})_{j \in J}$ with $K_j = \{k \in K : jk \in S\}$.

The subspace $\mathbb{R}_S^J$ is isomorphic to $\mathbb{R}_J$ (for the Euclidean structure).

Proof. $\langle \delta_S^j | \delta_S^{j'} \rangle = \langle \sum_{k \in K_j} f_k^j \delta_S^{jk} | \sum_{k \in K_{j'}} f_k^{j'} \delta_S^{j'k} \rangle = \sum_{k \in K_j} \sum_{k' \in K_{j'}} f_k^j f_{k'}^{j'} \langle \delta_S^{jk} | \delta_S^{j'k'} \rangle$

$$= \begin{cases} 0 & \text{if } j \neq j' \\ \sum_{k \in K_j} (f_k^j)^2 / f_{jk} = 1/f_j & \text{if } j = j' \end{cases} \quad \triangleleft$$

Theorem 2.8. *Given a measure* $u_K \in \mathbb{R}_K$ *and* $f_J^K(u_K)$ *its image by the transition* $f_J^K : \mathbb{R}_K \to \mathbb{R}_J$, *the orthogonal projection of* $i_S^K(u_K)$ *onto* $\mathbb{R}_S^J$ *is equal to* $i_S^J(f_J^K(u_K))$. *Similarly, the orthogonal projection of* $i_S^J(u_J)$ *onto* $\mathbb{R}_S^K$ *is equal to* $i_S^K(f_K^J(u_J))$.

Proof. $i_S^J(f_J^K(u_K))$ is the orthogonal projection of $i_S^K(u_K)$ on $\mathbb{R}_S^J$ if and only if $\forall j \in J \; : \; \langle i_S^J(f_J^K(u_K))|\delta_S^j\rangle = \langle i_S^K(u_K)|\delta_S^j\rangle$. One has:

$$\begin{aligned}\langle i_S^K(u_K)|\delta_S^j\rangle &= \sum_{j'k\in S}(u_k f_{j'}^k)\delta_{j'k}^j/f_{j'k} = \sum_{j'k\in S} u_k\delta_{j'k}^j/f_k \\ &= \sum_{k\in K_j} u_k f_k^j/f_k = \sum_{k\in K} u_k f_{jk}/(f_j f_k)\end{aligned}$$

and owing to the isomorphism between $\mathbb{R}^J$ and $\mathbb{R}_S^J$:

$$\langle i_S^J(f_J^K(u_K))|\delta_S^j\rangle = \langle f_J^K(u_K)|\delta_J^j\rangle = \sum_{k\in K} f_j^k u_k/f_j = \sum_{k\in K} u_k f_{jk}/(f_j f_k) \quad \triangleleft$$

Theorem 2.9. *Let* $z_{K\ell}$ *and* $z_{J\ell}$ *be the* ℓ*–th principal contrasts of norm 1 of the two clouds associated to the eigenvalue* $\lambda_\ell = \xi_\ell^2$; *then the angle* θ_ℓ *between the* ℓ*-th principal axes directed by* $i_S^J(z_{J\ell})$ *and* $i_S^K(z_{K\ell})$ *is such that* $\cos\theta_\ell = \xi_\ell$.

Proof. From Theorem 2.8, $i_S^J(f_J^K(z_{K\ell}))$ is the orthogonal projection of $i_S^K(z_{K\ell})$ onto $\mathbb{R}_S^K$, with, from the transition formulas (p.38), $i_S^J(f_J^K(z_{K\ell})) = \xi_\ell i_S^J(z_{J\ell})$. From the property of orthogonal projection, one has:

$$\langle i_S^J(z_{J\ell})|i_S^K(z_{K\ell})\rangle = \langle i_S^J(z_{J\ell})|i_S^J(f_J^K(z_{K\ell}))\rangle = \xi_\ell\|z_{J\ell}\|^2 = \xi_\ell \quad \triangleleft$$

This theorem provides the *geometrical interpretation of singular values.*

— *Remark.* Let $\mathcal{L}_J$ be the subspace of $\mathbb{R}_J$ spanned by the L unit–norm principal contrasts $(z_{K\ell})_{\ell=1,\dots L}$ and $\mathcal{L}_K$ the subspace of $\mathbb{R}_K$ spanned by the L unit–norm principal contrasts $(z_{J\ell})_{\ell=1,\dots L}$.

Let $\mathcal{L}_S^J = i_S^J(\mathcal{L}_J)$ and $\mathcal{L}_S^K = i_S^K(\mathcal{L}_K)$. The subspace $\mathcal{L}_S^J + \mathcal{L}_S^K \subseteq \mathbb{R}_S$ of the contrasts $a i_S^J(u_J) + b v_S^K(v_K)$ with $u_J \in \mathcal{L}_J$ and $v_K \in \mathcal{L}_K$ is $2L$–dimensional. Let us consider the bilinear form φ defined as follows.
$\forall u_S = i_S^J(u_J)$, $\forall u'_S = i_S^J(u'_J) \in \mathcal{L}_S^J$ and $\forall v_S = i_S^K(v_K)$, $\forall v'_S = i_S^K(v'_J) \in \mathcal{L}_S^K$, one has:

1. $\varphi(u_S, u'_S) = \langle u_J|u'_J\rangle = \sum_{j\in J} u_j\, u'_j/f_j = \sum_{j\in J} u^j\, u'_j$
2. $\varphi(v_S, v'_S) = \langle v_K|v'_K\rangle = \sum_{k\in K} v_k v'_k/f_k = \sum_{k\in K} v^k\, v'_k$
3. $$\begin{aligned}\varphi(u_S, v_S) &= \langle u_J|f_J^K(v_K)\rangle = \langle f_K^J(u_J)|v_K\rangle \\ &= \sum_{j\in J}\sum_{k\in K} f_{jk}u_j v_k/(f_j f_k) = \sum_{j\in J}\sum_{k\in K} f_{jk}u^j\, v^k\end{aligned}$$

Property 2.1. *For $1 \leq \ell \leq L$, one has $0 \leq \xi_\ell \leq 1$ and the bilinear form φ defines a scalar product on $\mathcal{L}_S^J + \mathcal{L}_S^K$.*

Proof. From the reconstitution formula $f_{jk}/(f_j f_k) = 1 + \sum_{\ell=1}^{L} \xi_\ell z_\ell^j z_\ell^k$, one deduces: $\varphi(u_S, v_S) = \sum_{\ell=1}^{L} \xi_\ell \sum_{j\in J} u_j z_\ell^j \sum_{k\in K} v_k z_\ell^k = \sum_{\ell=1}^{L} \xi_\ell u_\ell v_\ell$
with $u_\ell = \langle u_J | z_{J\ell}\rangle = \sum_{\ell=1}^{L} u_j z_\ell^j$ and $v_\ell = \langle v_K | z_{K\ell}\rangle = \sum_{\ell=1}^{L} v_k z_\ell^k$.

Any vector of $\mathcal{L}_S^J + \mathcal{L}_S^K$ is of the form $au_S + bv_S$ ($a \in \mathbb{R}$, $b \in \mathbb{R}$). One has $\varphi(au_S + bv_S, au_S + bv_S) = a^2 \sum_{j\in J} u_j^2/f_j + b^2 \sum_{k\in K} u_k^2/f_k + 2ab\varphi(u_S, v_S)$.

Now $\sum_{j\in J} u_j^2/f_j = \sum_{\ell=1}^{L} u_\ell^2$ and $\sum_{k\in K} v_k^2/f_k = \sum_{\ell=1}^{L} v_\ell^2$ hence one has
$\varphi(au_S + bv_S, au_S + bv_S) = a^2 \sum_{\ell=1}^{L} u_\ell^2 + b^2 \sum_{\ell=1}^{L} v_\ell^2 + 2ab \sum_{\ell=1}^{L} \xi_\ell u_\ell v_\ell$.

Since $0 \leq \xi_\ell \leq 1$, one has: $a^2 \sum_{\ell=1}^{L} u_\ell^2 + b^2 \sum_{\ell=1}^{L} v_\ell^2 + 2ab \sum_{\ell=1}^{L} \xi_\ell u_\ell v_\ell \geq$
$a^2 \sum_{\ell=1}^{L} \xi_\ell u_\ell^2 + b^2 \sum_{\ell=1}^{L} \xi_\ell v_\ell^2 + 2ab \sum_{\ell=1}^{L} \xi_\ell u_\ell v_\ell \geq \sum_{\ell=1}^{L} (a\sqrt{\xi_\ell} u_\ell + b\sqrt{\xi_\ell} v_\ell)^2 \geq 0$ ◁

Fundamental joint representation. Property 2.1 entails the joint representation of the $\mathcal{J}$ and $\mathcal{K}$ clouds as a "bi–cloud" in $\mathbb{R}_S$, which is defined by $\left(i_S^J(f_J^k - f_J)\right)_{k\in K}$ and $\left(i_S^K(f_K^j - f_K)\right)_{j\in J}$. To characterize this representation, we will describe the two–dimensional projected bi–cloud constituted by the two unidimensional ℓ–clouds, in the plane spanned by $i_S^J(z_{J\ell})$ and $i_S^K(z_{K\ell})$. In this plane, the projected $\mathcal{J}$–cloud lies on the axis directed by $i_S^K(z_{K\ell})$, and the projected $\mathcal{K}$–cloud lies on the axis directed by $i_S^J(z_{J\ell})$. The variance of each cloud is equal to λ_ℓ and the angle θ_ℓ between the two axes is such that $\cos\theta_\ell = \xi_\ell$. Orthogonal projections of the ideal points k of the $\mathcal{J}$–cloud onto the $\mathcal{K}$–cloud axis are the profile points k of the $\mathcal{K}$–cloud, and similarly the projections of the ideal points j of the $\mathcal{K}$–cloud are the profile points j of the $\mathcal{J}$–cloud (cf. Figure 2.6, p.46).

The fundamental joint representation is a genuine joint representation, since it provides a true distance between profile points j and k; the distance in the ℓ–plane between two profile points j and k is given by $d_\ell^2(j,k) = (y_\ell^j)^2 + (y_\ell^k)^2 - 2\xi_\ell y_\ell^j y_\ell^k = (y_\ell^j - y_\ell^k)^2 + 2\,(1-\xi_\ell) y_\ell^j y_\ell^k$. It is easily proved that the principal axes of the bi–cloud in the plane are the bissectors of the two ℓ-th principal axes with eigenvalues $\lambda_\ell \cos^2\theta_\ell/2$ and $\lambda_\ell \sin^2\theta_\ell/2$.

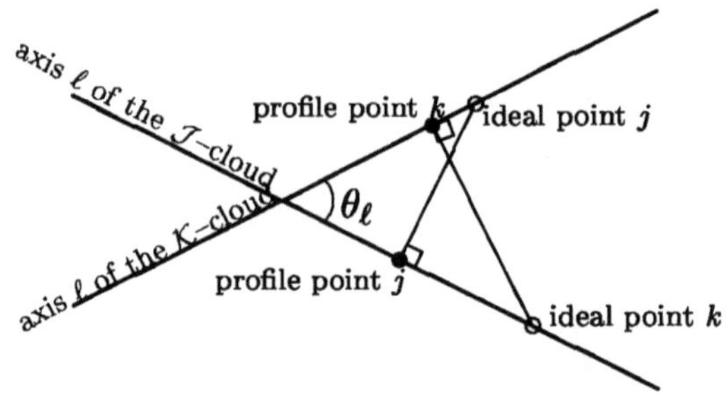

Figure 2.6. Bi–cloud in the ℓ–plane of $\mathbb{R}_S$.

MCA representation. An alternative joint representation is obtained by projecting ideal points on the two bissectors (Figure 2.7). As a result, on each bissector, one has a cloud of $J + K$ points whose variance is equal to $\cos^2 \theta_\ell/2 = (1 + \xi_\ell)/2$ for the first bissector cloud and $\sin^2 \theta_\ell/2 = (1 - \xi_\ell)/2$ for the second bissector cloud. This alternative joint representation coincides with one that is obtained when applying multiple correspondence analysis to a table with two questions J and K (see chapter *MULTIPLE*, exercise 5.3, p.243).

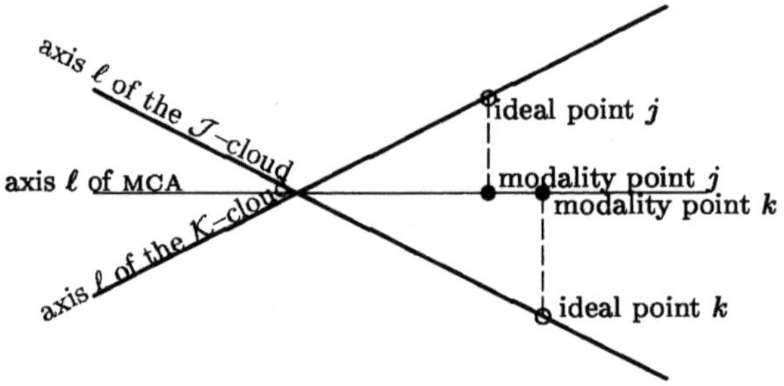

Figure 2.7. MCA representation in the ℓ–plane of $\mathbb{R}_S$.

Simultaneous representation. It is obtained from the fundamental joint representation by replacing, for each axis ℓ, one of the two clouds by its symmetric image with respect to the first bissector of axes, so as to bring the two ℓ–axes into coincidence. In simultaneous representation, each profile point j is the barycenter of profile points k, up to a product of affinities. The profile points j and k lie in different spaces that are simply brought into coincidence, the simultaneous representation is not a genuine joint representation, the apparent distance $|y_\ell^j - y_\ell^k|$ between points j and k on axis ℓ is not a genuine distance. In CA literature, the simultaneous representation is the usual symmetric representation and is often justified by a direct argument, as a compromise between the two dissymmetric barycentric representations (Benzécri, 1973, Vol 2, p.41). Simultaneous representation has been a long-standing tradition in CA, since Benzécri (1969), and has certainly been a major reason for the success of CA; it is essential for the interpretation of axes.

2.3.4 Methodology of Correspondence Analysis

The study of a contingency table begins with the examination of elementary statistics: marginal frequencies, graph of attractions, mean square contingency coefficient Φ^2, etc. Then the correspondence analysis of the table enables one to bring out the structures underlying data, to reduce dimensionality and to elaborate the interpretation of axes.

Hereafter we present the *aids to interpretation*, then we propose a strategy of interpretation based on the *method of the contributions of points and deviations* (Le Roux & Rouanet, 1998).

Aids to interpretation

Contributions of axes and points to the variance of cloud. The variance of cloud, that is, the Φ^2 coefficient of the contingency table, can be additively broken down according to points and axes.

$$\begin{cases} \Phi^2 &= \sum_{j\in J} f_j(\mathrm{GM}^j)^2 = \sum_{k\in K} f_k(\mathrm{GM}^k)^2 \\ &= \sum_{\ell=1}^{L} \lambda_\ell \end{cases} \tag{2.16}$$

The part of variance accounted for by point j is, by definition, the *absolute contribution of point j to overall variance*, and is denoted $\mathrm{Cta}_j = f_j(\mathrm{GM}^j)^2$. Similarly for point k: $\mathrm{Cta}_k = f_k(\mathrm{GM}^k)^2$. Dividing absolute contributions by Φ^2 yields the *relative contributions* denoted Ctr.

The *part of the variance* of the cloud accounted for by axis ℓ is equal to the eigenvalue λ_ℓ. The *proportion of variance* $\lambda_\ell/(\sum_{\ell=1}^{L}\lambda_\ell)$ accounted for by axis ℓ is a descriptive index of the importance of axis ℓ.

Contributions of points to the variance of an axis. The variance of each axis can be broken down according to the points:

$$\lambda_\ell = \sum_{j\in J} f_j(y_\ell^j)^2 = \sum_{k\in K} f_k(y_\ell^k)^2 \tag{2.17}$$

The *absolute contribution of point j to axis ℓ* is, by definition[5], the part of the variance of axis ℓ accounted for by point j, denoted $\text{Cta}_{j\ell}$,with

$$\text{Cta}_{j\ell} = f_j(y^j_\ell)^2; \quad \text{similarly} \quad \text{Cta}_{k\ell} = f_k(y^k_\ell)^2 \tag{2.18}$$

Contributions of deviations to the variance of an axis. Given two points j and j', with weights f_j and $f_{j'}$ respectively, and coordinates y^j_ℓ and $y^{j'}_\ell$ on axis ℓ; with the deviation between j and j', one can associate the weight $1/\big((1/f_j)+(1/f_{j'})\big)$, and the absolute contribution of deviation $\text{Cta}_{(j,j')\ell} = (y^j_\ell - y^{j'}_\ell)^2 \times (f_j f_{j'})/(f_j + f_{j'})$. This notion readily extends to two disjoint classes c and c' of points, with weights f_c and $f_{c'}$ respectively, and with average coordinates y^c_ℓ and $y^{c'}_\ell$ on axis ℓ; with the deviation between the mean points of classes c and c', one associates the weight $1/\big((1/f_c)+(1/f_{c'})\big)$, and the absolute contribution

$$\text{Cta}_{(c,c')\ell} = (y^c_\ell - y^{c'}_\ell)^2 \times (f_c f_{c'})/(f_c + f_{c'}) \tag{2.19}$$

Relative contributions and qualities of representation. Starting from the absolute contributions of points (or of deviations) to axes, several relative contributions can be defined.

• Dividing the absolute contribution by the variance λ_ℓ of axis yields the *relative contribution to axis* of the point (or of the deviation):

$$\text{Ctr}^\ell_j = \text{Cta}_{j\ell}\,/\lambda_\ell \quad \text{Ctr}^\ell_k = \text{Cta}_{k\ell}\,/\lambda_\ell \quad \text{Ctr}^\ell_{(c,c')} = \text{Cta}_{(c,c')\ell}\,/\lambda_\ell \tag{2.20}$$

• Dividing the absolute contribution by the overall absolute contribution of the point yields the *relative contribution* of axis ℓ to the point:

$$\text{Ctr}^j_\ell = \text{Cta}_{j\ell}\,/\,\text{Cta}_j \quad \text{Ctr}^k_\ell = \text{Cta}_{k\ell}\,/\,\text{Cta}_k \tag{2.21}$$

The contribution of axis ℓ to point j is the square of the cosine of the (acute) angle $\theta_{j\ell}$ between segment GM^j and axis ℓ:

$$\text{Ctr}^j_\ell = (\text{GA}^j_\ell)^2/(\text{GM}^j)^2$$

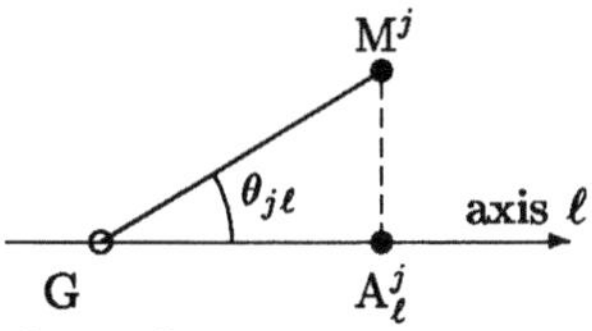

As such, this contribution characterizes the *quality of representation* of point j by axis ℓ and is denoted $\cos^2$. The relative contribution of point j to the subspace of the first L' axes, and its quality of representation by this subspace, can be written respectively:

$$\text{Ctr}^{L'}_j = \sum_{\ell=1}^{L'} \text{Cta}_{j\ell}\,/\sum_{\ell=1}^{L'} \lambda_\ell \qquad \text{Ctr}^j_{L'} = \sum_{\ell=1}^{L'} \cos^2\theta_{j\ell} = \sum_{\ell=1}^{L'} \text{Ctr}^j_\ell$$

[5]Cf. Benzécri (1992, p.340); Greenacre (1984, p.67).

Supplementary elements. Additional modalities defined by their profiles (on J or on K) can be put as *supplementary elements*; they are simply projected on principal axes or planes, without participating to their determination. Let j_s be a supplementary element added to J, with profile $f_K^{j_s} = (n_{j_sk}/n_{j_s})_{k\in K}$ ($n_{j_s} = \sum_{k\in K} n_{j_sk}$); it is represented by the point M^{j_s} such that $\mathrm{M}^{j_s} = \sum_{k\in K} f_k^{j_s}\mathrm{I}^k$, and its coordinate on axis ℓ is $y_\ell^{j_s} = \sum_{k\in K} f_k^{j_s} z_\ell^k$.

Methodology of interpretation

Overall analysis. In order to depict the overall data set, one checks that there is no modality having a predominant contribution to the variance of the cloud. If there is such a modality, a new analysis should be done putting this modality as a supplementary element. To determine the axes to be interpreted, one first considers the axes whose contributions exceed the average $\Phi^2/\min(J-1, K-1)$. Then one searches for a subset such that the proportion of variance accounted for is elevated enough, keeping in mind the difference between successive eigenvalues, preferably putting the cutoff point between well separated eigenvalues (the "elbow criterion"). Also most modalities should be reasonably well represented (a quality of representation greater than .5 as a rule of thumb) by this subspace.

Interpretation of axes. The interpretation of axes is an indispensable step of any geometric data analysis. One distinguishes *general axes*, that involve the whole of data and express oppositions, and *specific axes*, usually of remote ranks, that characterize particular points or subgroups of points. As said in §2.3.1, the axes of a correspondence analysis are defined by *contrasts*. As Benzécri (1992, p. 405) puts it: "Interpreting an axis amounts to finding out what is similar, on the one hand, between all the elements figuring on the right of the origin and, on the other hand between all that is written on the left; and expressing with conciseness and precision, the contrast (or opposition) between the two extremes."

In order to reach such an interpretation, we suggest the following strategy: (i) Look for points or groups of points whose contributions to the variance of axis are important (e.g. greater than the average contribution) and that account together for the major part of variance (c.g. 80%); to those points, add those that are well represented on the axis (to interpret an axis one cannot ignore a point located on the axis, even if its contribution is less than average). (ii) Divide the set of preceding points into two groups: those with positive coordinates as opposed to those with negative ones; calculate the contribution of the deviation between the mean points of the two groups, so as to appreciate the quality of the conclusion of opposition between the two groups. (iii) Make the study for both sets J and K, to assess the convergence and/or complementarity of conclusions; proceed to the simultaneous representation on the axis of the points retained for interpretation. For an illustration of this strategy, see e.g. Exercise 2.2.

When a modality is well represented by an axis with a predominant contribution, the axis is said to be *specific* of this modality. It may be appropriate to put the modality as a supplementary element so as to improve the interpretation of subsequent axes[6]. An axis is specific of a *group of modalities*, if there is a cluster of modalities on one side of the axis, with an important overall contribution and with qualities of representation on the axis markedly superior to those of the other points.

Proximities between points. When points of the *same cloud* are well represented in a subspace, one can evaluate approximately their distances from those of their projections. Detecting modalities that are close to each other is important in CA (and more generally in GDA), and enables one to proceed to grouping of modalities.

In the *barycentric representation* of the cloud M^j, the distances between profile points j and ideal points k are genuine distances in the whole space $\mathbb{R}_K$. But in projection in the first principal planes, ideal points are usually (for large size tables) poorly represented, so the apparent proximities will often not be useful. On the other hand, the barycentric relation, which is preserved in projection, may be used for the interpretation.

Simultaneous representation: Interpreting proximities between points j and points k (which do not lie in a same space) does require caution[7], because in the simultaneous representation, the apparent distance between a j point and a k point is not a genuine distance. Still, an interpretation in terms of attraction is possible, at least for points located at the periphery of the graph. Indeed, if j and k are preferentially associated (f_j^k and f_k^j elevated), point M^j is close to the ideal point I^k and consequently to $\widehat{\mathrm{M}}^k = \sum\limits_{j \in J} f_j^k \mathrm{M}^j$ (Example for *Hair & Eye Color* $j3$ and $k3$).

[6]One can give an upper bound of the rotation angle of axes due to the deletion of an element (Escofier, Le Roux, 1976), which enables one to assess the stability of axes without having to do the analysis again.

[7]The fundamental joint representation (p.45) helps to make sense of the reservations that can be expressed about this representation. Clearly, for a very small eigenvalue λ_ℓ, that is, for an angle θ_ℓ close to $\pi/2$, the distances of the joint representation are distorted in the simultaneous representation.

Two modalities j and k that appear as close on the first principal plane may have a low association rate (e.g. $j2$ and $k1$), and this for two reasons:

1. The location of j is determined not only by f_k^j but by the whole set $(f_k^j)_{k\in K}$, therefore point j may appear close to point k because it is strongly associated not to k itself but to its neighbors;

2. In the reconstitution formula $t^{jk} = \sum_{\ell=1}^{L} y_\ell^j y_\ell^k / \xi_\ell$, the positive contribution to t^{jk} brought by the positive terms $y_\ell^j y_\ell^k$ pertaining to the axes under study can be counterbalanced by negative contributions on other axes.

Other types of joint representations have been suggested in CA.

• Plotting $\sqrt{\xi_\ell}\, z_\ell^j$ and $\sqrt{\xi_\ell}\, z_\ell^k$ may be used to investigate in projection the sign of the association between (j,k) from the angle between the segments joining the origin of axes to the points j and k; yet this representation, which moves away from the basic geometric distance, cannot be recommended as a routine representation[8].

• When the sets J and K refer to comparable variables, MCA provides a genuine distance between points j and k (See chapter *MULTIPLE*).

2.3.5 The "Hair & Eye Color" Example

As an illustration we analyze in detail the data set *Hair and Eye Color*[9] briefly presented in chapter 1 (*OVERVIEW*, §1.1, p.2).

Data set and elementary statistics

The contingency table 2.2 (p.52) concerns 5387 Scottish children (Caithness county) cross–classified according to *Hair Color* $J = \{j1, j2, j3, j4, j5\}$ and *Eye Color* $K = \{k1, k2, k3, k4\}$[10].

The *association rates* (formula (2.4), p.32) between all modality pairs are given in Table 2.3 (for example between $j1$ and $k1$: $(326\times 5387)/(1455\times 718) - 1 = +0.681$). The *graph of attractions* is obtained by joining modalities with positive association rates[11]. We observe that *Light* and *Blue* eyes go with *Fair* and *Red* hair; *Dark* and *Black* eyes go with *Dark* hair.

[8]Some other graphs occasionally encountered are hardly worth mentioning, for instance plotting the column vectors **a** and **b** of the SVD of matrix **M**; such graphs, which amount to plot merely technical intermediate results in the geometric construction, have simply nothing to do with Correspondence Analysis.

[9]This example, presented to illustrate the procedure and the strategy of data analysis, is not meant to be typical of the large size tables usually analyzed by CA.

[10]Fisher (1940), Maung (1941) .

[11]For large size tables, only modalities with large positive association rates will be joined, using an appropriate cut point.

			Eye colors				
		Blue	*Light*	*Medium*	*Dark*	
		$k1$	$k2$	$k3$	$k4$	
	(Fair) $j1$	326	688	343	98	1455
	(Red) $j2$	38	116	84	48	286
Hair colors	(Medium) $j3$	241	584	909	403	2137
	(Dark) $j4$	110	188	412	681	1391
	(Black) $j5$	3	4	26	85	118
		718	1580	1774	1315	5387

Mean square contingency coefficient: $\Phi^2 = 0.2302$

Table 2.2. Joint and marginal absolute frequencies.

	$k1$	$k2$	$k3$	$k4$
$j1$	0.681	0.612	−0.284	−0.724
$j2$	−0.003	0.383	−0.108	−0.312
$j3$	−0.154	−0.068	0.292	−0.227
$j4$	−0.407	−0.539	−0.101	1.006
$j5$	−0.809	−0.884	−0.331	1.951
	Blue	*Light*	*Medium*	*Dark*

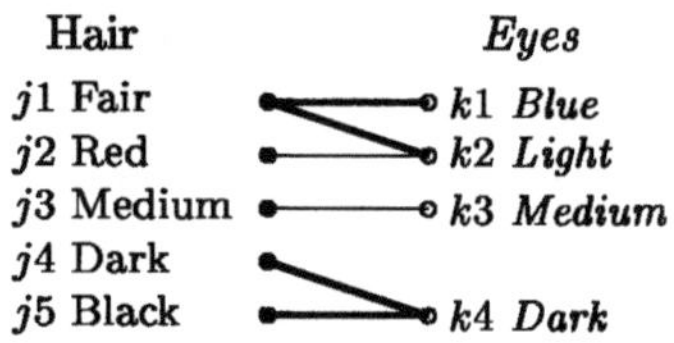

Table 2.3. Association rates and graph of attractions.

The calculation of the *conditional frequencies* of k if j (for instance, the frequency of $k3$ if $j1$ is $f^{j1}_{k3} = 343/1455 = 0.236$) and of the conditional frequencies of j if k, yield the two *transition tables*; cf. Table 2.4.

	$k1$	$k2$	$k3$	$k4$	
$j1$	.224	.473	.236	.067	[1]
$j2$	.133	.406	.294	.168	[1]
$j3$	.113	.273	.425	.189	[1]
$j4$	.079	.135	.296	.490	[1]
$j5$	.025	.034	.220	.720	[1]
	.133	.293	.329	.244	

Table 2.4. Transition from J to K.

	$k1$	$k2$	$k3$	$k4$	
$j1$	.454	.435	.193	.075	.270
$j2$	.053	.073	.047	.037	.053
$j3$	.336	.370	.512	.306	.397
$j4$	.153	.119	.232	.518	.258
$j5$	.004	.003	.015	.065	.022
	[1]	[1]	[1]	[1]	

Transition from K to J.

Let us comment on these results: i) (transition from J to K) for *Blue* and *Light eyes* (columns $k1$ and $k2$), conditional frequencies decrease from $j1$ to $j5$; whereas for *Dark eyes* (column $k4$), they increase; ii) (transition from K to J) for *Fair hair* (row $j1$), the conditional frequencies decrease from $k1$ to $k4$, whereas about the reverse holds for *Black hair* (row $j5$); the profiles of *Blue* and *Light eyes* (columns $k1$ and $k2$) are very close.

From formula $(\mathrm{GM}^j)^2 = \frac{n}{n_j^2}(\sum_{k\in K} \frac{n_{jk}^2}{n_k}) - 1$ one calculates the *distances* between profile points and mean point G (cf. Table 2.5): for example, $(\mathrm{GM}^{j2})^2 = \frac{5387}{286^2}\left(\frac{38^2}{718} + \cdots + \frac{48^2}{1315}\right) - 1 = 0.0707 = 0.266^2$. One verifies that the variance of each cloud $\sum_{j\in J} f_j(\mathrm{GM}^j)^2 = \sum_{k\in K} f_k(\mathrm{GM}^k)^2$ is equal to $0.230191 = \Phi^2$. From formula (2.18) (p.48), one deduces the absolute (Cta) and relative (Ctr) contributions of points to the variance (Table 2.6). We observe that the contribution of *Red hair* ($j2$) is very weak.

	$j1$	$j2$	$j3$	$j4$	$j5$
GM^j	0.571	0.266	0.213	0.598	1.132
f_j	.270	.053	.397	.258	.022

	$k1$	$k2$	$k3$	$k4$
GM^k	0.438	0.451	0.247	0.715
f_k	.133	.293	.329	.244

Table 2.5. Distances between mean point and profile points.

	$j1$	$j2$	$j3$	$j4$	$j5$
Cta_j	.08813	.00375	.01792	.09231	.02808
Ctr_j	.383	.016	.078	.401	.122

	$k1$	$k2$	$k3$	$k4$
Cta_k	.02555	.05956	.02015	.12493
Ctr_k	.111	.259	.088	.543

Table 2.6. Absolute and relative contributions of points to variance.

Results of correspondence analysis

The *eigenvalues* and *principal coordinates* (basic results of CA) are obtained by computer software (Table 2.7).

$\lambda_1 = .199245$, $\lambda_2 = .030087$, $\lambda_3 = .000859$

Hair y^j_ℓ	f_j	$\ell=1$	$\ell=2$	$\ell=3$
Fair $j1$	0.270	+0.544	+0.174	+0.013
Red $j2$	0.053	+0.233	+0.048	−0.118
Medium $j3$	0.397	+0.042	−0.208	+0.003
Dark $j4$	0.258	−0.589	+0.104	+0.010
Black $j5$	0.022	−1.094	+0.286	−0.046

Eyes y^k_ℓ	f_k	$\ell=1$	$\ell=2$	$\ell=3$
Blue $k1$	0.133	+0.400	+0.165	+0.064
Light $k2$	0.293	+0.441	+0.088	−0.032
Medium $k3$	0.329	−0.034	−0.245	+0.006
Black $k4$	0.244	−0.703	+0.134	−0.004

Table 2.7. Eigenvalues and principal coordinates.

Figure 2.8 (p.54) shows the two *barycentric representations*, Figure 2.9 (p.54) shows the *simultaneous representation*. From Table 2.7, one deduces absolute contributions. For example, for $j1$ and Axis 1, $\text{Cta}_{j1\ell1} = (1455/5387) \times (0.544)^2 = .0799$ (cf. formula (2.18), p.48); hence Table 2.8. One deduces:

• the *relative contributions of points to axes* (formula (2.20), p.48). For example, one has $\text{Ctr}^{\ell1}_{j2} = .00289/.19925 = .014$ (up to rounding–off errors).
• the *qualities of representation* (formula (2.21), p.48). The quality of representation of $j2$ by Axis 1 is $\text{Ctr}^{j2}_{\ell1} = \frac{.00289}{.00375} = .770$; for $j2$ by axis 2: $\text{Ctr}^{j2}_{\ell2} = .033$, and for $j2$ by plane 1-2: $\text{Ctr}^{j2}_{\ell1\_\ell2} = \text{Ctr}^{j2}_{\ell1} + \text{Ctr}^{j2}_{\ell2} = .803$.

Interpretation

The first axis contributes to 86.6% of variance, the second axis to 13.1% and the third to 0.4%. In spite of its dominant contribution, the first axis is not sufficient to summarize the data, because *Medium hair* ($j3$) and *Medium eye* ($k3$) are poorly represented on Axis 1 (.00070/.01792 = .039 and .00037/.02015 = .018 respectively). The first two axes provide a good summary of the data on two counts; on the one hand the contribution of plane 1-2 to the variance is equal to 99.6%, on the other hand all points (except $j2$) are very well represented in the plane (qualities of representation ≥ 0.979). Therefore we will interpret the first two axes, on the basis of coordinates (Table 2.7, p.53) and contributions (Table 2.8).

$\mathrm{Cta}_{j\ell}$	$\ell=1$	$\ell=2$	$\ell=3$	Cta_j
$j1$	.07993	.00816	.00004	.08813
$j2$	.00289	.00013	.00074	.00375
$j3$	.00070	.01721	.00000	.01792
$j4$	.08949	.00279	.00003	.09231
$j5$	.02623	.00180	.00005	.02808
λ_ℓ	.19924	.03009	.00086	.23019

$\mathrm{Cta}_{k\ell}$	$\ell=1$	$\ell=2$	$\ell=3$	Cta_k
$k1$	.02136	.00365	.00055	.02555
$k2$	.05697	.00230	.00029	.05956
$k3$	.00037	.01977	.00001	.02015
$k4$	.12055	.00438	.00000	.12493
λ_ℓ	.19924	.03009	.00086	.23019

Table 2.8. Absolute contributions of points and axes.

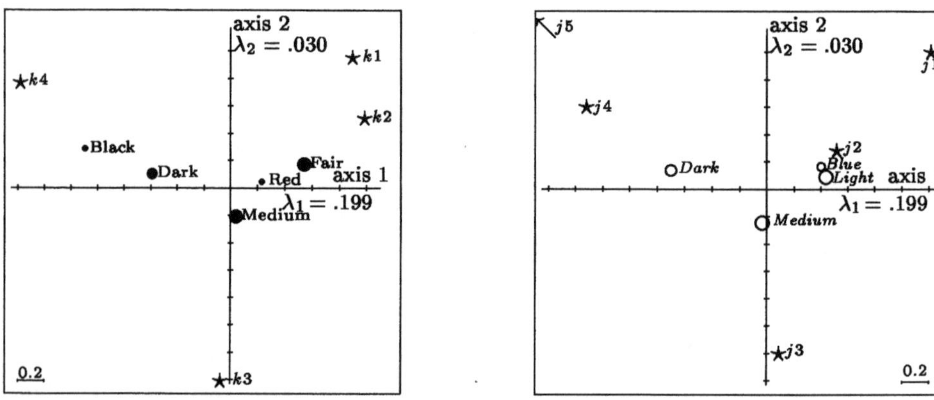

Figure 2.8. Barycentric diagrams (on these diagrams, ideal points are pictured by ⋆; "Hair color" points are pictured by •, "*Eye color*" points by ∘).

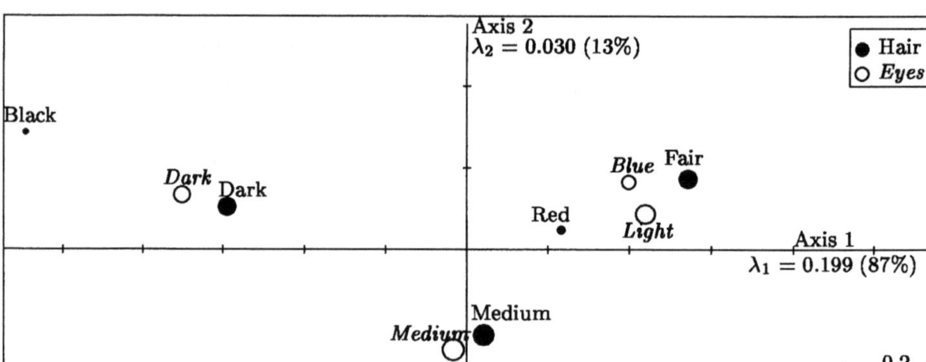

Figure 2.9. Simultaneous representation (the graphic scale is 3 times larger than the one of barycentric representations).

Interpretation of axis 1. The contributions of *Dark hair* ($j4$) and *Fair hair* ($j1$) to the variance of axis 1 are greater than average (respectively 45% and 40% > 20%). Besides, the quality of representation of *Black hair* ($j5$) is very high (0.934), therefore this modality is retained for interpretation. All together these three modalities contribute to 98% of the variance of axis 1, and the absolute contribution of the deviation between $j1$ vs the barycenter of $j4$ and $j5$ is equal to 0.189 (cf. formula (2.19)), that is, this opposition accounts for 95% of the variance of axis 1. Axis 1 opposes *Dark* ($j4$) and *Black* hair ($j5$) (on the left), to *Fair* hair ($j1$) (on the right).

One proceeds similarly for Eye colors. One retains *Dark* ($k4$) and *Light* ($k2$) eyes, to which one will add *Blue eyes* ($k1$), which on axis 1 is close to *Light eyes*, and whose quality of representation on this axis is good (.836). These three modalities contribute together to 99.8% of the variance of axis. Axis 1 opposes *Dark eyes* (on the left) to *Light* and *Blue* eyes (on the right); this opposition accounts for 99.6% of the variance of axis.

Axis 1 opposes Dark hair and eyes to Light ones.

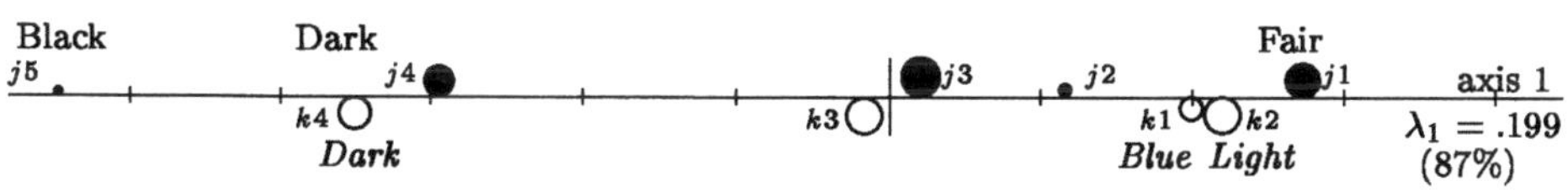

Figure 2.10. Simultaneous representation on axis 1 with the labels (in words) of the modalities used to interpret axis 1.

Interpretation of axis 2. *Medium hair* ($j3$) and *Medium eyes* ($k3$) have important contributions to axis 2 (57% and 66% respectively). In addition, these two modalities are very well represented on this axis (.96 and .98 respectively).

Axis 2 is specific of Medium hair and eyes, and it accounts for the special association between these two modalities.

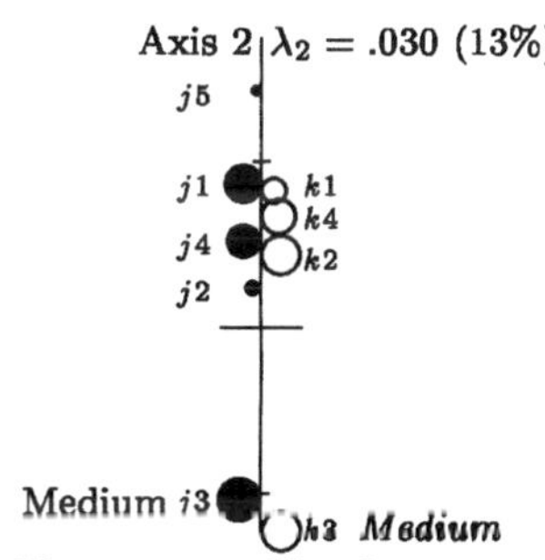

Figure 2.11: Simultaneous representation on axis 2

Study of plane 1-2. Since the points are well represented in the plane 1-2, the distances of the points of a same set can be accurately read on the plane diagram. Thus *Light eyes* ($k2$) is close to *Blue eyes* ($k1$)[12]; there is a large distance between *Black* ($j5$) and *Fair* ($j1$) hair, etc.

[12] For the modalities $k1$ and $k2$, there is quasi-distributional equivalence; if we pool together these two modalities, the plane 1-2 is stable (cf. chapter *STABILITY*, p.276).

The *simultaneous representation* suggests a pattern of three clusters corresponding to light, medium and dark colors. The proximity of *Medium hair* and *Medium eyes* is unambiguously interpreted in terms of attraction; these points are extreme on the clouds and stand apart from the others; similarly the proximity between *Dark eyes* vs *Black* and *Dark* hair reveals their attractions and again similarly for *Fair hair* vs *Blue* and *Light* eyes.

Also, the shape of clouds suggests an ordinal scale ("Guttman effect"[13]) corresponding to increasing coloration degrees for hair as well as for eyes. For eye colors, this ordering differs from that of the original contingency table (*Blue* and *Light* eyes are permuted). This ordering appears to be a property of data, since there is no ordinal assumption in CA.

Admittedly, the interpretation of the *Hair & Eye Color* table has only an illustrative value, because owing to the small table size, the conclusions brought by the analysis are not unexpected. With large size tables, the very same strategy of statistical analysis (performing CA followed by interpretation) will uncover structures (ordering, patterns, clusters, etc.) that may be unknown or only partly known before the analysis.

Table 2.9 shows the output of ADDAD software[14].

	$\mathrm{Ctr}^j_{\ell1\ell2}$	f_j	Ctr_j	$y^j_{\ell1}$	$\mathrm{Ctr}^j_{\ell1}$	$\mathrm{Ctr}^{\ell1}_j$	$y^j_{\ell2}$	$\mathrm{Ctr}^j_{\ell2}$	$\mathrm{Ctr}^{\ell2}_j$
	QLT	POID	INR	1#F	COR	CTR	2#F	COR	CTR
j1	1000	270	383	544	907	401	174	93	271
j2	803	53	16	233	770	14	48	33	4
j3	1000	397	78	42	39	4	−208	961	572
j4	999	258	401	−589	969	449	104	30	93
j5	998	22	122	−1094	934	132	286	64	60

	$\mathrm{Ctr}^k_{\ell1\ell2}$	f_k	Ctr_k	$y^k_{\ell1}$	$\mathrm{Ctr}^k_{\ell1}$	$\mathrm{Ctr}^{\ell1}_k$	$y^k_{\ell2}$	$\mathrm{Ctr}^k_{\ell2}$	$\mathrm{Ctr}^{\ell2}_k$
	QLT	POID	INR	1#F	COR	CTR	2#F	COR	CTR
k1	979	133	111	400	836	107	165	143	121
k2	995	293	259	441	956	286	88	39	76
k3	999	329	88	−34	18	2	−245	981	657
k4	1000	244	543	−703	965	605	134	35	145

• On the top of the output we recall the notation used in this book.
• QLT, quality of representation by axes 1 and 2; POID, (weight) relative frequency of point; INR, relative contribution to Φ^2 (variance of cloud);
• For each axis: ℓ#F, ℓ-th principal coordinate; COR, quality of representation; CTR, relative contribution of point to axis.

Table 2.9. Output of ADDAD software (all numbers are multiplied by 1000).

[13]Strictly speaking, there is a *Guttman effect* for axes 1 and 2 if there is a quadratic relation between the 2 principal variables $\forall j \in J : y^j_{\ell2} = a\big((y^j_{\ell1})^2 - \lambda_{\ell1}\big)$. See chapter *MULTIPLE*, p.220.

[14]A free version of the program can be obtained from the Brigitte Le Roux's homepage (see address p.335).

2.3.6 Statistical Characterizations of CA

Statistical characterizations of CA are found by seeking numerical codings of the basic sets J and K, that is, a *dual scaling* of J and K, so that an optimality criterion is satisfied. Several criteria can be chosen, that turn out to be equivalent and whose solutions are principal variables of CA[15]. Hereafter we present three optimality criteria, assuming throughout the positivity property: $\forall jk \in J \times K : f_{jk} \geq 0$. The statistical characterizations directly follow from the transition formulas and maximization properties.

Discriminant analysis

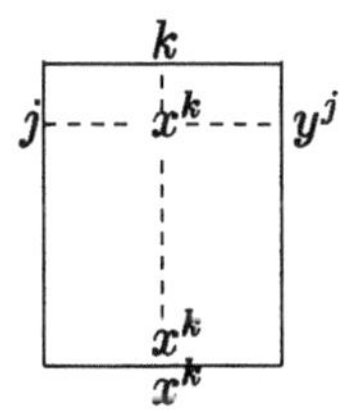

Let x^K be a variable over K with mean $\overline{x} = \sum_{k \in K} f_k x^k$. The classes $(\{j\} \times K)_{j \in J}$ constitute a partition of $J \times K$, with means $(y^j = \sum_{k \in K} x^k f_k^j)_{j \in J}$ and between–J variance $\operatorname{Var} y^J = \sum_{j \in J} f_j (y^j - \overline{x})^2$.

Problem: Find the variable x^K that best discriminates classes $(\{j\} \times K)_{j \in J}$, in the sense that the ratio $\operatorname{Var} y^J / \operatorname{Var} x^K$ (between–variance/total variance) is maximum (Fisher, 1940). That is, in Fisher's own words on the *Hair & Eye Color* example: "For what eye color scores are the five hair color classes most distinct?" The ratio $\operatorname{Var} y^J / \operatorname{Var} x^K$ is equal to $\|f_K^J(x_K^0)\| / \|x_K^0\|$ (with $x_k^0 = f_k(x^k - \overline{x})$, that is the variance of the cloud along the direction (G, x_K^0) (cf. theorem 2.4, p.36). From the theorem 2.6 (p.36), the solution is the principal variable associated with $\lambda_1 = \xi_1^2$.

If $x^J : J \to \mathbb{R}$, we denote $\check{x}^J = (x^j)_{jk \in J \times K}$ the "lifted" variable over $J \times K$ defined by its natural embedding in the space $\mathbb{R}^{JK}$ of variables over $J \times K$ weighted by f_{JK}; and similarly for $\check{x}^K = (x^k)_{jk \in J \times K}$.

Linearity of regressions

Problem: Find x^J and x^K such that the regression of $\check{x}^K$ onto $\check{x}^J$ and the regression of $\check{x}^J$ onto $\check{x}^K$ are both linear (Hirschfeld, 1935).

Let $x_0^j = x^j - \sum_{j \in J} f_j x^j$ and $x_0^k = x^k - \sum_{k \in K} f_k x^k$. The mean of $\check{x}_0^J = (x_0^j)_{jk \in J \times K}$ on $J \times \{k\}$ is $y_0^k = \sum_{j \in J} x_0^j f_j^k$; hence $y_0^K = (y_0^k)_{k \in K}$. The mean of $\check{x}_0^K$ on $\{j\} \times K$ is $y_0^j = \sum_{k \in K} x_0^k f_k^j$ hence $y_0^J = (y_0^j)_{j \in J}$. The regression of $\check{x}_0^J$

[15]There are also several ways to view CA as a principal component analysis (PCA), by constructing derived tables from the basic n_{JK} table and applying some form of biweighted PCA. These equivalences will be dealt with in chapter *PRINCIPAL*.

on $\overset{\vee}{x}{}_0^K$ is linear if and only if $\overset{\vee}{y}{}_0^K$ is proportional to $\overset{\vee}{x}{}_0^K$, therefore, if and only if $\exists\ \xi \in \mathbb{R}$ such that $\forall k \in K : \sum_{j \in J} x_0^j f_j^k = \xi x_0^k$. Similarly, the regression of $\overset{\vee}{x}{}_0^K$ on $\overset{\vee}{x}{}_0^J$ is linear if and only if $\exists \xi' \in \mathbb{R}$ such that $\forall j \in J : \sum_{k \in K} x_0^k f_k^j = \xi' x_0^j$. The two regressions are linear if and only if $\exists \xi$ and $\xi' \in \mathbb{R}$ such that $\forall k \in K : \sum_{j \in J} x_0^j f_j^k = \xi x_0^k$ and $\forall j \in J : \sum_{k \in K} x_0^k f_k^j = \xi' x_0^j$.

The solutions are therefore pairs of eigenvectors (and singular values) of homomorphisms f_J^K and f_K^J (transitions) (Definition 10.25, p.444). The solutions are L pairs of variables proportional to the standard variables $(z_\ell^K, z_\ell^J)_{\ell=1,\dots L}$. This property is known as *reciprocal averaging*.

Hair & Eye Color example. The picture shows the Cartesian diagram of standard principal variables $(\overset{\vee}{z}{}_1^K, \overset{\vee}{z}{}_1^J)$, with the two regressions (scaled principal variables) $y_1^j = \xi_1\, z_1^k$ (line $\mathcal{D}_1$) and $y_1^k = \xi_1\, z_1^j$ (line $\mathcal{D}_2$). The areas of circles are proportional to frequencies f_{jk}.

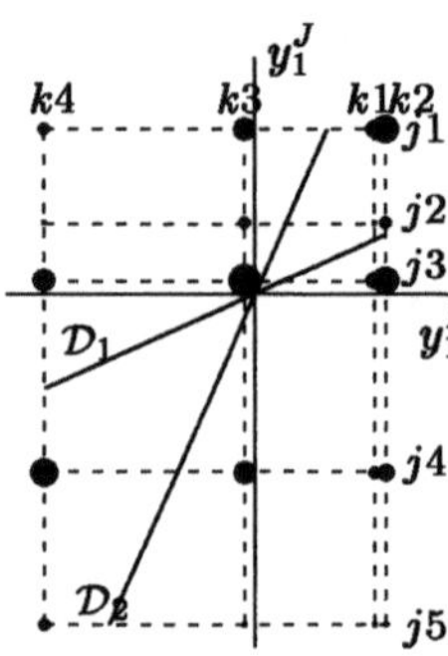

Canonical analysis

Problem: Find x^J and x^K such that the correlation between the variables $\overset{\vee}{x}{}^J$ and $\overset{\vee}{x}{}^K$ is maximum (Fisher, 1940).

For standard variables z^J and z^K, the maximum of $\text{Corr}(\overset{\vee}{z}{}^J | \overset{\vee}{z}{}^K) = \sum_{j \in J} \sum_{k \in K} f_{jk} z^j z^k$ is attained for the pair of variables $(z_{\ell 1}^K, z_{\ell 1}^J)$ associated with $\lambda_{\ell 1}$. The singular values $(\xi_\ell)_{\ell=1,2,\dots L}$ are called the *canonical correlations* of the contingency table. The reference to canonical analysis becomes clear if one goes back to the basic protocol $I \to J \times K$, where a set I of individuals is answering two questions J and K. As it is known, in the classic problem of canonical analysis, starting with two sets of numerical variables, one seeks two linear combinations of variables (one for each set) such that these two derived variables are of maximum correlation. If we denote $I{<}j{>}$ the set of individuals answering j to question J (nesting notation) the variable over I indicator of j is denoted δ_j^I with $\delta_j^i = 1$ if $i \in I{<}j{>}$ and 0 if not; similarly we define δ_k^I the variable over I indicator of k (disjunctive coding of J and K). The canonical analysis of the two sets of variables $(\delta_j^I)_{j \in J}$ and $(\delta_k^I)_{k \in K}$ amounts to seeking coefficients $(x^j)_{j \in J}$ and $(x^k)_{k \in K}$ such that the correlation between $\sum_{j \in J} x^j \delta_j^I$ and $\sum_{k \in K} x^k \delta_k^I$ is maximum.

Mean $\delta_j^I = f_j$ and $\text{Cov}(\delta_j^I|\delta_k^I) = f_{jk} - f_j f_k$; hence $\text{Cov}(\sum_{j\in J} x^j \delta_j^I | \sum_{k\in K} x^k \delta_k^I) = \sum_{j\in J}\sum_{k\in K}(f_{jk} - f_j f_k)x^j x^k = \sum_{j\in J}\sum_{k\in K} f_{jk} x_0^j x_0^k$ (with $x_0^j = x^j - \sum_{j\in j} f_j x^j$ and $x_0^k = x^k - \sum_{k\in K} f_k x^k$). One has $\sum_{j\in J}\sum_{k\in K}(f_{jk} - f_j f_k)x_0^j x_0^k = \sum_{j\in J} f_j x_0^j \sum_{k\in K} f_k^j x_0^k$, hence $\text{Corr}(\sum_{j\in J} x^j \delta_j^I | \sum_{k\in K} x^k \delta_k^I) = \frac{\langle x_0^J|^t f_K^J(x_0^K)\rangle}{\|x_0^K\|\times\|x_0^J\|}$ and the result (cf. Property 10.32, p.446).

Since the three statistical characterizations are equivalent, we will refer to all of them as the *canonical analysis of contingency tables*. As reflected in the foregoing references — and as already said in chapter OVERVIEW — the canonical analysis of contingency table was anterior to the formal–geometric approach; see Guttman (1959), Benzécri (1982a), Greenacre (1984). In the canonical approach, no geometry is involved; principal variables are determined up to a linear transformation, in other words, only the standard principal variables $(z_\ell^J, z_\ell^K)_{\ell\in L}$ are determined and there is no reason to rescale them as $(y_\ell^J, y_\ell^K)_{\ell\in L}$, as done in the geometric approach.

2.4 Extensions and Concluding Comments

In this section, we discuss the domain of the direct applicability of CA, and tackle the issue of its enlarged domain of applicability. Then we briefly compare CA to MDS. Lastly, we outline CA as a probabilistic model.

2.4.1 The direct Domain of Applicability of CA

The rough statement is often heard that "CA is applicable to any table of nonnegative numbers". The formal approach enables one to qualify this statement and to precisely delineate the conditions of applicability of CA. It makes it clear that beyond frequency tables, CA is directly applicable to any table of numbers whose marginal sums are strictly positive and which is a measure over $J \times K$, that is, for which *rows or columns are grouped by summation*, or in other words (Benzécri, 1992, p.155) refer to *extensive additive magnitudes*. As an example, consider a table where J is a set of exporting countries of some good, K a set of importing countries, and the number in cell (j, k) is the amount of goods (expressed in physical or monetary units) exported from country j to country k during a given period of time. Cf. Benzécri (1992, part V).

The positivity issue. A frequently asked question is: If some cells in a table are negative numbers, is CA still applicable?

To answer this question, it is crucial to dissociate CA, as a geometric method, from canonical correlation analysis. For canonical analysis, when some f_{jk} are negative, it is hard to conceive of a formula such as $\sum_{j \in J} \sum_{k \in K} f_{jk} z^j z^k$ as expressing a correlation. On the other hand, the fundamental joint representation of CA in the ℓ–subspace spanned by $z_{J\ell}$ and $z_{K\ell}$ remains feasible as long as the expression $\sum_{j \in J} \sum_{k \in K} f_{jk} z^j_\ell z^k_\ell = \xi_\ell < 1$; therefore, CA as a geometric construction remains applicable even when some cells are negative, as long as all eigenvalues are less than or equal to 1[16]. In this case, a negative (j, k) cell brings profile point M^j outside the simplex $(\mathrm{I}^k)_{k \in K}$ and profile point M^k outside the simplex $(\mathrm{I}^j)_{j \in J}$. An example of such a situation may arise when one analyzes the incomplete reconstitution of a contingency table (See Exercise 2.1).

Lexicographic paradigm and textual analysis. The lexicographic paradigm is defined as follows. Take a text (or a set of texts from one or several authors) divided into a set K of pieces or *chapters* $k \in K$, and a set J of verbal forms or *words* $j \in J$; consider the table $J \times K$ such that n_{jk} is the number of occurrences of word j in chapter k, and proceed to the CA of the table. The lexicographic paradigm has given birth to a fully fledged domain of direct application of CA. The set J will be chosen depending on the linguistic issues investigated in text K: the most frequent "full words" to investigate its *content*; "tool words" to investigate its *style*. An important area of textual analysis is the analysis of open questions in questionnaires; see Lebart & Salem (1988), Lebart & al (1998).

The book by Benzécri & Coll (1981) and the journal *Cahiers d'Analyse des Données* contain masterpiece analyses of Hebrew and Greek classics; as well as analyses pertaining to contemporary issues. As an example of the latter we may cite the study by Volochine (1995), motivated by the intriguing charge of plagiarism raised against Cholokhov (the famous Russian Nobel prize writer), in connection with his novel *And quiet flows the Don.* Volochine's paper presents a textual analysis (using tool words) of the eight chapters of the novel, completed by other texts by Cholokhov and also by Kriukov (a sometimes alleged concealed source for Cholokhov); the analysis shows convincingly i) that the first six chapters of the novel do differ (on the first axis) in style from the last two chapters and also from Cholokhov's

[16]Remember that negative cells do not necessarily entail eigenvalues greater than 1.

other works, therefore strengthening the suspicion of plagiarism; ii) but on the other hand that Kriukov's style differs (on the second axis) from Cholokov's, therefore weakening the Kriukov track. Such papers illustrate well the use of CA as a geometric model for investigating hypotheses.

2.4.2 The enlarged Domain of Applicability of CA

In many situations, CA can be applied after an appropriate *coding* of variables, guided by the sense of invariants provided by the formal approach. In that way CA has come to be applied to various *Individuals* × *Variables* tables. After Benzécri (1982a, p. 122), let us mention the following main types of extensions.

• *Intensive bipolar magnitudes*, such as scores belonging to an interval $[u, v]$. The standard coding is here the *doubling* of scores, replacing each score x by the two scores $(x - u, v - x)$, then proceeding to the correspondence analysis of the doubled table.

• *Binary variables.* The simplest coding is *logical coding* (1 presence, 0 absence of character); then the table as it stands is analyzed by CA; an alternative coding consists in doubling each binary variable beforehand, hence two $(1, 0)$ variables denoting presence vs absence respectively.

• *Categorized variables.* The standard coding is here the *disjunctive encoding* of variables, i.e. replacing each modality by a $(1, 0)$ variable; CA applied to the disjunctively coded table, known as *Multiple Correspondence Analysis* (MCA), is the most widespread extension of CA and will be presented in the *MULTIPLE* chapter.

2.4.3 CA vs MultiDimensional Scaling (MDS)

MultiDimensional Scaling (MDS), initiated by Shepard (1962) under the name *Proximity Analysis*, was developed by Shepard, then Carroll, Kruskal, etc. MDS, like CA, is concerned by Euclidean representations. See Benzécri (1965), Shepard (1966, 1980). The affinities between the two methods are well recognized today, and CA is sometimes put under the heading "MultiDimensional Scaling". It is certainly even more appropriate to put MDS under the heading "Geometric Data Analysis".

Similarity tables. The MDS paradigm is not the contingency table, but the *similarity table*, therefore for comparing CA and MDS two lines can be followed: (i) Investigating CA for analyzing similarity tables; (ii) investigating MDS for analyzing square contingency tables. A similarity table is a symmetric $K \times K$ table whose cells (k, k') contain the values of either a similarity or a dissimilarity index. For dissimilarities (e.g. distances), the greater the values, the more remote k and k' are; for similarities (also called proximities), the greater the values, the closer k and k' are.

The most genuine version of MDS is the *non-metric* one[17]. The nondiagonal $K(K-1)/2$ cells of the table are rank-ordered according to dissimilarities (or similarities), and a Euclidean representation of set K is constructed in a low-dimensional (most often one or two-dimensional) space, such that the ordering of dissimilarities is preserved as much as possible.

MDS can also be applied to moderately nonsymmetric tables, typically taking as a dissimilarity or similarity index between k and k' the half-sum of the (k, k') and (k', k) cell values. As a classical example of tables well handled by MDS, let us mention *confusion tables*, that is, square contingency tables $K \times K$, where $n_{kk'}$ is the number of times signal k has been judged identical to signal k'. The confusion table is first transformed into a similarity table, typically taking as a similarity index between k and k' (with $k \neq k'$) the half-sum of the two confusion frequencies $n_{kk'}$ and $n_{k'k}$, or the half-sum of the two association rates $t_{kk'}$ and $t_{k'k}$. For confusion tables, non-metric MDS and CA often lead to close results (See Exercise 2.3). Still, because MDS requires symmetrizing the table, it cannot allow studying the associations between the two sets (first and second signals). There is no provision for simultaneous representation in MDS.

Symmetric frequency tables. More generally, a symmetric table whose entries are similarity or dissimilarity indices can be analyzed by MDS. Now while for confusion tables (with dominant diagonal frequencies), cell values are naturally declared as similarities, for other sorts of tables the choice may be far from obvious. As an example, take the table referred to by Benzécri & Coll (1981 p.39, 1986a, 1986b) about the statistical patterns of consonants in Hebrew. From a set of "roots" (3 successive consonants), a $K \times K$ frequency table is constructed indexed by the $K = 23$ consonants of Hebrew; the frequency $n_{kk'}$ being the number of roots for which consonant k is in the first position and k' is the second position. Now, as it turns out, the diagonal cells in this table (two identical consonants in succession) are virtually null, that is, there are very few roots for which a same consonant k is in the first and the second positions. For CA, this particular feature is no problem; it simply entails that there is an opposition between each consonant in the first position and this same consonant in the second posi-

[17] For examples of nonmetric MDS, see Henley (1969). Metric MDS amounts to the principal analysis of a distance table. See chapter *CLOUD*, Exercice 3.1 p. 121.

tion; whereas in order to perform an MDS analyis, the association rates $t^{kk'}$ (with $k \neq k'$) must be declared as dissimilarities, not similarities.

In the remarkable note devoted to the comparison of MDS and CA, Benzécri (1986b) makes the point that, when using MDS, the status (similarity vs dissimilarity) of the association rates must be decided prior to analysis, therefore on linguistic grounds for that matter, whereas using CA requires no assumption of this sort and makes the similarity or dissimilarity structures emerge from data analyis. This property is an exemplary illustration of the Philosophy of Data Analysis embodied in CA. In his 1969 paper, in connection with the analysis of the Morse code data[18], Benzécri had already commented: "In non-artificial cases where analysis of proximities (i.e. MDS) has been successfully used...the data have a beautiful correspondence matrix structure which it is wiser to keep than to convert into proximities." This comment appears to be still applicable today.

2.4.4 CA as a Probabilistic Model

Around the time CA was developed as a description–oriented method, other methods, such as log–linear modeling or association models for contingency tables, were developed in the probabilistic (sampling–oriented) line; these methods have sometimes been opposed to CA, allegedly presumed to be a "descriptive only" method. Such an opposition is far–fetched. There is no doubt that CA, as well as association models, can be framed as a probabilistic model. It suffices to assume that the data are a sample from a population whose underlying factors, eigenvalues and principal coordinates are given the status of parameters. Taking all factors of the full–dimensional descriptive analysis defines the "saturated model" canonically associated with descriptive CA; then the least–square estimates of the parameters are simply the basic descriptive statistics of CA. Taking a more parsimonious number of factors defines an "unsaturated model", such as a one–factor model, a two–factor model, etc.; under an unsaturated model, the parameters need to be reestimated[19]. In effect, there has been quite a lot of statistical work developing CA as a probabilistic model. See e.g. Gilula & Haberman (1986). The existence of such work suffices to crack the legend of

[18]The analysis of confusion tables was one the earliest applications of CA, see Benzécri (1969, 1970).

[19]For the Hair & Eye table, the two–dimensional model yields results very close to descriptive analysis. For large size tables, investigations remain to be made, keeping in mind Caussinus's (1986) important remark: "When dealing with very large tables, the problem is no more with overparametrization but rather with underparametrization."

CA as being a "descriptive only" method with no pretense to probabilistic modeling. For those who like the sampling–oriented approach, the probabilistic modeling of CA provides a serious alternative to association models — with the advantage that the parameters of CA are easier to interpret, i.e. in terms of correlations rather than cross–product ratios.

This point being made, there is no reason to renounce the descriptive key idea. Starting the analysis of a complex data set with probabilistic modeling means facing all problems (validity and relevance) at the same time[20]. In this connection, what is perhaps the most at odds with the Philosophy of Data Analysis is the *fit–&–test technique* currently used in the sampling–oriented approach. "Put in the model as many terms as necessary until goodness–of–fit becomes nonsignificant". To realize the perverse effects of this technique, it suffices to consider the problem of the number of axes to be retained in a geometric analysis[21]. Not only the fit–&–test technique leaves no room for a descriptive appraisal of the importance of axes (i.e. proportion of inertia), but for large samples it detects trivial deviations from the model, and for small samples it fails to detect notable ones[22]. No serious model validation should rely on such an inconsistent technique. For the large size tables and large samples commonly used in GDA, testing for the significance of axes typically brings out plenty of significant axes, only a few of them are worth interpreting; therefore, using significance testing for deciding about the number of axes is of little interest. Whether the practice of GDA would gain in adopting such a questionable technique is more than dubious[23].

In his *Comment* to Goodman (1991), Benzécri (1991) lucidly reiterates the descriptive key idea in geometric data analysis; for the number of axes to be retained in CA, he emphasizes that significance testing should yield the way to interpretation. "The question is not, he says, whether the model passes or fails the test...(it is clear that with an appropriate number of individual instances, the fit shall not pass the test) but [the question goes]

[20]Perhaps this remark gives a clue to the question: How is it that the probabilistic model of CA is always exemplified on small size tables?

[21]A similar argument would run for loglinear, logistic models, etc.

[22]This inadequacy is a direct consequence of the "size–glued" property of significance testing: For any given (non–zero) deviation, the larger the sample, the more significant the fit — a consequence that has been pointed out for a long time, from Mahalanobis (1948) to Guttman (1977).

[23]The most common manifestation of the perverse effect of the fit–&–test technique is the *significance fallacy*: confusing significant deviations with important ones and non-significant deviations with negligible ones. At least the reluctance of CA users toward significance testing will have preserved their work from this pollution.

in the opposite direction: Granted that the data table differs significantly from the zero–hypothesis model of independence, how far can we go in building a scheme that involves only interpretable factors?"

Exercises of Chapter 2

Exercise 2.1 (Status)

In the n_{JK} contingency table given hereafter, J denotes the *origin* and K the *status* of the respondents to a questionnaire[24]. The Origin of respondent J is categorized in four modalities defined by the crossing of two factors: "status of father" $A = \{a1, a2\}$, with $a1$ = high and $a2$ = low, and "education level" of respondent $B = \{b1, b2\}$ with $b1$ = high and $b2$ = low.

One has $J = A \times B$ with $j1 = a1b1$, $j2 = a1b2$, $j3 = a2b1$ and $j4 = a2b2$. The Status of respondent K is categorized in three modalities $k1$ = high, $k2$ = medium and $k3$ = low. The mean square contingency of the $J \times K$ table is $\Phi^2 = 0.28089$.

	$k1$	$k2$	$k3$	
$j1 = a1b1$	106	88	8	202
$j2 = a1b2$	18	45	11	74
$j3 = a2b1$	106	776	133	1015
$j4 = a2b2$	27	1274	858	2159
	257	2183	1010	3450

1. Calculate the conditional frequencies f^j_k ("k if j") and the mean profile.

   Calculate the Φ^2–distances between the ideal points I^{k1}, I^{k2} and I^{k3}.

   Construct geometrically the simplex (here a triangle) of edges I^{k1}, I^{k2} and I^{k3}, and the cloud of the 4 points $(\mathrm{M}^j)_{j\in J}$ together with its mean point.

2. Eigenvalues and standard eigenvectors of the symmetric matrix $\mathbf{S}$ of general term $s_{kk'} = \frac{1}{\sqrt{n_k n_{k'}}} \sum_{j\in J} n_{jk} n_{jk'}/n_j$ (cf. opposite table).

	$\ell = 0$ $\lambda_0 = 1$	$\ell = 1$ $\lambda_1 = .24435$	$\ell = 2$ $\lambda_2 = .03654$
$k1$	.27293	.90327	−.33108
$k2$	.79546	−.01834	.60573
$k3$	.54107	−.42868	−.72352

   • What is the content of column ($\ell = 0$) of the above table ?

   • Determine the two standard principal variables over K. Deduce the scaled principal variables over J.

   • Draw the simultaneous representation.

3. Calculate the matrix $\mathbf{\Sigma}$ (product of the transition matrices $\mathbf{T}$ and $\mathbf{T}^\star$) of general term $\sigma_{kk'} = \sum_{j\in J} f^j_k f^{k'}_j = \left(\sum_{j\in J} n_{jk}\, n_{jk'}/n_j\right)/n_{k'}$, then determine its eigenvalues and eigenvectors. Deduce the standard principal contrasts.

4. Calculate the reconstitution of order 1 of the contingency table.

[24]Data adapted from Glass (1954).

Solution

1. Conditional frequencies of k if j: $f^j_k = n_{jk}/n_j$. Example: $f^{j1}_{k2} = 88/202 = .4356$, etc. See opposite table. Φ^2–distance between ideal points: $\mathrm{I}^k\mathrm{I}^{k'} = \sqrt{(n/n_k) + (n/n_{k'})}$.

	$k1$	$k2$	$k3$
$j1$	.5248	.4356	.0396
$j2$	.2432	.6081	.1486
$j3$	.1044	.7645	.1310
$j4$	.0125	.5901	.3974
	.0745	.6328	.2928

Hence $\mathrm{I}^{k1}\mathrm{I}^{k2} = 3.87$; $\mathrm{I}^{k1}\mathrm{I}^{k3} = 4.10$, and $\mathrm{I}^{k2}\mathrm{I}^{k3} = 2.24$.

One constructs the segment $\mathrm{I}^{k1}\mathrm{I}^{k3}$ of length $4.1u$ (u denoting the unit–distance), then a point I^{k2} at a $3.87u$ distance from I^{k1} and at a $2.24u$ distance from I^{k3}. From the barycentric relation $\mathrm{M}^j = \sum_{k\in K} f^j_k \mathrm{I}^k$, one constructs the profile points $j1, \ldots j4$.

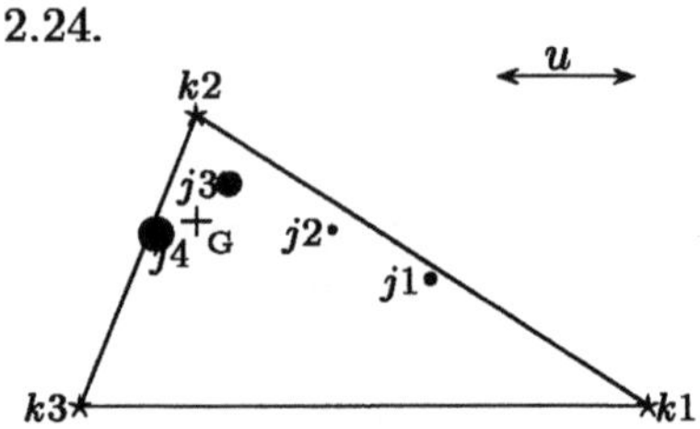

2. In column $\ell = 0$ we find the coordinates $(\sqrt{f_k})_{k\in K}$ of the mean point with respect to the orthonormal basis. Standard principal variables over K (cf. formula 2.13, p.40) are z^K_ℓ with $z^k_\ell = b_{k\ell}/\sqrt{n_k/n}$.

Example: $z^{k1}_{\ell 1} = .90327/\sqrt{\frac{257}{3450}} = 3.3095$, hence $y^{k1}_{\ell 1} = \sqrt{.24435} \times z^{k1}_{\ell 1} = 1.636$.

	$z^k_{\ell 1}$	$z^k_{\ell 2}$	$y^k_{\ell 1}$	$y^k_{\ell 2}$
$k1$	3.3095	−1.2130	1.636	−0.232
$k2$	−0.0231	0.7615	−0.011	0.146
$k3$	−0.7923	−1.3372	−0.392	−0.256

The computational formula for the scaled principal variables y^J_ℓ is: $\frac{\sqrt{n}}{n_j} \sum_{k\in K} \frac{n_{jk}}{\sqrt{n_k}} b_{k\ell}$,

that is, for $j1$: $y^{j1}_{\ell 1} = \frac{\sqrt{3450}}{202}\left(\frac{106}{\sqrt{257}}0.90327 - \frac{88}{\sqrt{2183}}0.01834 - \frac{8}{\sqrt{1010}}0.42868\right) = 1.695$.

Hence the principal coordinates of points j.

	y^J_1	y^J_2	f_J
$j1$	1.695	−0.358	.0586
$j2$	0.673	−0.031	.0214
$j3$	0.224	0.280	.2942
$j4$	−0.287	−0.097	.6258

(The scale of simultaneous representation is twice the one of the barycentric graph of question 1)

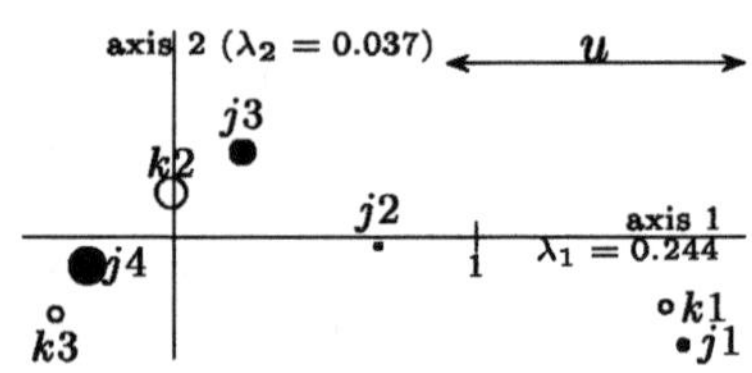

3. Computation of matrix $\mathbf{\Sigma}$ (product of the two transition matrices):

$$\begin{pmatrix} .5248 & .2432 & .1044 & .0125 \\ .4356 & .6081 & .7645 & .5901 \\ .0396 & .1486 & .1310 & .3974 \end{pmatrix} \begin{pmatrix} .4125 & .0403 & .0079 \\ .0700 & .0206 & .0109 \\ .4125 & .3555 & .1317 \\ .1051 & .5836 & .8495 \end{pmatrix} = \begin{pmatrix} 0.27788 & 0.07057 & 0.03117 \\ 0.59963 & 0.64624 & 0.61204 \\ 0.12254 & 0.28315 & 0.35678 \end{pmatrix}$$

By nullifying the determinant of matrix $\mathbf{\Sigma} - \lambda\mathbf{I}$, and developing it according to the first row, one gets: $(0.27788 - \lambda)(\lambda^2 - 1.00302\lambda + 0.05727) - 0.07057(0.113894 - 0.0.59963\lambda)$ $+0.03117(0.09059 + 0.12254\lambda) = \lambda^3 - 1.28089\lambda^2 + 0.28984\lambda - 0.00893 = 0$. Since 1 (trivial eigenvalue) is a root of the equation, one has: $(\lambda - 1)(\lambda^2 - 0.28089\lambda + 0.00893) = 0$, hence $\lambda_1 = 0.24434$ and $\lambda_2 = 0.03655$.

An eigenvector $\boldsymbol{\alpha} = (a_1, a_2, a_3)$ associated with eigenvalue λ_1 verifies the equation $\mathbf{\Sigma}\,\boldsymbol{\alpha} = \lambda_1\boldsymbol{\alpha}$, hence the system of 3 equations with 3 unknowns:

$$\begin{cases} 0.27788\,a_1 + 0.07057\,a_2 + 0.03117\,a_3 = 0.24434\,a_1 \\ 0.59963\,a_1 + 0.64624\,a_2 + 0.61204\,a_3 = 0.24434\,a_2 \\ 0.12254\,a_1 + 0.28315\,a_2 + 0.35678\,a_3 = 0.24434\,a_3 \end{cases}$$

Adding the 3 equations, one deduces $a_1 + a_2 + a_3 = 0$ (contrast on K). Replacing a_3 by $-(a_1+a_2)$ in the second equation, one gets $a_2 = 0.0591a_1$, and $a_3 = 0.9409a_1$; $\boldsymbol{\alpha} = (1, -0.0591, 0.9409)$ is an eigenvector of $\boldsymbol{\Sigma}$ associated with λ_1; the square of its norm is equal to $3450 \times \left(\frac{1}{257} + \frac{(0.0591)^2}{2183} + \frac{(0.9409)^2}{1010}\right) = (4.0563)^2$, hence the standard principal contrast on K: $\pm\boldsymbol{\alpha}/4.0563 = \pm(0.2465, -0.0146, -0.2319)$ (the results given above correspond to the + sign).
In the same way, vector $\boldsymbol{\beta} = (1, -5.3319, 4.3326)$ is an eigenvector of $\boldsymbol{\Sigma}$ associated with λ_2, with $||\boldsymbol{\beta}|| = 11.0668$; hence $z_{K\ell 2} = \pm(0.0904, -0.4818, +0.3915)$.
(the results given above correspond to the − sign).

4. Reconstitution of order 1: $n_j n_k/n + (n_j n_k/n) y_1^j z_1^k$. Let $\widehat{\mathbf{N}}$ be the matrix of general term $n_j n_k/n$.

$$\widehat{\mathbf{N}} = \frac{1}{3450}\begin{bmatrix}203\\74\\1015\\2159\end{bmatrix}\begin{bmatrix}257 & 2183 & 1010\end{bmatrix} = \begin{bmatrix}15.048 & 127.816 & 59.136\\5.512 & 46.824 & 21.664\\75.610 & 642.245 & 297.145\\160.830 & 1366.115 & 632.055\end{bmatrix}$$

$$\widehat{\mathbf{N}} \times \begin{bmatrix}1.695\\0.673\\0.224\\-0.287\end{bmatrix}\begin{bmatrix}3.3095 & -0.0231 & 0.7923\end{bmatrix} = \begin{bmatrix}84.411 & -3.005 & -79.417\\12.278 & -0.728 & -11.552\\56.052 & -3.323 & -52.736\\-152.761 & 9.057 & 143.723\end{bmatrix}$$

Hence the reconstitution of order 1:

99.46	122.81	−20.28
18.29	46.10	10.11
131.66	638.92	244.41
8.07	1375.17	775.78

Comments

Cf. question 1. The figure that is constructed from the distances between points is only defined up to orientation.

Cf. question 3. The (nonsymmetric) matrix $\boldsymbol{\Sigma}$ is the matrix of mapping $f_K^J \circ f_J^K$ in the canonical basis; its diagonalization directly yields the principal contrasts. In the calculation procedure (cf. p.41), one usually prefers the detour by the symmetric matrix $\mathbf{S}$.

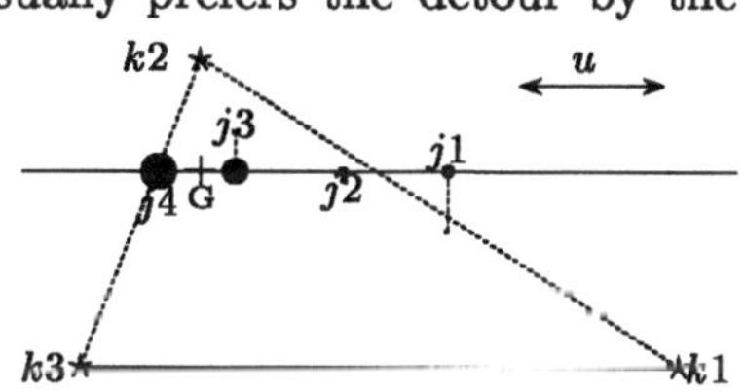

Cf. question 4. Applying CA to the reconstituted table of order 1 would yield the same first eigenvalue and principal variables as the initial table. The reconstituted table has a negative cell, namely $(j1, k3)$. The geometric construction is still valid since the eigenvalue is less than 1 (cf. 2.4); the projection of point $j1$ on axis 1 is seen to lie outside the simplex defined by the ideal points.

Exercise 2.2 (Social Representation)

In a research on social representation, the interviewees, belonging to 8 socio–professional *categories*, had to select among 9 *Qualities* the 3 that best describe a *nice person*. In the following $J \times K$ table, the figure in cell (j, k) is the number of times respondents in category j have mentioned quality k. For example, in cell $(j3, k4)$ (Salesmen, honest), 25 is the number of respondents in the *Salesmen* category who have mentioned the quality *honest* 25 times to describe a nice person.

- **Data table** (Maisonneuve, 1978).

*k*1 *conscientious*; *k*2 *generous*; *k*3 *cheerful*; *k*4 *honest*; *k*5 *intelligent*; *k*6 *helpful*; *k*7 *courageous*; *k*8 *understanding*; *k*9 *discreet*.

		k1	k2	k3	k4	k5	k6	k7	k8	k9	
Farmers	j1	19	9	9	26	10	16	19	4	8	120
Workers	j2	20	5	11	25	9	14	19	6	11	120
Salesmen	j3	20	3	9	25	15	13	10	13	12	120
Shopkeepers	j4	8	9	12	23	14	16	14	12	12	120
Employees	j5	10	5	8	26	19	13	11	13	15	120
Technicians	j6	10	5	12	24	22	13	11	13	10	120
Academics	j7	3	18	13	11	25	11	12	20	7	120
The Professions	j8	4	21	12	12	24	11	11	17	8	120
		94	75	86	172	138	107	107	98	83	960

- **Association rates**

	k1	k2	k3	k4	k5	k6	k7	k8	k9
j1	0.62	−0.04	−0.16	0.21	−0.42	0.20	0.42	−0.67	−0.23
j2	0.70	−0.47	0.02	0.16	−0.48	0.05	0.42	−0.51	0.06
j3	0.70	−0.68	−0.16	0.16	−0.13	−0.03	−0.25	0.06	0.16
j4	−0.32	−0.04	0.12	0.07	−0.19	0.20	0.05	−0.02	0.16
j5	−0.15	−0.47	−0.26	0.21	0.10	−0.03	−0.18	0.06	0.45
j6	−0.15	−0.47	0.12	0.12	0.28	−0.03	−0.18	0.06	−0.04
j7	−0.74	0.92	0.21	−0.49	0.45	−0.18	−0.10	0.63	−0.33
j8	−0.66	1.24	0.12	−0.44	0.39	−0.18	−0.18	0.39	−0.23

- $\Phi^2 = 0.127161$. **Eigenvalues** $\lambda_1 = 0.095218$, $\lambda_2 = 0.020730$, $\lambda_3 = 0.005934$.
- **Principal coordinates** and **Absolute contributions** of points and axes.

	n_j	y_1^j	y_2^j	y_3^j
j1	120	0.30040	−0.23055	−0.00751
j2	120	0.35808	−0.12874	0.00958
j3	120	0.24775	0.13489	0.14678
j4	120	0.00286	0.00488	−0.13787
j5	120	0.06659	0.20030	−0.06613
j6	120	0.00142	0.16001	−0.01176
j7	120	−0.49491	−0.03678	0.03661
j8	120	−0.48220	−0.10402	0.03030

1000 $\text{Cta}_{j\ell}$	$\ell = 1$	$\ell = 2$	$\ell = 3$	1000 Cta_j
j1	11.28	6.64	0.01	18.48
j2	16.03	2.07	0.01	18.73
j3	7.67	2.27	2.69	12.96
j4	0.00	0.00	2.38	3.00
j5	0.55	5.01	0.55	6.94
j6	0.00	3.20	0.02	4.73
j7	30.62	0.17	0.17	31.35
j8	29.06	1.35	0.11	30.98
1000 Cta_ℓ	95.22	20.73	5.93	127.16 = 1000Φ^2

	n_k	y_1^k	y_2^k	y_3^k
k1	94	0.52106	−0.08435	0.18420
k2	75	−0.58037	−0.30797	0.01480
k3	86	−0.10415	−0.03410	−0.01906
k4	172	0.25544	0.06732	−0.05226
k5	138	−0.29744	0.12772	0.03855
k6	107	0.09759	−0.03402	−0.06799
k7	107	0.13732	−0.19677	−0.06154
k8	98	−0.35073	0.16284	0.06833
k9	83	0.11872	0.16165	−0.07173

1000 $\text{Cta}_{k\ell}$	$\ell = 1$	$\ell = 2$	$\ell = 3$	1000 Cta_k
k1	26.58	0.70	3.32	30.63
k2	26.31	7.36	0.02	34.21
k3	0.97	0.10	0.03	2.28
k4	11.69	0.81	0.49	13.28
k5	12.72	2.34	0.21	16.15
k6	1.06	0.13	0.52	2.02
k7	2.10	4.32	0.42	7.31
k8	12.56	2.71	0.48	16.28
k9	1.22	2.26	0.44	5.01
1000 Cta_ℓ	95.22	20.73	5.93	127.16 = 1000Φ^2

Table 2.1. Data table and statistical results.

From the statistical results given in Table 2.1, answer the following questions.

1. Draw the graph of attractions (association rate > 0.35). Comment.
2. Graph the simultaneous representation of the cloud of Categories and the cloud of *Qualities* in plane 1-2.
3. How many axes should (at least) be retained for interpretation? (Indicate the adopted criterion).
4. Interpret the first and second axes by the method of point and deviation contributions.

5. By examining the cloud of categories, say whether the first two axes are sufficient for interpretation (explain why).

Solution

1. Graph of attractions (modalities have been ranked according to Axis 1).

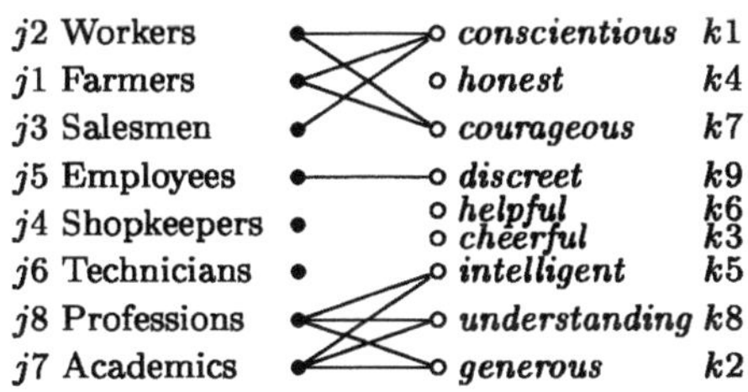

Two groups distinctly emerge from the graph: that of Workers, Farmers and Salesmen, in attraction with *conscientious* and *courageous*, and the one of The Professions and Academics, in attraction with *intelligent, understanding* and *generous*.

2. Simultaneous representation in plane 1-2

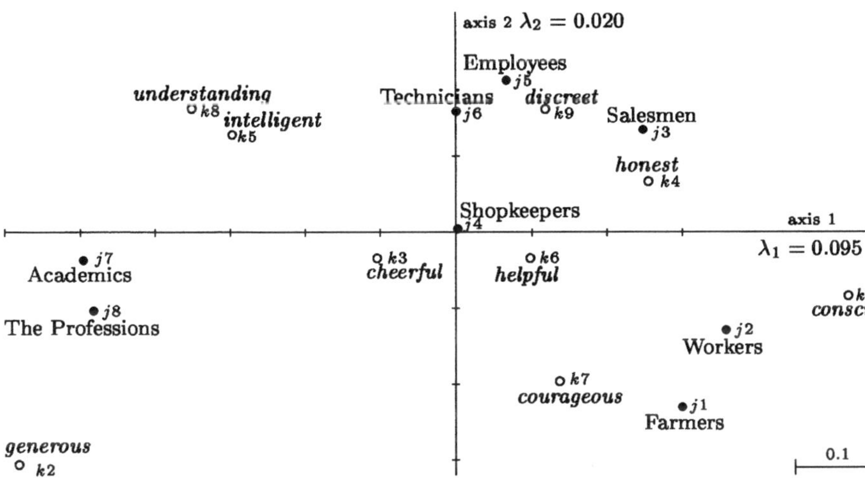

3. The contribution to the total variance of the first axis is 75%, that of the second axis is 16% (cf. §2.3.4, p.47). One will (at least) retain the first 2 axes, whose contributions are greater than the average contribution (1/7 = 14%).

4. We successively examine the cloud of Categories and the cloud of *Qualities*.

• *Interpretation of the first axis*

The three Categories $j7$, $j8$, $j2$ have contributions to Axis 1 (cf. Table 2.1, p.68) greater than the average contribution ($\text{Cta}_{j\ell 1} > \lambda_1/8 = 0.012$), to which we add $j1$ whose contribution is near average. Taken together these categories account for 0.08699/0.095218 = 91% of the variance of Axis 1; furthermore, they are well represented on Axis 1 (30.62/31.35 = .98, .94, .86, .61 respectively). Categories $j7$ (Academics) and $j8$ (The Professions) (on the left side of the graph) are opposed on Axis 1 to Categories $j1$ (Farmers) and $j2$ (Workers) (on the right side of the graph). The coordinates, weights and absolute contributions of the barycenters of the two clusters, denoted $c1$ and $c2$, are given in Table 2.2.

	Coordinate	Weight	Cta
$c1$	−0.48856	$2 \times 120/960 = .25$	0.05967
$c2$	0.32924	.25	0.02710

Table 2.2. Coordinates, weights and absolute contributions of $c1$ and $c2$

Absolute contribution of the deviation between $c1$ and $c2$ (cf. formula (2.19), p.48) is equal to $(0.32924 - (-0.48856))^2/(\frac{1}{0.25} + \frac{1}{0.25}) = 0.08360$. Points $c1$ and $c2$ together account for 91% of the variance of Axis 1, and their deviation for 88%; thus, for the cloud of Categories, $(c1, c2)$ constitutes a good summary of Axis 1.
The ***Qualities*** that contribute most to Axis 1 are $k1$, $k2$, $k5$, $k8$ and $k4$ ($\text{Cta}_k > \lambda_1/9 = .011$). Together they account for 94% of the variance of Axis 1 (the sum of their absolute contributions is 0.08987); furthermore, they are well represented on Axis 1 (.87, .77, .79, .77 and .88 respectively). Qualities $k2$ (***generous***), $k5$ (***intelligent***) and $k8$ (***understanding***) (group $d1$) are opposed on Axis 1 to qualities $k1$ (***conscientious***) and $k4$ (***honest***) (group $d2$) (cf. Table 2.3). ***Absolute contribution***

	Coordinate	Weight	Contribution
$d1$	−0.38246	$311/960\,(= 0.324)$	0.04739
$d2$	0.34930	$266/960\,(= 0.277)$	0.03381

Table 2.3. Coordinates, weights and absolute contributions of $d1$ and $d2$

of deviation (cf. formula (2.19), p.48): $(0.34930 + 0.38246)^2/(960(\frac{1}{266} + \frac{1}{311})) = 0.07997$. Points $d1$ and $d2$ together account for 85% of the variance of Axis 1, and their deviation for 84%. For ***Qualities***, $(d1, d2)$ constitutes a good summary of axis 1.

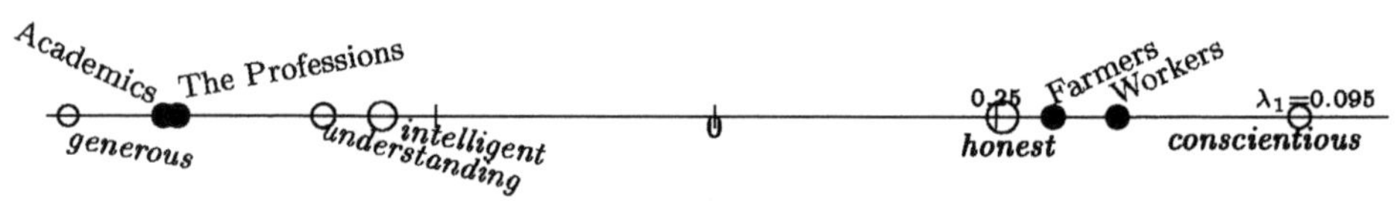

Simultaneous representation on Axis 1

To sum up: Axis 1 opposes Farmers and Workers, for whom a nice person is above all ***conscientious*** and ***honest***, to Academics and The Professions, for whom a nice person is above all ***generous***, ***intelligent*** and ***understanding***.

• ***Interpretation of the second axis***

The Categories that contribute most to Axis 2 ($\text{Cta}_{j\ell 2} > .00259$) are $j1$, $j5$ and $j6$. Together they contribute to 72% of the variance of axis. Category $j1$ (Farmers) is opposed on this axis to Categories $j5$ (Employees) and $j6$ (Technicians). The mean point of $\{j5, j6\}$ has coordinate +0.180, weight 0.25 and absolute contribution .00811. Point $j1$ and the mean point of $j5$ and $j6$ together account for 71% of the variance of axis, their deviation accounts for 68%.
Similarly, the ***Qualities*** which contribute most to Axis 2 ($\text{Cta}_{k\ell 2} > .0023$) are $k2$, $k7$, $k8$, $k5$ and $k9$. Together they contribute to 92% of the variance of Axis 2. The ***qualities*** $k2$ (***generous***) and $k7$ (***courageous***) are opposed on this axis to $k5$ (***intelligent***), $k8$ (***understanding***) and $k9$ (***discreet***). The mean point of $\{k2, k7\}$ has coordinate −0.2422, weight .190, and Cta .01112; the mean point of $\{k5, k8, k9\}$

has coordinate +0.1473, weight .332, and Cta .0072. Together they account for 88% of Axis 2, their deviation also accounts for 88% of Axis 2.

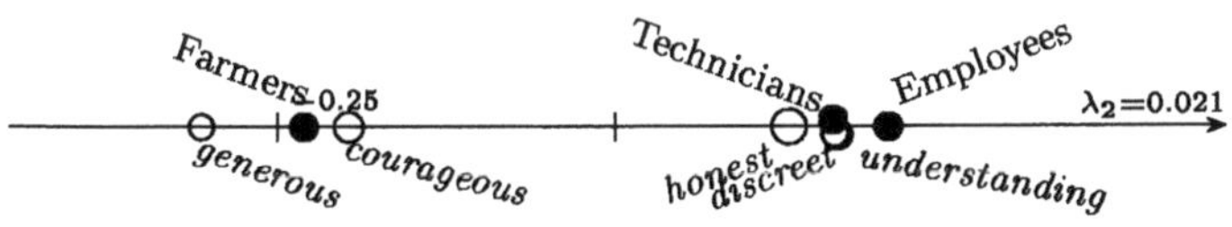

Simultaneous representation on Axis 2

To sum up: Axis 2 opposes Farmers to Employees and Technicians, and the ***Qualities courageous*** and ***generous*** to ***honest, discreet, understanding.***

5. Among Categories, only that of Shopkeepers, located near the mean point of the cloud, is poorly represented in plane 1-2; but it is well represented on Axis 3. Similarly, the quality ***cheerful***, close to the mean point of the cloud, is the only one that is poorly represented in plane 1–2 (quality of representation $0.47 < 0.5$). Therefore, it can be asserted that the first two axes, which account for nearly all Categories and ***Qualities***, are sufficient for interpretation.

Comment: The Φ^2 statistic is admittedly quite small, that is, the deviation of the table from independence is not large. Still, Correspondence Analysis, applied here to a table for which the two sets are without structure, nicely brings out the structures of the data table by providing a clear representation of oppositions.

Exercise 2.3 (Morse Data)

In Rothkopf's experiment (1957), subjects were presented on each trial with a pair of Morse signals (10 digits and 26 letters); subjects had to state whether the two signals of the pair were identical or different[25]. The table (confusion table) was constructed giving, for each pair of signals, a number proportional to the number of times the two signals of the pair had been judged identical.

In this exercise, the subtable involving the 10 digits (1...9,0) will be analyzed (the 10 Morse codes are reproduced hereafter). $j1, \ldots, j9, j0$ will denote the first signal presented, and $k1, \ldots k9, k0$ the second; e.g. $(j1, k1)$ refers to the pair "1 followed by 1", $(j1, k2)$ to the pair "1 followed by 2", etc.

• **Data Table** (J.B Kruskal & M. Wish, 1978).

	$k1$	$k2$	$k3$	$k4$	$k5$	$k6$	$k7$	$k8$	$k9$	$k0$	
$j1$	84	63	13	8	10	8	19	32	57	55	349
$j2$	62	89	54	20	5	14	20	21	16	11	312
$j3$	18	64	86	31	23	41	16	17	8	10	314
$j4$	5	26	44	89	42	44	32	10	3	3	298
$j5$	14	10	30	69	90	42	24	10	6	5	300
$j6$	15	14	26	24	17	86	69	14	5	14	284
$j7$	22	29	18	15	12	61	85	70	20	13	345
$j8$	42	29	16	16	9	30	60	89	61	26	378
$j9$	57	39	9	12	4	11	42	56	91	78	399
$j0$	50	26	9	11	5	22	17	52	81	94	367
	369	389	305	295	217	359	384	371	348	309	3346

1	• — — — —
2	• • — — —
3	• • • — —
4	• • • • —
5	• • • • •
6	— • • • •
7	— — • • •
8	— — — • •
9	— — — — •
0	— — — — —

[25]Subjects had no previous knowledge of Morse and did not receive information about the alphanumerical values of signals. Morse signals are made up of combinations of dots and dashes separated by spaces; the dash duration is three times the dot duration (and the space duration is equal to the dot duration).

• Correspondence Analysis results

$\Phi^2 = .6307$. *Eigenvalues*: $\lambda_1 = .3369$; $\lambda_2 = .1248$; $\lambda_3 = .0995$; $\lambda_4 = .0301$.
The first three principal coordinates for J and K are given in the following tables.

	n_j	y_1^j	y_2^j	y_3^j
$j1$	349	0.555	−0.297	0.121
$j2$	312	0.036	−0.654	−0.334
$j3$	314	−0.450	−0.499	−0.285
$j4$	298	−0.887	−0.044	0.199
$j5$	300	−0.975	0.074	0.600
$j6$	284	−0.441	0.465	−0.365
$j7$	345	−0.040	0.473	−0.393
$j8$	378	0.320	0.285	−0.111
$j9$	399	0.650	0.067	0.193
$j0$	367	0.686	0.070	0.318

	n_k	y_1^k	y_2^k	y_3^k
$k1$	369	0.467	−0.284	0.016
$k2$	389	0.096	−0.556	−0.270
$k3$	305	−0.524	−0.514	−0.291
$k4$	295	−0.851	−0.051	0.364
$k5$	217	−1.022	0.051	0.668
$k6$	359	−0.489	0.385	−0.274
$k7$	384	−0.105	0.494	−0.330
$k8$	371	0.389	0.312	−0.129
$k9$	348	0.749	0.080	0.277
$k0$	309	0.757	0.032	0.365

1. How many principal axes should be retained for interpretation ?
2. Draw the simultaneous diagram for the first principal axis, and interpret this axis. Do the same for the second axis.
3. Draw the simultaneous diagram in plane 1–2, and join segments ($j1, k1$), ($j2, k2$), etc. Interpret the plane representation. Do the same for plane 1–3.

Solution

1. The first 3 axes, whose eigenvalues exceed $\frac{\Phi^2}{9} = .07$, should at least be retained.
2. For each diagram (axis 1 and axis 2), digits of first signals are written on left of the axis; codes and digits of second signals on the right of the axis.
Axis 1 ranks signals in the order 5, 4, 3, 6, 7, 2, 8, 1, 9, 0; that is roughly, from 5, the shortest signal (• • • • •) to 0, the longest (— — — — —).
Axis 2 opposes the signals beginning with dashes (6, 7, 8, 9) to those beginning with dots (1, 2, 3, 4).

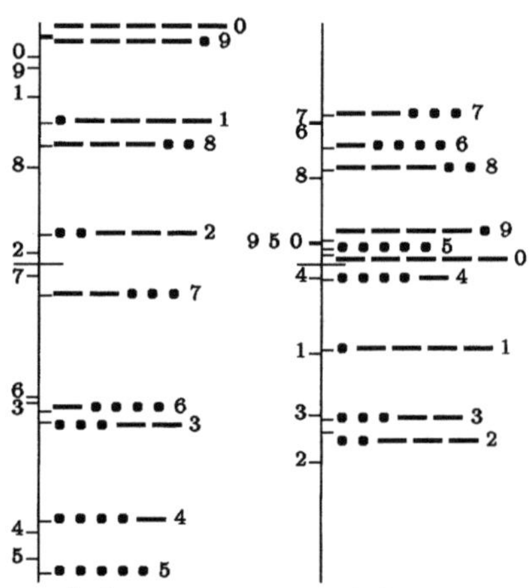

axis 1 axis 2

3. Plane 1-2 shows a circular order, which can be interpreted in terms of proximities between signals. Two neighboring signals only differ by one symbol, as for example 6 (— • • • •) and 7 (— — • • •). Besides, the two points j and k pertaining to the same signal are close to each other, which indicates strong attractions, and reflects the quasi–symmetry of the data table. In plane 1-3, the cloud has a parabolic shape which reflects the order on the first axis, that is, the order of durations. Axis 3 opposes medium durations to extreme (short and long) ones (Guttman effect, cf. *MULTIPLE*, p.220).

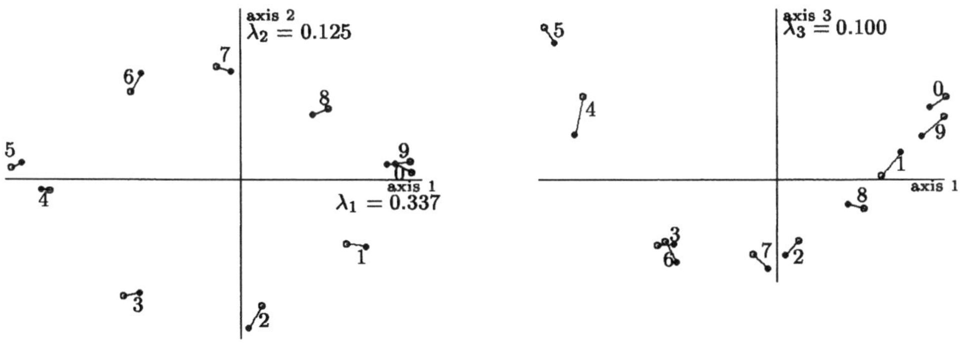

Bullets (•) represent first signals, empty circles (∘) second ones.

Comment: Correspondence Analysis and MultiDimensional Scaling.
In the literature, the Morse data are a classic example for the method of ***MultiDimensional Scaling*** (MDS).

The plane diagram yielded by nonmetric MDS for the Morse code subtable is shown on the opposite. As can be seen, the MDS cloud is strikingly close to those of CA. However, MDS yields a single cloud and therefore makes no provision for the simultaneous study of the two clouds provided by CA. See the discussion "CA vs MDS" in §2.4.3 (p.61).

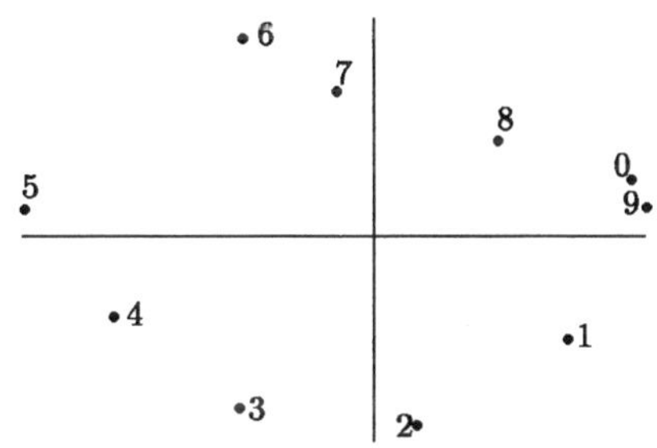

Exercise 2.4 (Block Structure)

Let J and K be two finite sets, with a partition of J in two classes denoted $J<c1>$ and $J<c2>$, and a partition of K in two classes denoted $K<c1>$ and $K<c2>$.

Let n_{JK} be a contingency table on $J \times K$ defined from two contingency tables, one on $J<c1> \times K<c1>$ with total mass n_{c1}, the other one on $J<c2> \times K<c2>$ of total mass n_{c2}, with $n = n_{c1} + n_{c2}$; and let n_{JK} be such that

$$n_{jk} = 0 \begin{cases} \text{for } j \in J<c1> \text{ et } k \in K<c2> \\ \text{for } j \in J<c2> \text{ et } k \in K<c1> \end{cases}$$

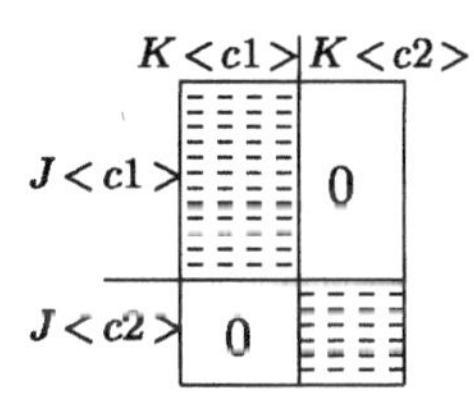

1. For the correspondence on $J<c1> \times K<c1>$, write the transition formulas corresponding to the trivial eigenvalue $\lambda_0 = 1$, and to eigenvalues $\lambda_{\ell<c1>}$.

2. Let δ^J_c be the variable on J indicator of $J<c>$ ($\delta^j_c = 1$ if $j \subset J<c>$ and 0 if $j \notin J<c>$). Show that $a^{c1}\delta^J_{c1} + a^{c2}\delta^J_{c2}$ and $a^{c1}\delta^K_{c1} + a^{c2}\delta^K_{c2}$ are principal variables of n_{JK} associated with eigenvalue 1. Determine a^{c1} and a^{c2} so that these variables are standard ones.

3. Let $(z^{J<c1>}_{\ell<c1>}, z^{K<c1>}_{\ell<c1>})$ be a pair of standard principal variables of the correspondence $n_{J<c1>K<c1>}$ associated with an eigenvalue $\lambda_{\ell<c1>} \neq 1$. Show that the variables t^J_ℓ and t^K_ℓ defined by

$$t^J_{\ell<c1>} = \begin{cases} z^{j<c1>}_\ell & \text{if } j \in J<c1> \\ 0 & \text{if } j \in J<c2> \end{cases} \qquad t^K_{\ell<c1>} = \begin{cases} z^{k<c1>}_\ell & \text{if } k \in K<c1> \\ 0 & \text{if } k \in K<c2> \end{cases}$$

are principal variables of n_{JK} associated with the eigenvalue $\lambda_{\ell<c1>}$.

Solution

1. Transition formulas for the trivial eigenvalue $\lambda_0 = 1$:

$$\begin{cases} \sum\limits_{j\in J<c1>} 1^j f^{K<c1>}_j = 1^{K<c1>} \\ \sum\limits_{k\in K<c1>} 1^k f^{J<c1>}_k = 1^{J<c1>} \end{cases} \qquad \begin{cases} \sum\limits_{j\in J<c2>} 1^j f^{K<c2>}_j = 1^{K<c2>} \\ \sum\limits_{k\in K<c2>} 1^k f^{J<c2>}_k = 1^{J<c2>} \end{cases}$$

Transition formulas for the nontrivial eigenvalue $\lambda_\ell \leq 1$:

$$\begin{cases} \sum\limits_{j\in J<c1>} z^j_{\ell<c1>} f^{K<c1>}_j &= \xi_{\ell<c1>} z^{K<c1>}_{\ell<c1>} \\ \sum\limits_{k\in K<c1>} z^k_{\ell<c1>} f^{J<c1>}_k &= \xi_{\ell<c1>} z^{J<c1>}_{\ell<c1>} \end{cases}$$

Similarly for the correspondence $n_{J<c2>K<c2>}$.

2. One has: $\sum\limits_{j\in J} \delta^j_{c1} f^K_j = \sum\limits_{j\in J<c1>} 1^j f^K_j = \begin{Bmatrix} 1 \text{ if } k \in K<c1> \\ 0 \text{ if } k \in K<c2> \end{Bmatrix} = \delta^K_{c1}$

Hence $$\begin{cases} \sum\limits_{j\in J} (a^{c1}\delta^j_{c1} + a^{c2}\delta^j_{c2}) f^K_j &= a^{c1}\delta^K_{c1} + a^{c2}\delta^K_{c2} \\ \sum\limits_{k\in K} (a^{c1}\delta^k_{c1} + a^{c2}\delta^k_{c2}) f^J_k &= a^{c1}\delta^J_{c1} + a^{c2}\delta^J_{c2} \end{cases}$$

The variables $a^{c1}\delta^J_{c1} + a^{c2}\delta^J_{c2}$ and $a^{c1}\delta^K_{c1} + a^{c2}\delta^K_{c2}$ are principal variables of n_{JK} associated with eigenvalue 1; these variables are standard ones for $a^{c1} = \sqrt{f_{c2}/f_{c1}}$ and $a^{c2} = -\sqrt{f_{c1}/f_{c2}}$.

3. One has: $\sum\limits_{j\in J} t^j_{\ell<c1>} f^K_j = \sum\limits_{j\in J<c1>} z^j_{\ell<c1>} f^K_j$.

Hence: $\sum\limits_{j\in J} t^j_\ell f^K_j = \begin{Bmatrix} \xi_{\ell<c1>} z^k_{\ell<c1>} & \text{for } k \in K<c1> \\ 0 & \text{for } k \in K<c2> \end{Bmatrix} = \xi_{\ell<c1>} t^K_{\ell<c1>}$.

Similarly $\sum\limits_{k\in K} t^k_{\ell<c1>} f^J_k = \xi_{\ell<c1>} t^J_{\ell<c1>}$. Since variables $t^K_{\ell<c1>}$ and $t^J_{\ell<c1>}$ have the same variance f_{c1}, the eigenvalue is $\lambda_{\ell<c1>}$.

Similarly for the principal variables of correspondence $n_{J<c2>K<c2>}$.

Comment

A correspondence n_{JK} has a ***block structure*** if there exists a partition $(J<c>)_{c\in C}$ and a partition $(K<c>)_{c\in C}$ such that $\forall c, c'$ with $c \neq c'$ one has $n_{jk} = 0$ for $j \in J<c>$ and $k \in K<c'>$. What has been shown in this exercise is that if a correspondence has a block structure, it has a nontrivial eigenvalue equal to 1. More generally, it can be shown that for a structure in C blocks, 1 is a nontrivial eigenvalue of multiplicity $C-1$.

Chapter 3

Euclidean Cloud

(*'CLOUD'*)

A cloud without metric is a cloud without shape.
J.P. Benzécri

Introduction

In this chapter, we study the central object of Geometric Data Analysis (GDA), namely, the *Euclidean cloud*, that is, a family of statistical observations conceptualized as points in a multidimensional Euclidean space[1].

Ready–made Euclidean clouds occur whenever observations are points in a plane (two–dimensional space) or in a physical three–dimensional space, structured as classical geometric Euclidean spaces, see chapter 10 (*MATH.BASES*) p.439. For instance, the impacts of bullets on a target, or the positions of the bees in a swarm define Euclidean clouds. In GDA, Euclidean clouds are constructed from numerical data sets, such as contingency tables in Correspondence Analysis, dissimilarity tables in metric or nonmetric MDS, Individuals×Variables tables in Principal Component Analysis for numerical variables and Multiple Correspondence Analysis for categorized ones.

The chapter is organized as follows. First of all, we define basic statistics (§3.1), followed by the orthogonal projection of a cloud onto a subspace and its breaking down into projected clouds (§3.2). We study the principal directions of a cloud, we present Specific Analysis (§3.3) and introduce principal hyperellipsoids (§3.4). Then we study the partition of a cloud into

[1] In this chapter, by a space (without qualification) we always mean a multidimensional geometric space whose elements are points, as opposed to a vector space whose elements are vectors.

subclouds, with the between–within breakdown (§3.5), and introduce Euclidean classification, that is, Agglomerative Hierarchical Clustering (AHC) with aggregation according to variance (§3.6). Finally, we move on "from points to numbers", choosing a Cartesian frame and working on coordinate variables (§3.7).

The Mathematical Bases relevant to this chapter are multidimensional geometry (*MATH.BASES*, §10.4) and the theorem of spectral analysis (§10.5).

3.1 Basic Statistics

In this chapter, we will study a Euclidean cloud, in the line of pure geometry, that is, using a coordinate–free approach. Without loss of generality, this study will be conducted on a cloud of ten points in a plane, the *Target example*, as represented on the opposite figure. When it comes to numerical computations, we will take a unit length u such that the frame of the picture is a $38u \times 38u$ square.

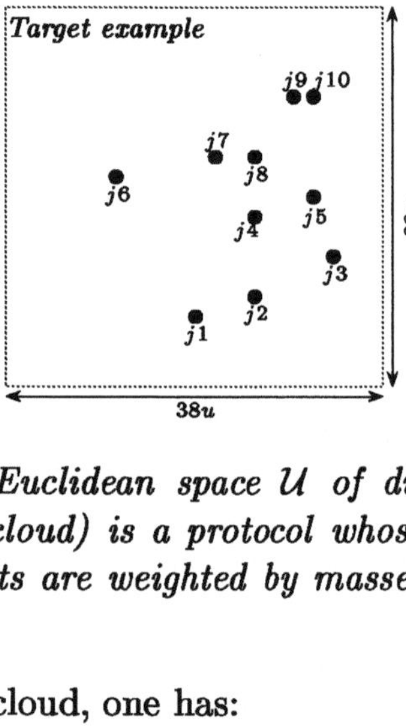

Definition 3.1 (Euclidean cloud). *Given a Euclidean space $\mathcal{U}$ of dimensionality K, a Euclidean cloud (in brief a cloud) is a protocol whose observations $(\mathrm{M}^j)_{j\in J}$ are points in $\mathcal{U}$. The points are weighted by masses $(\varpi_j)_{j\in J}$ (with $\forall j \in J : \varpi_j > 0$).*

Denoting J the set–theoretic support of the cloud, one has:

$$\mathrm{M}^J : \begin{array}{lll} J & \longrightarrow \mathcal{U} & \text{(Euclidean space)} \\ j & \longmapsto \mathrm{M}^j & \text{(point)} \end{array} \qquad \varpi_J : \begin{array}{lll} J & \longrightarrow \mathbb{R} & \\ j & \longmapsto \varpi_j & \text{(mass)} \end{array}$$

According to the *duality notation* (*CORRESPONDENCE*, p.26), points are indexed by upper indices and *weights* (or masses) by lower indices.

Weights are usually derived from an absolute frequency measure n_J (possibly elementary) with its relative frequency measure $f_j = n_j/n$, letting $n = \sum\limits_{j\in J} n_j$ (total absolute frequency of cloud). The affine *support* of the cloud (in brief support), denoted $\mathcal{M}\subseteq\mathcal{U}$, is the smallest subspace containing the points of the cloud; the *dimensionality of the cloud*, denoted L, is by definition the dimensionality of its support with $L \leq K$ ($\dim \mathcal{M} = L$). By *plane cloud* we mean a two–dimensional cloud ($L = 2$).

The basic statistics of a Euclidean cloud (mean point, sum of squares, variance and contributions) are the multidimensional extensions of the basic statistics of a numerical protocol.

Geometric notation. *Points* are denoted by capital letters: M, P, A, etc. *Geometric vectors*, elements of the vector space $\mathcal{V}$ underlying $\mathcal{U}$, are "arrowed": $\overrightarrow{\varepsilon}$, $\overrightarrow{\alpha}$, $\overrightarrow{0}$ (null–vector); the vector associated with the bipoint (P,M) is denoted $\overrightarrow{\mathrm{PM}}$, or M − P as the *deviation* of M from P ("endpoint − origin"). The *scalar product* on $\mathcal{V}$ is denoted $\langle\cdot|\cdot\rangle$ and the *Euclidean norm* is denoted $\|\cdot\|$; the *Euclidean distance* between two points M and P is denoted MP (or PM).

3.1.1 Mean Point

Definition 3.2 (Mean point). *Let* P *be a point in* $\mathcal{U}$; *the point* G *such that* $\overrightarrow{\mathrm{PG}} = \sum_{j\in J} f_j \overrightarrow{\mathrm{PM}^j}$ *does not depend on point* P; *it is called the mean point of the cloud and can be written* $\mathrm{G} = \sum_{j\in J} f_j\,\mathrm{M}^j$ (MATH.BASES, p.436).

Property 3.1 (Barycentric property). *The following equations define the barycentric characterization of the mean point.*

$$\sum_{j\in J} f_j \overrightarrow{\mathrm{GM}^j} = \overrightarrow{0} \quad \textit{or} \quad \sum_{j\in J} n_j \overrightarrow{\mathrm{GM}^j} = \overrightarrow{0}$$

Proof. If we let P = G in the preceding definition, we obtain the property. ◁

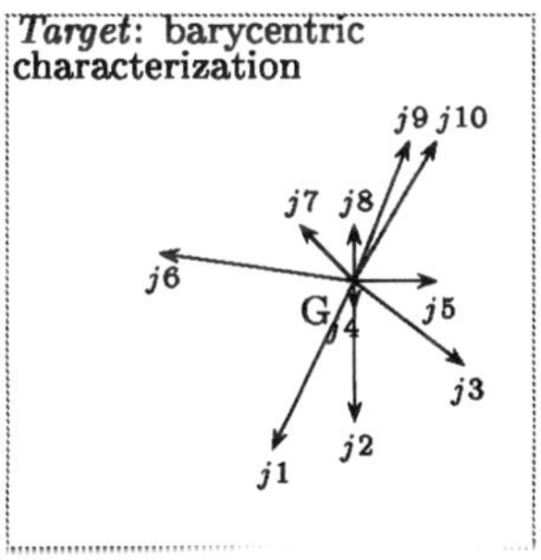

For a *cloud of two points* (M^j, n_j) and $(\mathrm{M}^{j'}, n_{j'})$, the mean point G belongs to the segment $\mathrm{M}^j\mathrm{M}^{j'}$; it is such that $n_j\overrightarrow{\mathrm{GM}^j} + n_{j'}\overrightarrow{\mathrm{GM}^{j'}} = \overrightarrow{0}$ (cf. opposite figure with $n_j = 3$, $n_{j'} = 2$; $\mathrm{GM}^j/\mathrm{GM}^{j'} = n_{j'}/n_j = 2/3$).
For $n_j = n_{j'}$, the mean point is the midpoint of the segment $\mathrm{M}^j\mathrm{M}^{j'}$.

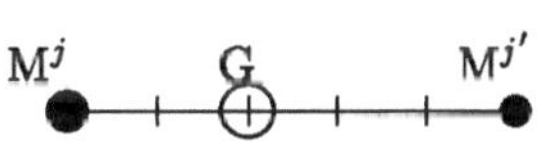

3.1.2 Inertia, Sum of Squares, Variance and Contributions

Definition 3.3 (Inertia). *Given a weighting measure* ϖ_J *and a reference point P, the inertia of the cloud with respect to point P is the* ϖ_J*–weighted sum of the squares of the distances of its points from point* P, *and is denoted* $\mathrm{In}^{\mathrm{P}}\mathrm{M}^J$, *with:*

$$\mathrm{In}^{\mathrm{P}}\mathrm{M}^J = \sum_{j\in J} \varpi_j\,(\mathrm{PM}^j)^2$$

The part of this sum accounted for by the point M^j is denoted $\mathrm{In}_j^{\mathrm{P}}$, with $\mathrm{In}_j^{\mathrm{P}} = \varpi_j(\mathrm{PM}^j)^2$.

The inertias most commonly used in GDA are the n_J–inertia or *Sum of Squares*, and the f_J–inertia or *Mean of Squares.*

Definition 3.4 (Variance). *The variance of the cloud* M^J *is the weighted mean of the squares of the distances of its points from the mean point* G.

$$\mathrm{Var}\,\mathrm{M}^J = \sum_{j\in J} f_j(\mathrm{GM}^j)^2$$

The variance of the cloud M^J can also be expressed as a function of the distances between pairs of points (cf. Exercise 3.1, p.121), that is:

$$\mathrm{Var}\,\mathrm{M}^J = \frac{1}{2}\sum_{j\in J}\sum_{j'\in J} f_j f_{j'}(\mathrm{M}^j\mathrm{M}^{j'})^2 = \sum_{jj'\in\mathcal{P}_2(J)} f_j f_{j'}(\mathrm{M}^j\mathrm{M}^{j'})^2 \qquad (3.1)$$

where $\mathcal{P}_2(J)$ denotes the set of $J(J-1)/2$ pairs of elements of J.

Definition 3.5 (Absolute contribution). *The absolute contribution of the point* M^j *to the variance of the cloud, denoted* Cta_j *("Ct" for Contribution, "a" for absolute), is the part of variance accounted for by this point:*

$$\mathrm{Cta}_j = f_j\,(\mathrm{GM}^j)^2$$

The breakdown of variance according to points is therefore:

$$\mathrm{Var}\,\mathrm{M}^J = \sum_{j\in J} \mathrm{Cta}_j$$

Definition 3.6 (Relative contribution). *The relative contribution of the point* M^j *to the variance of the cloud, denoted* Ctr_j *("Ct" for Contribution, "r" for relative), is the proportion of variance accounted for by the point* M^j*:*

$$\mathrm{Ctr}_j = \frac{\mathrm{Cta}_j}{\mathrm{Var}\,\mathrm{M}^J} = \frac{n_j(\mathrm{GM}^j)^2}{\sum_{j\in J} n_j(\mathrm{GM}^j)^2}$$

Contributions add up: According to the *duality notation*, index j is written below (subscript) in Cta_j and Ctr_j.

Target example. With the distance unit previously specified, the sum of squares with respect to the target center P is equal to 1280; the centered sum of squares is equal to 920, hence the variance is equal to 92.

Theorem 3.1 (First Huyghens' theorem). *The mean of the squares of the distances of the points of a cloud from a reference point* P *is equal to the variance of the cloud plus the square of the distance from the mean point of the cloud to the reference point* P.

$$\forall \mathrm{P} \in \mathcal{U} : \sum_{j\in J} f_j(\mathrm{PM}^j)^2 = (\mathrm{PG})^2 + \sum_{j\in J} f_j(\mathrm{GM}^j)^2$$

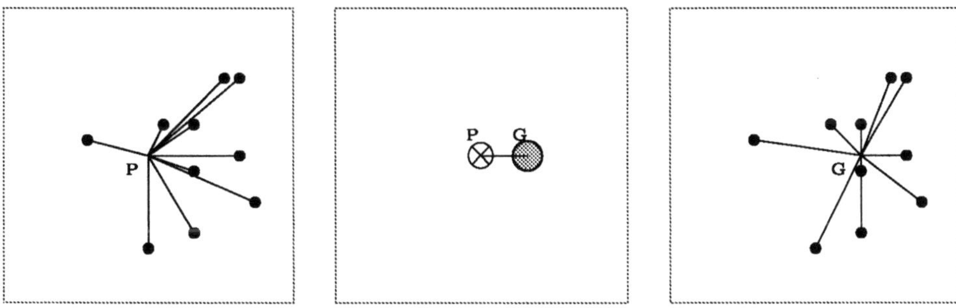

Target example: illustration of Huyghens' theorem.

Proof. From $\overrightarrow{\mathrm{PM}^j} = \overrightarrow{\mathrm{PG}} + \overrightarrow{\mathrm{GM}^j}$, one gets $(\mathrm{PM}^j)^2 = \mathrm{PG}^2 + (\mathrm{GM}^j)^2 + 2\langle\overrightarrow{\mathrm{PG}}|\overrightarrow{\mathrm{GM}^j}\rangle$, and $\sum_{j\in J} f_j\,(\mathrm{PM}^j)^2 = \mathrm{PG}^2 + \sum_{j\in J} f_j\,(\mathrm{GM}^j)^2 + 2\langle\overrightarrow{\mathrm{PG}}|\sum_{j\in J} f_j\,\overrightarrow{\mathrm{GM}^j}\rangle$. One has $\sum_{j\in J} f_j\overrightarrow{\mathrm{GM}^j} = \overrightarrow{0}$ (barycentric property), hence the theorem. ◁

From Huyghens' theorem, the *metric characterization of the mean point* follows: If to each point $\mathrm{P} \in \mathcal{U}$ there is assigned the quantity $\sum_{j\in J} f_j(\mathrm{PM}^j)^2$, this quantity is minimum when P coincides with the mean point G. In other words, the mean point is the *point of least squares*; among the clouds that reduce to a single point (i.e. zero–dimensional clouds), the one which is concentrated in point G is the one which best fits the observed cloud in the sense of least squares.

3.2 Projected Clouds

Let (M^J, n_J) be a Euclidean cloud, and let $\mathcal{H}$ be a subspace of $\mathcal{U}$. Let H^j be the orthogonal projection of point M^j onto $\mathcal{H}$ (MATH.BASES, p.439).

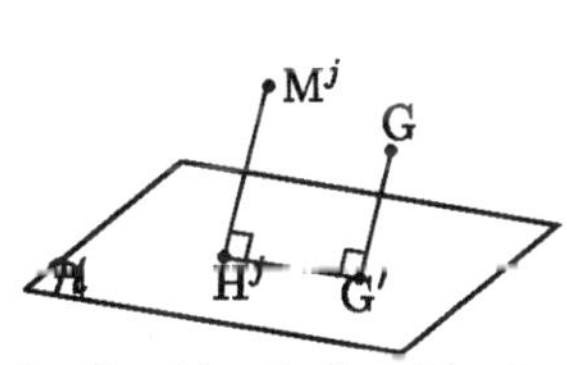

Figure 3.1: $\dim\mathcal{U}=3, \dim\mathcal{H}=2$

Definition 3.7 (Projected cloud). *The orthogonal projection onto* $\mathcal{H}$ *of the cloud* M^J, *or projected cloud, is the cloud* H^J *whose points are the orthogonal projections of points of the cloud* M^J *onto* $\mathcal{H}$.

Property 3.2. *The mean point of the cloud projected onto $\mathcal{H}$ is the projection onto $\mathcal{H}$ of the mean point of the cloud* M^J.

Proof. As an affine mapping, projection preserves barycenters, that is, from $\mathrm{M}^j \mapsto \mathrm{H}^j$ one gets $\sum\limits_{j\in J} f_j \mathrm{M}^j \mapsto \sum\limits_{j\in J} f_j \mathrm{H}^j$, therefore $\mathrm{G} \mapsto \mathrm{G}'$ (G' is the mean point of the projected cloud H^J). ◁

— *Remark.* In this chapter, all projections are orthogonal ones; consequently, we will often omit "orthogonal".

3.2.1 Variance in a Direction

Property 3.3 (Variance in a direction). *Projected clouds onto parallel subspaces of same dimensionalities have the same variances; their common value is called the variance of cloud in the direction of these subspaces.*

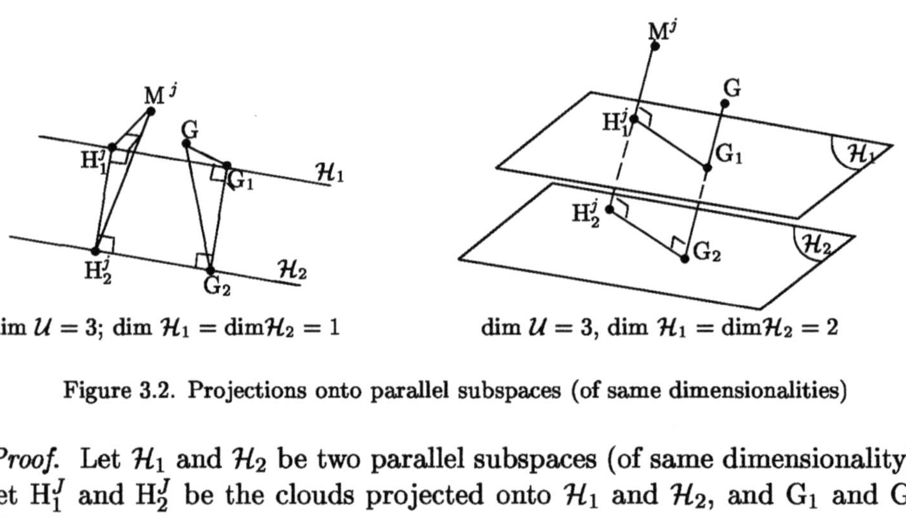

dim $\mathcal{U} = 3$; dim $\mathcal{H}_1 = \dim\mathcal{H}_2 = 1$ dim $\mathcal{U} = 3$, dim $\mathcal{H}_1 = \dim\mathcal{H}_2 = 2$

Figure 3.2. Projections onto parallel subspaces (of same dimensionalities)

Proof. Let $\mathcal{H}_1$ and $\mathcal{H}_2$ be two parallel subspaces (of same dimensionality); let H_1^J and H_2^J be the clouds projected onto $\mathcal{H}_1$ and $\mathcal{H}_2$, and G_1 and G_2 their mean points.

The cloud H_2^J is obtained from the cloud H_1^J by the translation of vector $\overrightarrow{\mathrm{G}_1\mathrm{G}_2}$; one has $\forall j \in J : \mathrm{H}_2^j = \mathrm{H}_1^j + \overrightarrow{\mathrm{G}_1\mathrm{G}_2}$ (Figure 3.2) and $\overrightarrow{\mathrm{G}_1\mathrm{H}}{}_1^j = \overrightarrow{\mathrm{G}_2\mathrm{H}}{}_2^j$, hence $\sum\limits_{j\in J} f_j\,(\mathrm{G}_1\mathrm{H}_1^j)^2 = \sum\limits_{j\in J} f_j\,(\mathrm{G}_2\mathrm{H}_2^j)^2$. ◁

A *line* (one–dimensional subspace) defined by a point and a vector $\overrightarrow{\alpha}$ ("direction vector") is called an *axis*; the variance of the projected cloud onto the line is called *variance in direction* $\overrightarrow{\alpha}$, or *variance of axis* (even though it does not depend on orientation).

Property 3.4. *The variance of the cloud* M^J *in direction* $\overrightarrow{\alpha}$ *is equal to:*

$$\sum_{j \in J} f_j \frac{\langle \overrightarrow{\mathrm{GM}^j} | \overrightarrow{\alpha} \rangle^2}{\| \overrightarrow{\alpha} \|^2}$$

Proof. The projected point of M^j onto the axis $(\mathrm{G}, \overrightarrow{\alpha})$ is the point A^j defined by the formula (*MATH.BASES*, p.439): $\overrightarrow{\mathrm{GA}}^j = \frac{\langle \overrightarrow{\mathrm{GM}}^j | \overrightarrow{\alpha} \rangle}{\| \overrightarrow{\alpha} \|^2} \overrightarrow{\alpha}$; hence: $\operatorname{Var} \mathrm{A}^J = \sum_{j \in J} f_j (\mathrm{GA}^j)^2 = \sum_{j \in J} f_j \frac{\langle \overrightarrow{\mathrm{GM}}^j | \overrightarrow{\alpha} \rangle^2}{\| \overrightarrow{\alpha} \|^2}$. ◁

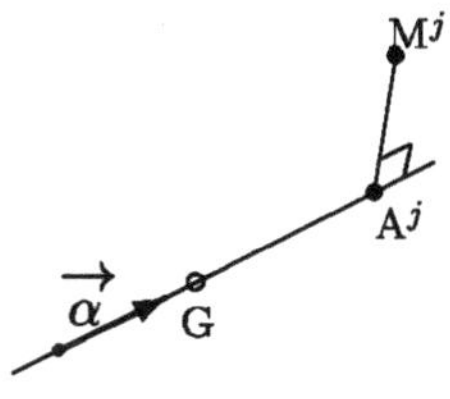

Definition 3.8 (Spherical cloud). *A cloud is spherical if its variance is the same in all the directions of its support.*

3.2.2 Residual Square Mean

Definition 3.9 (Residual square mean). *The mean of residual squares (or residual square mean) of a cloud with respect to a subspace is the weighted mean of the squares of the distances from the points of the cloud to this subspace.*

Let H^j be the orthogonal projection of M^j onto $\mathcal{H}$, hence the distance $\mathrm{M}^j\mathrm{H}^j$ of M^j to $\mathcal{H}$. The residual square mean of M^J with respect to $\mathcal{H}$ is equal to $\sum_{j \in J} f_j\, (\mathrm{M}^j\mathrm{H}^j)^2$.

Theorem 3.2 (General Huyghens' theorem). *The mean of residual squares of a cloud with respect to a subspace is equal to its residual square mean with respect to the parallel subspace of same dimensionality going through the mean point, plus the square of the distance from the mean point to this subspace.*

Let $\mathcal{A}$ be the subspace parallel to $\mathcal{H}$ going through the mean point G. Denoting A^j the projection of the point M^j onto $\mathcal{A}$, and denoting H^j and G' the respective orthogonal projections of points M^j and G onto $\mathcal{H}$ (cf. Figure 3.3, p.82), the theorem reads:

$$\sum_{j \in J} f_j\, (\mathrm{M}^j\mathrm{H}^j)^2 = \sum_{j \in J} f_j (\mathrm{M}^j\mathrm{A}^j)^2 + (\mathrm{GG}')^2$$

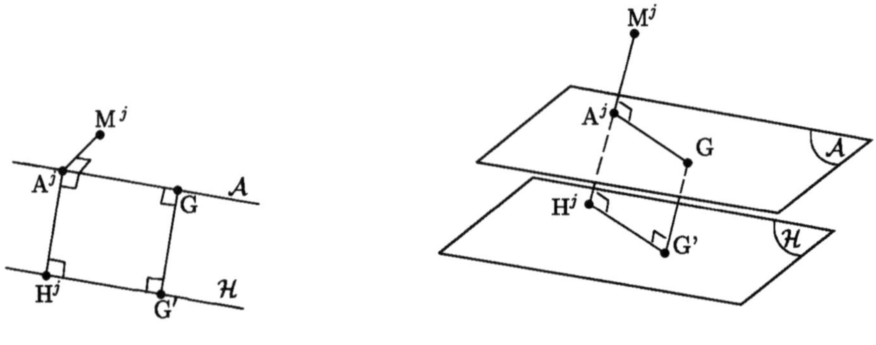

Figure 3.3. $\dim \mathcal{U} > \dim\mathcal{H} = 1$ and $\dim \mathcal{U} > \dim\mathcal{H} = 2$

Proof. One has: $\overrightarrow{\mathrm{M}^j\mathrm{H}^j} = \overrightarrow{\mathrm{M}^j\mathrm{A}^j} + \overrightarrow{\mathrm{A}^j\mathrm{H}^j}$, with $\overrightarrow{\mathrm{A}^j\mathrm{H}^j} = \overrightarrow{\mathrm{GG}'}$, therefore: $(\mathrm{M}^j\mathrm{H}^j)^2 = (\mathrm{M}^j\mathrm{A}^j)^2 + (\mathrm{GG}')^2 + 2\langle\overrightarrow{\mathrm{M}^j\mathrm{A}^j}|\overrightarrow{\mathrm{GG}'}\rangle$. From barycentric property (p.77) one has $\sum_{j\in J} f_j\,\overrightarrow{\mathrm{GM}^j} = \overrightarrow{0} = \sum_{j\in J} f_j\,\overrightarrow{\mathrm{GA}^j}$, hence $\sum_{j\in J} f_j\,\overrightarrow{\mathrm{M}^j\mathrm{A}^j} = \overrightarrow{0}$; it follows that $\sum_{j\in J} f_j\,(\mathrm{M}^j\mathrm{H}^j)^2 = \sum_{j\in J} f_j(\mathrm{M}^j\mathrm{A}^j)^2 + (\mathrm{GG}')^2$. ◁

3.2.3 Fitted and Residual Clouds

In the fitting perspective, whenever one replaces the cloud M^J by a cloud of lower dimensionality, the projected cloud A^J is called a *fitted cloud*, and $\forall j \in J$ the vector $\overrightarrow{\mathrm{A}^j\mathrm{M}^j}$ is called the *residual deviation*, where:

$$\forall j \in J \quad \mathrm{M}^j = \mathrm{A}^j + \overrightarrow{\mathrm{A}^j\mathrm{M}^j} \quad \text{with} \quad \overrightarrow{\mathrm{A}^j\mathrm{M}^j} \perp \mathcal{A}$$

observed point = fitted point + residual deviation

Property 3.5. *If* G′ *and* Hʲ *are the orthogonal projections of points* G *and* Mʲ *onto the subspace* $\mathcal{H}$*, the cloud* R^J *defined by* $\mathrm{R}^j = \mathrm{G}' + \overrightarrow{\mathrm{H}^j\mathrm{M}^j}$ $(j \in J)$ *only depends on the direction of* $\mathcal{H}$.

Proof.
$$\begin{aligned}\mathrm{R}^j &= \mathrm{G}' + \overrightarrow{\mathrm{H}^j\mathrm{M}^j}\\ &= (\mathrm{G}' + \overrightarrow{\mathrm{G}'\mathrm{G}}) + \overrightarrow{\mathrm{GG}'} + \overrightarrow{\mathrm{H}^j\mathrm{M}^j};\ \text{now}\ \ \overrightarrow{\mathrm{GG}'} = \overrightarrow{\mathrm{A}^j\mathrm{H}^j}\\ &= \quad \mathrm{G} \quad + \overrightarrow{\mathrm{A}^j\mathrm{H}^j} + \overrightarrow{\mathrm{H}^j\mathrm{M}^j} = \mathrm{G} + \overrightarrow{\mathrm{A}^j\mathrm{M}^j}\end{aligned}$$ ◁

Definition 3.10 (Residual cloud). *The residual cloud is defined by:*

$$\forall j \in J \ : \ \mathrm{R}^j = \mathrm{G} + \overrightarrow{\mathrm{A}^j\mathrm{M}^j}$$

residual point = mean point + residual deviation

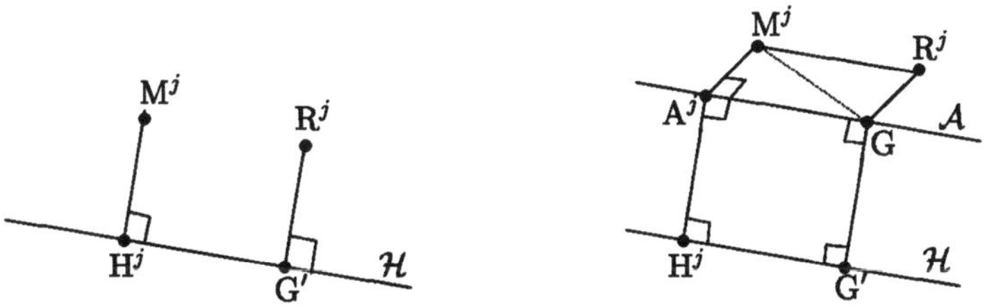

Figure 3.4. Residual cloud (dim $\mathcal{H} = 1$, dim $\mathcal{U} > 1$)

By construction, G is the mean point of the residual cloud R^J; the support of R^J is orthogonal to $\mathcal{H}$; hence the orthogonal breakdown:

$$\overrightarrow{\mathrm{GM}}^j = \overrightarrow{\mathrm{GA}}^j + \overrightarrow{\mathrm{GR}}^j \quad \text{with } \overrightarrow{\mathrm{GA}}^j \perp \overrightarrow{\mathrm{GR}}^j \text{ (cf. figure 3.4)}$$

The variance of the residual cloud is equal to the residual square mean of the cloud M^J with respect to $\mathcal{A}$: $\mathrm{Var}\,\mathrm{R}^J = \sum\limits_{j\in J} f_j(\mathrm{GR}^j)^2 = \sum\limits_{j\in J} f_j(\mathrm{A}^j\mathrm{M}^j)^2$.

Property 3.6 (Breakdown of variance of cloud). *The variance of a cloud is equal to its variance in the direction of $\mathcal{H}$ (variance of the projected cloud), plus the variance of the residual cloud.*

$$\mathrm{Var}\,\mathrm{M}^J = \mathrm{Var}\,\mathrm{A}^J + \mathrm{Var}\,\mathrm{R}^J \tag{3.2}$$

Total variance = fitted variance + residual variance

Proof. $\overrightarrow{\mathrm{GM}}^j = \overrightarrow{\mathrm{GA}}^j + \overrightarrow{\mathrm{GR}}^j$, with $\overrightarrow{\mathrm{GA}}^j \perp \overrightarrow{\mathrm{GR}}^j$, now $\mathrm{GR}^j = \mathrm{A}^j\mathrm{M}^j$, hence: $(\mathrm{GM}^j)^2 = (\mathrm{GA}^j)^2 + (\mathrm{A}^j\mathrm{M}^j)^2$. ◁

3.2.4 Variables Attached to an Axis

Let $(\mathrm{G}, \overrightarrow{\alpha})$ be an axis, one defines the projected cloud $\mathrm{A}^J = (\mathrm{A}^j)_{j\in J}$ onto this axis and the following variables.

The *covariant variable* $\alpha^J = (\alpha^j)_{j\in J}$ with $\alpha^j = \langle \overrightarrow{\mathrm{GM}}^j | \overrightarrow{\alpha} \rangle$; α^j is the *covariant coordinate* of the point M^j with respect to $(\mathrm{G}, \overrightarrow{\alpha})$. The α^J variable is centered and its variance is $\mathrm{Var}\,\mathrm{A}^J\, \|\overrightarrow{\alpha}\|^2$.

The *calibrated variable* y^J with $y^j = \alpha^j / \|\overrightarrow{\alpha}\|$. The variable y^J is centered and its variance is $\mathrm{Var}\,\mathrm{A}^J$.

The *standard variable* z^J with $z^j = y^j / \mathrm{SD}\, y^J = y^j / \mathrm{SD}\,\mathrm{A}^J$ (Mean $z^J = 0$ and $\mathrm{Var}\, z^J = 1$).

The *axial variable* t^J with $t^j = \langle \overrightarrow{\mathrm{GM}}^j | \overrightarrow{\alpha} \rangle / \|\overrightarrow{\alpha}\|^2$. The variable t^J is centered and its variance is $\mathrm{Var}\, t^J = \mathrm{A}^J / \|\overrightarrow{\alpha}\|^2$; t^j is the coordinate of A^j on axis $(\mathrm{G}, \overrightarrow{\alpha})$ (MATH.BASES, p.440).

If $\overrightarrow{\alpha}$ is multiplied by a coefficient κ, the covariant coordinate α^j of point M^j is multiplied by κ, hence the term "covariant" (nothing to do with the covariance of two variables). The calibrated variable y^J is the covariant variable associated with the unit–norm vector $\overrightarrow{\alpha}/\|\overrightarrow{\alpha}\|$. The covariant variable α^J is mainly of mathematical interest; the calibrated and the standard variables are *intrinsic* (up to orientation) and more directly interpretable statistically.

Dimensional considerations (*MATH.BASES*, p.442). The norm of a geometric vector is assimilable to a *length*, that is, it is of dimension 1; the scalar product of two vectors is of dimension 2. The covariant variable is therefore of dimension 2, the calibrated variable is of dimension 1, the standard variable is "without dimension". The properties of Euclidean clouds — being pure geometrical properties — are *homogeneous*, that is, they do not depend on unit length.

3.2.5 Linear Formalization

We now introduce the linear mappings that will intervene in the determination of principal directions; these mappings are defined in terms of the following vector spaces: the Euclidean vector space $\mathcal{V}$ underlying the geometric space $\mathcal{U}$; the vector space $\mathbb{R}^J$ of variables over J; and the vector space $\mathbb{R}_J$ of measures over J.

The linear algebra notions used in the rest of this chapter are recalled in chapter 10 (*MATH.BASES*).

Definition 3.11 (Linear mapping *Vac*). *Given a point* $\mathrm{P}\in\mathcal{U}$, *the mapping* Vac_{P} *("Va" for* **Variable**, *"c" for* **covariant***) is the linear mapping (homomorphism) (depending on* P*) such that:*

$$\begin{aligned} Vac_{\mathrm{P}} : \mathcal{V} &\longrightarrow \mathbb{R}^J \\ \overrightarrow{\alpha} &\longmapsto \alpha^J_{\mathrm{P}} \quad \text{with } \alpha^J_{\mathrm{P}} = \left(\langle \overrightarrow{\mathrm{PM}}^j | \overrightarrow{\alpha} \rangle\right)_{j\in J} \end{aligned}$$

Let Vac denote Vac_{P} when $\mathrm{P}=\mathrm{G}$. From $\overrightarrow{\mathrm{PM}}^j = \overrightarrow{\mathrm{PG}} + \overrightarrow{\mathrm{GM}}^j$ we deduce $\langle \overrightarrow{\mathrm{PM}}^j | \overrightarrow{\alpha} \rangle = \langle \overrightarrow{\mathrm{PG}} | \overrightarrow{\alpha} \rangle + \langle \overrightarrow{\mathrm{GM}}^j | \overrightarrow{\alpha} \rangle$, that is, $\alpha^j_{\mathrm{P}} = \langle \overrightarrow{\mathrm{PG}} | \overrightarrow{\alpha} \rangle + \alpha^j_{\mathrm{G}}$, hence:

$$\forall \overrightarrow{\alpha} \in \mathcal{V} \quad Vac_{\mathrm{P}}(\overrightarrow{\alpha}) = \langle \overrightarrow{\mathrm{PG}} | \overrightarrow{\alpha} \rangle 1^J + Vac(\overrightarrow{\alpha}) \tag{3.3}$$

Definition 3.12 (Linear mapping *Eff*). *Given a point* $\mathrm{P}\in\mathcal{U}$, *the mapping* $E\!f\!f^{\mathrm{P}}$ *(Eff for* **Effect***) is the linear mapping such that:*

$$\begin{aligned} E\!f\!f^{\mathrm{P}} : \mathbb{R}_J &\longrightarrow \mathcal{V} \\ u_J &\longmapsto \sum_{j\in J} u_j \overrightarrow{\mathrm{PM}}^j \end{aligned}$$

Let $u = \sum_{j\in J} u_j$. If mbox$u \neq 0$, $\mathit{Eff}^{\mathrm{P}}$ depends on P and $\mathit{Eff}^{\mathrm{P}}(u_J) = u\,\overrightarrow{\mathrm{PG}}$.
If $u = 0$ (u_J is a contrast), the vector $\sum_{j\in J} u_j\overrightarrow{\mathrm{PM}}^j$ does not depend on P, it is denoted $\sum_{j\in J} u_j\mathrm{M}^j$ (*MATH.BASES*, p.436) and called *vector–effect.*

From the orthogonal breakdown of u_J as the contrast $u_J^0 = (u_j - u f_j)_{j\in J}$ and the measure $u\,f_J$, one gets $\mathit{Eff}^{\mathrm{P}}(u_J) = \mathit{Eff}^{\mathrm{P}}(u_J^0) + u\,\mathit{Eff}^{\mathrm{P}}(f_J)$.

Letting Eff be the Effect mapping with G as a reference point, we have:
$\forall u_j \in \mathbb{R}_J$: $\mathit{Eff}(u_J) = \sum_{j\in J} u_j\overrightarrow{\mathrm{GM}}^j = \sum_{j\in J} u_j^0\mathrm{M}^j = \mathit{Eff}(u_J^0)$, hence:

$$\forall u_J \in \mathbb{R}_J \quad \mathit{Eff}^{\mathrm{P}}(u_J) = u\,\mathit{Eff}^{\mathrm{P}}(f_J) + \mathit{Eff}(u_J) \quad \text{with } \mathit{Eff}^{\mathrm{P}}(f_J) = \overrightarrow{\mathrm{PG}} \qquad (3.4)$$

— *Remark.* In Analysis of Variance, when u_J is a contrast, the vector $\sum u_j\mathrm{M}^j$ is interpreted as the *Effect* of the contrast on the cloud; hence the name *Effect* that we use for this mapping.

Duality scheme. Let us now recall (*CORRESPONDENCE*, p.27) that with each variable x^J the isomorphism $f_J : \mathbb{R}^J \to \mathbb{R}_J$ associates the measure $x_J = (f_j\,x^j)_{j\in J}$ such that x^J is the f_J–density of x_J. We are thus led to the following *duality scheme*:

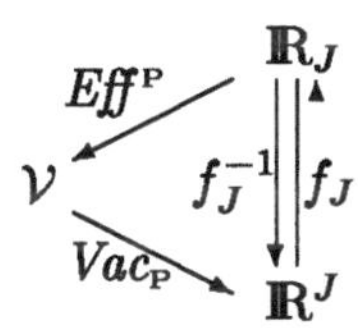

In connection with the duality scheme, there are the homomorphism $\mathit{Vac}_{\mathrm{P}}^* : \mathbb{R}^J \to \mathcal{V}$, adjoint of $\mathit{Vac}_{\mathrm{P}}$, as well as the two symmetric and positive homomorphisms $\mathit{Som}_{\mathrm{P}} : \mathcal{V} \to \mathcal{V}$ and $\mathit{Tom}_{\mathrm{P}} : \mathbb{R}^J \to \mathbb{R}^J$ (*MATH.BASES*, p.433).

Property 3.7. *The homomorphism $\mathit{Eff}^{\mathrm{P}} \circ f_J$ is the adjoint of $\mathit{Vac}_{\mathrm{P}}$.*

Proof. One has:
$\langle \mathit{Vac}_{\mathrm{P}}(\overrightarrow{\alpha})|x^J\rangle = \sum_{j\in J} f_j x^j \alpha_{\mathrm{P}}^j = \sum_{j\in J} f_j x^j \langle\overrightarrow{\mathrm{PM}}^j|\overrightarrow{\alpha}\rangle$
and $(\mathit{Eff}^{\mathrm{P}} \circ f_J)(x^J) = \sum_{j\in J} f_j x^j \overrightarrow{\mathrm{PM}}^j$. Hence
$\forall \overrightarrow{\alpha} \in \mathcal{V}\ \forall x^J \in \mathbb{R}^J$: $\langle \mathit{Vac}_{\mathrm{P}}(\overrightarrow{\alpha})|x^J\rangle = \langle \overrightarrow{\alpha}\,|\mathit{Eff}^{\mathrm{P}} \circ f_J(x^J)\rangle$
and the property of adjoint homomorphisms (*MATH.BASES*, Theorem 10.3, p.432). ◁

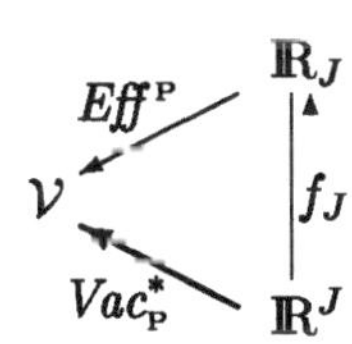

Linear mapping Vac^*. Let Vac^*_{P} be the homomorphism $Eff^{\mathrm{P}} \circ f_J$:

$$\begin{aligned} Vac^*_{\mathrm{P}} : \mathbb{R}^J &\longrightarrow \mathcal{V} \\ x^J &\longmapsto \sum_{j \in J} f_j\, x^j \overrightarrow{\mathrm{PM}}^j \end{aligned}$$

For P = G, we denote Vac the linear mapping. From $\overrightarrow{\mathrm{PM}}^j = \overrightarrow{\mathrm{PG}} + \overrightarrow{\mathrm{GM}}^j$ one deduces: $\sum_{j\in J} f_j x^j \overrightarrow{\mathrm{PM}}^j = \overline{x}\overrightarrow{\mathrm{PG}} + \sum_{j\in J} f_j x^j \overrightarrow{\mathrm{GM}}^j$ with $\overrightarrow{\mathrm{PG}} = Vac^*_{\mathrm{P}}(1^J)$ and $\overline{x} = \sum_{j\in J} f_j x^j$, hence:

$$\forall x^J \in \mathbb{R}^J\colon\ Vac^*_{\mathrm{P}}(x^J) = \overline{x}\ Vac^*_{\mathrm{P}}(1^J) + Vac^*(x^J) \tag{3.5}$$

Definition 3.13 (Endomorphism *Som*). *$Eff^{\mathrm{P}} \circ f_J \circ Vac_{\mathrm{P}}$ is the symmetric endomorphism on $\mathcal{V}$ associated with Vac_{P} and denoted Som_{P}.*

$$\begin{aligned} Som_{\mathrm{P}} : \mathcal{V} &\longrightarrow \mathcal{V} \\ \overrightarrow{\alpha} &\longmapsto \sum_{j\in J} f_j \langle \overrightarrow{\mathrm{PM}}^j | \overrightarrow{\alpha} \rangle \overrightarrow{\mathrm{PM}}^j \end{aligned}$$

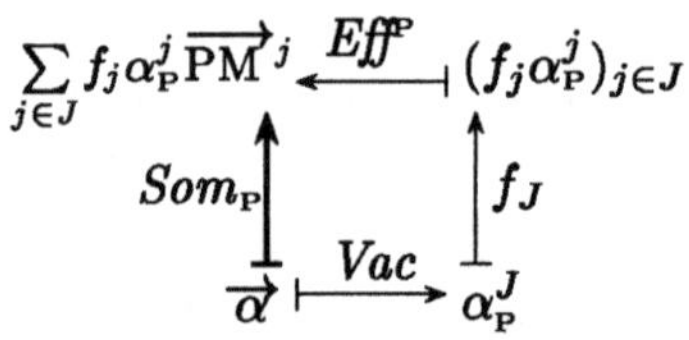

$Som_{\mathrm{P}} = Vac^*_{\mathrm{P}} \circ Vac_{\mathrm{P}}$ is a positive endomorphism of $\mathcal{V}$ (MATH.BASES, p.433).

Denoting *Som* the endomorphism with G as reference point, one has:

$$\forall \overrightarrow{\alpha} \in \mathcal{V} : Som_{\mathrm{P}}(\overrightarrow{\alpha}) = \langle \overrightarrow{\mathrm{PG}} \mid \overrightarrow{\alpha} \rangle \overrightarrow{\mathrm{PG}} + Som(\overrightarrow{\alpha}) \tag{3.6}$$

Property 3.8. *The variance of cloud in direction $\overrightarrow{\alpha}$ is given by:*

$$\frac{\| Vac(\overrightarrow{\alpha})\|^2}{\|\overrightarrow{\alpha}\|^2} = \frac{\langle\, Som(\overrightarrow{\alpha}) | \overrightarrow{\alpha} \rangle}{\|\overrightarrow{\alpha}\|^2}$$

Proof. One has: $\| Vac(\overrightarrow{\alpha})\|^2 = \operatorname{Var} \alpha^J = \operatorname{Var} \mathrm{A}^J \, \|\overrightarrow{\alpha}\|^2$ (cf. §3.2.4, p.83), and $\langle Som(\overrightarrow{\alpha}) | \overrightarrow{\alpha} \rangle = \| Vac(\overrightarrow{\alpha})\|^2$. ◁

Definition 3.14 (Endomorphism *Tom*). *$Vac_{\mathrm{P}} \circ Eff^{\mathrm{P}} \circ f_J$ is the symmetric endomorphism on $\mathbb{R}^J$ associated with Vac_{P} and denoted Tom_{P}.*

$$\begin{aligned} Tom_{\mathrm{P}} : \mathbb{R}^J &\to \mathbb{R}^J \\ x^J &\mapsto \Big(\sum_{j'\in J} f_{j'} x^{j'} \langle \overrightarrow{\mathrm{PM}}^{j'} | \overrightarrow{\mathrm{PM}}^j \rangle \Big)_{j\in J} \end{aligned}$$

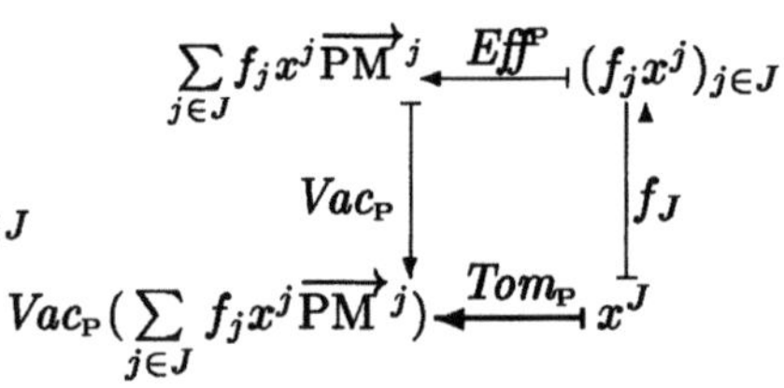

Bilinear and quadratic forms. We now take the mean point G as reference point.

- The bilinear form connected with *Vac* is the mapping $\mathcal{V} \times \mathbb{R}^J \to \mathbb{R}$ that associates with $(\overrightarrow{\alpha}, x^J)$ the scalar product $\langle Vac(\overrightarrow{\alpha}) | x^J \rangle$, i.e. the covariance between the covariant variable α^J and the variable x^J:
$\langle Vac(\overrightarrow{\alpha}) | x^J \rangle = \sum_{j \in J} f_j x^j \langle \overrightarrow{\mathrm{GM}}^j | \overrightarrow{\alpha} \rangle = \sum_{j \in J} f_j \alpha^j x^j = \mathrm{Cov}(\alpha^J | x^J)$
(since the covariant variable α^J is centered).

- The bilinear form connected with *Som* is the mapping $\mathcal{V} \times \mathcal{V} \to \mathbb{R}$ that associates with $(\overrightarrow{\alpha}, \overrightarrow{\beta})$ the scalar product $\langle \overrightarrow{\alpha} | Som(\overrightarrow{\beta}) \rangle$, i.e. the covariance of the covariant variables associated with the vectors $\overrightarrow{\alpha}$ and $\overrightarrow{\beta}$.
Indeed, $\langle \overrightarrow{\alpha} | Som(\overrightarrow{\beta}) \rangle = \langle \overrightarrow{\alpha} | \sum_{j \in J} f_j \langle \overrightarrow{\mathrm{GM}}^j | \overrightarrow{\beta} \rangle \overrightarrow{\mathrm{GM}}^j \rangle = \mathrm{Cov}(\alpha^J | \beta^J)$.
The quadratic form connected with *Som* is the mapping $\mathcal{V} \to \mathbb{R}$ that associates with $\overrightarrow{\alpha} \in \mathcal{V}$ the variance of α^J: $\langle Som(\overrightarrow{\alpha}) | \overrightarrow{\alpha} \rangle = \mathrm{Var}\, \alpha^J$.

- The bilinear form connected with *Tom* is the mapping $\mathbb{R}^J \times \mathbb{R}^J \to \mathbb{R}$ that associates with the pair of variables (x^J, y^J) the scalar product $\langle x^J | Tom(y^J) \rangle = \langle Eff(x_J) | Eff(y_J) \rangle$ (with $x_J = (f_j x^j)_{j \in J}$ and $y_J = (f_j y^j)_{j \in J}$). The quadratic form connected with *Tom* is the mapping form $\mathbb{R}^J \to \mathbb{R}$ that associates with the variable x^J the square of the norm of the effect of x_J, that is, $\| \sum_{j \in J} x_j \overrightarrow{\mathrm{GM}}^j \|$.

3.3 Principal Directions of a Cloud

The problem of principal directions. Given a high–dimensional cloud, one seeks to approximate it by a cloud in a lower dimensional space, such as a one–dimensional cloud (line) or a two–dimensional cloud (plane), etc. such that the approximated cloud is intuitively "of greatest elongation". When operationalized in terms of *variance*, this requirement leads to the *problem of principal directions*. For any given L' (with $1 \leq L' \leq L$), among the L'–dimensional projected clouds, we search the projected cloud having maximum variance. From the variance breakdown property (p.83), it is equivalent to search for the L'–dimensional subspace going through the mean point G for which the residual square mean is minimum. This subspace is called the *first L'-dimensional principal subspace*, and its direction, the *first L'-dimensional principal direction*. As will be seen, for any given

L', the first principal direction always exists and is generally unique. Thus there is in general a unique first principal line ($L' = 1$), then a unique first principal plane ($L' = 2$), etc., hence a *hierarchy of principal directions.* Furthermore, this hierarchy possesses the *heredity property,* that is, the first principal plane contains the first principal line, etc.

— *Remarks.* The "principal concept" extends to $L' = 0$, the mean point is the "principal zero–dimensional subspace" (first Huyghens' theorem, p.79).

The problem of principal directions is akin to that of *regression.* In both problems we search a fitting which is optimal in the sense of least squares. In regression, the least squares are *parallel* to a given direction and only involve affine structures; whereas in principal directions, they are "orthogonal least squares" (perpendicular to the sought direction) and involve the Euclidean distance.

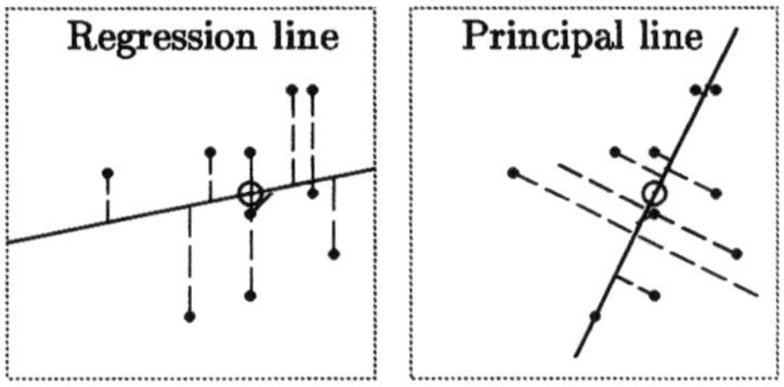

The case of a *plane cloud* is thoroughly treated by elementary algebra in Exercise 3.3 (p.123); the results are applied to the *Target example* in Exercise 3.4 (p.125).

3.3.1 Principal Axes

For the first principal line, the problem reads: Among all one–dimensional directions, find one along which the variance of cloud is maximum.

From Property 3.8 (p.86) the variance of the projected cloud on a line directed by $\overrightarrow{\alpha}$ is $(\| \mathit{Vac}(\overrightarrow{\alpha})\| / \|\overrightarrow{\alpha}\|)^2$. Therefore one looks for $\overrightarrow{\alpha}$ such that the ratio $\| \mathit{Vac}(\overrightarrow{\alpha})\| / \|\overrightarrow{\alpha}\|$ is maximum.

The solution is provided by the *spectral analysis* of the linear mapping *Vac* or of the endomorphism *Som*, which leads to the eigendirection and eigenvalue equation $\mathit{Som}(\overrightarrow{\alpha}) = \lambda\, \overrightarrow{\alpha}$. The maximum of $\| \mathit{Vac}(\overrightarrow{\alpha})\| / \|\overrightarrow{\alpha}\|$ is reached for any eigenvector of *Som* associated with its maximum eigenvalue λ_1 (MATH.BASES, p.446), and is equal to the maximum singular value $\xi_1 = \sqrt{\lambda_1}$.

Theorem 3.3 (Principal direction equation). *The direction vectors of principal lines are eigenvectors of the endomorphism Som.*

$$\forall \ell = 1, \ldots L \; : \; \mathit{Som}(\overrightarrow{\alpha_\ell}) = \lambda_\ell \overrightarrow{\alpha_\ell}$$

Proof. Any eigenvector $\overrightarrow{\alpha_1}$ of the endomorphism *Som* associated with the greatest eigenvalue λ_1 defines a *first principal one–dimensional direction* (or line direction); $\overrightarrow{\alpha_1}$ is called *first principal vector.*

If the eigenvalue λ_1 is simple, the first principal line direction is unique, the axis defined by $(\mathrm{G}, \overrightarrow{\alpha_1})$ is the first principal axis of the cloud M^J; the projected cloud A_1^J onto this axis is the first one–dimensional principal cloud, its variance (i.e. the variance along direction $\overrightarrow{\alpha_1}$) is equal to λ_1.

If the eigenvalue λ_1 is multiple of order p, there is a p–dimensional subspace for which every line direction is principal; a first principal line direction will be any line direction arbitrarily chosen among them, yielding a first one–dimensional principal cloud of variance λ_1.

With a first one–dimensional principal cloud A_1^J there is associated a residual cloud R_1^J such that $\overrightarrow{\mathrm{GR}_1^j} = \overrightarrow{\mathrm{GM}_1^j} - \overrightarrow{\mathrm{GA}_1^j}$ $(j \in J)$ (cf. Definition 3.10, p.82). If the residual cloud R_1^J is concentrated in the point G, that is, if M^J is one–dimensional, the search of principal direction is completed. If this is not the case, one determines the first principal direction of the residual cloud R_1^J, which leads to the *second principal line direction* of the cloud M^J. Let $Vac' : \mathcal{V} \to \mathrm{I\!R}^J$ be the linear mapping associated with the residual cloud R^J defined by $Vac'(\overrightarrow{\alpha}) = (\langle \overrightarrow{\mathrm{GR}_1^j} | \overrightarrow{\alpha} \rangle)_{j \in J}$. One has $Vac(\overrightarrow{\alpha}) = (\langle \overrightarrow{\mathrm{GM}^j} | \overrightarrow{\alpha} \rangle)_{j \in J}$ and $Vac_1(\overrightarrow{\alpha}) = (\langle \overrightarrow{\mathrm{GA}_1^j} | \overrightarrow{\alpha} \rangle)_{j \in J}$, therefore $Vac' = Vac - Vac_1$. Let likewise $Vac'^* = Vac^* - Vac_1^* : \mathrm{I\!R}^J \to \mathcal{V}$ such that $Vac_1^*(x^J) = \sum_{j \in J} f_j x^j \overrightarrow{\mathrm{GA}_1^j}$. It is easily proved that Som' $(= Vac'^* \circ Vac')$ is equal to $Som - Som_1$, with $Som_1 = Vac_1^* \circ Vac_1$. Now Som' has the same eigenvectors as Som (MATH.BASES, Property 10.28, p.443), and $\overrightarrow{\alpha_1}$ is an eigenvector of Som' associated with the eigenvalue 0, and the greatest eigenvalue of Som' is λ_2 (possibly equal to λ_1, when λ_1 is not simple). As a result, the maximum of $\| Vac'(\overrightarrow{\alpha}) \| / \| \overrightarrow{\alpha} \|$ is ξ_2 and it is attained by $\overrightarrow{\alpha_2}$, an eigenvector of Som associated with the second eigenvalue λ_2, that is, $Som(\overrightarrow{\alpha_2}) = \lambda_2 \overrightarrow{\alpha_2}$ and $\overrightarrow{\alpha_2} \perp \overrightarrow{\alpha_1}$ since the mapping Som is symmetric (MATH.BASES, Property 10.26, p.443).

From the residual cloud $(\overrightarrow{\mathrm{GM}^j} - \overrightarrow{\mathrm{GA}_1^j} - \overrightarrow{\mathrm{GA}_2^j})_{j \in J}$, one proceeds in the same way to determine a *third principal direction*, and so on, until the determination of an L-th one–dimensional principal cloud, with a zero–dimensional residual cloud. ◁

The construction of the L line directions entails the following property.

Property 3.9. *The principal vectors $(\overrightarrow{\alpha_\ell})_{\ell=1,\ldots L}$ are pairwise orthogonal.*

The line defined by $(\mathrm{G}; \overrightarrow{\alpha_\ell})$ is called the ℓ–th principal axis, or in brief *Axis ℓ*. The projected cloud A_ℓ^J onto Axis ℓ is called the *ℓ–th principal one–dimensional cloud.* The variance of the cloud A_ℓ^J is equal to λ_ℓ and is called variance of Axis ℓ.

Notation. The index ℓ is here a mere enumeration index, arbitrarily put as a lower index without reference to duality; as opposed to its position in the notation of contributions (cf. later §3.3.3)

Property 3.10 (Principal breakdown of variance). *The variance of the cloud is equal to the sum of eigenvalues:*

$$\mathrm{Var}\,\mathrm{M}^J = \sum_{\ell=1}^{L} \lambda_\ell$$

The breakdown of the variance of a cloud follows from the breakdown of the cloud into L one–dimension orthogonal principal clouds:

$$\forall j \in J : \mathrm{M}^j = \mathrm{G} + \sum_{\ell=1}^{L} \overrightarrow{\mathrm{GA}}{}_\ell^j \qquad (\ell \neq \ell' \Rightarrow \forall j \in J : \overrightarrow{\mathrm{GA}}{}_\ell^j \perp \overrightarrow{\mathrm{GA}}{}_{\ell'}^j)$$

Theorem 3.4. *A first L'–dimensional principal direction is spanned by L' first principal linearly independent vectors ($1 \leq L' \leq L$).*

Property 3.11 (Heredity property). *If $\lambda_{L'} > \lambda_{L'-1}$, a first principal L'–dimensional direction contains all the first principal L''–dimensional directions with $L'' < L'$.*

Thus, a first principal plane contains the first principal line, and so on. The theorem can be proved recursively, by showing that the L–dimensional subspace contains the $L-1$ one, then that it contains the principal line $\mathcal{D}_L$. Hereafter we sketch the proof by showing that the first principal plane is spanned by the first two principal lines (assuming $\lambda_1 > \lambda_2 > \lambda_3$).

Proof. Denoting $\mathcal{P}$ the first principal plane and $\mathcal{D}_1$ the first principal line, suppose that $\mathcal{D}_1 \not\subset \mathcal{P}$ (proof by contradiction).

Let $\mathcal{D}'_2 = \mathcal{P} \cap \mathcal{D}_1^\perp$ be the intersection of the first principal plane $\mathcal{P}$ with the hyperplane orthogonal to $\mathcal{D}_1$, and $\mathcal{D}'_1$ be the line of $\mathcal{P}$ orthogonal to $\mathcal{D}'_2$ ($\mathcal{D}'_1 \in \mathcal{P}$ and $\mathcal{D}'_1 \perp \mathcal{D}'_2$). If $\mathrm{A'}_1^J$ and $\mathrm{A'}_2^J$ are respectively the projected clouds onto $\mathcal{D}'_1$ and $\mathcal{D}'_2$, the variance of the projected cloud onto $\mathcal{P}$ is equal to $\mathrm{Var}\,\mathrm{A'}_1^J + \mathrm{Var}\,\mathrm{A'}_2^J$.

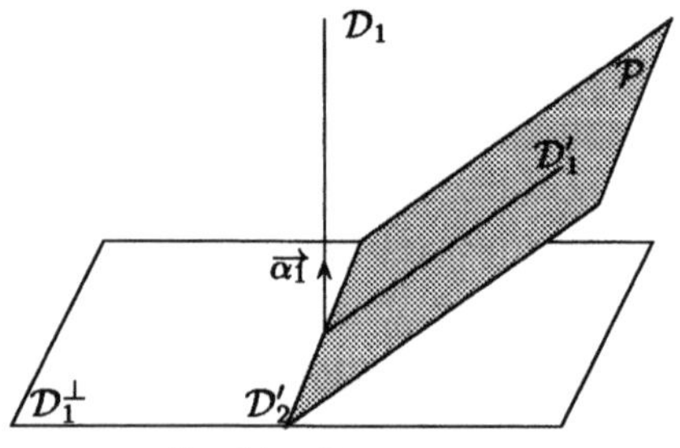

dim $\mathcal{U} = 3$

Since $\mathcal{D}_1$ is the first principal line, one has $\text{Var}\,\text{A'}_1^J \leq \text{Var}\,\text{A}_1^J = \lambda_1$, hence the cloud projected onto the plane spanned by $(\mathcal{D}_1, \mathcal{D}'_2)$ has a greater variance than the cloud projected onto the plane $\mathcal{P}$, which contradicts the hypothesis. Therefore the plane $\mathcal{P}$ contains the line $\mathcal{D}_1$.

Suppose now that $\mathcal{D}_2 \not\subset \mathcal{P}$. The intersection of the hyperplane orthogonal to $\mathcal{D}_2$ with $\mathcal{P}$ is $\mathcal{D}_1$; let $\mathcal{D}'_2$ denote the line of $\mathcal{P}$ orthogonal to $\mathcal{D}_1$. If A_1^J and A'^J_2 are respectively the projected clouds onto $\mathcal{D}_1$ and $\mathcal{D}'_2$, the variance of the projected cloud onto the plane $\mathcal{P}$ is equal to $\text{Var}\,\text{A}_1^J + \text{Var}\,\text{A'}_2^J = \lambda_1 + \text{Var}\,\text{A'}_2^J$.

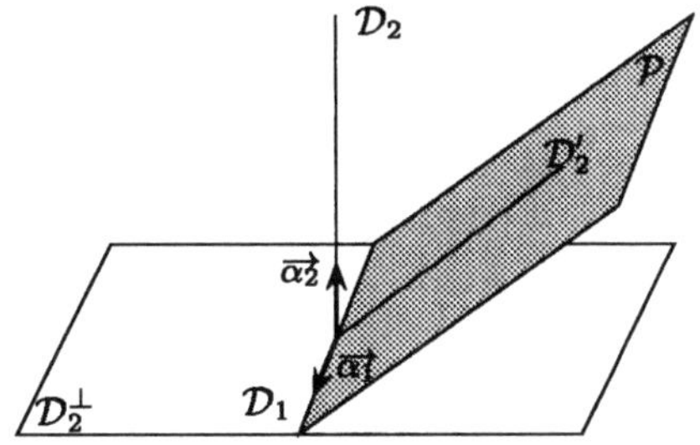

dim $\mathcal{U} = 3$

Since $\mathcal{D}'_2 \perp \mathcal{D}_1$, $\text{Var}\,\text{A'}_2^J \leq \lambda_2$, the variance of the projected cloud onto the plane $(\mathcal{D}_1, \mathcal{D}_2)$ is equal to $\lambda_1 + \lambda_2$; hence the variance of the projected cloud onto the plane $\mathcal{P}$ is lower than $\lambda_1 + \lambda_2$: the plane $\mathcal{P}$ contains $\mathcal{D}_2$. ◁

Multiple eigenvalues. The approach remains valid when some eigenvalues are multiple. With each multiple eigenvalue there is associated a principal subspace for which each direction is principal, and within which it is always possible to choose orthonormal directions.

Null eigenvalues. In the supplementary orthogonal subspace of the support of the cloud every direction is a principal one associated with a null eigenvalue; the corresponding principal clouds are concentrated in the point G and the associated principal variables are identically null.

3.3.2 Principal Variables

Definition 3.15 (Principal coordinates). *The coordinate y_ℓ^j on Axis ℓ in frame $(\text{G}, \overrightarrow{\alpha_\ell^\ell}/\|\overrightarrow{\alpha_\ell^\ell}\|)$ of the projected point A_ℓ^j of the point M^j onto Axis ℓ is called ℓ-th principal coordinate of the point M^j.*

$$y_\ell^j = \langle \overrightarrow{\text{GM}}^j | \overrightarrow{\alpha_\ell} \rangle / \|\overrightarrow{\alpha_\ell}\|$$

Property 3.12 (Principal breakdown of distances).

$$(\text{GM}^j)^2 = \sum_{\ell=1}^{L} (y_\ell^j)^2 \qquad (\text{M}^j\text{M}^{j'})^2 = \sum_{\ell=1}^{L} (y_\ell^j - y_\ell^{j'})^2$$

Proof. $\overrightarrow{\text{GM}}^j = \sum_{\ell=1}^{L} \overrightarrow{\text{GA}}_\ell^j = \sum_{\ell=1}^{L} y_\ell^j\, \overrightarrow{\alpha_\ell^\ell}/\|\overrightarrow{\alpha_\ell^\ell}\|$ where $(\overrightarrow{\alpha_\ell^\ell})_{\ell=1,\ldots L}$ is an orthogonal basis of the support of the cloud, hence the property. ◁

With the one-dimensional principal cloud referred to $(\mathrm{G}, \overrightarrow{\alpha_\ell})$ are attached the following two (centered and proportional) *principal variables*.
— The first *calibrated principal variable* y_1^J, in brief *principal variable*, is defined by $y_1^j = \langle \overrightarrow{\mathrm{GM}}^j | \overrightarrow{\alpha_1} \rangle / \| \overrightarrow{\alpha_1} \|$, i.e. the coordinate of point A_1^j in frame $(\mathrm{G}, \overrightarrow{\alpha_1}/\|\overrightarrow{\alpha_1}\|)$; one has $\operatorname{Var} y^J = \lambda_1$.
— The first *standard principal variable* z_1^J is defined by $z_1^j = y_1^j/\xi_1$, i.e. the coordinate of point A_1^j in frame $(\mathrm{G}, \xi_1\overrightarrow{\alpha_1}/\|\overrightarrow{\alpha_1}\|)$; one has $\operatorname{Var} z^J = 1$.

Property 3.13 (Passage formulas). *Between the principal vector $\overrightarrow{\alpha_\ell}$ and the standard principal variable z_ℓ^J, one has:*

$$\begin{cases} Vac(\overrightarrow{\alpha_\ell}/\|\overrightarrow{\alpha_\ell}\|) & = \xi_\ell \, z_\ell^J \\ Vac^*(z_\ell^J) & = \xi_\ell \, \overrightarrow{\alpha_\ell}/\|\overrightarrow{\alpha_\ell}\| \end{cases} \tag{3.7}$$

The passage formulas allow one to go from a principal vector to a principal variable and vice versa (*MATH.BASES*, Theorem 10.6, p.445).

Theorem 3.5. *The principal variables are eigenvectors of the endomorphism Tom.*

$$\forall \ell = 1, \dots L \quad Tom(z_\ell^J) = \lambda_\ell \, z_\ell^J \tag{3.8}$$

Proof. From passage formulas, it follows that $Tom(z_\ell^J) = Vac \circ Vac^\star(z_\ell^J) = \xi_\ell \, Vac(\overrightarrow{\alpha_\ell})/\|\overrightarrow{\alpha_\ell}\| = \lambda_\ell z_\ell^J$. ◁

Property 3.14. *The ℓ–th and ℓ'–th principal variables ($\xi_\ell \neq \xi_{\ell'}$) are uncorrelated.*

This property follows from the symmetry of the endomorphism *Tom*.

3.3.3 Contributions

Definition 3.16 (Absolute contribution of point and axis). *The absolute contribution of point* M^j *and Axis ℓ, denoted* $\mathrm{Cta}_{j\ell}$*, is the absolute contribution of* A_ℓ^j *to the variance of the projected cloud* A_ℓ^J *on Axis ℓ:*

$$\mathrm{Cta}_{j\ell} = f_j(\mathrm{GA}_\ell^j)^2 = f_j(y_\ell^j)^2$$

Since $\overrightarrow{\mathrm{GA}}_\ell^j = y_\ell^j \overrightarrow{\alpha_\ell}$, one has $\mathrm{Cta}_{j\ell} = f_j(y_\ell^j)^2$, $\sum_{j \in J} \mathrm{Cta}_{j\ell} = \lambda_\ell$ (variance of Axis ℓ), and $\sum_{\ell=1}^{L} \mathrm{Cta}_{j\ell} = f_j \sum_{\ell=1}^{L} (\mathrm{GA}_\ell^j)^2 = f_j(\mathrm{GM}^j)^2 = \mathrm{Cta}_j$ (absolute contribution of j to the variance of the cloud, cf. definition 3.5 p.78).

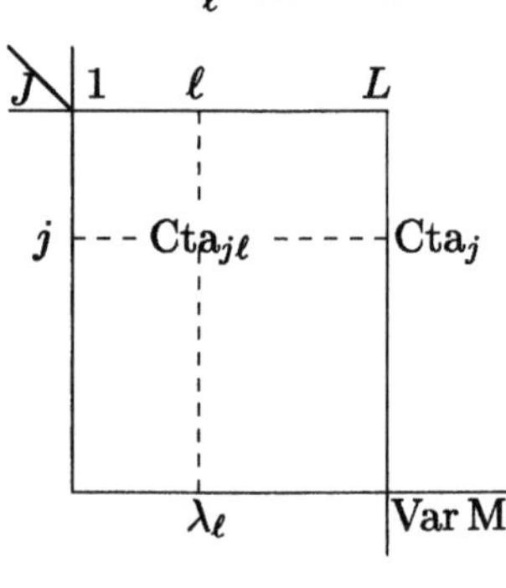

The *double breakdown of the variance of the cloud according to points and axes* follows:

$$\operatorname{Var} \mathrm{M}^J = \sum_{j\in J} \sum_{\ell=1}^{L} \mathrm{Cta}_{j\ell}$$

Absolute contributions add up; the lower positions of j and ℓ are meaningful in terms of duality notation. From absolute contributions, the following two relative contributions are defined.

Definition 3.17 (Relative contribution of point to axis).
The relative contribution of the point M^j *to Axis* ℓ*, denoted* Ctr_j^ℓ*, is the proportion of the variance of Axis* ℓ *accounted for by the point* M^j*:*

$$\mathrm{Ctr}_j^\ell = \mathrm{Cta}_{j\ell}/\lambda_\ell = f_j (y_\ell^j)^2/\lambda_\ell$$

More generally, the relative contribution of the point M^j to the L'–principal direction is defined by $\sum_{\ell=1}^{L'} \mathrm{Cta}_{j\ell} / (\sum_{\ell=1}^{L'} \lambda_\ell)$, which is the weighted mean (the weights being $(\lambda_\ell)_{\ell=1,\dots L'}$) of the relative contributions of M^j to axes 1 to L'. For $L' = L$, $\mathrm{Ctr}_j^L = f_j(\mathrm{GM}^j)^2/\operatorname{Var} \mathrm{M}^J$ is the relative contribution of the point M^j to the variance of the cloud (denoted Ctr_j, cf. p.78).

Definition 3.18 (Relative contribution of axis to point).
The relative contribution of Axis ℓ *to the point* M^j*, or quality of representation of* M^j *on Axis* ℓ*, denoted* Ctr_ℓ^j*, is equal to* $\mathrm{Cta}_{j\ell}/\mathrm{Cta}_j$.

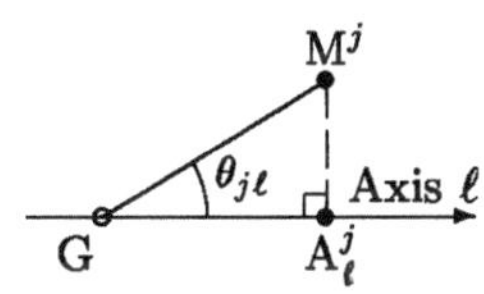

$\mathrm{Ctr}_\ell^j = (y_\ell^j)^2/(\mathrm{GM}^j)^2 = (\mathrm{GA}_\ell^j)^2/(\mathrm{GM}^j)^2$ is equal to the square of the cosine of the (acute) angle $\theta_{j\ell}$ between $\overrightarrow{\mathrm{GM}^j}$ with its projection onto Axis ℓ.

This contribution can be interpreted as the *quality of representation* of the point M^j on Axis ℓ: The closer to 1 its value is, the better the point is represented on axis. The quality of representation of the point M^j on the L'–principal space is the sum of the qualities of representation of this point on axes 1 through L': $\sum_{\ell=1}^{L'} \mathrm{Ctr}_\ell^j$. For $L' = L$ (complete reconstitution of the cloud), the quality of representation is equal to 1.

The duality notation applies here like for the conditional frequencies of a contingency table (§2.2.1, p.32): ℓ is an upper index in $\mathrm{Ctr}_j^\ell = \mathrm{Cta}_{j\ell}/\lambda_\ell$, and j is an upper index in Ctr_ℓ^j.

3.3.4 Specific Analysis of a Cloud

The *specific analysis* of a cloud consists in determining its principal axes under the constraint that they must belong to a given prespecified subspace.

Let a subspace $\mathcal{A} \subset \mathcal{U}$ going through the point G. Let us denote A^j the orthogonal projection of the point M^j onto $\mathcal{A}$ and B^j its projection onto $\mathcal{A}^\perp$, orthogonal supplementary subspace in $\mathcal{A}$ going through G. One has:

$$\forall j \in J \; : \; \overrightarrow{\mathrm{GM}}^j = \overrightarrow{\mathrm{GA}}^j + \overrightarrow{\mathrm{GB}}^j \quad \text{with } \overrightarrow{\mathrm{GA}}^j \perp \overrightarrow{\mathrm{GB}}^j$$

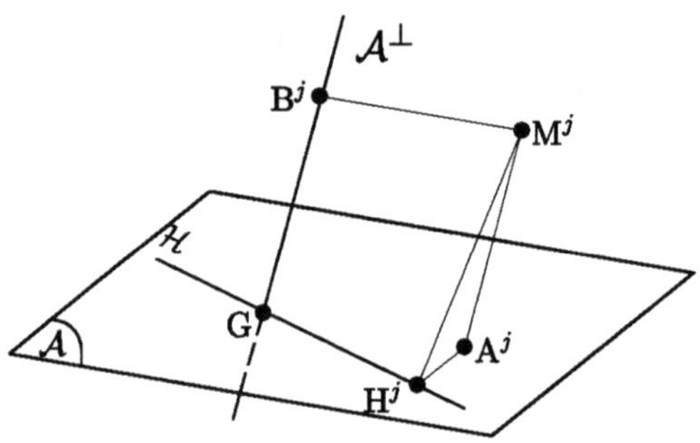

Figure 3.5. Specific analysis ($\dim \mathcal{U} = 3$, $\dim \mathcal{A} = 2$)

Let $\mathcal{H}$ be a subspace of $\mathcal{A}$. the point H^j, the orthogonal projection of M^j onto $\mathcal{H}$, is also the orthogonal projection of A^j onto $\mathcal{H}$, hence the relation: $(\mathrm{M}^j\mathrm{H}^j)^2 = (\mathrm{M}^j\mathrm{A}^j)^2 + (\mathrm{A}^j\mathrm{H}^j)^2$, and

$$\sum_{j\in J} f_j d^2(\mathrm{M}^j, \mathcal{H}) = \sum_{j\in J} f_j d^2(\mathrm{M}^j, \mathcal{A}) + \sum_{j\in J} f_j d^2(\mathrm{A}^j, \mathcal{H})$$

Therefore, the subspace $\mathcal{H}$ of $\mathcal{A}$ such that the residual square mean of the cloud M^J is minimum is the first principal subspace associated with the cloud A^J; the specific coordinates of the point M^j are therefore equal to the principal coordinates of the point A^j.

The projection of the bipoint (G,M^j) is (G,A^j), hence $\mathrm{GA}^j \leq \mathrm{GM}^j$: The variance of the specific cloud A^J is always less than or equal to that of cloud M^J. More precisely, $(\lambda'_\ell)_{\ell=1,\ldots L'}$ denoting the eigenvalues of the specific cloud ($L' \leq L$), then one has the property:

$$\sum_{\ell=1}^{L''} \lambda'_\ell \leq \sum_{\ell=1}^{L''} \lambda_\ell \quad (1 \leq L'' \leq L')$$

As will be seen in next chapters, subspaces involve in a specific analysis are usually supports of subcloud or derived cloud of methodological interest: e.g. the between–cloud associated with a partition, or, in MCA, a subcloud determined by some modalities of interest (*MULTIPLE*, §5.2).

3.4 Principal or Inertia Hyperellipsoids

This section is devoted to geometric characterizations that are intrinsically related to the principal directions of a cloud.

In the Euclidean space $\mathcal{U}$, we consider the geometric support of the cloud $\mathcal{M}\subseteq\mathcal{U}$ (see p.76), with the underlying vector subspace $\mathcal{L}\subseteq\mathcal{V}$ with its two orthogonal principal bases: the orthonormal one $(\overrightarrow{\alpha_\ell})_{\ell=1,\ldots L}$, and the "orthocalibrated" one $(\overrightarrow{\beta_\ell})_{\ell=1,\ldots L}$ with $\overrightarrow{\beta_\ell} = \xi_\ell\overrightarrow{\alpha_\ell}$.

In the space of variables, we consider the principal subspace $\mathcal{X}_0 = Vac(\mathcal{L})\subseteq \mathbb{R}^J_0$ (subspace of centered variables over J), also with its two orthogonal principal bases: the orthonormal one $(z^J_\ell)_{\ell=1,\ldots L}$, and the orthocalibrated one $(y^J_\ell)_{\ell=1,\ldots L}$.

Then we consider the bijections $\mathcal{L}\to\mathcal{X}_0$, $\mathcal{X}_0\to\mathcal{L}$, and $\mathcal{L}\to\mathcal{L}$ respectively associated with the homomorphisms Vac, Vac^* and Som, denoted with the same symbols and defined by

$$Vac\,(\overrightarrow{\alpha_\ell}) = y^J_\ell = \xi_\ell\, z^J_\ell \text{ and } Vac^*\,(z^J_\ell) = \overrightarrow{\beta_\ell} = \xi_\ell\,\overrightarrow{\alpha_\ell} \quad (\ell = 1,\ldots L)$$

$$Som\,(\overrightarrow{\alpha_\ell}) = Vac^*\circ Vac\,(\overrightarrow{\alpha_\ell}) = \lambda_\ell\,\overrightarrow{\alpha_\ell} \quad (\ell = 1,\ldots L)$$

3.4.1 Definitions and Properties

Let $\mathcal{H}\subset\mathcal{M}$ be the hyperplane of equation (in the orthonormal basis) $\sum\limits_{\ell=1}^{L} u_\ell y_\ell = \pm h$, with $\sum\limits_{\ell=1}^{L}(u_\ell)^2 = 1$; the unit–norm vector $\overrightarrow{u} = \sum\limits_{\ell=1}^{L} u_\ell\overrightarrow{\alpha_\ell}$ is a direction vector of the normal (perpendicular line) to $\mathcal{H}$, and h (>0) is the distance of the mean point G of the cloud to hyperplane $\mathcal{H}$ (MATH.BASES, p.440).

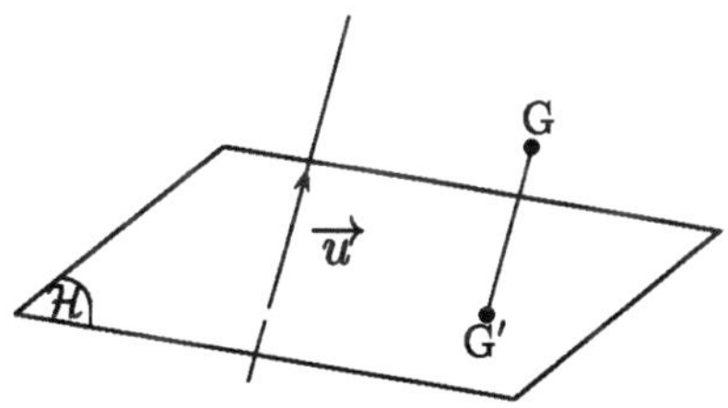

$\dim\mathcal{M} = 3$, $d(\mathrm{G},\mathcal{H}) = \mathrm{GG}' = h$

Let σ be the standard deviation (square root of variance) of the cloud along the direction $\overrightarrow{u}$ with $\sigma^2 = \|\,Vac(\overrightarrow{u})\|^2 = \sum\limits_{\ell=1}^{L}\lambda_\ell u^2_\ell$; and let $\kappa = h/\sigma$.

Definition 3.19. *The hyperellipsoid $\mathcal{E}_\kappa$ with principal axes of half-lengths $(\kappa\,\xi_\ell)_{\ell=1,\ldots L}$ is called the κ–principal, or κ–inertia hyperellipsoid of the cloud. For $\kappa = 1$, it is called the indicator hyperellipsoid.*

Relative to the orthonormal frame $(\mathrm{G}, (\overrightarrow{\alpha_\ell})_{\ell=1,\ldots L})$ of $\mathcal{M}$, the hyperellipsoid $\mathcal{E}_\kappa$ has Cartesian equation $\sum\limits_{\ell=1}^{L} y^2_\ell/\lambda_\ell = \kappa^2$, and tangential equation $\sum_{\ell=1}^{L} \lambda_\ell u_\ell^2 = h^2/\kappa^2$. In any orthonormal basis $(\overrightarrow{\delta_k})_{k\in K}$ of $\mathcal{U}$, the hyperellipsoid $\mathcal{E}_\kappa$ has tangential equation $\sum_{k\in K}\sum_{k'\in K} v^{kk'} u_k u_{k'} = h^2/\kappa^2$, where $v^{kk'} = \langle Som(\overrightarrow{\delta^k})|\overrightarrow{\delta^{k'}}\rangle$ denotes the covariance of the covariant variables associated with $\overrightarrow{\delta_k}$ and $\overrightarrow{\delta_{k'}}$ (see p.87), and $(u_k)_{k\in K}$ the coordinates of the unit–norm vector $\overrightarrow{u}$ (MATH.BASES, p.441)[2].

For the matrix formulations of those equations, see p.121.

Property 3.15. *The principal κ–hyperellipsoid is the envelope of the hyperplanes that are at the distance $\kappa\,\sigma$ of the mean point of the cloud.*

Proof. $h^2 = \kappa^2\sigma^2 \iff h^2 = \kappa^2 \sum_{\ell=1}^{L} \lambda_\ell u_\ell^2$, that is, $\sum_{\ell=1}^{L} \lambda_\ell u_\ell^2 = h^2/\kappa^2$ (tangential equation of κ-principal hyperellipsoid). ◁

Principal hyperellipsoids — also named *inertia hyperellipsoids* — are intrinsically attached to the cloud for which they provide a geometric summary. The distance between the two hyperplanes orthogonal to a given direction and tangent to the κ-hyperellipsoid is equal to 2κ times the standard deviation of the cloud along that direction. The contact points of the hyperplanes normal to a principal direction and tangent to the hyperellipsoid are the edges of the hyperellipsoid.

Property 3.16. *The indicator hyperellipsoid is the image by the mapping Vac* of the set of the standard variables (unit hypersphere) of $\mathcal{X}_0$.*

Proof. Let $x^J = \sum_{\ell=1}^{L} x_\ell z_\ell^J$ be a standard variable of $\mathcal{X}_0$ ($\sum_{\ell=1}^{L} x_\ell^2 = 1$). One has $Vac^*(x^J) = \sum_{\ell=1}^{L} x_\ell \xi_\ell \overrightarrow{\alpha}_\ell$. Letting $y_\ell = x_\ell \xi_\ell$, one has $\sum_{\ell=1}^{L} y_\ell^2/\lambda_\ell = 1$. When x^J goes through the hypersphere of unit radius of $\mathcal{X}_0$, its image $Vac^*(x^J)$ goes through the indicator hyperellipsoid. ◁

Property 3.17 (Projection of a hyperellipsoid). *If a cloud is orthogonally projected onto a subspace, the κ–hyperellipsoid of the projected cloud is the projection of the κ–hyperellipsoid of the cloud onto this subspace.*

[2] As a matter of fact, the tangential equation remains valid for a nonorthonormal basis, that is, κ is an affine index (independent from the initial metric on the whole space); of course, this invariance property does not hold if index κ is restricted to a strict principal subspace.

Proof. Let us refer the cloud to an orthonormal basis $(\overrightarrow{\delta_k})_{k\in K}$ such that $(\overrightarrow{\delta_k})_{k\in K'}$ is a basis of the subspace $\mathcal{L}'$ of $\mathcal{L}$ onto which the cloud is projected. The κ–hyperellipsoid $\mathcal{E}_\kappa$ has tangential equation $\sum\limits_{k\in K}\sum\limits_{k'\in K} u_k u_{k'} v^{kk'} = \frac{h^2}{\kappa^2}$. The cloud projected onto $\mathcal{L}'$ has variances and covariances $(v^{kk'})_{k\in K',k'\in K'}$, that is, the restriction to K' of the variances and covariances of the initial cloud; the κ–hyperellipsoid $\mathcal{E}'_\kappa$ of the projected cloud has therefore the equation in $\mathcal{L}'$: $\sum\limits_{k\in K'}\sum\limits_{k'\in K'} u_k u_{k'} v^{kk'} = h^2/\kappa^2$. On the other hand, the projection of $\mathcal{E}_\kappa$ onto $\mathcal{L}'$ is defined by the tangential equation of $\mathcal{E}_\kappa$ restricted to the hyperplanes that are orthogonal to $\mathcal{L}'$, which therefore verify $\forall k \in K - K' : u_k = 0$; which indeed comes back to $\mathcal{E}'_\kappa$. ◁

3.4.2 Hypersphere of the Contrast Space

Let $\mathcal{X}^0 = (f_J \circ Vac)(\mathcal{M})$ be the subspace of $\mathbb{R}_J$ spanned by the principal contrasts $(z_{J\ell})_{\ell=1,\dots L}$, with $z_{J\ell} = (f_j z^j_\ell)_{j\in J}$. On the one hand, the contrast $c^j_J = \sum\limits_{\ell=1}^{L} z^j_\ell z_{J\ell}$ is the projection of the measure δ^j_J onto $\mathcal{X}^0$; on the other hand, the effect of the contrast c^j_J on the cloud is $\overrightarrow{\mathrm{GM}}^j$. In $\mathcal{X}^0$, vector $\overrightarrow{\mathrm{GM}}^j$ will be represented by contrast c^j_J, and the cloud M^J will therefore be represented by a spherical cloud of unit variance in all directions. The image by the mapping $Eff^{-1} : \mathcal{L} \to \mathcal{X}^0$ of the κ–hyperellipsoid of equation $\sum\limits_{\ell=1}^{L} (y^j_\ell)^2/\lambda_\ell = \kappa^2$ is the κ–hypersphere of equation $\sum\limits_{\ell=1}^{L} (z^j_\ell)^2 = \kappa^2$.

3.4.3 Plane Cloud

Relative to the principal orthonormal basis $(\overrightarrow{\alpha_1}, \overrightarrow{\alpha_2})$, the principal κ–ellipse has equation:

$$y_1^2/\lambda_1 + y_2^2/\lambda_2 = \kappa^2 \tag{3.9}$$

Relative to any orthonormal basis $(\overrightarrow{\varepsilon}, \overrightarrow{\varepsilon}')$, where $v = \langle Som(\overrightarrow{\varepsilon})|\overrightarrow{\varepsilon}\rangle$, $v' = \langle Som(\overrightarrow{\varepsilon}')|\overrightarrow{\varepsilon}'\rangle$ and $c = \langle Som(\overrightarrow{\varepsilon})|\overrightarrow{\varepsilon}'\rangle$ (variances and covariance in the directions $\overrightarrow{\varepsilon}$, $\overrightarrow{\varepsilon}'$) and $r = c/\sqrt{v\,v'}$, the principal κ–ellipse has equation:

$$\frac{1}{1-r^2}\left(\frac{x^2}{v} - 2c\frac{xx'}{vv'} + \frac{x'^2}{v'}\right) = \kappa^2 \text{ or } \frac{1}{vv'-c^2}(v'x^2 - 2cxx' + vx'^2) = \kappa^2$$

For $-\pi \le \varphi \le +\pi$, let us consider the standard variable $t^J(\varphi)$ defined by $t^J(\varphi) = z^J_1 \cos\varphi + z^J_2 \sin\varphi$. When φ varies, the end of the vector $t^J(\varphi)$ goes through a unit circle.

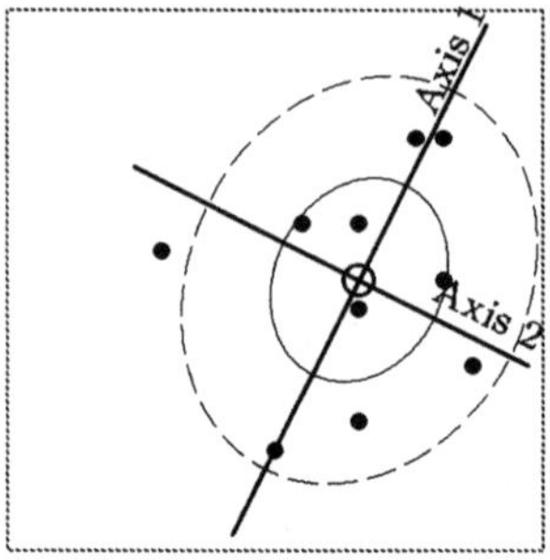

Figure 3.6. *Target example*: ($v = 56$, $v' = 36$, $c = 8$). Indicator ellipse ($\kappa = 1$, solid line) and concentration ellipse ($\kappa = 2$, dashed line).

One has: $Vac^*(t^J(\varphi)) = \cos\varphi\,\overrightarrow{\beta_1} + \sin\varphi\,\overrightarrow{\beta_2} = \cos\theta\,\overrightarrow{\alpha_1} + \sin\theta\,\overrightarrow{\alpha_2}$, with $\tan\theta = (\xi_2/\xi_1)\tan\varphi$; the indicator ellipse is the image of the unit radius circle of the variable space by the mapping *Vac** (see figure hereafter).

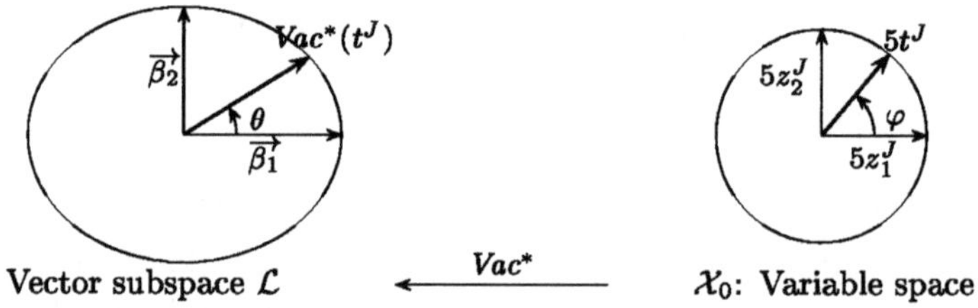

The standard deviation σ in direction $\overrightarrow{u}$ is equal to the length of segment GA (see figure 3.7), where A is the point where line (G, $\overrightarrow{u}$) intersects one of the tangents to the indicator ellipse perpendicular to this line. When $\overrightarrow{u}$ varies, the point A goes through a quartic curve (the "pedal curve" of ellipse); this curve has polar equation (in frame (G, $\overrightarrow{\alpha_1}$)) $\rho^2(\theta) = \lambda_1\cos^2\theta + \lambda_2\sin^2\theta$, and Cartesian equation (in frame (G, $\overrightarrow{\alpha_1}, \overrightarrow{\alpha_2}$)) $(y_1^2 + y_2^2)^2 = \lambda_1 y_1^2 + \lambda_2 y_2^2$. When the values ξ_1 and ξ_2 are clearly separated (which is not the case for the *Target example*), the pedal curve stands out well distinct from the ellipse (see figure 3.7 with $\xi_2/\xi_1 = 2/5$).

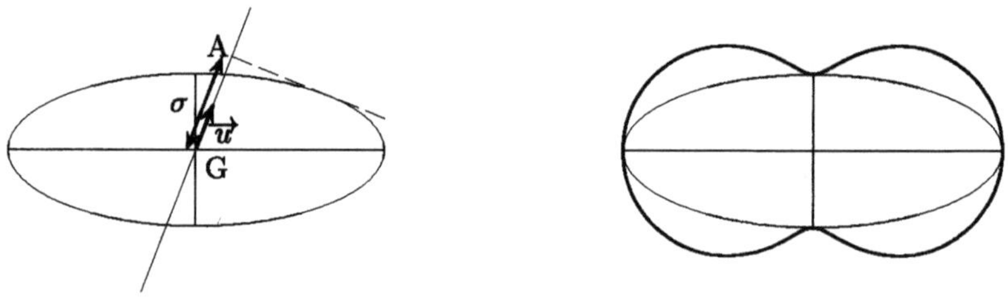

Figure 3.7. Standard deviation in a direction and pedal curve.

In any (principal or not) orthonormal basis $(\mathrm{O}, \overrightarrow{\varepsilon}, \overrightarrow{\varepsilon}')$, for any principal ellipse, the conjugate axis of $(\mathrm{O}, \overrightarrow{\varepsilon}')$ (ordinate axis x') is the *regression line* of x' on x, similarly the conjugate axis of $(\mathrm{O}, \overrightarrow{\varepsilon})$ (abscissa axis x) is the regression line of x on x' (for conjugate axes, cf. MATH.BASES, Definition 10.22, p. 441). The angles of those regression lines with the abscissa axis are θ and θ' defined by $\tan\theta = c/v$ and $\tan\theta' = v'/c$ respectively. The angle θ_1 between the first principal axis and the abscissa axis is such that $\tan\theta_1 = (v' - v + \sqrt{(v-v')^2 + 4c^2})/2c$. The first principal axis lies between the two regression lines: When $c > 0$, one has $\theta \leq \theta_1 \leq \theta'$, that is, $c/v \leq (v' - v + \sqrt{(v-v')^2 + 4c^2})/2c \leq v'/c$, a relation which follows from $c^2 \leq vv'$ i.e. $|r| \leq 1$.

When $v = v'$, the first principal axis is the bissector of axes, and the two regression lines are symmetrical with respect to this bissector.

Target example (Figure 3.8). One has: $\tan\theta' = 6.5$ ($\theta' = 81°3$), $\tan\theta_1 = 2$ ($\theta_1 = 63°4$), $\tan\theta = 0.20$ ($\theta = 11°3$).

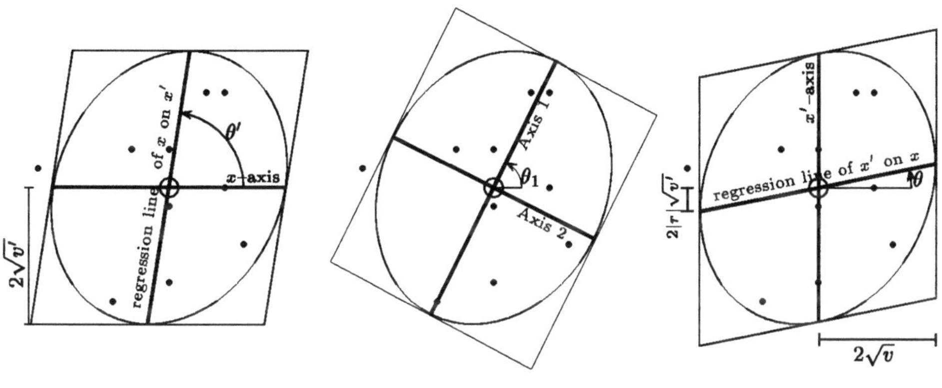

Figure 3.8. Target Example: Concentration ellipse and regression lines.

3.4.4 Concentration Hyperellipsoids

The concept of principal hyperellipsoids applies to *continuous distributions.* For the multidimensional normal and related distributions (such as the t–distribution), principal hyperellipsoids are equal density contours, and the density for every point inside a hyperellipsoid is higher than the density for every outside point. Given a cloud together with its κ–hyperellipsoid, one may consider a distribution of uniform density inside the hyperellipsoid; it can be shown that this distribution has the same principal axes as the cloud and that its eigenvalues are equal to $\left(\lambda_\ell \kappa^2/(L+2)\right)_{\ell=1,\dots L}$. Consequently, a uniform distribution on the κ–hyperellipsoid with $\kappa^2 = L+2$ has the same eigenvalues $(\lambda_\ell)_{\ell=1,\ldots L}$ as the cloud; this hyperellipsoid is called the *concentration hyperellipsoid* of the cloud (Cramér, 1946, p.283; Anderson, 1958, p.44; Malinvaud, 1980).

For the normal L–dimensional distribution, the proportion of the distribution inside the κ–hyperellipsoid is given by $P(\chi^2_L \leq \kappa^2)$, where χ^2_L denotes the classical variable χ^2 with L degrees of freedom. For a given value of κ, the higher the value of L, the smaller the corresponding proportion. For $L = 1$, the interval of half–length σ ($\kappa = 1$) ("indicator interval") contains 68.27% of distribution; that of half–length $2\,\sigma$ ("concentration interval") contains 95.45% of distribution.

For $L = 2$, the *indicator ellipse* ($\kappa = 1$) contains 39.35% of distribution; the *concentration ellipse* ($\kappa = 2$) contains 86.47% of distribution.

Magnitude of deviation. For the point M^j (with principal coordinates $(y^j_\ell)_{\ell=1,\ldots L}$), the coefficient $\kappa = \left(\sum_{\ell=1}^{L} (y^j_\ell)^2/\lambda_\ell\right)^{1/2}$ is an index of the *magnitude of the deviation* $\overrightarrow{\mathrm{GM}^j} = \mathrm{M}^j - \mathrm{G}$ that takes into account the direction of deviation. The κ–norm of $\overrightarrow{\mathrm{GM}^j}$, denoted $|\overrightarrow{\mathrm{GM}^j}|_\kappa$, is the multidimensional extension of standard score for numerical variables.

When the fit of the cloud by a distribution close to normal is acceptable, the κ–norm can be used as an *extremality index* of point M^j among the points of the cloud: the higher κ is, the more extreme the point is[3].

Target example. The two points M^{j2} and M^{j3} are both at the same geometric distance 10 from the mean point. Now for M^{j2}, one has $\kappa = 1.41$ and for M^{j3}, one has $\kappa = 1.66$; therefore the point M^{j3} (which is closer to Axis 2) is less extreme than M^{j2} (which is closer to Axis 1).

3.5 Between and Within Clouds

3.5.1 Subcloud

With a subset $J\langle c\rangle$ of J, there is associated the *subcloud* of points M^j such that $j \in J\langle c\rangle$. This subcloud will be denoted $\mathrm{m}^{J\langle c\rangle}$; its support $J\langle c\rangle$ is weighted by the restriction to $J\langle c\rangle$ of the absolute frequency–measure n_J, and its weight is $n_c = \sum_{j\in J\langle c\rangle} n_j$.

[3] The κ–index, applied to mean points of subclouds obtained by sampling, will be fundamental in statistical inference (chapter *INDUCTIVE*, §8.5).

Target example (opposite picture). Subcloud of three points $\{j1, j2, j6\}$ (diamonds).

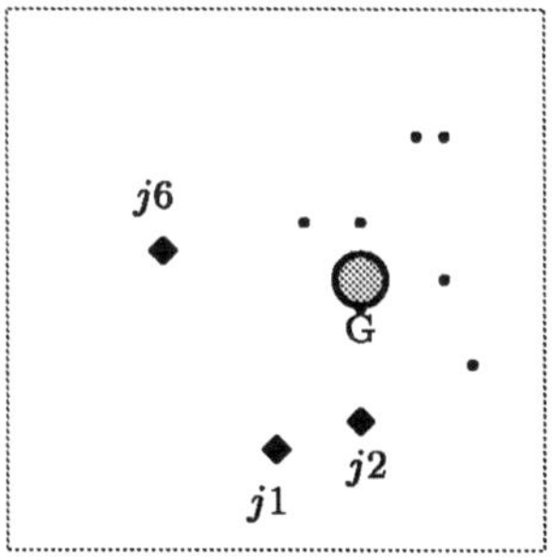

The *mean point of subcloud* is such that: $\mathrm{M}^c = \sum_{j \in J<c>} n_j \mathrm{M}^j / n_c$ (Definition 3.2, p.77), that is, letting $f_j^c = n_j/n_c$ ($j \in J<c>$) (duality notation), $\mathrm{M}^c = \sum_{j \in J<c>} f_j^c \mathrm{M}^j$. This point, called *center of class c*, is weighted by n_c.

The *variance of subcloud* is equal to $\sum_{j \in J<c>} f_j^c (\mathrm{M}^c\mathrm{M}^j)^2$ (Definition 3.4, p.78).

We will study statistics relevant to a subcloud by taking the cloud M^J as *reference cloud* and consequently its mean point G as *reference point* and its weight n as *reference weight*. In particular, the absolute contribution of M^c to the variance of M^J is equal to $\mathrm{Cta}_c = f_c(\mathrm{GM}^c)^2$, with $f_c = n_c/n$.

Definition 3.20. *The absolute contribution of the subcloud* $\mathrm{M}^{J<c>}$, *denoted* $\mathrm{Cta}_{J<c>}$, *is equal to the sum of the absolute contributions of its points:*

$$\mathrm{Cta}_{J<c>} = \sum_{j \in J<c>} \mathrm{Cta}_j = \sum_{j \in J<c>} f_j (\mathrm{GM}^j)^2$$

Definition 3.21. *The absolute within-contribution of the subcloud* $\mathrm{M}^{J<c>}$, *denoted* $\mathrm{Cta}_{J(c)}$, *is the* f_j*–weighted sum of the squares of distances of the points of the subcloud from its mean point.*

$$\mathrm{Cta}_{J(c)} = \sum_{j \in <c>} f_j (\mathrm{M}^c\mathrm{M}^j)^2$$

Within notation. In $J(c)$, "c" is between parentheses, not between chevrons (<>): $J(c)$ should be read "J within–c" (not "J in c").

Property 3.18. *The absolute within-contribution of a subcloud is equal to the product of its relative weight by its variance:*

$$\mathrm{Cta}_{J(c)} = f_c \operatorname{Var} \mathrm{M}^{J<c>}$$

Proof. $f_c \operatorname{Var} \mathrm{M}^{J<c>} = \sum_{j \in J<c>} f_c f_j^c (\mathrm{M}^c\mathrm{M}^j)^2 = \sum_{j \in J<c>} f_j (\mathrm{M}^c\mathrm{M}^j)^2 = \mathrm{Cta}_{J(c)}$. ◁

Definition 3.22. *The within–c cloud, denoted* $\mathrm{M}^{J(c)}$, *is obtained from the subcloud* $\mathrm{M}^{J<c>}$ *by the translation of vector* $\overrightarrow{\mathrm{M}^c\mathrm{G}}$.

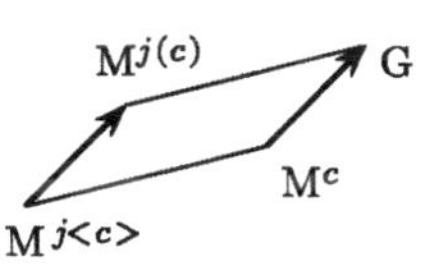

$$\forall j \in J\text{<}c\text{>} : \ \mathrm{M}^{j(c)} = \mathrm{M}^{j\text{<}c\text{>}} + \overrightarrow{\mathrm{M}^c\mathrm{G}}$$

Point $\mathrm{M}^{j(c)}$ is weighted by n_j. By construction, the within-c cloud has mean point G, and its variance is the variance of subcloud $\mathrm{M}^{J\text{<}c\text{>}}$.

Target example. Construction of within-c cloud, see opposite picture.

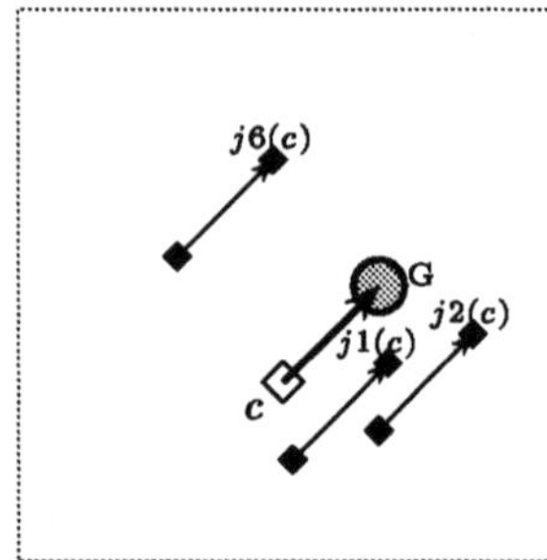

Property 3.19. *The absolute contribution of a subcloud is the sum of its mean point contribution and its within-contribution:*

$$\mathrm{Cta}_{J\text{<}c\text{>}} = \mathrm{Cta}_c + \mathrm{Cta}_{J(c)}$$

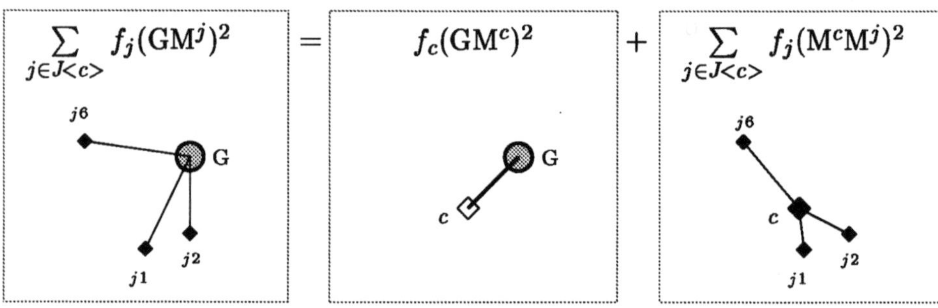

This property is obtained by applying the first Huyghens' theorem (p.79) to the subcloud taking G as reference point.

Property 3.20. *The variance and the absolute within-contribution of a subcloud of two points* $\{\mathrm{M}^j, \mathrm{M}^{j'}\}$ *are respectively given by*

$$\mathrm{Var}_{\{j,j'\}} = \frac{f_j f_{j'}}{(f_j + f_{j'})^2}(\mathrm{M}^j\mathrm{M}^{j'})^2; \qquad \mathrm{Cta}_{(j,j')} = \frac{(\mathrm{M}^j\mathrm{M}^{j'})^2}{\frac{1}{f_j} + \frac{1}{f_{j'}}}$$

Definition 3.23 (Dipole). *Dipole jj' is the bipoint* $(\mathrm{M}^j, \mathrm{M}^{j'})$ *weighted by* $\widetilde{n}_{jj'} = 1/(\frac{1}{n_j} + \frac{1}{n_{j'}}) = n_j n_{j'}/(n_j + n_{j'})$.

Property 3.21 (Dipole contribution). *The absolute contribution of the dipole jj' is equal to the absolute within-contribution of the subcloud of the two points* $\{\mathrm{M}^j, \mathrm{M}^{j'}\}$*:*

$$\mathrm{Cta}_{jj'} = \frac{\widetilde{n}_{jj'}}{n}(\mathrm{M}^j\mathrm{M}^{j'})^2 = \frac{(\mathrm{M}^j\mathrm{M}^{j'})^2}{\frac{1}{f_j} + \frac{1}{f_{j'}}} = \mathrm{Cta}_{(j,j')}$$

With the subcloud $\mathrm{M}^{J<c>}$, there are associated the three endomorphisms on $\mathcal{V}$ defined as follows:

$$\forall \overrightarrow{\alpha} \in \mathcal{V} : Som_{J<c>}(\overrightarrow{\alpha}) = \sum_{j\in J<c>} f_j \langle \overrightarrow{\mathrm{GM}}^j | \overrightarrow{\alpha} \rangle \overrightarrow{\mathrm{GM}}^j \ (f_j = n_j/n) \quad (3.10)$$

$$Som_{J(c)}(\overrightarrow{\alpha}) = \sum_{j\in J(c)} f_j^c \langle \overrightarrow{\mathrm{M}^c\mathrm{M}^j} | \overrightarrow{\alpha} \rangle \overrightarrow{\mathrm{M}^c\mathrm{M}^j} \ (f_j^c = n_j/n_c) \quad (3.11)$$

$$Som_c(\overrightarrow{\alpha}) = f_c \langle \overrightarrow{\mathrm{GM}}^c | \overrightarrow{\alpha} \rangle \overrightarrow{\mathrm{GM}}^c \quad (f_c = n_c/n) \quad (3.12)$$

Lemma 3.1. *Between these three endomorphisms, one has the relations:*

$$Som_{J<c>} = Som_c + f_c Som_{J(c)}$$

Proof. From relation 3.6 (p.86), applied to subcloud $\mathrm{M}^{J<c>}$, one has: $\sum_{j\in J<c>} f_j^c \langle \overrightarrow{\mathrm{GM}}^j | \overrightarrow{\alpha} \rangle \overrightarrow{\mathrm{GM}}^j = \langle \overrightarrow{\mathrm{GM}}^c | \overrightarrow{\alpha} \rangle \overrightarrow{\mathrm{GM}}^c + \sum_{j\in J<c>} f_j^c \langle \overrightarrow{\mathrm{M}^c\mathrm{M}}^j | \overrightarrow{\alpha} \rangle \overrightarrow{\mathrm{M}^c\mathrm{M}}^j$ $(\overrightarrow{\alpha} \in \mathcal{V})$ and thus, by multiplying f_c, the lemma. ◁

3.5.2 Partition of a Cloud, Between and Within Clouds

Let a partition of J into C classes (subsets of J) be denoted $J<C>$; hence the partition of M^J into subclouds (classes) $\mathrm{M}^{J<C>} = (\mathrm{M}^{J<c>})_{c\in C}$.

Property 3.22. *The variance of the cloud is the sum of the contributions of the classes (subclouds) of the partition:*

$$\mathrm{Var\,M}^J = \sum_{c\in C} \mathrm{Cta}_{J<c>}$$

Definition 3.24 (Between–cloud and variance). *The between–cloud is the derived cloud of the mean points of subclouds* $(\mathrm{M}^{J<c>})_{c\in C}$*; its variance is called between–variance.*

The between–cloud is denoted M^C. Its mean point is the mean point G of the total cloud and its variance is equal to $\sum_{c\in C} f_c(\mathrm{GM}^c)^2$.

Definition 3.25 (Within–cloud and variance). *The within–cloud, denoted* $\mathrm{M}^{J(C)}$*, is the union of the within–c clouds; its variance is called within–variance.*

Each point $\mathrm{M}^{j(c)}$ is weighted by n_j. The variance of the within–cloud is equal to the sum of the within–c contributions, or equivalently to the weighted mean of the variances of the subclouds $(\mathrm{M}^{J<c>})_{c\in C}$.

$$\mathrm{Var\,M}^{J(C)} = \sum_{c\in C} \mathrm{Cta\,M}^{J(c)} = \sum_{c\in C} f_c \,\mathrm{Var\,M}^{J<c>} \quad (3.13)$$

Target example. Consider the three–class partition: $J{<}c1{>}=\{j1, j2\}$ (■), $J{<}c2{>}=\{j6\}$ (◆) and $J{<}c3{>}=\{j3, j4, j5, j7, j8, j9, j10\}$ (•), hence the between–cloud M^C (▫, ◇, ○) and the within–cloud $\mathrm{M}^{J(C)}$.

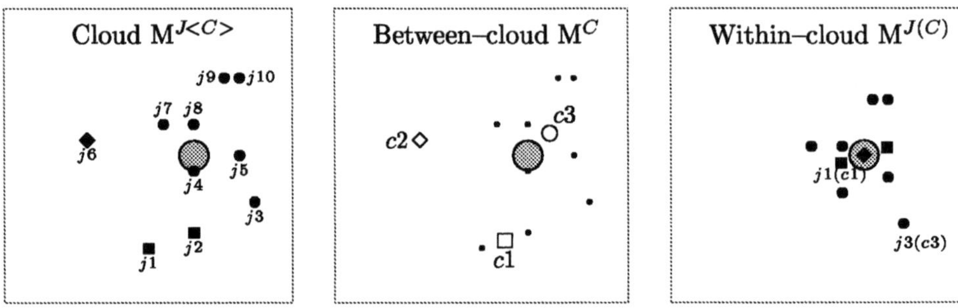

Property 3.23 (Between–within breakdown of variance). *The total variance (variance of the cloud) is the sum of the between and the within–variances:*

$$\mathrm{Var}\,\mathrm{M}^{J<C>} = \mathrm{Var}\,\mathrm{M}^{C} + \mathrm{Var}\,\mathrm{M}^{J(C)}$$

This property follows from Property 3.19 (p.102), by summing over all classes.

Definition 3.26. *The η^2–coefficient (or correlation ratio) of the partition is the ratio of the between–variance to the total variance.*

$$\eta^2 = \mathrm{Var}\,\mathrm{M}^{C} / \mathrm{Var}\,\mathrm{M}^{J}$$

Target example. One has: $\mathrm{Var}\,\mathrm{M}^J = 92$; $\mathrm{Var}\,\mathrm{M}^C = 57.429$; $\mathrm{Var}\,\mathrm{M}^{J(C)} = 34.571$, that is, $92 = 57.429 + 34.571$, and $\eta^2 = 0.624$.

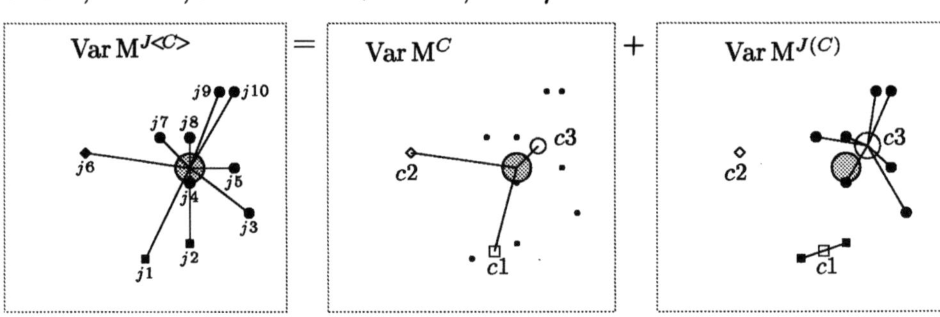

Definition 3.27 (Between–within cloud). *The between-within cloud is the union of the between and within–clouds.*

The between–within cloud has mean point G, its weight is equal to $2n$, therefore its variance is half of the variance of cloud M^J.

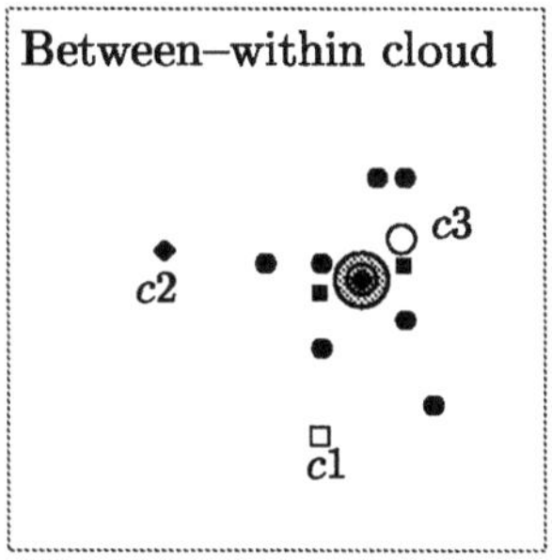

Theorem 3.6. *Let Som be the endomorphism of $\mathcal{V}$ for cloud M^J, Som_C the endomorphism for the between-cloud, and $Som_{J(C)}$ the endomorphism for the within-cloud:*

$$Som = Som_C + Som_{J(C)}$$

Corollary 3.1. *The between–within cloud has the same principal directions as the cloud M^J associated with eigenvalues $\lambda_\ell/2$ ($\ell = 1, \dots L$).*

Property 3.24 (Two–class partition). *For a partition with two classes $\{c, c'\}$, the following relation holds:*

$$G = \frac{n_c}{n} M^c + \frac{n_{c'}}{n} M^{c'}$$

One deduces the relations: $\overrightarrow{M^c G} = (n_{c'}/n)\overrightarrow{M^c M^{c'}} = (n_{c'}/n_c)\overrightarrow{G M^{c'}}$.

Property 3.25 (Grouping property). *If two classes of a partition are grouped together, the between–variance decreases from an amount equal to the contribution of the dipole defined by the centers of the grouped classes.*

Proof. Let $C{<}d{>} = \{c', c''\}$ be a pair of elements of C, and D the set obtained from C by grouping c' and c'' in a class d. From the between–within breakdown of variance (p.104), one deduces:

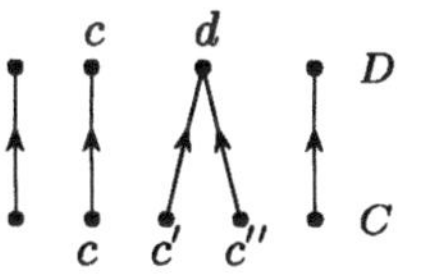

$\operatorname{Var} M^D = \operatorname{Var} M^C - \operatorname{Var} M^{C(D)}$;
$\operatorname{Var} M^{C(D)} = \sum_{\substack{c \in C \\ c \neq c'; c \neq c''}} f_c \times 0 + (f_{c'} + f_{c''}) \operatorname{Var}_{\{c', c''\}}$;
hence $\operatorname{Var} M^{C(D)} = \text{Cta}_{c'c''}$ which is the contribution of dipole $c'c''$, (cf. Property 3.21, p.102); hence $\operatorname{Var} M^D = \operatorname{Var} M^C - \text{Cta}_{c'c''}$ ◁

The grouping property is fundamental for Euclidean classification.

Target example. For the three–class partition (p.104), if the two classes c1 and c2 are grouped, one has $(M^{c1}M^{c2})^2 = 290$, hence $\text{Cta}_{(c1,c2)} = 290 \times \left(\frac{2}{10} \times \frac{1}{10}\right) / \left(\frac{2}{10} + \frac{1}{10}\right) = 19.333$; the new partition in two classes $d1 = \{c1, c2\}$ and $d2 = \{c3\}$ has between–variance 38.095, that is, the between–variance of the three–class partition (57.429) minus the contribution of the dipole $(c1, c2)$ (19.333).

3.6 Euclidean Classification

Automatic Classification — alias *Cluster analysis* — is a world in itself. Classification will be tackled here only within the framework of GDA, where the objects to classify are points of Euclidean clouds.

The overall objective of any classification method is intuitive. What is sought is to construct clusters of objects — in GDA subclouds of points — so that the objects within a same cluster are as close together as possible whereas those belonging to different clusters are as remote from one another as possible — even though, in some situations these two demands may not be easy to reconcile (see e.g. Benzécri's counterexample, 1992, p.572). At an early stage of the classificatory process, one may be content to delineate *clusters*, i.e. classes of points corresponding to high density zones, with the twofold requirement of class *separability* and of class *compactness*, in other words, of heterogeneity between classes and homogeneity within classes. At a more elaborate stage, one can seek a set of classes that constitute a *partition* of the cloud, that is, such that every point belongs to one and only one class. At an even more elaborate stage, what is sought is a *hierarchical classification*, that is, a *system of nested classes*, after the pattern of natural science, where animals are divided into vertebrates, molluscs, etc., the vertebrates being in turn subdivided into mammals, birds, etc.

Classification methods developed less rapidly than GDA, mainly owing to technical obstacles (the "algorithmic wall"!) that were only gradually overcome. Yet, Volume 1 of Benzécri & coll (1973)was already entirely devoted to classification ("taxinomy"). As for AHC, one can also refer to part V of Benzécri's book (1992); to chapter 4 of Lebart & al (1984), chapter 2 of Lebart & al (1995), and to the book by Jambu & Lebeaux (1983), etc.

The present section will be mainly devoted to the *agglomerative (or ascending) hierarchical classification* (AHC) method with variance as aggregation index; we call this method *Euclidean classification*, as it is the method which has definitely emerged as the one most in harmony with the mathematical structures of GDA.

3.6.1 Hierarchy, Hierarchical Tree

Definition 3.28. *Let J be a nonempty finite set; a set $\mathcal{H}$ of classes (nonempty subsets of J) is called a hierarchy on J if the classes of $\mathcal{H}$ possess the two following properties:*

P1 *(intersection): for any pair of classes of $\mathcal{H}$: either the two classes are disjoint, or one is included in the other.*

P2 *(union): every class of $\mathcal{H}$ is the union of the classes included in this class.*

In brief, hierarchy classes are obtained by grouping and are represented by a *hierarchical tree.*

Figure 3.9 shows such a tree for the *Target example.* At the basis of the tree there are 10 one–element classes, indexed from $\ell1$ to $\ell10$; thus class $\ell1$ has $j1$ as a single element, etc. Classes with several elements correspond to the *nodes* of the tree and are indexed from $\ell11$ to $\ell19$. The hierarchy is *dichotomous* (and represented by a binary tree), in the sense that every class indexed by a node is the union of two classes. For example, class $\ell13$ is the union of classes $\ell1$ and $\ell2$; class $\ell18$ is the union of classes $\ell13$ and $\ell6$, etc. A dichotomous hierarchy of J objects has $J-1$ nodes[4] (here 9 nodes numbered from $\ell11$ to $\ell19$). By "cutting" the tree at any level, one determines a partition of the set. For instance, cutting the tree between nodes $\ell17$ and $\ell18$ (Figure 3.9, p.107) generates the partition into three classes (cf. p.104) corresponding to the three nodes $\ell6$, $\ell13$ and $\ell17$.

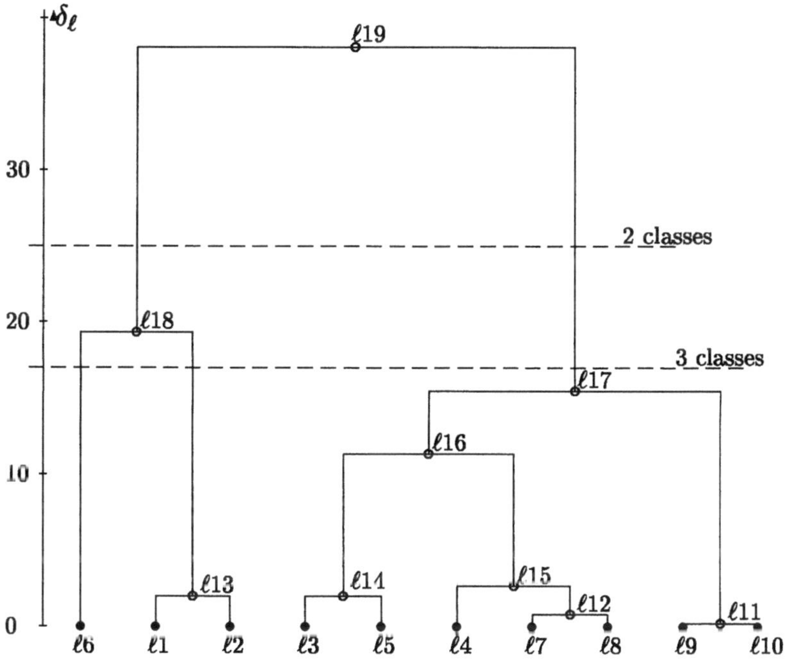

Figure 3.9. *Target example*: Hierarchical tree.

[4]Let us recall that we use capital letters to denote both finite sets and their cardinalities: e.g. the cardinality of the set J minus 1 is written $J-1$, etc.

3.6.2 Divisive vs Agglomerative Classification

In order to construct a hierarchy there are two main ways of : the divisive one and the agglomerative one.

In a *divisive classification* (or segmentation, or descending classification), one starts with the objects to be classified, regarded as a single class which is divided into classes; those classes are in turn subdivided, etc. A well–known example of divisive classification is the CART method (Classification And Regression Tree, Breiman & al., 1984).

In an *agglomerative (or ascending) classification*, one starts with the objects regarded as a set of one–element classes, from which one proceeds to successive aggregations, until all objects are grouped within a single class. At each step of the construction, one starts with a partition and groups two classes of this partition. In what follows, we will deal in detail with Agglomerative Hierarchical Classification (AHC).

3.6.3 Aggregation Indices

To render an AHC method operational, an aggregation index must be specified to select the classes to be aggregated at each step.

Given a set C of classes, an aggregation index $\delta(c, c')$ is an index defined on the two disjoint classes c and c' of C that can be interpreted either as a direct index of separability between classes c and c', or as an inverse index of compactness of the class grouping those classes. Several aggregation indices are commonly used in AGD. They are all defined from the geometric distance, and sometimes called "distance indices", even though most of them do not verify the triangle inequality: It may happen that two classes are (in the sense of the aggregation index) close to a same third one, while at the same time being distant from each other.

In any AHC method, one requirement must be satisfied, that is, the aggregation of two classes at some stage should not call into question previous aggregations. Specifically, suppose classes c' and c'' have already been aggregated into the new class d, which implies that $\forall c \in C - \{c', c''\}$: $\delta(c', c'') < \delta_{\min} = \min\{\delta(c, c')\,;\, \delta(c, c'')\}$. What we do not want is a class c such that $\delta(c, d) < \delta_{\min}$, since this would imply that class c should have been aggregated with d before aggregating c' and c''. Therefore the aggregation index must satisfy the following median inequality.

Definition 3.29 (Median inequality). *Aggregation index δ is said to verify the median inequality, if one has:*

$$\forall c \in C - \{c', c''\} : \delta(c', c'') \leq \delta_{\min} \Longrightarrow \delta(c, d) \geq \delta_{\min}$$

letting $\delta_{\min} = \min\{\delta(c,c') ; \delta(c,c'')\}$ *and d the pooled class of c' and c''.*

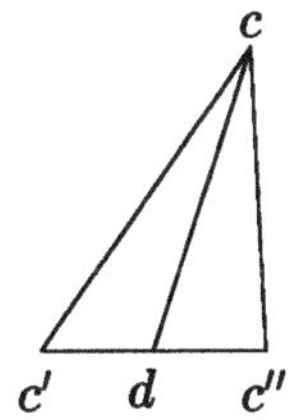

If c, c', c'' are pictured as vertices of a triangle, the median (c, d) opposed to the smallest side (c', c'') must be greater than (or equal to) the smaller of the other two sides; hence the name "median inequality". However, this pictorial figuration is metaphorical because the δ index is not a genuine distance.

Counterexample. As a separability index between classes, one might think of simply taking the geometric distance between the centers M^c and $\mathrm{M}^{c'}$ of classes c and c'. If classes c' and c'' are aggregated, it may happen that the center M^d of the new class d (lying between $\mathrm{M}^{c'}$ and $\mathrm{M}^{c''}$) is such that $\mathrm{M}^d\mathrm{M}^c < \min\{\mathrm{M}^c\mathrm{M}^{c'} ; \mathrm{M}^c\mathrm{M}^{c''}\}$, which contradicts the median inequality (see opposite figure).

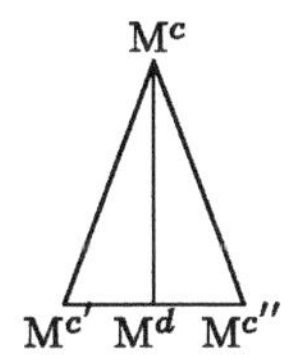

All aggregation indices used in GDA verify the median inequality.

Definition 3.30 (Variance index). *Given two classes c and c', the variance index is the contribution of the dipole (cf. p.102) of the class centers* M^c *and* $\mathrm{M}^{c'}$, *denoted* $\delta(c,c')$ *with*

$$\delta(c,c') = \frac{f_c f_{c'}}{f_c + f_{c'}}(\mathrm{M}^c\mathrm{M}^{c'})^2$$

This index is also called *inertia index*, or centered moment of order 2, or *Ward index* (see Ward, 1963).

Property 3.26 (Recurrence formula). *If class d is obtained by grouping the two classes c' and c'', the aggregation index verifies the formula:*

$\forall c \in C, c' \in C$ and $\forall c \in C - \{c', c''\}$:

$$\delta(c,d) = \frac{(n_c + n_{c'})\delta(c,c') + (n_c + n_{c''})\delta(c,c'') - n_c\delta(c',c'')}{n_c + n_{c'} + n_{c''}} \tag{3.14}$$

Proof. The sum of squares of the cloud made up of two points $\mathrm{M}^{c'}$ and $\mathrm{M}^{c''}$ with respect to point M^c is $n_{c'}(\mathrm{M}^c\mathrm{M}^{c'})^2 + n_{c''}(\mathrm{M}^c\mathrm{M}^{c''})^2$. From Huyghens' theorem (p.79), this sum of squares can be written $n_d(\mathrm{M}^c\mathrm{M}^d)^2 + n\,\delta(c',c'')$, where $\mathrm{M}^d = (n_{c'}\mathrm{M}^{c'} + n_{c''}\mathrm{M}^{c''})/n_d$ denotes the mean point of subcloud $\{\mathrm{M}^{c'}, \mathrm{M}^{c''}\}$ and $n_d = n_{c'} + n_{c''}$ its weight.
Therefore $n_{c'}(\mathrm{M}^c\mathrm{M}^{c'})^2 + n_{c''}(\mathrm{M}^c\mathrm{M}^{c''})^2 = n_d(\mathrm{M}^c\mathrm{M}^d)^2 + n\,\delta(c',c'')$. Replacing in this equality the squares of distances by their expression in terms of the aggregation index (cf. Definition 3.30), one gets $\frac{n_c+n_{c'}}{n_c}\delta(c,c') + \frac{n_c+n_{c''}}{n_c}\delta(c,c'') = \frac{n_c+n_d}{n_c}\delta(c,d) + \delta(c',c'')$, hence the formula. ◁

Corollary 3.2. *The variance index verifies the median inequality.*

Proof. This property follows from the recurrence formula. Indeed, by minoring $\delta(c, c')$ and $\delta(c, c'')$ by $\delta_{\min}$, and by majoring $\delta(c', c'')$ by $\delta_{\min}$

$$\delta(c, d) \geq \frac{n_c+n_{c'}}{n_c+n_d}\delta_{\min} + \frac{n_c+n_{c''}}{n_c+n_d}\delta_{\min} - \frac{n_c}{n_c+n_d}\delta_{\min} = \delta_{\min} \quad \triangleleft$$

The variance index does not satisfy the triangle inequality, as the following *counterexample* shows. Let three classes c, c', c'' with equidistant centers, with $f_c = \epsilon$, $f_{c'} = f_{c''} = (1-\epsilon)/2$. Hence $\delta(c, c') = \delta(c, c'')$ and $(\delta(c, c') + \delta(c, c''))/\delta(c', c'') = 8\epsilon/(1+\epsilon)$. If $\epsilon < 1/7$, one has $\delta(c, c') + \delta(c', c'') < \delta(c', c'')$. Class c is close to classes c' and c'' which are distant from each other.

In *Euclidean classification*, one takes as an aggregation index the *variance index*. Euclidean classification tends to aggregate firstly classes with small numbers of observations: In brief, the lighter classes are, the earlier they are aggregated.

Local and global viewpoints. In Euclidean classification, it follows from the *grouping property* (p.105) that at each stage the aggregated classes are those which lead to the minimal decrease of the between–variance of the partition, or equivalently to the minimal increase of the within–variance. Thus the variance index which, from a local point of view, is just one measure of deviation between two classes, among others, has good properties from a global viewpoint, since the two requirements of compactness within classes and separability between classes come down to the same thing.

Other aggregation indices in AHC. Beside the variance index, three AHC indices are also used in GDA. Let $j \in J{<}c{>}$ and $j' \in J{<}c'{>}$.

• *Minimal jump.* Let $D_{\min}(c, c') = \min\limits_{jj', j \neq j'} \mathrm{M}^j\mathrm{M}^{j'}$. Taking $D_{\min}(c, c')$ as aggregation index defines the minimal jump, or *single linkage clustering.*

• *Maximal jump.* Let $D_{\max}(c, c') = \max\limits_{jj'} \mathrm{M}^j\mathrm{M}^{j'}$. Taking $D_{\max}(c, c')$ as aggregation index defines the maximal jump, or *diameter index*, or *complete linkage clustering.*

• *Mean distance.* Let $D_{\text{mean}}(c, c') = \sum\limits_{j \in J<c>} \sum\limits_{j' \in J<c'>} f_j f_{j'} \mathrm{M}^j\mathrm{M}^{j'}/(f_c f_{c'})$ (weighted mean of distances between the points of classes c and c'). Taking $D_{\text{mean}}(c, c')$ as aggregation index defines the mean distance index, or *average linkage clustering.*

All three indices verify the median inequality. Only D_{mean} verifies the triangle inequality. The D_{min} index produces clusters that are clearly separated from each other but tend to induce "chaining effects"; the D_{max} index has the opposite features. The D_{mean} index is a compromise between D_{min} and D_{max}, since it takes into account all segments of the classes (the smallest as well as the biggest) like the variance index. For this reason, the D_{mean} index and the variance index are usually preferred; the latter has the global property stated earlier, which also entails technical advantages in term of memory space and algorithmic complexity.

3.6.4 Basic and Reciprocal Neighbor Algorithms

Basic algorithm. Once an aggregation index has been chosen, the *basic algorithm* of AHC is as follows:

Step 1. From the table of distances between the J points, calculate the aggregation index for the $J(J-1)/2$ pairs of one–element classes, then aggregate a pair of classes for which the index is minimum: hence a partition into $J-1$ classes.

Step 2. Calculate the aggregation indices between the new class and the $J-2$ others, and aggregate a pair of classes for which the index is minimum; this gives a second partition into $J-2$ classes in which the first partition is nested.

Step 3. Iterate the procedure until a single class is reached.

In *Euclidean classification*, at step ℓ, the partition C_ℓ is the between–cloud M^{C_ℓ}; the minimum of the aggregation index is called *level index* and denoted δ_ℓ, with $\delta_\ell = \operatorname{Var} \mathrm{M}^{C_{\ell-1}} - \operatorname{Var} \mathrm{M}^{C_\ell}$. At each step, it is sufficient to consider the between–cloud, without considering the points inside classes. In brief, "one thinks about classes, but one works on class centers".

Reciprocal neighbor algorithm. The calculation time for the basic algorithm is of the order of J^3, since at each step one seeks the smallest aggregation index of all node pairs. There are algorithms which lessen calculation time; for instance, the reciprocal neighbor algorithm, whose principle goes back to McQuitty (1966), introduces — without complicating the basic algorithm — an acceleration principle by aggregating at each step the node pairs that are nearest *reciprocal neighbors*.

Definition 3.31. *Two points* M^c *and* $\mathrm{M}^{c'}$ *are said to be reciprocal neighbors if* $(\delta(c, c''))_{\substack{c''\in C\\ c''\neq c}}$ *is minimum for* $c'' = c'$ *and if* $(\delta(c', c''))_{\substack{c''\in C\\ c''\neq c'}}$ *is minimum for* $c'' = c$.

In the basic algorithm, indeterminations may arise when pairs of nodes have equal aggregation indices. The reciprocal neighbor algorithm often enables one to remove such indeterminations. When there is no indetermination, the reciprocal neighbor algorithm yields the same result as the basic algorithm for any aggregation index that verifies the median inequality (see Benzécri, 1982b). The reciprocal neighbor algorithm will be illustrated on the *Target example.*

δ	$j1$	$j2$	$j3$	$j4$	$j5$	$j6$	$j7$	$j8$	$j9$
$j2$	**2**								
$j3$	11.6	4							
$j4$	6.8	3.2	4						
$j5$	14.4	6.8	**2**	2					
$j6$	13	17	27.4	10.6	20.2				
$j7$	13	10.6	12.2	2.6	5.8	5.2			
$j8$	14.6	9.8	8.2	1.8	2.6	10	**0.8**		
$j9$	29.2	20.8	13.6	8	5.2	19.4	5	2.6	
$j10$	31.4	21.8	13	9	5	23.2	6.8	3.6	**0.2**

Table 3.1. *Target example*: contributions of the $9 \times 10/2 = 45$ dipoles

• **Step 1.** From the distances between pairs of points, calculate the contributions of the 45 dipoles (Table 3.1). Four pairs of reciprocal neighbors are found with aggregation indices $\delta(\ell9, \ell10) = 0.2$ (class $\ell11$); $\delta(\ell7, \ell8) = 0.8$ (class $\ell12$); $\delta(\ell1, \ell2) = 2 = \delta(\ell3, \ell5)$ (classes $\ell13$ and $\ell14$).

• **Step 2.** Applying the recurrence formula, calculate the aggregation indices between the 4 new classes $\ell11, \ell12, \ell13, \ell14$ and $\ell4$, then between these 4 new classes and $\ell6$, then between the 4 new classes (opposite Table).

δ	$\ell4$	$\ell6$	$\ell11$	$\ell12$	$\ell13$
$\ell6$	10.6				
$\ell11$	11.267	28.334			
$\ell12$	**2.667**	9.867	8.5		
$\ell13$	6	19.334	50.5	22.6	
$\ell14$	3.334	31.067	17.3	13	16.4

Classes $\ell4$ and $\ell12$ are reciprocal neighbors hence the new class $\ell15 = \{j4, j7, j8\}$ at aggregation level 2.667.

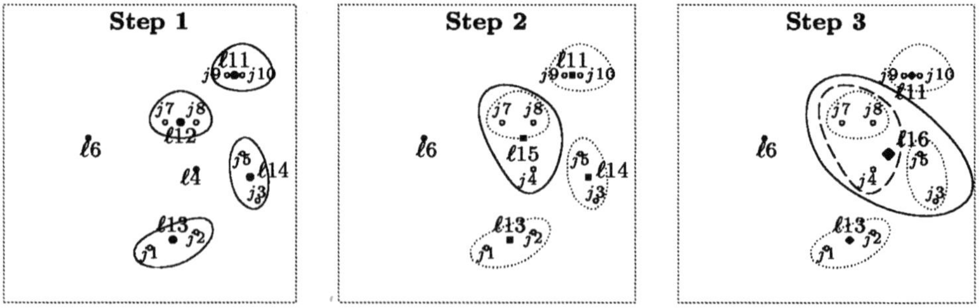

• **Step 3.** Calculate the aggregation indices between class $\ell 15$ and the 4 classes $\ell 6$, $\ell 11$, $\ell 13$, and $\ell 14$, hence the opposite table.

δ	$\ell 6$	$\ell 11$	$\ell 13$	$\ell 14$
$\ell 11$	28.334			
$\ell 13$	19.334	50.5		
$\ell 14$	31.067	17.3	16.4	
$\ell 15$	12.033	12.493	20.613	**11.333**

The two classes $\ell 14$ and $\ell 15$ are aggregated into the class $\ell 16$ at level 11.333.

• **Step 4.** Calculate the three aggregation indices between $\ell 16$ and $\ell 6$ $\ell 11$ and $\ell 13$. Hence the following table:

δ	$\ell 6$	$\ell 11$	$\ell 13$
$\ell 11$	28.333		
$\ell 13$	**19.333**	50.5	
$\ell 16$	20.666	**15.571**	20.857

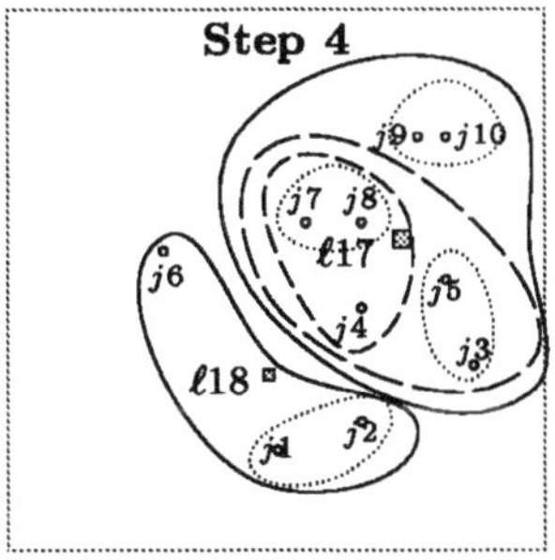

There are two pairs of reciprocal neighbors $(\ell 11, \ell 16)$ and $(\ell 6, \ell 13)$; they are respectively aggregated at level 15.572 into class $\ell 17 = \{j3, j4, j5, j7, j8, j9, j10\}$ and at level 19.333 into class $\ell 18 = \{j1, j2, j6\}$).

• **Step 5.** The aggregation of the two nodes $\ell 17$ and $\ell 18$ at level 38.095 (node $\ell 19$) achieves the classification.

Property 3.27 (Breakdown of variance for a class hierarchy). *The sum of all level indices is equal to the variance of the cloud.*

Proof. $\sum_{\ell=J+1}^{2J-1} \delta_\ell = \sum_{\ell=J+1}^{2J-1} \operatorname{Var} \mathrm{M}^{C_{\ell-1}} - \sum_{\ell=J+1}^{2J-1} \operatorname{Var} \mathrm{M}^{C_\ell} = \operatorname{Var} \mathrm{M}^J$ ◁

For each grouping, the level index is equal to the decrease of the between–variance, or else to the increase of the within–variance: When the nodes $\ell 11$ and $\ell 16$ (step 4) are grouped, the between–variance decreases by 15.571.

As one ascends the construction, the level indices form an *increasing sequence*; the higher in the hierarchy, the higher the heterogenity level where aggregation is made; this property follows from median inequality.

At each level, the between–variance of the corresponding partition is equal to the sum of the indices of higher levels (above), and the within–variance is equal to the sum of this level and those of lower levels (below). For instance, for the partition into three classes $\ell 6$, $\ell 13$ and $\ell 17$, the between–variance is the sum of the $C-1=2$ level indices running from $\ell = 2J - C + 1 = 18$ to $2J - 1 = 19$, that is $19.33 + 38.10 = 57.43$; the between–variance is equal to $0.2 + 0.8 + 2 + 2 + 2.67 + 11.33 + 15.67$, that is, 34.57. The sum of the nine level indices is equal to 92, i.e. the total variance. Table 3.2 (p.114) summarizes the 9 levels of the classification of the *Target example* (the classification process goes from bottom to top).

ℓ	δ_ℓ	classes		n	class description
$\ell19$	38.095	$\ell18$	$\ell17$	10	$j9\,j10\,j3\,j5\,j4\,j7\,j8\,j6\,j1\,j2$
$\ell18$	19.333	$\ell13$	$\ell6$	3	$j6\,j1\,j2$
$\ell17$	15.571	$\ell16$	$\ell11$	7	$j9\,j10\,j3\,j5\,j4\,j7\,j8$
$\ell16$	11.333	$\ell15$	$\ell14$	5	$j3\,j5\,j4\,j7\,j8$
$\ell15$	2.667	$\ell12$	$\ell4$	3	$j4\,j7\,j8$
$\ell14$	2.	$\ell5$	$\ell3$	2	$j3\,j5$
$\ell13$	2.	$\ell2$	$\ell1$	2	$j1\,j2$
$\ell12$	0.8	$\ell8$	$\ell7$	2	$j7\,j8$
$\ell11$	0.2	$\ell10$	$\ell9$	2	$j9\,j10$

Between Var		η^2_ℓ
$\ell19$	38.10	.414
$\ell18$	57.43	.624
$\ell17$	73.00	.793
$\ell16$	84.33	.917
$\ell15$	87.00	.957
$\ell14$	89.00	.967
$\ell13$	91.90	.989
$\ell12$	91.80	.998
$\ell11$	92.00	1

Table 3.2. *Target example*: steps of the AHC.

3.6.5 Reading a Classification, Aids to Interpretation

The hierarchical tree, once constructed, can be read in one of two ways, ascending or descending. The *ascending reading* allows one to single out, at the basis of the tree, the objects or classes of objects that are nearest; the *descending reading* is a better way for the overall interpretation, according to the maxim: "One constructs the hierarchy by ascending, one interprets it by descending".

When what is sought is a partition with a number of classes specified in advance, descending reading yields such a partition by cutting the tree as already shown in the *Target example* (Figure 3.9, p.107). More often, a hierarchy is sought with partitions whose numbers of classes are not fixed in advance. One will then look for a compromise between the number of classes — preferably small — and the proportion of variance — preferably high — accounted for by the finest partition of the retained hierarchy. The further one goes down the tree, the more classes there are, and so the accounted part of variance increases, but this increase gets smaller and smaller as level indices decrease. One will end the procedure at a degree of fineness beyond which a further subdividing of classes would not bring any appreciable and/or interpretable variance increase.

Target example. Cutting the tree between levels $\ell19$ and $\ell18$ generates the partition into two classes $\ell18$ and $\ell17$, with $\ell18 = \{j6, j1, j2\}$ and $\ell17 = \{j3, j5, j4, j7, j8, j9, j10\}$. This partition only accounts for 41.4% of the variance of the cloud; one will therefore cut the tree at a lower level. As a rule of thumb, the between–variance should be greater than the within–variance ($\eta^2 > 0.5$). Cutting between levels $\ell18$ and $\ell17$ determines the hierarchy with the previous three–class partition accounting for 62.4% of variance. Cutting the tree between levels $\ell17$ and $\ell16$ would generate a still finer hierarchy, with a four–class partition accounting for 79.3% of variance. Which of the foregoing hierarchies (with three or four–class partitions) is to be eventually retained will be a matter of appreciation. With real data, "one of the major roles of interpretation is to specify among the dichotomies of an AHC those which correspond to divisions that are geometrically well–defined and conceptually interpretable." (Benzécri, 1992, p. 573).

> *To sum up*: The end–product of the classification procedure is a hierarchy represented by a tree whose classes and partitions are interpreted.

3.6.6 Other Clustering Methods and Issues

Optimal partitions

An attractive feature of AHC is that it generates a set of nested partitions which permit differentiated interpretations. On the other hand, the AHC method does not possess the desirable property that each partition produced by the hierarchy has minimum variance among all the partitions with the same number of classes.

Target example. The η^2 of the partition into three classes $\{i1, i2, i3\}$, $\{i6\}$, $\{i4, i5, i7, i8, i9, i10\}$ is equal to 0.651, therefore greater than the η^2 ($=$ 0.624) of the three–class partition of the AHC.

Nevertheless, starting with a partition produced by AHC, one can always try to improve it in order to get (at least locally) an optimal partition with a fixed number of classes.

Among such methods let us mention the *method of moving centers*, that of *dynamic clustering* ("Nuées dynamiques"), and that of *stable clusterings* (Diday, 1971), which consists in constructing several partitions starting with different sets of centers, and looking for stable clusters, i.e. sets of individuals that remain in the same clusters throughout the process.

Let us also mention the *transfer algorithm*, which consists in moving one element from a class to another. Transferring element j from class c to class c' entails a variation of the between–variance equal to $(f_c f_j/(f_c - f_j)) \times (\mathrm{M}^c\mathrm{M}^j)^2 - (f_{c'} f_j/(f_{c'} + f_j)) \times (\mathrm{M}^{c'}\mathrm{M}^j)^2$. Given a partition, for each possible transfer we calculate the variation of between–variance, and choose the transfer for which the variance increase is greatest, then the procedure is iterated until the variance increase becomes negligible.

Joint use of GDA and Classification

In the analysis of a complex data set, it is recommended, as a rule, to use GDA and classification jointly. The two methods are complementary, especially at the stage of interpretation. In this connection, there is an issue that must be dealt with in each situation, and that is ***overall vs principal classification***. Since in GDA only a principal cloud is retained, the question arises whether the classification should be performed either on the full multidimensional cloud or on its principal projection.

3.7 From Points to Numbers: Matrix Formulas

By choosing a Cartesian frame and working on the coordinates of points, we will now go "from points to numbers", and get "numerical principal formulas", with their matrix expressions, for the direction and eigenvalues equations, and the passage and reconstitution formulas.

Letting $(\mathrm{O}, (\overrightarrow{\delta_k})_{k\in K})$ be a Cartesian frame of $\mathcal{U}$ with $q_{kk'} = \langle \overrightarrow{\delta_k} | \overrightarrow{\delta_{k'}} \rangle$, we denote $(x^{jk})_{k\in K}$ the coordinates of point M^j, that is: $\overrightarrow{\mathrm{OM}^j} = \sum\limits_{k\in K} x^{jk} \overrightarrow{\delta_k}$. The family $x^{jK} = (x^{jk})_{k\in K}$ is called the *initial profile* of point M^j ("initial" is opposed to "principal"). The family $x^{Jk} = (x^{jk})_{j\in J}$ defines the k–th *initial variable*; we denote $\overline{x}^k = \sum\limits_{j\in J} f_j x^{jk}$ its mean, $v^{kk} = \sum\limits_{j\in J} f_j (x^{jk} - \overline{x}^k)^2$ its variance, and $v^{kk'}$ the covariance between the initial variables x^{Jk} and $x^{Jk'}$. The K variables $(x^{Jk})_{k\in K}$ define the K–numerical protocol x^{JK}.

When individuals j are grouped, coordinates average up, that is, x^{Jk} is indeed a variable on J, therefore the upper position of index j is meaningful in terms of duality notation. On the other hand, the upper position of index k is here arbitrary.

When the mean point G of the cloud is taken as the origin, the initial variables are *centered*: $(x_0^{Jk})_{k\in K}$, with $x_0^{jk} = x^{jk} - \overline{x}^k$.

Matrix notation. We denote $\mathbf{X}$ the table $(J \times K)$ of protocol x^{JK}, $\mathbf{x}^j$ the K–row representing the profile of j and $\mathbf{x}^k$ the J–column representing variable x^{Jk}, $\mathbf{X}_0$ the centered protocol x_0^{JK}; we denote $\mathbf{f}_J$ the J–column of $(f_j)_{j\in J}$, and $\mathbf{F}_J$ the diagonal matrix with $(f_j)_{j\in J}$ diagonal terms. The following auxiliary matrices is used: the identity matrices $\mathbf{I}_J$ $(J \times J)$ and $\mathbf{I}_K$ $(K \times K)$, as well as the J–column $\mathbf{e}$ with all terms equal to 1.

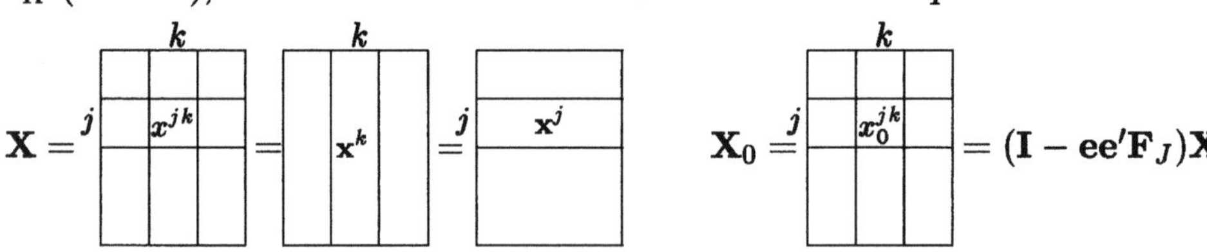

We denote $\mathbf{V}$ the $(K \times K)$ matrix of covariances between initial variables $(x^{Jk})_{k\in K}$, with $\mathbf{V} = \mathbf{X}_0'\mathbf{F}_J\mathbf{X}_0$, and $\mathbf{Q}$ the $(K \times K)$ matrix of scalar products between the basis vectors $(\overrightarrow{\delta_k})_{k\in K}$.

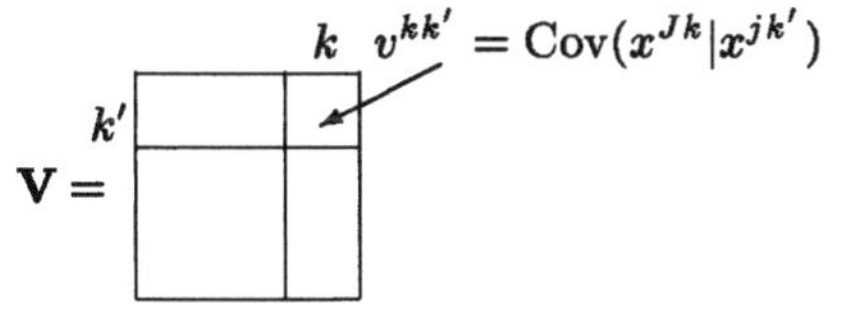

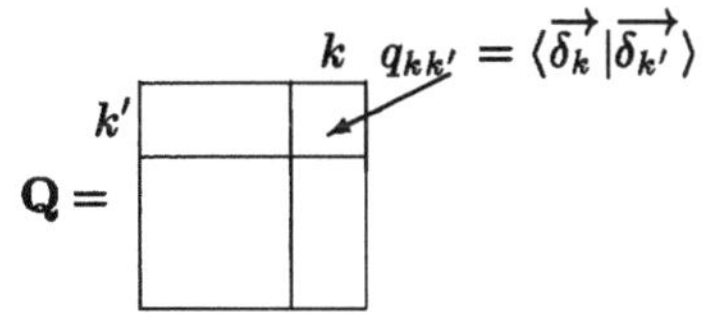

3.7.1 Basic Statistics

Property 3.28. *The Cartesian coordinates of the mean point of a cloud are the means of the initial variables.*

Proof. $\overrightarrow{\mathrm{OG}} = \sum_{j\in J} f_j \overrightarrow{\mathrm{OM}}^j = \sum_{j\in J} f_j \sum_{k\in K} x^{jk}\overrightarrow{\delta_k} = \sum_{k\in K}(\sum_{j\in J} f_j x^{jk})\overrightarrow{\delta_k}$. ◁

If $\overline{\mathbf{x}}^k$ denotes the K–row of means $(\overline{x}^k)_{k\in K}$, one has $\overline{\mathbf{x}}^k = \mathbf{f}_J'\mathbf{X}$.

Property 3.29. *The variance of cloud is equal to* $\sum_{k\in K}\sum_{k'\in K} v^{kk'}q_{kk'}$.

Proof. From $(\mathrm{GM}^j)^2 = \langle \sum_{k\in K}(x^{jk} - \overline{x}^k)\overrightarrow{\delta_k} | \sum_{k'\in K}(x^{jk'} - \overline{x}^{k'})\overrightarrow{\delta_{k'}}\rangle$, it follows:
$\sum_{j\in J} f_j\,(\mathrm{GM}^j)^2 = \sum_{k\in K}\sum_{k'\in K}\left(\sum_{j\in J} f_j(x^{jk} - \overline{x}^k)(x^{jk'} - \overline{x}^{k'})\right)\langle\overrightarrow{\delta_k}|\overrightarrow{\delta_{k'}}\rangle$. ◁

Matrix notation. $(\mathrm{GM}^j)^2 = \mathbf{x}_0'\mathbf{Q}\mathbf{x}_0^j$, $\mathrm{Var}\,\mathrm{M}^J = \mathrm{tr}\,\mathbf{VQ}$ (trace of $\mathbf{VQ}$). If the frame is orthonormal, one has $\mathbf{Q} = \mathbf{I}_K$, and the variance of cloud is equal to $\mathrm{tr}\mathbf{V} = \sum_{k\in K}\mathrm{Var}\,x^{jk}$ (sum of the variances of initial variables).

3.7.2 Principal Formulas

To get the principal formulas in matrix form, one starts from the linear mappings defined in §3.2.5 (p.84), and writes their matrices in the basis $(\overrightarrow{\delta_k})_{k\in K}$ of $\mathcal{V}$ and in the canonical bases $(\delta_j^J)_{j\in J}$ of $\mathbb{R}^J$ and $(\delta_J^j)_{j\in J}$ of $\mathbb{R}_J$.

Property 3.30. *The $(J \times K)$ matrix of Vac is* $\mathbf{X}_0\mathbf{Q}$.

Proof. From Definition 3.11 (p.84) $Vac(\overrightarrow{\delta_k}) = \sum_{j\in J} \langle \overrightarrow{\mathrm{GM}}^j | \overrightarrow{\delta_k} \rangle\, \delta_j^J$. One has $\langle \overrightarrow{\mathrm{GM}}^j | \overrightarrow{\delta_k} \rangle = \sum_{k'\in K} x_0^{jk} \langle \overrightarrow{\delta_{k'}} | \overrightarrow{\delta_k} \rangle$; hence the (j,k)–term of the matrix of *Vac* is equal to $\langle \overrightarrow{\mathrm{GM}}^j | \overrightarrow{\delta_k} \rangle = \sum_{k'\in K} q_{kk'} x_0^{jk'}$. ◁

Corollary 3.3. *The variance of cloud in direction* $\overrightarrow{\alpha} = \sum_{k\in K} a^k \overrightarrow{\delta_k}$ *is equal to* $(\mathbf{a}'\mathbf{QVQa})/(\mathbf{a}'\mathbf{Qa})$, *where* $\mathbf{a}$ *denotes the column of coefficients* $(a^k)_{k\in K}$.

Proof. From property 3.8 (p.86), the variance in direction $\overrightarrow{\alpha}$ is equal to $\|Vac(\overrightarrow{\alpha})\|^2/\|\overrightarrow{\alpha}\|^2$. The numerator is equal to $(\mathbf{X}_0\mathbf{Qa})'\mathbf{F}_J(\mathbf{X}_0\mathbf{Q})\mathbf{a} = \mathbf{a}'\mathbf{Q}(\mathbf{X}_0'\mathbf{F}_J\mathbf{X}_0)\mathbf{Qa}$, and the denominator to $\mathbf{a}'\mathbf{Qa}$, hence the formula. ◁

Property 3.31. *Let* $\mathbf{V}_L$ *be the* $(L \times L)$ *matrix of covariances of the* L *variables associated with a basis of the support of cloud, and* $\mathbf{Q}_L$ *the* $(L \times L)$ *matrix of the scalar products of the* L *vectors of this basis; if one has the property* $\mathbf{V}_L = \lambda \mathbf{Q}_L^{-1}$, *the cloud is spherical.*

Property 3.32. *The* $(K \times J)$ *matrix of Eff is* $\mathbf{X}_0'$.

Proof. $Eff(\delta_J^j) = \sum_{j'\in J} \delta_{j'}^j \overrightarrow{\mathrm{GM}}^{j'} = \overrightarrow{\mathrm{GM}}^j = \sum_{k\in K} x_0^{jk} \overrightarrow{\delta_k}$ (Definition 3.12, p.84). The (k,j)–term of the *Eff* matrix is therefore x_0^{jk}. ◁

The preceding two properties entail the following three corollaries.

Corollary 3.4. *The* $(K \times J)$ *matrix of* $Vac^\star = Eff \circ f_J$ *is* $\mathbf{X}_0'\mathbf{F}_J$.

Corollary 3.5. *The* $(K \times K)$ *matrix of* $Som = Vac^\star \circ Vac$ *is*

$$\mathbf{X}_0'\mathbf{F}_J\mathbf{X}_0\mathbf{Q} = \mathbf{VQ}$$

(Recall that the (k,k')–term of $\mathbf{V}$ is $v^{kk'} = \sum_j f_j x_0^{jk} x_0^{jk'}$).

Corollary 3.6. *The matrix of* $Tom = Vac \circ Vac^\star$ *is the* $(J \times J)$ *matrix* $\mathbf{X}_0\mathbf{Q}\mathbf{X}_0'\mathbf{F}_J = \mathbf{WF}_J$, *with* $\mathbf{W} = \mathbf{X}_0\mathbf{Q}\mathbf{X}_0'$, *with* $w_{jj'} = \langle \overrightarrow{\mathrm{GM}^j} | \overrightarrow{\mathrm{GM}^{j'}} \rangle$.

Relative to the orthonormal principal bases, the matrices of $Vac : \mathcal{L} \to \mathcal{X}_0$ and $Vac^* : \mathcal{X}_0 \to \mathcal{L}$ are both equal to the diagonal matrix $\mathbf{\Xi}$ of singular values $(\xi_\ell)_{\ell=1,\ldots L}$; the matrices of *Som* and *Tom* are diagonal and both equal to $\mathbf{\Lambda} = \mathbf{\Xi}^2$.

Property 3.33. *In matrix terms, the eigendirection and eigenvalue equation is* $\mathbf{VQa}_\ell = \lambda_\ell \mathbf{a}_\ell$, *with* $\mathbf{a}_\ell'\mathbf{Qa}_{\ell'} = 0$ *for* $\ell \neq \ell'$.

Proof. From Theorem 3.3 (p.88), one has: $\sum\limits_{k'\in K}\Big(\sum\limits_{k''\in K} v^{kk''} q_{k'k''}\Big) a^{k'} = \lambda\, a^k$; hence $\mathbf{VQa} = \lambda \mathbf{a}$ ◁

Le us denote $\mathbf{A}$ the $(K \times L)$ matrix of the L columns $(\mathbf{a}_\ell)_{\ell=1,\ldots L}$, and $\mathbf{\Lambda}$ the $(L \times L)$ diagonal matrix of eigenvalues $(\lambda_\ell)_{\ell=1,\ldots L}$, and $\mathbf{I}_L$ the $(L \times L)$ identity matrix; looking for unit–norm eigenvectors one has:

$$\mathbf{VQA} = \mathbf{A\Lambda} \quad \text{with } \mathbf{A'QA} = \mathbf{I}_L \tag{3.15}$$

Property 3.34 (Passage formulas). *In matrix terms, passage formulas between the ℓ–th principal direction (K–column $\mathbf{a}_\ell$) and the ℓ–th standard principal variable (J–column $\mathbf{z}_\ell$) read:*

$$\begin{cases} \mathbf{X}_0\mathbf{Q}\mathbf{a}_\ell &= \xi_\ell \mathbf{z}_\ell \quad (Vac) \\ \mathbf{X}_0{}'\mathbf{F}_J\mathbf{z}_\ell &= \xi_\ell \mathbf{a}_\ell \quad (Vac^\star) \end{cases} \quad \text{with} \quad \begin{cases} \mathbf{a}_\ell{}'\mathbf{Q}\mathbf{a}_\ell &= 1 \\ \mathbf{z}_\ell{}'\mathbf{F}_J\mathbf{z}_\ell &= 1 \end{cases} \tag{3.16}$$

In the *profile space*, the first equation corresponds to the passage from the initial profile x_0^{jK} of j to its principal profile $(y_\ell^j = \sum\limits_{k\in K} a_{k\ell} x_0^{jk})_{\ell=1,\ldots L}$, letting $a_{k\ell} = \sum\limits_{k'\in K} q_{kk'} a_\ell^{k'}$. In the *variable space*, the second equation corresponds to the passage from the k–th initial variable x^{Jk} to its regression coefficient $b_\ell^k = \sum\limits_{j\in J} f_j z_\ell^j x_0^{jk}$ with the ℓ–th standard principal variable.

If we denote $\mathbf{Z}$ the $(J \times L)$ matrix of the L standard principal variables $(z_\ell^J)_{\ell=1,\ldots L}$, and $\mathbf{\Xi}$ the $(L \times L)$ diagonal matrix of the singular values $(\xi_\ell)_{\ell=1,\ldots L}$, the passage formulas read:

$$\begin{cases} \mathbf{X}_0\mathbf{QA} &= \mathbf{Z\Xi} \quad (Vac) \\ \mathbf{X}_0{}'\mathbf{F}_J\mathbf{Z} &= \mathbf{A\Xi} \quad (Vac^\star) \end{cases} \quad \text{with} \quad \begin{cases} \mathbf{A'QA} = \mathbf{I}_L \\ \mathbf{Z'F}_J\mathbf{Z} = \mathbf{I}_L \end{cases} \tag{3.17}$$

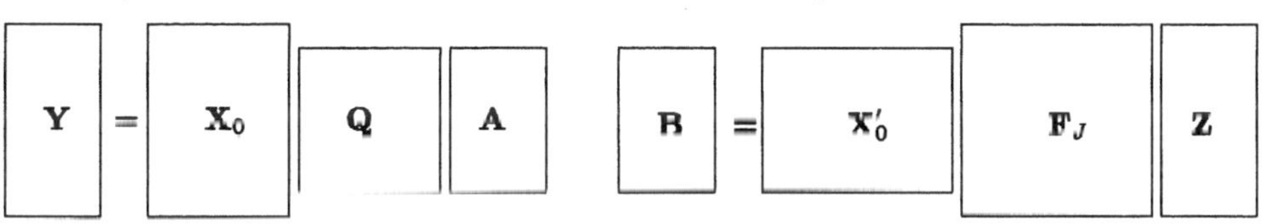

Corollary 3.7 (Principal variable equation).
$\mathbf{X}_0\mathbf{QX'}_0\mathbf{F}_J\mathbf{z}_\ell = \lambda_\ell \mathbf{z}_\ell$ *with* $\mathbf{z'}_\ell\mathbf{F}_J\mathbf{z}_\ell = 1$, *that is* $\mathbf{WF}_J\mathbf{z}_\ell = \lambda_\ell \mathbf{z}_\ell$

Property 3.35 (Reconstitution of order 1). *With the first principal one–dimensional cloud there is associated the reconstitution $\widetilde{\mathbf{X}}_1$ of order 1 of $\mathbf{X}_0$:*

$$\widetilde{\mathbf{X}}_1 = \mathbf{y}_1\mathbf{a}_1' = \xi_1\mathbf{z}_1\mathbf{a}_1{}' = \mathbf{z}_1\mathbf{b}_1'$$

Proof. $\overrightarrow{\mathrm{GA}}{}^j_\ell = y^j_\ell \overrightarrow{\alpha_\ell} = y^j_\ell \sum_{k\in K} a^k_\ell \overrightarrow{\delta_k} = \sum_{k\in K} \tilde{x}^{jk}_\ell \overrightarrow{\delta_k}$; hence $\tilde{x}^{Jk}_\ell = a^k_\ell y^J_\ell$. ◁

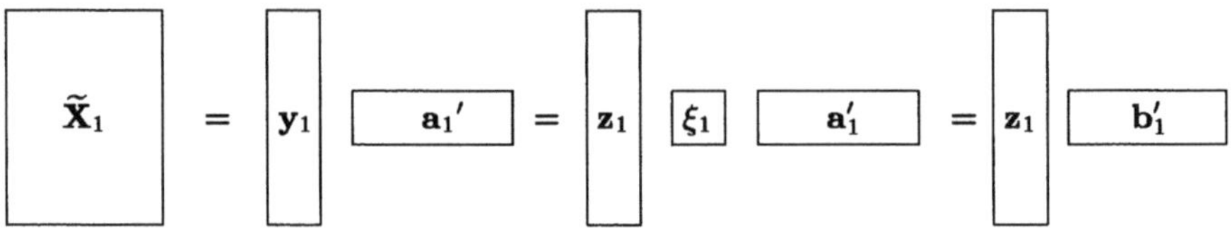

More generally, with the principal L'–dimensional cloud there is associated the *reconstitution of order* L':

$$x^{Jk}_{L'} = \sum_{\ell=1}^{L'} a^k_\ell y^J_\ell = \sum_{\ell=1}^{L'} b^k_\ell z^J_\ell \tag{3.18}$$

hence the reconstitution of $\widetilde{\mathbf{X}}_{L'}$ of order L' of $\mathbf{X}_0$:

$$\widetilde{\mathbf{X}}_{L'} = \sum_{\ell=1}^{L'} \mathbf{y}_\ell \mathbf{a}_{\ell'} = \sum_{\ell=1}^{L'} \xi_\ell \mathbf{z}_\ell \mathbf{a}'_\ell \quad \text{with} \quad \mathbf{a}'_\ell \mathbf{Q} \mathbf{a}_\ell = 1 \text{ and } \mathbf{z}'_\ell \mathbf{F}_J \mathbf{z}_\ell = 1$$

Proof. $\overrightarrow{\mathrm{GA}}{}^j_{L'} = \sum_{\ell=1}^{L'} \overrightarrow{\mathrm{GA}}{}^j_\ell = \sum_{k\in K} \big(\sum_{\ell=1}^{L'} y^j_\ell a^k_\ell\big) \overrightarrow{\delta_k} = \sum_{k\in K} \tilde{x}^{jk}_{L'} \overrightarrow{\delta_k}$. ◁

For $L' = L$, one gets the *formula of complete reconstitution*:

$$\mathbf{X}_0 = \mathbf{Y}\mathbf{A}' = \mathbf{Z}\boldsymbol{\Xi}\mathbf{A}' = \mathbf{Z}\mathbf{B}'$$

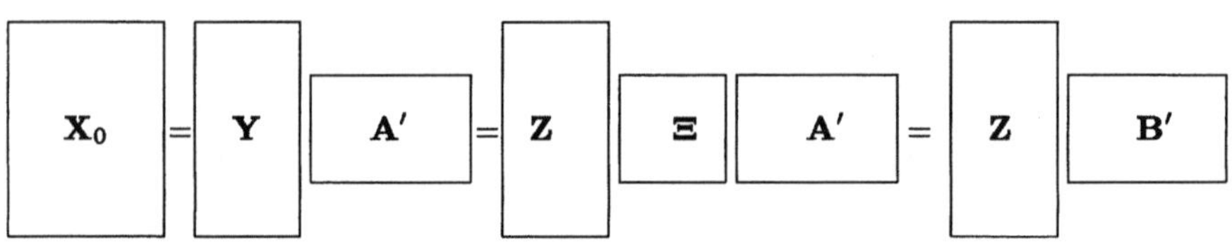

From the reconstitution formula, the other principal formulas can be derived. For instance, postmultiplying both sides of the equation $\mathbf{X}_0 = \mathbf{Z}\boldsymbol{\Xi}\mathbf{A}'$ by $\mathbf{QA}$, one gets: $\mathbf{X}_0\mathbf{QA} = \mathbf{Z}\boldsymbol{\Xi}\mathbf{A}'\mathbf{QA} = \mathbf{Z}\boldsymbol{\Xi}$, etc. To the reconstitution of $\mathbf{X}_0$ there is attached the *reconstitution of matrix* $\mathbf{V}$ *of covariances*: that of order 1 is $\lambda_1 \mathbf{a}_1 \mathbf{a}'_1$; that of order L' (with $L' \leq L$) is such that $\sum_{\ell=1}^{L'} \lambda_\ell \mathbf{a}_\ell \mathbf{a}'_\ell$.

The complete reconstitution of matrix $\mathbf{V}$ is $\sum_{\ell=1}^{L} \lambda_\ell \mathbf{a}_\ell \mathbf{a}'_\ell$, that is $\mathbf{V} = \mathbf{A}\boldsymbol{\Lambda}\mathbf{A}'$ with $\mathbf{A}'\mathbf{QA} = \mathbf{I}_L$.

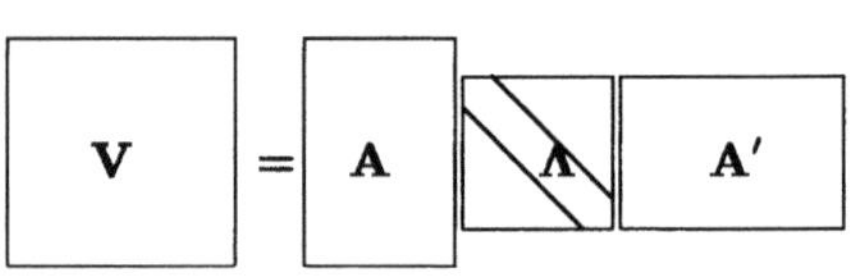

If the initial frame is orthonormal ($\mathbf{Q} = \mathbf{I}_K$), letting $d_1^{kk'} = v^{kk'} - \lambda_1 a_1^k a_1^{k'}$ be the residual deviation, and $\mathbf{D}_1$ the $K \times K$ matrix of residual deviations, one has: $\sum\limits_{k\in K}\sum\limits_{k'\in K}(d_1^{kk'})^2 = \operatorname{tr}\mathbf{D}_1'\mathbf{D}_1 = \sum\limits_{\ell=2}^{L}\lambda_\ell^2$.

The reconstitution of order L' of the variance of the cloud is $\sum\limits_{\ell=1}^{L'}\lambda_\ell$; the complete reconstitution is $\operatorname{tr}\mathbf{VQ} = \sum\limits_{\ell=1}^{L}\lambda_\ell = \operatorname{tr}\mathbf{\Lambda} = \operatorname{Var}\mathrm{M}^J$.

3.7.3 Principal Hyperellipsoids

In any basis, the κ–hyperellipsoid has Cartesian equation $\mathbf{x}'\mathbf{V}^{-1}\mathbf{x} = \kappa^2$ and tangential equation $\mathbf{u}'\mathbf{V}\mathbf{u} = 1/\kappa^2$. With respect to principal basis, those equations become $\mathbf{y}'\mathbf{\Lambda}^{-1}\mathbf{y} = \kappa^2$ and $\mathbf{u}'\Lambda\mathbf{u} = 1/\kappa^2$ respectively.

Exercises of Chapter 3

Exercise 3.1 (Analysis of a distance table)
Let M^J a Euclidean cloud on weighted support (J, n_J) with mean point G. One lets $\Delta^{jj'} = (\mathrm{M}^j\mathrm{M}^{j'})^2$ the square of distance between M^j and $\mathrm{M}^{j'}$; $\Delta^j = \sum\limits_{j'\in J} f_{j'}\Delta^{jj'}$;

$\Delta = \sum\limits_{j\in J}\sum\limits_{j'\in J} f_j f_{j'}\Delta^{jj'}$ and $w^{jj'} = \langle\overrightarrow{\mathrm{GM}^j}|\overrightarrow{\mathrm{GM}^{j'}}\rangle$.

1. Prove that $\Delta^j = (\mathrm{GM}^j)^2 + \operatorname{Var}\mathrm{M}^J$. Deduce: $\Delta = 2\operatorname{Var}\mathrm{M}^J$.

2. Prove that $\Delta^{jj'} = (\mathrm{GM}^j)^2 + (\mathrm{GM}^{j'})^2 - 2w^{jj'}$.

3. From the two preceding questions deduce *Torgerson's formula*:
$$w^{jj'} = -(\Delta^{jj'} - \Delta^j - \Delta^{j'} + \Delta)/2$$

4. For the elementary cloud of 10 points of the *Target example*, one gives the table of the squares of distances ($\Delta^{jj'}$) between the 45 pairs of points. Deduce $\operatorname{Var}\mathrm{M}^J$.
Given $\mathrm{OG} = 6$, calculate the mean of the squares of distances from the points of the cloud to the point O.

Δ	$j1$	$j2$	$j3$	$j4$	$j5$	$j6$	$j7$	$j8$	$j9$
$j2$	40								
$j3$	232	80							
$j4$	136	64	80						
$j5$	288	136	40	40					
$j6$	260	340	548	212	404				
$j7$	260	212	244	52	116	104			
$j8$	292	196	164	36	52	200	16		
$j9$	584	416	272	160	104	388	100	52	
$j10$	628	436	260	180	100	464	136	72	4

5. For the partition into 3 classes with $J{<}c1{>} = \{j1, j2\}$, $J{<}c2{>} = \{j6\}$ and $J{<}c3{>} = \{j3, j4, j5, j7, j8, j9, j10\}$, one gives the squares of distances from the 3 class centers M^{c1}, M^{c2}, M^{c3} to point G: $(\mathrm{GM}^{c1})^2 = 130$, $(\mathrm{GM}^{c2})^2 = 200$ and $(\mathrm{GM}^{c3})^2 = 16.32654$. Deduce the between–variance then the within–

variance. Using the above table, calculate the variances of the 3 subclouds and deduce the contributions to the variance of the 3 subclouds.

6. From the 10×10 table of question 4, deduce (using Torgerson's formula) the matrix $\mathbf{W}$ of scalar products. Verify numerically that the variables $(y_\ell^J)_{\ell=1,2}$ found in the *Target example* exercise (question 3, p.126) are solutions of the principal variables equation $\mathbf{W}\mathbf{F}_J\mathbf{y}_\ell = \lambda_\ell \mathbf{y}_\ell$ (p.119).

7. *Application to* CA. Given a contingency table n_{JK}, consider the $J \times J$ table of square distances $(\mathrm{M}^j\mathrm{M}^{j'})^2 = \sum\limits_{k \in K} \frac{(f_k^j - f_k^{j'})^2}{f_k} = \delta^{jj'}$. Prove that the principal analysis of the table of distances yields the principal coordinates $(y_\ell^J)_{\ell \in L}$ of CA. Verify the property on the *Eye & Hair Color* (CORRESPONDENCE), by starting with the square distances $\delta^{jj'}$, then constructing the scalar products $w^{jj'}$.

Solution

1. From Huyghens' theorem (p.79) taking M^j as a reference point, one has, : $\sum\limits_{j' \in J} f_j(\mathrm{M}^j\mathrm{M}^{j'})^2 = (\mathrm{GM}^j)^2 + \mathrm{Var}\,\mathrm{M}^J$; hence Δ^j and $\Delta = \sum\limits_{j \in J} f_j \Delta^j$.

2. $(\mathrm{M}^j\mathrm{M}^{j'})^2 = \|\overrightarrow{\mathrm{GM}}^{j'} - \overrightarrow{\mathrm{GM}}^{j}\|^2 = (\mathrm{GM}^j)^2 + (\mathrm{GM}^{j'})^2 - 2\langle \overrightarrow{\mathrm{GM}}^{j} | \overrightarrow{\mathrm{GM}}^{j'} \rangle$; hence $\Delta^{jj'}$.

3. From question 1: $(\mathrm{GM}^j)^2 + (\mathrm{GM}^{j'})^2 = \Delta^j + \Delta^{j'} - \Delta$, hence $w^{jj'}$ by replacing in the relation obtained in question 2.

4. The sum of squares of the distances between the 45 pairs of points is equal to 9 200, hence the variance $\frac{1}{10} \times \frac{1}{10} \times 9\,200 = 92$. From Huyghens' theorem (p.79), one has: $\sum\limits_{j \in J} f_j(\mathrm{OM}^j)^2 = 6^2 + 92 = 128$.

5. Between–variance: $\frac{2}{10} \times 130 + \frac{1}{10} \times 200 + \frac{7}{10} \times 16.32654 = 57.429$. From the between–within breakdown of the variance (property 3.23, p.104), one deduces the within–variance: $\mathrm{Var}\,\mathrm{M}^{J(C)} = \mathrm{Var}\,\mathrm{M}^J - \mathrm{Var}\,\mathrm{M}^C = 92 - 57.429 = 34.571$. One has $\mathrm{Var}\,\mathrm{M}^{J<c1>} = \frac{1}{2} \times \frac{1}{2} \times 40 = 10$, hence $\mathrm{Cta}_{J(c1)} = \frac{2}{10} \times \mathrm{Var}\,\mathrm{M}^{J<c1>} = 2$, and $\mathrm{Cta}_{c1} = \frac{2}{10} \times 130 = 26$; from the property 3.19 (p.102), one deduces $\mathrm{Cta}_{J<c1>} = 26 + 2 = 28$. Likewise, one has $\mathrm{Cta}_{J<c2>} = 20$, and $\mathrm{Cta}_{J<c3>} = 44$.

Comment

Analysis of distance tables. When a data set is a $J \times J$ table of distances between J objects, together with a weighting f_J, an Euclidean cloud can be constructed such that the distances between points are reconstituted, at least approximately, using Torgerson formula then the principal variable equation. The analysis is also known as "principal coordinate analysis". It amounts to *metric* MDS.

Analysis of dissimilarity tables. When a data set is a *dissimilarity* table[5], approximate Euclidean clouds can be still be constructed. One method is the *additive constant*, where a constant is added to each dissimilarity, such that the resulting table is a distance table. Another method is *nonmetric* MDS, which produces

[5] A dissimilarity index satisfies the positiveness and symmetry axioms, but not necessarily the triangle axiom.

an approximate multidimensional representation from the mere rank ordering of dissimilarities (Shepard, 1962, 1966). Nonmetric as well as metric MDS are GDA methods, since they produce Euclidean clouds[6]. For further discussion, see Benzécri (1964, 1965), Benzécri & al (1973, Vol. 2, p. 93-95), Saporta (1990).

Exercise 3.2 (Euclidean cloud)

One considers a cloud of 4 points O, A, B, C, such that OA = OB = OC = r and AB = AC = BC = a; point O has weight $1-m$, and each of the points A, B and C has weight $m/3$, with $0 < m < 1$. Let O' be the mean point of the subcloud of the three points A, B, C.

1. Express distances O′A, O′B and O′C as a function of a. Which relation a and r must verify so that the four points A, B, C and O define a Euclidean cloud? Express distance OO′ as a function of a and r. What is the dimensionality of the cloud? In which case does the cloud reduce to a plane cloud?

   Prove that the mean point G of the cloud is such that $\overrightarrow{OG} = m\overrightarrow{OO'}$.

2. Express the variance of the subcloud of the three points A, B, C as a function of a; deduce that the within–contribution of this subcloud to the cloud of the 4 points is equal to $ma^2/3$.

3. One considers the partition in two classes $\mathcal{O} = \{\mathrm{O}\}$, and $\mathcal{A} = \{\mathrm{A,B,C}\}$. Express the *within–variance* as a function of a and m.

   What are the two points of the *between–cloud* associated with this partition, and what are their weights? Deduce the *between–variance* and the *variance* of the 4–point cloud as a functions of r, a and m.

4. What is the variance of the cloud in the direction OO′? of the plane ABC?

5. Show that OO′ is the principal line of the 4–point cloud associated with the eigenvalue $m(1-m)(r^2-a^2/3)$, and that every line going through G parallel to plane ABC is principal with the double eigenvalue $ma^2/6$.

Exercise 3.3 (Principal directions of a plane cloud)

Let (M^J, n_J) be a plane cloud referred to an orthonormal Cartesian frame with origin G (mean point of the cloud) and basis vectors $(\vec{\varepsilon}, \vec{\varepsilon}\,')$.

One lets $\overrightarrow{\mathrm{GM}}^j = x_0^j\vec{\varepsilon} + x'^j_0\vec{\varepsilon}\,'$. One denotes respectively v, v' the variances of centered variables x_0^J and x'^J_0, c their covariance and r their correlation.[7]

1. Express the variance of the cloud M^J as a function of v and v'.

---

[6] MDS algorithms do not satisfy the heredity property, that is, the principal coordinates of the one–dimensional representation may differ from the first principal coordinates in the two–dimensional representation, etc.

[7] In this exercise, one solves, by elementary algebra, the problem of principal directions and variables of a plane cloud.

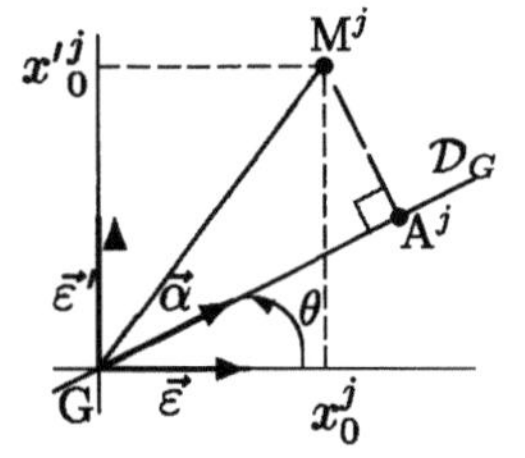

2. Let $\vec{\alpha} = \cos\theta\,\vec{\varepsilon} + \sin\theta\,\vec{\varepsilon}\,'$. Show that the variance of the cloud in direction $\vec{\alpha}$ can be written: $V(\theta) = v\cos^2\theta + v'\sin^2\theta + 2c\cos\theta\sin\theta$.
3. Show that, in general, $V(\theta)$ is maximum for $\theta = \theta_1$ and minimum for $\theta = \theta_2$ (determined up to a $k\pi$ factor) verifying equation $c\tan^2\theta + (v - v')\tan\theta - c = 0$.

   Show that $\theta_2 = \theta_1 \pm \pi/2$. Write the solutions $\tan\theta_1$ and $\tan\theta_2$ of this equation and discuss the sign of $\tan\theta_1$ in relation to the sign of c (particular cases $c = 0$ and $\theta = \pm\pi/2$ will be studied in question 8).

   Express unit–norm principal vectors $\vec{\alpha_1}$ and $\vec{\alpha_2}$ as a function of θ_1, $\vec{\varepsilon}$ and $\vec{\varepsilon}\,'$ $(-\pi/2 \leq \theta_1 \leq +\pi/2, \theta_2 = \theta_1 + \pi/2)$.
4. For $\ell = 1, 2$, one lets: $\lambda_\ell = v + c\tan\theta_\ell$. Show that θ_ℓ and λ_ℓ (ℓ–th eigenvalue) are solutions of the following system of two equations:
   $$v\cos\theta + c\sin\theta = \lambda\cos\theta \qquad \text{and} \qquad c\cos\theta + v'\sin\theta\lambda\sin\theta$$
5. Write the two solutions of equation $\lambda^2 - (v + v')\lambda + vv' - c^2 = 0$.
6. Show that $\lambda_\ell = v + c\tan\theta_\ell$ is the variance of the cloud in the direction of angle θ_ℓ ($\ell = 1, 2$).
7. Prove that if $\lambda_1 \neq \lambda_2$ the 2 principal variables are uncorrelated.
8. Study the following particular cases: $c = 0$ and $v < v'$; $c = 0$ and $v > v'$; $v = v'$ and $c \neq 0$; $c/\sqrt{vv'} = r = \pm 1$.

Solution

1. $\operatorname{Var} M^J = \sum_{j\in J} f_j (GM^j)^2 = \sum_{j\in J} f_j\left((x^j_0)^2 + (x'^j_0)^2\right) = v + v'$.

2. Point A^j, projection of point M^j on line $\mathcal{D}_G$, is such that $\overrightarrow{GA}^j = y^j\,\vec{\alpha}$, with $y^j = x^j_0\cos\theta + x'^j_0\sin\theta$, hence the variance of cloud A^J is written:

$$V(\theta) = \sum_{j\in J} f_j(x^j_0\cos\theta + x'^j_0\sin\theta)^2 = v\cos^2\theta + v'\sin^2\theta + 2c\sin\theta\cos\theta \qquad (3.19)$$

3. $V(\theta)$ can be derived everywhere on interval $(-\pi/2\,;\pi/2)$ and has derivative: $V'(\theta) = 2(v' - v)\cos\theta\sin\theta + 2c(\cos^2\theta - \sin^2\theta)$; or else (for $\theta \neq \pm\pi/2$): $V'(\theta) = -2\cos^2\theta\big(c\tan^2\theta + (v - v')\tan\theta - c\big)$. The angles θ for which $V(\theta)$ is extremum are the solutions of $V'(\theta) = 0$. Hence the ***principal direction equation***:

$$c\tan^2\theta + (v - v')\tan\theta - c = 0 \qquad (\text{with} \quad -\pi/2 < \theta < +\pi/2) \qquad (3.20)$$

This equation of unknown $\tan\theta$ has two solutions verifying:

$$\tan\theta = (v' - v \pm \sqrt{(v - v')^2 + 4c^2})/2c \qquad (c \neq 0)$$

(for cases $v = v'$, $c = 0$ and $\theta = \pm\pi/2$, cf. question 8).
The study of the sign of $V'(\theta)$ $(-\pi/2 \leq \theta \leq \pi/2)$ shows that $V(\theta)$ is maximum for $\tan\theta_1 > 0$ if $c > 0$, and for $\tan\theta_1 < 0$ if $c < 0$. The angle θ_1 for which $V(\theta)$

is maximum has the sign of c, and it is given by taking the + sign in the above formula. The angle θ_2 rendering $V(\theta)$ minimum is such that $\theta_2 = \theta_1 \pm \pi/2$ since $\tan\theta_1 \times \tan\theta_2 = -1$ (product of the roots of Equation (3.20)). Hence unit–norm principal vectors are $\vec{\alpha_1^*} = \cos\theta_1 \vec{\epsilon} + \sin\theta_1 \vec{\epsilon}\,'$ and $\vec{\alpha_2^*} = -\sin\theta_1 \vec{\epsilon} + \cos\theta_1 \vec{\epsilon}\,'$.

— *Remark.* $\tan 2\theta = 2\tan\theta/(1-\tan^2\theta)$, then Equation (3.20) can be written $\tan 2\theta = 2c/(v-v')$.

4. The relation $v + c\tan\theta = \lambda$ can be written: $v\cos\theta + c\sin\theta = \lambda\cos\theta$. In Equation (3.20), if one replaces $v + c\tan\theta$ by λ, one has: $\tan\theta(\lambda - v') = c$, hence $c\cos\theta + v'\sin\theta = \lambda\sin\theta$.

Let $\mathbf{a}' = \boxed{\cos\theta \quad \sin\theta}$, and $\mathbf{V}$ the variance–covariance matrix, these equations can be written $\mathbf{Va} = \lambda\mathbf{a}$ (diagonalization of $\mathbf{V}$, cf. *MATH.BASES*, p.443).

5. In Equation (3.20), if one replaces $\tan\theta$ by $(\lambda - v)/c$ $(c \neq 0)$, one obtains the *eigenvalue equation*, whose solutions are:

$$\lambda_1 = \tfrac{v+v'}{2} + \tfrac{1}{2}\sqrt{(v-v')^2 + 4c^2} \qquad \text{and} \qquad \lambda_2 = \tfrac{v+v'}{2} - \tfrac{1}{2}\sqrt{(v-v')^2 + 4c^2}$$

6. If one replaces $v\cos\theta_\ell + c\sin\theta_\ell$ by $\lambda_\ell\cos\theta_\ell$ then $c\cos\theta_\ell + v'\sin\theta_\ell$ by $\lambda_\ell\sin\theta_\ell$ in Equation 3.19, one obtains: $V(\theta_\ell) = \lambda_\ell$.

7. One has:
$$\begin{aligned} \alpha_1^j &= \langle \overrightarrow{\mathrm{GM}}^j | \overrightarrow{\alpha_1} \rangle \text{ therefore } & y_1^j &= x_0^j\cos\theta_1 + x'^j_0\sin\theta_1 \\ \alpha_2^j &= \langle \overrightarrow{\mathrm{GM}}^j | \overrightarrow{\alpha_2} \rangle \text{ therefore } & y_2^j &= -x_0^j\sin\theta_1 + x'^j_0\cos\theta_1 \end{aligned}$$

One has $\mathrm{Cov}(\alpha_1^J|\alpha_2^J) = (v'-v)\cos\theta_1\sin\theta_1 + c(\cos^2\theta_1 - \sin^2\theta_1) = \frac{1}{2}V'(\theta_1)$, hence (question 3): $V'(\theta_1) = 0$, so $\mathrm{Cov}(\alpha_1^J|\alpha_2^J) = 0$.

8. If $c = 0$ and $v > v'$ one has $\theta_1 = 0$ and $\theta_2 = \pi/2$, if $c = 0$ and $v < v'$ one has $\theta_1 = \pi/2$ and $\theta_2 = 0$: initial axes are principal.

If $v = v'$ and $c \neq 0$, then $\tan^2\theta_\ell = 1$, hence $\theta_\ell = \pm\pi/4$ (the principal lines are the bissectors of the plane) and $\lambda = v(1 \pm r)$. If $v = v'$ and $c = 0$, $V(\theta)$ does not depend on θ, then every direction of the plane is principal, and is associated with the eigenvalue of multiplicity 2: $\lambda_1 = \lambda_2 = v = \mathrm{Var}\,\mathrm{M}^J/2$.

For a *one–dimensional cloud*, one has $vv' = c^2$ (or $r = \pm 1$), $\lambda_1 = v + v'$ and $\lambda_2 = 0$.

If $c > 0$ and $v > v'$, then $0 < \theta_1 < \pi/4$. If $c > 0$ and $v' > v$, then $\pi/4 < \theta_1 < \pi/2$.

If $c < 0$ and $v > v'$, then $-\pi/4 < \theta_1 < 0$. If $c < 0$ and $v' > v$,

$-\pi/2 < \theta_1 < -\pi/4$. That is, to sum up: $|\theta_1| < \pi/4 \iff v > v'$; $|\theta_1| = \pi/4 \iff v = v'$; $|\theta_1| > \pi/4 \iff v < v'$.

Exercise 3.4 (Target example)

If one refers the plane (p.76) to two rectangular axes (horizontal and vertical) going through point O (center of the target), the abscissas and ordinates of points M^j are shown in the following table (the graduations of the figure below correspond to $2u$). Let $(\mathrm{O}, \vec{\varepsilon_1}, \vec{\varepsilon_2})$ be the orthonormal Cartesian frame and x^{Jk1} and x^{Jk2} the two initial variables.

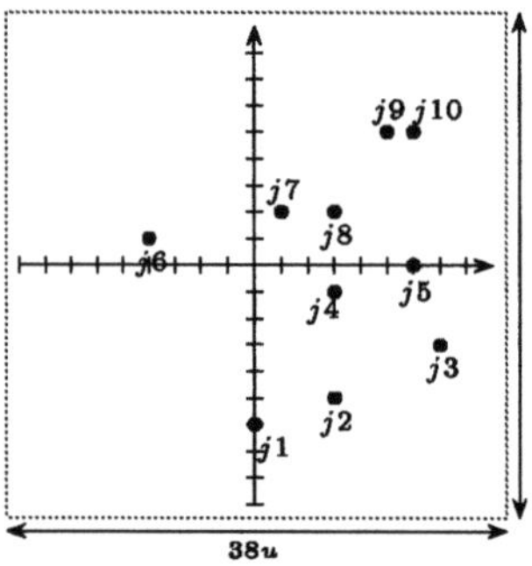

	x^{Jk1}	x^{Jk2}
$j1$	0	−12
$j2$	6	−10
$j3$	14	−6
$j4$	6	−2
$j5$	12	0
$j6$	−8	2
$j7$	2	4
$j8$	6	4
$j9$	10	10
$j10$	12	10

1. Calculate the coordinates of the mean point; deduce the table $\mathbf{X}_0$ of coordinates of points in frame $(G, \vec{\varepsilon_1}, \vec{\varepsilon_2})$. Write the variance–covariance matrix (denoted $\mathbf{V}$). Deduce the variance of the cloud.

2. Diagonalize covariance matrix $\mathbf{V}$ by solving the matrix equation $\mathbf{Va} = \lambda\,\mathbf{a}$, with $\mathbf{a}' = \boxed{\cos\theta \quad \sin\theta}$ and write (2×2) matrix $\mathbf{A}$ verifying $\mathbf{VA} = \mathbf{A\Lambda}$ (with $\mathbf{AA}' = \mathbf{I}$). Draw the principal lines and the two principal clouds.

3. Write the passage formula from the centered initial variables to the calibrated principal variables; calculate the calibrated principal variables and present the matrix diagram going with them. Plot the projected clouds on principal axes 1 and 2 (pictured both horizontally), as well as the plane cloud referred to principal axes (Axis 1 horizontal, Axis 2 vertical).

4. Calculate the regression coefficients (b_1^{k1}, b_2^{k1}) of initial variable x^{Jk1} on the two standard principal variables z_1^J and z_2^J, and the correlation coefficients (r_1^{k1}, r_2^{k2}). Calculate the regression and correlation coefficients for x^{Jk2}. Express the centered initial variables as a function of the standard principal variables; write the reconstitution of matrix $\mathbf{X}_0$. Draw the centered initial variables in the principal plane.

5. Write the Cartesian equations of the concentration ellipse in the initial and the principal frames and plot the ellipse.

Solution

1. Coordinates of the mean point: $(6, 0)$.
$\operatorname{Var} x^{Jk1} = 40, \operatorname{Var} x^{Jk2} = 52$ and $\operatorname{Cov}(x^{Jk1}|x^{Jk2}) = +8$.

Variance–covariance matrix: $\mathbf{V} = \begin{bmatrix} 40 & 8 \\ 8 & 52 \end{bmatrix}$

Variance of the cloud: $\operatorname{Var} \mathrm{M}^J = 40 + 52 = 92$.

$\mathbf{X}_0 =$

−6	−12
0	−10
8	−6
0	−2
6	0
−14	2
−4	4
0	4
4	10
6	10

2. Direction and eigenvalue equations ($\mathbf{Va} = \lambda\mathbf{a}$):

$$\begin{cases} 40\cos\theta + 8\sin\theta = \lambda\cos\theta \\ 8\cos\theta + 52\sin\theta = \lambda\sin\theta \end{cases} \text{ or } \begin{cases} 8\tan\theta = \lambda - 40 \\ (\lambda - 52)\tan\theta = 8 \end{cases}$$

From $\frac{\lambda-40}{8} = \frac{8}{\lambda-52}$, one deduces: $\lambda^2 - 92\lambda + 2016 = 0$.
The solutions of this equation are $\lambda_1 = 56$ and $\lambda_2 = 36$. From the preceding system of two equations, one deduces: $\tan\theta_1 = 2$, hence $\theta_1 = +63°4$ and $\theta_2 = -26°6$.

Thus one has: $\cos^2\theta_1 = 1/(1+\tan^2\theta) = 1/5$, hence $\cos\theta_1 = 1/\sqrt{5}$, $\sin\theta_1 = 2/\sqrt{5}$; and $\cos\theta_2 = -2/\sqrt{5}$, $\sin\theta_2 = 1/\sqrt{5}$.

***Passage matrix* A** from initial basis vectors $(\overrightarrow{\varepsilon_1}, \overrightarrow{\varepsilon_2})$ to principal basis vectors $(\overrightarrow{\alpha_1}, \overrightarrow{\alpha_2})$ is equal to:

$$\mathbf{A} = \begin{bmatrix} \cos\theta_1 & \cos\theta_2 \\ \sin\theta_1 & \sin\theta_2 \end{bmatrix} = \begin{bmatrix} 1/\sqrt{5} & -2/\sqrt{5} \\ 2/\sqrt{5} & 1/\sqrt{5} \end{bmatrix} = \begin{bmatrix} \mathbf{a}_1 & \mathbf{a}_2 \end{bmatrix}$$

Principal vectors:

$$\overrightarrow{\alpha_1} = (\overrightarrow{\varepsilon_1} + 2\overrightarrow{\varepsilon_2})\sqrt{5}$$
$$\overrightarrow{\alpha_2} = (-2\overrightarrow{\varepsilon_1} + \overrightarrow{\varepsilon_2})/\sqrt{5}$$

3. Passage formula from centered initial variables to calibrated principal variables (cf. opposite scheme):

$$(1/\sqrt{5})\, x_0^{Jk1} + (2/\sqrt{5})\, x_0^{Jk2} = y_1^J$$
$$(-2/\sqrt{5})\, x_0^{Jk1} + (1/\sqrt{5})\, x_0^{Jk2} = y_2^J$$

		y_1^J $\overrightarrow{\alpha_1}$	y_2^J $\overrightarrow{\alpha_2}$	
x_0^{Jk1}	$\overrightarrow{\varepsilon_1}$	$1/\sqrt{5}$	$-2/\sqrt{5}$	$= \mathbf{A}$
x_0^{Jk2}	$\overrightarrow{\varepsilon_2}$	$2/\sqrt{5}$	$1/\sqrt{5}$	

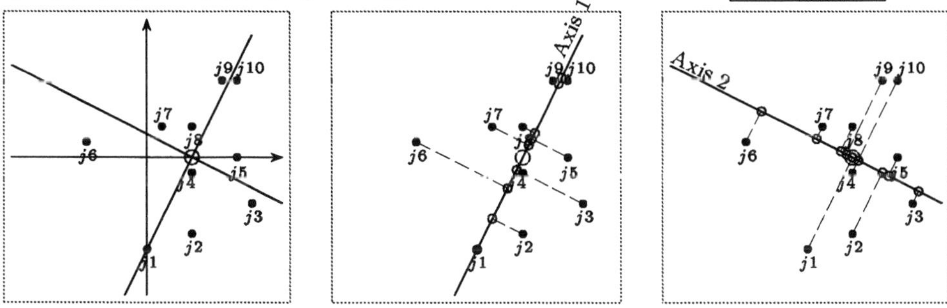

Graphical display of the principal lines and the principal clouds (empty circles)

Matrix formula: $\mathbf{X}_0\mathbf{A} = \mathbf{Y}$ (cf. p 119).

$$\begin{bmatrix} -6 & -12 \\ 0 & -10 \\ 8 & -6 \\ 0 & -2 \\ 6 & 0 \\ -14 & 2 \\ -4 & 4 \\ 0 & 4 \\ 4 & 10 \\ 6 & 10 \end{bmatrix} \begin{bmatrix} \frac{1}{\sqrt{5}} & -\frac{2}{\sqrt{5}} \\ \frac{2}{\sqrt{5}} & \frac{1}{\sqrt{5}} \end{bmatrix} = \frac{1}{\sqrt{5}} \begin{bmatrix} -30 & 0 \\ -20 & -10 \\ -4 & -22 \\ -4 & -2 \\ 6 & -12 \\ -10 & 30 \\ 4 & 12 \\ 8 & 4 \\ 24 & 2 \\ 26 & -2 \end{bmatrix} = \begin{bmatrix} -13.4 & 0 \\ -8.9 & -4.5 \\ -1.8 & -9.8 \\ -1.8 & -0.9 \\ 2.7 & -5.4 \\ -4.5 & 13.4 \\ 1.8 & 5.4 \\ 3.6 & 1.8 \\ 10.7 & 0.9 \\ 11.6 & -0.9 \end{bmatrix}$$

Diagrams with graph scale — 1.5u (graduations every 5u).

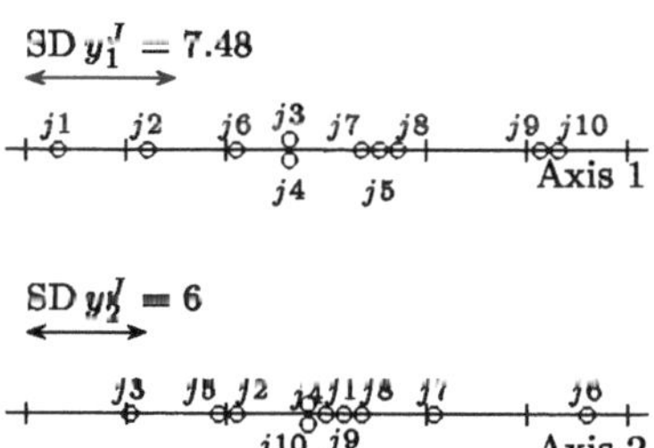

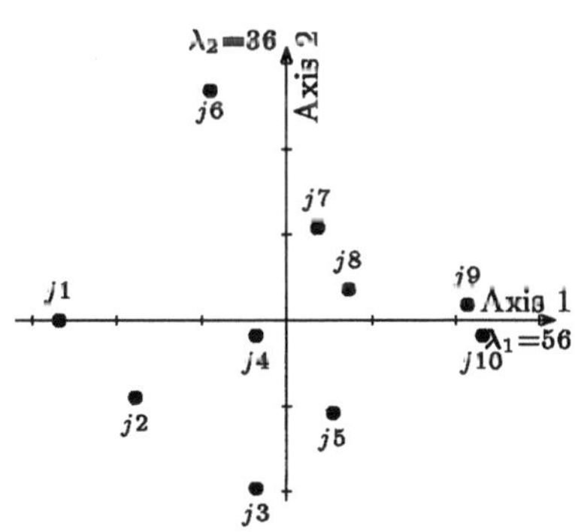

4. Regression coefficients:
$b_1^{k1} = \sqrt{56} \times \frac{1}{\sqrt{5}} = 3.35 \quad b_1^{k2} = \sqrt{56} \times \frac{2}{\sqrt{5}} = 6.69$
$b_2^{k1} = \sqrt{36} \times \frac{-2}{\sqrt{5}} = -5.37 \quad b_2^{k2} = \sqrt{36} \times \frac{1}{\sqrt{5}} = 2.68$
$x_0^{Jk1} = 3.35\, z_1^J - 5.37\, z_2^J$ and $x_0^{Jk2} = 6.69\, z_1^J + 2.68\, z_2^J$
Correlation coefficients:
$r_1^{k1} = .529,\ r_2^{k1} = -.849,\ r_1^{k2} = .928,\ r_2^{k2} = .372.$
Graphical display in variable space (see opposite figure, with graphical scale equal to $3u$); the coordinates of the initial variables are equal to the regression coefficients.

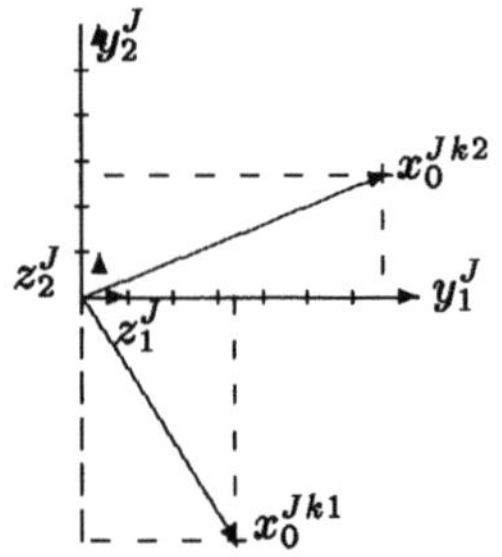

Reconstitution of centered protocol:

$$x_0^{Jk1} = (1/\sqrt{5})\, y_1^J - (2/\sqrt{5})\, y_2^J$$
$$x_0^{Jk2} = (2/\sqrt{5})\, y_1^J + (1/\sqrt{5})\, y_2^J$$

↙	y_1^J $\vec{\alpha_1}$	y_2^J $\vec{\alpha_2}$
x_0^{Jk1} $\vec{\epsilon_1}$	$1/\sqrt{5}$	$-2/\sqrt{5}$
x_0^{Jk2} $\vec{\epsilon_2}$	$2/\sqrt{5}$	$1/\sqrt{5}$

Matrix formulation: $\mathbf{X}_0 = \mathbf{YA}'$ (cf. Equations (3.17), p 119).

$$\begin{bmatrix} -6 & -12 \\ 0 & -10 \\ 8 & -6 \\ 0 & -2 \\ 6 & 0 \\ -14 & 2 \\ -4 & 4 \\ 0 & 4 \\ 4 & 10 \\ 6 & 10 \end{bmatrix} = \frac{1}{\sqrt{5}} \begin{bmatrix} -30 & 0 \\ -20 & -10 \\ -4 & -22 \\ -4 & -2 \\ 6 & -12 \\ -10 & +30 \\ 4 & +12 \\ 8 & +4 \\ 24 & +2 \\ 26 & -2 \end{bmatrix} \times \begin{bmatrix} \frac{1}{\sqrt{5}} & \frac{2}{\sqrt{5}} \\ \frac{-2}{\sqrt{5}} & \frac{1}{\sqrt{5}} \end{bmatrix} = \begin{bmatrix} -6 & -12 \\ -4 & -8 \\ -0.8 & -1.6 \\ -0.8 & -1.6 \\ 1.2 & 2.4 \\ -2 & -4 \\ 0.8 & 1.6 \\ 1.6 & 3.2 \\ 4.8 & 9.6 \\ 5.2 & 10.4 \end{bmatrix} + \begin{bmatrix} 0 & 0 \\ 4 & -2 \\ 8.8 & -4.4 \\ 0.8 & -0.4 \\ 4.8 & -2.4 \\ -12 & 6 \\ -4.8 & 2.4 \\ -1.6 & 0.8 \\ -0.8 & 0.4 \\ 0.8 & -0.4 \end{bmatrix}$$

5. Cartesian equation of the concentration ellipse (cf. §3.4.3, p.97):
in frame $(O, \vec{\epsilon_1}, \vec{\epsilon_2})$: $\frac{1}{1-0.175}\left(\frac{(x^{k1})^2}{40} - 6\frac{x^{k1}x^{k2}}{2080} + \frac{(x^{k2})^2}{52}\right) = 4;$
in frame $(O, \vec{\alpha_1}, \vec{\alpha_2})$: $(y_1)^2/56 + (y_2)^2/36 = 4.$
Graphical display: cf. p.98.

Chapter 4

Principal Component Analysis

(*'PRINCIPAL'*)

Introduction

In this chapter, we start with numerical data and we go "from numbers to points", by representing data geometrically as a cloud of points. More precisely, we study a finite set K of numerical variables over a finite set J of individuals. In applications, individuals, that is, statistical individuals, can refer to living persons, as well as to firms, experimental units, etc. To deal with such data, a classical method is ***Principal Component Analysis*** (PCA): The K variables (***initial variables***) are replaced by L $(1 \leq L \leq K)$[1] new variables (***principal variables***) that are linear combinations of initial variables and constitute the best summary of these variables in the least–square sense. In traditional Multivariate Analysis, PCA is essentially concerned with the associations (covariances or correlations) between variables.

In this chapter, we recast PCA as a geometric method: Starting from the $J \times K$ data table, the basic step of PCA consists in constructing a Euclidean cloud of J points. The following step consists in reducing the dimensionality of the cloud by determining its principal directions and variables, as done in chapter 3 (*CLOUD*). Thus the geometric approach leads not only to investigating the associations between variables, but also to describing and interpreting the proximities between individual points.

[1] Recall that we denote the cardinality (number of elements) of a finite set by the same letter as the set itself.

The origin of PCA goes back to Karl Pearson (1901), at least. But for a long time its application to real data sets was hindered by the computational burden; the routine practice of PCA has only become possible at the computer age. Prior to that, a related method was (and still is) used, namely *Factor Analysis*, that was extensively applied to psychometric data, following Spearman (1904) then Thurstone (1947). The aim of Factor Analysis, just like that of PCA, is to summarize a set of variables by a smaller number of variables; the main difference — to put it briefly — is that while in PCA the new (principal) variables are linear combinations of the initial ones, in Factor Analysis the new variables have the status of underlying "factors": See Exercise 4.4 (*Spearman*, p.171)[2]. The thrust of Factor Analysis is on variables, not on individuals (like PCA in traditional Multivariate Analysis).

For any $J \times K$ table studied in this chapter, we assume throughout that each element k (column of the table) indexes a variable over J in the formal sense defined in chapter 2 (*CORRESPONDENCE*, §2.1, p.24), that is, there is an absolute measure frequency n_J over J and each variable satisfies the basic *averaging property*: For each k, whenever individuals j are grouped, the values are n_J-averaged[3]. With $j \in J$, there is associated the family of the K values of variables called the *profile of* j (row of the table).

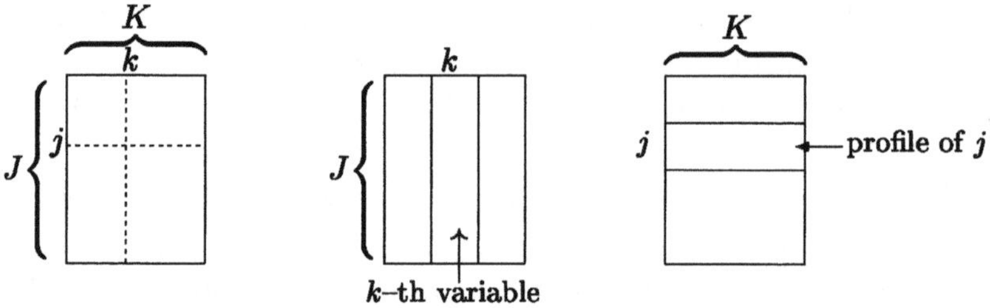

Figure 4.1. $J \times K$ data table, table of K variables, table of J profiles.

In this chapter we will study two types of tables, whether the profiles are *variables over* K (i.e. grouping by averaging over K), or *measures over* K (i.e. grouping by summing over K). Like in CA, the geometric representation is determined by the distances between profiles. The choice of variables which define these distances is methodologically crucial and refers to the frame model of data (cf. *OVERVIEW*, §1.5, p.14). Unlike transition

[2]The principal variables yielded by a PCA are often called "factors" like in classical Factor Analysis.

[3]Data tables for which the J and/or K sets are equipped with further properties will be studied by the procedures of the Analysis of Structured Data; see chapter *STRUCTURED*.

variables in CA, however, the variables over J in the tables we study here may be defined on different scales[4]. Before defining the distance between profiles, it is therefore appropriate to inquire whether the K scales are comparable. If this is not the case, the variables can be brought to a common scale by some *scaling procedure.* Scaling procedures may or may not involve an external norm; the paradigm is *standardization*, i.e. subtracting mean then dividing by standard deviation.

When the K variables over J are *on a common scale* on the one hand and the J *profiles are variables* over K on the other hand, we refer to the values in the table as *scores* denoted $(x^{jk})_{jk \in J \times K}$ (j and k being both upper indices); the $J \times K$ table is called a protocol of scores and denoted x^{JK}. Some data tables, such as the performances at several tests or exams, are ready–made protocols of scores. In a protocol of scores, the relative importances of the K variables are specified by a weighting measure over K $\varpi_K = (\varpi_k)_{k \in K}$ (with $\varpi_k > 0$ for all $k \in K$), which (unlike in CA) is usually not induced by the data themselves; we call (x^{JK}, n_J, ϖ_K) a *biweighted protocol of scores*, since it involves two weights, one over individuals (as is the rule) and one over variables. Therefore the *Euclidean distance between two profiles* x^{jK} and $x^{j'K}$ is defined by

$$\Big(\sum_{k \in K} \varpi_k (x^{jk} - x^{j'k})^2 \Big)^{1/2} \tag{4.1}$$

More generally, the distance between two profiles can be defined from a definite positive matrix $\mathbf{Q} = [q_{kk'}]$ as $\Big(\sum_{k \in K} \sum_{k' \in K} q_{kk'} (x^{jk} - x^{j'k})(x^{jk'} - x^{j'k'}) \Big)^{\frac{1}{2}}$. For a biweighed protocol of scores as above, matrix $\mathbf{Q}$ is diagonal.

When the *profiles are measures* over K, they are denoted $x^j_K = (x^j_k)_{k \in K}$ (lower index k); the table is called a *table of measures* and denoted x^J_K. Therefore, given a weighting measure ϖ_K, the *Euclidean distance between two profiles* x^j_K and $x^{j'}_K$ is defined by

$$\Big(\sum_{k \in K} (x^j_k - x^{j'}_k)^2 / \varpi_k \Big)^{1/2} \tag{4.2}$$

The chapter is organized as follows. We deal in detail with a biweighted protocol of scores (§4.1), with two particular cases, namely simple PCA (or analysis of covariances) (§4.2), and standard PCA (or analysis of correlations) (§4.3). We then sketch the general case (§4.4). We proceed by

[4]This is typically the case with *tables of measurements* for which the variables may refer to heterogeneous magnitudes evaluated on various units.

studying a biweighted protocol of measures (§4.5). Lastly, we give some methodological principles and guidelines for interpretation (§4.6).

The mathematical notions used in this chapter are essentially those of multidimensional geometry; cf. chapter 10 (*MATH.BASES*), §10.4 (p.435). The definitions of the basic statistics concerning Euclidean clouds can be found in chapter 3 (*CLOUD*), §3.1 (p.76). For the determination of principal axes and variables of the cloud of individuals, the reader can refer to chapter *CLOUD*, §3.3 (p.87). As throughout the book, the duality between measures and variables is pervasive; cf. chapter 2 (*CORRESPONDENCE*), §2.1 (p.24).

4.1 Biweighted PCA of a Protocol of Scores

A protocol x^{JK}, equipped with the absolute frequency measure n_J and with the weighting ϖ_K ($\forall k \in K : \varpi_k > 0$), is called a ***biweighted protocol of scores*** if it satisfies the following two properties:

Property 1: The K variables $(x^{Jk})_{k\in K}$ over (J, f_J) (letting $f_j = n_j/n$, and $n = \sum_{j\in J} n_j$) are on the same numerical scale; the variables are weighted by $(\varpi_k)_{k\in K}$, the weights reflect the importance granted to each variable.

Property 2: The J profiles $(x^{jK})_{j\in J}$ are variables over (K, ϖ_K), that is, grouping on K entails averaging over K according to the weighting ϖ_K.

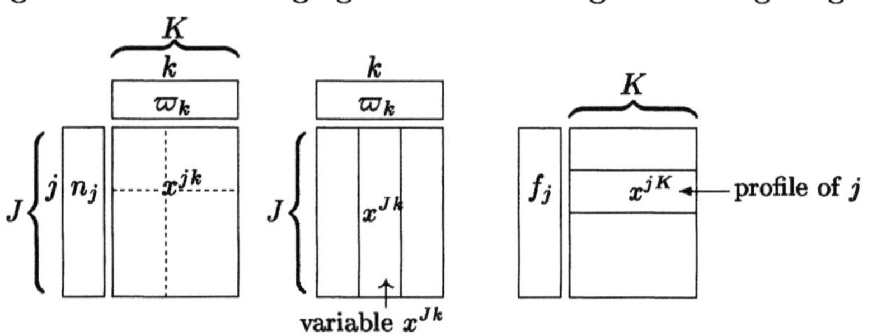

Figure 4.2. Biweighted protocol (x^{JK}, n_J, ϖ_K), set of K variables $(x^{Jk})_{k\in K}$ over (J, f_J), set of J profiles $(x^{jK})_{j\in J}$ over (K, ϖ_K).

Space of variables. The space of variables over J is equipped as a vector space with the Euclidean structure induced by f_J and denoted $\mathbb{R}^J$; the canonical basis of indicator variables $(\delta_j^J)_{j\in J}$ is orthogonal, with $\|\delta_j^J\| = \sqrt{f_j}$ (*CORRESPONDENCE*, §2.1, p.28). We denote $\overline{x}^k = \sum_{j\in J} f_j x^{jk}$ the mean of the variable x^{Jk}, hence the centered variable $x_0^{Jk} = (x^{jk} - \overline{x}^k)_{j\in J}$. We denote $v^{kk'} = \sum_{j\in J} f_j(x^{jk} - \overline{x}^k)(x^{jk'} - \overline{x}^{k'})$ the covariance between x^{Jk}

and $x^{Jk'}$, and $\operatorname{Var} x^{Jk} = v^{kk}$ the variance of x^{Jk}. With each individual $j \in J$ there is associated the ϖ_K–average of scores $\overline{x}^j = \sum\limits_{k \in K} \varpi_k x^{jk}/\varpi$ (letting $\varpi = \sum\limits_{k \in K} \varpi_k$). The family $(\overline{x}^j)_{j \in J}$ defines the ϖ_K*–Average variable* over J, denoted $\overline{x}^J$.

Space of profiles. The space of profiles over K is equipped as a vector space with the Euclidean metric induced by the weighting ϖ_K (CORRESPONDENCE, formula (2.1), p.28). The distance $d(j, j')$ between profiles x^{jK} and $x^{j'K}$ is such that $d^2(j, j') = \sum\limits_{k \in K} \varpi_k (x^{jk} - x^{j'k})^2$; therefore, two individuals are all the more distant as they differ on "heavily weighted" variables. We denote $\overline{x}^K = \sum\limits_{j \in J} f_j x^{jK}$ the *mean profile*, and $\overline{x} = \sum\limits_{k \in K} \varpi_k \overline{x}^k / \varpi = \sum\limits_{j \in J} f_j \overline{x}^j$ the *overall mean*.

— *Remark.* Although the profiles of individuals are also variables (over K), in PCA the status of profiles differs from the one of variables over J.

Matrix notation. Let $\mathbf{X} = [x^{jk}]$ denote the $J \times K$ data table, $\mathbf{F}$ the J–diagonal matrix with diagonal $(f_j)_{j \in J}$, and $\mathbf{\Omega}$ the K–diagonal matrix with diagonal $(\varpi_k)_{k \in K}$. The K–column of means $(\overline{x}^k)_{k \in K}$ is $\overline{\mathbf{x}} = \mathbf{X}'\mathbf{F}\mathbf{e}$ ($\mathbf{e}$ denotes the J–column of ones), hence the $J \times K$ matrix of the centered protocol $\mathbf{X}_0 = \mathbf{X} - \mathbf{e}\overline{\mathbf{x}}'$. The $K \times K$ covariance matrix is $\mathbf{V} = \mathbf{X}_0'\mathbf{F}\mathbf{X}_0$, and the $J \times J$ matrix of general term $w^{jj'} = \sum\limits_{k \in K} \varpi_k (x^{jk} - \overline{x}^k)(x^{j'k} - \overline{x}^k)$ is $\mathbf{W} = \mathbf{X}_0 \mathbf{\Omega} \mathbf{X}_0'$.

4.1.1 Cloud of Individuals

The geometric affine space[5] within which the profiles of individuals are represented is called the *space of individuals* (also called space of observables).

Construction of the Euclidean cloud

In a K–dimensional geometric space $\mathcal{U}$, referred to the Cartesian frame $(\mathrm{O}, (\vec{\delta_k})_{k \in K})$, with the profile $x^{jK} = (x^{jk})_{k \in K}$ of j weighted by f_j there is associated the point M^j defined by $\overrightarrow{\mathrm{OM}^j} = \sum\limits_{k \in K} x^{jk} \vec{\delta_k}$ and weighted by f_j.

[5]Recall the geometric notations (CLOUD, p.77): *Points* are denoted by capital letters, *geometric vectors* are "arrowed"; the *scalar product* is denoted by $\langle \cdot | \cdot \rangle$, the *Euclidean norm* by $\| \cdot \|$ and the distance between points M and P by PM.

The set of J profiles, that is $(x^{jK}, f_j)_{j\in J}$, is thus geometrically represented by the weighted Euclidean cloud $(\mathrm{M}^j, f_j)_{j\in J}$. The distance $\mathrm{M}^j\mathrm{M}^{j'}$ between points M^j and $\mathrm{M}^{j'}$ is equal to the distance (previously defined by formula (4.1)) between the profiles of j and j'.

$$(\mathrm{M}^j\mathrm{M}^{j'})^2 = \sum_{k\in K} \varpi_k (x^{jk} - x^{j'k})^2 \tag{4.3}$$

The basis $(\overrightarrow{\delta_k})_{k\in K}$ of the vector space $\mathcal{V}$ underlying the geometric space $\mathcal{U}$ is orthogonal, but not orthonormal (except if $\varpi_K = 1_K$): $\|\overrightarrow{\delta_k}\| = \sqrt{\varpi_k}$. Relative to the Cartesian frame $(\mathrm{O}, (\overrightarrow{\delta_k})_{k\in K})$, the coordinates of the *mean point* $\mathrm{G} = \sum\limits_{j\in J} f_j\mathrm{M}^j$ of the cloud M^J (cf. Definition 3.2, p.77) are the means $(\overline{x}^k)_{k\in K}$ of the initial variables: $\overrightarrow{\mathrm{OG}} = \sum\limits_{k\in K} \overline{x}^k \overrightarrow{\delta_k}$, hence $\overrightarrow{\mathrm{GM}^j} = \sum\limits_{k\in K} x_0^{jk} \overrightarrow{\delta_k}$ (with $x_0^{jk} = x^{jk} - \overline{x}^k$). The distance GM^j of point M^j from the mean point G is thus equal to $\left(\sum\limits_{k\in K} \varpi_k (x^{jk} - \overline{x}^k)^2\right)^{1/2}$.

The dimensionality of the affine support of the cloud — i.e. the smallest subspace of $\mathcal{U}$ containing the cloud — is equal to the rank L $(\leq K)$ of the covariance matrix $\mathbf{V}$.

Property 4.1 (Variance of cloud). *The variance of the cloud is equal to the ϖ_K–weighted sum of the variances of the initial variables.*

$$\mathrm{Var}\,\mathrm{M}^J = \sum_{k\in K} \varpi_k \,\mathrm{Var}\, x^{Jk} \tag{4.4}$$

Proof. From $\mathrm{Var}\,\mathrm{M}^J = \sum\limits_{j\in J} f_j (\mathrm{GM}^j)^2$ (CLOUD, Definition 3.4, p.78), one deduces $\mathrm{Var}\,\mathrm{M}^J = \sum\limits_{j\in J}\sum\limits_{k\in K} \varpi_k f_j (x^{jk} - \overline{x})^2 = \sum\limits_{k\in K} \varpi_k \,\mathrm{Var}\, x^{Jk}$. ◁

The part of variance of the cloud accounted for by the variable x^{Jk}, denoted Cta_k, defines the ***absolute contribution*** of the variable x^{Jk} (weighted by ϖ_k) to the variance of the cloud: $\mathrm{Cta}_k = \varpi_k \,\mathrm{Var}\, x^{Jk}$.

In order to determine the principal directions of the cloud M^J, we express the four fundamental linear mappings $Vac : \mathcal{V} \to \mathbb{R}^J$, $Vac^* : \mathbb{R}^J \to \mathcal{V}$, $Som : \mathcal{V} \to \mathcal{V}$ and $Tom : \mathbb{R}^J \to \mathbb{R}^J$ (CLOUD, §3.2.5, p.84-86 and p.117) as functions of (x^{JK}, n_J, ϖ_K). The images of the basis vectors $(\overrightarrow{\delta_k})_{k\in K}$ (canonical basis of $\mathcal{V}$) and $(\delta_j^J)_{j\in J}$ (canonical basis of $\mathbb{R}^J$) by the linear mappings are the following.

$$\begin{array}{llrr}
Vac(\overrightarrow{\delta_k}) & = \left(\langle\overrightarrow{\mathrm{GM}^j}|\overrightarrow{\delta_k}\rangle\right)_{j\in J} = \sum\limits_{j\in J}(x_0^{jk}\varpi_k)\delta_j^J & \text{in matrix terms} & \mathbf{X_0\Omega} \\
Vac^*(\delta_j^J) & = f_j\overrightarrow{\mathrm{GM}^j} = \sum\limits_{k\in K}(f_j x_0^{jk})\overrightarrow{\delta_k} & '' & \mathbf{X_0'F} \\
Som(\overrightarrow{\delta_k}) & = \sum\limits_{k'\in K}\varpi_k v^{kk'}\overrightarrow{\delta_{k'}} & '' & \mathbf{V\Omega} \\
Tom(\delta_j^J) & = \sum\limits_{j'\in J} f_j w^{jj'}\delta_{j'}^J & '' & \mathbf{WF}
\end{array}$$

Theorem 4.1 (Passage formulas). *The direction coefficients* $(a_\ell^k)_{k\in K}$ *of the* ℓ*–th principal line and the* ℓ*–th standard principal variable* z_ℓ^J *satisfy the following system of equations (letting* $a_{k\ell} = \varpi_k a_\ell^k$ *and* $z_{j\ell} = f_j z_\ell^j$*):*

$$\begin{cases} \sum\limits_{k\in K} x_0^{Jk} a_{k\ell} & = \xi_\ell z_\ell^J \\ \sum\limits_{j\in J} x^{jK} z_{j\ell} & = \xi_\ell a_\ell^K \end{cases} \quad \text{with} \quad \begin{cases} \sum\limits_{k\in K} \varpi_k (a_\ell^k)^2 & = 1 \\ \sum\limits_{j\in J} f_j (z_\ell^j)^2 & = 1 \end{cases} \tag{4.5}$$

Proof. The principal directions and variables of the cloud M^J are obtained from the singular value decomposition (SVD) of *Vac* and *Vac**, with the passage formulas (3.7) (CLOUD, p.92). ◁

Let us recall that the ℓ–th principal axis — in brief *Axis* ℓ — goes through the mean point G, admits $\overrightarrow{\alpha_\ell} = \sum\limits_{k\in K} a_\ell^k \overrightarrow{\delta_k}$ as a unit–norm director vector, and has an *arbitrary orientation* (CLOUD, §3.3.1, p.88).

Property 4.2 (Principal coordinates). *The principal coordinate* y_ℓ^j *of the point* M^j *on Axis* ℓ *is equal to* $\xi_\ell z_\ell^j$*.*

Proof. By definition (CLOUD, p.91), the principal coordinate of the point M^j on Axis ℓ is equal to $\langle\overrightarrow{\mathrm{GM}^j}|\overrightarrow{\alpha_\ell}\rangle$, that is, $y_\ell^j = \sum\limits_{k\in K}\sum\limits_{k'\in K} x_0^{jk} a_\ell^k \langle\overrightarrow{\delta_k}|\overrightarrow{\delta_{k'}}\rangle = \sum\limits_{k\in K} x_0^{jk}(a_\ell^k \varpi_k)$. From the first equation (4.5), one deduces $y_\ell^j = \xi_\ell z_\ell^j$. ◁

Property 4.3. *If the Average variable* $\overline{x}^J$ *is constant, then the measure* $a_{K\ell}$ *is a contrast over* K.

Proof. $\sum\limits_{k\in K} a_{k\ell} = \sum\limits_{k\in K}\varpi_k a_\ell^k = \frac{1}{\xi_\ell}\sum\limits_{j\in J}(\sum\limits_{k\in K}\varpi_k x^{jk}) z_{j\ell}$ (equation (4.5)) $= \frac{1}{\xi_\ell}\sum\limits_{j\in J}\varpi\,\overline{x}\,z_{j\ell} = 0$ (since $\overline{x}^J = \overline{x}1^J$ and $z_{J\ell}$ is a contrast over J). ◁

From Equations (4.5) we directly deduce the following property.

Property 4.4 (Equation for principal directions). *Let* $a_{k\ell} = \varpi_k a_\ell^k$, *the principal direction* $\overrightarrow{\alpha_\ell} = \sum_{k\in K} a_\ell^k \overrightarrow{\delta_k}$ *is such that*

$$\forall k \in K \qquad \sum_{k'\in K} v^{kk'} a_{k'\ell} = \lambda_\ell a_\ell^k \tag{4.6}$$

Property 4.5. *If axes* ℓ *and* ℓ' *correspond to distinct eigenvalues, they are orthogonal:* $\lambda_\ell \neq \lambda'_\ell \Rightarrow \overrightarrow{\alpha_\ell} \perp \overrightarrow{\alpha_{\ell'}}$, *i.e.* $\sum_{k\in K} \varpi_k a_\ell^k a_{\ell'}^k = 0$.

Proof. Cf. Property 3.9 (CLOUD, p.89). If eigenvalues are multiple, orthogonal principal axes can be chosen in corresponding eigenspaces. ◁

Reconstitution formulas

The reconstitution of distances defined p.131 by formula (4.1) is:

$$d^2(j,j') = \sum_{\ell\in L} (y_\ell^j - y_\ell^{j'})^2 \tag{4.7}$$

The reconstitution of order $L' \leq L$ of the table of centered variables (cf. CLOUD, formula (3.18), p.120) is:

$$\forall k \in K \quad \tilde{x}_{L'}^{Jk} = \sum_{\ell=1}^{L'} a_\ell^k y_\ell^J \quad \text{with} \quad \sum_{k\in K} \varpi_k (a_\ell^k)^2 = 1 \tag{4.8}$$

$$\text{Reconstitution of covariances:} \quad \forall k \in K\ \forall k' \in K\ \tilde{v}_{L'}^{kk'} = \sum_{\ell=1}^{L'} \lambda_\ell a_\ell^k a_\ell^{k'} \tag{4.9}$$

Contributions of points to the variance of axes

Let us recall the definitions of contributions (CLOUD, §3.3.3, p.92).

The *absolute contribution* of the point M^j and Axis ℓ is $\text{Cta}_{j\ell} = f_j(y_\ell^j)^2$, with $\sum_{\ell=1}^{L} \text{Cta}_{j\ell} = \text{Cta}_j$ (absolute contribution of j to the total variance). The *double breakdown of the variance* of the cloud according to points and axes follows:

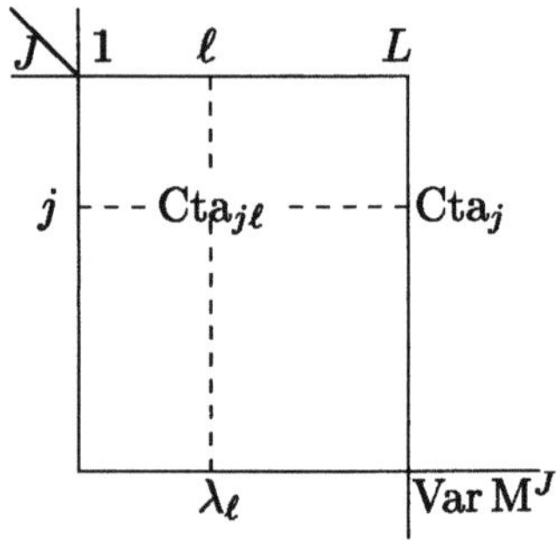

$$\text{Var}\,M^J = \sum_{j\in J} \sum_{\ell=1}^{L} \text{Cta}_{j\ell}$$

The *relative contribution* of the point M^j to Axis ℓ is $\text{Ctr}_j^\ell = \text{Cta}_{j\ell}/\lambda_\ell$.

The *quality of representation* of the point M^j on Axis ℓ, denoted Ctr_ℓ^j, is equal to $\mathrm{Cta}_{j\ell} / \mathrm{Cta}_j$, that is, $\mathrm{Ctr}_\ell^j = (y_\ell^j)^2/(\mathrm{GM}^j)^2$; it is the square of the cosine of the angle between $\overrightarrow{\mathrm{GM}^j}$ and its projection onto Axis ℓ. The quality of representation of the point M^j on the principal subspace generated by the first L' principal axes is equal to $\sum\limits_{\ell=1}^{L'} \mathrm{Ctr}_\ell^j$, and is equal to 1 for $L' = L$ (complete reconstitution).

Matrix formulas. If $\mathbf{a}_\ell$ denotes the K–column of coefficients $(a_\ell^k)_{k\in K}$ and $\mathbf{z}_\ell$ the J–column of values $(z_\ell^j)_{j\in J}$, the passage formulas (4.5) read:

$$\begin{cases} \mathbf{X}_0\mathbf{\Omega a}_\ell &= \xi_\ell \mathbf{z}_\ell \\ \mathbf{X}_0\mathbf{F z}_\ell &= \xi_\ell \mathbf{a}_\ell \end{cases} \quad \text{with} \quad \begin{cases} \mathbf{a}'_{\ell'}\mathbf{\Omega a}_\ell &= 1 \\ \mathbf{z}'_{\ell'}\mathbf{F z}_\ell &= 1 \end{cases} \text{if } (\ell = \ell'), = 0 \text{ if not} \tag{4.10}$$

Equation 4.6 (p.136) reads: $\mathbf{V\Omega a}_\ell = \lambda_\ell \mathbf{a}_\ell$

Let us denote $\mathbf{y}_\ell$ the J–column of the principal coordinates of points M^j on Axis ℓ, one has $\mathbf{y}_\ell = \xi_\ell \mathbf{z}_\ell$. The reconstitutions of order L' of matrix $\mathbf{X}_0$ and of covariance matrix $\mathbf{V}$ read:

$$\begin{cases} \widetilde{\mathbf{X}}_{L'} &= \sum\limits_{\ell=1}^{L'} \mathbf{a}'_\ell \mathbf{y}_\ell \\ \widetilde{\mathbf{V}}_{L'} &= \sum\limits_{\ell=1}^{L'} \lambda_\ell \mathbf{a}'_\ell \mathbf{a}_\ell \end{cases} \quad \text{with} \quad \begin{cases} \mathbf{a}'_{\ell'}\mathbf{\Omega a}_\ell &= 1 \\ \mathbf{y}'_{\ell'}\mathbf{F y}_\ell &= \lambda_\ell \end{cases} \text{if } (\ell = \ell'), = 0 \text{ if not} \tag{4.11}$$

4.1.2 Space of Variables

In this section, we study the properties of the PCA of a biweighted protocol of scores in the space $\mathbb{R}^J$ of variables over (J, f_J).

The following property is directly deduced from Equations (4.5).

Property 4.6 (Equation for principal variables). *Letting* $z_{j\ell} = f_j z_\ell^j$, *the ℓ–th principal variable z_ℓ^J satisfies the equation*

$$\forall j \in J \qquad \sum_{j' \subset J} w^{jj'} z_{j'\ell} = \lambda_\ell z_\ell^j \qquad (\ell = 1, \ldots L) \tag{4.12}$$

Property 4.7. *The principal variables are pairwise uncorrelated.*

Proof. $\mathrm{Cov}(y_\ell^J | y_{\ell'}^J) = \mathrm{Cov}\left(\sum\limits_{k\in K} a_{k\ell} x^{Jk} | \sum\limits_{k'\in K} a_{k'\ell'} x^{Jk'}\right) = \sum\limits_{k\in K}\sum\limits_{k'\in K} a_{k\ell}\, a_{k'\ell'} v^{kk'}$; we have $\sum\limits_{k'\in K} v^{kk'} a_{k'\ell'} = \lambda_{\ell'} a_{\ell'}^k$ (Property 4.6 (p.136), then $\mathrm{Cov}(y_\ell^J | y_{\ell'}^J) =$

$\lambda_{\ell'} \sum\limits_{k \in K} a_{k\ell} a^k_{\ell'} = \lambda_{\ell'} \langle \overrightarrow{\alpha_\ell} | \overrightarrow{\alpha_{\ell'}} \rangle$. Therefore principal variables associated with orthogonal principal axes of the support of cloud are uncorrelated (if an eigenvalue is multiple, orthogonal axes are chosen in the corresponding eigenspace). ◁

The L standard principal variables $(z^J_\ell)_{\ell \in L}$ define an orthonormal basis of the subspace $\mathcal{X}_0 \subset \mathbb{R}^J$ spanned by the centered initial variables $(x_0^{Jk})_{k \in K}$.

Statistical characterization

The statistical characterization of biweighted PCA can be formulated as follows: Among the variables $y^J = \sum\limits_{k \in K} a_k \, x^{Jk}$ ($a_k \in \mathbb{R}$), seek the one that verifies: $\operatorname{Var} y^J / (\sum\limits_{k \in K} a_k^2 / \varpi_k)$ maximum.

Property 4.8. *Seeking* $(a_k)_{k \in K}$ *such that* $\operatorname{Var}(\sum\limits_{k \in K} a_k \, x^{Jk}) / (\sum\limits_{k \in K} a_k^2 / \varpi_k)$ *is maximum is equivalent to determining the first principal axis.*

Proof. Let $\overrightarrow{\alpha} = \sum\limits_{k \in K} a^k \overrightarrow{\delta_k}$ (with $a^k = a_k / \varpi_k$), $\langle \overrightarrow{\mathrm{GM}^j} | \overrightarrow{\alpha} \rangle = \sum\limits_{k \in K} x_0^{jk} \varpi_k a^k = \sum\limits_{k \in K} x_0^{jk} a_k$, then $\sum\limits_{j \in J} f_j \langle \overrightarrow{\mathrm{GM}^j} | \overrightarrow{\alpha} \rangle^2 = \sum\limits_{j \in J} f_j \big(\sum\limits_{k \in K} a_k x_0^{jk} \big)^2 = \operatorname{Var}\big(\sum\limits_{k \in K} a_k x^{Jk} \big)$. From Property 3.4 (CLOUD, p.81), the variance of the projected cloud onto the line $(\mathrm{G}, \overrightarrow{\alpha})$ is equal to $\sum\limits_{j \in J} f_j \langle \overrightarrow{\mathrm{GM}^j} | \overrightarrow{\alpha} \rangle^2 / \| \overrightarrow{\alpha} \|^2$, therefore is equal to $\operatorname{Var}(\sum\limits_{k \in K} a_k \, x^{Jk}) / (\sum\limits_{k \in K} a_k^2 / \varpi_k)$. ◁

Principal variables of orders 2, 3, etc. are then determined. For the second variable, one seeks the variable $\sum\limits_{k \in K} a'_k x^{Jk}$ uncorrelated with the first principal variable such that $\operatorname{Var}(\sum\limits_{k \in K} a'_k x^{Jk}) / (\sum\limits_{k \in K} (a'_k)^2 / \varpi_k)$ is maximum. One has $\operatorname{Cov}(\sum\limits_{k \in K} a'_k x^{Jk} | \sum\limits_{k' \in K} a_{k'} x^{Jk'}) = 0 \iff \sum\limits_{k \in K} \sum\limits_{k' \in K} a'_k v^{kk'} a_{k'} = 0$; now one has $\sum\limits_{k' \in K} v^{kk'} a_{k'} = \lambda a^k$, hence $\sum\limits_{k \in K} a'_k a^k = 0$; therefore the principal variable is associated with an axis orthogonal to the first principal axis (or taken to be orthogonal if $\lambda_1 = \lambda_2$), that is, associated with the second principal axis, etc.

Regressions of the initial variables on principal variables

Property 4.9. *The coefficient* b^k_ℓ *of the regression of the initial variable* x^{Jk} *on the* ℓ*–th standard principal variable is equal to* $\xi_\ell a^k_\ell$.

Proof. From the 2nd Equation (4.5), one deduces $\mathrm{Cov}(x^{Jk}|z_\ell^J) = \xi_\ell a_\ell^k$. ◃

We are going to express the *regression* of initial variables x^{Jk} on principal variables. The reconstitution formula (4.8) (p.136) applied to the variable x_0^{Jk} with $L' = L$, reads $x_0^{Jk} = \sum_{\ell=1}^{L} a_\ell^k y_\ell^J$, with $a_\ell^k = (1/\xi_\ell) b_\ell^k$ and $y_\ell^J = \xi_\ell z_\ell^J$, therefore one has:

$$x_0^{Jk} = \sum_{\ell=1}^{L} b_\ell^k z_\ell^J \qquad (4.13)$$

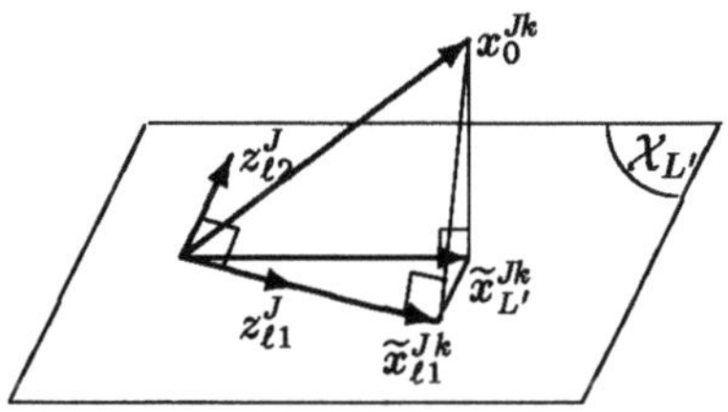

Let $\mathcal{X}_{L'}$ be the subspace of $\mathbb{R}^J$ generated by the first L' principal variables ($L' \leq L$).

The variable $\widetilde{x}_{L'}^{Jk} = \sum_{\ell=1}^{L'} b_\ell^k z_\ell^J$ is the multiple regression of x_0^{Jk} on $\mathcal{X}_{L'}$ since $x_0^{Jk} - \widetilde{x}_{L'}^{Jk} = \sum_{\ell=L'+1}^{L} b_\ell^k z_\ell^J$ is a linear combination of the uncorrelated variables $(z_\ell^J)_{\ell=1,\dots L'}$ orthogonal to $\mathcal{X}_{L'}$; see above figure for $L' = 2$.

Since the principal variables are uncorrelated, the partial regression coefficient of x_0^{Jk} on z_ℓ^J coincides with the simple regression coefficient, and $\mathrm{Var}\, x^{Jk} = \sum_{\ell=1}^{L} (b_\ell^k)^2$.

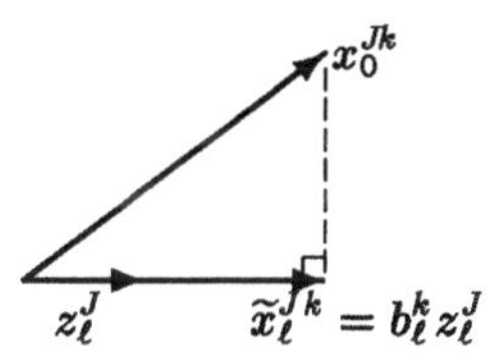

Correlations between initial and principal variables

Let r_ℓ^k be the correlation coefficient between the initial variable x^{Jk} and ℓ–th principal variables and $\mathrm{SD}\, x^{Jk}$ the standard deviation of x^{Jk}, one has:

$$r_\ell^k = b_\ell^k / \mathrm{SD}\, x^{Jk}, \quad \text{hence } \sum_{\ell=1}^{L} (r_\ell^k)^2 = 1$$

The correlation coefficient r_ℓ^k is equal to the cosine of the angle between the variables (vectors) x_0^{Jk} and z_ℓ^J.

The multiple correlation coefficient $R_{L'}^k$ of x^{Jk} with the first L' principal variables is given by $(R_{L'}^k)^2 = \sum_{\ell=1}^{L'} (r_\ell^k)^2$; therefore, the closer $(R_{L'}^k)^2$ is to 1, the better the quality of the fit of the initial variable by the first L' principal variables.

Regressions of principal variables on initial variables

It follows from Equation (4.6) (p.136), that coefficients $(a_{k\ell})_{k\in K}$ verify the normal equations $\sum\limits_{k'\in K} v^{kk'} a_{k'\ell} = \text{Cov}(y_\ell^J | x^{Jk})$ $(k \in K)$, since $\lambda_\ell a_\ell^k = \xi_\ell b_\ell^k = \text{Cov}(\xi_\ell z_\ell^J | x^{Jk})$, therefore the formula $y_\ell^J = \sum\limits_{k\in K} x_0^{Jk} a_{k\ell}$ (cf. first equation (4.5), p.135) expresses the multiple regression of the calibrated principal variable y_ℓ^J on the K centered initial variables $(x_0^{Jk})_{k\in K}$. The partial regression coefficient of y_ℓ^J on x_0^{Jk} is equal to $a_{k\ell}$; the simple regression coefficient is equal to $\lambda_\ell a_\ell^k / \text{Var}\, x^{Jk} = \lambda_\ell a_{k\ell} / \text{Cta}_k$ (with $\text{Cta}_k = \varpi_k \text{Var}\, x^{Jk}$).

Contributions of variables to the variance of axes

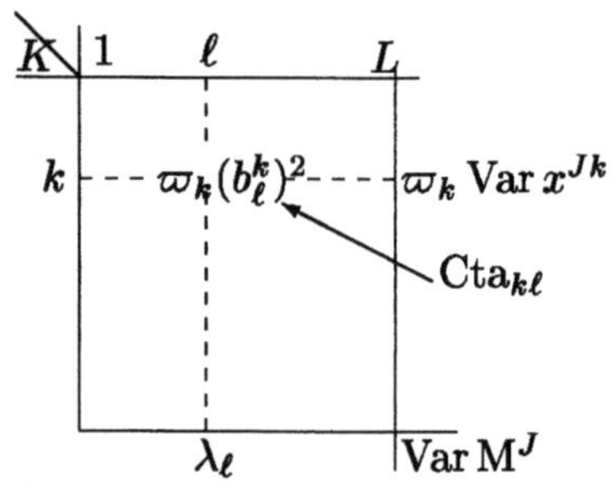

• The *absolute contribution of the initial variable x^{Jk} to the variance of the ℓ-th principal variable* is $Cta_{k\ell} = \varpi_k (b_\ell^k)^2$.

One has:

$\text{Cta}_k = \sum\limits_{\ell=1}^{L} \varpi_k (b_\ell^k)^2 = \varpi_k \text{Var}\, x^{Jk}$

$\text{Cta}_\ell = \sum\limits_{k\in K} \varpi_k (b_\ell^k)^2 = \lambda_\ell$.

• The *relative contribution of the variable x^{Jk} to the variance of the ℓ-th principal variable* is denoted Ctr_k^ℓ, with

$$\text{Ctr}_k^\ell = \varpi_k (b_\ell^k)^2 / \lambda_\ell = \varpi_k (a_\ell^k)^2$$

• The relative contribution $\text{Ctr}_\ell^k = \text{Cta}_{k\ell} / \text{Cta}_k$ is called *quality of fit of x^{Jk} by the ℓ-th principal variable*, and is equal to $(r_\ell^k)^2$, square of the correlation between the initial variable x^{Jk} and the ℓ-th principal variable.

Average variable

The interest of the properties of the Average variable $\overline{x}^J = \sum\limits_{k\in K} \varpi_k x^{Jk} / \varpi$ has been shown in Rouanet & Le Roux (1993). The mean and the variance of $\overline{x}^J$ are equal to

$$\text{Mean}\, \overline{x}^J = \sum_{k\in K} \varpi_k \overline{x}^k / \varpi = \overline{x} \text{ (overall mean)} \tag{4.14}$$

$$\text{Var}\, \overline{x}^J = \sum_{k\in K} \sum_{k'\in K} v^{kk'} \varpi_k \varpi_{k'} / \varpi^2 \tag{4.15}$$

Property 4.10. $\text{Cov}(x^{Jk} | \overline{x}^J) = \sum\limits_{k'\in K} v^{kk'} \varpi_{k'} / \varpi$

Proof. $\text{Cov}(x^{Jk} | \overline{x}^J) = \sum\limits_{k'\in K} \varpi_{k'} \text{Cov}(x^{Jk} | x^{Jk'}) / \varpi = \sum\limits_{k'\in K} v^{kk'} \varpi_{k'} / \varpi$ ◁

Corollary 4.1. $\forall k \in K \ \sum_{k' \in K} v^{kk'} \varpi_{k'} = 0 \iff \overline{x}^J = \overline{x} 1^J$.

Proof. $\forall k \in K : \sum_{k' \in K} v^{kk'} \varpi_{k'} = 0 \iff \text{Cov}(x^{Jk} | \overline{x}^J) = 0$

$$\iff x^{Jk} = a_k \overline{x}^J + b_k 1^J \ (a_k, b_k \in \mathbb{R})$$

Taking the ϖ_K–weighted sum of the K preceding equations, one has $\varpi \overline{x}^J = \big(\sum_{k \in K} \varpi_k a_k\big) \overline{x}^J + \big(\sum_{k \in K} \varpi_k b_k\big) 1^J$, therefore $\overline{x}^j$ does not depend on j.

Conversely, $\overline{x}^J = \overline{x} 1^J \Rightarrow \forall k \in K : \text{Cov}(x^{Jk} | \overline{x}^J) = 0$; from the preceding Property 4.10, we deduce $\sum_{k' \in K} v^{kk'} \varpi_{k'} / \varpi = 0$. ◁

Property 4.11. *The Average variable is the axial variable associated with the axis* $(\mathrm{O}, \sum_{k \in K} \overrightarrow{\delta_k})$.

Proof. Let us recall (CLOUD, §3.2.4, p.83) that the values of the axial variable associated with the line $(\mathrm{O}, \sum_{k \in K} \overrightarrow{\delta_k})$ are equal to the coordinates of the orthogonal projection of the points $(\mathrm{M}^j)_{j \in J}$ onto this line, that is, are equal to $(\langle \overrightarrow{\mathrm{OM}^j} | \sum_{k \in K} \overrightarrow{\delta_k} \rangle / \| \sum_{k \in K} \overrightarrow{\delta_k} \|^2)_{j \in J}$. Since the basis $(\overrightarrow{\delta_k})_{k \in K}$ is orthogonal with $\|\overrightarrow{\delta_k}\| = \sqrt{\varpi_k}$, one has $\| \sum_{k \in K} \overrightarrow{\delta_k} \|^2 = \sum_{k \in K} \|\overrightarrow{\delta_k}\|^2 = \sum_{k \in K} \varpi_k = \varpi$, and $\langle \overrightarrow{\mathrm{OM}^j} | \sum_{k \in K} \overrightarrow{\delta_k} \rangle = \sum_{k \in K} \langle \overrightarrow{\mathrm{OM}^j} | \overrightarrow{\delta_k} \rangle = \sum_{k \in K} x^{jk} \|\overrightarrow{\delta_k}\|^2 = \sum_{k \in K} x^{jk} \varpi_k$, hence $\langle \overrightarrow{\mathrm{OM}^j} | \sum_{k \in K} \overrightarrow{\delta_k} \rangle / \| \sum_{k \in K} \overrightarrow{\delta_k} \|^2 = \sum_{k \in K} \varpi_k x^{jk} / \varpi = \overline{x}^j$. ◁

Corollary 4.2. *The centered Average variable is the axial (centered) variable associated with* $(\mathrm{G}, \sum_{k \in K} \overrightarrow{\delta_k})$.

Theorem 4.2. *The centered (nonnull) Average variable is principal if and only if* $\sum_{k' \in K} v^{kk'} \varpi_{k'}$ *does not depend on* k*; the common value is the associated eigenvalue* $\lambda = \varpi \operatorname{Var} \overline{x}^J (\neq 0)$.

Proof. $\overline{x}^J$ is a principal variable if and only if $\sum_{k \in K} \overrightarrow{\delta_k}$ is a principal vector, therefore (applying Equation (4.6) (p.136) with $a^k = 1$ and $a_k = \varpi_k$) if and only if for all $k \in K$ one has $\sum_{k' \in K} v^{kk'} \varpi_{k'} = \lambda$. Taking the ϖ_K–average of these K equations yields: $\lambda = \sum_{k \in K} \sum_{k' \in K} \varpi_k v^{kk'} \varpi_{k'} / \varpi = \varpi \operatorname{Var} \overline{x}^J$ (cf. formula (4.15), p.140). ◁

Definition 4.1 (Thurstone condition). *The protocol* (x^{JK}, n_J, ϖ_K) *is said to satisfy Thurstone condition if and only if* $\sum\limits_{k' \in K} v^{kk'} \varpi_{k'}$ *does not depend on* k.

In matrix notation, if $\boldsymbol{\varpi} = \boldsymbol{\Omega}\mathbf{e}$ denotes the K–column of weights $(\varpi_k)_{k \in K}$ and $\mathbf{e}$ the K–column of ones, Thurstone condition reads $\mathbf{V}\boldsymbol{\varpi} = \lambda\mathbf{e}$.

Corollary 4.3. *Under Thurstone condition, all regression coefficients of the initial variables on the (nonconstant) Average variable are equal to 1.*

Proof. It follows from Theorem 4.2 (p.141) that $\sum\limits_{k' \in K} v^{kk'} \varpi_{k'} = \varpi \operatorname{Var} \overline{x}^J$ and from Property 4.10 (p.140) that $\operatorname{Cov}(x^{Jk} | \overline{x}^J) = \sum\limits_{k' \in K} v^{kk'} \varpi_{k'} / \varpi$; hence the regression coefficient $\operatorname{Cov}(x^{Jk} | \overline{x}^J) / \operatorname{Var} \overline{x}^J = 1$. ◁

Property 4.12. *Under Thurstone condition, the variable* $\sum\limits_{k \in K} c_k x^{Jk}$ *is uncorrelated with the (nonconstant) Average variable if and only if* c_K *is a contrast over* K.

Proof. $\operatorname{Cov}(\sum\limits_{k \in K} c_k x^{Jk} | \overline{x}^J) = \sum\limits_{k \in K} c_k \operatorname{Cov}(x^{Jk} | \overline{x}^J) = \sum\limits_{k \in K} c_k \sum\limits_{k' \in K} v^{kk'} \varpi_{k'}$ (cf. Property 4.10, p.140), then $\operatorname{Cov}(\sum\limits_{k \in K} c_k x^{Jk} | \overline{x}^J) = \sum\limits_{k \in K} c_k \lambda$ (Thurstone condition). $\sum\limits_{k \in K} c_k \lambda = 0 \iff \sum\limits_{k \in K} c_k = 0$, since $\lambda = \varpi \operatorname{Var} \overline{x}^J \neq 0$ (the Average variable $\overline{x}^J$ being assumed to be nonconstant). ◁

— *Particular case:* $\sum\limits_{k' \in K} v^{kk'} \varpi_{k'} = 0$. Any eigenvector $\sum\limits_{k \in K} c_k \overrightarrow{\delta_k}$ associated with a nonnull eigenvalue λ verifies Equations (4.6) (p.136), that is, $\sum\limits_{k' \in K} v^{kk'} c_{k'} = \lambda c^k \ (k \in K)$.

From $\sum\limits_{k \in K} \varpi_k v^{kk'} = 0$, we deduce that the ϖ_K–average of these K equations is equal to 0: $\lambda \sum\limits_{k \in K} \varpi_k c^k / \varpi = \sum\limits_{k' \in K} (\sum\limits_{k \in K} \varpi_k v^{kk'}) c_{k'} / \varpi = 0$.

Property 4.13. *If the Average variable* $\overline{x}^J$ *is principal, then any principal variable associated with eigenvalues* $\lambda_\ell \neq \varpi \operatorname{Var} \overline{x}^J$ *is a variable derived by contrast among the initial centered variables.*

Proof. The other principal variables are uncorrelated with $\overline{x}^J$ (Property 4.5, p.136), and they are derived by contrast (Property 4.12). ◁

— *Remark.* When $\lambda = \varpi \operatorname{Var} \overline{x}^J$ is a multiple eigenvalue, one may always choose, in the associated eigensubspace, a basis of uncorrelated variables consisting in the Average variable and variables derived by contrasts.

Property 4.14. *Under Thurstone condition, the regression coefficients of initial variables on a variable derived by a contrast among the K initial centered variables have a null ϖ_K-average.*

Proof. Let $\gamma^k = \operatorname{Cov}(\sum\limits_{k' \in K} c_{k'} x^{Jk'} | x^{Jk}) = \sum\limits_{k' \in K} c_{k'} v^{kk'}$.
Then $\sum\limits_{k \in K} \varpi_k \gamma^k = \sum\limits_{k' \in K} c_{k'} (\sum\limits_{k \in K} v^{kk'} \varpi_k) = \lambda \sum\limits_{k' \in K} c_{k'} = 0$. ◁

— *Remark.* Under Thurstone condition, the principal axes other than the one with director vector $\sum\limits_{k \in K} \overrightarrow{\delta_k}$ express oppositions among initial variables, that is, they refer to *bipolar factors*, in Thurstone's words.

4.1.3 Computation Procedure and Matrix Formulas

The calculation method is based on the diagonalization of a symmetric matrix. The principal directions satisfy Equation (4.6) (p.136), that is, are determined by the eigenvectors of the symmetric endomorphism *Som*.

With respect to basis $(\overrightarrow{\delta_k})_{k \in K}$, the matrix of *Som* is not symmetric, its general term is $v^{kk'} \varpi_{k'}$.

With respect to the orthonormal basis $(\overrightarrow{\varepsilon_k} = \overrightarrow{\delta_k} / \sqrt{\varpi_k})_{k \in K}$, the matrix of *Som*, denoted **S**, is symmetric, its general term is $s_{kk'} = \sqrt{\varpi_k \varpi_{k'}}\, v^{kk'}$. Let us denote $(c_{k\ell})_{k \in K}$ the coordinates of a unit–norm eigenvector $\overrightarrow{\alpha_\ell}$ of *Som* in the orthonormal basis $(\vec{\varepsilon_k})_{k \in K}$, one has $\sum\limits_{k' \in K} s_{kk'} c_{k'\ell} = \lambda_\ell c_{k\ell}$.

In the basis $(\overrightarrow{\delta_k})_{k \in K}$, the coordinates of $\overrightarrow{\alpha_\ell}$ are $(a_\ell^k = c_{k\ell} / \sqrt{\varpi}_k)_{k \in K}$.

Let $\mathbf{\Omega}^{1/2}$ denote the $K \times K$ diagonal matrix $(\sqrt{\omega_k})_{k \in K}$; $\mathbf{\Lambda}$ the $L \times L$ diagonal matrix of eigenvalues $(\lambda_\ell)_{\ell = 1 \ldots L}$; $\mathbf{c}_\ell$ the K–column of coefficients $(c_{k\ell})_{k \in K}$; $\mathbf{C}$ the $K \times L$ matrix of K–columns $(\mathbf{c}_\ell)_{\ell = 1 \ldots L}$; $\mathbf{y}_\ell$ the J–column of the principal coordinates $(y_\ell^j)_{j \in J}$ of points $(\mathrm{M}^j)_{j \in J}$ on Axis ℓ; $\mathbf{b}_\ell$ the K–column of the regression coefficients of initial variables on the ℓ–th standard principal variable; hence the following computation procedure[6].

Step 1. Compute the matrix $\mathbf{S} = [s_{kk'}]$ with $s_{kk'} = \sqrt{\varpi_k \varpi_{k'}}\, v^{kk'}$, that is, $\mathbf{S} = \mathbf{\Omega}^{1/2} \mathbf{V} \mathbf{\Omega}^{1/2}$.

[6]A program of biweighted PCA, written by P. Bonnet and B. Le Roux, is available, see p.417.

Step 2. For $\ell = 1, \ldots L$, calculate the eigenvalues λ_ℓ, and the eigenvectors $(c_{k\ell})_{k \in K}$, with $\sum\limits_{k \in K} c_{k\ell}^2 = 1$ of the matrix $\mathbf{S}$, that is, determine $\mathbf{C}$ verifying $\mathbf{SC} = \mathbf{C\Lambda}$ with $\mathbf{C'C} = \mathbf{I}$.

Step 3. For $\ell = 1, \ldots L'$ $(\leq L)$ and $j \in J$, compute $y_\ell^j = \sum\limits_{k \in K} \sqrt{\varpi_k}\, c_{k\ell}\, x_0^{jk}$, that is, compute $\mathbf{y}_\ell = \mathbf{X}_0 \mathbf{\Omega}^{1/2} \mathbf{c}_\ell$.

Step 4. For $\ell = 1, \ldots L'$ and $k \in K$, compute $b_\ell^k = \sqrt{\lambda_\ell}\, c_{k\ell} / \sqrt{\varpi_k}$, thats is, compute $\mathbf{b}_\ell = \sqrt{\lambda_\ell}\, \mathbf{\Omega}^{-1/2} \mathbf{c}_\ell$.

Steps 3 and 4 are obtained by replacing a_ℓ^k by $c_{k\ell}/\sqrt{\varpi_k}$ and $a_{k\ell}$ by $c_{k\ell}\sqrt{\varpi_k}$ in the passage formulas (p.135).

Synopsis. Let us denote $\mathbf{A} = [a_\ell^k]$ the $K \times L$ matrix of L columns $\mathbf{a}_\ell$, $\mathbf{A}^\star = \mathbf{A}'\mathbf{\Omega}$, hence, letting $a_{k\ell} = \varpi_k\, a_\ell^k$, $\mathbf{A}^{\star\prime} = \mathbf{\Omega A} = [a_{k\ell}]$; let us denote $\mathbf{Y} = [y_\ell^j]$ the $J \times L$ matrix of L columns $\mathbf{y}_\ell$.

- Equation for principal directions: $\mathbf{VA}^{\star\prime} = \mathbf{A\Lambda}$ with $\mathbf{A}^\star\mathbf{A} = \mathbf{I}$.

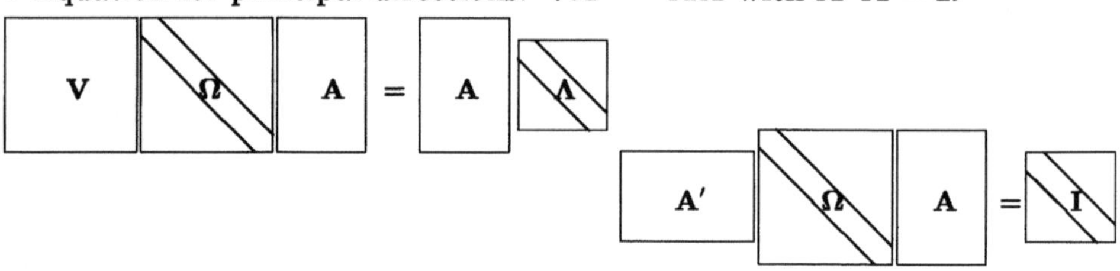

- Principal coordinates $\mathbf{Y} = \mathbf{X}_0\mathbf{A}^{\star\prime}$

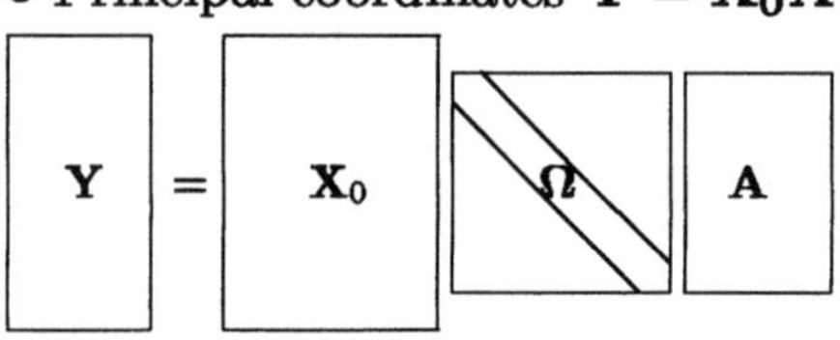

- Reconstitution: $\mathbf{X}_0 = \mathbf{YA}'$

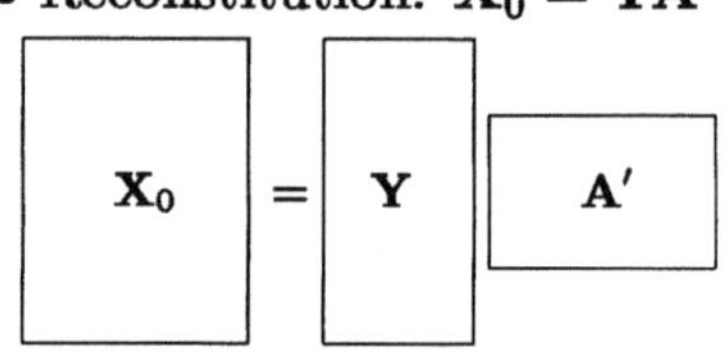

- Regression coefficients: $\mathbf{B} = \mathbf{A\Xi}$

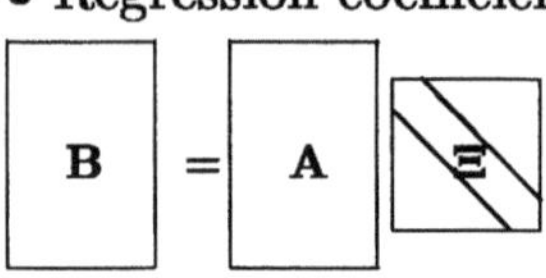

- Reconstitution $\mathbf{V} = \mathbf{A\Lambda A}'$

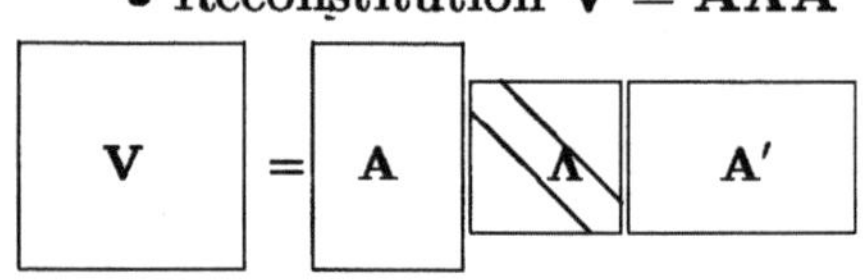

Figure 4.3. Synopsis of biweighted PCA

Main results of a biweighted PCA. Eigenvalues (λ_ℓ) i.e. *variances of axes*; *principal coordinates* (y_ℓ^j) of points; *regression coefficients* (b_ℓ^k) of initial variables on standard principal variables.

A biweighted PCA is thus summarized by the diagram of Figure 4.4. Together with the tables of this figure, there are associated the histogram of eigenvalues, the representations in the space of individuals (cloud of points) and in the space of variables (cf. below §4.1.4). Furthermore, as aids to interpretation, there are the tables of contributions of points (§4.1.1, p.136) and of variables (§4.1.2, p.140).

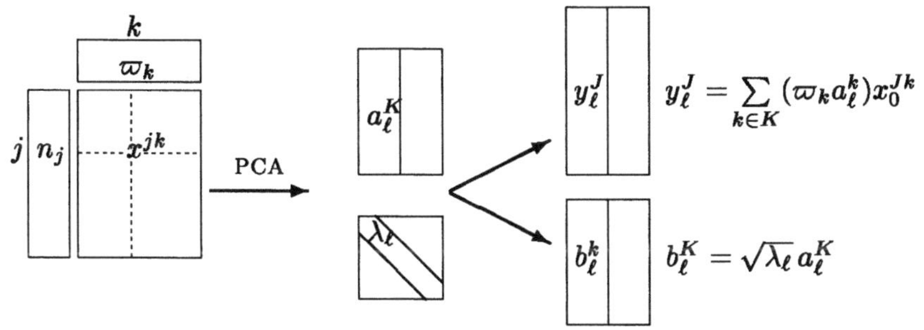

Figure 4.4. Summary of biweighted PCA

4.1.4 Graphical Representations

In the *space of individuals* (geometric space), one represents the projection of the cloud of individuals on the first principal axes, or on principal planes 1-2, 1-3, 2-3, etc. Thus on the principal plane 1-2 referred to its orthonormal principal frame $(\vec{\alpha}_1, \vec{\alpha}_2)$, the point M^j is represented by the point $\widetilde{\mathrm{M}}^j$ that has abscissa y_1^j and ordinate y_2^j, etc. On this figure, the approximate distances between the points of the cloud are read with the help of their qualities of representation on the principal axes and planes.

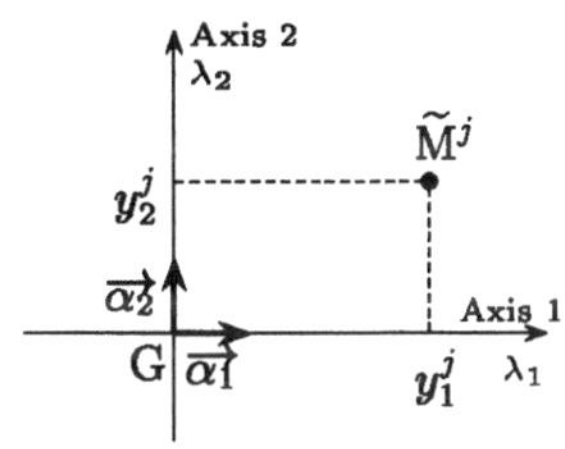

In the *space of variables* (Euclidean vector space), the regressions of the K centered initial variables $(x_0^{Jk})_{k\in K}$ on principal variables 1, 2, 3, etc., or on two principal variables (1-2, 2-3, 1-3, etc.) are represented by vectors.

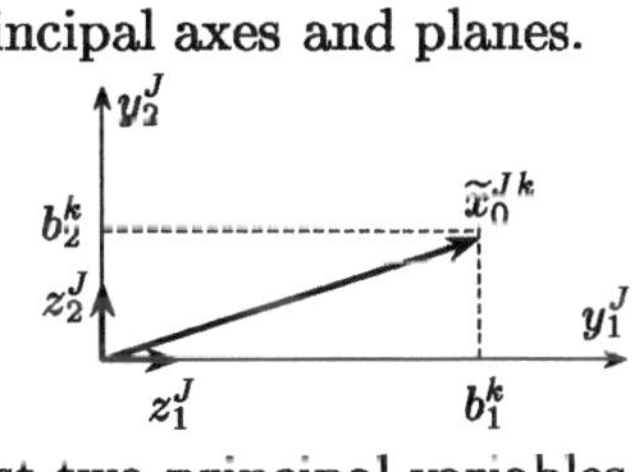

Thus the regression of the variable x^{Jk} on the first two principal variables is represented by vector $\widetilde{x}_0^{Jk}$ that has coordinates (b_1^k, b_2^k) in the principal basis of standard variables (z_1^J, z_2^J) (cf. above figure). The correlations between regressed variables, that is, the approximate correlations of the initial variables can be read on the figure as cosines between the vectors representing those variables.

"Cloud of variables". Sometimes the regressed variables are merely represented by the endpoints of vectors, hence a "sort of cloud" (to speak like Benzécri) analogous to the cloud of individuals. In general, however, the symmetry between the two clouds is imperfect. The "cloud of variables" admits as a mean point the endpoint of the vector representing the Average variable, and this point only coincides with the origin of axes when the Average variable has a null variance (cf. Exercise *Scotland Yard*, p.166).

4.1.5 Supplementary Elements

Starting with an Individuals×Variables table, one may on the one hand define the distance between individuals from a selected subset of variables, which are then called *active variables*; on the other hand, one may determine the principal axes from a selected subset of individuals, which are called *active individuals.* By *supplementary elements*, one designates individuals that do not participate in the determination of principal axes or variables that are not used for the distance calculation. In the analysis of data, supplementary variables[7] are regressed on principal variables, and supplementary individuals are projected on principal axes.

Supplementary individuals. Let $(x^{j_s k})_{k\in K}$ be the profile of a supplementary individual j_s, this profile is represented by the point M^{j_s} defined by $\overrightarrow{\mathrm{OM}}^{j_s} = \sum\limits_{k\in K} x^{j_s k}\vec{\delta}_k$.

The ℓ–th principal coordinate of the point M^{j_s} is:

$$y_\ell^{j_s} = \sum_{k\in K} a_{k\ell}(x^{j_s k} - \overline{x}^k) = \frac{1}{\sqrt{\lambda_\ell}}\sum_{k\in K} \varpi_k b_\ell^k (x^{j_s k} - \overline{x}^k) \tag{4.16}$$

where $a_{k\ell}$ and $\overline{x}^k = \operatorname{Mean} x^{Jk}$ are determined by active individuals.

The quality of representation of the supplementary individual j_s on Axis ℓ is equal to $(y_\ell^{j_s})^2/(\mathrm{GM}^{j_s})^2 = (y_\ell^{j_s})^2/(\sum\limits_{k\in K} \varpi_k(x^{j_s k} - \overline{x}^k)^2)$. If the point M^{j_s} belongs to the affine support of the cloud of active individuals, then $(\mathrm{GM}^{j_s})^2 = \sum\limits_{\ell=1}^{L}(y_\ell^{j_s})^2$ and its quality of representation on the L–principal subspace (support of the cloud) is equal to 1; if not, its quality of representation is less than 1; if $\overrightarrow{\mathrm{GM}}^{j_s}$ is orthogonal to the support of the cloud, its quality of representation is equal to 0.

[7]Supplementary variables, also called "illustrative", may be quite important methodologically.

Supplementary variables. A supplementary variable x^{Jk_s} is given by its values $(x^{jk_s})_{j\in J}$ on active individuals. The regression coefficient $b_\ell^{k_s}$ of x^{Jk_s} on the ℓ-th standard principal variable is $b_\ell^{k_s} = \mathrm{Cov}(x^{Jk_s}|z_\ell^J)$; if the supplementary variable is standardized, $b_\ell^{k_s}$ is the correlation coefficient between x^{Jk_s} and the ℓ-th principal variable. From the first equation (4.5) (p.135), one deduces $\mathrm{Cov}(x^{Jk_s}|z_\ell^J) = \mathrm{Cov}(x^{Jk_s}|(1/\xi_\ell)\sum_{k\in K} x_0^{Jk}a_{k\ell}$, hence

$$b_\ell^{k_s} = (1/\xi_\ell)\sum_{k\in K} a_{k\ell}v^{kk_s} \quad \text{with } v^{kk_s} = \mathrm{Cov}(x^{Jk}|x^{Jk_s}) \tag{4.17}$$

This formula shows that a supplementary variable can be given by its covariances with the initial variables. Since principal variables are uncorrelated, the variable $\sum_{\ell=1}^{L'} b_\ell^{k_s} z_\ell^J$ $(L' \leq L)$ is the multiple regression of x^{Jk_s} on the first L' standard principal variables, and is called *principal regression of order L'*. If $L' = L$, the principal regression is said to be complete, and it amounts to the multiple regression of the supplementary variable x^{Jk_s} on the K initial variables.

Proof. The regression of x^{Jk_s} on the L principal variables is $\widetilde{x}_L^{Jk_s} = \sum_{\ell=1}^{L} b_\ell^{k_s} z_\ell^J$. From $\xi_\ell z_\ell^J = \sum_{k\in K} x_0^{Jk}a_{k\ell}$ we deduce $\widetilde{x}_L^{Jk_s} = \sum_{k\in K}(\sum_{\ell=1}^{L} a_{k\ell}b_\ell^{k_s}/\xi_\ell)x_0^{Jk}$: the partial regression coefficients of the supplementary variable on initial variables are equal to $(\sum_{\ell=1}^{L} a_{k\ell}b_\ell^{k_s}/\xi_\ell)_{k\in K}$. ◁

The *quality of fit* of the supplementary variable by the L principal variables is $(R_L^{k_s})^2 = \mathrm{Var}(\sum_{\ell=1}^{L} b_\ell^{k_s} z_\ell^J)/\operatorname{Var} x^{Jk_s} = \sum_{\ell=1}^{L}(b_\ell^{k_s})^2/\operatorname{Var} x^{Jk_s}$. If $x_0^{Jk_s} \in \mathcal{X}_0$ (subspace spanned by centered initial variables), then $R^2 = 1$, else $R^2 < 1$.

> *To sum up*: Putting a variable as a supplementary element amounts to a multiple regression on principal variables; when all principal variables are taken, the procedure amounts to the multiple regression of the supplementary variable on the K initial variables.

— *Particular case.* Let δ_c^J be the *indicator variable* of class J<c>, i.e. such that $\delta_c^j = 1$ if $j \in J$<c>, 0 elsewhere. Let $n_c = \sum_{j\in J<c>} n_j$, $f_c = n_c/n$ and $\overline{z}_\ell^c = \sum_{j\in<c>} n_j z_\ell^j/n_c$, the regression coefficient $b_{c\ell}$ of variable δ_c^J on the standard principal variable z_ℓ^J is equal to $f_c\overline{z}_\ell^c$.

Toward structured data. The method of supplementary elements is the first step toward the analysis of structured data, to be presented in chapter *STRUCTURED* and applied in chapter *CASE STUDIES*.

For example, consider a class of individuals $J\langle c\rangle$ (nonempty subset of J), weighted by $n_c = \sum_{j\in J\langle c\rangle} n_j$. The mean point of the subcloud $\mathrm{M}^{J\langle c\rangle} = (\mathrm{M}^j)_{j\in J\langle c\rangle}$ is $\mathrm{M}^c = \sum_{j\in J\langle c\rangle} f_j^c \mathrm{M}^j$ (with $f_j^c = n_j/n_c$), and the ℓ-th principal coordinate of M^c is $\overline{y}_\ell^c = \sum_{j\langle c\rangle} f_j^c y_\ell^j$ (mean of the coordinates of the individuals belonging to class c). The following properties hold for each principal Axis ℓ:

1. $\overline{y}_\ell^c = \sum_{k\in K} a_{k\ell}(\overline{x}^{ck} - \overline{x}^k)$ is the coordinate of the supplementary profile $\overline{x}^{cK} = \sum_{j\in J\langle c\rangle} f_j^c x^{jK}$, mean of the profiles of individuals in class c.

2. $\overline{y}_\ell^c = \xi_\ell b_\ell^c = \xi_\ell b_{c\ell}/f_c$, where $b_{c\ell}$ denotes the regression coefficient of δ_c^J (indicator variable of c) on the ℓ–th standard principal variable z_ℓ^J.

3. The correlation coefficient of the indicator variable δ_c^J with the ℓ–th principal variable is $r_\ell^c = (\sqrt{f_c/(1-f_c)}\,\overline{y}_\ell^c/\xi_\ell$.

4.1.6 Biweighted PCA and Correspondence Analysis

Property 4.15. *The correspondence analysis of the contingency table* n_{JK} *is equivalent to the biweighted* PCA *of the protocol* $f^{JK} = (f^{jk})_{j\in J,k\in K}$, *letting* $f^{jk} = f_{jk}/(f_j f_k)$, *with weightings* n_J *and* f_K.

Proof. The f_K–mean of f^{JK} is equal to 1^J, its f_J–mean is equal to 1^K. The centered protocol is that of association rates $(t^{jk} = f^{jk} - 1)_{j\in J,k\in K}$, with $\mathrm{Cov}(f^{Jk}|f^{Jk'}) = \sum_{j\in J} f_{jk} f_{jk'}/(f_j\, f_k\, f_{k'}) - 1$. The Average variable over J is constant, therefore, applying Property (4.3) (p.135), the measure $a_{K\ell}$ defining the ℓ–th principal line is a contrast over K that we denote $z_{K\ell}$. The passage formulas (4.5) read:

$$\begin{cases} \sum_{k\in K} z_{k\ell} t^{Jk} & = \xi_\ell\, z_\ell^J \\ \sum_{j\in J} z_{j\ell} t^{jK} & = \xi_\ell\, z_\ell^K \end{cases} \quad \text{with} \quad \begin{cases} z_{k\ell} & = f_k z_\ell^k \\ z_{j\ell} & = f_j z_\ell^j \end{cases} \tag{4.18}$$

Replacing in these equations t^{jk} by $f_{jk}/(f_j\, f_k)$, one obtains the transition formulas of CA (*CORRESPONDENCE*, p.38). ◁

4.2 Simple PCA (Analysis of Covariances)

Definition 4.2 (Simple PCA). *Given a protocol of scores, the* PCA *performed by taking all weights ϖ_k equal to 1 is called a simple* PCA *(the weighting over J still being measure n_J, possibly elementary).*

For a simple PCA, the square of the *distance between two profiles* is the sum of the squares of the deviations corresponding to each variable:

$$d^2(j,j') = (\mathrm{M}^j\mathrm{M}^{j'})^2 = \sum_{k\in K}(x^{jk} - x^{j'k})^2$$

The Euclidean metric defined on $\mathcal{U}$ is elementary and the canonical basis $(\overrightarrow{\delta_k})_{k\in K}$ is orthonormal ($\mathbf{Q} = \mathbf{I}_K$). The isomorphism, that with a measure over K associates its density, reduces here to $\forall k \in K : a_{k\ell} = a^k_\ell$ (still distinguishing upper and lower indices in accordance with duality notation). The *variance of cloud* is equal to the sum of the variances of initial variables:

$$\operatorname{Var} \mathrm{M}^J = \sum_{k\in K} \operatorname{Var} x^{Jk} \tag{4.19}$$

Passage formulas (4.5) (p.135) read:

$$\begin{cases} \sum\limits_{k\in K} a_{k\ell} x_0^{Jk} = \xi_\ell z_\ell^J & \text{with } \sum\limits_{k\in K}(a^k_\ell)^2 = \sum\limits_{k\in K}(a_{k\ell})^2 = 1 \\ \sum\limits_{j\in J} x^{jK}(f_j z^j_\ell) = \xi_\ell a^K_\ell & \text{with } \sum\limits_{j\in J} f_j (z^j_\ell)^2 = 1 \end{cases} \tag{4.20}$$

The *equation for principal directions* (cf. Equations (4.6), p.136) reads:

$$\sum_{k'\in K} v^{kk'} a_{k'\ell} = \lambda_\ell a^k_\ell \qquad \text{with } a^k_\ell = a_{k\ell} \tag{4.21}$$

The properties of variables, stated at §4.1.2 for biweighted PCA, are valid for simple PCA.

> *To sum up:* In simple PCA, principal directions and eigenvalues are obtained by determining eigenvalues and eigenvectors of the matrix $\mathbf{V}$ of variances–covariances between initial variables; that is in brief:
>
> *Simple* PCA *is the analysis of covariances.*

Property 4.16 (Biweighted vs simple PCA). *If the weights $(\varpi_k)_{k\in K}$ of a protocol of scores are integers, the biweighted* PCA *is equivalent to the simple* PCA *obtained by duplicating ϖ_k times each initial variable x^{Jk} in the sense that both analyses yield the same standard principal variables and eigenvalues.*

4.2.1 Computation Procedure for Simple PCA

1. Calculate the covariance matrix **V**.

2. Diagonalize **V**, hence $(\lambda_\ell, (a_{k\ell})_{k\in K})_{\ell=1,\ldots L}$ with $\sum_{k\in K} (a_{k\ell})^2 = 1$.

3. For $\ell = 1, 2, \ldots L'$ $(\leq L)$, calculate $(y_\ell^j = \sum_{k\in K} a_{k\ell} x_0^{jk})_{j\in J}$.

4. For $\ell = 1, 2, \ldots L'$ $(\leq L)$, calculate $(b_\ell^k = \sqrt{\lambda_\ell}\, a_\ell^k)_{k\in K}$.

If $\mathbf{a}_\ell$ denotes the K–column of coefficients $(a_{k\ell} = a_\ell^k)_{k\in K}$ and $\mathbf{z}_\ell$ denotes the J–column of values of the ℓ–th standard principal variable, one has:

$$\mathbf{X}_0\mathbf{a}_\ell = \xi_\ell \mathbf{z}_\ell \qquad \mathbf{X}_0'\mathbf{F}\mathbf{z}_\ell = \xi_\ell \mathbf{a}_\ell \qquad \mathbf{V}\mathbf{a}_\ell = \lambda_\ell \mathbf{a}_\ell$$

Let **A** denote the $K \times L$ matrix of columns $\mathbf{a}_\ell$, one has:

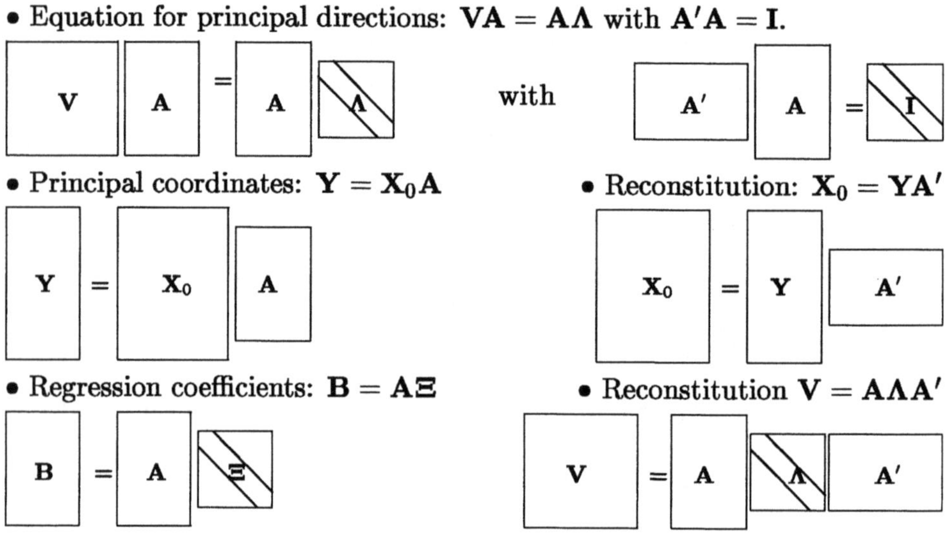

Figure 4.5. Synopsis of simple PCA

4.3 Standard PCA (Analysis of Correlations)

When the K variables $(x^{Jk})_{k\in K}$ over (J, n_J) are heterogeneous, one way of bringing them to a common scale is the standardization procedure, that is, $(x^{jk} \mapsto (x^{jk} - \overline{x}^k)/\operatorname{SD} x^{Jk})$, hence a protocol of scores with the standard deviation as a common unit scale, on which a simple PCA can be performed.

Definition 4.3 (Standard PCA). *Given a protocol $(x^{Jk})_{k\in K}$, the simple* PCA *performed on the associated standardized variables is called the standard* PCA *of the protocol.*

Therefore performing a standard PCA on a family of variables amounts to taking the standard deviation as a common scale on the one hand, and to giving to the standardized variables weights equal to 1 on the other hand.

If $r^{kk'}$ denotes the correlation between variables x^{Jk} and $x^{Jk'}$, the *equation for principal directions* reads:

$$\sum_{k' \in K} r^{kk'} a_{k'\ell} = \lambda_\ell a_\ell^k \qquad \text{with} \quad a_\ell^k = a_{k\ell} \tag{4.22}$$

> *To sum up*: In standard PCA, eigenvalues and principal directions are obtained by diagonalizing the correlation matrix of initial variables:
>
> *Standard* PCA *is the analysis of correlations.*

Property 4.17 (Standard vs weighted PCA). *Given a family of variables $x^{JK} = (x^{Jk})_{k \in K}$ over (J, n_J), and the associated standardized variables $(x_0^{Jk} / \text{SD}\, x^{Jk})_{k \in K}$:*

(i) The standard PCA *of x^{JK} is equivalent to the biweighted* PCA *of x^{JK} taking as weights the reciprocals of the variances: $(\varpi_k = 1/ \text{Var}\, x^{Jk})_{k \in K}$.*

(ii) The simple PCA *of x^{JK} is equivalent to the biweighted* PCA *of the standardized protocol taking as weights the variances: $\varpi_k = \text{Var}\, x^{Jk}$.*

Proof. (i) In the standard PCA, the distance between j and j' is given by $d^2(j, j') = \sum_{k \in K} (x^{jk} - x^{j'k})^2 / \text{Var}\, x^{Jk}$.

(ii) In the simple PCA of x^{JK}, the distance between j and j' is given by $d^2(j, j') = \sum_{k \in K} (x^{jk} - x^{j'k})^2 = \sum_{k \in K} \text{Var}\, x^{Jk} \left((x^{jk} - x^{j'k}) / \text{SD}\, x^{Jk} \right)^2$. ◁

4.3.1 Properties of Standard PCA

The properties of standard PCA are obtained by replacing, in the formulas of simple PCA, the covariances by the correlations, and the variances by 1; the most important properties are listed hereafter.

- The *variance of cloud* is equal to the number of variables. All variables contribute to the variance of cloud equally.

$$\text{Var}\, \text{M}^J = K \qquad \forall k \in K \,:\, \text{Cta}_k = 1$$

- The *regression coefficient* b_ℓ^k of $x_0^{Jk} / \text{SD}\, x^{Jk}$ (standardized initial variable) on z_ℓ^J (ℓ–th standard principal variable) is equal to the correlation coefficient r_ℓ^k between x^{Jk} and z_ℓ^J.

The *partial regression* coefficients of the standardized initial variable $x_0^{Jk} / \mathrm{SD}\, x^{Jk}$ on the first L' standard principal variables $(z_\ell^J)_{\ell=1,\dots L'}$ are equal to the correlation coefficients $(r_\ell^k)_{l=1,\dots L'}$ $(L' \leq L)$.

The *multiple correlation coefficient* $R_{L'}^k$ of the initial variable x^{Jk} with the first L' principal variables is given by

$$(R_{L'}^k)^2 = \sum_{\ell=1}^{L'} (r_\ell^k)^2 \quad \text{and } (R_L^k)^2 = \sum_{\ell=1}^{L} (r_\ell^k)^2 = 1 \text{ (L'=L)}$$

- The coefficients $a_\ell^k = r_\ell^k / \xi_\ell$ verify the two following properties:

  (i) $\forall \ell = 1, \dots L : \sum_{k \in K} (a_\ell^k)^2 = 1;$

  (ii) $\forall k \in K : \sum_{\ell=1}^{L} \lambda_\ell (a_\ell^k)^2 = 1.$

- The relative contribution of the initial variable x^{Jk} to the variance of Axis ℓ is equal to $(a_\ell^k)^2$; the *quality of fit* of the variable x^{Jk} by the ℓ–th principal variable is equal to $(r_\ell^k)^2 = \lambda_\ell (a_\ell^k)^2$.

- In the *multiple regression* of z_ℓ^J on the K standardized initial variables, the partial regression coefficient on $x_0^{Jk} / \mathrm{SD}\, x^{Jk}$ is equal to r_ℓ^k / λ_ℓ, whereas the simple regression coefficient is r_ℓ^k. The ratio of the partial regression coefficient to the simple regression coefficient is $1/\lambda_\ell$; this ratio is less than one for the eigenvalue(s) exceeding 1.

- The regression coefficient of a supplementary variable x^{Jk_s} on the ℓ–th standard principal variable z_ℓ^J is equal to $(1/\xi_\ell) \sum_{k \in K} r^{kk_s} a_{k\ell}$, where r^{kk_s} is the correlation coefficient between x^{Jk} and x^{Jk_s}.

4.3.2 Graphical Representations

The *graphical representations of individuals* are those described in §4.1.4 (p.145) with the following specific features.

In the *graphical representation of variables*, the length of the vector representing a variable in plane (ℓ, ℓ') is equal to the multiple correlation coefficient of this variable with the principal variables ℓ and ℓ'. If in this plane we draw a circle of radius 1 — the so–called *circle of correlations* — the endpoints of the vectors representing standard variables are inside the circle. If two variables are perfectly represented in plane (ℓ, ℓ'), the endpoints lie on the circle, and the cosine of the angle between the two vectors is equal to their correlation coefficient.

4.3.3 Computation Procedure for Standard PCA

1. Calculate the correlation matrix $\mathbf{R}$.
2. Diagonalize $\mathbf{R}$, hence $(\lambda_\ell, (a_{k\ell})_{k\in K})_{\ell=1,\dots L}$ with $\sum\limits_{k\in K}(a_{k\ell})^2 = 1$.
3. For $\ell = 1,\dots L'$ $(\leq L)$ calculate: $(r_\ell^k = \sqrt{\lambda_\ell}\, a_{k\ell})_{k\in K}$.
4. For $\ell = 1,\dots L'$ $(\leq L)$ calculate $(y_\ell^j = \sum\limits_{k\in K} a_{k\ell} x_0^{jk}/\sqrt{v^{kk}})_{j\in J}$.

In the following synopsis, $\mathbf{A}$ denotes the matrix with general term $a_\ell^k = a_{k\ell}$, and $\mathbf{V}_K$ the diagonal matrix whith diagonal terms $v^{kk} = \operatorname{Var} x^{Jk}$.

• Equation for principal directions: $\mathbf{RA} = \mathbf{A\Lambda}$ with $\mathbf{A'A} = \mathbf{I}$

• Principal coordinates: $\mathbf{Y} = \mathbf{X}_0\mathbf{V}_K^{-\frac{1}{2}}\mathbf{A}$ • Reconstitution: $\mathbf{X}_0\mathbf{V}_K^{-\frac{1}{2}} = \mathbf{YA'}$

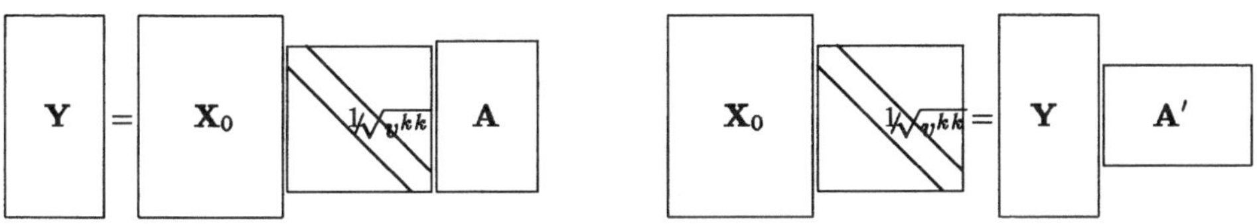

• Correlation coefficients: $\mathbf{B} = \mathbf{A\Xi}$ • Reconstitution $\mathbf{R} = \mathbf{A\Lambda A'}$

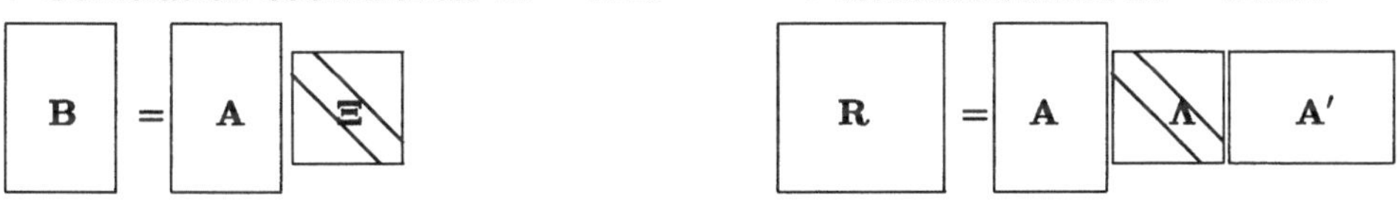

Figure 4.6. Synopsis of standard PCA

4.4 General PCA of a Protocol of Scores

In the general case, starting with a protocol of scores x^{JK}, the definition of a Euclidean distance between profiles requires specifying the $K \times K$ positive definite matrix $\mathbf{Q} = [q_{kk'}]$ of scalar products between the basis vectors, i.e. $q_{kk'} = \langle \delta_k^K | \delta_{k'}^K \rangle$.

The distance between the profiles of j and j' is defined by

$$\sum_{k\in K}\sum_{k'\in K} q_{kk'}(x^{jk} - x^{j'k})(x^{jk'} - x^{j'k'}) = (\mathrm{M}^j\mathrm{M}^{j'})^2 \tag{4.23}$$

The mean point G of the cloud M^J (with $\overrightarrow{\mathrm{OM}^j} = \sum_{j\in J} x^{jk}\overrightarrow{\delta_k}$) is such that $\mathrm{G} = \sum_{k\in K} \overline{x}^k \overrightarrow{\delta_k}$ and the variance is equal to $\sum_{k\in K}\sum_{k'\in K} v^{kk'} q_{kk'}$. Then the PCA of x^{JK} again consists in determining the principal axes and variables of the cloud so constructed, as done in chapter 3 (*CLOUD*). Let us denote $\mathbf{a}_\ell$ the K–column of coefficients $(a_\ell^k)_{k\in K}$, $\mathbf{A}$ the matrix of the L columns $\mathbf{a}_\ell$ and $\mathbf{A}^* = \mathbf{A}'\mathbf{Q}$; the main results can be summarized as follows.

- Equation for principal directions:

$$\forall k \in K \;:\; \sum_{k''\in K}\Big(\sum_{k'\in K} a_\ell^{k'} q_{k'k''}\Big) v^{kk''} = \lambda_\ell a_\ell^k \quad \text{with} \sum_{k\in K}\sum_{k'\in K} q_{kk'} a_\ell^k a_\ell^{k'} = 1$$

$$\mathbf{VQA} = \mathbf{A\Lambda} \text{ with } \mathbf{A'QA} = \mathbf{I}, \text{ or } \mathbf{VA}^{*\prime} = \mathbf{A\Lambda} \text{ with } \mathbf{A^*A} = \mathbf{I}$$

- Principal variables (principal coordinates of points):

$$y_\ell^J = \sum_{k\in K}\Big(\sum_{k'\in K} a_\ell^{k'} q_{kk'}\Big) x_0^{Jk}, \quad \text{in matrix notation} \quad \mathbf{y}_\ell = \mathbf{X}_0\mathbf{Q}\mathbf{a}_\ell$$

- Regression coefficient of the initial variable x^{Jk} on ℓ–th standard principal variable z_ℓ^J:

$$b_\ell^k = \xi_\ell a_\ell^k, \quad \text{in matrix notation } \mathbf{b}_\ell = \xi_\ell \mathbf{a}_\ell$$

- Reconstitution of data table:

$$\forall k \in K \; x_0^{Jk} = \sum_{\ell=1}^{L} a_\ell^k y_\ell^J, \quad \text{in matrix notation} \quad \mathbf{X}_0 = \mathbf{YA}'$$

- Reconstitution of covariances:

$$\forall k \in K\; \forall k' \in K \;:\; v^{kk'} = \sum_{\ell=1}^{L} \lambda_\ell a_\ell^k a_\ell^{k'} \qquad \text{in matrix notation } \mathbf{V} = \mathbf{A\Lambda A}'$$

In a general PCA, the *properties of variables* are exactly the same as those of the PCA of a biweighted protocol of scores (§4.1.2, p.137) letting $a_{k\ell} = \sum_{k'\in K} q_{kk'} a_\ell^{k'}$.

Computation procedure. The procedure is based on the diagonalization of the symmetric endomorphism *Som*. Since the matrix of a symmetric endomorphism relative to an orthonormal basis is symmetric (*MATH.BASES*, p.433), and owing to the robustness of diagonalization procedures for symmetric matrices, it is recommended as a rule to work with an orthonormal basis (constructed e.g. by a procedure of the Gram–Schmidt type).

4.5 PCA of a Table of Measures

In this section, we consider a $J \times K$ data table for which the profile of each individual j is a *measure over* K, denoted $x^j_K = (x^j_k)_{k\in K}$. Still assuming a frequency measure n_J and a weighting ϖ_K, we write the table (x^J_K, n_J, ϖ_K)[8]. The profiles of individuals $(x^j_K)_{j\in J}$ are vectors of the space $\mathbb{R}_K$ of measures over (K, ϖ_K), with the scalar product defined by $\forall u_K \in \mathbb{R}_K\ \forall v_K \in \mathbb{R}_K$: $\langle u_K | v_K \rangle = \sum\limits_{k\in K} u_k\, v_k / \varpi_k$.

The distance between profiles is defined by

$$d^2(j,j') = \sum_{k\in K} \frac{(x^j_k - x^{j'}_k)^2}{\varpi_k} \tag{4.24}$$

Matrix notation. Let us denote $\mathbf{X} = [x^j_k]$ the $J \times K$ data table, $\mathbf{F}$ the diagonal matrix of relative frequencies $(f_j)_{j\in J}$, and $\boldsymbol{\Omega}$ the diagonal matrix of weights $(\varpi_k)_{k\in K}$. The K–column of means $(\overline{x}_k)_{k\in K}$ is $\overline{\mathbf{x}} = \mathbf{X}'\mathbf{F}\mathbf{e}$; the table of centered variables is $\mathbf{X}_0 = \mathbf{X} - \mathbf{e}\overline{\mathbf{x}}'$; the $K \times K$ covariance matrix $\mathbf{V} = [v_{kk'}]$ with $v_{kk'} = \mathrm{Cov}(x^J_k | x^J_{k'})$) is $\mathbf{V} = \mathbf{X}'_0\mathbf{F}\mathbf{X}_0$; the $J \times J$ matrix with general term $w^{jj'} = \sum\limits_{k\in K} (x^j_k - \overline{x}_k)(x^{j'}_k - \overline{x}_k)/\varpi_k$ is $\mathbf{W} = \mathbf{X}_0\boldsymbol{\Omega}^{-1}\mathbf{X}'_0$.

4.5.1 Cloud of Individuals

The Euclidean cloud associated with the J profiles $(x^j_K)_{j\in J}$ is constructed in a way similar to that of §4.1.1 (p.133). The norm of the geometric vector $\overrightarrow{\delta}^k$ associated with Dirac measure δ^k_K is now equal to $\sqrt{1/\varpi_k}$. Relative to the Cartesian frame $(\mathrm{O}, (\overrightarrow{\delta}^k)_{k\in K})$, the point M^j is such that $\overrightarrow{\mathrm{OM}^j} = \sum\limits_{k\in K} x^j_k \overrightarrow{\delta}^k$; the coordinates of the *mean point* G of the cloud are equal to $(\overline{x}_k)_{k\in K}$ (where $\overline{x}_k = \sum\limits_{j\in J} f_j x^j_k$ is the mean of the variable x^J_k). The *variance of the cloud* is equal to $\sum\limits_{k\in K} \mathrm{Var}\, x^J_k/\varpi_k$, and the absolute contribution of variable x^J_k is $\mathrm{Cta}_k = \mathrm{Var}\, x^J_k/\varpi_k$.

By expressing in terms of data (x^J_K, f_J, ϖ_K) the images of the canonical basis vectors, namely $(\overrightarrow{\delta}^k)_{k\in K}$ of $\mathcal{V}$ and $(\delta^J_j)_{j\in J}$ of $\mathbb{R}^J$, by the linear mappings $Vac : \mathcal{V} \to \mathbb{R}^J$, $Vac^* : \mathbb{R}^J \to \mathcal{V}$, $Som : \mathcal{V} \to \mathcal{V}$ and $Tom : \mathbb{R}^J \to \mathbb{R}^J$ (CLOUD, p.84-86 and p.117), one obtains:

[8] A table of measures x^J_K will not be confused with a "table of measurements" x^{JK}, cf. footnote 4, p.131.

$$
\begin{array}{llll}
Vac(\overrightarrow{\delta}^k) & = \sum\limits_{j\in J} \dfrac{(x_k^j - \overline{x}_k)}{\varpi_k} \delta_j^J & \text{in matrix notation} & \mathbf{X}_0\mathbf{\Omega}^{-1} \\
Vac^*(\delta_j^J) & = \sum\limits_{k\in K} f_j(x_k^j - \overline{x}_k)\overrightarrow{\delta}^k & '' & \mathbf{X}_0'\mathbf{F} \\
Som(\overrightarrow{\delta}^k) & = \sum\limits_{k'\in K} (v_{kk'}/\varpi_k)\overrightarrow{\delta}^{k'} & '' & \mathbf{V}\mathbf{\Omega}^{-1} \\
Tom(\delta_j^J) & = \sum\limits_{j'\in J} f_j w^{jj'} \delta_{j'}^J & '' & \mathbf{WF}
\end{array}
$$

Theorem 4.3 (Passage formulas). *The direction coefficients* $(a_{k\ell})_{k\in K}$ *of the* ℓ*–th principal line and the* ℓ*–th principal variable* z_ℓ^J *satisfy the following system of equations (letting* $a_\ell^k = a_{k\ell}/\varpi_k$ *and* $z_{j\ell} = f_j z_\ell^j$*):*

$$
\begin{cases}
\sum\limits_{k\in K} (x_k^J - \overline{x}_k) a_\ell^k & = \xi_\ell z_\ell^J \\
\sum\limits_{j\in J} x_K^j z_{j\ell} & = \xi_\ell a_{K\ell}
\end{cases}
\quad \text{with} \quad
\begin{cases}
\sum\limits_{k\in K} (a_{k\ell})^2/\varpi_k & = 1 \\
\sum\limits_{j\in J} f_j (z_\ell^j)^2 & = 1
\end{cases}
\tag{4.25}
$$

Proof. The result is obtained by applying Equations (3.7) (CLOUD, p.92) to the table of measures. ◁

Let us recall that *Axis* ℓ goes through point G and that its direction is $\overrightarrow{\alpha}_\ell = \sum\limits_{k\in K} a_{k\ell}\overrightarrow{\delta}^k$.

Property 4.18. *The principal coordinate* y_ℓ^j *of the point* M^j *on the principal Axis* ℓ *going through point* G *and directed by the unit–norm vector* $\overrightarrow{\alpha_\ell} = \sum\limits_{k\in K} a_{k\ell}\overrightarrow{\delta}^k$ *is equal to* $\xi_\ell z_\ell^j$*.*

Proof. $y_\ell^j = \langle \overrightarrow{\mathrm{GM}^j} | \overrightarrow{\alpha_\ell} \rangle = \sum\limits_{k\in K} (x_k^j - \overline{x}_k) a_{k\ell} \| \overrightarrow{\delta_k} \| = \sum\limits_{k\in K} x_0^{jk} (a_{k\ell}/\varpi_k)$. From the first equation (4.25), one deduces $y_\ell^j = \xi_\ell z_\ell^j$. ◁

The following property is deduced from Equations (4.25).

Property 4.19 (Equation for principal directions). *Letting* $a_\ell^k = \frac{a_{k\ell}}{\varpi_k}$*, the principal direction* $\overrightarrow{\alpha_\ell} = \sum\limits_{k\in K} a_{k\ell}\overrightarrow{\delta}^k$ *is such that*

$$
\forall k \in K \qquad \sum_{k'\in K} v_{kk'} a_\ell^{k'} = \lambda_\ell a_{k\ell} \tag{4.26}
$$

Reconstitution formulas

- The reconstitution of *distances* reads

$$d^2(j,j') = \sum_{\ell \in L} (y_\ell^j - y_\ell^{j'})^2 \tag{4.27}$$

- The reconstitution of order $L' \leq L$ of *initial variable* x_k^J is $\overline{x}_k + \sum_{\ell=1}^{L'} a_{k\ell}\, y_\ell^J$ (where L is the dimensionality of the cloud).
- The reconstitution of order L' of *covariances* is $\sum_{\ell=1}^{L'} \lambda_\ell a_{k\ell} a_{k'\ell}$.

4.5.2 Space of Variables

The properties in the space of variables are analogous to those of the biweighted PCA of a protocol of scores.

Property 4.20 (Equation for principal variables). *The ℓ-th principal variable z_ℓ^J satisfies the equation:*

$$\forall j \in J \qquad \sum_{j' \in J} w^{jj'}(f_{j'} z_\ell^{j'}) = \lambda_\ell z_\ell^j \qquad (\ell = 1 \dots L) \tag{4.28}$$

- Regression of initial variable x_k^J on principal variable z_ℓ^J:

$$x_k^J - \overline{x}_k = \sum_{\ell=1}^{L} b_{k\ell} z_\ell^J \qquad \text{with} \quad b_{k\ell} = \xi_\ell a_{k\ell}$$

- Regression of principal variable y_ℓ^J on initial variable x_k^J:

$$y_\ell^J = \sum_{k \in K} a_\ell^k (x_k^J - \overline{x}_k 1^J) \qquad \text{with} \quad a_\ell^k = a_{k\ell}/\varpi_k$$

Total variable

Since profiles are measures over K, with each profile x_K^j one can associate the total of its values denoted $\underline{x}^j$, with $\underline{x}^j = \sum_{k \in K} x_k^j$, hence the Total variable $\underline{x}^J = (\underline{x}^j)_{j \in J}$.

The mean of the Total variable is $\underline{x} = \sum_{j \in J} f_j \underline{x}^j = \sum_{k \in K}(\sum_{j \in J} f_j x_k^j) = \sum_{k \in K} \overline{x}_k$.

Property 4.21. *If the Total variable $\underline{x}^J$ is constant, then the measure $a_{K\ell}$ is a contrast over K.*

Proof. If $\underline{x}^J = \underline{x}1^J$, from the second equation 4.25 (p.156), one deduces: $\xi_\ell \sum_{k\in K} a_{k\ell} = \sum_{j\in J}(\sum_{k\in K} x_k^j) z_{j\ell} = \sum_{j\in J} \underline{x} z_{j\ell} = 0$, hence $\sum_{k\in K} a_{k\ell} = 0$ ◁

Property 4.22. *The Total variable $\underline{x}^J$ is the axial variable associated with $(\mathrm{O}, \sum_{k\in K} \varpi_k \overrightarrow{\delta}^k/\varpi)$.*

Proof. Let us recall (CLOUD, §3.2.4, p.83) that the values of the axial variable associated with the line $(\mathrm{O}, \sum_{k\in K} \varpi_k \overrightarrow{\delta}^k/\varpi)$ are equal to the coordinates of the orthogonal projection of the points $(\mathrm{M}^j)_{j\in J}$ onto this line, that is, are equal to $(\langle\overrightarrow{\mathrm{OM}}^j | \sum_{k\in K} \varpi_k \overrightarrow{\delta}^k/\varpi\rangle / \| \sum_{k\in K} \varpi_k \overrightarrow{\delta}^k/\varpi\|^2)_{j\in J}$.
$\| \sum_{k\in K} \varpi_k \overrightarrow{\delta}^k/\varpi\|^2 = \sum_{k\in K} \varpi_k^2 \|\overrightarrow{\delta}^k\|^2/\varpi^2 = 1/\varpi$.
$\langle\overrightarrow{\mathrm{OM}}^j | \sum_{k\in K} \varpi_k \overrightarrow{\delta}^k/\varpi\rangle = \sum_{k\in K} \varpi_k (x_k^j/\varpi_k)/\varpi = \underline{x}^j/\varpi$. Hence the coordinate $\langle\overrightarrow{\mathrm{OM}}^j | \sum_{k\in K} \varpi_k \overrightarrow{\delta}^k/\varpi\rangle / \| \sum_{k\in K} \varpi_k \overrightarrow{\delta}^k/\varpi\|^2 = \underline{x}^j$. ◁

4.5.3 Biweighted PCA of a Table of Measures and CA

Property 4.23. *If the Total variable is constant and equal to $\underline{x}1^J$, assuming $\forall k \in K : \underline{x}\,\overline{x}_k > 0$, then the* PCA *of $(x_K^J, n_J, \underline{x}\,\overline{x}_K)$ is equivalent to the correspondence analysis of the measure $u_{JK} = (n_j x_k^j)_{j\in J, k\in K}$*

Proof. $u_k = \sum_{j\in J} u_{jk} = n\,\overline{x}_k$; $u_j = \sum_{k\in K} u_{jk} = \underline{x}\,n_j$; $u = \sum_{j\in J} u_j = n\,\underline{x}$.

Let P^J be the cloud associated with the measure u_{JK} by CA (CORRESPONDENCE, §2.2.2, p.33). The Φ^2–distance between points P^j and $\mathrm{P}^{j'}$ is such that $(\mathrm{P}^j\mathrm{P}^{j'})^2 = \sum_{k\in K} \left((u_{jk}/u_j) - (u_{j'k}/u_{j'})\right)^2/(u_k/u)$. Therefore, by replacing (u_{jk}/u_j) by $(x_k^j/\underline{x})$ and (u_k/u) by $(\overline{x}_k/\underline{x})$, one has

$$(\mathrm{P}^j\mathrm{P}^{j'})^2 = \frac{\left((x_k^j/\underline{x}) - (x_k^{j'}/\underline{x})\right)^2}{\overline{x}_k/\underline{x}} = \sum_{k\in K} \frac{(x_k^j - x_k^{j'})^2}{\underline{x}\,\overline{x}_k}$$

Letting $\varpi_k = \underline{x}\,\overline{x}_k$, one is back to the distance $d^2(j,j')$ defined by formula (4.24) (p.155). The weighted cloud (P^J, p_J) (with $p_j = f_j$) associated with the CA of the measure u_{JK} is thus identical to the weighted cloud (M^J, f_J) associated with the PCA of the biweighted protocol $(x_K^J, f_J, \underline{x}\,\overline{x}_K)$. ◁

Corollary 4.4. *The biweighted* PCA *of the transition table* $f_K^J = (f_k^J)_{k\in K}$ *associated with contingency table* n_{JK}*, with weightings* n_J *and* f_K *(letting* $f_k = n_k/n$*), is equivalent to the* CA *of* n_{JK}*.*

Proof. One has the relations $f_k^j = n_{jk}/n_j$, $\sum\limits_{k\in K} f_k^J = 1^J$, $\overline{x}^k = \sum\limits_{j\in J} f_j f_k^j = f_k$ and $\underline{\underline{x}} = \sum\limits_{k\in K} \overline{x}_k = 1$, therefore the PCA of the table of measures (f_K^J, n_J, f_K) is equivalent to the CA of the contingency table $n_{JK} = (n_j f_k^j)_{j\in J, k\in K}$. ◁

Matrix formulas

Let $\mathbf{a}_\ell = (a_{k\ell})_{k\in K}$ (K–column), the passage formulas (4.25) read:

$$\mathbf{X}_0 \boldsymbol{\Omega}^{-1} \mathbf{a}_\ell = \xi_\ell \mathbf{z}_\ell \qquad \text{and} \qquad \mathbf{X}_0' \mathbf{F} \mathbf{z}_\ell = \xi_\ell \mathbf{a}_\ell$$

and the equations (4.26) and (4.28) read in matrix notation:

$$\mathbf{V}\boldsymbol{\Omega}^{-1}\mathbf{a}_\ell = \lambda_\ell \mathbf{a}_\ell \quad \text{and} \quad \mathbf{W}\mathbf{F}\mathbf{z}_\ell = \lambda_\ell \mathbf{z}_\ell$$

with $\mathbf{a}_\ell' \boldsymbol{\Omega}^{-1} \mathbf{a}_\ell = 1 = \mathbf{z}_\ell' \mathbf{F} \mathbf{z}_\ell$ and $\mathbf{a}_{\ell'}' \boldsymbol{\Omega}^{-1} \mathbf{a}_\ell = 0 = \mathbf{z}_{\ell'}' \mathbf{F} \mathbf{z}_\ell$ $(\ell' \neq \ell)$.

4.5.4 Computation Procedure and Synopsis

Step 1. Calculate the symmetric matrix $\mathbf{S}= [s_{kk'}]$ with $s_{kk'} = \frac{v_{kk'}}{\sqrt{\varpi_k \varpi_{k'}}}$.

Step 2. For $\ell = 1, \dots L$, calculate the eigenvalues λ_ℓ and the eigenvectors $(c_{k\ell})_{k\in K}$ of the matrix $\mathbf{S}$, with $\sum\limits_{k\in K} c_{k\ell}^2 = 1$.

Step 3. For $\ell = 1, \dots L' (\leq L)$ and $k \in K$, compute $b_{k\ell} = \sqrt{\lambda_\ell}\, c_{k\ell} \sqrt{\varpi_k}$.

Step 4. For $\ell = 1, \dots L' \leq L)$ and $j \in J$, compute $y_\ell^j = \sum\limits_{k\in K} (x_k^j - \overline{x}_k) \frac{c_{k\ell}}{\sqrt{\varpi_k}}$.

— *Remark.* This algorithm applied to protocol (f_K^J, n_J, f_K) leads to putting the matrix of general term $(1/\sqrt{n_k n_{k'}}) \sum\limits_{j\in J} \frac{n_{jk} n_{jk'}}{n_j} - \sqrt{f_k f_{k'}} = s_{kk'} - \sqrt{f_k f_{k'}}$ to diagonal form, where $\mathbf{S} = [s_{kk'}]$ is the symmetric matrix associated with CA (*CORRESPONDENCE*, p.40).

The formula of step 3 gives the principal contrast $y_{K\ell} = b_{K\ell}$ hence the calibrated principal variable $y_\ell^K = (y_{k\ell}/f_k)_{k\in K}$ of CA.

By the formula of step 4, one obtains the principal variable of CA, that is, $y_\ell^J = \sum\limits_{k\in K} z_\ell^k f_k^J$ (with $z_\ell^k = y_\ell^k / \xi_\ell$).

Synopsis and main results. Let $\mathbf{A}$ denote the $K \times L$ matrix of general term $a_{k\ell}$, $\mathbf{A}^{\star\prime} = \boldsymbol{\Omega}^{-1}\mathbf{A}$, with $\mathbf{A}^{\star\prime} = [a^k_\ell]$ (letting $a^k_\ell = a_{k\ell}/\varpi_k$), and $\mathbf{Y} = [y^j_\ell]$ the $J \times L$ matrix of L columns $\mathbf{y}_\ell$.

• Equation for principal directions: $\mathbf{V}\mathbf{A}^{\star\prime} = \mathbf{A}\boldsymbol{\Lambda}$ with $\mathbf{A}^\star\mathbf{A} = \mathbf{I}$.

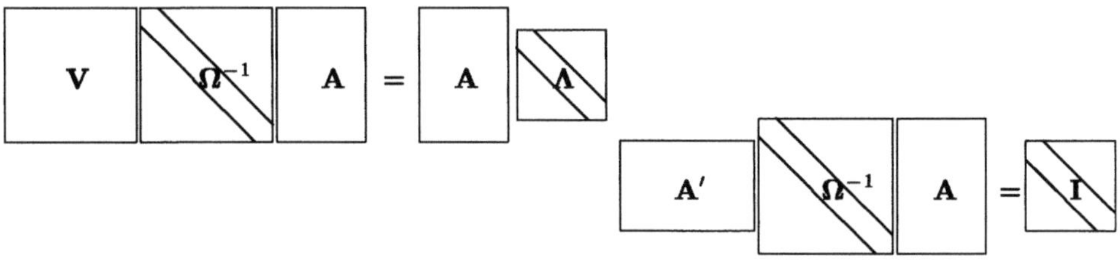

• Principal coordinates $\mathbf{Y} = \mathbf{X}_0\mathbf{A}^{\star\prime}$

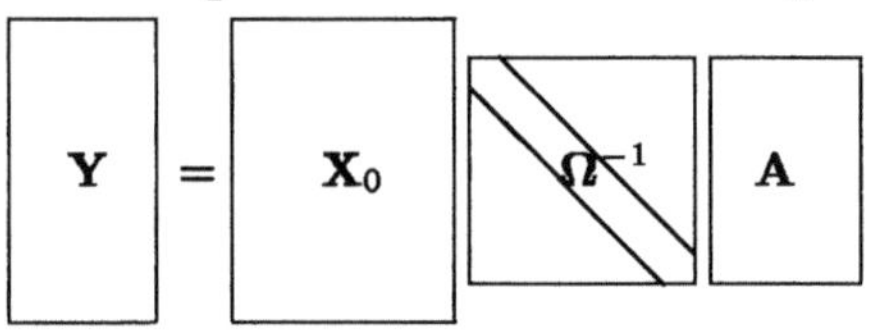

• Reconstitution: $\mathbf{X}_0 = \mathbf{Y}\mathbf{A}'$

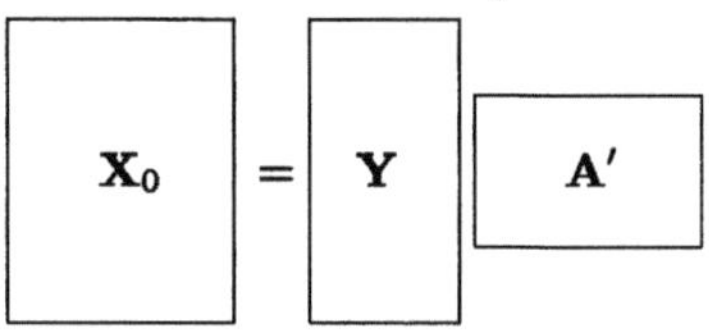

• Regression coefficients: $\mathbf{B} = \mathbf{A}\boldsymbol{\Xi}$

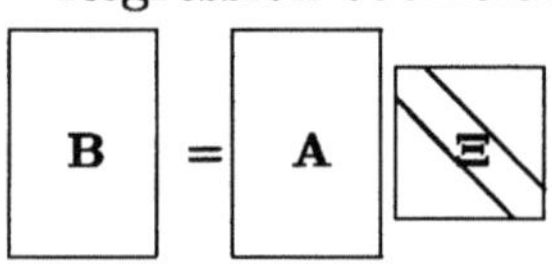

• Reconstitution $\mathbf{V} = \mathbf{A}\boldsymbol{\Lambda}\mathbf{A}'$

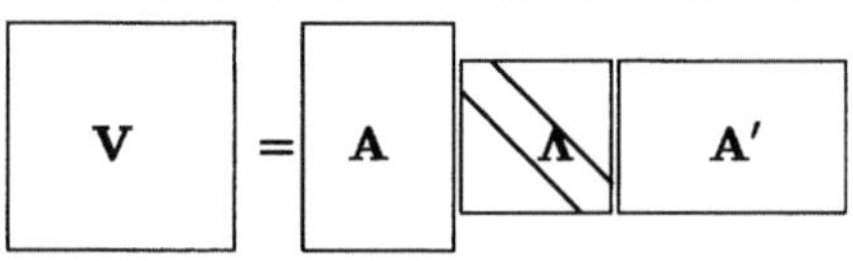

Figure 4.7. Synopsis of biweighted PCA of table of measures

4.6 Methodology of PCA

In a research study, the first stage consists in specifying the set of variables (in what follows, numerical variables) and the set of individuals relevant to the field under study.

Statistical analysis should always begin with the examination of elementary statistics: means, variances, correlations, diagrams for each of the variables (histograms) and pairs of variables (matched clouds, Cartesian diagrams), etc. Then a PCA may be performed to reduce dimensionality, both by summarizing the set of variables and by fitting to the cloud of individuals a lower–dimensional cloud.

Among variables, one will specify those that will serve to define the distances between individuals (active variables); among individuals, one will specify those that will serve to define the principal axes of the cloud (active individuals).

The choice of the distance between individuals presupposes a common numerical scale among variables, together with a weighting on variables. This provides a clue for selecting the most appropriate type of PCA; simple and standard PCA being the most common choices. As already said, when the numerical variables are heterogeneous — such as measurements referring to various magnitudes evaluated on different units — simple PCA may not make sense, whereas standard PCA always offers a solution. When the variables are on a ready–made scale — such as achievement scores from 0 to 20, or various lengths expressed in centimeters, etc. —, both a simple PCA and a standard PCA may be considered. If the variances of variables are of the same order of magnitude, the two analyses will yield close results[9]. If the variances differ widely, in simple PCA the variables with large variances will be predominant to the detriment of the others; a standard PCA may be preferred, which amounts to changing scales (i.e. SD) and giving equal weights to the variables (in new scales) (cf. Exercise 4.3).

This section deals with the methodology of the interpretation of a PCA. Biweighted PCA will be discussed, simple and standard PCA being particular cases. To begin with, a global analysis of the cloud can be done, based on the contributions of axes and on the contributions of individuals and variables to the overall variance; then the interpretation of axes will be carried out, based on the contributions of variables and of individuals to axes. Lastly, the exploration of the cloud and analyses involving supplementary elements will be undertaken.

4.6.1 Global Analysis of Cloud

Contributions of individuals and of variables. In order for the analysis to provide a representation of the *data as a whole*, on the one hand the contributions of individuals to total variance and on the other hand those of variables to total variance should be of the same orders of magnitude. Sometimes there are individuals, or groups of individuals, whose contributions are much larger than those of others. If such individuals really pertain to the research field under study (assuming no errors in data)[10], rather than discarding them from the analysis, one will make a new analysis by putting these individuals as supplementary ones. Other procedures can be used, for instance proceeding to a rank encoding and performing a PCA on ranked data. Similarly, finding a variable with a predominant contribution may

[9]Cf. *STABILITY*, §7.3, p.281.

[10]Unusually large contributions may simply point to outliers. In this connection, let us mention the exceptional ability of GDA for detecting outliers.

lead to a new analysis with this variable as a supplementary one, or to another procedure, such as MCA after categorization of variables[11].

Contributions of axes. At least, all axes whose contributions exceed the mean contribution ($\text{Var}\, \text{M}^J/K$) should be retained. To make sure that the first L' axes appropriately summarize the data one will check the following properties:

1. The first L' axes account for an important proportion of the variance of the cloud (e.g. more than 75%).
2. All (or almost all) variables are well fitted by the first L' principal variables (as an example see the *Parkinson* case study p.344); if this is not the case, another principal variable (at least) should be added (as an example, see exercise *Scotland Yard*, p.166).
3. Most individuals are well represented by the first L' axes (cf. *Parkinson* case study, p.344).

— *Particular cases*

• *Uniform* n_J. If the measure n_J is uniform (particular case $\forall j : n_j = 1$), the contribution of individual j to the variance of the cloud is proportional to $(\text{GM}^j)^2$, and the contribution to the variance of Axis ℓ is proportional to $(y^j_\ell)^2$. More generally, the contribution of a point to a principal subspace is proportional to the square of the distance of its projection from the origin. Consequently, ordering the individuals by their contributions amounts to ordering them by their distances from the origin.

• *Uniform* ϖ_K. Similarly, if the weighting ϖ_K is uniform (particular case $\forall k \in K : \varpi_k = 1$), the contribution of variable k to the variance of the cloud is proportional to its variance and its contribution to Axis ℓ is proportional to $(b^k_\ell)^2$ (square of regression coefficient).

4.6.2 Interpretation of Axes

The interpretation of axes is essentially based on the contributions of variables and points to axes. We reproduce hereafter the tables of pages 136 and 140 that give the absolute contributions of variables and individuals to axes and from which relative contributions are obtained.

The interpretation of axes is done firstly from variables, secondly from individuals.

[11]See e.g. the EPGY case study p. 394.

1. One selects the *variables* whose correlations (and/or contributions) with the axis are, in absolute value, the most important; then from the signs of correlations one gives an interpretation of the axis. When the correlations of all initial variables with a principal variable are of the same sign (the + sign is generally chosen, as the orientation of an axis is arbitrary), this principal variable is often called a size factor; on the correlation circle, the variables are all located on the same side, and the correlation of the Average variable with the "size factor" is close to 1 (cf. exercises 4.3 and 4.4).

There is a common situation in which the initial variables are indicators of some achievement performance, some indicators being "direct" and positively correlated with the "size factor", whereas "inverse" indicators are negatively correlated (cf. *Parkinson* case study). The oppositions expressed by this axis are then more apparent than real, since they basically reflect the orientations of indicators.

2. The interpretation of an axis is made more precise by studying the *profiles of the individuals* — or groups of individuals near one another on the axis — that contribute most to the variance of axis. For an application see *Parkinson* study p.344.

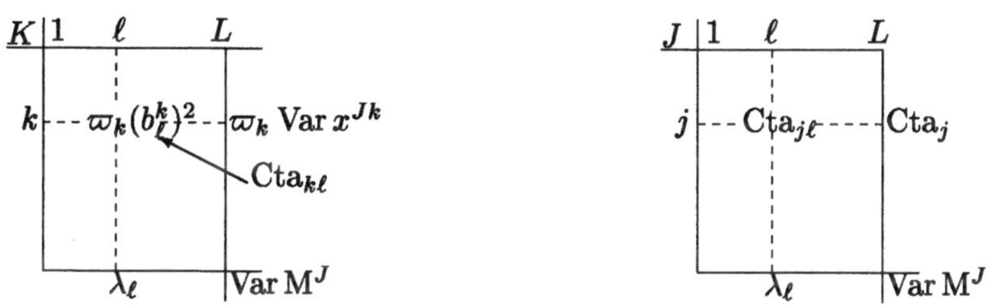

Table 4.1. Tables of absolute contributions.

4.6.3 Complementary Analyses

If a *classification* compatible with the Euclidean structure of the cloud is performed, i) one represents the classes and their mean points on principal axes; ii) one effectuates a *double breakdown of variance* according to axes and nodes; iii) for each class of individuals, one draws the concentration ellipse that constitutes a geometric summary of the class.

The interpretation of axes is also made more precise by supplementary variables. If, in addition, data are structured, the cloud of individuals can be investigated using these structures and the variance can be broken down according to axes and sources of variation (cf. STRUCTURED, §6.2.5, p.260).

Exercises of Chapter 4

Exercise 4.1 (Biweighted PCA)

Let $x^{IK} = (x^{Ik1}, x^{Ik2})$ be a bivariate protocol of scores on the elementary support I, with the weighting ($\varpi_{k1} = 9, \varpi_{k2} = 4$) on variables.

x^{Ik}	ϖ_k	$i1$	$i2$	$i3$	$i4$	$i5$	$i6$	$i7$	$i8$	$i9$	$i10$	$\overline{x}^k$
$k1$	9	16	13	13	6	16	11	13	17	15	10	13
$k2$	4	15	5	12	11	10	12	9	7	15	4	10

1. Relative to two orthogonal vectors $\vec{\delta}_{k1}$ and $\vec{\delta}_{k2}$, with norms proportional to $\sqrt{\varpi_{k1}} = 3$ and $\sqrt{\varpi_{k2}} = 2$ respectively, represent the cloud of points.
2. Effectuate the biweighted PCA of protocol x^{IK}. Calculate the eigenvalues λ_1 and λ_2, the regression coefficients b_1^k and b_2^k of the initial variables on the two standard principal variables and the correlation coefficients r_1^k and r_2^k; the principal coordinates (y_1^i, y_2^i) of the 10 individuals and their qualities of representation on the two principal axes. Draw the two principal axes on the graph of question 1.
3. Let $\overline{x}^I = (9\,x^{Ik1} + 4\,x^{Ik2})/13$ be the Average variable. Calculate the regression coefficients of $\overline{x}^I$ on the standard principal variables.
4. Relative to two the standard principal variables, represent graphically the centered *initial variables*, the scaled principal variables and the centered Average variable. Relative to the principal axes, represent the *cloud of individuals* with its indicator ellipse.
5. Check that variables $\sqrt{\varpi_k}x_0^{Ik1}$ and $\sqrt{\varpi_k}x_0^{Ik2}$ are conjugate diameters of the principal ellipse of equation $(b_1^2/\lambda_1) + (b_2^2/\lambda_2) = 1$.

Solution

1.

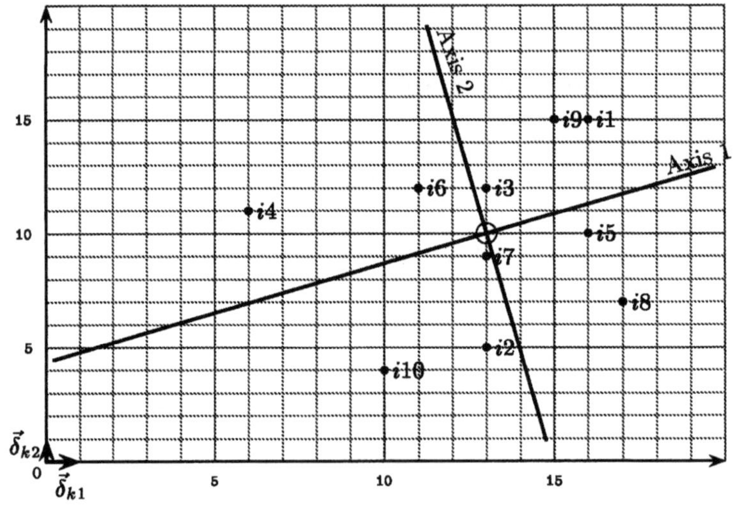

2. Covariance matrix $\mathbf{V} = \begin{bmatrix} 10 & 2 \\ 2 & 13 \end{bmatrix}$ Scalar product matrix $\mathbf{\Omega} = \begin{bmatrix} 9 & 0 \\ 0 & 4 \end{bmatrix}$ Matrix $\mathbf{V\Omega}$ $\mathbf{V\Omega} = \begin{bmatrix} 90 & 8 \\ 18 & 52 \end{bmatrix}$

The eigenvalues of $\mathbf{V\Omega}$ are $\lambda_1 = 93.4722$ and $\lambda_2 = 48.5278$; hence $\xi_1 = 9.7$ and $\xi_2 = 7.0$. The following tables give the two $\mathbf{\Omega}$–normed eigenvectors $(a_\ell^K)_{\ell=\ell1,\ell2}$ of $\mathbf{V\Omega}$, the regression coefficients $b_\ell^k = \sqrt{\lambda_\ell} a_\ell^k$ and the correlation coefficients $r_\ell^k = b_\ell^k / \operatorname{SD} x^{Ik}$.

	$\ell=1$	$\ell=2$
a_ℓ^{k1}	0.3202	−0.0927
a_ℓ^{k2}	0.1390	0.4803

	$\ell=1$	$\ell=2$
b_ℓ^{k1}	3.0957	−0.6454
b_ℓ^{k2}	1.3437	3.3459

	$\ell=1$	$\ell=2$
r_ℓ^{k1}	.979	−.204
r_ℓ^{k2}	.373	.923

One verifies $(b_{\ell1}^{k1})^2 + (b_{\ell2}^{k1})^2 = 10$ and $(b_{\ell1}^{k2})^2 + (b_{\ell2}^{k2})^2 = 13$.

Passage formulas (cf. Equations (4.5), p.135): $y_\ell^I = \sum_{k\in K} a_{k\ell} x_0^{Ik}$, with $a_{k\ell} = \varpi_k a_\ell^k$.

$y_1^I = (0.3202\times 9)x_0^{Ik1} + (0.1390\times 4)x_0^{Ik2}$; $y_2^I = -(0.0927\times 9)x_0^{Ik1} + (0.4803\times 4)x_0^{Ik2}$.

	$i1$	$i2$	$i3$	$i4$	$i5$	$i6$	$i7$	$i8$	$i9$	$i10$
y_1^I	11.4	−2.8	1.1	−19.6	8.6	−4.7	−0.6	9.9	8.5	−12.0
y_2^I	7.1	−9.6	3.8	7.8	−2.5	5.5	−1.9	−9.1	7.9	−9.0

Table of coordinates $(y_1^i, y_2^I)_{i\in I}$ (1 decimal place).

	$i1$	$i2$	$i3$	$i4$	$i5$	$i6$	$i7$	$i8$	$i9$	$i10$
Axis 1	.72	.08	.08	.87	.92	.42	.08	.54	.54	.64
Axis 2	.28	.92	.92	.13	.08	.58	.92	.46	.46	.36

Table of qualities of representation $(y_\ell^i)^2/((x_0^{ik1})^2 + (x_0^{ik2})^2)$.

3. The regression coefficients of the Average variable are $\bar{b}_\ell = \sum_{k\in K} \varpi_k b_\ell^k / \sum_{k\in K} \varpi_k$; hence $\bar{b}_1 = 2.557$; $\bar{b}_2 = 0.583$.

4.

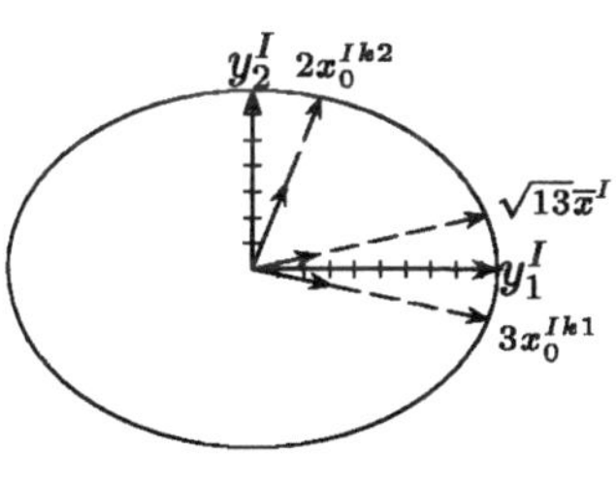

Space of variables.

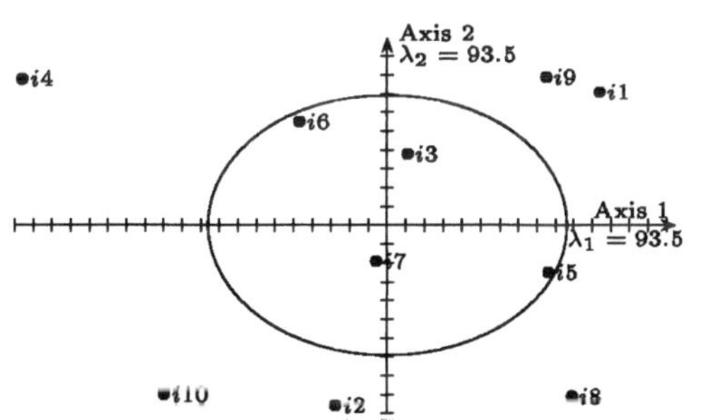

Cloud of individuals and indicator ellipse.

5. The coordinates of vectors $\sqrt{\varpi_{k1}} x_0^{Ik1}$ and $\sqrt{\varpi_{k2}} x_0^{Ik2}$ verify the ellipse equation, as well as the relation $(-0.6454/3.0957) \times (3.3459/1.3437) = -.519 = -\lambda_2/\lambda_1$, hence the conjugate diameter property (cf. *MATH.BASES*, Property 10.25, p.442).

Exercise 4.2 (Non–diagonal metric for two variables)

Consider two variables (x^J, x'^J), letting $v = \operatorname{Var} x^J$, $v' = \operatorname{Var} x'^J$ and $c = \operatorname{Cov}(x^J | x'^J)$, and assuming $c > 0$, $v > c$, and $v' > c$. Let the matrix of scalar products $\mathbf{Q} = \begin{pmatrix} v' - c - \gamma & \gamma \\ \gamma & v - c - \gamma \end{pmatrix}$, whith $\gamma \in [0, \gamma_0]$ and $\gamma_0 = \dfrac{(v-c)(v'-c)}{v + v' - 2c}$.

Prove the following properties:

• For all γ, the largest eigenvalue is $\lambda_1 = vv' - c^2$; it is associated with the Sum vector $\vec{\delta} + \vec{\delta'}$ and with the weighted Average variable $\overline{x}^J = ((v' - c)x^J + (v - c)x'^J)/(v + v' - 2c)$.

• The second eigenvalue is $\lambda_2 = (v - c)(v' - c) - \gamma(v + v' - 2c)$; it is associated with the Difference vector $(v - c)\vec{\delta} - (v' - c)\vec{\delta}'$ and with the Difference variable $x^J - x'^J$, which is uncorrelated with the Average variable (for any γ).

• When γ increases from 0 (biweighted PCA) to γ_0, λ_2 decreases from $(v-c)(v'-c)$ to 0; the ratio λ_2/λ_1 decreases from $(v - c)(v' - c)/(v + v' - c^2)$ to 0.

Comment

For $\gamma = 0$, $\mathbf{Q}$ is diagonal and the analysis is a biweighted PCA satisfying the Thurstone condition, with the property $\mathbf{VQe} = (vv' - c^2)\mathbf{e}$.
This exercise illustrates, for $K = 2$, the extension of the Thurstone condition to a non–diagonal metric, namely $\mathbf{VQe}$ proportional to $\mathbf{e}$; and shows how, under this condition, the main properties of biweighted PCA extend to the non–diagonal case.

Exercise 4.3 (Scotland Yard)

Table 1 provides the correlations between 7 anthropometric measurement variables carried out on a sample of 3000 criminals randomly selected from Scotland Yard files (London, 1902).

	*k*1	*k*2	*k*3	*k*4	*k*5	*k*6	*k*7	
*k*1	1							*Head Length*
*k*2	.40163	1						*Head Breadth*
*k*3	.39454	.61779	1					*Face Breadth*
*k*4	.30071	.15040	.32097	1				*Left Middle Finger*
*k*5	.30539	.13515	.28869	.84638	1			*Left Forearm (Cubit)*
*k*6	.33886	.20614	.36322	.75871	.79699	1		*Left Foot*
*k*7	.33993	.18308	.34527	.66084	.79986	.73636	1	*Height*

Table 1. Correlations between 7 anthropometric measurement variables (*Source*: Macdonell (1902)).

1. Comment on the correlation table.
2. Using Table 2, calculate the main results of the PCA, namely the proportion of variance accounted for by each principal variable, by the first two and by the first three, and calculate the correlations of the 7 initial variables with the first 3 principal ones. Calculate the qualities of fit of the initial variables by the first principal variable, by the first two and by the first three.

   How many axes should be retained for interpretation?
3. Calculate the variance of the Average variable (defined as the average of the 7 initial standardized variables) and the regression coefficients of this variable on the first 3 standardized principal variables.
4. Represent graphically the regressions of the standardized initial variables and of the Average variable on the standard principal variables 1 and 2, then on 1 and 3. In planes (1,2) and (1,3), draw the circle of correlations.

	λ_1 3.79939	λ_2 1.50127	λ_3 0.65097	λ_4 0.36029	λ_5 0.33920	λ_6 0.23529	λ_7 0.11360
$k1$	.27612	.36423	.88234	−.08652	.06936	.00506	−.01562
$k2$	.21189	.63957	−.25561	.68598	−.09318	.03391	.01763
$k3$	.29502	.51249	−.38255	−.69591	.11312	.03499	−.07530
$k4$	.43757	−.23486	−.07039	.11304	.61831	.31500	.50409
$k5$	.45577	−.27667	−.03578	.11498	.03918	.29001	−.78459
$k6$	.45028	−.17825	−.05932	.05167	.02937	−.87076	.01432
$k7$	.43574	−.17936	−.00642	−.09476	−.76747	.23676	.35196

Table 2. Eigenvalues (λ_ℓ) and normed eigenvectors $(a_\ell^k)_{k\in K}$ of the correlation matrix.

5. Interpret the first three principal variables.
6. *Thurstone weighting.* The results of biweighted PCA with weights $(\varpi_k)_{k\in K}$: (1.43955, 1.93108, 0.71061, 1.18068, 0.62429, 0.23542, 0.87837) are given in Table 3 (see *Comment*, p.170). Calculate the variance of the cloud. Effectuate a graphical representation of variables (cf. question 4). Proceed to the intepretation of these results as for the standard PCA. Calculate the correlation between the weighted Average variable $\widehat{x}$ and the plain Average variable $\overline{x}$ of standard PCA.

λ_1=3.41955; λ_2=1.84284; λ_3=0.91034; λ_4=0.3457; λ_5=0.33080; λ_6=0.08595; λ_7=0.06476

	r_1^k	r_2^k	r_3^k
$k1$ *Head Length*	.699	.145	.700
$k2$ *Head Breadth*	.699	.647	−.262
$k3$ *Face Breadth*	.699	.282	−.178
$k4$ *Left Middle F.*	.699	−.599	−.157
$k5$ *Left Forearm*	.699	−.622	−.135
$k6$ *Left Foot*	.699	−.482	−.111
$k7$ *Height*	.699	−.511	−.091
Weighted Average	.699	.000	.000
W. sum of squares	3.419	1.843	0.910

	$(R_1^k)^2$	$(R_{1,2}^k)^2$	$(R_{1,2,3}^k)^2$
$k1$	.489	.510	.999
$k2$	.489	.907	.975
$k3$	.489	.568	.600
$k4$	.489	.847	.872
$k5$	.489	.875	.893
$k6$	.489	.721	.734
$k7$	.489	.750	.758
$\widehat{k}$	1	1	1

Table 3. Eigenvalues, correlations and qualities of fit for the first three principal variables.

Solution

1. All correlations are positive (cf. Table 1). The 6 highest correlations ($.661 < \cdot < .847$) are those of the measurements of limbs and of *Height* ($k4, k5, k6, k7$). The other correlations are much inferior to the preceding ones ($< .405$), with the exception of the correlation between *Head* ($k2$) and *Face* ($k3$) (.618).

2. The percentages of variance accounted for by each principal variable are: 54.3% ; 21.4% ; 9.3% ; 5.1% ; 4.8% ; 3.4% ; 1.6%. The percentage accounted for by the first two is 75.7%; and by the first three 85.0%. To summarize the data (cf. §4.6.1, p.161), at least the first two principal variables, whose variances (eigenvalues) exceed 1 (average of eigenvalues), must be retained; they account together for 75.7% of the total variance. Still, two principal variables are not sufficient to ensure a satisfactory fit of *Head Length* ($k1$, $R^2 = .489$). Adding the third principal variable makes the quality of fit rise to .996 for $k1$; for the *Height* variable ($k7$, $R^2 = .770$), however, it does not improve the fit; getting a good fit for $k7$ would require adding the fifth principal variable (0.970).

	r_1^k	r_2^k	r_3^k
*k*1 *Head Length*	.538	.446	.712
*k*2 *Head Breadth*	.413	.784	−.206
*k*3 *Face Breadth*	.575	.628	−.309
*k*4 *Left Middle Finger*	.853	−.288	−.057
*k*5 *Left Forearm (Cubit)*	.888	−.339	−.029
*k*6 *Left Foot*	.878	−.218	−.048
*k*7 *Height*	.849	−.220	−.005
Average	.713	.113	.008
Sum of squares	3.799	1.501	0.651

	$(R_1^k)^2$	$(R_{1,2}^k)^2$	$(R_{1,2,3}^k)^2$
*k*1	.290	.489	.996
*k*2	.171	.785	.827
*k*3	.331	.725	.820
*k*4	.727	.810	.813
*k*5	.789	.904	.905
*k*6	.770	.818	.820
*k*7	.721	.770	.770
$\overline{k}$	.975	.999	1.000

Table 4. Correlations $r_\ell^k = \xi_\ell a_\ell^k$ between the 7 initial variables and the first 3 principal ones, and qualities of fit (squares of simple and multiple correlations) by the first one $(r_1^k)^2$, by the first two $(R_{1,2}^k)^2 = (r_1^k)^2 + (r_2^k)^2$, and by the first 3 principal variables $(R_{1,2,3}^k)^2 = (r_1^k)^2 + (r_2^k)^2 + (r_3^k)^2$.

3. The variance of the Average variable is equal to .52208 (the average of the 49 correlations, see Formula (4.15), p.140). The regression coefficients of the Average variable on the first three standard principal variables ($\sum_{k \in K} r_\ell^k/7$) are .713, .113, .008 (see Table 2); the correlation with the first principal variable: $.713/\sqrt{0.52208} = .987$, hence an angle of 9°1 (see Figure 1 of question 4).

4.

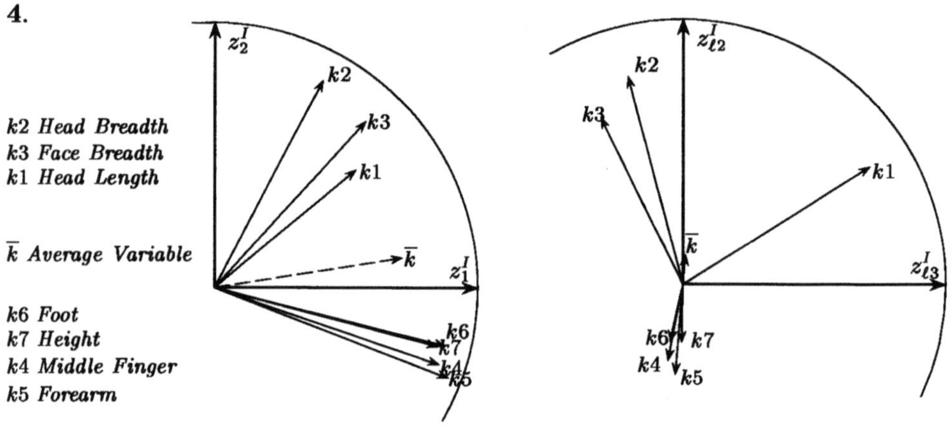

Figure 1. Circles of correlation (planes 1-2 and 2-3)

5. The correlations of the initial variables with the first principal variable are all positive, the average correlation being .713; and the correlation between the average variable and the first principal variable is high, therefore the ***first principal variable*** is interpreted as a "size factor"; in the cloud of the individuals (data not available), the 3000 individuals would tend to be ordered along Axis 1 from low to high measurement values.

For the ***second principal variable***, 3 correlations are positive and 4 are negative, the average being .113; this variable is as a "bipolar factor" that opposes "head" variables ($k2, k3, k1$) to "limb" ones ($k4, k5, k6$) and to *Height* ($k7$).

The *third principal variable* is also bipolar (average correlation .008), it opposes *Head Length* ($k1$) to all other variables, especially *Head Breadth* ($k2$) and *Face Breadth* ($k3$); the correlations of $k4$ to $k7$ with the 3-th principal variable are virtually null.

6. With Thurstone weighting, the percentages of variance accounted for by the first three principal variable are: 48.9%, 26.3% and 13.0% respectively; hence in cumulative terms: 48.9%, 75.2% and 88.2%. Three principal axes must be retained. The overall interpretation is very much the same as for the standard PCA, but the message is clearer. See Figure 2.

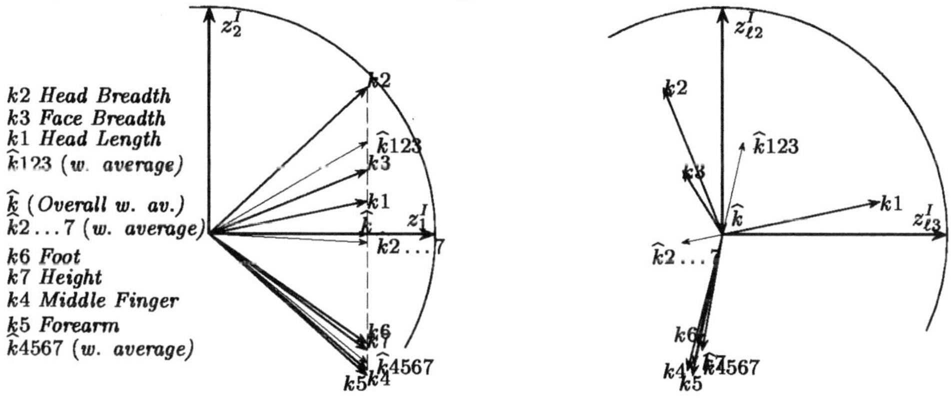

Figure 2. Variables in planes 1-2 and 2-3 (Thurstone weighting)

The weighted Average variable $\widehat{x}$ is exactly proportional to the first principal variable ("size factor") $\widehat{x} = (1.43955x^{k1} + \ldots + 0.87837x^{k7})/7 = 0.699z_1$, and it is exactly orthogonal to the subsequent principal variables. The correlations r_1^k of all variables x^k with the first principal variable are all equal to 0.699.

For the second principal variable ("bipolar factor") we have exactly $\sum \varpi_k r_2^k = 0$: $1.43955 \times 0.1448 + \ldots - 0.87837 \times 0.5108 = 0$; hence $4.08124r_2^{123} + 2.91876r_2^{4567} = 0$, letting $r_2^{123} = (1.43955 \times 1.1448 + \ldots + 0.71061 \times 0.2815)/4.08124 = 0.4062$ (weighted average of regression coefficients of variables $k1$, $k2$ and $k3$); for variables $k4$, $k5$, $k6$ and $k7$: $r_2^{4567} = (-1.18068 \times 0.5988 + \ldots - 0.87837 \times 0.5108)/2.91876 = -0.5678$.

Similarly, for the third principal variable we have exactly $\sum \varpi_k r_3^k = 0$, and we have $1.43955\, r_3^1 + 5.56045\, r_3^{234567} = 0$, with $r_3^1 = 0.6998$ and $r_3^{234567} = -0.1812$.

The weighted Average variables $\widehat{k}$, $\widehat{k}123$, $\widehat{k}4567$ and $\widehat{k}234567$ are shown in plane 1 2 and in plane 1-3. In the latter plane, the vectors $\widehat{k}123$ and $\widehat{k}4567$ are collinear and opposite in direction, as well as $k1$ and $\widehat{k}2\ldots7$. In plane 2-3, the variables may also be represented as a cloud of weighted points. See opposite Figure.

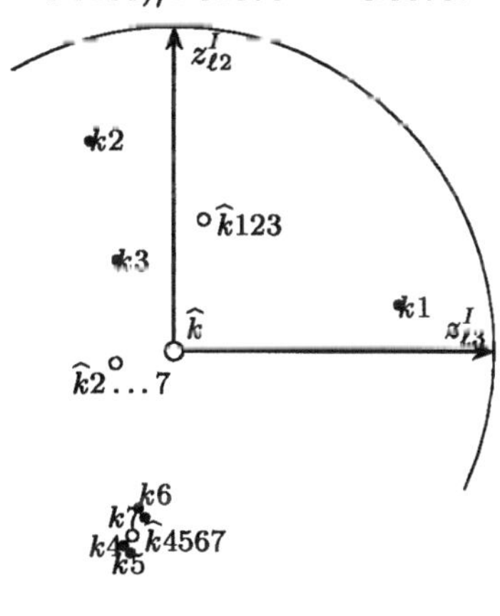

The correlation between the weighted Average variable $\widehat{x}$ and the plain average $\overline{x}$ of the standard PCA is found to be .967.

Comments and complements

1. The data of this exercise are those of the historical paper by Macdonell (1902), one of Karl Pearson's collaborators at the Galton Laboratory. Using the available anthropometric measurements of criminals as a database, Macdonell sought "the best manner in which these measurements can be applied for the identification of criminals."
After performing various regressions, Macdonell declares that according to K. Pearson, "the ideal index characters would be given if we calculated ... the principal axes of the correlation ellipsoid." The effective realization of this project was deferred by Macdonell to an unspecified future.

2. *Thurstone weighting.* Given any non-singular correlation matrix $\mathbf{R}$, one may calculate coefficients $(\varpi_k)_{k\in K}$ such that (in matrix notation) the column $\boldsymbol{\varpi}$ is proportional to $\mathbf{R}^{-1}\mathbf{e}$ (where $\mathbf{e}$ is the column of ones). Whenever $\varpi_k > 0$ for all $k \in K$, these coefficients can be taken as basic weights; Thurstone condition is then satisfied, and the associated principal variable is proportional to the weighted Average variable $\widehat{x}$ of the standard variables $(x^k)_{k\in K}$: $\widehat{x} = (\sum \varpi_k\, x^k)/\varpi$.
One further has $\mathrm{Var}\,\widehat{x} = 1/\mathbf{e}'\mathbf{R}^{-1}\mathbf{e} = \mathrm{Cov}(\widehat{x}|x^k)$ for all $k \in K$, and the correlation between $\widehat{x}$ and x^k is $1/(\mathbf{e}'\mathbf{R}^{-1}\mathbf{e})^{-1/2}$ for all k.
If more precisely we take $\boldsymbol{\varpi} = \frac{K}{\mathbf{e}'\mathbf{R}^{-1}\mathbf{e}}\mathbf{R}^{-1}\mathbf{e}$, then the sum of weights ϖ is equal to the number of variables, like in standard PCA: $\varpi = \boldsymbol{\varpi}'\mathbf{e} = K$, and the associated eigenvalue is $\lambda = K/(\mathbf{e}'\mathbf{R}^{-1}\mathbf{e})$.
The high correlation between the weighted average and the plain Average variables illustrates the definite imperviousness of the average with respect to differential weighting, a well–known fact among psychometricians; see e.g. Guttman (1941).

3. The means and standard deviations (SD) of the 7 measurements (in centimeters) are also reported in Macdonell's paper:

	$k1$	$k2$	$k3$	$k4$	$k5$	$k6$	$k7$
Means	19.1663	15.0442	13.6500	11.5474	45.0586	25.6877	166.4572
SD	0.6046	0.5014	0.5060	0.5479	1.9627	1.1792	6.4540

The total variance is equal to 48.069; the contribution of $k7$ (Height) to the total variance is equal to 86.7%.
From the SD and correlations, the variances and covariances can be reconstituted, and a PCA of the covariance matrix can be performed. If the analysis is done, the first principal variable is found to represent 94.1% of the total variance of the cloud (sum of the seven variances), that is, the cloud is practically one–dimensional, and Variable *Height* ($k7$) has overwhelming contribution (91.8%) and quality of representation (.997). Such results are expected from the data, since Variable *Height* (hence its SD) largely exceeds all the others in magnitude.
As a conclusion, the physical measurement scale (centimeters) is simply not appropriate for yielding a non–trivial summary of the variables.

Exercise 4.4 (Spearman data)

The table p.171 gives the correlations between 6 performance variables of English pupils in the following subjects: *Literature* ($k1$), *French* ($k2$), *English* ($k3$), *Mathematics* ($k4$), *Auditive discrimination* ($k5$), *Music* ($k6$). The eigenvalues (λ_ℓ) and vectors $(a_\ell^k)_{k\in K}$ of the correlation matrix are given in the following table.

	$k1$	$k2$	$k3$	$k4$	$k5$	$k6$
$k1$	1					
$k2$	.83	1				
$k3$	.78	.67	1			
$k4$	.70	.67	.64	1		
$k5$	.66	.65	.54	.45	1	
$k6$	.63	.57	.51	.51	.40	1

Source: Spearman (1904)

	λ_1	λ_2	λ_3	λ_4	λ_5	λ_6
	4.103	0.619	0.512	0.357	0.270	0.139
$k1$	.462	−.033	−.027	−.128	−.266	−.835
$k2$	.441	−.119	.007	.228	−.734	.448
$k3$	.416	−.003	−.342	−.765	.191	.298
$k4$	.397	.255	−.558	.567	.380	.019
$k5$	.367	−.712	.388	.157	.425	.059
$k6$	.356	.643	.648	−.018	.171	.099

1. Calculate the proportions of variance accounted for by each principal axis. From the first eigenvalue and the 6 coefficients of the first principal variable $y = \sum a_k\, x^k$, calculate the correlation coefficients $(b^k)_{k\in K}$ between the initial variables $(x^k)_{k\in K}$ and the first principal variable y. How many axes should be retained?

2. One considers the Average variable $\overline{x}$ of the 6 (standard) initial variables x^k; calculate the variance of the Average variable, and its covariance and correlation with the first principal variable y. Interpret the first axis.

3. From the first principal variable, reconstitute the correlation table of the 6 initial variables. Calculate the residual deviations, and the sum of squares of the deviations, both for diagonal and off–diagonal terms.

4. A classical Factor Analysis (see *Comment* below) has been performed on data, yielding a single common factor, denoted g, with the following six factor loadings $(c^k)_{k\in K}$ (i.e. correlations between factor g and the six initial variables): (.96, .89, .80, .75, .67, .64). From the loadings $(c^k)_{k\in K}$, calculate the correlation between variables g and y (first principal variable of PCA), using the formula $\text{corr}\,(g|y) = \sum c^k\, a_k/\sqrt{\lambda}$; then reconstitute the correlation table by the multiplicative formula $\widehat{r}^{kk'} = c^k\, c^{k'}$, and compare the reconstitution to the one of the preceding question.

5. In order to predict variable g from the observed $(x^k)_{k\in K}$ variables, construct the predictor variable $\widetilde{g} = \sum \beta_k x^k$, from the preceding 6 factor loadings and the correlation table. Calculate $\text{Var}\,\widetilde{g}$ (quality of representation of variable $\widetilde{g}$), the correlation between $\widetilde{g}$ and the first principal variable y, and the correlation between variable $\widetilde{g}$ and the Average variable $\overline{x}$.

Solution

1. Only the first eigenvalue exceeds 1, by a large amount. The percentages of variance of the first 3 axes are 68.4%, 10.3% and 8.5%. The correlation between variable x^k and the first principal variable is $b^k = \sqrt{\lambda_1}\, a_k$ (example for $k1$: $\sqrt{4.103}\times 0.462 = .936$), hence the 6 correlations $(b^k)_{k\in K}$: (.936, .893, .843, .804, .743, .721). These results suggest retaining one axis to summarize data.

2. The variance of the Average variable $\overline{x}$ is equal to .6783 (average of the 36 correlations); its covariance with the first standard principal variable is .8233 (average of the 6 preceding correlations); hence the correlation $.8233/\sqrt{.6783} = 1.000$. Thus not only the first principal variable of PCA is interpreted as a size factor, but it is virtually proportional to the Average variable $\overline{x}$.

3. From the correlations $(b^k)_{k\in K}$ between the initial variables and the first principal variable, one reconstitutes the correlations by the formula $\widetilde{r}^{kk'} = b^k b^{k'}$, hence the following reconstitution table of correlations and residual deviation table.

	$k1$	$k2$	$k3$	$k4$	$k5$	$k6$
$k1$	.876					
$k2$	.836	.798				
$k3$	.789	.753	.710			
$k4$	.753	.718	.678	.647		
$k5$	.696	.664	.626	.598	.553	
$k6$	.675	.644	.608	.580	.536	.520

$k1$	$k2$	$k3$	$k4$	$k5$	$k6$
+.124					
−.006	+.202				
−.009	−.083	+.290			
−.053	−.048	−.038	+.353		
−.036	−.014	−.086	−.148	+.447	
−.045	−.074	−.098	−.070	−.136	+.480

The sum of squares of the 36 residual terms is $\sum_{\ell=2}^{6} \lambda_\ell^2 = 0.865$, that is, 0.695 (diagonal terms) +0.170 (off–diagonal).

4. The correlation between variables g and y is $\text{corr}(g|y) = (.462\times.96+\ldots+.356\times .64)/\sqrt{4.102} == 0.96$. Factor g of Factor Analysis is seen to be highly correlated with the first principal variable y of PCA.

The reconstitution of correlations from the factor loadings of Factor Analysis and the residual deviations are given by the following tables:

	$k1$	$k2$	$k3$	$k4$	$k5$	$k6$
$k1$	.922					
$k2$	.854	.792				
$k3$	.768	.712	.640			
$k4$	.720	.668	.600	.562		
$k5$	.643	.596	.536	.503	.449	
$k6$	.614	.570	.512	.480	.429	.410

$k1$	$k2$	$k3$	$k4$	$k5$	$k6$
+.078					
−.024	+.208				
+.012	−.042	+.360			
−.020	+.002	+.040	+.438		
+.017	+.054	+.004	−.052	+.551	
+.016	+.000	−.002	+.030	−.029	+.590

The sum of squares of the 36 residual terms is 1.047, that is, 1.022 (diagonal terms)+0.025 (off–diagonal). The overall fit is not as good as the previous one; but the fit of the off–diagonal terms (correlations) is much better. The correlations are systematically underestimated by PCA — all residuals are negative, one of them reaches .148 in absolute value —, whereas with Factor Analysis some residuals are negative and others positive, all of them lying between −.052 and +.054.

5. The coefficients $(\beta_k)_{k\in K}$ verify the regression equation $\sum r^{kk'}\beta_{k'} = c^k$, that is, in matrix notation: $\mathbf{R}\boldsymbol{\beta} = \mathbf{c}$.

Solving the equation yields the β_k coefficients (.585, .258, .089, .079, .021, .035). One finds $\text{Var}\,\widetilde{g} = .958$. The correlation between $\widetilde{g}$ and the first principal variable is found to be equal to .978, and the correlation with the Average variable is .974.

Comments and complements

1. The data of this exercise are those of the historical example of Spearman (1904), they are the first illustration of ***classical Factor Analysis***, that is, of the analysis with common and specific factors, that has been widely used to analyze psychometric data.

2. In Spearman's data, the variables can be arranged so that in each column of the correlation table, the values tend to decrease when one goes away from the diagonal — the sole exception being variable $k4$ which is less correlated with $k5$ than with $k6$. Such a correlation table is said to be ***hierarchical***.

3. In Factor Analysis, hierarchical correlation tables correspond to a single common factor structure. Each one of the initial standard variables $(x^k)_{k\in K}$ is broken down into two terms as follows: $\forall k : x^k = c^k\, g + e^k$. In the first term, g (standard variable) denotes the (here unique) ***common factor***; the second term $(e^k)_{k\in K}$ denotes the ***specific factor*** associated with variable k. Specific factors are centered variables and are supposed to be uncorrelated among themselves and with the g factor. Coefficient c^k is the correlation between x^k and g and is called the ***factor loading*** of variable k in factor g. The reconstitution of the off–diagonal correlations ($k \neq k'$) is done from the factor loadings by the formula $\widetilde{r}^{kk'} = c^k\, c^{k'}$ for $k \neq k'$; whereas the square of the loading of variable k, i.e. $(c^k)^2$ is not meant to reconstitute 1, but the proportion of variance of variable k accounted for by the common factor. Conditionally on factor g, the correlations between the initial variables vanish, that is, factor g accounts for the observed (unconditional) correlations between the initial variables, hence the status of a "general factor" conferred to factor g in psychometric theory.

4. From Burt to Thurstone, Factor Analysis has been extended to several common factors, and in the psychometric tradition, it is commonplace to contrast Principal Component Analysis to Factor Analysis. Thus according to Maxwell (1977), PCA should be preferred for data with low error measurements, such as anthropological measurements (e.g. exercise 4.3), and Factor Analysis, for data where individual variability does not reflect measurement errors, such as achievement scores (e.g. the present exercise).

Exercise 4.5 (Doubling technique)

Let $(x^I_q)_{q\in Q}$ be a family of Q numerical variables over a finite set I (with elementary weighting), such that each variable q is bounded on both sides by a pair of numbers (u_q, v_q) with $v_q > u_q$, i.e. for each $\forall i\ :\ u_q \leq x^i_q \leq v_q$. The doubling technique consists in replacing each variable x^I_q by the two variables $x^i_{tq} = x^i_q - u_q$ and $x^i_{t'q} = v_q - x^i_q$; hence the doubled table x^I_K, with $K = T \times Q$ letting $T = \{t, t'\}$. The doubled table will be analyzed as a table of measures (cf. *PRINCIPAL*, §4.5, p.155); or equivalently, letting $m_{ik} = x^i_k$ (since $\forall i \in I : n_i = 1$), the associated m_{IK} table will be analyzed by CA as a measure over $I \times K$.
The following notation will be used:
$e_q = v_q - u_q$; $\overline{e} = \frac{1}{Q} \sum\limits_{q\in Q} e_q$; $\overline{x}_q = \sum\limits_{i\in I} x^i_q / n$ (mean of variable x^I_q).

1. Determine $m_i = \sum\limits_{k\in K} m_{ik}$, $m_k = \sum\limits_{i\in I} m_{ik}$ (for $k \in T \times Q$), and the total mass m of m_{IK}, as fonctions of $\overline{e}$, u_q, v_q, e_q and $\overline{x}_q$.

   Show that the square of Φ^2–distance between the profiles of individuals i and i' is $(\mathrm{M}^i\mathrm{M}^{i'})^2 = \sum\limits_{q\in Q} \frac{1}{\pi_q}(x^i_q - x^{i'}_q)^2$, with $\pi_q = \frac{Q\overline{e}}{e_q}(v_q - \overline{x}_q)(\overline{x}_q - u_q)$.

2. Show that the CA of the doubled table is equivalent to the biweighted PCA of the table of measures $(x_Q^I, 1_Q, \pi_Q)$, in the sense that the eigenvalues λ_ℓ and the ℓ–th principal variables over I are equal.

   Show that $\forall q : (\overline{x}_q - u_q)\overrightarrow{\mathrm{GM}}^{tq} + (v_q - \overline{x}_q)\overrightarrow{\mathrm{GM}}^{t'q} = \overrightarrow{0}$, and therefore that points M^{tq}, $\mathrm{M}^{t'q}$ and G are collinear. Show further that $\overrightarrow{\mathrm{GM}}^{tq}$ represents the variable $\frac{1}{\overline{x}_q - u_q} x_{0q}^I$, hence that $\mathrm{GM}^{tq} = \frac{1}{\overline{x}_q - u_q}\ \mathrm{SD}\, x_q^I$, $\mathrm{GM}^{t'q} = \frac{1}{v_q - \overline{x}_q}\ \mathrm{SD}\, x_q^I$, and $\mathrm{M}^{tq}\mathrm{M}^{t'q} = \frac{e_q}{(v_q - \overline{x}_q)(\overline{x}_q - u_q)}\ \mathrm{SD}\, x_q^I$. Express the variables y_ℓ^K of CA as functions of the regression coefficients $b_\ell^q = \mathrm{Cov}(x_q^I | z_\ell^I)$ of the PCA.

3. Examine the particular case where the x_q^I variables are ***rank variables***, taking values from 1 to n.

4. Examine the particular case where the x_q^I variables are ***dichotomous variables*** encoded in 1 and 0. Show that the CA of the doubled table is equivalent, for the principal variables over I, to the standard PCA of the $I \times Q$ table, with the eigenvalues of the doubled CA being equal to $1/Q$ times those of PCA. Compare the two analyses in the space of variables.

Solution

1. $m_i = Q\overline{e}$; $m_{tq} = n\,(\overline{x}_q - u_q)$; $m_{t'q} = n\,(v_q - \overline{x}_q)$; $m = nQ\overline{e}$. Hence $(\mathrm{M}^i\mathrm{M}^{i'})^2 = \sum\limits_{k\in K} \frac{\left((m_{ik}/m_i) - (m_{i'k}/m_{i'})\right)^2}{m_k/m} = \frac{1}{Q\overline{e}} \sum\limits_{k\in K} \frac{(m_{ik} - m_{i'k})^2}{m_k/n} = \frac{1}{Q\overline{e}} \sum\limits_{k\in K} \frac{(x_k^i - x_k^{i'})^2}{m_k/n} = \frac{1}{Q\overline{e}} \sum\limits_{q\in Q} (x_q^i - x_q^{i'})^2 \left(\frac{1}{v_q - \overline{x}_q} + \frac{1}{\overline{x}_q - u_q}\right) = \sum\limits_{q\in Q} \frac{1}{\pi_q}(x_q^i - x_q^{i'})^2$, with $\pi_q = \frac{Q\overline{e}}{e_q}(v_q - \overline{x}_q)(\overline{x}_q - u_q)$.

2. Since the distances between the points representing individuals in the biweighted PCA and the CA of the doubled table are equal, the principal directions are the same. The two clouds are weighted by $(1/n)_{i\in I}$, the standard principal variables are equal.

Applying the transition formula $y_\ell^k = \sum\limits_{i\in I} z_\ell^i \frac{m_{ik}}{m_k}$, one has $y_\ell^{tq} = \sum\limits_{i\in I} z_\ell^i \frac{x_{tq}^i}{n(\overline{x}_q - u_q)} = \frac{1}{\overline{x}_q - u_q}\mathrm{Cov}(x_{tq}^I | z_\ell^I)$, that is, $y_\ell^{tq} = b_\ell^q/(\overline{x}_q - u_q)$, and $y_\ell^{t'q} = -b_\ell^q/(v_q - \overline{x}_q)$. These relations can also be obtained from the representation of $\overrightarrow{\mathrm{GM}}^{tq}$ as $\frac{1}{\overline{x}_q - u_q} x_{0q}^I$, by projecting onto the ℓ–th axis.

3. *Rank data* (assuming no ties). One has $u_q = 0$, $v_q = n + 1 = e_q = \overline{e}$, $x_{tq}^i = x_q^i$, $x_{t'q}^i = n + 1 - x_q^i$, $\overline{x}_q = (n+1)/2$, $\mathrm{Var}\, x_q^I = (n^2 - 1)/12$, $\pi_q = Q(n+1)^2/4$. The CA of the doubled table is equivalent (up to a multiplicative factor) to a simple PCA, or to a standard PCA (since the variances are equal).

4. *Dichotomous variables.* One has: $u_q = 0$, $v_q = 1$, $\overline{x}_q = f_q$ (relative frequency of value 1), $\mathrm{Var}\, \delta_q^I = f_q(1 - f_q)$. Therefore $\pi_q = Qf_q(1 - f_q)$.

The distance $d(i, i')$ of the CA of the doubled table, i.e. of the PCA of the table of measures $(\delta_Q^I, 1^I, \pi_Q)$, is defined by $d^2(i, i') = \frac{1}{Q} \sum\limits_q \frac{(\delta_q^i - \delta_q^{i'})^2}{f_q(1 - f_q)}$. Let us now consider the standardized variable associated with δ_q^I, that is, the variable $\frac{\delta_{0q}^I}{\sqrt{f_q(1 - f_q)}}$;

the square of distance of the associated PCA (i.e. of the standard PCA of δ^I_Q) is $\sum_q \frac{(\delta^i_q-\delta^{i'}_q)^2}{f_q(1-f_q)}$. As a conclusion, the principal variables over I defined by the CA of the doubled table are proportional to those of the standard PCA of the $I \times Q$ table of indicator variables.
One further has $\mathrm{GM}^{tq} = \sqrt{\frac{1-f_q}{f_q}} = \frac{1}{f_q}\,\mathrm{SD}\,\delta^I_q$, $\mathrm{GM}^{t'q} = \sqrt{\frac{f_q}{1-f_q}} = \frac{1}{1-f_q}\,\mathrm{SD}\,\delta^I_q$, and $\mathrm{M}^{tq}\mathrm{M}^{t'q} = 1/\sqrt{f_q(1-f_q)} = \frac{1}{\mathrm{SD}\,\delta^I_q}$. Thus in the CA of the doubled table, variable δ^I_q is represented by $\frac{1}{f_q}\delta^I_{0q}$, a vector proportional to the standard PCA representation (standard variable) by a proportionality factor equal to $\sqrt{(1-f_q)/f_q}$.

Comment

The doubling technique — together with the disjunctive coding leading to MCA — has played a major role in the extension of the CA to various types of data.
The measure vs variable duality helps to make sense of this technique. Thus in the doubled table x^I_K, grouping of t and t' entails for each q summing on T; if furthermore, given t, grouping on Q entails summing on x^i_{tq} (and symmetrically on $x^i_{t'q}$), this suggests assimilating each profile x^i_K to a measure over K.
Two important applications of the doubling technique are the analysis of score data (Exercise 4.6), and the geometric study of Guttman scales (see *MULTIPLE*, Exercise 5.4, p.245).

Exercise 4.6 (Doubling scoring)

In this exercise, the doubling technique is applied to a biweighted protocol of scores, denoted here $(x^{IQ}, 1_I, \varpi_Q)$; with all scores lying on the same interval $[0, a]$. The double scoring procedure consists in constructing the associated m_{IK} doubled table and analyzing this table by CA. Let $\forall q \in Q$: $\mathrm{Mean}\, x^{Iq} = \overline{x}^q$; $x^I_q = \varpi_q\, x^{Iq}$, $u_q = 0$, $v_q = \varpi_q\, a$, hence $e_q = \varpi_q\, a$, $\overline{e} = \frac{a}{Q}\varpi$ with $\varpi = \sum \varpi_q$.

1. Show that the doubling procedure is equivalent to a biweighted PCA on $(x^{IQ}, 1_I, \varpi'_Q)$ (in the sense that the eigenvalues and the principal variables over I coincide), with the new weighting $\varpi'_q = \frac{\varpi_q}{\varpi}\frac{1}{(a-\overline{x}^q)\overline{x}^q}$.

   Show that $\overline{x}_q\,\overrightarrow{\mathrm{GM}}^{tq} + (a-\overline{x}_q)\,\overrightarrow{\mathrm{GM}}^{t'q} = \overrightarrow{0}$, hence that $\mathrm{GM}^{tq} = \frac{1}{\overline{x}_q}\ \mathrm{SD}\, x^I_q$, $\mathrm{GM}^{t'q} = \frac{1}{a-\overline{x}_q}\ \mathrm{SD}\, x^I_q$, and $\mathrm{M}^{tq}\mathrm{M}^{t'q} = \frac{a}{(a-\overline{x}_q)\overline{x}_q}\ \mathrm{SD}\, x^I_q$.

   Examine the particular case where $\forall q \in Q$: $\overline{x}_q = \overline{x}$.

2. *Application.* Consider the data table of Exercise 4.1 with two variables denoted $(x^{Iq}, x^{Iq'})$, with all scores in the interval $[0, 20]$.

   Starting with the weights $(\varpi_q = 9, \varpi_{q'} = 4)$ like in Exercise 4.5, construct the associated doubled table, and calculate the new weights $(\varpi'_q, \varpi'_{q'})$ of the equivalent PCA.

x^{Iq}	$i1$	$i2$	$i3$	$i4$	$i5$	$i6$	$i7$	$i8$	$i9$	$i10$	$\overline{x}^q$	Var – Cov	
q	16	13	13	6	16	11	13	17	15	10	13	10	2
q'	15	5	12	11	10	12	9	7	15	4	10	2	13

3. The CA of the doubled table yields the following results:

	$i1$	$i2$	$i3$	$i4$	$i5$	$i6$	$i7$	$i8$	$i9$	$i10$
$y^I_{\ell 1}$	.321	−.068	.027	−.579	.254	−.142	−.014	.298	.237	−.335
$y^I_{\ell 2}$	.205	−.269	.108	.203	−.064	.150	−.054	−.246	.226	−.259

	tq	$t'q$	tq'	$t'q'$
$y^K_{\ell 1}$	.240	−.445	.123	−.123
$y^K_{\ell 2}$	−.042	.077	.339	−.339

Eigenvalues $\lambda_1 = 0.07851$; $\lambda_2 = 0.03757$.

Perform the equivalent biweighted PCA and verify those results. Calculate the regression coefficients b^q_ℓ and the correlation coefficients r^q_ℓ. In the space of variables, compare the representation of PCA to the one of the doubled CA. Represent graphically the vector variables $(x^{Iq}, x^{Iq'})$ of PCA, and the points $(\mathrm{M}^{tq}, \mathrm{M}^{t'q}, \mathrm{M}^{tq'}, \mathrm{M}^{t'q'})$ of CA; in the latter graph, plot the individuals as a simultaneous representation.

Solution

1. From Exercise 4.5, one has $\pi_q = \frac{Q\,\overline{e}}{e_q}(v_q - \overline{x}_q)(\overline{x}_q - u_q)$ hence $\pi_q = \varpi(a - \overline{x}^q)\varpi_q\overline{x}^q$. Then one has $(\mathrm{M}^i\mathrm{M}^{i'})^2 = \sum\limits_{q\in Q} \varpi'_q(x^{iq} - x^{i'q})^2 = \sum\limits_{q\in Q} \frac{1}{\pi_q}(x^i_q - x^{i'}_q)^2 = \sum\limits_{q\in Q} \frac{\varpi_q^2}{\pi_q}(x^{iq} - x^{i'q})^2$ therefore $\varpi_q^2/\pi_q = \varpi'_q = (\varpi_q/\varpi)\frac{1}{(a-\overline{x}^q)\overline{x}^q}$. In the particular case $\forall q \in Q : \overline{x}_q = \overline{x}$, one has $\varpi'_q = \frac{\varpi_q}{\varpi}\frac{1}{(a-\overline{x})\overline{x}}$, that is, ϖ'_q is proportional to ϖ_q. The relations involving M^{tq} and $\mathrm{M}^{t'q}$ follow from the preceding exercise.

2. *Application*: $x^I_q = 9x^{Iq}$, $u_q = 0$, $v_q = 180$, $x^I_{q'} = 4x^{Iq'}$, $u_{q'} = 0$, $v_{q'} = 80$, hence the doubled table:

	$i1$	$i2$	$i3$	$i4$	$i5$	$i6$	$i7$	$i8$	$i9$	$i10$
tq	144	117	117	54	144	99	117	153	135	90
$t'q$	36	63	63	126	36	81	63	27	45	90
tq'	60	20	48	44	40	48	36	28	60	16
$t'q'$	20	60	32	36	40	32	44	52	20	64
	260	260	...						...	260

Weights of the equivalent PCA: $\varpi'_q = \frac{9}{13}\frac{1}{(20-13)\times 13} = \frac{9}{1183} = 0.76078 \times 10^{-2}$, $\varpi'_{q'} = \frac{4}{13}\frac{1}{(20-10)\times 10} = \frac{4}{1300} = 0.30769 \times 10^{-2}$. We remark that the ratio of the new weights $\frac{\varpi'_q}{\varpi q'} = 2.47$ is not far from the ratio of the initial ones $\frac{\varpi_q}{\varpi_{q'}} = 2.25$.

3. *Equivalent biweighted* PCA.

Covariance matrix — new scalar product matrix (new weights) — matrix $\mathbf{V}\widetilde{\mathbf{\Omega}}$

$$\mathbf{V} = \begin{bmatrix} 10 & 2 \\ 2 & 13 \end{bmatrix} \qquad \widetilde{\mathbf{\Omega}} = \begin{bmatrix} \frac{9}{1183} & 0 \\ 0 & \frac{4}{1300} \end{bmatrix} \qquad \mathbf{V}\widetilde{\mathbf{\Omega}} = \begin{bmatrix} \frac{90}{1183} & \frac{8}{1300} \\ \frac{18}{1183} & \frac{52}{1300} \end{bmatrix}$$

Eigenvalues. The eigenvalues of $\mathbf{V}\widetilde{\mathbf{\Omega}}$ are found to be λ_1=0.07851 and λ_2=0.03757, that is, they are equal to those of the CA of the doubled table.

Principal variables over I.

In the line of Exercise 4.1 (p.164), we compute the $\widetilde{\mathbf{\Omega}}$–normed eigenvectors of $\mathbf{V}\widetilde{\mathbf{\Omega}}$ that are found to be

	$\ell = 1$	$\ell = 2$
a^q_ℓ	11.1192	−2.7940
$a^{q'}_\ell$	4.3934	17.4842

Then we have $y_\ell^I = \sum_k a_\ell^q \varpi_q x_0^{Iq}$.

Therefore $y_{\ell 1}^I = (11.1192 \times \frac{9}{1183}) x_0^{Iq} + (4.3934 \times \frac{4}{1300}) x_0^{Iq'}$. Thus for $i1$:
$y_{\ell 1}^{i1} = (11.1192 \times \frac{9}{1183}) \times (16 - 13) + (4.3934 \times \frac{4}{1300}) \times (15 - 10) = .321$.
This value is seen to be equal to the value in the CA of the doubled table.
And similarly for $y_{\ell 2}^I$.

Regression and correlation coefficients.
We further compute the regression coefficients $b_\ell^q = \sqrt{\lambda_\ell} a_\ell^q$, and the correlation coefficients $r_\ell^q = b_\ell^q / \mathrm{SD}\, x^{Iq}$; hence the following tables:

	$\ell = 1$	$\ell = 2$
b_ℓ^q	3.1155	−0.5416
$b_\ell^{q'}$	1.2310	3.3889

and

	$\ell = 1$	$\ell = 2$
r_ℓ^q	.985	−.171
$r_\ell^{q'}$	.341	.940

Verifications: ${b_{\ell 1}^q}^2 + {b_{\ell 2}^q}^2 = 10 = \mathrm{Var}\, x^{Iq}$; and ${b_{\ell 1}^{q'}}^2 + {b_{\ell 2}^{q'}}^2 = 13 = \mathrm{Var}\, x^{Iq'}$.

Comparisons in the space of variables.

In PCA, we have for the x^{Iq} variable: $b_{\ell 1}^q = 3.1155, b_{\ell 2}^q = -0.5416$. Hence for the doubled CA: $y_{\ell 1}^{tq} = \frac{1}{\bar{x}^q} b_{\ell 1}^q = \frac{1}{13} \times 3.1155 = .240$, and $y_{\ell 2}^{tq} = \frac{1}{13} \times (-.5416) = -.042$, which agrees with the numerical results of CA. And similarly for $x^{Iq'}$.

In the space of variables the two representations are not equivalent, since $\overrightarrow{\mathrm{GM}}^{tq}$ corresponds to $\frac{1}{13} x_0^{Iq}$, and $\overrightarrow{\mathrm{GM}}^{tq'}$ to $\frac{1}{10} x_0^{Iq'}$. The figure hereafter shows the two graphs in the space of variables for PCA and for CA; the differences between the two representations are apparent.

In the CA graph, we have simultaneously plotted the cloud of individuals, whose principal coordinates are obtained by the transition formula from those of the four points ($\mathrm{M}^{tq}, \mathrm{M}^{t'q}, \mathrm{M}^{tq'}, \mathrm{M}^{t'q'}$). Thus for point $i1$ and Axis $\ell 1$ we have:
$\frac{1}{260} \times (144 \times 0.240 - 36 \times 0.445 + 60 \times 0.123 - 20 \times 0.123) \times \frac{1}{\sqrt{0.07851}} = .321 = y_{\ell 1}^{i1}$.
The (expected) similarity between this cloud of individuals and the one of Exercise 4.1 is apparent.

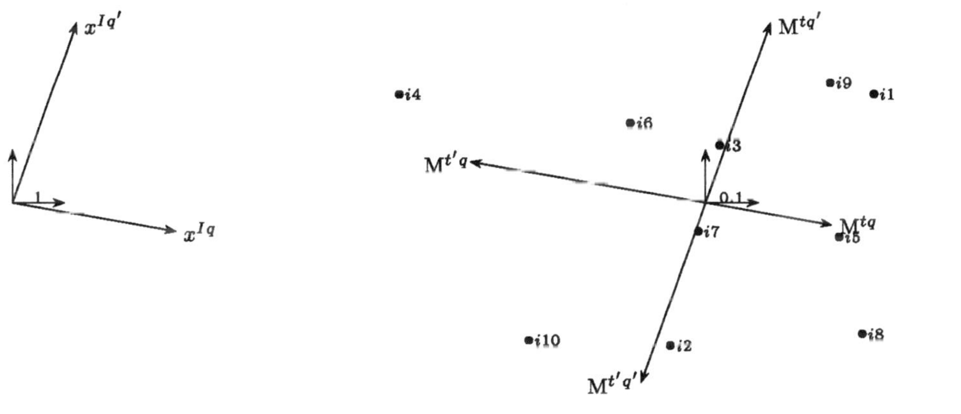

Space of variables: PCA (left) and doubled CA (right)

Comment

This investigation of variants of CA and PCA shows the crucial importance of "the explicit choice of masses and of distance formula" (Benzécri, 1992, p.55). It also shows that variants that are equivalent for the representation of individuals may differ in the space of variables. For bounded scores, like those of tests or exams — as opposed e.g. to unbounded physical measurements — double scoring offers an alternative to standard PCA. The interest of CA lies in the mutual representation of individuals and variables in terms of barycentric properties.

Chapter 5

Multiple Correspondence Analysis (MCA)

(*'MULTIPLE'*)

Social reality is multidimensional.
Pierre Bourdieu

Introduction

In this chapter, we consider a set a of *categorized variables* (i.e. variables with a finite number of categories) defined on a finite set of *individuals*, i.e. a protocol taking values in the Cartesian product of several categorized variables, and we analyze this protocol by the GDA method of Multiple Correspondence Analysis (MCA). For the presentation of MCA, it will be convenient to adopt the *language of questionnaire*, where individuals give responses to a set of questions[1], more precisely, where for each question, each individual gives one and only one response chosen among a set of categories, or *response modalities* (in brief modalities); such a questionnaire is said to be in *"standard format"*. Denoting I the set of individuals and Q the set of questions, the basic data table analyzed by MCA is thus an $I \times Q$ (Individuals×Questions) table, with in cell (i, q) the modality of question q chosen by individual i.

MCA provides a geometric model of data, that is, it represents individuals by points, and summarizes the relations between the categorized

[1]The language of questionnaire is proper in surveys where individuals are people responding to questions; it is metaphorical when individuals (people, or states, or firms, etc.) are described by properties or attributes.

variables. Methodologically, MCA thus appears to be the counterpart of PCA for categorized variables.

From a historical perspective, the reference Guttman (1941) is a precursor paper of MCA as an optimal scaling (in very much the same way as Fisher's 1940 paper is for CA): see Exercise 5.5 (p.248). The paper by Burt (1950) must also be mentioned, if only because the table attached to Burt's name in MCA is found in this paper. In the early seventies, MCA emerged as an extension of CA to a table of Individuals×Categorized variables after disjunctive coding: see Benzécri (1977), Lebart (1975)[2]. In the late seventies, MCA software for large format questionnaires[3] was developed (see Lebart & al, 1977), then MCA became a major GDA method for the analysis of questionnaires, and has been constantly used since, especially in Bourdieu's sociological school[4].

In MCA, as a GDA method, a Euclidean cloud of points representing the individuals is constructed, and principal axes and variables are determined. A restricted number of axes are interpreted which provides a summary of data. The first and crucial step of an MCA is the specification of the distance between individuals, defined from their responses. Given two individuals i and i' and a question $q \in Q$, we first define the part of distance between i and i' due to question q, that we denote $d_q(i, i')$.

• If for question q, individuals i and i' choose the same response, we say that question q is an *agreement question* between i and i' and the part of distance due to question q is null:

$$d_q(i, i') = 0 \tag{5.1}$$

• If for question q, individual i chooses modality k and individual i' modality $k' \neq k$, we say that question q is a *disagreement question*, and the part of distance due to question q is defined by

$$d_q^2(i, i') = \frac{1}{f_k} + \frac{1}{f_{k'}} \tag{5.2}$$

---

[2]The paper by Benzécri reproduces a mimeographed note of 1972 ("BinMult"), in which he credits Lebart for his shrewd thinking about applying CA to multiway tables and for introducing the "Burt table" name. The phrase "Analyse des Correspondances Multiples" appears for the very first time in Lebart's paper which is devoted to MCA as a method in its own right.

[3]A questionnaire is considered to have a small–size format when its number of questions is small, regardless of the number of individuals.

[4]See e.g. Bourdieu & Saint–Martin (1978), Bourdieu (1984), Bourdieu (1989), Bourdieu (1999), Lebaron (2001).

where $f_k = \frac{n_k}{n}$, $n = \sum\limits_{k \in K} n_k$ is the total number of individuals and n_k is the number of individuals choosing k to question q (and similarly $f_{k'} = \frac{n_{k'}}{n}$). Then the overall distance $d(i, i')$ between i and i' is defined as the quadratic average of the q–distances $\left(d_q(i, i')\right)_{q \in Q}$ defined by formulas (5.1) and (5.2).

$$d^2(i, i') = \frac{1}{Q} \sum_{q \in Q} d_q^2(i, i') \tag{5.3}$$

One notices that only "disagreement questions" create distance between individuals, and that the smaller the frequencies, the greater the distance between individuals. This property tends to enhance the importance of infrequent modalities, which is a desirable property up to a certain point; very infrequent modalities need to be pooled with others or to be put as "passive" ones (cf. specific MCA in §5.2).

The chapter is organized as follows. We present the multiple correspondence of a questionnaire in standard format and the statistical characterization of MCA (§5.1). Then we present the two kinds of specific MCA (§5.2). Then we outline a methodology for the interpretation of MCA (§5.3). Lastly, we illustrate the procedure and the strategy of interpretation on the *Culture* example (§5.4).

The mathematical notions used in this chapter are again those of multidimensional geometry; see §10.4 in chapter 10 (*MATH.BASES*). The basic definitions about a Euclidean cloud are those of §3.1 in chapter 3 (*CLOUD*). For the determination of principal axes and variables, see §2.3 in chapter 2 (*CORRESPONDENCE*). As everywhere in the book, the measure vs variable duality (cf. *CORRESPONDENCE*, §2.1) is omnipresent in this chapter; it is the key to the equivalence properties between MCA with CA and PCA.

5.1 Multiple Correspondence Analysis of a Questionnaire (standard MCA)

In this section, we firstly construct and study the cloud of individuals (§5.1.1), secondly the cloud of modalities (§5.1.2). Then we study modality mean–points and Burt cloud (§5.1.3). Then we present the statistical characterizations of MCA (§5.1.4). Lastly, we give a synopsis (§5.1.5) and a computation procedure (§5.1.6).

Let I be the set of the n individuals answering the questionnaire, and Q the set of questions. We denote[5] $K\!<\!q\!>$ the set of the response modalities

[5] *Nesting notation* (cf. p.25). Read "K in q".

of question q. The sets of modalities of two different questions are disjoint: $K\langle q\rangle \cap K\langle q'\rangle = \emptyset$ for $q \neq q'$. Letter K denotes the overall set of modalities (taking all Q questions): $K = \bigcup_{q\in Q} K\langle q\rangle$. The basic protocol of a standard format questionnaire is defined by $I \to \prod\limits_{q\in Q} K\langle q\rangle$ (Cartesian product of the Q sets $K\langle q\rangle$); we assume the elementary frequency measure over I: $(\forall i \in I : n_i = 1)$[6]. A Q–tuple of modalities $(k\langle q\rangle)_{q\in Q} \in \prod\limits_{q\in Q} K\langle q\rangle$ is called a *response pattern*[7]. The subset of modalities chosen by i (response pattern of individual i) is denoted $K\langle i\rangle$. In such a questionnaire, each question induces a *partition* of the set I of individuals.

5.1.1 Cloud of Individuals

In this subsection, we first give a direct geometric definition of the cloud of individuals and a definition of the multiple correspondence measure. We state the basic properties and we determine principal axes and variables. We proceed with the cloud of response patterns, then we present the contributions of individuals and supplementary individuals.

Cloud construction

We construct the cloud of individuals, denoted M^I, such that the Euclidean distance $\mathrm{M}^i\mathrm{M}^{i'}$ is equal to $d(i,i')$ (cf. formula (5.3), p.181)[8]. Let $\mathcal{U}_K$ be a K–dimensional geometric space with affine frame $\big(\mathrm{O}, (\mathrm{I}^k)_{k\in K}\big)$; the point O is the origin–point and points I^k are called *ideal points*. The point M^i is defined as the *equibarycenter* of the Q ideal points I^k corresponding to the response pattern $K\langle i\rangle$ and is weighted by $n_i = 1$.

$$\mathrm{M}^i = \sum_{k\in K\langle i\rangle} \mathrm{I}^k/Q \qquad (5.4)$$

Equivalently, $\mathrm{M}^i = \frac{1}{Q}\sum\limits_{k\in K} \delta_{ik}\,\mathrm{I}^k$ with $\delta_{ik} = \begin{cases} 1 & \text{if } i \text{ has chosen } k \\ 0 & \text{if not} \end{cases}$.
For a standard questionnaire, one has $\forall q \in Q : \sum\limits_{k\in K\langle q\rangle} \delta_{ik} = 1$. Therefore, the dimensionality of the affine *support of cloud* is at most equal to $K - Q$.

---

[6]The assumption $n_I = 1_I$ (elementary measure) is convenient though not theoretically essential.

[7]In this chapter, letter J will be reserved for response patterns; response pattern j is weighted by the number n_j of individuals having j as a response pattern.

[8]*Recall of notation.* Roman capitals designate points, such as M^i, I^k; italic capitals designate sets such as I (individuals), J (response patterns), and their cardinalities.

Notation synopsis.

I	: set of n individuals
Q	: set of active questions
$K{<}q{>}$	: set of modalities of question q
K	: overall set of modalities
$K{<}i{>}$	: subset of modalities chosen by i (response pattern)
$I{<}k{>}$	: subset of n_k individuals who choose modality k
δ_{ik}	: $\begin{cases} 1 & \text{if } i \text{ has chosen } k \\ 0 & \text{if not} \end{cases}$
$\sum_{i\in I} \delta_{ik} = n_k$	: number of individuals who have chosen k
$\sum_{k\in K<q>} \delta_{ik} = 1 \quad (q \in Q)$	
$\sum_{k\in K} \delta_{ik} = Q$	: number of active questions
$\sum_{i\in I}\sum_{k\in K} \delta_{ik} = nQ$ hence $\sum_{k\in K} n_k = nQ$	
$\sum_{i\in I} \delta_{ik}\delta_{ik'} = n_{kk'}$	: number of individuals who have chosen both k and k'

Let $\mathcal{V}_K$ be the vector space underlying $\mathcal{U}_K$ with basis $(\overrightarrow{\mathrm{OI}^k})_{k\in K}$. With respect to this Cartesian frame, one has:

$$\overrightarrow{\mathrm{OM}^i} = (1/Q) \sum_{k\in K} \delta_{ik} \overrightarrow{\mathrm{OI}^k} \tag{5.5}$$

The *Cartesian coordinates of the point* M^i are $(\delta_{ik}/Q)_{k\in K}$. The measure $p^i_K = (\delta_{ik}/Q)_{k\in K}$ over K is called the *numerical profile of individual i*. In the geometric modeling, the weights $(\delta_{ik})_{ik\in I\times K}$ play the role of a measure over the Cartesian product $I \times K$, wich leads to the following definition[9].

Definition 5.1 (Multiple Correspondence). *Measure δ_{IK} over $I \times K$ is called a multiple correspondence if the following relations hold: $K{<}Q{>}$ (modalities are nested in questions) and $\forall q \in Q$: $I{<}K{<}q{>}{>}$ (each question determines a partition of individuals).*

Example. Consider a questionnaire with three questions $q1$, $q2$ and $q3$ with $K{<}q1{>}= \{k1, k2\}$ (2 modalities), $K{<}q2{>}= \{k3, k4\}$ (2 modalities), $K{<}q3{>}= \{k5, k6, k7\}$ (3 modalities), hence $Q = 3$ and $K = 2+2+3 = 7$. There are $2 \times 2 \times 3 = 12$ possible response patterns. Table 5.1 shows the

[9] δ_{IK} is the analog of n_{JK} in CA; in MCA we have $\mathrm{M}^i = \sum_{k\in K} \delta_{ik}\, \mathrm{I}^k / \sum_{k\in K} \delta_{ik}$; whereas in CA we have $\mathrm{M}^j = \sum_{k\in K} n_{jk}\mathrm{I}^k / \sum_{k\in K} n_{jk}$ (cf. chapter 2, formula (2.5), p.34).

basic questionnaire and the multiple correspondence table δ_{IK}, exhibiting an individual i with response pattern $\{k2, k4, k6\}$.

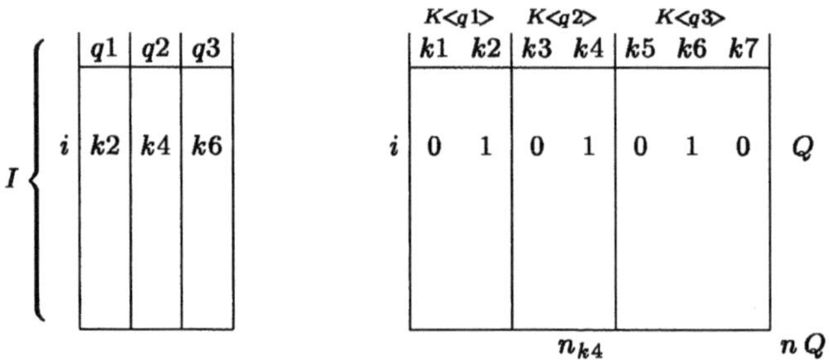

Figure 5.1. Example with $Q = 3$ and $K = 2 + 2 + 3 = 7$. Basic questionnaire ($I \times Q$ table), and multiple correspondence δ_{IK} (zero–one $I \times K$ table).

Since $Q = 3$, each point can be represented in the subspace spanned by the ideal points associated with its response pattern. Figure 5.2 depicts the construction of the point $\mathrm{M}^i = \frac{1}{3}(\mathrm{I}^{k2} + \mathrm{I}^{k4} + \mathrm{I}^{k6})$, together with the relation $\overrightarrow{\mathrm{OM}}^i = \frac{1}{3}(\overrightarrow{\mathrm{OI}}^{k2} + \overrightarrow{\mathrm{OI}}^{k4} + \overrightarrow{\mathrm{OI}}^{k6})$.

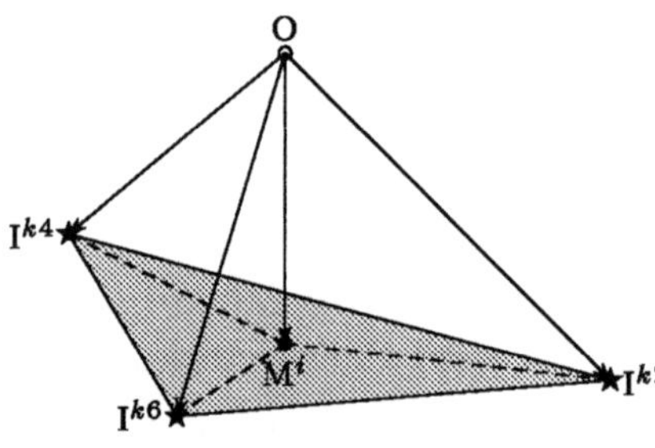

Figure 5.2. Example with $Q = 3$ and $K = 2 + 2 + 3 = 7$. Geometric representation of individual i with response pattern $(k2, k4, k6)$ by point M^i.

If all twelve response patterns are observed, the cloud is $7 - 3 = 4$–dimensional, and cannot be represented in a three–dimensional space; but some projected clouds can be represented. Figure 5.3 illustrates the construction of the projected cloud in the three–dimensional subspace framed by the four ideal points I^{k1}, I^{k2} (question $q1$), I^{k3}, I^{k4} (question $q2$), in the case where for each question the marginal frequencies are equal. The projected cloud consists in four points, each of which is the projection of three response pattern points. For instance, H^{246} is the orthogonal projection of the three points associated with $(k2, k4, k6)$, $(k2, k4, k5)$ and $(k2, k4, k7)$; in the same way, H^{236} is the projection of $(k2, k3, k6)$, $(k2, k3, k5)$ and $(k2, k3, k7)$, etc. The four projected points lie in the plane spanned by the four midpoints $I^{24} = (I^2 + I^4)/2$, I^{13}, I^{14} and I^{23}; they exhibit a parallelogram shape, in accordance with the equipollence property: see below Property 5.2.

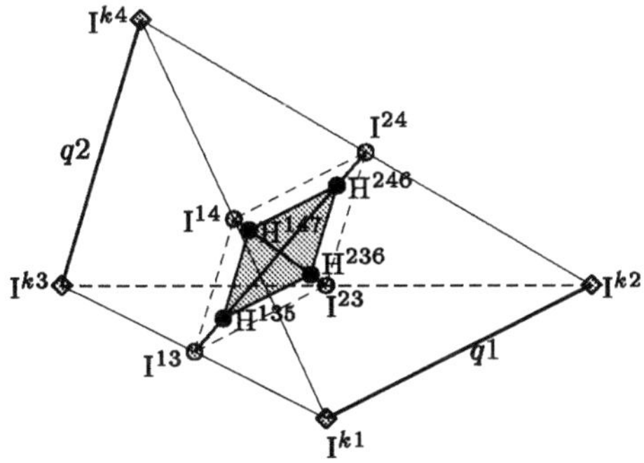

Figure 5.3. Example with $Q = 3$ and $K = 2+2+3 = 7$. Ideal points and projected cloud associated with questions $q1$ and $q2$.

Basic properties

Property 5.1 (Mean point). *The Cartesian coordinates of the mean point of the cloud are equal to* $(f_k/Q)_{k\in K}$.

Proof. From Definition 3.2 (CLOUD, p.77), one has $\overrightarrow{\mathrm{OG}} = \sum_{i\in I} \overrightarrow{\mathrm{OM}^i}/n$, hence $\overrightarrow{\mathrm{OG}} = \sum_{k\in K} (\sum_{i\in I} \delta_{ik}/Q)\overrightarrow{\mathrm{OI}^k}/n = \sum_{k\in K} n_k\overrightarrow{\mathrm{OI}^k}/(nQ) = \sum_{k\in K} \frac{f_k}{Q}\overrightarrow{\mathrm{OI}^k}$. ◁

Property 5.2 (Equipollence property). *If individuals i and i' choose the same modalities, except for one question for which i has chosen k and i' has chosen k', one has:* $\overrightarrow{\mathrm{M}^i\mathrm{M}^{i'}} = \overrightarrow{\mathrm{I}^k\mathrm{I}^{k'}}/Q$.

Metric on $\mathcal{U}_K$

The metric on $\mathcal{U}_K$ is taken such that the distance between the points M^i and $\mathrm{M}^{i'}$ is equal to the distance $d(i,i')$ defined by formula (5.3), p.181. One has $\overrightarrow{\mathrm{M}^i\mathrm{M}^{i'}} = (1/Q)\sum_{k\in K}(\delta_{i'k} - \delta_{ik})\overrightarrow{\mathrm{OI}^k}$, and $(\mathrm{M}^i\mathrm{M}^{i'})^2 = (1/Q^2)\sum_{k\in K}\sum_{k'\in K}(\delta_{i'k} - \delta_{ik})(\delta_{i'k'} - \delta_{ik'})\langle\overrightarrow{\mathrm{OI}^k}|\overrightarrow{\mathrm{OI}^{k'}}\rangle$.

If one takes $\overrightarrow{\mathrm{OI}^k} \perp \overrightarrow{\mathrm{OI}^{k'}}$, then $(\mathrm{M}^i\mathrm{M}^{i'})^2 = (1/Q^2)\sum_{k\in K}(\delta_{i'k} - \delta_{ik})^2(\mathrm{OI}^k)^2$, with $(\delta_{i'k} - \delta_{ik})^2 = 1$ if k is a disagreement modality between i and i' and 0 if not. Hence, letting $(\mathrm{OI}^k)^2 = Q/f_k$, one gets $(\mathrm{M}^i\mathrm{M}^{i'})^2 = \frac{1}{Q}\sum_{k\in K}\frac{(\delta_{i'k} - \delta_{ik})^2}{f_k} =$

$\frac{1}{Q}\sum\limits_{q\in Q}\sum\limits_{k\in K<q>}\frac{(\delta_{i'k}-\delta_{ik})^2}{f_k}=\frac{1}{Q}\sum\limits_{q\in Q}(\mathrm{M}^i\mathrm{M}^{i'})_q^2$, with $(\mathrm{M}^i\mathrm{M}^{i'})_q^2=\sum\limits_{k\in K<q>}\frac{(\delta_{i'k}-\delta_{ik})^2}{f_k}$.
Now if for question q, individual i has chosen modality k and individual i' modality k', we have $(\mathrm{M}^i\mathrm{M}^{i'})_q^2=\frac{1}{f_k}+\frac{1}{f_{k'}}$; if both i and i' have chosen k, we have $d_q(i,i')=0$; hence the following formula equivalent to formula (5.3).

$$(\mathrm{M}^i\mathrm{M}^{i'})^2=\frac{1}{Q}\sum_{q\in Q}(\mathrm{M}^i\mathrm{M}^{i'})_q^2=\frac{1}{Q}\sum_{q\in Q}d_q^2(i,i')=\frac{1}{Q}\sum_{k\in K}\frac{(\delta_{ik}-\delta_{i'k})^2}{f_k} \quad (5.6)$$

Definition 5.2 (Metric on $\mathcal{U}_K$). *The metric on $\mathcal{U}_K$ is such that the vectors $\overrightarrow{\mathrm{OI}}^k$ are pairwise orthogonal with $(\mathrm{OI}^k)^2=Q/f_k$.*

In particular, one has $(\mathrm{I}^k\mathrm{I}^{k'})^2=Q(\frac{1}{f_k}+\frac{1}{f_{k'}})$ for $k\neq k'$.

Property 5.3. *$\overrightarrow{\mathrm{OG}}$ is orthogonal to the support of cloud, and* $\mathrm{OG}=1$.

Proof. Let us show that $\forall\, i\in I:\ \overrightarrow{\mathrm{OG}}\perp\overrightarrow{\mathrm{GM}}^i$, i.e. $\langle\overrightarrow{\mathrm{OG}}|\overrightarrow{\mathrm{GM}}^i\rangle=0$.
$\langle\overrightarrow{\mathrm{OG}}|\overrightarrow{\mathrm{GM}}^i\rangle=\sum\limits_{k\in K}\frac{f_k}{Q}(\frac{\delta_{ik}}{Q}-\frac{f_k}{Q})(\mathrm{OI}^k)^2=\sum\limits_{k\in K}\frac{f_k}{Q}(\frac{\delta_{ik}}{Q}-\frac{f_k}{Q})\frac{Q}{f_k}=\frac{1}{Q}\sum\limits_{k\in K}(\delta_{ik}-f_k)$.
From $\sum\limits_{k\in K}\delta_{ik}=Q$ and $\sum\limits_{k\in K}f_k=Q$, one deduces $\langle\overrightarrow{\mathrm{OG}}|\overrightarrow{\mathrm{GM}}^i\rangle=0$.
$(\mathrm{OG})^2=\sum\limits_{k\in K}(f_k/Q)^2(\mathrm{OI}^k)^2=\sum\limits_{k\in K}f_k/Q=1$ ◁

Corollary 5.1. $\sum\limits_{k\in K}a_k\overrightarrow{\mathrm{OI}}^k\perp\overrightarrow{\mathrm{OG}}\iff\sum\limits_{k\in K}a_k=0.$

Proof. $\langle\overrightarrow{\mathrm{OG}}|\sum\limits_{k\in K}a_k\overrightarrow{\mathrm{OI}}^k\rangle=0\iff\sum\limits_{k\in K}(\frac{f_k}{Q})a_k(\mathrm{OI}^k)^2=\sum\limits_{k\in K}(\frac{f_k}{Q})a_k(\frac{Q}{f_k})=\sum\limits_{k\in K}a_k=0.$ ◁

Property 5.4. *The distance* GM^i *of the point* M^i *from the mean point* G *is such that* $(\mathrm{GM}^i)^2=\Big((1/Q)\sum\limits_{k\in K<i>}(1/f_k)\Big)-1$.

Proof. $(\mathrm{OM}^i)^2=\sum\limits_{k\in K}(\delta_{ik}/Q)^2(\mathrm{OI}^k)^2=(1/Q)\sum\limits_{k\in K}\delta_{ik}/f_k$ and $\mathrm{OG}^2=1$.
$\overrightarrow{\mathrm{OG}}\perp\overrightarrow{\mathrm{GM}}^i\Rightarrow(\mathrm{OM}^i)^2=\mathrm{OG}^2+(\mathrm{GM}^i)^2$, hence the result. ◁

— *Remark.* The more an individual chooses infrequent modalities (small f_k), the more remote this individual is from the mean point. Individuals that choose infrequent modalities are located at the periphery of the cloud.

Property 5.5 (Variance of cloud). *The variance of the cloud is equal to* $(K/Q)-1$ *(mean number of modalities per question minus 1).*

Proof. $\operatorname{Var} M^I = \sum_{i\in I}(GM^i)^2/n = \big(1/(nQ)\big)\big(\sum_{i\in I}\sum_{k\in K}\delta_{ik}/f_k\big) - \sum_{i\in I} 1/n$, hence $\operatorname{Var} M^I = \big(1/(nQ)\big)\big(\sum_{k\in K} n_k/f_k\big) - 1 = (K/Q) - 1$. ◁

Corollary 5.2. *If all questions have the same number p of modalities, the variance of the cloud is equal to $p-1$. In the particular case of dichotomous questions ($p=2$) the variance is equal to 1.*

— *Remark.* The variance of the cloud does not depend on data, but only on the questionnaire format. There is a similar property for standard PCA: the variance of cloud is equal to the number of variables (*PRINCIPAL*, §4.3).

Principal axes and variables

The following theorem provides a further motivation for calling measure δ_{IK} a "multiple correspondence".

Theorem 5.1 (MCA fundamental theorem). *The principal axes and variables of the cloud of individuals are given by the Correspondence Analysis of measure δ_{IK}.*

Proof. The cloud M^I (defined above) is identical to the cloud P^I constructed by CA from measure δ_{IK} over the Cartesian product $I \times K$, that is, for every pair (i, i') the equality $P^iP^{i'} = M^iM^{i'}$ holds, and the points P^i and M^i have the same weights. Indeed the profile of the point P^i is $p^i_K = (p^i_k)_{k\in K}$ with $p^i_k = \delta_{ik}/(\sum_{k\in K}\delta_{ik}) = \delta_{ik}/Q$, and its weight is $\sum_{k\in K}\delta_{ik}/(nQ) = 1/n$ (cf. chapter 2, §2.2.2 p.33).

Let $p_k = \sum_{i\in I}\delta_{ik}/nQ = f_k/Q$; the Φ^2–distance of center p_K between points P^i and $P^{i'}$ is such that $(P^iP^{i'})^2 = \sum_{k\in K}\frac{(p^i_k - p^{i'}_k)^2}{f_k/Q} = \frac{1}{Q}\sum_{k\in K}\frac{(\delta_{ik}-\delta_{i'k})^2}{f_k} = (M^iM^{i'})^2$ (cf. formula (5.6), p.186). ◁

—*Remark.* We can also prove this theorem by applying the results of §3.3 (*CLOUD*, p.87) to the cloud M^I: cf. Exercise 5.1, p.241.

> *To sum up*: To proceed to the *Multiple Correspondence Analysis* (MCA) of questionnaire $I \to \prod_{q\in Q} K\langle q\rangle$ (Cartesian product of the Q sets $K\langle q\rangle$) is to proceed to the simple Correspondence Analysis (CA) of the multiple correspondence δ_{IK}.

Corollary 5.3 (Transition formulas). *The ℓ-th standard principal variable z_ℓ^K over $(K, f_K/Q)$ and the ℓ-th standard principal variable z_ℓ^I over $(I, (1/n)_{i\in I})$ of the measure δ_{IK} verify the equations:*

$$\begin{cases} \forall i \in I & \sum\limits_{k\in K<i>} z_\ell^k/Q &= \xi_\ell z_\ell^i \\ \forall k \in K & \sum\limits_{i\in I<k>} z_\ell^i/n_k &= \xi_\ell z_\ell^k \end{cases} \quad \textit{with} \quad \begin{cases} \sum\limits_{i\in I}(z_\ell^i)^2/n &= 1 \\ \sum\limits_{k\in K}(z_\ell^k)^2/(f_k/Q) &= 1 \end{cases} \tag{5.7}$$

Proof. Transition formulas (2.8) (chapter 2, p.38) of the CA of δ_{IK} read: $\sum\limits_{k\in K} p_k^I z_\ell^k = \xi_\ell z_\ell^I$ and $\sum\limits_{i\in I} p_i^K z_\ell^i = \xi_\ell z_\ell^K$, with $p_k^i = \delta_{ik}/Q$ and $p_i^k = \delta_{ik}/n_k$; hence the property. ◁

Property 5.6 (Equation for principal directions). *Let $\lambda_\ell = \xi_\ell^2$ and $f_k^{k'} = n_{kk'}/n_{k'}$, the ℓ-th normed principal contrast over $(K, f_K/Q)$ verifies*

$$\forall k \in K \qquad (1/Q)\sum_{k'\in K} f_k^{k'} z_{k'\ell} = \lambda_\ell z_{k\ell} \tag{5.8}$$

Proof. From preceding Equations (5.7), one gets $\xi_\ell z_\ell^k = \sum\limits_{i\in I} p_i^k \frac{1}{\xi_\ell} \sum\limits_{k'\in K} p_{k'}^i z_\ell^{k'}$;

now $\sum\limits_{i\in I} p_i^k p_{k'}^i = \sum\limits_{i\in I} \frac{\delta_{ik}\delta_{ik'}}{n_k Q} = \frac{n_{kk'}}{n_k Q} = \frac{f_{k'}^k}{Q}$, hence the property. ◁

The ℓ-*th principal axis* (Axis ℓ) is defined by $(\mathrm{G}, \overrightarrow{\alpha_\ell})$ with $\overrightarrow{\alpha_\ell} = \sum\limits_{k\in K} z_{k\ell}\overrightarrow{\mathrm{OI}^k}$.

Property 5.7 (Principal coordinates). *The principal coordinate of the ideal point* I^k *on Axis* ℓ *is equal to* $z_\ell^k = z_{k\ell}/(f_k/Q)$; *the principal coordinate of the point* M^i *on Axis* ℓ *is equal to* $y_\ell^i = \xi_\ell z_\ell^i$.

Proof. By Definition 3.15 (CLOUD, p.91), the principal coordinate of the point I^k on Axis ℓ is $\langle \overrightarrow{\mathrm{GI}^k} | \overrightarrow{\alpha_\ell}\rangle = \langle \overrightarrow{\mathrm{OI}^k} | \overrightarrow{\alpha_\ell}\rangle = z_{k\ell}(\mathrm{OI}^k)^2 = z_{k\ell}\frac{Q}{f_k} = z_\ell^k$.
The relation $\mathrm{M}^i = \sum\limits_{k\in K<i>} \mathrm{I}^k/Q$, projected on Axis ℓ, reads $y_\ell^i = \sum\limits_{k\in K<i>} z_\ell^k/Q$; and from the first equation (5.7) we get $y_\ell^i = \xi_\ell z_\ell^i$. ◁

Corollary 5.4 (Reconstitution of distances). *The distance $d(i, i')$ defined by formula (5.3) is obtained from the principal coordinates $(y_\ell^i)_{\ell=1,\dots L}$ (L being the cloud dimensionality), one has:*

$$d^2(i, i') = \sum_{\ell=1}^{L} (y_\ell^i - y_\ell^{i'})^2$$

Proof. One has $\overrightarrow{\mathrm{M}^i\mathrm{M}^{i'}} = \sum_{\ell=1}^{L}(y_\ell^{i'} - y_\ell^i)\overrightarrow{\alpha_\ell}$, the set $(\overrightarrow{\alpha_\ell})_{\ell=1,\cdots L}$ is an orthonormal basis of the support of the cloud, hence the property. ◁

Definition 5.3 (Disjunctive protocol and indicator variables). *Given a questionnaire $I \to \prod_{q\in Q} K{<}q{>}$, with each modality k there is associated the indicator variable δ_k^I with $\delta_k^i = 1$ if individual i chooses k and 0 if not. The numerical protocol $\delta_K^I = (\delta_k^I)_{k\in K}$ is called the disjunctive protocol of the questionnaire, or the questionnaire is said to be in complete disjunctive format.*

One has: $\sum_{k\in K} \delta_k^I = \sum_{q\in Q}\sum_{k\in<q>} \delta_k^I = \sum_{q\in Q} 1^I = Q\,1^I$. As is well known for indicator variables, one has $\mathrm{Mean}\,\delta_k^I = f_k$, and $\mathrm{Var}\,\delta_k^I = f_k(1 - f_k)$.

— *Remark.* The disjunctive protocol δ_K^I and the multiple correspondence δ_{IK} are represented by the same $I \times K$ *disjunctive table*; but for the disjunctive protocol, grouping on J entails averaging, whereas for the multiple correspondence, it entails summing.

Theorem 5.2 (MCA as a PCA). *Given a questionnaire $I \to \prod_{q\in Q} K{<}q{>}$, the* MCA *of the questionnaire is equivalent to the biweighted* PCA *of the disjunctive protocol $(\delta_k^I)_{k\in K}$ weighted by $n_I = 1_I$ and $\varpi_K = Q\,f_K$.*

Proof. This theorem follows from Property 4.23 (PRINCIPAL, p.158), since the Total variable $\underline{x}^I = \sum_{k\in K}\delta_k^I$ is constant ($\underline{x}^I = Q\,1^I$). ◁

Property 5.8. *The regression coefficient of the indicator variable δ_k^I on the ℓ-th standard principal variable z_ℓ^I is equal to $Q\xi_\ell z_{k\ell}$; the correlation coefficient is equal to $\xi_\ell z_\ell^k\sqrt{f_k/(1-f_k)}$.*

Proof. $\mathrm{Cov}(\delta_k^I|z_\ell^I) = \sum_{i\in I}\delta_k^i z_\ell^i/n = n_k\sum_{i\in I}(\delta_{ik}/n_k)z_\ell^i/n = f_k\sum_{i\in I}p_i^k z_\ell^i = \xi_\ell f_k z_\ell^k$ (from the second equation (5.7)), hence the regression coefficient is equal to $\xi_\ell Q(f_k/Q)z_\ell^k = Q\xi_\ell z_{k\ell}$. From $\mathrm{SD}\,\delta_k^I = \sqrt{f_k(1-f_k)}$, one deduces the correlation coefficient: $\xi_\ell f_k z_\ell^k/\sqrt{f_k(1-f_k)} = \sqrt{f_k/(1-f_k)}(\xi_\ell z_\ell^k)$. ◁

Property 5.9. *The regression of the ℓ-th standard principal variable z_ℓ^I on the indicator variables $(\delta_k^I)_{k\in K<q>}$ of the modalities of question q is equal to $\sum_{k\in K<q>}\xi_\ell z_\ell^k\delta_k^I$, and the square of the multiple correlation is equal to $\lambda_\ell\sum_{k\in K<q>} f_k(z_\ell^k)^2$.*

Proof. One has $\sum_{k\in K<q>} \delta^I_k = 1^I$, therefore the regression of z^I_ℓ on $(\delta^I_k)_{k\in K<q>}$ is equal to $\sum_{k\in K<q>} b^k_\ell \delta^I_k$, with $b^k_\ell = \frac{\langle z^I_\ell|\delta^I_k\rangle}{\|\delta^I_k\|^2}$. One has $\|\delta^I_k\|^2 = \sum_{i\in I}\delta^i_k/n = f_k$ and $\langle z^I_\ell|\delta^I_k\rangle = \sum_{i\in I} z^i_\ell\delta^i_k/n = \sum_{i\in I<k>} z^i_\ell/n = \xi_\ell f_k z^k_\ell$ (2nd equation (5.7), p.188), hence $b^k_\ell = \xi_\ell z^k_\ell$. By definition, the square of the multiple correlation is equal to $\mathrm{Var}(\sum_{k\in K<q>} b^k_\ell\delta^I_k)/\operatorname{Var} z^I_\ell$ which is equal to $\mathrm{Cov}(\sum_{k\in K<q>} b^k_\ell\delta^I_k|z^I_\ell) = \sum_{k\in K<q>} b^k_\ell \mathrm{Cov}(\delta^I_k|z^I_\ell) = \sum_{k\in K<q>} b^k_\ell\xi_\ell f_k z^k_\ell = \lambda_\ell \sum_{k\in K<q>} f_k(z^k_\ell)^2$. ◁

Cloud of response patterns

Let J be a set indexing the observed response patterns; we will denote $I<j>$ the set of individuals whose response pattern is j, and n_j their number (with $\forall j : n_j \geq 1$). The set J of observed response patterns induces a partition of individuals denoted $I<J>$[10]; hence the protocol of response patterns $J \to \prod_{q\in Q} K<q>$, each response pattern j being weighted by n_j. The cloud M^J is obtained from the cloud M^I by replacing the n_j points $(\mathrm{M}^i)_{i\in I<j>}$ (all on the same location) by a single point $\mathrm{M}^j = \mathrm{M}^i$ with weight n_j.

Definition 5.4 (Multiple correspondence of response patterns). *The measure m_{JK} over $J \times K$ defined by $m_{jk} = \sum_{i\in I<j>} \delta_{ik}$, that is, the measure obtained by summing δ_{IK} over the classes $(I<j>)_{j\in J}$, is called a multiple correspondence of response patterns.*

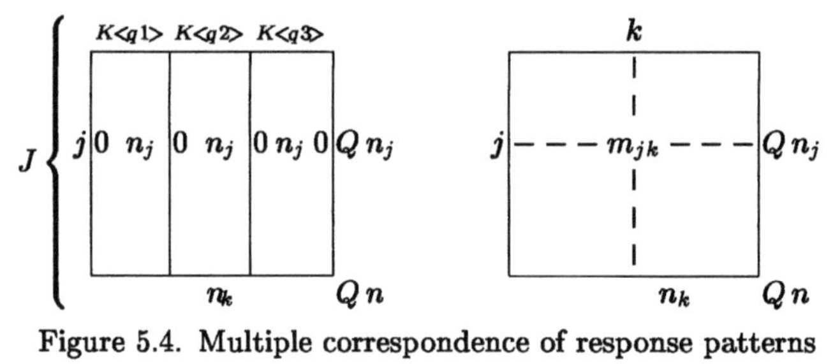

Figure 5.4. Multiple correspondence of response patterns

One has $m_j = \sum_{k\in K} m_{jk} = \sum_{i\in I<j>}\sum_{k\in K}\delta_{ik} = Qn_j$ and $m_k = \sum_{j\in J} m_{jk} = \sum_{j\in J}\sum_{i\in I<j>}\delta_{ik} = \sum_{i\in I}\delta_{ik} = n_k$, and $m = \sum_{k\in K} m_k = \sum_{k\in K} n_k = Qn$.

[10]The relation $I<J>$ is a nesting relation, i.e. with each individual there is associated one and only one response pattern.

Property 5.10. *The* CA *of the multiple correspondence* m_{JK} *is equivalent to the* CA *of* δ_{IK}.

This property results from the principle of distributional equivalence (*CORRESPONDENCE*, §2.2, p.34).

Definition 5.5 (Disjunctive protocol of response patterns). *The disjunctive protocol of response patterns, denoted* δ_K^J, *is defined by* $\delta_k^j = 1$ *if* k *is a modality of response pattern* j *and* $\delta_k^j = 0$ *if not.*

One has the relation $\delta_k^j = m_{jk}/n_j$; the variable δ_k^J is the n_J–density of the measure m_{Jk} over J. The indicator variables $(\delta_k^J)_{k\in K}$ verify the property $\sum_{k\in K} \delta_k^j = Q$. The MCA of the questionnaire $J \to \prod_{q\in Q} K{<}q{>}$ is equivalent to the biweighted PCA of the protocol of indicator variables $(\delta_k^J)_{k\in K}$ with weightings n_J and $Q\,f_K$.

— *Remark.* The disjunctive protocol of response patterns δ_K^J and the multiple correspondence m_{JK} are both represented by $J \times K$ numerical tables, but the disjunctive table is made of zeros and ones, whereas in the multiple correspondence the ones are replaced by the frequencies $(n_j)_{j\in J}$.

Contributions of individuals

The absolute contribution of individual i to the variance of the cloud is

$$\mathrm{Cta}_i = (\mathrm{GM}^i)^2/n = \Big(\tfrac{1}{Q}\sum_{k\in K<i>} 1/n_k\Big) - 1/n$$

This contribution is an increasing function of the distance of the point M^i from the mean point of the cloud.

The absolute contribution of individual i and Axis ℓ is: $\mathrm{Cta}_{i\ell} = (y_\ell^i)^2/n$.

For response pattern j, the absolute contribution to total variance is $\mathrm{Cta}_j = \sum_{i\in I<j>} \mathrm{Cta}_i = n_j\,\mathrm{Cta}_i - \big(\frac{f_j}{Q}\sum_{k\in K<j>} 1/f_k\big) - f_j$, and the absolute contribution to Axis ℓ is $\mathrm{Cta}_{j\ell} = f_j(y_\ell^j)^2$ (with $f_j = n_j/n$).

Supplementary individuals

Let i_s be a *supplementary individual*, that is, an individual that has not taken part in the determination of axes, defined by the set of modalities $K{<}i_s{>}$. The method of supplementary elements, like in CA and PCA, consists in applying to $K{<}i_s{>}$ the first transition formula, so as to assign principal coordinates. As for active individuals, the point M^{i_s} is the barycenter of the ideal points $(\mathrm{I}^k)_{k\in K<i_s>}$:

$$\mathrm{M}^{i_s} = \sum_{k\in K<i_s>} \mathrm{I}^k/Q$$

and its principal coordinate on Axis ℓ, denoted $y_\ell^{i_s}$ is equal to $\sum\limits_{k\in K<i_s>} z_\ell^k/Q$.

5.1.2 Cloud of Modalities

We construct the cloud of modalities of the multiple correspondence δ_{IK}, by applying the results of the chapter CORRESPONDENCE, especially those of §2.2 (p.31); then we relate the cloud of modalities so constructed to the cloud of individuals (§5.1.3).

Recall (see p.183) that $\sum\limits_{k\in K} \delta_{ik} = Q$; $\sum\limits_{i\in I} \delta_{ik} = n_k$ and $\sum\limits_{i\in I}\sum\limits_{k\in K} \delta_{ik} = nQ$.

Cloud of modality points

Given a modality $k \in K$, the CA of the measure δ_{IK} over $I \times K$ associates its profile $p_I^k = (\delta_{ik}/n_k)_{i\in I}$. The profiles are elements of the space $\mathbb{R}_I$ of measures over (I, f_I), with the fundamental measure $f_I = (1/n)_{i\in I}$.

In the associated affine space $\mathcal{U}_I$, the profile p_I^k of the modality k is represented by the *modality point* M^k weighted by $\sum\limits_{i\in I} \delta_{ik}/(nQ) = f_k/Q$; the Dirac measure δ_I^i is represented by the ideal point I^i, and the null measure by the origin–point O. By the barycentric relation (§2.2.2, p.33), every point M^k is the barycenter of ideal points $(\mathrm{I}^i)_{i\in I}$ weighted by $(p_i^k)_{i\in I}$:

$$\mathrm{M}^k = \sum_{i\in I} p_i^k\, \mathrm{I}^i = \sum_{i\in I<k>} \mathrm{I}^i/n_k \tag{5.9}$$

Therefore the modality point M^k is the equibarycenter of ideal points $(\mathrm{I}^i)_{i\in I<k>}$ corresponding to the subset $I<k>$ of individuals who have chosen modality k. Every point M^k is weighted by f_k/Q $\left(= \sum\limits_{i\in I} \delta_{ik}/(\sum\limits_{i\in I}\sum\limits_{k\in K} \delta_{ik})\right)$. Hence the cloud M^K of modality points.

Basic properties

Mean point. The mean profile $f_I = (1/n)_{i\in I}$ is represented by the mean point G of the cloud M^K; one has:

$$\mathrm{G} = \sum_{k\in K} (f_k/Q)\mathrm{M}^k = \sum_{i\in I} \mathrm{I}^i/n$$

With the question q there is associated the subcloud of its modalities: $\mathrm{M}^{K<q>} = (\mathrm{M}^k)_{k\in K<q>}$.

Property 5.11. *The mean point of the subcloud of modalities of any question is the mean point of the cloud:* $\forall q \in Q : \sum\limits_{k\in K<q>} f_k \mathrm{M}^k = \mathrm{G}$.

Proof. $\sum\limits_{k\in K<q>} f_k \mathrm{M}^k = \sum\limits_{k\in K<q>} f_k \sum\limits_{i\in I<k>} \mathrm{I}^i/n_k = \sum\limits_{k\in K<q>} \sum\limits_{i\in I} \delta_{ik}\mathrm{I}^i/n = \sum\limits_{i\in I} \mathrm{I}^i/n$
since $\forall i \in I : \sum\limits_{k\in K<q>} \delta_{ik} = 1$. ◁

— *Remark.* From this property, it follows that the projections onto an axis of the modality points of a same question are located on both sides of the origin. In particular, for a question with two modalities k and k', the points M^k and $\mathrm{M}^{k'}$ are collinear with the point G, one has: $f_k\overrightarrow{\mathrm{GM}}^k = -f_{k'}\overrightarrow{\mathrm{GM}}^{k'}$. This property is called "lever principle" by Benzécri.

Metric on $\mathcal{U}_I$

The vector space $\mathbb{R}_I$ of measures over I is equipped with the Φ^2–metric of center $f_I = (1/n)_{i\in I}$ (CORRESPONDENCE, p.29). In the associated geometric space $\mathcal{U}_I$, vectors $(\overrightarrow{\mathrm{OI}}^i)_{i\in I}$ form an orthogonal basis, with $\mathrm{OI}^i = \sqrt{n}$.

One easily verifies that $\overrightarrow{\mathrm{OG}}$ is perpendicular to the support of the cloud M^K, and that $\mathrm{OG} = 1$.

Property 5.12. *The distance of the modality point* M^k *from the mean point* G *is* $\mathrm{GM}^k = \sqrt{(1/f_k) - 1}$.

Proof. $(\mathrm{OM}^k)^2 = \sum\limits_{i\in I<k>} (\mathrm{OI}^i)^2/(n_k)^2 = \sum\limits_{i\in I<k>} n/n_k^2 = n\, n_k/(n_k)^2 = 1/f_k$.
$\overrightarrow{\mathrm{OG}} \perp \overrightarrow{\mathrm{GM}}^k \Rightarrow (\mathrm{GM}^k)^2 = (\mathrm{OM}^k)^2 - \mathrm{OG}^2 = (1/f_k) - 1$. ◁

Property 5.13. *Recall that* $n_{kk'}$ *is the number of individuals who have chosen both modalities* k *and* k'*; letting* $f_{kk'} = n_{kk'}/n$ *the distance between modality points* M^k *and* $\mathrm{M}^{k'}$ *is such that*

$$(\mathrm{M}^k\mathrm{M}^{k'})^2 = (1/f_k) + (1/f_{k'}) - 2\, f_{kk'}/(f_k\, f_{k'})$$

Proof. $(\mathrm{M}^k\mathrm{M}^{k'})^2 = \frac{1}{1/n} \sum\limits_{i\in I} (\frac{\delta_{ik}}{n_k} - \frac{\delta_{ik'}}{n_{k'}})^2 = n(\sum\limits_{i\in I} \frac{\delta_{ik}}{n_k^2} + \sum\limits_{i\in I} \frac{\delta_{ik'}}{n_{k'}^2} - 2\sum\limits_{i\in I} \frac{\delta_{ik}\delta_{ik'}}{n_k n_{k'}})$
$- (n/n_k) + (n/n_{k'}) - 2n_{kk'}/(n_k n_{k'}/n)$ ◁

— *Remarks.* The less frequent a modality k, the farther the point M^k is from the mean point. The distance $\mathrm{M}^k\mathrm{M}^{k'}$ is all the larger as the number of individuals that have chosen k and non–k' or k' and non–k is large.

Principal axes and variables

Property 5.14. *The principal axes and variables of the cloud of modalities verify the transition equations of the multiple correspondence* δ_{IK}.

Proof. This property results from the construction of the cloud M^K from the CA of the measure δ_{IK} over $I \times K$ (cf. equations (5.7), p.188). ◁

The connection between the cloud of modalities and the cloud of individuals will be made later on (§5.1.3).

Property 5.15 (Principal coordinates). *The principal coordinate of the ideal point* I^i *on Axis* ℓ *is equal to* z^i_ℓ; *the principal coordinate of the modality point* M^k *on Axis* ℓ *is* $y^k_\ell = \xi_\ell z^k_\ell$.

Proof. By Definition 3.15 (CLOUD, p.91), the coordinate of I^i on Axis ℓ (defined by $(\mathrm{G}, \overrightarrow{\alpha_\ell})$, with $\|\overrightarrow{\alpha_\ell}\| = 1$) is $\langle \overrightarrow{\mathrm{GI}^i} | \overrightarrow{\alpha_\ell} \rangle = \langle \overrightarrow{\mathrm{OI}^i} | \overrightarrow{\alpha_\ell} \rangle = n z_{i\ell} = z^i_\ell$.

The second transition formula (p.188) expresses that the projection onto Axis ℓ of the modality point M^k is the equibarycenter of the projection onto Axis ℓ of the ideal points $(\mathrm{I}^i)_{i \in I<k>}$, hence $y^k_\ell = \xi_\ell z^k_\ell$. ◁

Property 5.16. *The regression coefficient of indicator variable* δ^I_k *on the* ℓ*-th standard principal variable* z^I_ℓ *is equal to* $Q\, y_{k\ell} = f_k y^k_\ell$.

Proof. From Property 5.8 (p.189), this coefficient is equal to $Q\xi_\ell z_{k\ell}$, therefore to $Q\, y_{k\ell} = Q(f_k/Q) y^k_\ell$. ◁

Contributions of modalities

Contributions to the variance of the cloud. One applies the definitions of the chapter CLOUD (p.78).

• The *absolute contribution of the modality point* M^k to the variance of the cloud, denoted Cta_k, is its weight multiplied by its squared distance to the mean point: $\mathrm{Cta}_k = (f_k/Q)(\mathrm{GM}^k)^2 = (f_k/Q)(1/f_k - 1) = (1 - f_k)/Q$. Its *relative contribution* is $\mathrm{Ctr}_k = \mathrm{Cta}_k \,/\, \mathrm{Var}\, \mathrm{M}^K$, hence:

$$\mathrm{Ctr}_k = (1 - f_k)/(K - Q) \tag{5.10}$$

• The nesting structure $K<Q>$ invites one to study the contributions of questions and the relative contributions of modalities to questions. The *absolute contribution of the question* q, denoted Cta_q, is the sum of the contributions of its modalities:

$$\mathrm{Cta}_q = \sum_{k \in K<q>} \mathrm{Cta}_k = (K_q - 1)/Q$$

The variance v_q of the subcloud $\mathrm{M}^{K<q>}$ of modalities of question q is equal to $\sum_{k \in K<q>} f_k(\mathrm{GM}^k)^2 = Q\,\mathrm{Cta}_q = K_q - 1$.

The *relative contribution* of the question q to the variance of cloud is

$$\mathrm{Ctr}_q = (K_q - 1)/(K - Q) \tag{5.11}$$

The *relative contribution of a modality to its question* is:
$\mathrm{Ctr}_k^q = \mathrm{Cta}_k \,/\, \mathrm{Cta}_q = (1 - f_k)/(K_q - 1) \quad (k \in K<q>)$

• The *absolute contribution of the dipole* $\{k, k'\}$, or absolute contribution of the deviation between the two modality points M^k and $\mathrm{M}^{k'}$, is $(\mathrm{M}^k\mathrm{M}^{k'})^2/(Q/f_k + Q/f_{k'})$ (cf. CLOUD, property 3.21, p.102), hence:

$$\mathrm{Cta}_{(k,k')} = (1/Q)\Big(1 - 2\,n_{kk'}/(n_k + n_{k'})\Big) \tag{5.12}$$

— *Remark.* This contribution is the aggregation index used for aggregating modalities in an ascending hierarchical classification according to variance (cf. CLOUD, §3.6). The index $1 - Q\,\mathrm{Cta}_{(k,k')} = n_{kk'}/\frac{n_k+n_{k'}}{2}$ is the Dice and Sørensen' similarity index.

Property 5.17. *The cosine of the angle $\theta_{kk'}$ between $\overrightarrow{\mathrm{GM}}^k$ and $\overrightarrow{\mathrm{GM}}^{k'}$ is equal to the correlation between the indicator variables δ_k^I and $\delta_{k'}^I$.*

Proof. $\langle\overrightarrow{\mathrm{OM}}^k|\overrightarrow{\mathrm{OM}}^{k'}\rangle = \langle\overrightarrow{\mathrm{OG}} + \overrightarrow{\mathrm{GM}}^k|\overrightarrow{\mathrm{OG}} + \overrightarrow{\mathrm{GM}}^{k'}\rangle = \mathrm{OG}^2 + \langle\overrightarrow{\mathrm{GM}}^k|\overrightarrow{\mathrm{GM}}^{k'}\rangle$; $\mathrm{OG}^2 = 1$, $\langle\overrightarrow{\mathrm{OM}}^k|\overrightarrow{\mathrm{OM}}^{k'}\rangle = \sum_{i \in I} n\,(\delta_{ik}/n_k)(\delta_{ik'}/n_{k'}) = n\,n_{kk'}/(n_k\,n_{k'})$, hence $\langle\overrightarrow{\mathrm{GM}}^k|\overrightarrow{\mathrm{GM}}^{k'}\rangle = f_{kk'}/(f_k\,f_{k'}) - 1$ and $\cos\theta_{kk'} = \frac{(f_{kk'} - f_k f_{k'})}{\sqrt{f_k\,f_{k'}(1-f_k)(1-f_{k'})}}$. ◁

— *Remarks.* The less frequent a modality, the more it contributes to the variance of cloud (the contribution is always less than $1/Q$). In the limit ($f_k = 0$), one would have a point M^k at an infinite distance from the mean point, with a maximum contribution ($1/Q$). In this case, one will attempt to group an infrequent modality with another one of the same question, and if this proves uneasy one will proceed to a specific MCA (cf. §5.2).

The contribution of a question to the variance of the cloud does not depend on data; it is all the larger as the number of modalities of this question is larger. When all questions have the same numbers of modalities, they contribute to the variance of cloud equally. In the MCA of a questionnaire, one often seeks to give equal importances to questions, which may lead one to group some modalities of a same question.

The cloud of modalities is entirely defined by the joint frequencies of pairs of modalities, in other words by the set of all contingency tables between pairs of questions ("Burt table", see hereafter §5.1.3). The cloud of modalities provides a synthetic representation of all these tables.

Contributions to the variance of an axis. One applies the definitions of §3.3.3 (CLOUD, p.92).

• The *absolute contribution of modality k to Axis ℓ*, denoted $\text{Cta}_{k\ell}$, is, by definition, equal to its weight multiplied by the square of its coordinate: $\text{Cta}_{k\ell} = (f_k/Q)(y_\ell^k)^2$. Hence the *relative contribution* $\text{Ctr}_k^\ell = (f_k/Q)(z_\ell^k)^2$.

The angle $\theta_{k\ell}$ $(0 \leq \theta_{k\ell} \leq \pi)$ between $\overrightarrow{\text{GM}^k}$ and Axis ℓ is defined by $\cos\theta_{k\ell} = y_\ell^k/\text{GM}^k = \sqrt{f_k/(1-f_k)}\, y_\ell^k$, hence the *quality of representation* of k on Axis ℓ: $\cos^2\theta_{k\ell} = f_k(y_\ell^k)^2/(1-f_k) = Q\,\text{Cta}_{k\ell}/(1-f_k)$.

The coefficient $\cos\theta_{k\ell}$ is equal to the *correlation* between the ℓ–th principal variable and the variable δ_k^I over I indicator of k (cf. property 5.8).

• The *absolute contribution of the question q to Axis ℓ* is the sum of the contributions of its modalities: $\text{Cta}_{q\ell} = \sum_{k\in K<q>} \text{Cta}_{k\ell} = \sum_{k\in K<q>} (f_k/Q)(y_\ell^k)^2$.

If $v_{q\ell} = \sum_{k\in K<q>} f_k(y_\ell^k)^2$ denotes the variance of the modalities of question q on Axis ℓ, one has the relation $v_{q\ell} = Q\,\text{Cta}_{q\ell}$.

The *relative contribution of the question q to Axis ℓ* is denoted Ctr_q^ℓ, with $\text{Ctr}_q^\ell = \text{Cta}_{q\ell}/\lambda_\ell$. The relative contribution of the modality k to its question q on Axis ℓ is denoted $\text{Ctr}_{k\ell}^q$:

$$\text{Ctr}_{k\ell}^q = \text{Cta}_{k\ell}\,/\,\text{Cta}_{q\ell} = f_k(y_\ell^k)^2/v_{q\ell} \qquad (5.13)$$

• For *two modalities k and k′ of the same question*, the absolute contribution of the dipole $\{k, k'\}$, or the absolute contribution of the deviation between the two modality points M^k and $\text{M}^{k'}$ to Axis ℓ, denoted $\text{Cta}_{(k,k')\ell}$, is such that, letting $p_{(k,k')} = 1/\left(\frac{1}{f_k} + \frac{1}{f_{k'}}\right)$

$$\text{Cta}_{(k,k')\ell} = \tfrac{1}{Q} \times p_{(k,k')} \times (y_\ell^k - y_\ell^{k'})^2 \qquad (k \,\&\, k' \in K<q>) \qquad (5.14)$$

The angle $\theta_{(k,k')\ell}$ $(0 \leq \theta_{(k,k')\ell} \leq \pi)$ between $\overrightarrow{\text{M}^k\text{M}^{k'}}$ and Axis ℓ, for two modalities of the same question is defined by

$$\cos\theta_{(k,k')\ell} = \frac{y_\ell^{k'} - y_\ell^{k'}}{\text{M}^k\text{M}^{k'}} = \frac{y_\ell^{k'} - y_\ell^k}{\sqrt{1/f_k + 1/f_{k'}}} \qquad \left(\begin{smallmatrix}k\\ k'\end{smallmatrix} \in K<q>\right)$$

hence its *quality of representation*: $\cos^2\theta_{(k,k')\ell} = \frac{(y_\ell^k - y_\ell^{k'})^2}{1/f_k + 1/f_{k'}} = Q\,\text{Cta}_{(k,k')\ell}$.

— *Remark.* For a modality k of a question q and the grouping, denoted $\widetilde{k}$, of the modalities of this question other than k, the points M^k, G and $\mathrm{M}^{\widetilde{k}}$ are collinear; one has $f_k\overrightarrow{\mathrm{GM}}^k = -(1-f_k)\overrightarrow{\mathrm{GM}}^{\widetilde{k}}$ and

$$\cos^2\theta_{k\ell} = \cos^2\theta_{\widetilde{k}\ell} = \cos^2\theta_{(k,\widetilde{k})\ell} = Q\,\mathrm{Cta}_{(k,\widetilde{k})\ell}$$

Supplementary modalities and questions

Let us consider a modality k_s that has not served to define the distance (construction of cloud) and consequently to determine the principal axes, that is, a supplementary (or illustrative) modality. Such a modality can be represented by the point M^{k_s} $\sum\limits_{i\in I<k_s>} \mathrm{I}^i/n_{k_s}$, barycenter of the ideal points $(\mathrm{I}^i)_{i\in I<k_s>}$ associated with the individuals that have chosen this modality. The coordinate $y_\ell^{k_s}$ of the point M^{k_s} on Axis ℓ is: $y_\ell^{k_s} = \sum\limits_{i\in I<k_s>} z_\ell^i/n_{k_s}$.

5.1.3 Modality Mean–points in the Cloud of Individuals

Returning to the cloud of individuals, let us now consider the partition of individuals induced by a question (active or supplementary) (cf. p.182).

Modality mean–points. With modality k, there is associated the subset $I<k>$ of n_k individuals who have chosen this modality; hence the *subcloud* $\mathrm{M}^{I<k>}$. The mean point of this subcloud, denoted $\widehat{\mathrm{M}}^k$, is called a *modality mean–point*; it is weighted by n_k, and defined by $\widehat{\mathrm{M}}^k = \sum\limits_{i\in I<k>} \mathrm{M}^i/n_k$.

Property 5.18. *The Cartesian coordinates of the modality mean–point* $\widehat{\mathrm{M}}^k$ *are equal to* $(n_{kk'}/(Q\,n_k))_{k'\in K}$, *with* $n_{kk'} = \sum\limits_{i\in I}\delta_{ik}\delta_{ik'}$ *and* $n_k = \sum\limits_{i\in I}\delta_{ik}$.

Proof. $$\overrightarrow{\mathrm{O}\widehat{\mathrm{M}}}^k = \sum_{i\in I}\delta_{ik}\overrightarrow{\mathrm{OM}}^i/n_k = \sum_{i\in I}(\delta_{ik}/n_k)\sum_{k'\in K}(\delta_{ik'}/Q)\overrightarrow{\mathrm{OI}}^{k'}$$
$$= \sum_{k'\in K}\Big(\sum_{i\in I}\frac{\delta_{ik}\,\delta_{ik'}}{n_k\,Q}\Big)\overrightarrow{\mathrm{OI}}^{k'} = (1/Q)\sum_{k'\in K}(n_{kk'}/n_k)\overrightarrow{\mathrm{OI}}^{k'} \quad \triangleleft$$

Property 5.19. *The ℓ–th principal coordinate of modality mean–point* $\widehat{\mathrm{M}}^k$ *on Axis ℓ, denoted* $\widehat{y}_\ell^k$, *is equal to* $\xi_\ell y_\ell^k$ *where* y_ℓ^k *is the coordinate of modality point* M^k *on Axis ℓ.*

$$\widehat{y}_\ell^k = \xi_\ell y_\ell^k \tag{5.15}$$

Proof. $\widehat{y}^k_\ell = \sum\limits_{i\in I<k>} y^i_\ell/n_k = \xi_\ell \sum\limits_{i\in I<k>} z^i_\ell/n_k = \xi_\ell \sum\limits_{i\in I} \delta_{ik} z^i_\ell/n_k = \xi_\ell \sum\limits_{i\in I} p^k_i z^i_\ell$ with, from equations (5.7), $\sum\limits_{i\in I} p^k_i z^i_\ell = y^k_\ell$. ◁

— *Remark.* Equivalently, $\widehat{y}^k_\ell/\xi_\ell = y^k_\ell$, that is, y^k_ℓ expresses the deviation in SD units of the point $\widehat{\mathrm{M}}^k$ from the mean point G of the cloud on Axis ℓ.

Partition of the cloud of individuals by a question

Any question $q \in Q$ induces a partition of the cloud of individuals into K_q (number of modalities of question q) subclouds $(\mathrm{M}^{I<k>})_{k\in K<q>}$.

Definition 5.6 (Between–q cloud). *The cloud* $(\widehat{\mathrm{M}}^k, n_k)_{k\in K<q>}$ *of the mean points of the* K_q *subclouds induced by question* q *is called between–*q *(or inter–*q*) cloud and denoted* $\widehat{\mathrm{M}}^{K<q>}$.

Property 5.20. *i) For all* q*, the mean point of the between–*q *cloud is the mean point* G *of the cloud. ii) Let* $(\Phi^2)^{qq'}$ *be the* Φ^2 *coefficient*[11] *of the contingency table crossing questions* q *and* q'*, then the between–*q *variance is* $(1/Q) \sum\limits_{q'\in Q} (\Phi^2)^{qq'}$.

Proof. See Exercise 5.2. ◁

Recall (cf. CLOUD, p.104) that the ratio of the between–variance to the total variance is called *correlation ratio*; for question q and Axis ℓ, this ratio, denoted $\eta^2_{q\ell}$, is also called the *discrimination coefficient* of question q for Axis ℓ.

Property 5.21. *On Axis* ℓ*, the correlation ratio* $\eta^2_{q\ell}$ *associated with question* q *is equal to the variance* $v_{q\ell}$ *of the subcloud of the modalities of question* q *in projection on Axis* ℓ.

Proof. $\eta^2_{q\ell} = \sum\limits_{k\in K<q>} f_k (\widehat{y}^k_\ell)^2/\lambda_\ell = \sum\limits_{k\in K<q>} f_k (y^k_\ell)^2 = v_{q\ell}$. ◁

Corollary 5.5. *The correlation ratio* $\eta^2_{q\ell}$ *is equal to the square of the multiple correlation of* z^I_ℓ *with* $(\delta^I_k)_{k\in K<q>}$.

Proof. This corollary follows from Property (5.9) (p.189). ◁

[11] i.e. $(\Phi^2)^{qq'} = \sum\limits_{k\in K<q>} \sum\limits_{k'\in K<q'>} (f_{kk'} - f_k f_{k'})^2/(f_k f_{k'})$.

Definition 5.7 (Burt table). *The Burt table associated with the multiple correspondence δ_{IK} is the symmetric table $K \times K$ defined by*

$$b_{kk'} = \sum_{i \in I} \delta_{ik}\, \delta_{ik'} = \begin{cases} n_k & \text{if } \; k = k' \\ 0 & \text{if } \; k<q> \neq k'<q> \\ n_{kk'} & \text{if } \; k<q> \text{ et } k'<q'> \text{ and } q \neq q' \end{cases}$$

Letting $b_k = \sum_{k' \in K} b_{kk'}$, one has: $\sum_{k \in <q>} b_k = n\,Q$, and $b = \sum_{k \in K} b_k = n\,Q^2$.

The Burt table is made of Q^2 subtables. The Q diagonal blocks are square diagonal with terms equal to n_k. The $Q\,(Q-1)$non–diagonal blocks are the contingency tables between pairs of questions (each one appearing twice). Lastly, the $K<q> \times K$ subtable constituted by the rows pertaining to question q is called the *Burt q–subtable.*

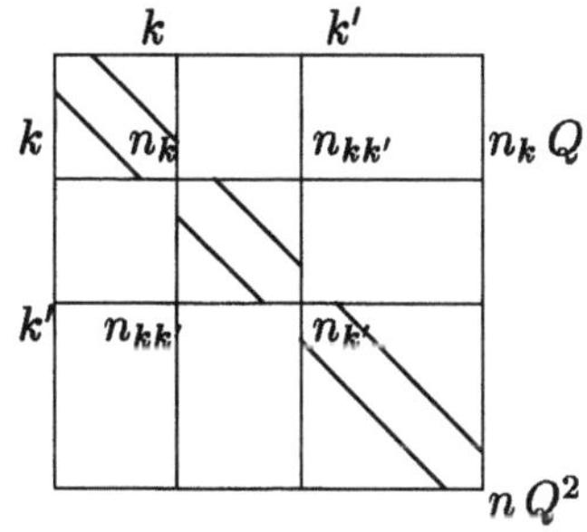

Φ^2 of Burt table. Denoting $(\Phi^2)^{qq'}$ the Φ^2 coefficient of the contingency table crossing questions q and q', one has the following properties:

$$\begin{cases} (\Phi^2)^{qq} = K_q - 1 \\ (\Phi^2)^{qq'} = \sum\limits_{k \in K<q>} \sum\limits_{k' \in K<q'>} \dfrac{(n_{kk'})^2}{n_k\, n_{k'}} - 1 \qquad \text{for} \quad q' \neq q \end{cases}$$

The Φ^2 of the *Burt q–subtable* denoted $(\Phi^2)^q$, is the average of the Φ^2 of the Q subtables that constitute this subtable: $(\Phi^2)^q = (1/Q) \sum_{q' \in Q} (\Phi^2)^{qq'}$. Consequently, it is also the between–q variance (Property 5.20).

The Φ^2 of the *Burt table* is the average of the Φ^2 of the Q Burt q–subtables, or else the average of the Φ^2 of all Q^2 subtables of Burt table:

$$\Phi^2_{\text{Burt}} = \frac{1}{Q} \sum_{q \in Q} (\Phi^2)^q = \frac{1}{Q^2} \sum_{q \in Q} \sum_{q' \in Q} (\Phi^2)^{qq'} = \frac{1}{Q}\left(\frac{K-Q}{Q} + 2\frac{1}{Q} \sum_{qq' \in \mathcal{P}_2(Q)} (\Phi^2)^{qq'} \right)$$

where $\mathcal{P}_2(Q)$ denotes the set of all $Q(Q-1)/2$ pairs of questions.

For the proofs of these properties, see Exercise 5.2.

Definition 5.8 (Burt cloud). *The cloud of the K points associated with Burt table, that is, the union of the Q clouds $(\widehat{\mathrm{M}}^{K<q>})_{q \in Q}$, is called the Burt cloud and denoted $(\widehat{\mathrm{M}}^K, f_K/Q)$.*

The variance of the Burt cloud is equal to $\frac{1}{Q} \sum_{q \in Q} \operatorname{Var} \widehat{\mathrm{M}}^{K<q>}$, that is,

$\operatorname{Var} \widehat{\mathrm{M}}^K = \frac{1}{Q} \sum_{q \in Q} (\Phi^2)^q = \Phi^2_{\text{Burt}}$.

Property 5.22 (CA of Burt table and MCA). *The standard principal variables of the Burt cloud are the standard principal variables of the multiple correspondence δ_{IK}, associated with the squares of eigenvalues.*

Proof. From equations (5.7) (p.188), one deduces $z^i_\ell = \frac{1}{\xi_\ell} \sum\limits_{k' \in K} p^i_{k'} z^{k'}_\ell$ and $\sum\limits_{i \in I} p^k_i z^i_\ell = \frac{1}{\xi_\ell} \sum\limits_{i \in I} p^k_i \sum\limits_{k' \in K} p^i_{k'} z^{k'}_\ell = \xi_\ell z^k_\ell$, hence $\frac{1}{Q n_k} \sum\limits_{k' \in K} \big(\sum\limits_{i \in I} \delta_{ik'} \delta_{ik} \big) z^{k'}_\ell = \xi^2_\ell z^k_\ell$, that is, $\sum\limits_{k' \in K} (b_{kk'}/b_k) z^{k'}_\ell = \lambda_\ell z^k_\ell$: the transition formula (unique because the correspondence b_{KK} is symmetrical) of Burt table[12]. ◁

Binary questionnaire. In the case of a *binary questionnaire*, that is, a questionnaire with two questions q and q', let $A = K\langle q\rangle$ and $B = K\langle q'\rangle$ (with $K = A \cup B$), three analyses can be contemplated, namely

i) the analysis of the $I \times K$ disjunctive table,

ii) the analysis of the $K \times K$ Burt table,

iii) the analysis of the $A \times B$ contingency table.

It can be shown (Exercise 5.3, p.243) that the three analyses are equivalent, in the sense that they satisfy the following equivalence property.

Property 5.23 (Equivalence property). *Let μ_ℓ denote the ℓ-th eigenvalue of the contingency table n_{AB} (with $\ell \leq \min(A, B) - 1$), then the ℓ-th eigenvalue λ_ℓ of the disjunctive table is equal to $\frac{1}{2}(1 + \sqrt{\mu_\ell})$, and the standard principal variables over A and B of n_{AB} determine the standard principal variables over $K = A \cup B$ of the disjunctive table.*

Proof. See Exercise 5.3 (p.243). ◁

Variance rates and contributions

Variance rates or inertia rates, modified rates. The variance rates (τ_ℓ) of the *disjunctive table* are $\lambda_\ell / (\sum\limits_{\ell \in L} \lambda_\ell) = \lambda_\ell / (\frac{K}{Q} - 1)$; those of the *Burt table* are $\lambda^2_\ell / (\sum\limits_{\ell \in L} \lambda^2_\ell)$. The eigenvalues of the Burt table are *more separated* than those of the disjunctive table, since $\lambda_\ell / \lambda_{\ell+1} > 1$ entails $\lambda^2_\ell / \lambda^2_{\ell+1} > \lambda_\ell / \lambda_{\ell+1}$, therefore for the Burt table the *decrease trend* of eigenvalues is steeper than for the disjunctive table.

To appreciate the relative importances of principal axes, Benzécri has proposed to calculate *modified rates*[13] (τ'_ℓ) obtained as follows.

---

[12] For a symmetric table, the transition formula can associate with a principal variable either the same variable (direct factor), or the variable with opposite sign (inverse factor). For Burt table, all principal variables are direct factors: see Benzécri (1992, p.412).

[13] See Benzécri (1992, p.412).

Calculate the mean $\overline{\lambda}$ of the (nontrivially null) eigenvalues, that is, $\overline{\lambda} = \frac{\mathrm{Var}\,\mathrm{M}^K}{K-Q} = \frac{1}{Q}$; calculate $\lambda'_\ell = (\frac{Q}{Q-1})^2(\lambda_\ell - \overline{\lambda})^2$; determine $L' \leq L$ such that $\lambda_{L'} \geq \overline{\lambda}$ and $\lambda_{L'+1} < \overline{\lambda}$; then the modified rates are equal to

$$\tau'_\ell = \lambda'_\ell / (\sum_{\ell=1}^{L'} \lambda'_\ell) \quad (\ell \leq L') \tag{5.16}$$

When Q is large, λ'_ℓ differs little from λ^2_ℓ. Modified rates can be interpreted as an index of the deviation of the cloud from sphericity[14].

Let us call *modified Burt table* the table obtained by replacing the Q diagonal blocks of the Burt table by square subtables of zeros. It can be shown that the eigenvalues of the modified Burt table are equal to λ'_ℓ, and that the only direct principal variables are those whose eigenvalues are greater than $1/Q$. The modified rates can thus be interpreted in terms of relative contributions to the direct axes of the modified Burt table. The modified rates of the MCA of a binary questionnaire are equal to the rates of the CA of the corresponding contingency table n_{AB}.

Contributions of modalities to axes. Recall (§5.1.2, p.194) that $\mathrm{Cta}_{k\ell}$ denotes the absolute contribution of the modality point $(\mathrm{M}^k, (f_k/Q))$ to Axis ℓ, with $\mathrm{Cta}_{k\ell} = (f_k/Q)(y^k_\ell)^2$, and that $\mathrm{Ctr}^\ell_k = \mathrm{Cta}_{k\ell}/\lambda_\ell$ denotes the corresponding relative contribution.

In the *cloud of individuals*, the absolute contribution of a modality mean–point $(\widehat{\mathrm{M}}^k, f_k)$ to Axis ℓ is $\mathrm{Cta}_{\widehat{k}\ell} = f_k(\widehat{y}^k_\ell)^2 = \lambda_\ell Q\,\mathrm{Cta}_{k\ell}$. Its relative contribution $\mathrm{Ctr}^\ell_{\widehat{k}}$ to Axis ℓ is equal to $\lambda_\ell Q\,\mathrm{Ctr}^\ell_k$, and therefore is greater than the relative contribution Ctr^ℓ_k of modality mean–point M^k, for all principal axes such that $\lambda_\ell > \overline{\lambda} = 1/Q$.

The *relative contribution of a modality to its question* ($\mathrm{Ctr}^q_{k\ell}$, cf. formula (5.13), p.196), is the same, axis by axis, whether calculated from the cloud of modalities or from the cloud of individuals; it is equal to the relative contribution of this modality to the variance of its question.

5.1.4 Statistical Characterizations of MCA

Like for CA, the principal variables of MCA can be characterized as *optimal codings* of a pair of sets, the set I of individuals and the set K of modalities. These statistical characterizations can be obtained as special cases of CA, or they can be derived directly, as was done by Guttman (1941), on the data

[14] For a detailed study see Durand (1998).

table that we call a multiple correspondence[15]. Hereafter we summarize optimality criteria for the discriminant and the canonical analyses.

Discriminant analysis. Let $x^K = (x^k)_{k\in K}$ br a numerical coding of modalities, one can calculate the "score" $\widehat{x}^i = \frac{1}{Q}\sum_{k\in K<i>} x^k$ of individual i, that is, the average of the Q values of the modalities chosen by i. If one seeks x^K so as to maximize the ratio $\operatorname{Var}\widehat{x}^I / \operatorname{Var} x^K$, the solution is the first principal variable over K, with $\operatorname{Var}\widehat{x}^I / \operatorname{Var} x^K = \lambda_1$. In brief, the optimal coding of K best discriminates the individuals.

Optimal codings are evidently defined up to a linear transformation. If ones takes $x^K = z^K$ (first standard principal variable over K), then $\widehat{x}^I = y^I$ (first principal variable over I), with $y^i = \sum_{k\in K<i>} z^k/Q$ (transition formula). If ones takes $x^K = y^K$, then $\widehat{x}^I = \widehat{y}^I$, with $\widehat{y}^i = \sum_{k\in K<i>} y^k/Q$.

The construction is symmetrical, the roles of I and K can be interchanged. The optimal coding of I best discriminates the modalities.

Canonical analysis. If one seeks numerical codings x^I and x^K so as to maximize the correlation between the corresponding variables over $I\times K$ with respect to the basic measure δ_{IK}, the solution is the pair of principal variables (y^I, y^K). A related criterion is that of *generalized canonical analysis* in the sense of Carroll (1968): One seeks a variable x^I such that if η_q^2 denotes the correlation ratio between x^I and the categorized variable q, the sum $\sum_{q\in Q} \eta_q^2$ is maximum; the solution is again the first principal variable.

For Q dichotomous variables, the MCA amounts to the standard PCA of a protocol of Q indicator variables: cf. Exercise 4.5 (p.173).

5.1.5 Synopsis

•*Points and coordinates*

	Overall	Principal	The link
Cloud of Individuals	$\mathrm{M}^i = \frac{1}{Q}\sum_{k\in K<i>} \mathrm{I}^k$	$y_\ell^i = \frac{1}{Q}\sum_{k\in K<i>} z_\ell^k$	$y_\ell^i = \xi_\ell z_\ell^i$
Cloud of Modalities	$\mathrm{M}^k = \frac{1}{n_k}\sum_{i\in I<k>} \mathrm{I}^i$	$y_\ell^k = \frac{1}{n_k}\sum_{i\in I<k>} z_\ell^i$	$y_\ell^k = \xi_\ell z_\ell^k$
Cloud of Individuals	$\widehat{\mathrm{M}}^k = \frac{1}{n_k}\sum_{i\in I<k>} \mathrm{M}^i$	$\widehat{y}_\ell^k = \frac{1}{n_k}\sum_{i\in I<k>} y_\ell^i$	$\widehat{y}_\ell^k = \xi_\ell y_\ell^k$
Cloud of Modalities	$\widehat{\mathrm{M}}^i = \frac{1}{Q}\sum_{k\in K<i>} \mathrm{M}^k$	$\widehat{y}_\ell^i = \frac{1}{Q}\sum_{k\in K<i>} y_\ell^k$	$\widehat{y}_\ell^i = \xi_\ell y_\ell^i$

[15]Like Fisher for CA, Guttman was concerned with optimal scaling, not geometry.

• *Contributions of modality points to axes*

Cloud of modalities	$(\mathrm{M}^k, \frac{f_k}{Q})$	$\mathrm{Cta}_{k\ell} = \frac{f_k}{Q}(y_\ell^k)^2$	$\mathrm{Ctr}_k^\ell = \frac{f_k}{Q}(z_\ell^k)^2$	$\mathrm{Ctr}_{k\ell}^q = f_k(y_\ell^k)^2/v_{q\ell}$
Cloud of individuals	$(\widehat{\mathrm{M}}^k, f_k)$	$\lambda_\ell Q\,\mathrm{Cta}_{k\ell}$	$Q\,\mathrm{Cta}_{k\ell}$	$\mathrm{Ctr}_{k\ell}^q$

5.1.6 Computation Procedure

In terms of a symmetric matrix, one calculates the eigenvalues λ_ℓ and the normed eigenvectors $(c_{k\ell})_{k\in K}$ ($\sum\limits_{k\in K} c_{k\ell}^2 = 1$) of the $K \times K$ symmetric matrix $\mathbf{S} = [s_{kk'}]$, with

$$s_{kk'} = \frac{1}{Q} \times \frac{n_{kk'} - n_k\, n_{k'}/n}{\sqrt{n_k\, n_{k'}}} \qquad \begin{cases} k \in K \\ k' \in K \end{cases}$$

One then deduces the coordinates of modalities on Axis ℓ:

$$y_\ell^k = \sqrt{\lambda_\ell}\sqrt{Q}\frac{c_{k\ell}}{\sqrt{n_k/n}} \qquad (k \in K)$$

The coordinates of individuals on Axis ℓ follow:

$$y_\ell^i = \frac{1}{\sqrt{Q}} \sum_{k\in K<i>} \frac{c_{k\ell}}{\sqrt{n_k/n}} \qquad (i \in I)$$

— *Remark.* $(\sqrt{n_k})_{k\in K}$ is eigenvector of the matrix $\mathbf{S}$ associated to the eigenvalue 0; very often, one calculates the symmetric matrix with general term $(1/Q)n_{kk'}/\sqrt{n_k\, n_{k'}}$, then (cf. *MATH.BASES*, Property 10.28, p.443) one gets the trivial eigenvalue equal to 1.

5.2 Specific MCA

Specific analysis is especially important in MCA. Two kinds of specific MCA will be presented in what follows.

• *Specific* MCA *by restriction to a subset of modalities.*

A case in point for specific MCA is the problem of infrequent modalities. As we have seen, points representing infrequent modalities are remote from the center of the cloud; they contribute heavily to the variance of their question, and they can be too influential for the determination of axes. What should be done in the first place is attempting to pool infrequent modalities with some other modalities of the same question. When this is not naturally feasible, it is necessary, in order to preserve the constitutive properties of multiple correspondence, to build up a variant of MCA that, without deleting the modality out of the analysis, simply ignores it in the calculation of distances between individuals. This method so designed that it preserves the structural regularities of MCA we call *specific* MCA (Le Roux, 1999).

In specific MCA, such modalities are called *passive modalities* (as opposed to active ones) of active questions. Beside infrequent modalities, specific MCA will also be usefully applied to modalities of "no–interest", such as the "Others" category of a question when this category is a heterogeneous mixture. Specific MCA will be performed in the *Political Space Study* in Chapter 9 (*CASE STUDIES*).

• *Specific* MCA *of a subcloud of individuals.* This kind of specific MCA arises when it is wished to study a subcloud of individuals, while preserving the distances defined for the whole cloud. Specific MCA differs from the simple MCA applied to the relevant subtable, in which the distances between points and the weights of points are defined from the margins of the data subtable, and consequently differ from those of the MCA of the overall data table. By contrast, in a specific analysis, both distances and weights are defined from the initial complete data table. This kind of specific MCA will be put to work in Chapter 9, when we will construct the *concentration ellipses of subclouds.*

5.2.1 Restriction to a Subset of Modalities

Let $K' \subset K$ be the subset of response modalities of the Q questions used to define the distance between individuals, that is the Q *active questions.* These K' response modalities are *active*, as opposed to the $K'' = K - K'$ other response modalities of these questions, that are *passive*[16].

The specific distance $d'(i,i')$ between individuals i and i' is defined by:

$$d'^2(i,i') = \frac{1}{Q}\sum_{q\in Q} d'^2_q(i,i') = \frac{1}{Q}\sum_{q\in Q}\Big(\sum_{k\in K'<q>} \frac{(\delta_{ik} - \delta_{i'k})^2}{f_k}\Big) \tag{5.17}$$

• $d'_q(i,i') = d_q(i,i')$ (cf. p.181) if both individuals i and i' have chosen active modalities of question q (agreement question or not);

• $d'_q(i,i') = 1/\sqrt{f_k}$ if q is a disagreement question for which one individual has chosen an active modality k and the other a passive one.

[16] "Passive modalities" of active questions will be distinguished from "supplementary modalities", i.e. the modalities of supplementary questions.

— *Remarks*

(i) For two individuals i and i' who have not chosen passive response modalities, the distance is unchanged: $d'(i,i') = d(i,i')$. For two individuals who disagree about a single question for which one has chosen an active response modality $k \in K'$ and the other a passive modality $k' \in K''$, the square of distance decreases by the amount $1/(Qf_{k'})$.

(ii) If one had proceeded to the CA of the measure $\delta_{IK'}$, the distance between individuals i and i' who have chosen only active modalities would have decreased and been equal to $d(i,i')\sqrt{\sum\limits_{k\in K'} n_k/(nQ)}$. But if two individuals i and i' disagree only about one question q for which one has chosen an active modality and the other a passive one, then $\sum\limits_{k\in K'} \delta_{ik} \neq \sum\limits_{k\in K'} \delta_{i'k}$; hence, owing to normalization, even active agreement questions would generate distance, which is clearly an undesirable property.

Cloud of individuals

From a geometric viewpoint, the specific analysis of the cloud M^I amounts to studying the cloud R^I which is the orthogonal projection of the cloud M^I (defined p.182) onto the subspace, denoted $\mathcal{R}$, generated by $(\mathrm{O}, (\mathrm{I}^k)_{k\in K'})$: $\forall i \in I \;:\; \overrightarrow{\mathrm{OR}}^i = (1/Q) \sum\limits_{k\in K'} \delta_{ik}\overrightarrow{\mathrm{OI}}^k$. The projected cloud $(\mathrm{R}^I, 1_I)$ is called *specific cloud*. One has $\mathrm{R}^i\mathrm{R}^{i'} = d'(i,i')$.

Support, mean point and variance. The dimensionality of the *support of the specific cloud* is at most equal to $K' - (Q - Q'')$, where Q'' denotes the number of questions for which at least one modality has been put as passive. If there is exactly one passive modality per question, then $K' = K - Q$ and $Q'' = Q$, the dimensionality is at most $K - Q$ (like in standard MCA).

Property 5.24. *The Cartesian coordinates of the mean point* G' *of the specific cloud* R^I *are equal to* f_k/Q *for* $k \in K'$ *and 0 for* $k \in K''$. *One has* $(\mathrm{OG}')^2 = \sum\limits_{k\in K'} f_k/Q$.

Proof. $\overrightarrow{\mathrm{OG}}' = \sum\limits_{i\in I} \overrightarrow{\mathrm{OR}}^i/n = (1/Qn) \sum\limits_{k\in K'} (\sum\limits_{i\in I} \delta_{ik})\overrightarrow{\mathrm{OI}}^k = \sum\limits_{k\in K'} (f_k/Q)\overrightarrow{\mathrm{OI}}^k$.
$\mathrm{OG}'^2 = \sum\limits_{k\in K'} (f_k/Q)^2(\mathrm{OI}^k)^2 = \sum\limits_{k\in K'} (f_k/Q)^2(Q/f_k) - \sum\limits_{k\in K'} f_k/Q$. ◁

— *Remark.* $\overrightarrow{\mathrm{OG}}'$ is not orthogonal to the support of the cloud: if two individuals i and i' have not chosen the same number of active modalities, one has $\sum\limits_{k\in K'} \delta_{ik} \neq \sum\limits_{k\in K'} \delta_{i'k}$ and $\langle \overrightarrow{\mathrm{OG}}' | \overrightarrow{\mathrm{R}^{i'}\mathrm{R}^{i}} \rangle = \sum\limits_{k\in K'} (\delta_{ik} - \delta_{i'k})/Q \neq 0$.

Property 5.25 (Specific variance). *The variance of the specific cloud is called specific variance and is equal to* $(K'/Q) - \sum\limits_{k\in K'} f_k/Q$.

Proof. By definition, $\operatorname{Var} \mathrm{R}^I = \frac{1}{n} \sum\limits_{i\in I} (\mathrm{G}'\mathrm{R}^i)^2$. From first Huyghens' theorem (CLOUD, p.79), one has $\frac{1}{n} \sum\limits_{i\in I} (\mathrm{G}'\mathrm{R}^i)^2 = \frac{1}{n} \sum\limits_{i\in I} (\mathrm{OR}^i)^2 - \mathrm{OG}'^2$, with $(\mathrm{OR}^i)^2 = \sum\limits_{k\in K'} (\delta_{ik}/Q)^2 (\mathrm{OI}^k)^2 = \sum\limits_{k\in K'} \delta_{ik}/(Q f_k)$ and $\mathrm{OG}'^2 = \sum\limits_{k\in K'} f_k/Q$. One deduces $\operatorname{Var} \mathrm{R}^I = \sum\limits_{k\in K'} \left(\frac{1}{n} \sum\limits_{i\in I} \delta_{ik}\right)/(Q f_k) - \frac{1}{n} \sum\limits_{i\in I} \sum\limits_{k\in K'} \frac{f_k}{Q} = \frac{K'}{Q} - \sum\limits_{k\in K'} \frac{f_k}{Q}$. ◁

— *Remark.* From $\mathrm{G}'\mathrm{R}^i \leq \mathrm{GM}^i$ (contracting property of orthogonal projection) one deduces $\operatorname{Var} \mathrm{R}^i \leq \operatorname{Var} \mathrm{M}^I$.

Specific principal axes and variables. The principal direction of the specific cloud R^I are called *specific principal directions* of the cloud M^I. Let $\vec{\alpha}_\ell = \sum\limits_{k\in K} a_{k\ell} \overrightarrow{\mathrm{OI}}^k$ a unit–norm vector, with $\sum\limits_{k\in K} a_{k\ell}^2/(f_k/Q) = 1$; the family $(a_{k\ell})_{k\in K}$ is called the *direction coefficients* of $\vec{\alpha}_\ell$. For the specific cloud R^I, the direction coefficients of a principal line are such that $a_{k\ell} = 0$ for $k \in K''$.

Theorem 5.3. *The ℓ–th principal direction coefficients $(a_{k\ell})_{k\in K'}$ and the ℓ-th standard principal variable z'^I_ℓ of the specific cloud verify the following system of equations, letting $a^k_\ell = a_{k\ell}/(f_k/Q)$ and $\underline{a}_\ell = \sum\limits_{k\in K'} a_{k\ell}$:*

$$\begin{cases} \forall i \in I & \sum\limits_{k\in K'} \delta_{ik}\, a^k_\ell/Q - \underline{a}_\ell & = \xi'_\ell z'^i_\ell \\ \forall k \in K' & \sum\limits_{i\in I} \delta_{ik}\, z'^i_\ell/n_k & = \xi'_\ell a^k_\ell \end{cases} \tag{5.18}$$

Proof. With the specific cloud R^I there is associated the biweighted protocol $(x^I_{K'}, 1_I, f_{K'}/Q)$ of measures over K', with $(x^i_k = \delta_{ik}/Q)_{k\in K'}$. For $k \in K'$ one has $\operatorname{Mean} x^I_k = \sum\limits_{i\in I} x^i_k/n = \sum\limits_{i\in I} \delta_{ik}/(nQ) = f_k/Q$. The direction coefficients of principal axes and the standard principal variables over I associated with this protocol are obtained from the passage formulas (4.25) (PRINCIPAL, p.156) by replacing j by i, $\overline{x}_k$ by f_k/Q and by letting $z'^i_\ell = z'_{i\ell}/(1/n)$; one gets $\sum\limits_{k\in K'} (\frac{\delta_{ik}}{Q} - \frac{f_k}{Q}) a^k_\ell = \xi_\ell z^i_\ell$ and $\sum\limits_{i\in I} \frac{\delta_{ik}}{Q} z'_{i\ell} = \xi_\ell a_{k\ell}$. ◁

Corollary 5.6 (Equation for specific principal directions). *The direction coefficients $(a_{k\ell})_{k\in K'}$ of the ℓ–th specific principal line verify the equation, letting $f_k^{k'} = n_{kk'}/n_{k'}$ and $\lambda'_\ell = (\xi'_\ell)^2$:*

$$\frac{1}{Q}\sum_{k'\in K'}(f_k^{k'} - f_k)a_{k'\ell} = \lambda'_\ell a_{k\ell} \qquad (k \in K') \tag{5.19}$$

Proof. Equation (5.19) follows from Equations (5.18). ◁

The *specific Axis* ℓ is defined by $(\mathrm{G}', \overrightarrow{\alpha}'_\ell)$, with $\overrightarrow{\alpha}'_\ell = \sum_{k\in K'} a_{k\ell}\overrightarrow{\mathrm{OI}}^k$.

Property 5.26 (Specific principal coordinates). *The ℓ–th specific principal coordinate of the ideal point I^k $(k \in K')$ is equal to $a_\ell^k - \underline{a}_\ell$; the ℓ–th specific principal coordinate y'^i_ℓ of the point M^i is equal to $\xi'_\ell z'^i_\ell$.*

Proof. The specific principal coordinate of the ideal point I^k on axis $(\mathrm{G}', \overrightarrow{\alpha}'_\ell)$ is equal to $\langle \overrightarrow{\mathrm{G'I}}^k | \overrightarrow{\alpha}'_\ell\rangle$ with $\langle \overrightarrow{\mathrm{G'I}}^k | \overrightarrow{\alpha}'_\ell\rangle = \langle \overrightarrow{\mathrm{OI}}^k | \overrightarrow{\alpha}'_\ell\rangle - \langle \overrightarrow{\mathrm{OG}}' | \overrightarrow{\alpha}'_\ell\rangle = a_{k\ell}(\mathrm{OI}^k)^2 - \sum_{k\in K'}(f_k/Q)a_{k\ell}(\mathrm{OI}^k)^2 = a_\ell^k - \sum_{k\in K'} a_{k\ell} = a_\ell^k - \underline{a}_\ell$ (since $(\mathrm{OI}^k)^2 = Q/f_k$ and $a_\ell^k = a_{k\ell}/(f_k/Q)$).
The specific principal coordinate of the point M^i is the principal coordinate of the specific point R^i. One has $\overrightarrow{\mathrm{G'R}}^i = \overrightarrow{\mathrm{OR}}^i - \overrightarrow{\mathrm{OG}}'$ and $\overrightarrow{\mathrm{OR}}^i = \sum_{k\in K'}\delta_{ik}\overrightarrow{\mathrm{OI}}^k/Q$, hence $y'^i_\ell = \langle \overrightarrow{\mathrm{G'R}}^i | \overrightarrow{\alpha}'_\ell\rangle = \sum_{k\in K'}\frac{\delta_{ik}}{Q}a_{k\ell}(\mathrm{OI}^k)^2 - \underline{a}_\ell$, that is, $y'^i_\ell = \sum_{k\in K'}\frac{\delta_{ik}}{Q}a_\ell^k - \underline{a}_\ell = \xi'_\ell z'^i_\ell$ (first equation (5.18)). ◁

Specific cloud of modalities

The specific cloud of modalities is the subcloud of the cloud M^K restricted to the active modalities $k \in K'$ with weights and distances unchanged.

For the specific cloud $(\mathrm{M}^{K'}, \frac{1}{Q}f_{K'})$, the *mean point* G of the cloud M^K is a reference point. Recall (cf. §5.1.2, p.192) that $\overrightarrow{\mathrm{GM}}^k = \sum_{i\in I}(p_i^k - 1/n)\overrightarrow{\mathrm{OI}}^i$, with $p_i^k = \delta_{ik}/n_k$.

Property 5.27. *The absolute contribution of the subcloud $\mathrm{M}^{K'}$ is equal to the specific variance.*

Proof. The absolute contribution of the point M^k (as a point of the cloud M^K) is $\mathrm{Cta}_k = (1 - f_k)/Q$ (cf. formula (5.10), p.194). Hence the contribution of the subcloud $\mathrm{M}^{K'}$ is $\sum_{k\in K'} \mathrm{Cta}_k = (K'/Q) - \sum_{k\in K'} f_k/Q$. ◁

The absolute contribution of a question q with no passive modality ($K'\langle q\rangle = K\langle q\rangle$) is equal to $(K_q - 1)/Q$; whereas the contribution of a question with passive modalities is equal to $(K'_q - 1)/Q + \sum_{k\in K''\langle q\rangle} f_k/Q$.

Theorem 5.4. *With respect to the reference point* G *the principal axes and variables of the subcloud* $(\mathrm{M}^k, \frac{f_k}{Q})_{k\in K'}$ *are defined by equations (5.18).*

Proof. Let us recall (CLOUD, §3.2.5, p.84) that, taking P = G as reference point, the mappings $Vac'_{\mathrm{G}} : \mathcal{V}_I \to \mathbb{R}^K$ and $Vac'^*_{\mathrm{G}} : \mathbb{R}^K \to \mathcal{V}_I$ attached to cloud $\mathrm{M}^{K'}$ are defined by

$$Vac'_{\mathrm{G}}(\overrightarrow{\mathrm{OI}^i}) = (\langle\overrightarrow{\mathrm{GM}^k}|\overrightarrow{\mathrm{OI}^i}\rangle)_{k\in K'} \text{ and } Vac'^*_{\mathrm{G}}(\delta_k^K) = (f_k/Q)\overrightarrow{\mathrm{GM}^k}\ (k\in K')$$

Recall that $\overrightarrow{\mathrm{GM}^k} = \sum_{i\in I}(p_k^i - \frac{1}{n})\overrightarrow{\mathrm{OI}^i}$, with $p_k^i = \delta_{ik}/n_k$. From $\langle\overrightarrow{\mathrm{GM}^k}|\overrightarrow{\mathrm{OI}^i}\rangle = \sum_{i\in I}(p_i^k - 1/n)(\mathrm{OI}^i)^2$ and $(\mathrm{OI}^i)^2 = n$, one deduces $Vac'_{\mathrm{G}}(\overrightarrow{\mathrm{OI}^i}) = np_i^{K'} - 1^{K'}$ and $Vac'^*_{\mathrm{G}}(\delta_k^K) = \sum_{i\in I}(p_k^i - \frac{f_k}{Q})\overrightarrow{\mathrm{OI}^i}/n$ for $k \in K'$ and $\overrightarrow{0}$ for $k\in K''$.

Let $\overrightarrow{\alpha}'_\ell = \sum_{i\in I} a_{i\ell}\overrightarrow{\mathrm{OI}^i}$ a unit–norm vector ($\sum_{i\in I} a_{i\ell}^2 n = 1$), and $t_\ell^{K'}$ a variable on K' such that $\sum_{k\in K'}(f_k/Q)(t_\ell^k)^2 = 1$. The passage formulas pertaining to reference point G read

$$Vac'_{\mathrm{G}}(\overrightarrow{\alpha}'_\ell) = \xi'_\ell t_\ell^K \text{ and } Vac'^*(t_\ell^K) = \xi_\ell \overrightarrow{\alpha}'_\ell$$

Hence their expressions for cloud $\mathrm{M}^{K'}$:

$$\begin{cases} \forall k \in K' & \sum_{i\in I} a_{i\ell}(n\,p_i^k - 1) = \xi'_\ell t_\ell^k \\ \forall i \in I & \sum_{k\in K'}(p_k^i - \frac{f_k}{Q})t_\ell^k = \xi'_\ell a_{i\ell} \end{cases} \tag{5.20}$$

From the second equation (5.20), one deduces that $\sum_{i\in I} a_{i\ell} = 0$.

Indeed, $\xi'\sum_{i\in I} a_{i\ell} = \sum_{k\in K'}\left(\sum_{i\in I}(p_k^i - \frac{f_k}{Q})\right)t_\ell^k$. Now $\forall k \in K$ $\sum_{i\in I} p_k^i = n_k/Q$, therefore $\sum_{i\in I} p_k^i - \sum_{i\in I} f_k/Q = (n_k/Q) - n(f_k/Q) = 0$, hence $\sum_{i\in I} a_{i\ell} = 0$. Letting $a_\ell^k = t_\ell^k$ and $z'^i_\ell = a_\ell^i = na_{i\ell}$, one finds again Equations (5.18). ◁

Property 5.28 (Specific principal coordinates). *The specific principal coordinate of point* M^k *($k \in K'$) on specific Axis ℓ, denoted* y'^k_ℓ, *is such that* $y'^k_\ell = \xi'_\ell a^k_\ell$.

Proof. By definition, the specific coordinate of point M^k on specific Axis ℓ is equal to $\langle \overrightarrow{\mathrm{GM}}^k | \vec{\alpha}_\ell \rangle$, with $\overrightarrow{\mathrm{GM}}^k = \sum\limits_{i\in I}(p^k_i - \frac{1}{n})\overrightarrow{\mathrm{OI}}^i$; hence $y'^k_\ell = \sum\limits_{i\in I}(p^k_i - \frac{1}{n})a_{i\ell}(\mathrm{OI}^i)^2 = \sum\limits_{i\in I}(np^k_i - 1)a_{i\ell} = \xi'_\ell t^k_\ell$ (first Equation 5.20). ◁

Properties of specific MCA

• The modified rates are calculated by formula (5.16) p.201, where $\overline{\lambda}$ is the ratio of the specific variance (Property 5.25, p.206) to the dimensionality of the support of the specific cloud (cf. p.205), hence one has:

$$\overline{\lambda} = \Big(\frac{K'}{Q} - \sum_{k\in K'} \frac{f_k}{Q}\Big) / (K' - (Q - Q''))$$

• The specific principal variables over I are centered, and their variances are equal to the eigenvalues.

$$\mathrm{Mean}\, y'^I_\ell = 0 \quad \text{and} \quad \mathrm{Var}\, y'^I_\ell = \xi'^2_\ell = \lambda'_\ell$$

• From the second equation (5.18, p.206), the coordinate y'^k_ℓ of a modality k on specific Axis ℓ is equal to $(1/\xi'_\ell)\widehat{y}'^k_\ell$, where $\widehat{y}'^k_\ell = \sum\limits_{i\in I<k>} y'^i_\ell / n_k$ is the elementary mean of the coordinates of the points $(\mathrm{M}^i)_{i\in I<k>}$ associated with the individuals who have chosen modality k. Thus, one can calculate the coordinates of any modality whether passive or supplementary.

• The active and passive modalities of an active question q induce a partition of individuals; one has therefore $\sum\limits_{k\in K<q>} n_k \widehat{y}'^k_\ell = 0$, and $\sum\limits_{k\in K} n_k \widehat{y}'^k_\ell = 0$, hence $\sum\limits_{k\in K} n_k y'^k_\ell = 0$.

For any question q, the mean of the specific coordinates of all K_q (active and passive) modalities of the question is null.

• $\sum\limits_{k\in K'} \frac{f_k}{Q}(y'^k_\ell)^2 = \lambda'_\ell = \sum\limits_{k\in K'} \mathrm{Cta}_k$, and $\mathrm{Var}\, y'^K_\ell = \lambda'_\ell + \sum\limits_{k\in K''} \frac{f_k}{Q}(y'^k_\ell)^2$.

The raw sum of squares, weighted by f_k/Q, of the coordinates of active modalities is equal to λ'_ℓ.

• The coordinate of a supplementary individual i_s having chosen modalities $K<i_s>$ is given by the first equation (5.18):

$$y^{i_s}_\ell = \sum_{k\in K'<i_s>} a^k_\ell / Q - \sum_{k\in K'} a_{k\ell}$$

Computation procedure

In terms of a symmetric matrix, one calculates the eigenvalues λ'_ℓ and the normed eigenvectors $(c_{k\ell})_{k\in K'}$ $(\sum_{k\in K'} c^2_{k\ell} = 1)$ of the $K' \times K'$ symmetric matrix $\mathbf{T} = [t_{kk'}]$, with

$$t_{kk'} = \frac{1}{Q} \times \frac{n_{kk'} - n_k n_{k'}/n}{\sqrt{n_k\, n_{k'}}} \qquad \begin{cases} k \in K' \\ k' \in K' \end{cases}$$

One then deduces the coordinates of modalities on specific axis ℓ:

$$y'^k_\ell = \sqrt{\lambda'_\ell}\sqrt{Q}\frac{c_{k\ell}}{\sqrt{n_k/n}} \qquad (k \in K')$$

The coordinates of individuals on specific axis ℓ follow:

$$y'^i_\ell = \frac{1}{\sqrt{Q}}\Big(\sum_{k\in K'<i>} \frac{c_{k\ell}}{\sqrt{n_k/n}} - \sum_{k\in K'} (\sqrt{n_k/n})\, c_{k\ell}\Big) \qquad (i \in I)$$

— *Remark.* $(\sqrt{n_k})_{k\in K'}$ is not an eigenvector of the matrix $\mathbf{T}$.

5.2.2 Analysis of a Subcloud of Individuals

In this section we study a subcloud of individuals $\mathrm{M}^{I'}$ $(I' \subset I)$ by determining its principal axes and variables.

Starting from the cloud of modalities associated to the whole cloud M^I, we define a specific cloud of modalities associated to the subcloud. For that, we associate with the cloud a table of measures, and perform the biweighted PCA of this table (cf. *PRINCIPAL*, §4.5, p.155). This method makes it possible to interpret the importance of modalities for the subcloud.

Special notation for the present subsection. From now on, the letter n will denote the number of individuals of the subcloud (i.e. the number of elements of I'), and N will denote the total number of individuals (number of elements of I). We denote $n_k = \sum_{i\in I'} \delta_{ik}$ the number of individuals in I' choosing modality k, $f_k = n_k/n$ the associated relative frequency, and $n_{kk'} = \sum_{i\in I'} \delta_{ik}\delta_{ik'}$ the number of individuals in I' choosing both k and k'. We denote $N_k = \sum_{i\in I} \delta_{ik}$ the number of individuals in I choosing modality k, and $F_k = N_k/N$ the associated relative frequency.

— *Remark.* The distance between two individuals $i \in I'$ and $i' \in I'$ is the one defined from the whole cloud, more precisely if for question q, i chooses modality k and i' chooses modality k', one has (using the notation of this subsection) $d_q^2(i,i') = (1/F_k) + (1/F_{k'})$, while from the sub–questionnaire $I' \to \prod_{q \in Q} K\langle q\rangle$ we would have obtained $(1/f_k) + (1/f_{k'})$.

Subcloud of individuals

We consider a subcloud $\mathrm{M}^{I'} \subset \mathrm{M}^I$. The origin–point O and the K ideal points $(\mathrm{I}^k)_{k \in K}$ define the frame of $\mathcal{U}_K$, and the metric on $\mathcal{U}_K$ is unchanged, that is, $\overrightarrow{\mathrm{OI}}^k \perp \overrightarrow{\mathrm{OI}}^{k'}$ for $k \neq k'$ and $(\mathrm{OI}^k) = \sqrt{\frac{Q}{F_k}}$.

One has $\overrightarrow{\mathrm{OM}}^i = \sum_{k \in K} (\delta_{ik}/Q)\overrightarrow{\mathrm{OI}}^k$ $(i \in I')$, therefore the biweighted protocol associated with the cloud $\mathrm{M}^{I'}$ is $\left(p_K^{I'}, (1/n), (F_k/Q)\right)$, with $p_k^i = \delta_{ik}/Q$.

Support, mean point and variance. The *mean point* G′ of the subcloud $\mathrm{M}^{I'}$ is $\mathrm{G}' = \sum_{i \in I'} \mathrm{M}^i/n$ (CLOUD, Definition 3.2, p.77).
From $\overrightarrow{\mathrm{OM}}^i = \sum_{k \in K} \delta_{ik}\overrightarrow{\mathrm{OI}}^k/Q$, one deduces $\overrightarrow{\mathrm{OG}}' = \sum_{k \in K} f_k \overrightarrow{\mathrm{OI}}^k/Q$, then one has $(\mathrm{OG}')^2 = \frac{1}{Q}\sum_{k \in K} f_k^2/F_k$.

Property 5.29 (Specific variance). *The variance of the subcloud, called specific variance, is equal to* $\frac{1}{Q}\sum_{k \in K} f_k(1-f_k)/F_k$

Proof. From Definition (3.4) (CLOUD, p.78), the variance of the subcloud $\mathrm{M}^{I'}$ is equal to $\sum_{i \in I'} (\mathrm{G'M}^i)^2/n = \sum_{i \in I'} (\mathrm{OM}^i)^2/n - \mathrm{OG}'^2$ (first Huyghens' theorem, CLOUD, p.79) . Hence $\operatorname{Var} \mathrm{M}^{I'} = \sum_{k \in K} \frac{1}{n} \sum_{i \in I'} (\frac{\delta_{ik}}{Q})^2 (\mathrm{OI}^k)^2 - \frac{1}{Q}\sum_{k \in K} \frac{f_k^2}{F_k} = \frac{1}{Q}\sum_{k \in K}\left(\frac{f_k}{F_k} - \frac{f_k^2}{F_k}\right) = \frac{1}{Q}\sum_{k \in K} \frac{f_k(1-f_k)}{F_k}$. ◁

Principal axes and variables of the subcloud. The vector $\overrightarrow{\mathrm{OG}}$ is orthogonal to the support of cloud M^I and consequently to the support of subcloud $\mathrm{M}^{I'}$, therefore any vector $\overrightarrow{\alpha} = \sum_{k \in K} a_k \overrightarrow{\mathrm{OI}}^k$ in the support of the subcloud $\mathrm{M}^{I'}$ is associated with a contrast over K:

$$\overrightarrow{\mathrm{OG}} \perp \overrightarrow{\alpha} \Rightarrow \sum_{k \in K} (F_k/Q) a_k (\overrightarrow{\mathrm{OI}}^k)^2 = \sum_{k \in K} a_k = 0$$

As a special case, the principal axes of the subcloud are associated with contrasts over K.

Theorem 5.5 (Principal variables). *The ℓ–th principal variables over K and over I' verify the following equations:*

$$\begin{cases} \frac{1}{Q}\sum\limits_{k\in K}(\delta_{ik}-f_k)z'^k_\ell & = \xi'_\ell z'^i_\ell \ (i\in I') \\ \frac{1}{F_k}\sum\limits_{i\in I'}\delta_{ik}z'^i_\ell/n & = \xi'_\ell z'^k_\ell \ (k\in K) \end{cases} \text{with} \begin{cases} \sum\limits_{k\in K}\frac{F_k}{Q}(z'^k_\ell)^2 = 1 \\ \sum\limits_{i\in I'}(z'^i_\ell)^2/n = 1 \end{cases} \tag{5.21}$$

Proof. The unit–norm principal vector $\overrightarrow{\alpha'_\ell} = \sum\limits_{k\in K} z'_{k\ell}\overrightarrow{\mathrm{OI}}^k$ and the standard principal variable $z'^{I'}_\ell$ are obtained by the passage formulas (4.25) for a protocol of measures (PRINCIPAL, p.156) replacing J by I', x^J_k by $p^{I'}_k$ and $\overline{x}_k$ by f_k/Q (since Mean $p^{I'}_k = \sum\limits_{i\in I'}(\delta_{ik}/Q)/n = f_k/Q$); hence the equations: $\sum\limits_{k\in K}(p^{I'}_k - f_k/Q)z'^k_\ell = \xi'_\ell z'^{I'}_\ell$ and $\sum\limits_{i\in I'} p^i_K z'_{i\ell} = \xi'_\ell z'_{K\ell}$. ◁

Corollary 5.7 (Equation for principal directions). *The principal axes of the subcloud $\mathrm{M}^{I'}$ defined by the principal contrasts z'_K verify the following equation:*

$$\frac{1}{Q}\times\sum_{k'\in K}(f_{kk'}-f_kf_{k'})z'_{k'}/F_{k'} = \xi'^2 z'_k \tag{5.22}$$

Proof. Recall the equation (4.26): $\sum\limits_{k'\in K} v_{kk'}(a_{k'\ell}/\varpi_{k'}) = \lambda_\ell a_{k\ell}$ (PRINCIPAL, p.156). Equation (5.22) is obtained by replacing in the preceding equation $v_{kk'}$ by $\mathrm{Cov}(p^{I'}_k|p^{I'}_{k'}) = \sum\limits_{i\in I'}(1/n)(\delta_{ik}-f_k)(\delta_{ik'}-f_kf'_k)/Q^2$, that is, $\mathrm{Cov}(p^{I'}_k|p^{I'}_{k'}) = (n_{kk'}/n - f_kf_{k'})/Q^2$ and ϖ'_k by $F_{k'}/Q$. ◁

— *Remark.* F_K is an eigenvector associated with the eigenvalue 0.

Specific cloud of modalities

The cloud of modalities associated with the cloud M^I is defined in space $\mathcal{U}_I$ (with $(\mathrm{OI}^i)^2 = N$) by points $\mathrm{M}^k = \sum\limits_{i\in I\langle k\rangle}\mathrm{I}^i/N_k$. The cloud of modalities associated with the subcloud $\mathrm{M}^{I'}$ is defined in space $\mathcal{U}_{I'}$ by points $(\mathrm{R}^k, F_k/Q)_{k\in K}$; it is constructed as follows.

1. Define the point M'^k such that $\overrightarrow{\mathrm{OM}}'^k = \sum\limits_{i\in I'}\delta_{ik}\overrightarrow{\mathrm{OI}}^i/(nF_k)$.

2. With each point M'^k, associate the point G^k defined by the orthogonal projection of M'^k onto $\sum\limits_{i\in I'} \overrightarrow{\mathrm{OI}}^i$: $\overrightarrow{\mathrm{OG}}^k = u^k \sum\limits_{i\in I'} \overrightarrow{\mathrm{OI}}^i$, with $u^k = \frac{1}{n}\frac{f_k}{F_k}$.

   One has $u^k = \langle \sum\limits_{i\in I'} \overrightarrow{\mathrm{OI}}^i | \overrightarrow{\mathrm{OM}}'^k \rangle / \| \sum\limits_{i\in I'} \overrightarrow{\mathrm{OI}}^i \|^2$ and $\langle \sum\limits_{i\in I'} \overrightarrow{\mathrm{OI}}'^i | \overrightarrow{\mathrm{OM}}'^k \rangle = \frac{1}{nF_k} \sum\limits_{i\in I'} \delta_{ik} (\mathrm{OI}^i)^2 = \frac{n_k N}{nF_k} = N \frac{f_k}{F_k}$ and $\| \sum\limits_{i\in I'} \overrightarrow{\mathrm{OI}}^i \|^2 = \sum\limits_{i\in I'} (\mathrm{OI}^i)^2 = nN$, hence $u^k = \frac{1}{n} f_k / F_k$.

3. The point R^k is defined by $\overrightarrow{\mathrm{OR}}^k = \overrightarrow{\mathrm{G}^k\mathrm{M}'^k} = \frac{1}{n} \sum\limits_{i\in I'} \frac{\delta_{ik} - f_k}{F_k} \overrightarrow{\mathrm{OI}}^i$, and is weighted by F_k/Q; by construction, its mean point is O.

Theorem 5.6. *The principal axes and variables of the cloud $(\mathrm{R}^K, (F_k/Q))$ are defined by the equations (5.21).*

Proof. The biweighted protocol $(x_{I'}^K, \frac{F_K}{Q}, (\frac{1}{n})_{i\in I'})$ is associated with the cloud $(\mathrm{R}^K, (F_k/Q))$. One has $x_i^k = (\delta_{ik} - f_k)/(nF_k)$ and $\overline{x}_i = 0$. Replacing in the passage formulas (4.25) of a protocol of measures (PRINCIPAL, p.156) K by I', J by K and $\overline{x}_k$ by $\overline{x}_i = 0$, one obtains the following two equations: $\sum\limits_{i\in I'} (\delta_{ik} - f_k)/(nF_k) z'^i_\ell = \xi'_\ell z'^k_\ell$ $(k \in K)$ and $\sum\limits_{k\in K} (\delta_{ik} - f_k) z'_{k\ell}/(nF_k) = \xi_\ell z'_{i\ell}$ $(i \in I')$. Hence letting $z'_{i\ell} = (1/n) z'^i_\ell$ and $z'_{k\ell} = (F_k/Q) z'^k_\ell$, the equations (5.21). ◁

Computation procedure

In terms of a symmetric matrix, one calculates the eigenvalues λ'_ℓ and the normed eigenvectors $(c_{k\ell})_{k\in K}$ ($\sum\limits_{k\in K} c_{k\ell}^2 = 1$) of the $K \times K$ symmetric matrix $\mathbf{T} = [t_{kk'}]$, with

$$t_{kk'} = \frac{1}{Q} \times \frac{N}{n} \times \frac{n_{kk'} - n_k n_{k'}/n}{\sqrt{N_k N_{k'}}} \qquad \begin{cases} k \in K \\ k' \in K \end{cases}$$

One then deduces the coordinates of modalities on Axis ℓ:

$$y'^k_\ell = \sqrt{\lambda'_\ell} \sqrt{Q} \frac{c_{k\ell}}{\sqrt{N_k/N}} \qquad (k \in K)$$

The coordinates of the individuals of the subcloud on Axis ℓ follow:

$$y'^i_\ell = \frac{N}{\sqrt{Q}} \sum_{k\in K} \frac{(\delta_{ik} - n_k/n) c_{k\ell}}{\sqrt{N_k}} \qquad (i \in I')$$

— *Remark.* $(\sqrt{N_k})_{k\in K}$ is an eigenvector of $\mathbf{T}$ associated with eigenvalue 0.

5.3 Methodology of MCA

The domain of direct application of MCA is constituted by the tables arising from large questionnaires with numerous individuals (say more than a hundred), where questions pertaining to various themes are categorized variables, that is, the Individuals×Questions tables. A good indication for MCA occurs when within each question, there is no structure on the set of possible response modalities, as is the case e.g. for questions $q1$ through $q4$ of the *Culture Example* (§5.4.1, p.221). This does not preclude a fruitful use of MCA for questionnaires comprising questions with ordered (say) modalities; see e.g. the *Political Space* questionnaire in chapter 9. As a matter of fact, it is usual to treat by MCA a mix of questions with unordered and ordered modalities.

When contemplating an MCA, there are several basic choices to make, namely weights of individuals, active questions and structuring factors, then the coding of active questions; we discuss these choices hereafter.

5.3.1 Structuration of Questions

• *Weighting of individuals.* In a questionnaire where individuals are persons, equal weights are usually taken. When the statistical individuals are not equivalent, a differential weighting may be in order, e.g. the countries of Europe will be weighted by their populations, etc.

• *Choice of active questions.* Active questions are those that determine the geometric space, that is, they serve to define the distance between individuals. For active questions, a further choice has to be made when a *specific* MCA is contemplated, namely the choice of active modalities vs passive ones (see below p.216).

The contribution of a question to the variance of a cloud is an increasing function of the number of its modalities: cf. formula (5.10), p.194. To avoid giving disproportionate importances to some questions, it is recommended to construct questions having about equal numbers of modalities, possibly after grouping modalities of a same question; or at least, it is advisable to balance the various sets of questions referring to the same headings. For instance, in the *Political Space* study (p.365), one has tried to balance the importance of questions relevant to ethnocentrism, to social issues, etc.

• *Structuring factor.* In a questionnaire, there are questions that do not serve to construct the geometric space, but are assigned the role of a *structuring factor.* As a first example, in an educational study where a knowledge space is constructed, the interrogation naturally arises: "Is there a difference between boys and girls in this space?" This interrogation leads to treat the question about Gender in the questionnaire, not as an active question, but as a structuring factor; accordingly, the two subclouds for boys and girls can be studied, comparing not only their mean points, but also their dispersions (e.g. concentration ellipses), etc.

As another example, in a political study where a space of people's values is constructed, the question "For which candidate did you vote?" will be put as a structuring factor, and allow investigating how electorates differ in the space of values, by studying the subclouds corresponding to electorates, calculating between and within variances, etc. If the Vote question had been put as active, it would create distance between individuals voting differently, then it would be no wonder to find differences among electorates!

A structuring factor can also be put as a *supplementary question*, which provides the principal coordinates of its modalities on axes in the space of modalities, in other words the standardized deviations of the corresponding modality mean–points in the space of individuals.

Descriptive conclusions about structuring factors can further be extended by inductive analyses, as will be discussed in chapter INDUCTIVE, and applied in the *Parkinson* and *Political Space* studies (chapter 9).

Structuration of active questions

It is always advantageous to structure the set of active questions in connection with interrogations and hypotheses related with the gathering of data. Two common structures are nesting and crossing[17].

When each question refers to one and only one heading, the questions are *nested* within the headings. For instance, in the *Political space* study, the questions fall under six headings (ethnocentrism, politics, etc.).

When the same questions are repeated, for instance in a longitudinal survey, questions are *crossed* with time. The EPGY case study (p.394) provides a different instance of a crossing structure; each subject matter (geometry, logic, etc.) is crossed with latency times, error ratess, and numbers of exercises necessary to master a concept.

5.3.2 Coding Data for MCA

Hereafter, we mention problems of data coding, and we suggest solutions that extend the range of application of MCA to data that are not directly relevant of the standard format (cf. p.179).

[17]The topic of the structuration of data will be further studied in chapter 6.

"Junk modalities". In forced–choice questionnaires, there are often modalities of the "Others" sort which put together disparate properties. Such "junk modalities", which are not genuine categories, are better discarded from the construction of the space; this can be done by performing a specific MCA, in which they are put as *passive modalities.*

Infrequent modalities. As we have seen, infrequent response modalities may greatly influence the determination of principal axes. Therefore, before performing an MCA, it is mandatory to systematically search for all "rare modalities" — as a rule of thumb, one may apply the *five percent rule*: "no frequency less than 5%" — and dispose of them. A first solution is to *group* an infrequent modality with a semantically related modality of the same question. See e.g. question $q4$ of the *Culture Example*, Table 5.4 (p.222).

It may happen, however, that there is no defensible grouping; alternative solutions must then be found. For instance, in the Culture Example, we chose to discard from the analysis 152 individuals (out of 2872), who had not responded to at least one of the first four questions, before proceeding to a standard MCA. Again, in the *Political Space* study, we discarded 30 individuals (out of 3010), who did not give answers to more than 5 questions (out of 20 active questions); still, for each question, there remained non-response modalities (NR) with low frequencies (less than 4.2%); a specific MCA was then performed, putting these low frequency NR modalities as *passive elements.*

Multiple choice questions and barycentric coding. In actual questionnaires, there may be questions (not in standard format) where more than one option ("modality") may be chosen. For instance, there is a set $A = \{a1, \ldots, a6\}$ of options and the individual is invited to choose 3 options. To deal with this situation, a solution is *barycentric coding.*

If for example an individual chooses $a1$, $a3$ and $a6$, these three modalities are coded 1/3, the other modalities being coded 0, hence the pattern (1/3, 0, 1/3, 0, 0, 1/3, 0). If the three modalities are chosen with an order of preference, one can proceed to a coding with differential weights (for instance if an individual chooses $a3$, then $a1$, then $a6$, the coding may be $(.2, 0, .7, 0, 0, .1, 0)$).

Numerical variables grouped in classes and fuzzy coding. Very commonly, numerical variables, such as scores of various sort, are incorporated into the MCA format, after the grouping of scores into a small number of classes, that may either be taken as intervals of equal lenghts, or constructed from the frequency distributions; then each individual is allocated to a class. The EPGY case study exemplifies this well–tried procedure.

More sophisticated procedures can be used, such as the *fuzzy coding*, whose motivation runs as follows. An individual located at the boundary of two classes can be said to belong half to one class and half to the other, whereas an individual in the central part of a class 100% belongs to this class. The fuzzy coding procedure is depicted in the following figure (with $u \leq (a_\ell + a_{\ell+1})/p$ and $p = 4$ in the graph). Let class $c_\ell = [a_\ell; a_{\ell+1}]$.

An individual with score x such that $a_\ell+u \leq x \leq a_{\ell+1}-u$ is given the value 1 in class c_ℓ and 0 in classes $c_{\ell-1}$ and $c_{\ell+1}$; an individual with score x such that $a_\ell \leq x \leq a_\ell + u$ is given the value $0.5\times(x-a_\ell)/u$ in c_ℓ and $1 - 0.5 \times (x - a_\ell)/u$ in class $c_{\ell-1}$.

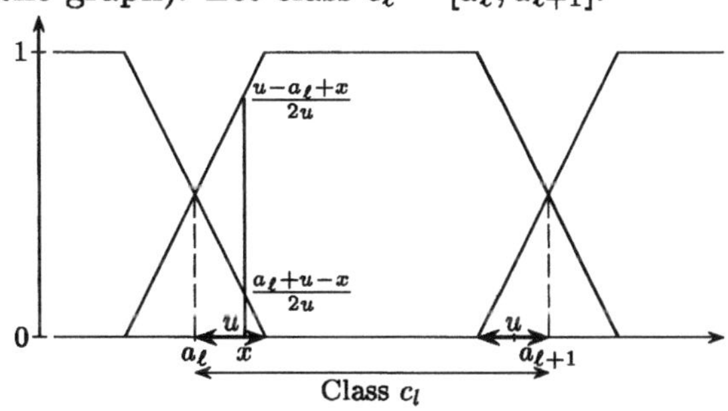

Figure 5.5. Fuzzy coding of a numerical variable.

Barycentric and fuzzy codings pave the way to a *generalized* MCA, applicable to an Individuals$\times$Questions table, such that with each individual there is associated, for each question, a *positive normalized measure* over the modality set of the question (see Cazes, 1990).

5.3.3 Strategy of Interpretation

Hereafter, "interpretation" means *statistical interpretation*, referring to the analyzed data, that should be done before proceeding to comments that make use of external information.

The strategy of interpretation in MCA can be conducted in three phases:

1) Looking at the two basic clouds (modalities and individuals) in several principal planes, in first place in plane 1-2.

2) Interpretation of axes using the method of the Contributions of points and deviations.

3) Exploration of the cloud of individuals and Structured Data Analysis. The *method of contributions of points and deviations* (Le Roux & Rouanet, 1998) is a guide along the line of Benzécri's quotation reproduced in *CORRESPONDENCE*, (p.49). Recall (cf. p.196) that the contribution of a point to an axis is the part of variance of axis accounted for by the point, that the contribution of a *deviation* between points (an extension of this notion) is

the part of variance accounted for by a contrast among groups of points. This method can be summarized as follows.

Step 1. *Important questions* (or groups of questions).

In the cloud of modalities, *rank questions* according to decreasing contributions to Axis ℓ, or equivalently $\eta^2_{q\ell} = Q\lambda\,\mathrm{Ctr}^\ell_q$. The questions that account together for a substantive amount of the variance of axis (say normally 75%) provide a first overall interpretation of axis. A question q whose $\mathrm{Ctr}^\ell_q > 1/Q$, i.e. whose $\eta^2_{q\ell}$ exceeds λ_ℓ is surely important.

Step 2. *Important modalities.*

One takes the modalities whose contributions exceed the average contribution (*basic criterion*): i.e. whose relative contributions (Ctr) are greater than $1/K$ (K being the number of active modalities). In addition, one retains groups of modalities of a same question that are close on axis and related semantically, if their barycenters have a contribution exceeding the basic criterion. Then one ranks these modalities and barycenters according to decreasing contributions to axis, and one makes sure that they provide a good summary of axis (say 75% of variance). If the cumulated amount is not sufficient, take a less stringent criterion.

Step 3. *Contributions of modalities to questions.*

For each question retained at Step 1, take the modalities retained in Step 2. If the modalities are divided into two groups on both sides of the origin, determine the barycenters of groups, then calculate the percentage of inertia of the question (on axis) accounted by the deviation between barycenters. *Remark:* More than 2 groups may at times be needed.

Step 4. *Landmark Patterns in the cloud of individuals.*

From the modalities brought out at Step 3, construct patterns involving those modalities. Those patterns will serve as landmarks in the cloud of individuals, e.g. to describe *typical patterns.*

Step 5. *Geometric summary.*

Simplified diagrams will be presented, substantiated by the quantitative appraisals of Step 3.

A first application of this strategy will be presented in §5.4.3 (p.229).

5.3.4 Joint Use of MCA and AHC

A Euclidean classification, performed on the results of a MCA, especially the cloud of individuals, is often a useful complement for interpretation. The representation of the classes of the partition stemming from the class hierarchy in the principal planes, together with their concentration ellipses, enables one to deepen the caracterizations of classes.

It is often advantageous to perform an AHC from the principal coordinates. Rather that confining oneself to the first two or three axes, which provide a coarse summary of data, one will eliminate the last axes (which presumably express mainly "noise") to get more stable results. The joint use of MCA and AHC will be illustrated in chapter *CASE STUDIES*.

5.3.5 Shapes of Clouds and Data Structures

In MCA, one encounters various shapes of clouds. Finding a cloud with a special shape often reveals an underlying structure in the data that may be rendered apparent in the data table by rearranging rows and columns.

Triangular shape

The points of the cloud appear to be scattered within a triangle, with a first vertex, or "pole", on one side of the first axis (say), and the other two poles on the two sides on the second axis. This structure is often encountered in evaluation questionnaires, with poles corresponding to "agree", "disagree" and "don't know". See e.g. the questionnaire on TV Variety shows reported in Benzécri (1992, p.508).

Block Structure

Suppose an opposition between two clusters of points is found on Axis 1, and that in plane 1-2, one cluster appears to be opposed to a subcloud elongated along the direction of Axis 2; then the underlying structure is (approximately) a two–block structure (cf. *CORRESPONDENCE*, Exercise 2.4, p.73). If the rows and columns of the data table are reordered according to the dichotomy of the first axis, two diagonal blocks appear on the table, with virtually null entries outside the blocks.

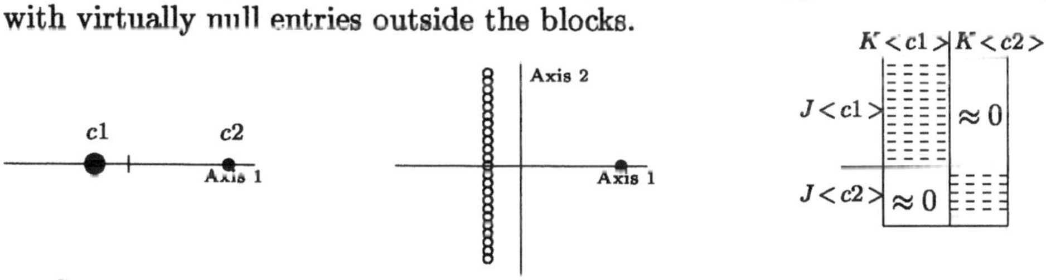

Similarly, finding three clusters in plane 1-2 would point to a three–block structure; in plane 1-3, the cloud would exhibit a shape corresponding to the one in plane 1-2 in the figure above. Etc.

In the case of two blocks, if one block has few modalities, an analysis may be carried out by putting these modalities as supplementary elements; if the two blocks have comparable sizes, they can be analyzed separately.

Guttman effect

The noncorrelation between principal variables excludes linear dependence, but not nonlinear dependence. A cloud may exhibit a quasi-functional relationship (evoking a "parabolic shape") from Axis 2 to Axis 1; this finding suggests an underlying approximate total ordering of modalities and individuals. If the rows and columns of the table are ranked along the order of the first axis, a central strip of high frequencies appears, with corners of low frequencies. The "parabolic shape" of a cloud is thus the geometric expression of a hierarchical scale, or *Guttman scale* (see below); accordingly it has been named by Benzécri *Guttman effect* (sometimes translated in English by "horseshoe effect"). Strictly speaking, there is a Guttman effect (following Benzécri, 1992, p.94) for two axes ℓ and ℓ' if the quadratic relation holds between the principal variables $\forall i \in I : y^i_{\ell'} = a\big((y^i_\ell)^2 - \lambda_\ell\big)$.

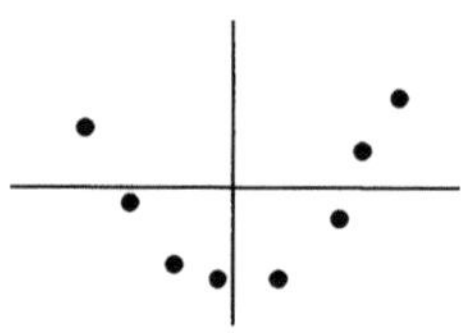

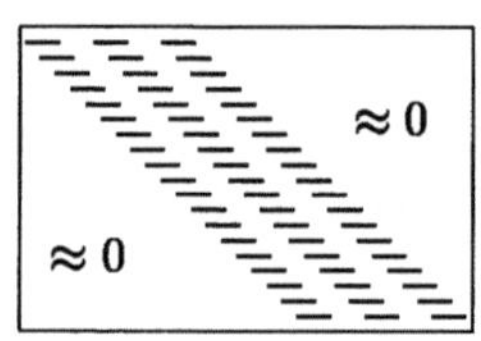

Guttman scale. Let a $I \times Q$ table with Q dichotomous questions or items (+ −, e.g. checks vs fails); if the items and the individuals can be ordered so that among the 2^Q possible patterns, only the $Q+1$ corresponding to increasing numbers of + are observed ("Guttman triangle"), the $I \times Q$ table defines a *Guttman scale.* If the doubling technique is applied to the $I \times Q$ table, putting all − items on the right of + ones, the hierarchical structure is expressed in the $I \times 2Q$ table by a "parallelogram" array, known as a *scalogram.* In Exercise 5.4 (p.245), the CA of a scalogram, yielding a Guttman scale, is studied in detail.

Triangle

	q1+	q2+	q3+	q4+
4	1	1	1	1
3	0	1	1	1
2	0	0	1	1
1	0	0	0	1
0	0	0	0	0

Doubling ⟶

Scalogram

	q1+	q2+	q3+	q4+	q1−	q2−	q3−	q4−
4	1	1	1	1	0	0	0	0
3	0	1	1	1	1	0	0	0
2	0	0	1	1	1	1	0	0
1	0	0	0	1	1	1	1	0
0	0	0	0	0	1	1	1	1

As a conclusion, the MCA procedure, which does not assume any ordinal structures on the data set — unlike so many statistical procedures — does bring out these structures whenever they exist in the data. The extraction of various scales from a questionnaire has been a long–time topic in social psychology and sociology; MCA provides an efficient tool for carrying out this search.

5.4 The Culture Example

Owing to the importance of MCA in the analysis of survey data, we have chosen to illustrate the procedure and strategy of data analysis on a real data set (even though of a small size, that is, with few questions, cf. footnote 3, p.180): the *Culture example.* This example has been devised as a compromise between a mere illustration and a real case study; it should give a fair idea of how a large questionnaire can be analyzed by MCA[18].

The *Culture example* has been constructed from a 1997 survey on the cultural practices of French people, sponsored by the Department of Studies and Prospective of the Ministry of Culture and Communication. In the survey, a sample of 3002 individuals aged 15 or more, representative of the French population, answered a very comprehensive questionnaire (125 questions): cf. Donnat (1998)[19].

For the purpose of illustrating MCA, we selected a set of $Q = 6$ questions pertaining to *leisure activities*, and we confined the study to the set I of 2720 individuals aged 18 or more who answered fully the six questions. Standard MCA was applied to the $I \times Q$ table.

We successively present the six active questions and elementary statistics (§5.4.1). We discuss the data set and the main results of MCA (5.4.2). We proceed to the interpretation of axes (§5.4.3). We examine three supplementary questions (§5.4.4). Then we explore the cloud of individuals and study three structuring factors (§5.4.5).

5.4.1 Active Questions and Elementary Statistics

The six questions, with their sets of modalities, are given below, together with their marginal response frequencies (absolute and relative).

Questions $q1$ through $q4$ are reproduced directly from the questionnaire, for each question each individual gives one and only one response.

(*q*1). As a general rule, do you prefer *Leisure activities* that you can do

---

[18]This example has been constructed and studied in collaboration with W. Ackermann.

[19]We thank O. Donnat for transmitting the data file of the survey to us.

q1	*Leisure activity*	freq.	%
q1r1	*alone*	434	16.0
q1r2	with *partner*	507	18.6
q1r3	with *friends*	1060	39.0
q1r4	with *family*	719	26.4
	Total	2720	100.0

Table 5.1. Question *q1* (*Leisure activity*) with marginal response frequencies.

(***q2***). As a general rule, would you say that during your *Free time*

q2	*Free time*	freq.	%
q2r1	you *lack time* to do all that you would like to do	1121	41.2
q2r2	you don't lack time but you have *always something to do*	1157	42.5
q2r3	sometimes you have *nothing particular to do*	241	8.9
q2r4	often you *do nothing in particular*	201	7.4
	Total	2720	100.0

Table 5.2. Question *q2* (*Free time*) with marginal response frequencies.

(***q3***). If you had *More time*, your first choice of activity would be

q3	*More time*	freq.	%
q3r1	to *rest*, not to do anything in particular	304	11.2
q3r2	to *take courses* to improve your work situation	262	9.6
q3r3	to discover or practice more *physical activities*	573	21.1
q3r4	to discover or practice more *artistic activities* (music, painting, etc.)	394	14.5
q3r5	to develop your *general knowledge*	449	16.5
q3r6	to *take care* of your *family*	316	11.6
q3r7	to do some *home DIY* (gardening, etc.)	422	15.5
	Total	2720	100.0

Table 5.3. Question *q3* (*More time*) with marginal response frequencies.

(***q4***). When you *Go out* in the evening, do you usually go

q4	*Going out*	freq.	%
q4r1	*alone*	202	7.4
q4r2	with your *partner*	911	33.5
q4r3	with *family*, children, parents, etc.	546	20.1
q4r4a	with *friends*	538	19.8
q4r4b	with a *group* (workers' council, club, etc.)	54	1.4
q4r5	you *don't go out* in the evening.	469	17.2
	Total	2720	100.0

Table 5.4. Question *q4* (*Going out*) with marginal response frequencies.

In addition to the four preceding questions, we have constructed two further questions $q5$ and $q6$.

(**$q5$**). Question $q5$ was built from questions about the time of watching *TV*, yielding the following five categories (hours per week).

(**$q6$**). Question $q6$ was built from questions concerning the number of *Books* and comic strips read during the last 12 months, yielding the following five categories:

$q5$	*TV*	freq.	%
$q5r1$	never	257	9.4
$q5r2$	less than 10h	435	16.0
$q5r3$	10-19h	794	29.2
$q5r4$	19-30h	705	25.9
$q5r5$	over 30h	529	19.4
	Total	2720	100.0

Table 5.5. Question $q5$ (*TV*) with marginal response frequencies.

$q6$	*Books*	freq.	%
$q6r1$	no books	603	22.2
$q6r2$	1-4 books	482	17.7
$q6r3$	5-12 books	641	23.6
$q6r4$	13-39 books	563	20.7
$q6r5$	40 books or more	431	15.8
	Total	2720	100.0

Table 5.6. Question $q6$ (*Books*) with marginal response frequencies.

— *Remark.* The numbering of the modalities of questions $q1$ to $q4$ does not imply any underlying order; the categories of questions $q5$ and $q6$ are ordered, but recall that the order structure is not used in MCA.

In addition to the foregoing variables, identification variables (such as *Gender*, *Age* and *Education*) will be examined, firstly as supplementary elements (§5.4.4, p.233) in the cloud of modalities, then as structuring factors (in the sense of the chapter *STRUCTURED*) in the cloud of individuals.

5.4.2 Data Set and Results of MCA

The data table concerns 2720 individuals and 6 questions ($Q = 6$). Like the authors of the survey, for question $q4$ (*Going out*) we grouped the modality "as part of a group" of low frequency (1.4%) with modality "with friends"; hence the modality $q4r4$ ("friends & group"). So we end up with $K = 30$ active modalities.

The variance of the cloud is equal to $(K - Q)/Q = (30 - 6)/6 = 4$. The number of (nontrivially null) eigenvalues is $K - Q = 24$, hence the mean of eigenvalues $\overline{\lambda} = 4/24 = .1667$.

The numbers of modalities per question vary between 4 and 7. Recall that the relative contribution of question q to the variance of the cloud is equal to $(K_q - 1)/(K - Q)$ (cf. p.192). The contributions of $q4$, $q5$ and $q6$ are equal to 16.7%, and those of $q1$ and $q2$ are 12.5%; that of question $q3$ (with 7 modalities) is equal to 25%; as it turns out, question $q3$ is the central question of this study of leisure activities.

There are 2012 observed response patterns to be compared with the number of individuals (2720); which points out the diversity of responses.

Eigenvalues and inertia rates

Ten eigenvalues exceed $\overline{\lambda} = .1667$: $\lambda_1 = .2925$, $\lambda_2 = .2415$, $\lambda_3 = .2248$, $\lambda_4 = .2073$, $\lambda_5 = .1950$, $\lambda_6 = .1832$, $\lambda_7 = .1790$, $\lambda_8 = .1758$, $\lambda_9 = .1733$, $\lambda_{10} = .1688 > .1667$. By dividing eigenvalues by the variance of cloud, one obtains inertia rates (τ_ℓ); by applying formula (5.16), p.201, one gets the modified inertia rates (τ'_ℓ)[20].
$\tau_1 = .073$, $\tau'_1 = .569$; $\tau_2 = .060$, $\tau'_2 = .201$; $\tau_3 = .056$, $\tau'_3 = .122$; $\tau_4 = .052$, $\tau'_4 = .059$.

Weights and coordinates of modalities

Table 5.7 gives the weights of the 30 modalities (column 2) and the coordinates of the 30 modalities on the first three axes (columns 3-5). The weight of modality k is f_k/Q (relative frequency divided by the number of active questions); for $k1 = q1r1$ (Table 5.1, p.222), one gets $\frac{434}{2720}/6 = .0266$.

Variances of axes: $\lambda_1 = .2925$; $\lambda_2 = .2415$; $\lambda_3 = .2248$

K	$Q\&R$	weights	Coordinates Axis 1	Coordinates Axis 2	Coordinates Axis 3	K	$Q\&R$	weights	Coordinates Axis 1	Coordinates Axis 2	Coordinates Axis 3
$k1$	$q1r1$	.0266	−0.540	−0.645	0.885	$k16$	$q4r1$	.0124	0.060	−0.428	0.435
$k2$	$q1r2$	.0311	−0.033	0.537	0.885	$k17$	$q4r2$	.0558	0.242	0.421	0.632
$k3$	$q1r3$	.0650	0.575	−0.563	−0.460	$k18$	$q4r3$	.0335	−0.296	0.926	−0.829
$k4$	$q1r4$	.0441	−0.498	0.841	−0.481	$k10$	$q4r4$	.0363	0.873	−0.928	−0.680
$k5$	$q2r1$	.0687	0.392	0.396	0.061	$k20$	$q4r5$	.0287	−1.253	−0.540	1.409
$k6$	$q2r2$	.0709	−0.236	−0.091	0.136	$k21$	$q5r1$	.0157	0.035	−0.519	0.553
$k7$	$q2r3$	.0148	0.071	−0.707	−0.547	$k22$	$q5r2$	.0267	0.628	0.117	0.275
$k8$	$q2r4$	.0123	−0.914	−0.840	−0.471	$k23$	$q5r3$	.0487	0.258	0.241	−0.104
$k9$	$q3r1$	.0186	−0.843	−0.741	−0.086	$k24$	$q5r4$	.0432	−0.046	0.225	−0.044
$k10$	$q3r2$	.0161	0.099	−0.155	−0.777	$k25$	$q5r5$	.0324	−0.859	−0.507	−0.279
$k11$	$q3r3$	.0351	0.619	−0.010	−0.532	$k26$	$q6r1$	.0369	−0.891	−0.299	−0.355
$k12$	$q3r4$	.0241	0.695	−0.003	0.517	$k27$	$q6r2$	.0295	−0.258	0.325	−0.317
$k13$	$q3r5$	.0275	0.187	−0.171	0.777	$k28$	$q6r3$	.0393	0.199	0.182	0.042
$k14$	$q3r6$	.0194	−0.571	1.076	−0.418	$k29$	$q6r4$	.0345	0.667	0.053	0.094
$k15$	$q3r7$	.0259	−0.715	0.022	0.271	$k30$	$q6r5$	.0264	0.366	−0.286	0.666

Table 5.7. Weights and coordinates of the 30 modalities on the first three axes.

Coordinates of individuals

We do not give the table of coordinates of individuals (2012 response patterns) on the first three axes, but it is easy to calculate them from Table 5.7 (this page). Indeed, we know that the point representing an individual is defined as the equibarycenter of the Q ideal points corresponding to its response pattern. From Property 5.7 (p.188), the coordinate y^i_ℓ of individual i on Axis ℓ is equal to $\xi_\ell z^i_\ell$, with $\xi_\ell z^i_\ell = \sum_{k \in K<i>} z^k_\ell / Q$ (transition formula

[20] See an example of calculation p.225.

(5.7), p.188). For example, from Table 5.7, we deduce the coordinates of the individual with response pattern (314424): its coordinate on Axis 1 is equal to $(1/6) \times (0.575 + 0.392 + 0.695 + 0.873 + 0.628 + 0.667)/\sqrt{.29252} = 1.180$.

Aids to interpretation: contributions of modalities and questions

Table 5.8 gives the relative contributions of the 30 modalities to the variance of the cloud (column 2) and to the first three axes (columns 3 to 5) as well as the relative contributions of questions.

Variances of axes: $\lambda_1 = .2925$; $\lambda_2 = .2415$; $\lambda_3 = .2248$

Q	Q&R	Cloud	Axis 1	Axis 2	Axis 3
	q1r1	.035	.027	**.046**	**.093**
	q1r2	.034	.000	**.037**	**.108**
	q1r3	.025	**.073**	**.085**	**.061**
	q1r4	.031	**.037**	**.129**	**.045**
q1		.125	.137	.297	.307
	q2r1	.024	**.036**	**.045**	.001
	q2r2	.024	.013	.002	.006
	q2r3	.038	.000	**.031**	.020
	q2r4	.039	**.035**	**.036**	.012
q2		.125	.085	.114	.039
	q3r1	.037	**.045**	**.042**	.001
	q3r2	.038	.001	.002	**.043**
	q3r3	.033	**.046**	.000	**.044**
	q3r4	.036	**.040**	.000	**.029**
	q3r5	.035	.003	.003	**.074**
	q3r6	.037	.022	**.093**	.015
	q3r7	.035	**.045**	.000	.008
q3		.250	.202	.140	.214

Q	Q&R	Cloud	Axis 1	Axis 2	Axis 3
	q4r1	.039	.000	.009	.010
	q4r2	.028	.011	**.041**	**.099**
	q4r3	.033	.010	**.119**	**.102**
	q4r4	.033	**.095**	**.129**	**.074**
	q4r5	.034	**.154**	**.035**	.021
q4		.167	.270	.333	.308
	q5r1	.038	.000	.018	.021
	q5r2	.035	**.036**	.002	.009
	q5r3	.030	.011	.012	.002
	q5r4	.031	.000	.009	.000
	q5r5	.034	**.082**	**.034**	.011
q5		.167	.129	.074	.044
	q6r1	.032	**.100**	.014	**.021**
	q6r2	.034	.007	.013	**.013**
	q6r3	.032	.005	.005	.000
	q6r4	.033	**.053**	.000	.001
	q6r5	.035	.012	.009	**.052**
q6		.167	.177	.041	.088

Table 5.8. Relative contributions of modalities and questions to the cloud and to the first three axes; in bold, contributions of modalities retained for the interpretation of axes.

Examples of calculations

Hereafter we illustrate the calculation of the aids to interpretation of MCA, starting from marginal frequencies of questions (Tables 5.1 to 5.6, p.222-223) and eigenvalues and modality coordinates (Table 5.7, p.224).

• *Modified rates* (cf. formula (5.16), p.201)

For each eigenvalue λ_ℓ exceeding the average $\overline{\lambda} = 1/Q$ (in standard MCA), one computes $(\lambda_\ell - \overline{\lambda})^2$; that is here, for the 10 eigenvalues greater than $\overline{\lambda} = 1/6 = 0.166667$, we compute : .015840, .005599, .003384, .001648, .000802, .000274, .000151, .000084, .000043, .000004; hence the sum $S = .015840 + \ldots + .000004 = 0.027830$. The modified rate for Axis ℓ is then given by $\tau'_\ell = (\lambda_\ell - \overline{\lambda})^2/S$, that is, for Axis 1: $\tau'_1 = .015840/0.027830 = .5692$, etc.

• *Contribution of a modality to cloud* (Ctr, 2nd column of Table 5.8)

The relative contribution of modality k to the cloud (cf. formula (5.10), p.194) is $\mathrm{Ctr}_k = (1 - f_k)/(K - Q)$.

Example for $k1 = q1r1$: $\mathrm{Ctr}_{k1} = (2720 - 434)/(2720 \times 24) = .035$.

• *Contributions of modalities to axes* (columns 3 to 5 of Table 5.8)

The absolute contribution of modality k to Axis ℓ (cf. §5.1.2, p.196) is $\mathrm{Cta}_{k\ell} = (f_k/Q)\,(y_\ell^k)^2$. Example for $k1 = q1r1$ and Axis 1: $\mathrm{Cta}_{k1\ell1} = (\frac{434}{2720}/6) \times (-.540)^2 = .0078$; hence, by dividing by the variance of Axis 1 ($\lambda_1 = .2925$), its relative contribution to Axis 1,that is, $.0078/.2925 = .027$.

• *Contributions of questions to axes*

The contribution of a question is the sum of the contributions of its modalities. Thus, for question $q1$, its absolute contribution to Axis 1 is equal to $.0078 + \ldots + .0109 = .0402$, and its relative contribution to Axis 1 is equal to $.027 + .000 + .073 + .037 = .137 = .0402/.2925$.

• *Contributions of deviations to axes*

The contribution of a deviation to an axis is equal to the contribution of the dipole composed of the barycenters of the two groups of modalities which are opposed on this axis (cf. formula (5.14), p.196).

Example: for question $q4$ and Axis 1, there is an opposition between the two modalities $q4r4$ and $q4r5$, the calculations of the contribution of the corresponding deviation are summarized in the following table.

Modalities	$q4r4$	$q4r5$
Absolute frequency	592	469
Coordinates	+0.873	−1.253
Dipole		
Abs. freq.	$\tilde{n}_d = 1/(\frac{1}{592} + \frac{1}{469}) = 261.685$	
Deviation	$d = 0.873 - (-1.253) = 2.126$	
Absolute contribution of deviation: $\frac{1}{6} \times \frac{261.685}{2720}(2.126)^2 = .0725$		

From the relative contribution of $q4$ ($\mathrm{Ctr}_{q4}^{\ell 1} = .270$, see Table 5.8, p.225), one deduces its absolute contribution $\lambda_{\ell 1} \times \mathrm{Ctr}_{q4}^{\ell 1} = .2925 \times .270 = .079$, hence the relative contribution of the deviation between modalities $q4r4$ and $q4r5$ to question $q4$: $.0725/.079 = .92$.

For question $q3$ and Axis 1, the opposition is between $q3r3$ and $q3r4$ on the one hand, $q3r1$ and $q3r7$ on the other hand. We apply the general rule: Weights add up by grouping, whereas coordinates average up. The deviation between $q3r3$ and $q3r4$ on the one hand, $q3r1$ and $q3r7$ on the other hand, has a relative contribution to question $q3$ equal to $0.0511/(.2925 \times .202) = .866$. Hence the following table.

Modalities	$q3r3$	$q3r4$	$q3r1$	$q3r7$
Abs. freq.	573	394	304	422
Coordinates	+0.619	+0.695	−0.843	−0.715
Barycenters	$q3r3\&q3r4$		$q3r1\&q3r7$	
Abs. freq.	$573+394=967$		$304+422=726$	
Coordinates	$\frac{573\times0.619+394\times0.695}{573+394}=+0.650$		$-\frac{304\times0.843+422\times0.715}{304+422}=-0.768$	
Deviation				
Abs. freq.	$\tilde{n}_d=1/\left(\frac{1}{967}+\frac{1}{726}\right)=414.673$			
Dipole	$d=+0.650-(-0.768)=1.418$			
	Absolute contribution of deviation: $\frac{1}{6}\times\frac{414.673}{2720}(1.418)^2=.0511$			

Cloud of modalities and cloud of individuals in plane 1-2

We immediately give the graphical representation of the cloud of 30 modalities in the plane of the first two axes (Figure 5.6), as well as that of the cloud of 2720 individuals (Figure 5.7, p.228).

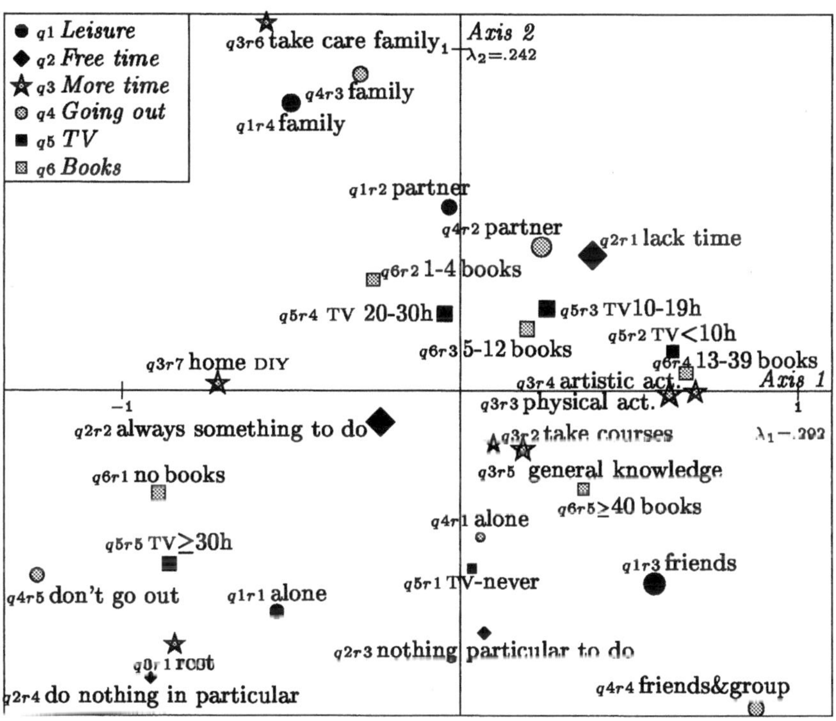

Figure 5.6. Cloud of the 30 modalities in plane 1-2. The sizes of markers are proportional to response frequencies.

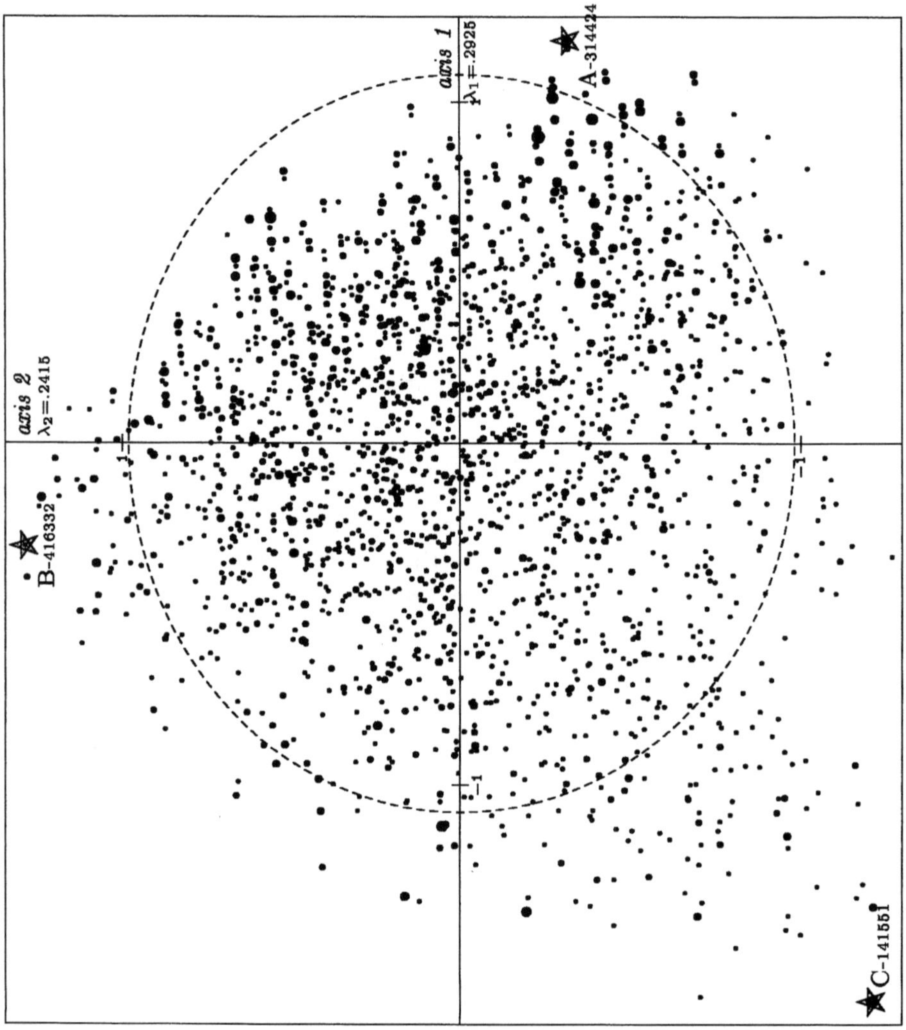

Figure 5.7. Cloud of individuals in plane 1-2 with its concentration ellipse and 3 typical patterns (points A, B, C).

First comments

The first eigenvalue is predominant; its modified rate (57%) is nearly three times that of the second one (20%), then the eigenvalues decrease slowly.

The *visual examination* of the cloud of modalities in plane 1-2 (Figure 5.6, p.227) shows that modalities are organized around three poles: friends (bottom right), family (top), and withdrawal and rest (bottom left). By examining the cloud of individuals (Figure 5.7), one observes its triangular shape, with a wider scattering of points in the left bottom quadrant.

5.4.3 Interpretation

How many axes should be interpreted?

The first eigenvalue is well separated from the second one ($\lambda_1 - \lambda_2 = .0510$); its modified rate is equal to 57%. By adding the second axis, one arrives to a cumulated rate of 77%. But the deviation $\lambda_2 - \lambda_3 = .0166$ is smaller than $\lambda_3 - \lambda_4 = .0176$, so we add the third axis, then the cumulated modified rate reaches 89%. For the following eigenvalues the deviations are small and decrease regularly, we do not attempt to interpret the subsequent axes.

The examination of the contributions of questions (cf. Table 5.8, p.225) shows that Axis 1 involves the six questions, whereas Axis 2 mainly involves the two questions $q4$ (*Going out*) and $q1$ (*Leisure activities*) whose contributions (33% and 30%) greatly exceed the average contribution (17% $= 1/Q$); three questions have important contributions to Axis 3, i.e. $q4$ (*Going out*, 31%), $q1$ (*Leisure activities*, 31%) and $q3$ (*More time*, 21%).

Interpretation of axes

For interpreting an axis, we will use the *method of contributions of points and deviations* (cf. §5.3.3, p.217). We will base the interpretation of an axis on the modalities which verify the following *basic criterion*: their contributions to axis exceed the average contribution (here $100/30 = 3.3\%$).

Interpretation of Axis 1 ($\lambda_1 = .2925$). The following Table 5.9, constructed from Tables 5.7 and 5.8 (p.224-225), gives the 14 modalities whose contributions exceed $100/30 = 3.3\%$.

			left	right	deviation
$q1$ ● *Leisure*	$r3$	friends		7.3	
	$r4$	family	3.7		75.2
$q2$ ◇ *Free time*	$r1$	lack time		3.6	
	$r4$	do nothing in particular	3.5		71.6
$q3$ ★ *More time*	$r3$	physical activities		4.6	
	$r7$	home DIY	4.5		
	$r1$	rest	4.5		
	$r4$	artistic activities		4.0	86.6
$q4$ *Going out*	$r5$	don't go out	15.4		
	$r4$	friends&group		9.5	91.8
$q5$ ■ TV	$r5$	≥ 30h	8.2		
	$r2$	< 10h		3.6	85.6
$q6$ *Books*	$r1$	no books	10.0		
	$r4$	13-39 books		5.3	83.7
			49.8	37.9	

Table 5.9. Interpretation of Axis 1. Relative contribution (percentage) of each modality to axis, written either in column "left" or in column "right" according to its position on graph; relative contribution of deviation to question.

These 14 modalities are depicted on Figure 5.8 (this page); they are divided into two groups, one group on the left (negative coordinates), the other one on the right (positive coordinates). For each question, the contribution of the deviation between the barycenters of the two groups has been calculated by formula 5.14 p.196 (see examples p.226).

As we can see in Table 5.9 (p.229), these 14 modalities contribute together to 88% of the variance of Axis 1, and for each question the contribution of the deviation is greater than 71%: it can be said that these 14 modalities provide a good summary of Axis 1. Furthermore this axis is a general one because all questions intervene in the interpretation of Axis 1.

One will now interpret Axis 1 by examining these modalities in detail.

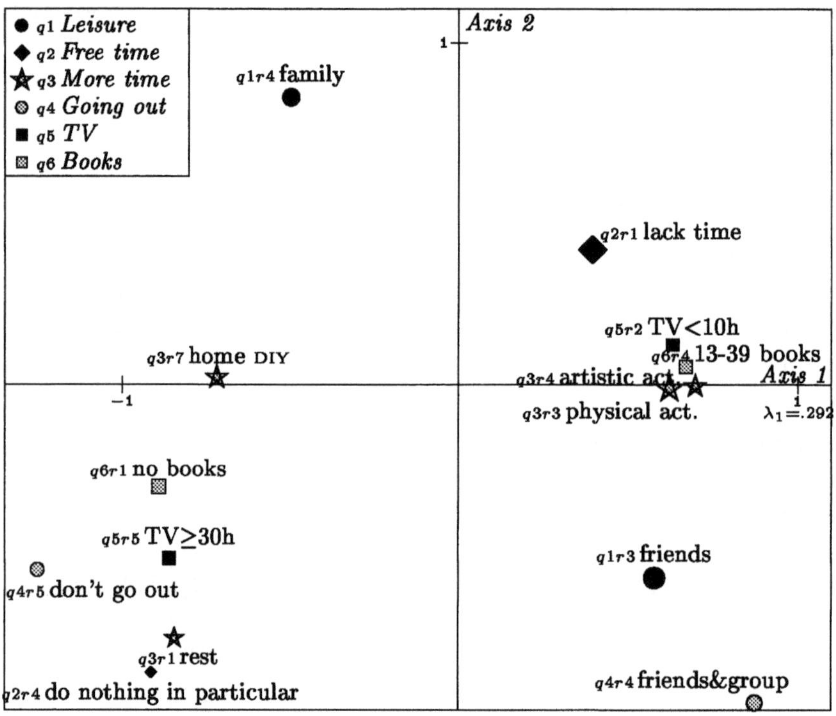

Figure 5.8. Interpretation of Axis 1: 14 modalities most contributing to axis.

• On the right, one finds going out with friends, physical and artistic activities, reading many books, little time on watching TV, lack of time.

• On the left, one finds leisure activities in family, doing nothing in particular, resting, not going out, no book reading, watching a lot of TV.

To sum up: Axis 1 opposes leisure activities linked with social life to more individually oriented ones.

Interpretation of Axis 2 ($\lambda_2 = .2415$). In the same way, one constructs Table 5.10 and Figure 5.9.

Question		Modality		bottom	top	deviation
*q*1	●	*r*4	family		12.9	
		*r*3	friends	8.5		
		*r*1	alone	4.6		
Leisure		*r*2	partner		3.7	97.5
*q*2	◇	*r*1	lack time		4.5	
		*r*4	do nothing in particular	3.6		95.8
Free time		*r*3	nothing particular to do	3.1		
*q*3	★	*r*6	take care family		9.3	
More time		*r*1	rest	4.2		92.6
*q*4	●	*r*4	friends & group	12.9		
		*r*3	family		11.9	
		*r*2	partner		4.1	
Going out		*r*5	don't go out	3.5		87.3
*q*5 *TV*	■	*r*5	≥ 30h	3.4		
				43.8	46.4	

Table 5.10. Interpretation of Axis 2. Relative contribution (percentage) of each modality to axis, written either in column "bottom" or in column "top" according to its position on graph; relative contribution of deviation to question.

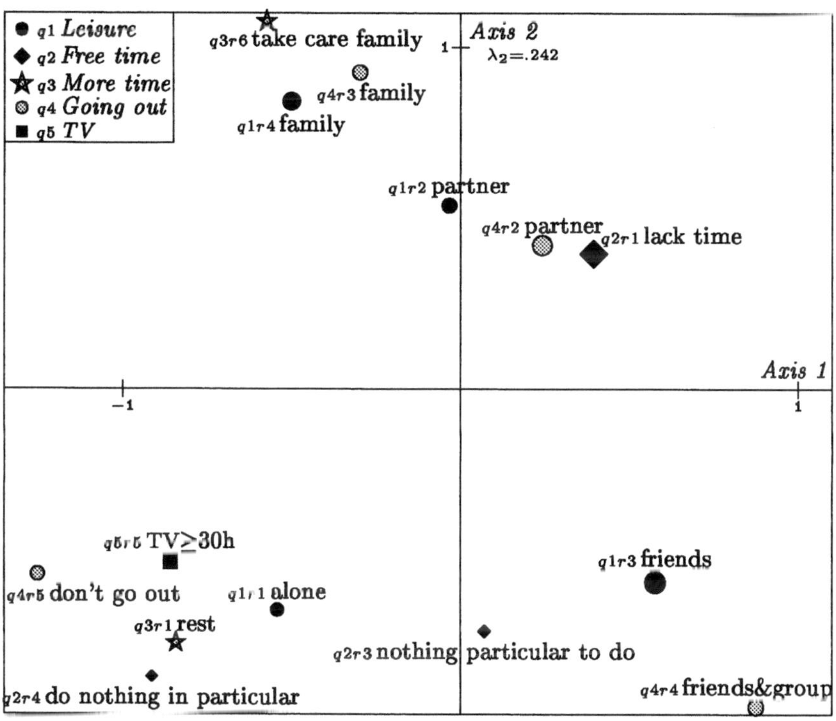

Figure 5.9. Interpretation of Axis 2: 14 modalities most contributing to axis.

Questions $q4$ (*Going out*) and $q1$ (*Leisure*) contribute together to 63% of the variance of Axis 2. They highlight the clear–cut opposition between the modalities Family or Partner ($q1r4$, $q1r2$, $q3r6$, $q4r3$, $q4r2$) associated with "lack time" ($q2r1$) on the one hand (top), and the modalities Friends & group or Alone ($q1r3$, $q4r4$, $q1r1$) associated with "rest" ($q3r1$) on the other hand (bottom). We add the modalities $q2r3$ and $q2r4$ (bottom) which are close on Axis 2 and close semantically ("do nothing in particular"); furthermore the contribution of their barycenter (coordinate -0.768, weight .027) is equal to 6.6% > 3.3%. The modality $q5r5$ ("over 30 hours " on TV) has also been retained, since it meets the basic criterion (3.4% > 3.3%), even though there are no other modalities of this question that can be opposed to it (owing to their weak contributions < 1.8%).

> *To sum up*: Axis 2 opposes activities practiced in family to doing nothing in particular or going out with friends.

Interpretation of Axis 3 ($\lambda_3 = .2248$). Proceeding similarly, for Axis 3, Table 5.11 and Figure 5.10 (p.233) are constructed.

		bottom	top	deviation
$q1$ ●	$r2$ partner		10.8	
	$r1$ alone		9.3	
	$r3$ friends	6.1		
Leisure	$r4$ family	4.5		100.0
$q3$ ★	$r5$ general knowledge		7.4	
	$r3$ physical act.	4.4		
	$r2$ take courses	4.3		71.4
More time	$r4$ artistic act.		2.9	
$q4$ ●	$r3$ family	10.2		
	$r2$ partner		9.9	
Going out	$r4$ friends & group	7.4		85.7
$q6$ ■	$r5$ 40 books or more		5.2	
	$r1$ no books	2.1		
Books	$r2$ 1-4 books	1.3		96.7
		40.3	45.5	

Table 5.11. Interpretation of Axis 3. Relative contribution (percentage) of each modality to axis, written either in column "bottom" or in column "top" according to its position on graph; relative contribution of deviation to question.

Questions $q4$ and $q1$ contribute together to 62% of axis. There are 11 modalities that verify the basic criterion; to which we have added modality $q3r4$ ("Artistic activities"), whose contribution (2.9%) is not far from the average contribution (3.3%) and is furthermore superior to those (< 1.5%) of the modalities of this question not retained for interpretation. We have also retained the modalities $q6r1$ ("no books") and $q6r2$ ("less than 4" books) that are close on Axis 3 (coordinates -0.355 and -0.317 respectively) and whose barycenter (coordinate -0.338, weight .066) has a contribution equal to 3.4% exceeding criterion. So we arrive at 14 modalities that highlight the opposition "alone" and "partner" (on the same side of axis), going with a preference for "cultural activities" – in the classic sense: reading, artistic activities, etc. — on the one hand, versus activities such as physical activities, etc. with family and friends on the other hand. Observe that "artistic" and "physical" activities, that were close in plane 1-2, are opposed along Axis 3.

To sum up: On Axis 3, there is an opposition between leisure activities practiced in family, with friends or in group, and more "cultural" activities practiced individually or with partner.

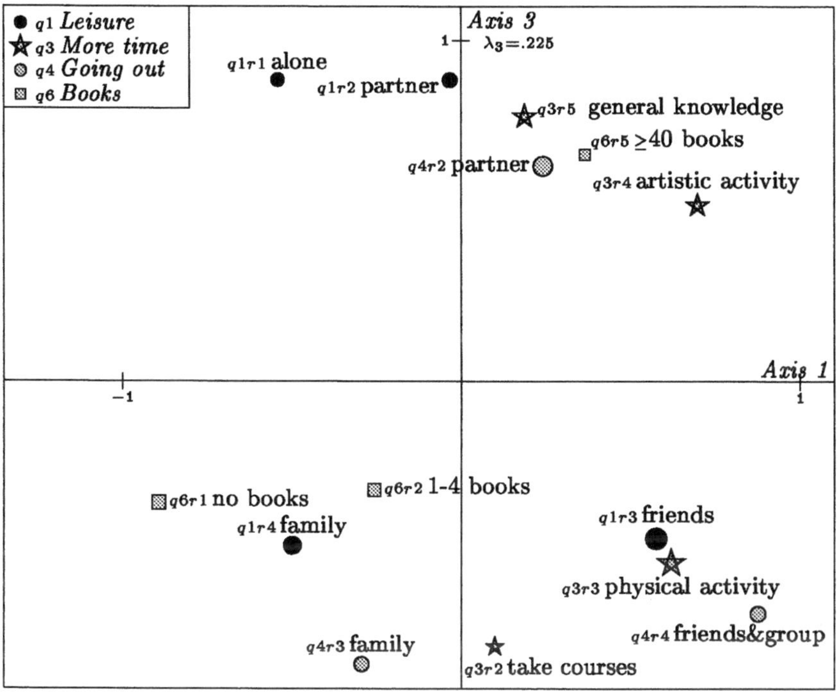

Figure 5.10. Interpretation of Axis 3: 14 modalities most contributing to axis

5.4.4 Supplementary Questions

To illustrate the procedure of supplementary questions (cf. p.197 and p.215), we will study three questions: Gender, Age (6 classes) and Education (8 classes[21]). The coordinates of the modalities of these questions on the first three axes are reported in Table 5.12 (p.234). On Figures 5.11 to 5.14 (p.234-235) these modalities are represented in planes 1-2 and 1-3 (graphical scale half of preceding figures).

[21]The eight classes correspond to the highest grade attained: no degree; CEP (elementary school); BEPC (ninth grade); CAP-BEP (professional certificates); Bac (Baccalauréat, high–school diploma); Bac+2 (two years after Bac); >Bac+2 (more than two years after Bac); University students.

Gender	Freq.	Axis 1	Axis 2	Axis 3
Men	1284	0.005	−0.077	0.023
Women	1436	−0.005	0.069	−0.021

Age	Freq.	Axis 1	Axis 2	Axis 3
$18 < \cdots \leq 25$	449	0.701	−0.498	−0.421
$25 < \cdots \leq 35$	574	0.211	0.214	−0.189
$35 < \cdots \leq 45$	520	0.060	0.451	−0.180
$45 < \cdots \leq 55$	394	−0.021	0.123	0.266
$55 < \cdots \leq 65$	317	−0.342	−0.080	0.344
$\cdots > 65$	466	−0752	−0.336	0.381

Education	Freq.	Axis 1	Axis 2	Axis 3
no degree	433	−0.668	−0.177	−0.246
CEP	458	−0.666	−0.067	0.129
BEPC	201	0.072	0.200	−0.041
CAP-BEP	652	0.021	0.151	−0.126
Bac	344	0.324	0.026	0.129
Bac+2	210	0.500	0.180	0.215
>Bac+2	250	0.647	0.156	0.451
Students	172	1.094	−0.680	−0.375

Table 5.12. Coordinates of supplementary modalities on axes 1, 2 and 3.

As we know (cf. remark p.197), the difference of coordinates between modalities along an axis in the cloud of modalities is equal to the deviation between the corresponding modality mean–points in the cloud of individuals expressed in standard deviation units: a deviation greater than 1 will be regarded as "large", a deviation less than 0.5 as "small".

Gender (Table 5.12, Figures 5.11 and 5.12). Differences between Men and Women are quite small on each of the three axes (coordinate deviations: < 0.01 on Axis 1, < 0.15 on Axis 2, < 0.04 on Axis 3)[22].

Age (Table 5.12, Figures 5.11 and 5.12). On Axis 1, the Age classes are ordered. The deviation between the extreme classes (18-25 vs >65 years) is equal to 1.45, therefore large. On Axis 2, there is an opposition between 36-45 years (age class linked with "family") versus firstly 18-25 years (deviation 0.95) and secondly over 65 years (deviation 0.79). On Axis 3, the age classes are ordered, with a cleavage between <45 years and >45 years; the deviation 18-25 vs >65 is equal to 0.80.

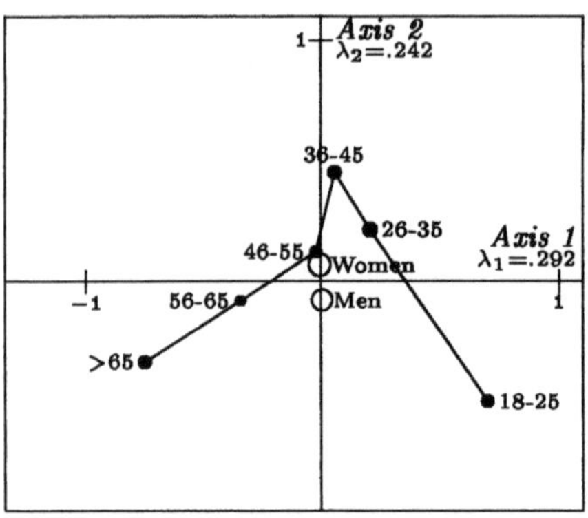

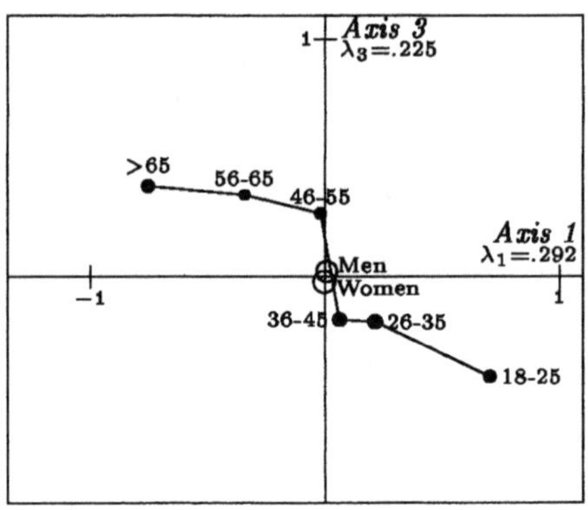

Cloud of modalities: Supplementary questions *Gender* and *Age*

Figure 5.11. in plane 1-2. Figure 5.12. in plane 1-3.

[22]The stability study done in chapter *STABILITY*(p.273) confirms this result, showing that the first three axes of the within–cloud are close to those of the overall cloud.

Education (Table 5.12, p.234). On Figures 5.13 and 5.14, one has represented the points (gray circles) of the 8 Education modalities. We also grouped together the following points: no degree&CEP, BEPC&CAP-BEP, Bac+2&>Bac+2, hence 3 new points (black circles) to which we add Bac grade. These 4 points are joined by a line (leaving aside the category Students); they are well ordered on Axis 1, with a large deviation between extreme categories (1.25). On Axis 2, no deviation exceeds 0.45. On Axis 3 the deviation between no–degree and >Bac+2 reaches 0.70.

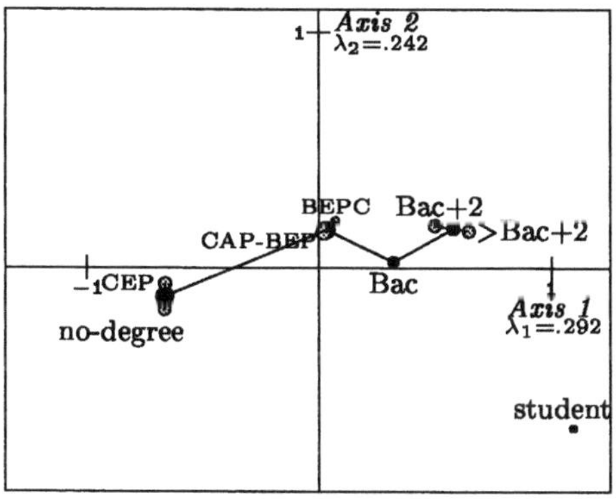

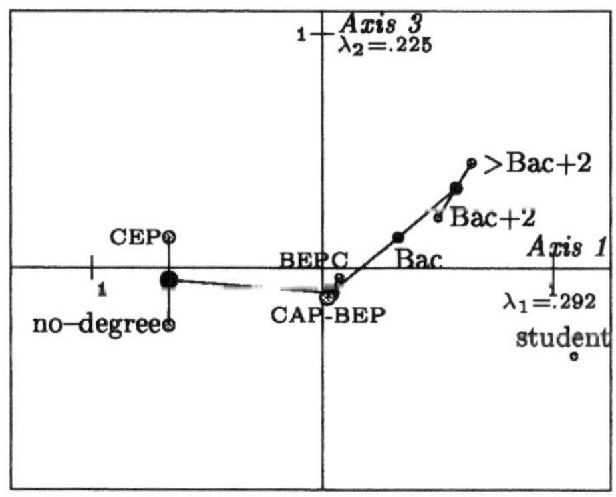

Cloud of modalities: Supplementary question *Education*

Figure 5.13. in plane 1-2.

Figure 5.14. in plane 1-3.

5.4.5 Exploration of the Cloud of Individuals

Shape of the cloud

In *plane 1-2*, the cloud of individuals has roughly a triangular shape (cf. Figure 5.7, p.228). To illustrate the three vertices, we create three "typical patterns". Pattern 314424 is constructed by taking, for each question, the modality which is extreme on the right of Axis 1 and has a high contribution, namely $q1r3$ (leisure activity with friends), $q2r1$ (lack time), $q3r4$ (artistic activity), $q4r4$ (going out with friends & group); $q5r2$ (TV <10h); $q6r4$ (13-39 books); we label this pattern *"sociability +"*. Pattern 416332 is constructed by taking, for each question, the more extreme modality on the top of Axis 2, namely $q1r4$ (leisure activity with family), $q2r1$ (lack time), $q3r6$ (take care family), $q4r3$ (going out with family); $q5r3$ (TV 10-19h); $q6r2$ (1-4 books); we label this pattern *"family +"*. Pattern 141551 is constructing by choosing the modalities with negative coordinates on both axes 1 and 2, namely $q1r1$ (alone), $q2r4$ (do nothing particular), $q3r1$ (rest), $q4r5$ (don't go out); $q5r5$ (TV≥30h); $q6r1$ (no books); we label this pattern *"withdrawal and rest"*. These three typical patterns[23] are represented on Figure 5.7 (p.228): A (pattern 314424, right), B (pattern 416332, top), C (pattern 141551, left bottom).

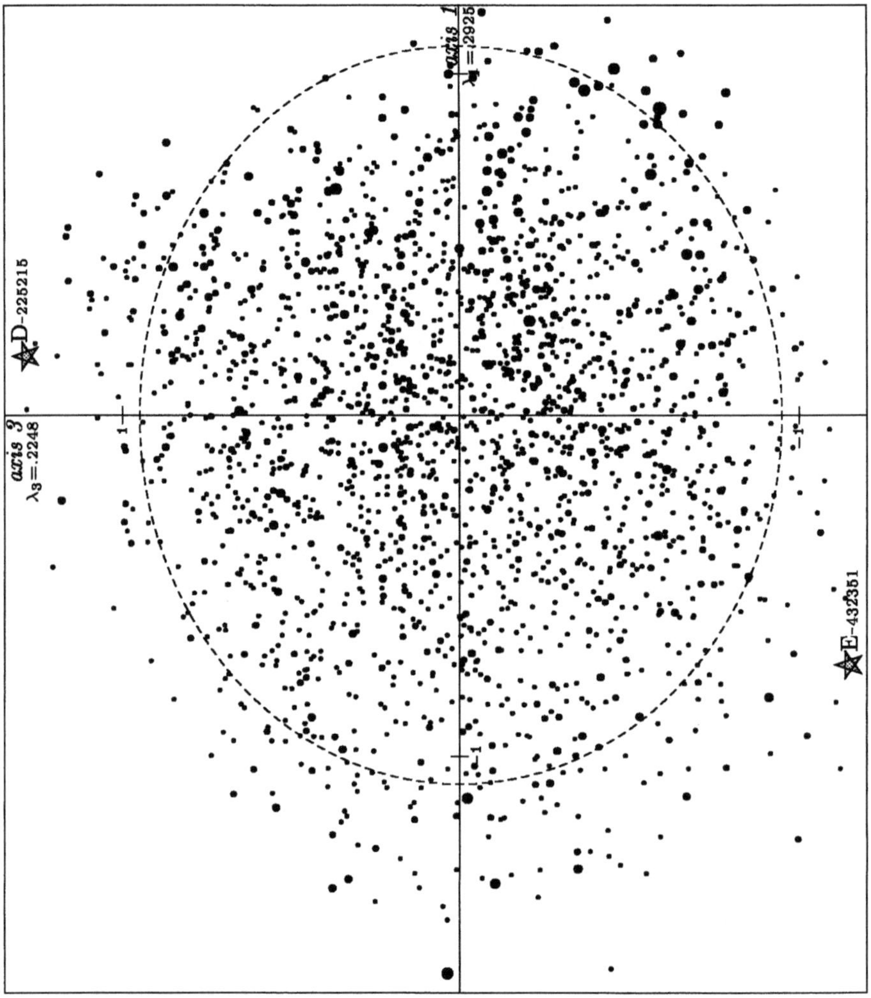

Figure 5.15. Cloud of individuals in plane 1-3 with two typical patterns (points D and E) characterizing Axis 3 (concentration ellipse in dotted line).

In order to better interpret Axis 3, we construct two typical patterns, namely 225215 *"cultural activities with partner"* and 432351 *"leisure activities in family"*, hence the points D and E on Figure 5.15.

[23]Here, the typical patterns are observed 3, 2 and 9 times respectively. For calculating coordinates, see §5.4.2, p.225.

Structured data analysis: study of Age and Education

A deeper study of supplementary questions consists in treating them as *structuring factors* of the cloud of individuals, that is, in studying the subclouds of individuals induced by these questions (cf. p.214 and chapter STRUCTURED). We illustrate the approach on the two questions *Age* and *Education*, studying them first separately, then jointly.

Study of Age. Each age class defines a subcloud with its mean point (called modality mean–point) and its variance, hence six subclouds of individuals. Table 5.13 gives, on the first three axes, the coordinates of the modality mean–points: $\widehat{y}^k_\ell = \xi_\ell y^k_\ell$, (formula (5.15), p.197). For example: for Axis 1, the coordinate of the modality 18-25 years is 0.701 (Table 5.12, p.234), hence the one of the modality mean–point $\sqrt{.2925} \times 0.701 = .379$.

Age	Abs. freq.	Coordinates Axis 1	Axis 2	Axis 3	Qualities of repr. Axis 1	Axis 2	Axis 3	Variances Axis 1	Axis 2	Axis 3
18-25	449	0.379	−0.245	−0.200	.519	.217	.144	.1931	.1884	.1938
26-35	574	0.114	0.105	0.090	.255	.216	.157	.2102	.2400	.2157
36-45	520	0.032	0.222	−0.085	.012	.544	.080	.2057	.2331	.2101
46-55	394	−0.011	0.061	0.126	.004	.115	.497	.2730	.2013	.2183
56-65	317	−0.185	−0.040	0.163	.360	.016	.280	.2626	.1910	.2074
> 65	466	−0.407	−0.165	0.181	.622	.103	.123	.2789	.2145	.1793
							within	.2335	.2146	.2042
							between	.0591	.0270	.0207
							total (λ)	.2925	.2415	.2248

Table 5.13. Age. Coordinates and qualities of representation of the modality mean–points on the first three axes, variances of the 6 subclouds and double breakdown of variances.

To the partition induced by the six Age classes, there is associated the between and the within variance (cf. CLOUD, §3.5.2). Table 5.13 gives the double breakdown of the variance according to axes and partition. To study the dispersion and shape of subclouds, we have represented the concentration ellipse of each subcloud in plane 1–2 (Figure 5.16, p.238), with its eccentricity parameter and its area (cf. MATH.BASES, §10.21, p.441).

In the *cloud of modalities* (§5.4.4, p.233), we had already noticed that the age modalities were ordered on Axis 1, with an important deviation between the extreme modalities, and that the age class 36-45 was opposed to the extreme classes on Axis 2. In the *cloud of individuals*, we now observe that the within–class dispersion is very large, and that the between–class variance is only well marked on Axis 1. In plane 1-2, we further observe a shift in the orientation of ellipses with increasing age, confirming the difference in leisure practices between age groups.

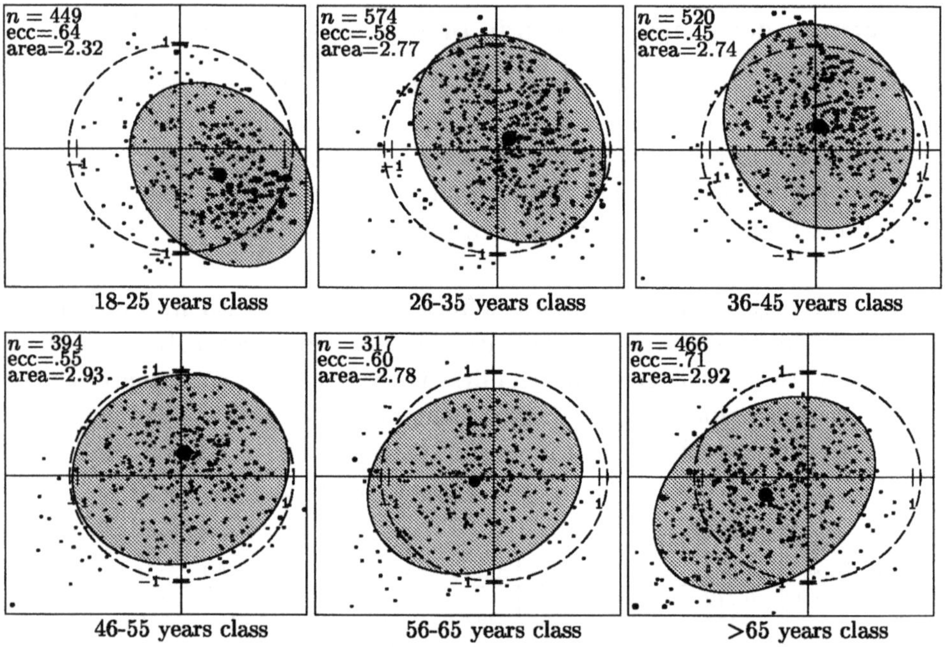

Figure 5.16. Six subclouds of individuals according to age class in plane 1-2 with their concentration ellipses (global ellipse in dotted line).

Study of Education. We proceed in the same way to study the partition in 7 classes induced by the Education factor; hence the Table 5.14 giving the coordinates of the modality mean–points, their qualities of representation and their variances on the first three axes, the within and between variances according to axes (leaving aside students). The within–variance, already strong on Axis 1, becomes even stronger on axes 2 and 3.

		Coordinates			Qualities of repr.			Variances		
Education	freq	Axis 1	Axis 2	Axis 3	Axis 1	Axis 2	Axis 3	Axis 1	Axis 2	Axis 3
no degree	433	−0.361	−0.087	−0.116	.772	.045	.080	.303	.264	.196
CEP	458	−0.360	−0.033	0.061	.831	.007	.024	.248	.240	.200
BEPC	201	0.039	0.098	−0.020	.028	.176	.007	.188	.238	.229
CAP-BEP	652	0.011	0.074	−0.060	.005	.242	.159	.195	.242	.215
Bac	344	0.175	0.013	0.061	.656	.004	.080	.194	.239	.238
Bac+2	210	0.270	0.089	0.102	.588	.064	.084	.156	.220	.239
>Bac+2	250	0.350	0.077	0.214	.535	.026	.200	.130	.187	.239
Students	172	0.592	−0.334	−0.178	.641	.204	.058	.124	.117	.155

Table 5.14. Education. For the first three axes: coordinates and qualities of representation of modality mean–points; variances of the subclouds and double breakdown of variances.

	Variances		
	Axis 1	Axis 2	Axis 3
within	.212	.237	.218
between	.066	.004	.010
total (λ)	.2925	.2415	.2248

To study the dispersions and shapes of subclouds in plane 1-2, we have represented (Figure 5.17) the concentration ellipses of the subclouds, by showing in the same picture the two classes "no degree" and CEP (education level $d1$), then the two classes BEPC and CAP-BEP (level $d2$), then the class Bac (level $d3$), last the two classes Bac+2 and >Bac+2 (level $d4$).

For the two subclouds "no degree" and "CEP", one observes not only that the mean–points are very close but that the ellipses are almost superposed. The same remark holds for BEPC and CAP-BEP. One observes a sharp contrast between the wide subclouds "no degree and CEP" and the narrow subclouds Bac+2 and Bac>+2.

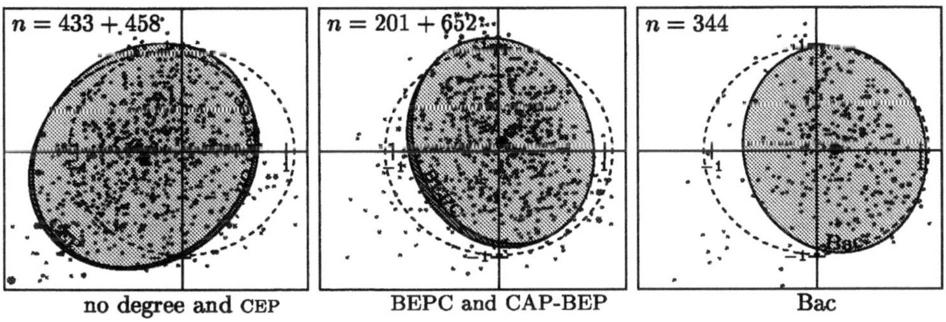

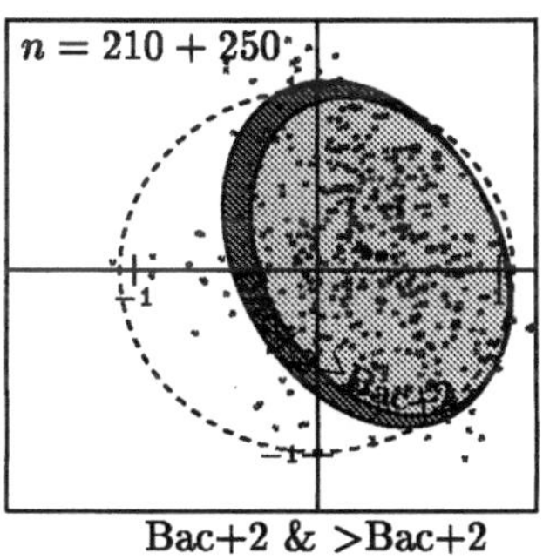

	ecc.	area plane 1-2
no degree	.54	3.51
CEP	.65	2.96
BEPC	.55	2.63
CAP-BEP	.50	2.71
Bac	.46	2.70
Bac+2	.69	2.24
>Bac+2	.69	1.89

Figure 5.17. Seven subclouds of individuals according to Education modalities in plane 1-2 with their concentration ellipses (global ellipse in dotted line).

Crossing Age and Education. To analyze the crossing Age and Education, we take the following categories: for Age, $a1$: $18-35$, $a2$: $36-55$, $a3$:>55 years; for Education, we take the 4 preceding levels $d1$ through $d4$ (still leaving students aside); hence $n = 2548$ individuals in $3 \times 4 = 12$ composite classes, and we denote A the factor *Age* with 3 modalities and D the factor *Education* with 4 modalities. The coordinates of the 12 modality mean–points with their absolute frequencies are given in Table 5.15 (p.240).

	freq.	Axis 1	Axis 2	Axis 3
*a*1*d*1	155	−0.071	−0.004	−0.289
*a*2*d*1	260	−0.293	0.076	−0.110
*a*3*d*1	476	−0.492	−0.152	0.107
*a*1*d*2	351	0.077	0.038	−0.186
*a*2*d*2	335	0.011	0.178	−0.033
*a*3*d*2	167	−0.094	−0.030	0.202

	freq.	Axis 1	Axis 2	Axis 3
*a*1*d*3	142	0.281	−0.057	−0.104
*a*2*d*3	128	0.209	0.175	0.079
*a*3*d*3	74	−0.086	−0.134	0.346
*a*1*d*4	205	0.386	0.014	0.075
*a*2*d*4	189	0.301	0.198	0.180
*a*3*d*4	66	0.124	−0.038	0.387

Table 5.15. Crossing Age in 3 classes with Education in 4 levels. Coordinates of 12 composite modality mean– points.

The opposite table shows two additive breakdowns of the variance of the crossing $A \star D$ on the first three axes: D and A-within D (denoted $A(D)$); then the additive protocol (denoted $A + D$), and the interaction (denoted $A.D$).

	Variances		
	Axis1	Axis2	Axis3
$A \star D$	.0800	.0140	.0270
D	.0659	.0041	.0062
$A(D)$	.0141	.0099	.0208
$A+D$	.0783	.0135	.0268
$A.D$	.0017	.0005	.0003

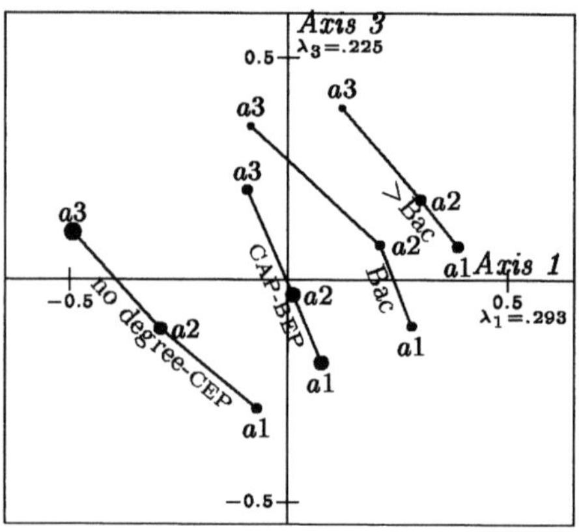

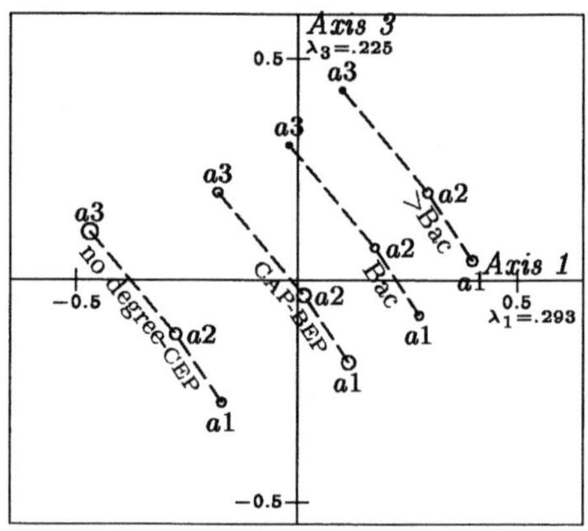

Figure 5.18. Cloud of the 12 composite modality mean–points in plane 1-3.

Figure 5.19. Additive cloud of the 12 composite modality mean–points in plane 1-3

Since the variances on Axis 3 of $A*D$, D and $A(D)$ are greater than the corresponding ones on Axis 2, we show, as a graphical illustration (Figure 5.18) the 12 composite modality mean–points in plane 1-3, joining the 3 age points for each of the 4 education levels. The 4 lines are not far from being parallel, which suggests that an *additive cloud* can be adequately fitted (see §6.3.1, *STRUCTURED*, p.261); this additive cloud is shown on Figure 5.19.

Lastly, Figure 5.20 shows, in plane 1-3, the concentration ellipses of the four extreme groups; on the left figure, it is shown $a1d1$ and $a3d1$, i.e. "young" vs "old" for the lowest education level $d1$, and on the right one $a1d4$ and $a3d4$, i.e. "young" vs "old" for the highest education level $d4$. The $a1d4$ ellipse, with its main dispersion along Axis 3, differs noticeably from the three other ones.

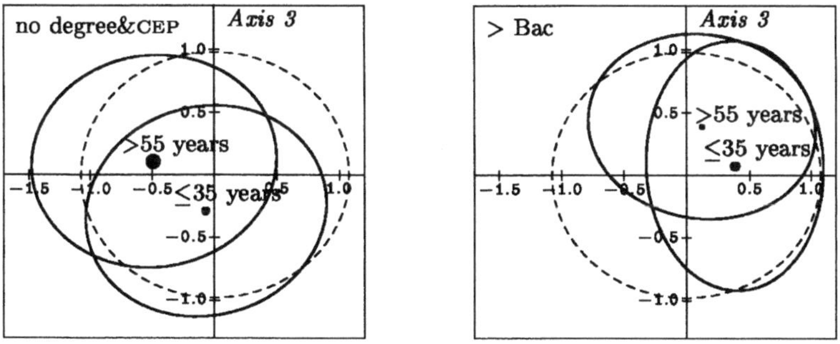

Figure 5.20. Plane 1-3. Concentration ellipses of extreme groups. On the left *a1d1* ($n = 155$) and *a3d1* ($n = 476$); on the right *a1d4* ($n = 205$) and *a3d4* ($n = 66$), together with global ellipse; the graphical scale is half that of Figures 5.18 and 5.19.

Conclusion

As said at the beginning of this section, the Culture Example was devised for illustrative purposes. Owing to the small number of questions, the geometric space that we have constructed is far from a genuine sociological "space of leisure activities". In spite of its limitations, this space has enabled us to investigate structuring factors as Age and Education (problem of the first kind); and it might also serve for predicting the location of other cultural practices (problem of the second kind). For instance, for museum visiting, we read in Donnat (1998) that the most frequent response modalities are "with children" and "with partner"; these modalities — and the corresponding subclouds of individuals – are naturally located in the geometric space of the Culture Example. In spite of their tentative character, the conclusions also nicely concur with the studies concerning sociability in the French society: see e.g. Héran (1988).

Exercises of Chapter 5

Exercise 5.1 (Proof of theorem 5.1)

The aim of this exercise is to determine the principal directions and variables of a cloud M^I stemming from a questionnaire (see §5.1.1) by applying theorems and properties of the *CLOUD* chapter (§3.3, p.87).

1. Consider the linear mappings $Vac : \mathcal{V} \to \mathbb{R}^I$ and $Vac^* : \mathbb{R}^I \to \mathcal{V}$ (cf. chapter 3, p.84 and p.85). Give the images, by those linear mappings, of the vectors $(\overrightarrow{\mathrm{OI}^k})_{k\in K}$ (basis of $\mathcal{V}$) and $(\delta^I_i)_{i\in I}$ (canonical basis of $\mathbb{R}^I$) .
2. Write the passage formulas; deduce that the principal axes and variables of the cloud M^I are given by the CA of the measure δ_{IK}.

Solution

1. From definition 3.11 (CLOUD, p.84): $Vac(\overrightarrow{\mathrm{OI}}^k) = \left(\langle\overrightarrow{\mathrm{GM}}^i|\overrightarrow{\mathrm{OI}}^k\rangle\right)_{i\in I}$. One has here $\langle\overrightarrow{\mathrm{GM}}^i|\overrightarrow{\mathrm{OI}}^k\rangle = \langle\overrightarrow{\mathrm{OM}}^i|\overrightarrow{\mathrm{OI}}^k\rangle - \langle\overrightarrow{\mathrm{OG}}|\overrightarrow{\mathrm{OI}}^k\rangle$, with $\langle\overrightarrow{\mathrm{OM}}^i|\overrightarrow{\mathrm{OI}}^k\rangle = p^i_k(\mathrm{OI}^k)^2 = Q\,p^i_k/f_k$, and $\langle\overrightarrow{\mathrm{OG}}|\overrightarrow{\mathrm{OI}}^k\rangle = (\mathrm{OG})^2 = 1$.
From definition 3.7 (CLOUD, p.85): $Vac^*(\delta^I_i) = f_i\overrightarrow{\mathrm{GM}}^i$. One has here $f_i = 1/n$ and $\overrightarrow{\mathrm{GM}}^i = \sum\limits_{k\in K}(p^i_k - (f_k/Q))\overrightarrow{\mathrm{OI}}^k = \sum\limits_{k\in K}(\delta_{ik} - f_k)\overrightarrow{\mathrm{OI}}^k/(Q)$.
To sum up: $Vac(\overrightarrow{\mathrm{OI}}^k) = \frac{Q}{f_k}p^I_k - 1^I$ and $Vac^*(\delta^I_i) = \frac{1}{nQ}\sum\limits_{k\in K}(\delta_{ik} - f_k)\overrightarrow{\mathrm{OI}}^k$.

2. Let $\overrightarrow{\alpha_\ell} = \sum\limits_{k\in K} a_{k\ell}\overrightarrow{\mathrm{OI}}^k$ be a vector belonging to the support of cloud ($\overrightarrow{\mathrm{OG}} \perp \overrightarrow{\alpha_\ell} \Rightarrow \sum\limits_{k\in K} a_{k\ell} = 0$) and having unit–norm ($\sum\limits_{k\in K} a^2_{k\ell}/(f_k/Q) = 1$).
One has $Vac(\overrightarrow{\alpha_\ell}) = \sum\limits_{k\in K} a_{k\ell}\,Vac(\overrightarrow{\mathrm{OI}}^k) = \sum\limits_{k\in K}\frac{Q}{f_k}a_{k\ell}\,p^I_k - (\sum\limits_{k\in K}a_{k\ell})1^I = \sum\limits_{k\in K}\frac{Q}{f_k}a_{k\ell}\,p^I_k$.
Let z^I_ℓ be a standard variable on I: $\sum\limits_{i\in I} z^i_\ell = 0$ and $\sum\limits_{i\in I}(z^i_\ell)^2/n = 1$.

$$Vac^*(z^I_\ell) = \sum_{i\in I} z^i_\ell\,Vac^*(\delta^I_i) = \sum_{k\in K}\Big(\sum_{i\in I}\delta_{ik}\,z^i_\ell/(nQ)\Big)\overrightarrow{\mathrm{OI}}^k$$

Applying the passage formulas (3.7) : $Vac(\overrightarrow{\alpha_\ell}) = \xi_\ell z^I_\ell$ and $Vac^*(z^I_\ell) = \xi_\ell\overrightarrow{\alpha_\ell}$ to cloud M^I (CLOUD, p.92), letting $z_{k\ell} = a_{k\ell}$ and $z^k_\ell = a_{k\ell}/(f_k/Q)$, one gets the two following equations: $\sum\limits_{k\in K} p^i_k z^k_\ell = \xi_\ell z^i_\ell$ ($i \in I$) and $\sum\limits_{i\in I} z^i_\ell\delta_{ik}/n_k = \xi_\ell a_{k\ell}/(f_k/Q) = \xi_\ell z^k_\ell$ ($k \in K$). Hence letting $p^k_i = \delta_{ik}/n_k$ one arrives at the two equations:

$$\begin{cases}\sum\limits_{k\in K} p^I_k z^k_\ell &= \xi_\ell z^I_\ell = y^I_\ell\\ \sum\limits_{i\in I} p^K_i z^i_\ell &= \xi_\ell z^K_\ell = y^K_\ell\end{cases}\quad\text{with}\quad\begin{cases}\sum\limits_{i\in I}(z^i_\ell)^2/n &= 1\\ \sum\limits_{k\in K}(z^k_\ell)^2/(f_k/Q) &= 1\end{cases}$$

that are the transition equations of the correspondence analysis of measure δ_{IK} (CORRESPONDENCE, equations (2.8), p.38).

Exercise 5.2 (Φ^2 of Burt table)

1. 2. 3. Prove the three properties of the Φ^2 of Burt table enunciated p.199.

4. Show that the variance between–q is equal to $(1/Q)\sum\limits_{q'\in Q}(\Phi^2)^{qq'}$.

Solution

1. $(\Phi^2)^{qq'} = \sum\limits_{k\in K\langle q\rangle}\sum\limits_{k'\in K\langle q'\rangle}\frac{(n_{kk'})^2}{n_k\,n_{k'}} - 1 = Q^2\sum\limits_{k\in K\langle q\rangle}\sum\limits_{k'\in K\langle q'\rangle}\frac{(b_{kk'})^2}{b_k\,b_{k'}} - 1.$

$(\Phi^2)^{qq} = Q^2\sum\limits_{k\in K\langle q\rangle}\sum\limits_{k'\in K\langle q\rangle}\frac{(b_{kk'})^2}{b_k\,b_{k'}} - 1 = \sum\limits_{k\in K\langle q\rangle}\frac{(n_k)^2}{n_k\,n_k} - 1 = K_q - 1.$

2. $(\Phi^2)^q = \sum\limits_{k'\in K}\sum\limits_{k\in K\langle q\rangle}\frac{(n_{kk'})^2}{b_k\,n_{k'}} - 1 = \sum\limits_{q'\in Q}\sum\limits_{k'\in K\langle q'\rangle}\sum\limits_{k\in K\langle q\rangle}\frac{(b_{kk'})^2}{b_k\,n_{k'}} - 1$
$= \sum\limits_{q'\in Q} Q\sum\limits_{k'\in K\langle q'\rangle}\sum\limits_{k\in K\langle q\rangle}\frac{(b_{kk'})^2}{b_k\,b_{k'}} - 1 = \sum\limits_{q'\in Q} Q\frac{(\Phi^2)^{qq'}+1}{Q^2} - 1 = \frac{1}{Q}\sum\limits_{q'\in Q}(\Phi^2)^{qq'}$

3. $\Phi^2_{\text{Burt}} = \sum\limits_{k\in K}\sum\limits_{k'\in K}\frac{b^2_{kk'}}{b_k\, b_{k'}} - 1 = \sum\limits_{q\in Q}\sum\limits_{k\in K<q>}\sum\limits_{q'\in Q}\sum\limits_{k'\in K<q'>}\frac{b^2_{kk'}}{b_k\, b_{k'}} - 1$
$= \sum\limits_{q\in Q}\frac{(\Phi^2)^q+1}{Q} - 1 = \frac{1}{Q}\sum\limits_{q\in Q}(\Phi^2)^q = \frac{1}{Q^2}\sum\limits_{q\in Q}\sum\limits_{q'\in Q}(\Phi^2)^{qq'}$

4. By definition, $\operatorname{Var}\widehat{\mathrm{M}}^{K<q>} = \sum\limits_{k\in K<q>} f_k(\mathrm{G}\widehat{\mathrm{M}}^k)^2$. Since $\overrightarrow{\mathrm{OG}}$ is orthogonal to the support of cloud, one has $\overrightarrow{\mathrm{OG}} \perp \overrightarrow{\mathrm{G}\widehat{\mathrm{M}}^k}$, and $(\mathrm{O}\widehat{\mathrm{M}}^k)^2 = \mathrm{OG}^2 + (\mathrm{G}\widehat{\mathrm{M}}^k)^2$. Owing to property 5.18, one has $\overrightarrow{\mathrm{O}\widehat{\mathrm{M}}^k} = (1/Q)\sum\limits_{k'\in K}(f_{kk'}/f_k)\overrightarrow{\mathrm{OI}^{k'}}$.
Therefore $(\mathrm{O}\widehat{\mathrm{M}}^k)^2 = \frac{1}{Q^2}\sum\limits_{k'\in K}(f_{kk'}/f_k)^2(\mathrm{OI}^{k'})^2 = \frac{1}{Q f_k}\sum\limits_{k'\in K} f^2_{kk'}/(f_k\, f_{k'})$. Hence $\operatorname{Var}\widehat{\mathrm{M}}^{K<q>} = \sum\limits_{k\in K<q>} f_k(\mathrm{G}\widehat{\mathrm{M}}^k)^2 - 1 = (1/Q)\Big(\sum\limits_{k\in K<q>}\sum\limits_{k'\in K} f^2_{kk'}/(f_k\, f_{k'})\Big) - 1 =$
$(1/Q)\sum\limits_{q'\in Q}\Big(\sum\limits_{k'\in K<q'>}\sum\limits_{k\in K<q>}(f_{kk'})^2/(f_k\, f_{k'}) - 1\Big) = (1/Q)\sum\limits_{q'\in Q}(\Phi^2)^{qq'} = (\Phi^2)^q$.

Exercise 5.3 (Binary correspondence)

Given a questionnaire with 2 questions q and q', with $A = K<q>$, $B = K<q'>$ and $K = A\cup B$; three tables can be built: disjunctive table, contingency table and Burt table (with two diagonal blocks and two rectangular ones).

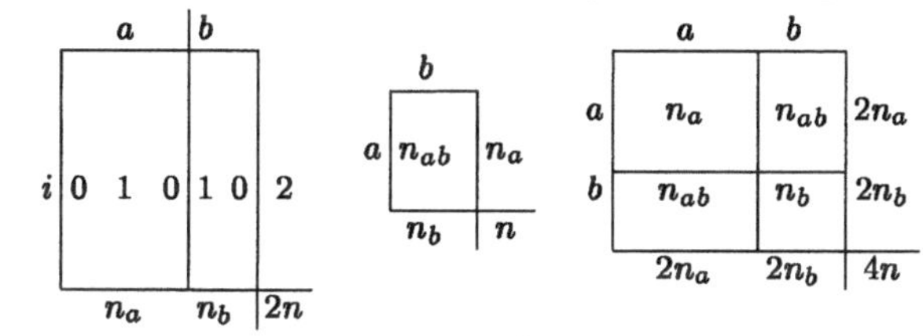

• Show that the correspondence analyses of these three tables are equivalent, in the sense that knowing the principal variables on modalities for one analysis[24] entails their knowledge for the other ones.
Suggested method: Start from the standard principal variables (z^A_ℓ, z^B_ℓ) of the contingency table, and use the transition formulas to obtain the standard principal variables of the Burt table.
• Show that the modified rates of the disjunctive table are equal to the variance rates of the contingency table n_{AB}.

Solution

The equivalence between the MCA of δ_{IK} and the CA of b_{KK} follows from the property 5.22 (p.200), not confined to the binary case. Both analyses lead to the same standard principal variables $(z^K_\ell)_{\ell\in L}$; if $(\lambda_\ell)_{\ell\in L}$ are the eigenvalues of δ_{IK}, then $(\lambda^2_\ell)_{\ell\in L}$ are eigenvalues of b_{KK}; if $(y^K_\ell)_{\ell\in L}$ are the principal variables of δ_{IK}, then $\widehat{y}^k_\ell = \sqrt{\lambda_\ell}\, y^k_\ell$ are principal variables of b_{KK}.
In what follows we show the equivalence between the CA of n_{AB} and that of b_{KK}. Let $(z^A_\ell, z^B_\ell)_{\ell=1\ldots R}$ be the standard principal variables of contingency table n_{AB} and $(\mu_\ell)_{\ell=1\ldots R}$ the R (nontrivial and nonnull) eigenvalues. Let (z^A_ℓ, z^B_ℓ) and $(z^A_\ell, -z^B_\ell)$ be variables on K obtained by concatening z^A_ℓ and z^B_ℓ, and z^A_ℓ and

[24] The principal variables on individuals can only be obtained from the CA of δ_{IK}.

$-z_\ell^B$ respectively. We will show that these variables on K are standard principal variables of the Burt table.

For Burt table, the transition formula reads $\sum_{k'\in K} b_{k'}^k z_\ell^{k'} = \lambda_\ell z_\ell^k$, letting λ_ℓ the eigenvalue of the disjunctive table (cf. p.189). For $k \in A$, one has $b_{k'}^k = \frac{1}{2}$ for $k' = a$; $b_{k'}^k = 0$ for $k' \in A$ and $k' \neq a$; and $b_{k'}^k = \frac{1}{2} f_b^a$ for $k' = b \in B$. For $k \in B$, one has $b_{k'}^k = \frac{1}{2}$ for $k' = b$; $b_{k'}^k = 0$ for $k' \in B$ and $k' \neq b$; and $b_{k'}^k = \frac{1}{2} f_a^b$ for $k' = a \in A$. Let us now write the transition formulas for the contingency table:

$$\begin{cases} \sum_{a\in A} f_a^b z_\ell^a = \sqrt{\mu_\ell}\, z_\ell^b \\ \sum_{b\in B} f_b^a z_\ell^b = \sqrt{\mu_\ell}\, z_\ell^a \end{cases} \quad \text{therefore} \quad \begin{cases} \frac{1}{2}(z_\ell^a + \sum_{b\in B} f_b^a z_\ell^b) = \frac{1}{2}(1+\sqrt{\mu_\ell}) z_\ell^a \\ \frac{1}{2}(z_\ell^b + \sum_{a\in A} f_a^b z_\ell^a) = \frac{1}{2}(1+\sqrt{\mu_\ell}) z_\ell^b \\ \frac{1}{2}(z_\ell^a - \sum_{b\in B} f_b^a z_\ell^b) = \frac{1}{2}(1-\sqrt{\mu_\ell}) z_\ell^a \\ \frac{1}{2}(z_\ell^b - \sum_{a\in A} f_a^b z_\ell^a) = \frac{1}{2}(1-\sqrt{\mu_\ell}) z_\ell^b \end{cases}$$

The preceding relations express the transition formula for the Burt table, with $z_\ell^k = z_\ell^a$ for $k \in A$ and $z_\ell^k = z_\ell^b$ for $k \in B$ (the first two lines, with $\lambda_\ell > 1/2$) and $z_\ell^k = z_\ell^a$ for $k \in A$ and $z_\ell^k = -z_\ell^b$ for $k \in B$ (the last two lines, with $\lambda_\ell < 1/2$).
The expressions of the eigenvalues λ_ℓ of the disjunctive table (binary correspondence) follow. There are R (nontrivial) eigenvalues greater than 1/2, namely $\lambda_\ell = \frac{1}{2}(1+\sqrt{\mu_\ell})$, R (nonnull) eigenvalues smaller than 1/2, namely $\lambda_\ell = \frac{1}{2}(1-\sqrt{\mu_\ell})$, and $A+B-2-2R$ eigenvalues equal to 1/2 (thus when $B \leq A$ and $R = B-1$, there are $A+B-2$ eigenvalues, $B-1$ greater than 1/2, $B-1$ less than 1/2, and $A-B$ equal to 1/2). The principal variables associated with $\lambda_\ell = 1/2$ are null on one of the two sets (A or B).

Modified rates. For $\lambda_\ell > 1/2$, one has $\lambda_\ell - 1/2 = \sqrt{\mu_\ell}/2$, hence (cf. p.201) $\lambda'_\ell = 4(\lambda_\ell - 1/2)^2 = \mu_\ell$. The sum of the λ'_ℓ corresponding to $\lambda_\ell > 1/2$ is therefore $\sum_{\ell=1}^{R} \mu_\ell = \Phi^2$ of n_{AB} table. That is, the modified rate of the binary correspondence is equal to μ_ℓ/Φ^2, i.e. the variance rate of the contingency table n_{AB}.

Comments

i) The standard principal variables are the same in the three analyses.

ii) *Geometrically*, the MCA is linked with the fundamental joint representation of CA (cf. chapter 2, §2.3.3, p.45). Letting θ_ℓ the acute angle such that $\mu_\ell = \cos^2\theta_\ell$, one has $\cos^2\frac{\theta}{2} = \frac{1}{2}(1+\cos\theta_\ell)$ and $\sin^2\frac{\theta}{2} = \frac{1}{2}(1-\cos\theta_\ell)$. With each $\ell = 1\ldots R$, there is in MCA a plane associated with two orthogonal axes such that the projected points M_ℓ^k and N_ℓ^k have coordinates $z_\ell^k \cos\frac{\theta}{2}$ and $z_\ell^k \sin\frac{\theta}{2}$ respectively. The opposite figure is essentially that of chapter 2 (Figure 2.7, p.46), up to notation.

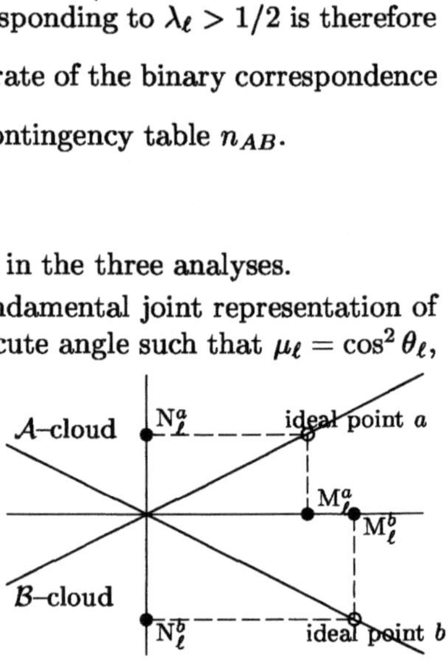

Exercise 5.4 (CA of scalogram)

In this exercise, one proceeds to the CA of a scalogram, i.e. the disjunctive table associated with a hierarchical scale ("Guttman scale"). The Q dichotomous questions being ordered from 1 to p (positive integer), let $K = \{1, 2, \ldots 2p\}$ and $J = \{0, 1, 2, \ldots p\}$; the scalogram is the correspondence n_{JK} defined by:

$$n_{jk} = 1 \text{ if } j + 1 \leq k \leq p + j \text{ and } 0 \text{ if not}$$

1. Write down the table for $p = 4$. For any p, express the marginal sums $n_j = \sum_{k=1}^{2p} n_{jk}$, $n_k = \sum_{j=0}^{p} n_{jk}$, and the total n as functions of p.
2. Let a variable on J be of the form $(a\,j + b)_{j \in J}$. Write the expression of b as a function of p and a so that this variable be f_J–centered; one will denote x_1^J this variable. Give the expression of variable $x_1^K = \sum_{j=0}^{p} n_{jk} x_1^j / n_k$ (f_J^K–averaging). Show that x_1^J is a principal variable of correspondence n_{JK}, and give the expression of the associated eigenvalue λ_1.
3. Let $x_2^J = (c\,j^2 + d\,j + e)_{j \in J}$ a variable over J. Determine d and e so that x_2^J be centered and orthogonal to variable x_1^J. Determine the eigenvalue λ_2. Express x_2^J as a function of x_1^J, and x_2^K as a function of x_1^K.
4. For $p = 4$, calculate the principal variables, and proceed to the simultaneous representation in plane 1-2.
5. PCA *vs* CA *of scalogram.* How do the results of the standard PCA of the $J \times Q$ table (Guttman triangle) (eigenvalues and principal variables over J and over Q) compare with those of the CA of scalogram? Same question if one proceeds to the PCA of the $J \times 2Q$ table (scalogram).

Recall: $\sum_{i=1}^{p} i = p(p+1)/2$; $\sum_{i=1}^{p} i^2 = p(p+1)(2p+1)/6$; $\sum_{i=1}^{p} i^3 = (p(p+1)/2)^2$.

Solution

1.

	k1	k2	k3	k4	k5	k6	k7	k8	
j0	1	1	1	1	0	0	0	0	4
j1	0	1	1	1	1	0	0	0	4
j2	0	0	1	1	1	1	0	0	4
j3	0	0	0	1	1	1	1	0	4
j4	0	0	0	0	1	1	1	1	4
	1	2	3	4	4	3	2	1	20

$n_j = \sum_{k=1}^{2p} n_{jk} = \sum_{k=j+1}^{j+p} 1 = (j+p) - j = p$;

hence $n = \sum_{j=0}^{p} n_j = p(p+1)$.

One can also write the correspondence:

for $k \leq p$, $n_{jk} = 1$ if $0 \leq j \leq k-1$ and 0 if not, with $n_k = \sum_{j=0}^{p} n_{jk} = \sum_{j=0}^{k-1} 1 = k$;

for $k > p$, $n_{jk} = 1$ if $k - p \leq j \leq p$ and 0 if not, with $n_k = \sum_{j=0}^{p} n_{jk} = 2p - k + 1$.

2. The mean of the variable $(a\,j + b)_{j \in J}$ is $ap/2 + b$, hence the centered variable $x_1^J = \big(a(j - p/2)\big)_{j \in J}$. The centered variable $x_1^K = (\sum_{j=0}^{p} x_1^j \frac{n_{jk}}{n_k})_{k \in K}$ is such that:

for $k \leq p$, $x_1^k = \sum_{j=0}^{p} a(j-\frac{p}{2})\frac{n_{jk}}{n_k} = \frac{a}{k}\sum_{j=0}^{k-1}(j-\frac{p}{2}) = \frac{a}{k}\left(\frac{(k-1)k}{2} - \frac{p}{2}k\right) = \frac{a}{2}(k-p-1)$;

for $k > p$, $\sum_{j=0}^{p} a(j-\frac{p}{2})\frac{n_{jk}}{n_k} = \frac{a}{2p-k+1}\sum_{j=k-p}^{p}(j-\frac{p}{2}) = \frac{a}{2}(k-p) = x_1^k$.

If one proceeds to the f_K^J–averaging of x_1^K, one has:

$$\sum_{k=1}^{2p} n_{jk}\, x_1^k/n_j = \frac{1}{p}\sum_{k=1}^{2p} n_{jk}\, x_1^k = \frac{1}{p}\sum_{k=j+1}^{j+p}\frac{a}{2}(k-p) - \sum_{k=1}^{p}\frac{n_{jk}}{p}\frac{a}{2}$$
$$= \frac{a}{2p}\left(\frac{j(j+1)}{2} - \frac{(j-p)(j+1-p)}{2} - (p-j)\right) = a\,\frac{p+1}{2p}\,(j-\frac{p}{2}) = \frac{p+1}{2p}\,x_1^j.$$

The variable $x_1^J = \left(a(j-\frac{p}{2})\right)_{j\in J}$ is therefore principal associated with the eigenvalue $\lambda_1 = (p+1)/2p$; similarly, variable x_1^K is principal associated with λ_1.

3. The mean of the variable $(c\,j^2 + d\,j + e)_{j\in J}$ is equal to $\sum_{j=0}^{p}\frac{c\,j^2+d\,j+e}{p+1}$, that is, $\frac{1}{p+1}\left(c\,\frac{p(p+1)(2p+1)}{6} + d\,\frac{p(p+1)}{2} + e\,(p+1)\right) = c\,\frac{p(2p+1)}{6} + d\,\frac{p}{2} + e$.

The scalar product of variable $(c\,j^2 + d\,j + e)_{j\in J}$ with variable $(j)_{j\in J}$ is equal to $\frac{1}{p+1}\sum_{j=0}^{p}(c\,j^3 + d\,j^2 + e\,j) = \frac{p}{2}\left(c\,\frac{p(p+1)}{2} + d\,\frac{2p+1}{3} + e\right)$.

The variable $(c\,j^2 + d\,j + e)_{j\in J}$ is orthogonal to variable $x_1^J = (j-\frac{p}{2})_{j\in J}$ if $\frac{p}{2}\left(c\,\frac{p(p+1)}{2} + d\,\frac{2p+1}{3} + e\right) - \frac{p}{2}\left(c\,\frac{p(2p+1)}{6} + d\,\frac{p}{2} + e\right) = 0$, that is $c\,p + d = 0$, hence $e = c\,\frac{p(p-1)}{6}$. The variable $(c\,j^2 + d\,j + e)_{j\in J}$ centered and orthogonal to x_1^J can therefore be written $x_2^J = \left(c(j^2 - p\,j + \frac{p(p-1)}{6})\right)_{j\in J}$.

By f_J^K–averaging of the variable x_2^J, one gets $x_2^K = (\sum_{j=0}^{p} x_2^j\frac{n_{jk}}{n_k})_{k\in K}$, such that, for

$$k < p\colon\ x_2^k = \sum_{j=0}^{p} c(j^2 - p\,j + \frac{p(p-1)}{6})\frac{n_{jk}}{n_k} = \frac{c}{k}\sum_{j=0}^{k-1}(j^2 - p\,j + \frac{p(p-1)}{6})$$
$$= \frac{c}{k}\left(\frac{(k-1)k(2k-1)}{6} - p\frac{(k-1)k}{2} + \frac{p(p-1)}{6}k\right)$$
$$= \frac{c}{6}\left(2(p+1-k)^2 - (p+1)(p+1-k)\right);$$

$$p < k \leq 2p\colon\ x_2^k = \sum_{j=0}^{p} c(j^2 - pj + \frac{p(p-1)}{6})\frac{n_{jk}}{n_k} = \frac{c}{2p-k+1}\sum_{j=k-p}^{p}(j^2 - p\,j + \frac{p(p-1)}{6})$$
$$= \frac{c}{2p-k+1}\sum_{j=k-p}^{p}\left((p-j)^2 - p(p-j)\right) + c\frac{p(p-1)}{6}$$
$$= \frac{c}{2p-k+1}\sum_{j=0}^{2p-k}(j^2 - pj) + c\frac{p(p-1)}{6} = \frac{c}{6}\left(2(k-p)^2 - (p+1)(k-p)\right)$$

By f_K^J–averaging of this variable x_2^K, one gets: $\sum_{k=1}^{2p}\frac{n_{jk}}{n_j}x_2^k = \frac{1}{p}\sum_{k=j+1}^{j+p}x_2^k$

$$= \frac{c}{6p}\left(\sum_{k=j+1}^{p}\left(2(p+1-k)^2 - (p+1)(p+1-k)\right) + \sum_{k=p+1}^{j+p}\left(2(k-p)^2 - (p+1)(k-p)\right)\right)$$
$$= \frac{c}{6p}\sum_{k=j}^{j+p}2(k-p)^2 - c\frac{p+1}{p}\left(\sum_{k=1}^{p-j}k + \sum_{k=1}^{j}k\right) = c\frac{p+1}{6p}(j^2 - pj + \frac{p(p-1)}{6}) = \frac{p+1}{6p}x_2^j.$$

Hence the eigenvalue $\lambda_2 = \frac{p+1}{6p}$.

One has the relations: $x_2^j = c\big(\frac{1}{a^2}(x_1^j)^2 - \frac{p(p+2)}{12}\big)$; for $k \le p$: $x_2^k = \frac{c}{3a}\big(\frac{4}{a}(x_1^k)^2 + (p+1)x_1^k\big)$ and for $k > p$: $x_2^k = \frac{c}{3a}\big(\frac{4}{a}(x_1^k)^2 - (p+1)x_1^k\big)$.

4. $y_1^J = (\sqrt{5}/4)(-2,-1,0,1,2)$ and $y_2^J = (\frac{5}{4}/\sqrt{21})(2,-1,-2,-1,2)$.

$y_1^K = \frac{1}{2\sqrt{2}}(-4,-3,-2,-1,1,2,3,4)$ and $y_2^K = \frac{1}{6}\sqrt{\frac{5}{14}}(12,3,-2,-3,-3,-2,3,12)$

Between principal variables, one has the relations: $y_2^j = \frac{1}{2\sqrt{21}}\big(8(y_1^j)^2 - 5\big)$.

For $k \le p$: $y_2^k = \frac{\sqrt{5}}{(3\sqrt{7}}\big(4\sqrt{2}(y_1^k)^2 + 5y_1^k\big)$; for $k > p$: $y_2^k = \frac{\sqrt{5}}{(3\sqrt{7}}\big(4\sqrt{2}(y_1^k)^2 - 5y_1^k\big)$.

The simultaneous representation of the two sets J and K follows, with (left figure) one parabola arc for the 5 j–points, and (right figure) two parabola arcs for the 8 k–points.

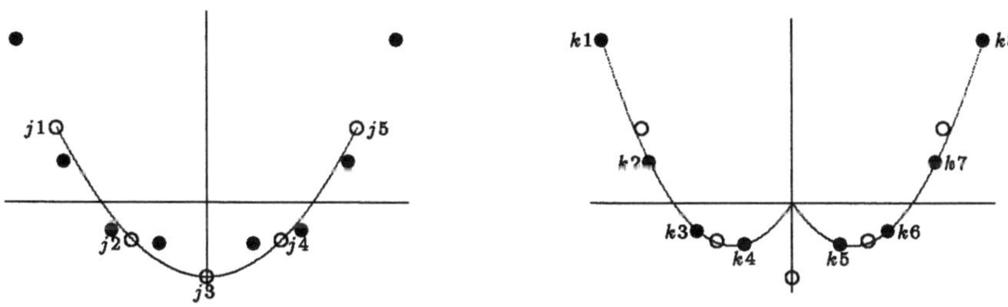

5. By application of Exercise 4.5 (question 4, dichotomous variables), the standard PCA of the $J \times Q$ table (Guttman triangle) leads to the following results.
— For all ℓ, the eigenvalues λ'_ℓ of PCA are proportional to the eigenvalues λ_ℓ of CA, i.e. $\lambda'_\ell = p\lambda_\ell$; therefore $\lambda'_1 = (p+1)/2$ and $\lambda'_2 = (p+1)/6$ (hence for $p = 4$: $\lambda'_1 = 2.5$ and $\lambda'_2 = 0.833$). — The ***cloud of individuals*** is homothetic to the one of CA. — In the ***cloud of modalities***, the representation of variable δ_k^J is proportional to the corresponding one of CA by the factor $\sqrt{\frac{f_k}{1-f_k}} = \sqrt{\frac{k}{p+1-k}}$ (since Mean $\delta_k^J = f_k = \frac{k}{p+1}$). In PCA, the ℓ–th principal coordinate of modality k is therefore $b_\ell^k = \sqrt{\frac{k}{p+1-k}}\, y_\ell^k$ (where y_ℓ^k is the principal coordinate of k in CA). We therefore have b_1^k proportional to $\sqrt{k(p+1-k)}$ and b_2^k proportional to $b_1^k(\frac{p+1}{2} - k)$. If we express k as a function of b_1^k, we get (after some calculation) the result: b_2^k proportional to $b_1^k\sqrt{\frac{9(p+1)^2}{p(p+2)} - b_1^{k\,2}}$, that is, $A^2 b_2^{k\,2} = b_1^{k\,2}(B^2 - b_1^{k\,2})$, where A and B are coefficients depending on p. Thus in PCA, the points $k1$ to $k4$ lie on a quartic curve, instead of the parabola arc of CA.

Comment

Whereas the CA of the $J \times Q$ table (triangle) differs from the CA of the $J \times 2Q$ table (scalogram), the PCA of the $J \times 2Q$ table leads to essentially the same results as the PCA of the $J \times Q$ table, for modalities as well as for individuals. For $p = 4$, the points representing modalities $k6$ to $k8$ are symmetrical to those of $k1$ to $k4$ with respect to the origin of axes. The underlying quartic curve goes through the origin; its shape (lemniscate type) is more apparent for large values of p: See Figure for $p = 50$.

PCA of scalogram for $p = 4$ and $p = 50$

Exercise 5.5 (Burt's data)

The Burt's data table below, reproduced from Burt (1950), gives, for $n = 100$ individuals (men living in Liverpool), the observed patterns and their absolute frequencies for 4 attributes ("Questions"), that is, *A Hair* (*a*1: fair, *a*2: red, *a*3: dark), *B Eyes* (*b*1: light, *b*2: mixed, *b*3: brown), *C Head* (*c*1: narrow, *c*2: wide), *D Stature* (*d*1: tall, *d*2: short).

K	Abs.freq
*a*1*b*1*c*1*d*1	8
*a*1*b*1*c*1*d*2	4
*a*1*b*1*c*2*d*1	2
*a*1*b*2*c*1*d*1	1
*a*1*b*2*c*1*d*2	1
*a*1*b*2*c*2*d*1	2
*a*1*b*2*c*2*d*2	2
*a*1*b*3*c*2*d*2	2

K	Abs.freq
*a*2*b*1*c*1*d*1	6
*a*2*b*1*c*2*d*1	2
*a*2*b*2*c*1*d*1	2
*a*2*b*2*c*1*d*2	1
*a*2*b*2*c*2*d*2	2
*a*2*b*3*c*1*d*2	2

K	Abs.freq
*a*3*b*1*c*1*d*1	9
*a*3*b*1*c*2*d*1	2
*a*3*b*2*c*1*d*1	3
*a*3*b*2*c*1*d*2	12
*a*3*b*2*c*2*d*1	2
*a*3*b*2*c*2*d*2	8
*a*3*b*3*c*1*d*1	1
*a*3*b*3*c*1*d*2	19
*a*3*b*3*c*2*d*1	3
*a*3*b*3*c*2*d*2	4

*a*1	*a*2	*a*3	*b*1	*b*2	*b*3	*c*1	*c*2	*d*1	*d*2
22	15	63	33	36	31	69	31	53	47

The procedure of MCA applied to Burt's data leads to results given below.

- $\lambda_1 = .48939$, $\lambda_2 = .29869$, $\lambda_3 = .25429$. $\lambda_4 = .20550$, $\lambda_5 = .17866$, $\lambda_6 = .07347$.
- *Coordinates and contributions* (in bold: the contributions of the modalities retained for interpretation).

Q	*K*	weight	Ctr	Coordinates Axis 1	Coordinates Axis 2	Ctr Axis 1	Ctr Axis 2
Hair	*a*1 fair	.055	.130	−0.896	−0.495	**.090**	.045
	*a*2 red	.038	.142	−0.860	0.049	**.057**	.000
	*a*3 dark	.158	.062	0.518	0.161	**.086**	.014
A			.333			**.233**	.059
Eyes	*b*1 light	.083	.112	−1.252	0.285	**.264**	.022
	*b*2 mixed	.090	.107	0.402	−0.938	.030	**.265**
	*b*3 brown	.078	.115	0.866	0.785	**.119**	**.160**
B			.333			**.413**	**.447**
Head	*c*1 narrow	.173	.052	−0.098	0.515	.003	**.153**
	*c*2 wide	.078	.115	0.219	−1.145	.008	**.340**
C			.167			.011	**.493**
Stature	*d*1 tall	.108	.095	−0.943	−0.016	**.195**	.000
	*d*2 short	.143	.072	0.712	0.012	**.147**	.000
D			.167			**.343**	.000

1. Show that it is appropriate to retain two axes. Proceed to a brief interpretation of the MCA results and to geometric representations.

2. The "factorial method" devised by Burt was applied to Burt's data and led to a "size factor" followed by subsequent "bipolar factors", whose saturations are given in the following table reproduced from Burt (1950).

	Size factor	Bipolar factor 1	Bipolar factor 2
*a*1	.469	−.420	−.231
*a*2	.387	−.333	.019
*a*3	.794	.411	.128
*b*1	.574	−.719	.164
*b*2	.600	.241	−.563
*b*3	.557	.482	.437
*c*1	.831	−.081	.048
*c*2	.557	.121	−.638
*d*1	.656	−.619	−.010
*d*2	.755	.537	.009

Compare the results of Burt's factorial method with those of MCA; more specifically, verify that Burt's saturations can be obtained from the coordinates of the MCA.

Solution

1. There are $Q = 4$ questions, $K = 10$ modalities, 24 observed patterns, out of $3 \times 3 \times 2 \times 2 = 36$ possible patterns. The variance of the cloud is $\Phi^2 = (K - Q)/Q = (10 - 4)/4 = 1.5$. The first eigenvalue is predominant; then the eigenvalues decrease slowly from λ_2 to λ_5. Three eigenvalues exceed the average $1/Q = 0.25$, with inertia rates $\tau_1 = .48939/1.5 = .326$, $\tau_2 = .199$, $\tau_3 = .113$. Computing $S = (0.48939-0.25)^2+(0.29869-0.25)^2+(0.25429-0.25)^2 = 0.0547+0.0024+0.0000 = 0.0597$, we get the modified rates $\tau'_1 = 0.0573/.0597 = 96\%$; $\tau'_2 = 4\%$; $\tau'_3 = 0\%$. These results suggest to retain one or two axes.
The examination of the Ctr of questions to axes (average contribution $1/Q = 25\%$) shows that Axis 1 involves only 3 questions, namely ***Eyes*** (41%), ***Stature*** (34%), and ***Hair*** (23%); whereas Axis 2 involves ***Head*** (49%) and ***Eyes*** (45%). In order to take into account all four questions, Axis 2 must be retained.
Now examining the contributions of modalities (average $1/K = 10\%$): The modalities most contributing to Axis 1 are *b*1 (***light*** eyes 26%) versus *b*3 (***brown*** eyes 12%); *d*1 (***tall*** 20%) versus *d*2 (***short*** 15%); then *a*1 with *a*2 (***fair*** and ***red*** hair together 15%)[25] versus *a*3 (***dark*** hair 9%). The modalities most contributing to Axis 2 are *c*2 (***wide*** head 34%) versus *c*1 (***narrow*** head 15%); *b*2 (***mixed*** eyes 27%) versus *b*3 (***brown*** eyes 16%).

[25] The barycenter of *a*1 and *a*2 is 0.881; hence Ctr $= .(0.881)^2 \times .093/.48939 = .1475$.

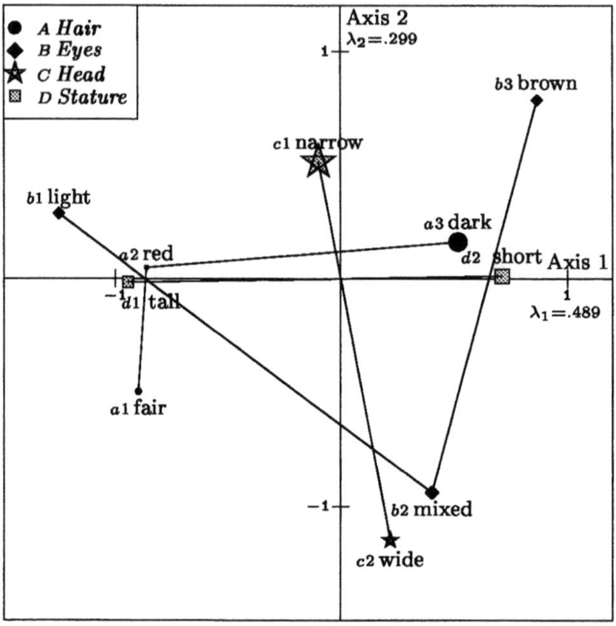

Modalities in plane 1-2. Markers are proportional to response frequencies.

2. *Burt's factorial method vs* MCA. Burt's size factor corresponds to the trivial factor of MCA ($\ell = 0$) constant equal to 1, with saturations $s_0^k = \sqrt{f_k}$. For instance, for $a1$, we have $f_k = 0.22$, hence $s_0^k = .469$. For subsequent factors, the saturations s_ℓ^k of Burt's factorial method are related to coordinates y_ℓ^k of MCA by the relation $s_\ell^k = \sqrt{f_k}\, y_\ell^k$. For instance, for $a1$ and $\ell = 1$, one has $y_{\ell 1}^k = .896$, and $s_{\ell 1}^k = .469 \times .896 = .420$.

Comment

The data of this exercise are those of the historical paper by Burt (1950), another precursor paper of MCA, that deserves special mention, because it purports to show that "qualitative data", as well as quantitative ones, are amenable to factor analysis; and also because it contains the first appearance of what came to be known as the "Burt table". However, Burt's factorial method definitely differs from MCA; Burt's saturations differ from the principal coordinates of MCA by the coefficient $\sqrt{f_k}$; and (like in Guttman 1941 paper), no geometric representation of either modalities or individuals is presented or even intended.

Chapter 6

Structured Data Analysis

(*'STRUCTURED'*)

There is what we know,
and there is what we would like to know.

Introduction

In the present chapter, we show how, starting from the space constructed by Geometric Data Analysis (GDA), the structures of the basic sets of the data table can be integrated into the analyses to perform a *Structured Data Analysis*.

Let us consider the basic object of GDA, namely the *two-way table*, as in the chapters *CORRESPONDENCE*, *PRINCIPAL* and *MULTIPLE*. A geometric data analysis brings out the structures of the table, that is, the relations between the two sets indexing rows and columns; it does not take into account, however, the structures with which the sets themselves may be equipped. By *structuring factors*, we mean relevant variables describing the two basic sets that do not serve to construct the geometric space[1]; by *structured data*, we designate data tables whose basic sets are equipped with structuring factors; and by *Structured Data Analysis*, we mean the embedding of structuring factors into GDA, in the line of Analysis of Variance including its multivariate extension (MANOVA) —while preserving the principles of the GDA construction (Le Roux & Rouanet, 1984). Structured Data Analysis is thus the synthesis between GDA and Analysis of Variance.

[1](i) As will become clear, "structuring factors" have nothing to do with principal variables (often called "factors"). (ii) See also Note 14 (p.265) on a more extensive use of "structuring factor".

Clearly, structured data constitute the rule rather than the exception. More than that, the questions of interest connected with structuring factors may not only be relevant, but central to the study that has led to the geometric construction. For example, in the textual research by Volochine (1995) mentioned in chapter *CORRESPONDENCE* (p.60), the set of texts analyzed by CA is structured by works, authors and years, which are structuring factors in the sense defined here, and are the main object of research. The need of an approach dealing explicitly with structured data has been felt for a long time among specialists of GDA[2].

In fact, we have already put to work Structured Data Analysis earlier in this book, for instance in the *Culture Example* (*MULTIPLE*, §5.4, p.221), where we examined subclouds of individuals induced by Age or Education. The present chapter provides a systematic presentation of this approach, focused on Individuals×Variables tables (analyzed by MCA or PCA), and more specifically on the structuring factors on individuals who carry all the information. In fact, the set of individuals may itself be constructed from basic structuring factors, as will be seen below[3].

The chapter is organized as follows. We start with an overview discussion of structuring factors and questions of interest (§6.1). We outline the formal theory of *Analysis of Comparisons*, with the concepts of contrast and comparison, nesting and crossing structures, and double breakdown (decomposition) of inertias (§6.2). We proceed with additive cloud and interaction (§6.3). Related topics and comments follow, together with a further example of Structured Data Analysis (§6.4).

As we proceed in the chapter, we give references to structured data analyses to be found in this book (especially in the *CASE STUDIES* chapter), and in various articles.

6.1 Structuring Factors and Questions of Interest

The experimental paradigm. In the experimental paradigm, there is a clear distinction between *experimental factors*, alias *independent variables* (treatments under experimenter's control, etc.), and *dependent variables*, alias response variables, recorded on *statistical units* (an often–used term in experimental settings to designate "statistical individuals"). Statistical

[2]See Escofier & Pagès (1988); see also Lebart & al (1995), who warn about the risk of proliferation of *ad hoc* procedures to cope with all sorts of structures.

[3]A similar discussion would apply to the set of variables, which may also be described by relevant characteristics not used to construct the geometric space.

analysis aims at studying the *effects* of experimental factors on dependent variables. When there are several dependent variables, a GDA can be performed on them, and the resulting geometric space takes on the status of a "dependent variable". The experimental factors on individuals become structuring factors, and their effects become *vector–effects* in this space[4].

For the formal presentation in subsequent sections, we will take the experimental paradigm as a leading case, and borrow from the classical tool of experimental data analysis, namely Analysis of Variance (ANOVA), with its multivariate extensions (MANOVA), as well as regression[5].

Observational data. Consider an educational questionnaire: for each student (individual), knowledge variables on various subject matters are recorded; suppose that a multidimensional knowledge space is constructed by some method of GDA, carried out from the Individuals×Variables table[6].

In addition to the knowledge variables used as active ones to "create distance in knowledge" between individuals, the educational questionnaire may include other information about individuals, such as identification characteristics: gender, age, etc. Even though such variables do not serve to construct the knowledge space, they are of obvious relevance, as reflected in the natural question of interest: "How are boys vs girls scattered in the knowledge space?" Gender and Age of students are structuring factors of individuals. Carrying over the experimentally–minded language, we may speak of the "effect of Gender" (assimilated to an independent variable) on the knowledge space ("geometric dependent variable"). For observational data, this language may be farfetched, and the more neutral and versatile language of *prediction* may be preferred, leading to state the question of interest of the first kind: Knowing the gender of a student ("predictor variable"), we would like to know about ("to predict") the position of this student in the knowledge space ("geometric variable to be predicted").

As another kind of a question of interest, suppose a variable giving the results of students at some final exam is available. This variable triggers once again a question of interest: "How are the successes at the exam scattered in the cloud of individuals?" Taking the exam variable as a structuring factor on the set of individuals, the knowledge space is now the predictor, and the structuring factor is the variable to be predicted, leading

[4]The *Parkinson study* in chapter 9 (§9.1) provides such an example.

[5]ANOVA vs *regression*. The distinction is mainly technical: ANOVA factors are categorized variables, whereas regression factors are numerical. With the advent of loglinear and logistic techniques, the distinction has become rather tenuous...

[6]The EPGY *study* in chapter 9 (§9.3) provides an example of this sort.

to the question of interest of the second kind: Knowing the position of a student in the knowledge space ("geometric predictor"), we would like to predict the success of this student at the exam ("variable to be predicted")[7].

Explanatory schemes. Questions of interest are often stated in "explanatory" terms. Dependent variables are "variables to be explained", and independent variables are "explanatory variables". This language, explicitly or implicitly, implies a reference to some underlying *explanatory scheme.* Thus in the educational example, one tries to figure out to what extent gender, age, etc. explain the position of students in the knowledge space, and also to figure out to what extent the position in the space explains the success at exams.

Explanatory schemes can be more or less elaborate, and they typically stem from non–statistical considerations[8]. Thus in a questionnaire leading to construct a social space in Bourdieu's perspective, the two kinds of questions of interest can be formulated (paraphrasing Bourdieu) as follows:

i) How are individual positions in the social space explained by the class fractions (structuring factor) to which individuals belong?

ii) How do in turn individual positions in the social space explain ("field effect") the position–takings of individuals (structuring factor) on various issues (politics, etc.)?

Such questions lead to an *explanatory use of* GDA, enlarged with structured data analyses.

Supplementary variables vs structuring factors. Actually, there is a technique traditional in GDA, namely that of *supplementary variables*, that already deals with structuring factors. This technique has been presented in the preceding chapters. We used it, for instance, in the Culture Example, where we represented the modalities of identification variables (gender, age, etc.) as supplementary points in the space of modalities: cf. e.g. Figure 5.11, p.234. Recall that the concept of "supplementary elements" also applies to individuals: see the Parkinson case study, p.336.

---

[7]In the *Political Space* study in chapter *CASE STUDIES* (§9.2) there is a question of interest of the second kind: The vote at the elections is studied in the geometric space of values and attitudes of electors.

[8]Statistical techniques by themselves had better viewed to be neutral as regards "explanation" (Statistics does not explain anything!). Thus taking the case of regression methods, the length of a metallic bar can be regressed on temperature, and temperature can be regressed on length as well. Stating that "Temperature explains the lengthening of bar" (not the reverse) implies that an *explanatory use of regression* is made, referring to the physical theory of dilatation. Similar considerations apply to the explanatory use of Geometric Data Analysis methods.

Users of GDA have widely used the technique of supplementary variables both in a prediction and in an explanatory perspective. For instance, Bourdieu (1979), building the lifestyle space of *La Distinction*, points out that he has put the status variables of age, father's profession, education level, and income, not as active variables, but as supplementary ones, in order to "give its full strength" to the demonstration that differences in lifestyle can be explained by those status variables.

The limitations of the technique of supplementary variables become apparent, however, when it is realized that in the space of individuals, the modalities of questions induce subclouds of individuals, and that considering supplementary modalities amounts to confining attention to the mean points of subclouds (modality mean–points), leaving aside their dispersion. For this reason, in the Culture Example, identification variables were further studied as structuring factors in the cloud of individuals, summarizing subclouds by concentration ellipses, etc. Compare e.g. Figure 5.11 (p.234) with Figure 5.16 (p.238)[9]. For further uses of supplementary variables as structuring factors in this book, see e.g. the EPGY case study (§9.3). For a review of the technique of supplementary elements, see Cazes (1982).

Relationships between structuring factors. In experimental data, the relationships between factors take the form of a more or less complex *factorial design*, involving simple relations such as nesting and crossing of factors, or more complex ones (latin squares, etc.)[10]. The factorial design can itself be constructed from basic factors, a typical example being the $S \times T$ design where the basic factor S ("Subjects") is crossed with a Treatment (or Time) factor T. Thus in the Parkinson study, there are 15 patients (factor P) observed twice, hence $15 \times 2 = 30$ statistical units associated with the Patients factor.

In observational data, even in the absence of prior design, there are also relationships between structuring factors. For instance, the individuals in a survey are *nested* within Age classes, since any individual belongs to one and only one age class; Gender and Age Classes are *crossed factors*

[9]Taking again *La Distinction*, this concern led Bourdieu, for studying the fractions of the upper class — "the most powerful explanatory factor", as he puts it — to examine the corresponding subclouds of individuals, that is, to treat the class fractions as what we now call a structuring factor. See Diagrams 11 and 12 of *La Distinction*. The examination of subclouds revealed e.g. that the individuals of "professions libérales" are concentrated in the center of the space, whereas "cadres" are scattered throughout the whole plane. As supplementary elements, these two class fractions would have been merely represented by two points, both located near the center of diagram. Cf. Rouanet & al (2000).

[10]For presentations of complex designs in the same line as the present chapter, see Rouanet & Lépine (1977), Hoc (1983), Lecoutre (1984), and Bernard (1994).

whenever there are individuals of both genders in every age class. In many observational data sets, nesting and crossing are sufficient to characterize the whole factorial structure, and all effects of interest are of the following types: main effects, between effects, and within–effects (also called conditional effects in regression), and interaction effects. On the other hand, when factors happen to be crossed in observational data, the crossing relation is usually not orthogonal (as opposed to experimental data, where orthogonality is often ensured by design), that is, structuring factors are typically *correlated*. As a consequence, conditional effects must be carefully distinguished from main effects. For instance, if Age and Gender are correlated, the effect of Age conditional upon Gender may differ from the main effect of Age[11]. This phenomenon is known in demography and econometrics under the name of "structural effect", and is often stated in regression terms: When regressor variables are correlated, the regression coefficients of a variable in a multiple regression differ from those in a simple regression.

6.2 Analysis of Comparisons

Analysis of Variance, like Geometric Data Analysis, can be tackled by the formal approach. The resulting construction, that we call ***Analysis of Comparisons***, is founded on set theory, to formalize factors and factorial designs, and on abstract linear algebra, to formalize ANOVA concepts, such as Sums of Squares (SS), etc., around the formalized concept of comparison. In short, a factor is a (nonempty) finite set, and a comparison is a vector space of contrasts on a factor. Measure vs variable duality is fundamental in Analysis of Comparisons. This method, formulated for numerical variables by Rouanet & Lépine (1976, 1977), was extended to Euclidean clouds by Le Roux & Rouanet (1983, 1984, 1998). In what follows, we outline the Analysis of Comparisons in view of its use in Structured Data Analysis, taking the experimental paradigm as a leading case.

6.2.1 Descriptors, Factors and Relations between Factors

Descriptors and factors. Let us consider a (nonempty) finite set of observations indexed by a set J; in GDA, the observations are the points of a cloud, and J is the set–theoretic support of cloud (cf. *CLOUD*, p.76), for instance a set of individuals. Formally, a ***descriptor*** of observations is a set

[11]Both effects may even be opposite to each other: the so–called ***reversal paradox***. See below p.264.

D associated with a mapping $J \to D$, that expresses the description of the observations by D. The image set C of the description mapping is a finite set called *factor* C, associated with the surjective mapping $J \to C$, that is, each modality $c \in C$ of the factor is the description of at least one observation. In short, a factor is a surjective descriptor, indexing the classes of a partition of the observations. The set J, support of observations, is itself a factor, associated with a *bijective* (i.e. surjective and injective) description, i.e. indexing the partition in one–element classes of the observations.

Compound factor, composite modalities, nesting and crossing. Given two factors A and B, the Cartesian product $A \times B$ is a compound descriptor. With this compound descriptor there is associated a compound factor, denoted $A \otimes B$, which is the subset of $A \times B$ made of the composite modalities (pairs $(a, b) \in A \times B$) corresponding to at least one observation; one has $A \otimes B \subseteq A \times B$.

• *Nesting relation.* If the compound factor $A \otimes B$ is such that for every modality $a \in A$ there is one and only one modality $b \in B$, factor A is said to be nested in factor B. We then write the compound factor $A \otimes B = A{<}B{>}$ (read "A in B"), and the set of composite modalities $a \in A$ such that $(a, b) \in A \otimes B$ is written $A{<}b{>}$. The notation $A{<}B{>}$ will also be used to denote the relation "A nested in B".

• *Crossing relation.* If the compound factor $A \otimes B = A \times B$, then for each pair $(a, b) \in A \times B$, there is at least one observation; in this case factors A and B are said to be crossed. We write the crossing relation $A \star B$, or $B \star A$ (the relation is symmetric).

Properties

The compound factor $A \otimes B$ is canonically nested in factor A, and also in factor B.

The support J is canonically nested in any factor, simple or compound; hence the notation $j{<}c{>}$ ("j in c") (nesting notation, cf. p.25), which makes explicit that observation j belongs to class $c \in C$.

6.2.2 Contrasts and Comparisons

Let J denote a factor equipped with a strictly positive measure n_J ($\forall j \in J : n_j > 0$). Let $\mathbb{R}_J$ denote the vector space of measures over J (cf. §2.1.4, p.27), with the Euclidean structure induced by n_J; recall (cf. formula (2.2), p.28) that the scalar product on $\mathbb{R}_J$ is defined by:

$$\forall u_J, v_J \in \mathbb{R}_J : \langle u_J | v_J \rangle = \sum_{j \in J} u_j \, v_j / n_j$$

Recall that a *contrast over J* is a measure c_J whose total mass (sum of coefficients) is null: $\sum_{j\in J} c_j = 0$. Then, by definition, a *comparison over J* is a subspace of contrasts over J; the dimensionality of this subspace is by definition the number of degrees of freedom (d.f.) of the comparison. A non–zero contrast generates a 1 d.f. comparison.

Let M^J be a cloud of points in a Euclidean space $\mathcal{U}$, with underlying vector space $\mathcal{V}$, and let G be its mean point. Recall that the overall *sum of squares* (or *inertia*) is by definition $\sum_{j\in J} n_j (\mathrm{GM}^j)^2$, where GM^j denotes the Euclidean distance between points G and M^j.

With any contrast c_J one associates: firstly, a *vector–effect*, or effect of the contrast over cloud M^J, which is the geometric vector $\sum_{j\in J} c_j \mathrm{M}^j \in \mathcal{V}$ (cf. CLOUD, definition 3.12, p.84); secondly, a *sum of squares* SS, or inertia, which is by definition the ratio of the square norm of the vector effect (in $\mathcal{V}$) to the square norm of contrast c_J (in $\mathbb{R}_J$), that is, $\mathrm{SS} = \| \sum_{j\in J} c_j \mathrm{M}^j \|^2 / (\sum_{j\in J} c_j^2/n_j)$. This SS is associated with any proportional contrast; therefore it also defines the *SS of the 1 d.f. comparison* generated by the contrast; the SS of a comparison with p d.f. is the sum of the inertias of p orthogonal contrasts generating the comparison.

An application of the preceding notions has already been presented in this book, namely the *contribution of a deviation*, which is the part of variance accounted for by a contrast among groups of points; see the *method of contributions of points and deviations* (chapter 5, §5.3.3, p.217), and its application to the Culture Example (§5.4, p.221)[12]. We are now going to apply these basic notions to the two structures of nesting and crossing between factors.

6.2.3 Nesting Structure

As a typical example of nesting structure, we take a set J of individuals with a partition into classes indexed by C. In ANOVA, it is classical to break down the source of variation $J\langle C\rangle$ into the two sources between–C (or inter–C) and within–C (or intra–C); we denote the latter source $J(C)$ (read "J within–C"). We first define between and within contrasts.

- *Between–C contrasts and comparisons.* The contrast e_J is a between–C contrast if $\forall c \in C$ and $\forall j \in J\langle c\rangle$: $e_j/n_j = e_c/n_c$ (with $e_c = \sum_{j\in J\langle c\rangle} e_j$

[12]See also the exercise "Social representation" (chapter 2, p.67).

and $n_c = \sum_{j \in J\langle c \rangle} n_j$). The between–$C$ contrasts generate the between–C comparison with $(C-1)$ d.f.

- *J within–C contrast and comparisons.* The contrast e_J is a J within–C contrast if $\forall c \in C$: $\sum_{j \in J\langle c \rangle} e_j = 0$. The within–$C$ contrasts generate the within–C comparison with $(J-C)$ d.f.

In geometric terms, the cloud M^J (weighted by n_J) is broken down into the *between-cloud* (M^C, n_C) and the *within-cloud* (residual cloud) $(\mathrm{M}^{J(C)}, n_J)$ (cf. §3.5.2, p.103). The breakdown is orthogonal in the sense that the covariant variables of the clouds M^C and $\mathrm{M}^{J(C)}$ associated with any vector are noncorrelated; as a consequence, there is an additive breakdown of the variance of the cloud into between–variance and within variance (cf. property 3.23, p.104).

6.2.4 Crossing Structure

When two factors A and B are crossed, it is customary to break down the source of variation $A \star B$ into three sources, namely the two main sources A and B and interaction $A.B$. In Analysis of Comparisons, these operations are formalized in terms of between, within and interaction contrasts and comparisons.

- *Between contrast and comparison.* Between A contrasts are defined by the nesting of $A \star B$ in A; between B contrasts are defined by the nesting of $A \star B$ in B; hence the between comparisons A and B, and their direct sum (in the sense of vector spaces) denoted $A \oplus B$.

- *Within contrast and comparison.* A contrast c_{AB} is a B–within–A contrast if for each $a \in A$, c_{AB} is a contrast over B, i.e. if $\forall a \in A$: $\sum_{b \in B} c_{ab} = 0$.

  A B–within–A comparison is generated by a family of B–within–A contrasts. Within–B contrasts and comparisons are similarly defined.

- *Interaction contrast and comparison.* A contrast c_{AB} over $A \star B$ is an interaction contrast if it is both a within–A contrast and a within–B contrast. The product $c_A \otimes c_B = (c_a\, c_b)_{a \in A, b \in B}$ of a contrast over A by a contrast over B is an interaction contrast. The $(A-1) \times (B-1)$ product–contrasts obtained from a basis of contrasts over A and a basis of contrasts over B constitute a basis of the overall interaction comparison.

Let n_{AB} denote the weighting measure over $A \star B$, n_A and n_B the marginal measures (with $n_a = \sum_{b \in B} n_{ab}$, $n_b = \sum_{a \in A} n_{ab}$), and $n = \sum_{a \in A} \sum_{b \in B} n_{ab}$. For any design (whether orthogonal or not), the interaction comparison $A.B$ is orthogonal to comparisons A and B (and consequently to $A \oplus B$). When the crossing is orthogonal ($\forall ab \in A \star B : n_{ab} = n_a n_b / n$), comparisons A and B are orthogonal, and the three inertias associated with A, B and $A.B$ add up. Remember that with observational data, the crossing is generally nonorthogonal.

In geometric terms, two main derivations are associated with the cloud $(\mathrm{M}^{AB}, n_{AB})$.

• *Derivation by transition.*

A transition τ_B^A from A to B is a family of positive measures over B indexed by A, that is, $\forall a \in A : \tau_b^a \geq 0$ and $\sum_{b \in B} \tau_b^a = 1$. For instance, the fundamental measure n_{AB} induces the transition f_B^A (from A to B) defined by $f_b^a = n_{ab}/n_a$, and the transition f_A^B (from B to A) defined by $f_a^b = n_{ab}/n_b$. With the transition τ_B^A there is associated the derived cloud on A defined by

$$\forall a \in A \; : \; \mathrm{M}^a = \sum_{b \in B} \tau_b^a \mathrm{M}^{ab}$$

The point M^a is the barycenter of points $(\mathrm{M}^{ab})_{b \in B}$ with weights $(\tau_b^a)_{b \in B}$. Below, $\mathrm{M}^A = (\mathrm{M}^a)_{a \in A}$ will designate the derived cloud of mean points associated with the transition f_B^A, and M^B the derived cloud of mean points associated with the transition f_A^B.

• *Derivation by contrast.*

Given a contrast γ_B (non–null) over B, one defines the vector $\overrightarrow{V}^a$ such that $\overrightarrow{V}^a = \sum_{b \in B} \gamma_b \mathrm{M}^{ab}$ called the vector–effect of contrast γ_B within–a, or conditional to a. Hence the derived protocol of vector–effects $(\overrightarrow{V}^a)_{a \in A}$. The derivation by contrast extends the notion of protocol of differences.

6.2.5 Double Breakdown of Variances

Questions of interest often lead to breaking down the overall source of variation into several sources, and to calculating the variances associated with each of these sources. When calculations are made for each principal axis, a *double breakdown*, by axes and sources of variation, results. Nonorthogonality is no obstacle to breaking down sources of variation, even though the breakdown of SS is no longer additive.

Two sorts of relative contributions follow. The *relative contribution of a source of variation to an axis* is obtained by dividing the variance of the source on axis by the variance of axis (eigenvalue). The *contribution of an axis to the source of variation* is obtained by dividing the variance of the source on axis by the total variance of the source. The first contribution extends the relative contribution of a point to an axis, the second extends the quality of representation of a point on an axis.

From the double breakdown table, in the case of a nesting (possibly following a classification), one may seek the axes for which the opposition between certain groups is important; in the case of a crossing, one may seek the axes for which the interaction is important. The bulk of interaction may be carried by axes of remote ranks, which may lead to investigating axes beyond the first ones.

An example of double breakdown has been shown in the *Culture Example* (p. 237), where individuals are nested within Age classes. Another example will be found in the *Political Space* study, where individuals are nested in the electorates (p.381).

6.3 Additive and Interaction Clouds

Given a Euclidean cloud defined on the crossing of two factors A and B: $(\mathrm{M}^{AB}, n_{AB})$, we now study in detail the links between the two crossed factors A and B, relying on the characterization of interaction given above (i.e. in terms of interaction contrasts, p.259), that is always valid[13]. We firstly characterize the *additive cloud*, then the interaction cloud. Then we discuss the "structural effect".

6.3.1 Additive Cloud

Definition 6.1. *A cloud is without interaction if the effect on the cloud of any interaction contrast is null.*

For any quadruple (a, a', b, b'), the effect of contrast $(\delta_A^a - \delta_A^{a'}) \otimes (\delta_B^b - \delta_B^{b'})$ on the cloud M^{AB} is the vector $\mathrm{M}^{ab} - \mathrm{M}^{a'b} - \mathrm{M}^{ab'} + \mathrm{M}^{a'b'}$; the cloud is without interaction if and only if this vector is null, i.e. if the figure formed by the four points M^{ab}, $\mathrm{M}^{a'b}$, $\mathrm{M}^{ab'}$, $\mathrm{M}^{a'b'}$ is a parallelogram. Equivalently, the cloud is without interaction if and only if for any contrast γ_B, the vector

[13] The elementary property of interaction effect, as a simple deviation from the sum of main effects, does not extend to the nonorthogonal case.

protocol $\overrightarrow{V}^A$ is constant over A. It can be shown that a cloud is without interaction if and only if it satisfies the following *additive property*.

Lemma 6.1. *Let φ_{AB} be a positive normalized measure over $A \times B$, with marginal measures $\varphi_a = \sum\limits_{b\in B} \varphi_{ab}$ and $\varphi_b = \sum\limits_{a\in A} \varphi_{ab}$. For any cloud M^{AB} without interaction, one has $\mathrm{P} = \sum\limits_{a\in A}\sum\limits_{b\in B} \varphi_{ab}\mathrm{M}^{ab} = \sum\limits_{a\in A}\sum\limits_{b\in B} \varphi_a\varphi_b\mathrm{M}^{ab}$.*

Proof. Indeed, measure $(\varphi_{ab} - \varphi_a\varphi_b)_{ab\in A\times B}$ is an interaction contrast since $\forall b \in B : \sum\limits_{a\in A}(\varphi_{ab} - \varphi_a\varphi_b) = 0$ and $\forall a \in A : \sum\limits_{b\in B}(\varphi_{ab} - \varphi_a\varphi_b) = 0$; therefore the effect of this contrast on the additive cloud M^{AB} is null, one has: $\sum\limits_{a\in A}\sum\limits_{b\in B}(\varphi_{ab}-\varphi_a\varphi_b)M^{ab} = \overrightarrow{0}$, hence $\sum\limits_{a\in A}\sum\limits_{b\in B} \varphi_{ab}\mathrm{M}^{ab} - \sum\limits_{a\in A}\sum\limits_{b\in B} \varphi_a\varphi_b\mathrm{M}^{ab} = \vec{0}$, hence $\mathrm{P} = \sum\limits_{a\in A}\sum\limits_{b\in B} \varphi_{ab}\mathrm{M}^{ab} = \sum\limits_{a\in A}\sum\limits_{b\in B} \varphi_a\varphi_b\mathrm{M}^{ab}$. ◁

Property 6.1 (Additive property). *Let φ_{AB} be a positive normalized measure over $A \times B$ with marginal measures $\varphi_a = \sum\limits_{b\in B} \varphi_{ab}$ and $\varphi_b = \sum\limits_{a\in A} \varphi_{ab}$. For any cloud M^{AB} without interaction one has the additive property:*

$$\mathrm{M}^{ab} = \mathrm{P}^a + \mathrm{P}^b - \mathrm{P} \quad \text{with} \quad \mathrm{P}^a = \sum_{b\in B} \varphi_b\mathrm{M}^{ab}, \quad \mathrm{P}^b = \sum_{a\in A} \varphi_a\mathrm{M}^{ab}$$

Proof. From parallelogram property $\mathrm{M}^{ab} - \mathrm{M}^{a'b} - \mathrm{M}^{ab'} + \mathrm{M}^{a'b'} = \overrightarrow{0}$, one deduces: $\sum\limits_{a'\in A}\sum\limits_{b'\in B} \varphi_{a'b'}(\mathrm{M}^{ab} - \mathrm{M}^{a'b} - \mathrm{M}^{ab'} + \mathrm{M}^{a'b'}) = \overrightarrow{0}$.
One has: $(\sum\limits_{a'\in A}\sum\limits_{b'\in B} \varphi_{a'b'})\mathrm{M}^{ab} = \mathrm{M}^{ab}$; $\sum\limits_{a'\in A}(\sum\limits_{b'\in B} \varphi_{a'b'})\mathrm{M}^{a'b} = \sum\limits_{a'\in A} \varphi_{a'}\mathrm{M}^{a'b}$; $\sum\limits_{b'\in B}(\sum\limits_{a'\in A} \varphi_{a'b'})\mathrm{M}^{ab'} = \sum\limits_{b'\in B} \varphi_{b'}\mathrm{M}^{ab'}$; $\sum\limits_{a'\in A}\sum\limits_{b'\in B} \varphi_{a'b'}\mathrm{M}^{a'b'} = \mathrm{P}$; hence the additive property. ◁

Property 6.2. *Given a cloud $(\mathrm{M}^{AB}, n_{AB})$, there is a unique additive cloud fitted to the cloud that has the same marginal derived clouds M^A and M^B, with therefore $\mathrm{M}^a = \sum\limits_{b\in B} f^a_b\mathrm{M}^{ab}$ et $\mathrm{M}^b = \sum\limits_{a\in A} f^b_a\mathrm{M}^{ab}$.*

We denote $\mathrm{M}^{A\oplus B}$ the fitted additive cloud, or *additive reference cloud.*

When the basic cloud is without interaction, the additive reference cloud coincides with the basic cloud; in other words, there are interaction effects only if there are deviations of the observed cloud from the additive reference cloud. When factors A and B are orthogonal ($n_{ab} = n_a n_b/n$), one has

$M^a = P^a$ and $M^b = P^b$, and the point $M^{a\oplus b}$ can be directly constructed by the vector sum $\overrightarrow{GM}^{a\oplus b} = \overrightarrow{GM}^a + \overrightarrow{GM}^b$; the cloud $M^{A\oplus B}$ is such that $M^{a\oplus b} = M^a + M^b - G$. When the crossing is not orthogonal, the construction is not straightforward; terms must be added that depend on the canonical correlations associated with the n_{AB} measure (Le Roux, 1991).

An example of additive cloud has been presented in the *Culture Example* (p.240); see also Bernard, Le Roux, Rouanet, Schiltz (1989).

6.3.2 Interaction Cloud

Definition 6.2. *A cloud is an interaction cloud if the effect (on this cloud) of any within contrast is null.*

Define the residual cloud $M^{A.B}$ as $M^{a.b} = G + M^{ab} - M^{a\oplus b}$. The residual cloud satisfies the following two properties.

Property 6.3. *(i) The residual cloud* $M^{A.B}$ *is doubly centered at point* G*:*

$$\forall a \in A : \sum_{b\in B} f_b^a M^{a.b} = G \quad \text{et} \quad \forall b \in B : \sum_{a\in A} f_a^b M^{a.b} = G$$

(ii) The cloud $M^{A.B}$ *is an interaction cloud.*

The breaking down of cloud M^{AB} into an additive and an interaction clouds is *orthogonal*, in the sense that for any vector $\overrightarrow{a} \in \mathcal{U}$, the covariant variables $\langle \overrightarrow{a} | \overrightarrow{GM}^{a\oplus b} \rangle_{ab\in A\times B}$ and $\langle \overrightarrow{a} | \overrightarrow{GM}^{a.b} \rangle_{ab\in A\times B}$ are noncorrelated. This suggests the following strategy for studying interaction:

- Construct the derived clouds that are attached with the structures of data.
- Proceed to the double breakdown of inertias on the principal axes of the initial cloud, on those of the additive cloud, and on those of the interaction cloud.
- Explore the interactions by studying the deviations $M^{ab} - M^{a\oplus b}$ between observed and fitted points, in the principal axes of the initial cloud, putting the additive cloud as supplementary; or directly study the interaction cloud in its own sytem of principal axes.

6.3.3 "Structural Effect"

When the crossing of two factors is nonorthogonal, a "structural effect", as defined above (p.256), may arise.

Again taking a cloud $(\mathrm{M}^{AB}, n_{AB})$ and a contrast γ_B, consider the *main effect of* γ_B defined as the effect of γ_B on the derived cloud M^B, i.e. as the vector $\sum\limits_{b\in B} \gamma_b \mathrm{M}^b = \sum\limits_{b\in B}\sum\limits_{a\in A} \gamma_b f_a^b \mathrm{M}^{ab}$.

When the crossing is orthogonal, one has $f_a^b = f_a$, hence $\sum\limits_{b\in B} \gamma_b \mathrm{M}^b = \sum\limits_{a\in A} f_a \sum\limits_{b\in B} \gamma_b \mathrm{M}^{ab} = \sum\limits_{a\in A} f_a \overrightarrow{V}^a$, where $\overrightarrow{V}^a = \sum\limits_{b\in B} \gamma_b \mathrm{M}^{ab}$ is the within–a effect of γ_B. That is, the main effect of γ_B is the weighted average of the within–a effects; in particular, when the latter are equal for all $a \in A$ (no interaction), the main effect is equal to the within–effect.

When the crossing is not orthogonal, there is no such property. For instance, for a factor B with two modalities b and b', the main effect is the vector $\overrightarrow{\mathrm{M}^b\mathrm{M}^{b'}}$, and the within–$a$ effects are the vectors $\overrightarrow{\mathrm{M}^{ab}\mathrm{M}^{ab'}}$; even when these vectors are equal for all $a \in A$ (no interaction), the main vector–effect is not equal to the within vector–effect. That is, the "structural effect" is a weighting phenomenon that has nothing to do with interaction.

In Rouanet & al (2002), the interest of studying "structural effects" in connection with GDA is shown; since the ratio of effects is preserved by projection, conditional effects can be compared to main effects in principal planes.

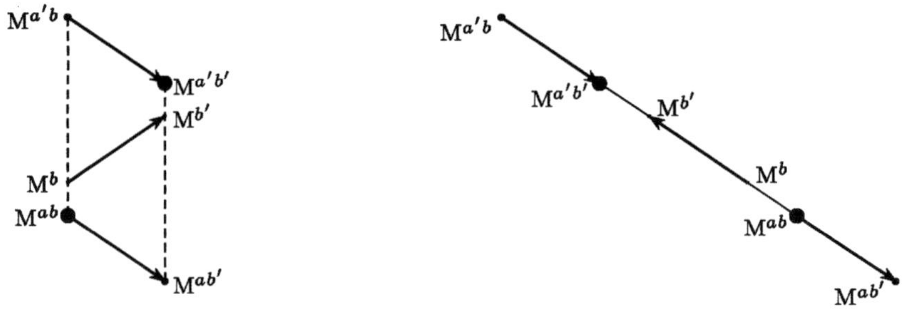

Figure 6.1. "Structural effect" for a cloud of four points $(\mathrm{M}^{ab}, \mathrm{M}^{ab'}, \mathrm{M}^{a'b}, \mathrm{M}^{a'b'})$ and weights $n_{ab} = n_{a'b'} = 5, n_{a'b} = n_{ab'} = 1$. Owing to nonorthogonal weighting, the vector–effect $\mathrm{M}^{b'} - \mathrm{M}^b$ (main effect of B) is not equal to the vector $\mathrm{M}^{ab'} - \mathrm{M}^{ab} = \mathrm{M}^{a'b'} - \mathrm{M}^{a'b}$ (within effect, there is no interaction) are not equal. They can even be opposite to each other (Figure on the right)

6.4 Related Topics and Comments

6.4.1 The Rehabilitation of Individuals

When Individuals ×Variables tables are analyzed by PCA or MCA, geometric representations are often confined to variables: circle of correlations in PCA, or cloud of modalities in MCA. The interest of representing the cloud of individuals is felt when the individuals are "known persons", such as university professors in Bourdieu (1984) (*Homo Academicus*), much less when individuals are anonymous, as is the case in most opinion surveys. As a matter of fact, if individual points are not distinguished, this cloud is no more than an undifferentiated collection of points. But when there are structuring factors of individuals, the interest of exploring the cloud of individuals and examining subclouds of individuals is soon realized. In this way, Structured Data Analysis should contribute to "rehabilitate individuals" in statistical analysis. In the *CASE STUDIES* chapter of this book, the clouds of individuals will be systematically studied in the line of Structured Data Analysis[14].

An especially enlightening analysis of a cloud of individuals, reported hereafter, is that of the *Basketball study*.

The Basketball study (Wolff, Rouanet, Grosgeorge, 1998). This study concerns the judgments of basketball experts in the activity of selecting high–level potential players. For 8 young players (structuring factor P), video sequences were constructed concerning typical game situations. These sequences were submitted to 9 experts (structuring factor E), who for each player expressed free verbal judgments about the potentialities of this player. The compound factor $I = P \times E$, made of the $8 \times 9 = 72$ (player, expert) combinations, was taken as the set of "statistical individuals" (or "units"), on which the expert judgments were recorded. A content analysis of the 72 records was carried out, leading to construct 11 judgement variables about four aspects of performance relating namely to *upper*

[14] *Extended use of "structuring factor".* In MCA, at least, individuals can also be partitioned according to their responses to active questions, exactly like structuring factors, and this property can be used for exploring the cloud of individuals. See e.g. Bonnet, Le Roux & Lemaine (1994), where an extended use of the phrase "structuring factor" is made. Of course, when it comes to interpretation, especially when there is an explanatory scheme, it will matter whether the partition of individuals under study is induced by an active question, or by a *bona fide* structuring factor (non–active question). In particular, the inference procedures presented in chapter *INDUCTIVE* are only valid for structuring factors in the strict sense.

body (4 variables: statics, hand and arm dynamics, arm strenght), *lower body* with four variables (statics, dynamics, leg strength and overall speed), *global judgment* (1 variable), *play strategy* (2 variables: attack and defense). The following weights were attributed to the 11 variables, following experts' advice: 1.5, 3.5, 1.5, 0.5; 1.5, 1.0, 0.5, 0.5; 2.5; 1.5, 1.5 respectively, hence a total weight equal to 16. A weighted PCA was performed on the 72 × 11 table of standardized variables. Two eigenvalues ($\lambda_1 = 9.298$ and $\lambda_2 = 2.104$) were found to exceed the average $\overline{\lambda} = 16/11 = 1.455$. Axis 1 corresponds to a size factor interpreted as "dexterity", and Axis 2 is related to "strategy" (attack and defense).

Then in the space of individuals, from the basic cloud of 72 points, three clouds were derived: the cloud of the 8 mean points of players (indexed by factor P), the one of experts (factor E) and the interaction cloud. The double breakdown of variance according to these three sources of variation and the first two axes is shown in Table 6.1.

	Variances	
	Axis1	Axis2
$I = P \star E$	9.298	2.104
P	8.913	1.731
E	0.051	0.124
$P.E$	0.335	0.250
$P \oplus E$	8.964	1.855

Table 6.1. Basketball study: double breakdown of variance according to players (P), experts (E), interaction $P.E$, and variances for the additive cloud ($P \oplus E$).

Table 6.1 shows the large individual differences among players, as contrasted to an overall fair homogeneity of experts, and also to a moderate interaction ($P.E$) between players and experts. Figure 6.2 (p.267) depicts the basic cloud in plane 1-2, structured by the 8 players mean points; Figure 6.3 (p.267) shows the fitted additive cloud, i.e. the fitted cloud without interaction.

6.4.2 Structured Data Analysis and Regression

• ANOVA and regression are closely connected methods, and properties stated in terms of ANOVA can be stated equivalently in terms of regression. For instance, when in an ANOVA with several factors all interaction terms are discarded, the analysis is equivalent to a multiple linear regression on the associated indicator variables, where regression coefficients are interpreted as conditional effects.

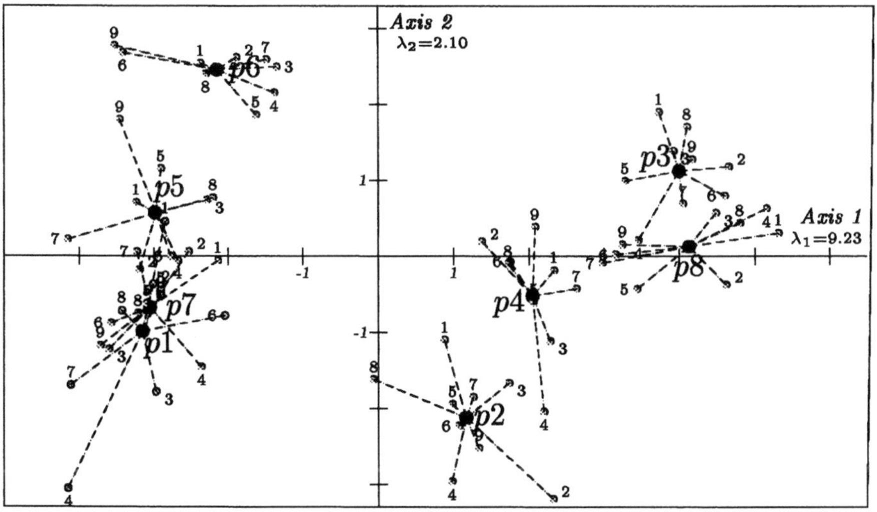

Figure 6.2. Basic cloud in plane 1-2: 8 × 9 = 72 (player, expert) points, joined to the 8 player mean points.

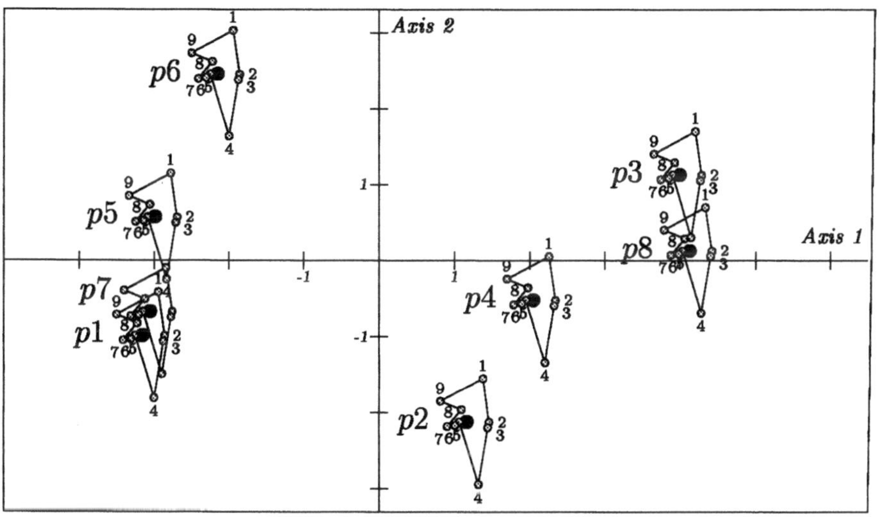

Figure 6.3. Additive cloud in plane 1-2: 8 × 9 = 72 (player, expert) points; for each player, lines have been drawn joining expert points (numbered arbitrarily from 1 to 9) to exhibit the parallelism property of additive cloud

• Regression and GDA, especially PCA, are also related. As seen in earlier chapters of this book, the technique of supplementary variables amounts to a regression onto a principal space. For instance in PCA, constructing a supplementary variable over the whole principal space amounts to performing a multiple linear regression of this variable on the set of all initial variables. The quality of representation of the supplementary variable thus provides an upper bound for the multiple R^2 coefficients of all multiple regressions that can be performed on any subset of initial variables.

• Regression techniques are traditionally used to investigate the influence of several independent variables on a dependent variable. In such situations, constructing a geometric space from the independent variables and embedding the regression models into GDA representations, at least in the space of variables, sheds light on the results of regressions, especially in the context of collinearity and "structural effects". As an example, see the *Cookie study* in Rouanet & al (2002).

6.4.3 Concluding Comment

Confronted with an Individuals×Variables table, some GDA users just perform a straight PCA or MCA, and juxtapose to it a traditional ANOVA or regression analysis, conducted and interpreted separately. In Structured Data Analysis, we start from a geometric model of data, and we "graft" ANOVA or regression procedures, taking into account whenever necessary the special properties (esp. nonorthogonality) of observational data. This approach does not require amending the basic geometric model e.g. modifying the definition of distance; thus it enriches the analysis of data without having to "invent a new method" everytime a new structure is encountered. To sum up, it avoids both the rigidity of a stereotyped technique and the proliferation of ad hoc procedures.

Chapter 7

Stability of a Euclidean Cloud

(*'STABILITY'*)

Introduction

In this chapter, we present stability problems in Geometric Data Analysis. The aim is to study the sensitivity of the methods of determination of the principal axes and variables of a Euclidean cloud to some data modifications, or *perturbations*, so as to bring out their mechanism of variation.

Stability studies have been tackled along several lines.

The line followed by Escofier & Le Roux (1972, 1975, 1976, 1977) consists in comparing the analysis of a *reference Euclidean cloud* to that of a cloud obtained by some modification. The influence of the perturbation is evaluated by an *interval for eigenvalues and inertia rates* and by *upper bounds for the rotation angles of principal subspaces*; intervals and upper bounds are expressed as functions of the results of the analysis of the reference cloud: eigenvalues, contributions and qualities of representation of points. This approach refers to the following linear algebra problems: comparing the eigenvalues and studying the rotation of invariant subspaces of a symmetric endomorphism by the addition of a symmetric endomorphism and/or multiplication by a positive–definite symmetric one.

Another line of study is that of *sensitivity analysis* and influence functions, in the sense of functional analysis[1]. This method has been mainly applied to PCA: see Krzanowski (1984), Critchley (1985), Tanaka (1988);

[1] See e.g. Hampel (1974) or Lecoutre & Tassi (1987, chapter 2).

for its application to CA, see Pack & Joliffe (1992), Bénasséni (1993).

Still another line is that of *bootstrap*, discussed e.g. by Diaconis & Efron (1983), Daudin & al (1988), Lebart & al (1995), Lebart & al (2003).

In this chapter, we present four stability problems, following the first line, which is the most congenial to the geometric approach followed elsewhere in this book. To begin with, we study the effect of the coding according to a partition (§7.1). Then, we deal with the influence of a group of points (§7.2). Then we study the effect of a change of metric (§7.3). We proceed with the study of the influence of a variable or a modality (§7.4). For each problem, we firstly present the results for a Euclidean cloud, then we particularize to the clouds stemming from PCA, CA, and MCA. The chapter closes with the linear algebra theorems on which our stability studies are based, namely, the comparison of eigenvalues and the rotation of the invariant subspaces of a symmetric endomorphism by a perturbation (§7.5).

7.1 Effect of Coding according to a Partition

In GDA, one often proceeds to the grouping of elements. It is a common saying that in CA, since the distance satisfies the principle of distributional equivalence, grouping modalities, whose profiles are similar, entails "little change" in the outcome of analysis. In this section we state mathematical results that formalize this assertion.

Grouping problems may concern either individuals or modalities.

— When grouping individuals according to a partition, the stability study permits to identify the principal variables that are little modified and therefore reflect differences among groups.

— In a questionnaire analysis, one attempts to balance the contributions of questions (at least within a same heading), which motivates the search of groupings of modalities that leave the analysis stable.

To begin with, we will recall some definitions and properties pertaining to the partition of a cloud (§7.1.1). Then we will present the effect of grouping into C classes ($C \geq 2$), into two classes (§7.1.3), and the grouping of two elements, e.g. quasi–distributional equivalence in CA (§7.1.4).

7.1.1 Partition of a Cloud

Let M^J be a cloud (reference cloud) in a Euclidean space $\mathcal{U}$; we denote G its mean point and Var M^J its variance. Let $\mathcal{V}$ be the underlying vector space of $\mathcal{U}$. With the cloud M^J there is associated the endomorphism $Som : \mathcal{V} \to \mathcal{V}$

(see CLOUD, Definition 3.13, p.86). Let $\mathcal{M}\subseteq\mathcal{U}$ be the geometric support of the cloud M^J with the underlying vector subspace $\mathcal{L}\subseteq\mathcal{V}$, and L be its dimensionality.

From now on all endomorphisms are restricted to $\mathcal{L}$.

$$\forall \overrightarrow{\alpha} \in \mathcal{L} : Som(\overrightarrow{\alpha}) = \sum_{j\in J} f_j \langle \overrightarrow{\mathrm{GM}^j} | \overrightarrow{\alpha} \rangle \overrightarrow{\mathrm{GM}^j} \in \mathcal{L} \tag{7.1}$$

Consider the grouping $J{<}c{>}$ of elements of J in class c, and the associated subcloud $\mathrm{M}^{J<c>}$ with the three endomorphisms on $\mathcal{L}$ — namely $Som_{J<c>}$, Som_c and $Som_{J(c)}$ — which are defined in chapter CLOUD (p.103), and which verify the between–within relation $Som_{J<c>} = Som_c + f_c Som_{J(c)}$ (cf. Lemma 3.1, p.103). For a partition into C classes, one has the between–within breakdown of the endomorphism Som (cf. Theorem 3.6, p.105):

$$Som = Som_C + Som_{J(C)} \tag{7.2}$$

The endomorphism $Som_C = \sum_{c\in C} Som_c$ is associated with the between–cloud M^C and the endomorphism $Som_{J(C)} = \sum_{c\in C} f_c Som_{J(c)}$ with the within–cloud $\mathrm{M}^{J(C)}$ (see Definitions 3.24 and 3.25, p.103). Let $(\lambda_\ell)_{\ell=1,\dots L}$ denote the eigenvalues of Som, $(\lambda'_\ell)_{\ell=1,\dots L}$ the ones of $Som_{J(C)}$ and $(\mu_\ell)_{\ell=1,\dots L}$ the ones of Som_C. Comparing the between–cloud M^C to the reference cloud M^J amounts to studying the influence of the perturbation $-Som_{J(C)}$ on the eigenvalues and the eigenvectors of Som. In the same way, comparing the within–cloud $\mathrm{M}^{J(C)}$ to the reference cloud M^J— that is, studying the effect of grouping into classes — amounts to comparing the eigenvalues and eigenvectors of Som to those of $Som_{J(C)}$ under the perturbation $-Som_C$.

The study of stability by grouping does not entail any change of metric: the clouds to be compared lie in the same Euclidean space.

7.1.2 Stability and Partition into C Classes ($C \geq 2$)

Property 7.1 (Eigenvalues). *Between the eigenvalue λ_ℓ of the cloud M^J and the eigenvalue λ'_ℓ of the within–cloud $\mathrm{M}^{J(C)}$, one has the inequalities:*

$$\lambda_\ell - \mu_1 \leq \lambda'_\ell \leq \lambda_\ell - \mu_L$$

This property follows from Theorem 7.1 (relation (2b), p.292).

If $\mathrm{Var}\,\mathrm{M}^C$ denotes the between–variance, one has $\mu_1 \leq \mathrm{Var}\,\mathrm{M}^C$ and $\mu_L \geq 0$, hence $\lambda_\ell - \mathrm{Var}\,\mathrm{M}^C \leq \lambda'_\ell \leq \lambda_\ell$.

The inertia rates pertaining to the first ℓ axes are denoted $\tau_{1\to\ell} = \sum_{\ell'=1}^{\ell} \lambda_{\ell'} / \operatorname{Var} \mathrm{M}^J$ for the reference cloud M^J, $\tau'_{1\to\ell} = \sum_{\ell'=1}^{\ell} \lambda'_{\ell'} / \operatorname{Var} \mathrm{M}^{J(C)}$ for the within-cloud $\mathrm{M}^{J(C)}$, and $\upsilon_{1\to\ell} = \sum_{\ell'=1}^{\ell} \mu_{\ell'} / \operatorname{Var} \mathrm{M}^C$ for the between-cloud M^C. Then letting $\eta^2 = \operatorname{Var} \mathrm{M}^C / \operatorname{Var} \mathrm{M}^J$ (correlation ratio), one has the following property.

Property 7.2 (Inertia rates). *Between the inertia rate $\tau_{1\to\ell}$ of the cloud M^J and the inertia rate $\tau'_{1\to\ell}$ of the within-cloud, pertaining to the first ℓ principal axes, one has the following inequalities:*

$$\frac{\tau_{1\to\ell} - \eta^2 \upsilon_{1\to\ell}}{1-\eta^2} \leq \tau'_{1\to\ell} \leq \frac{\tau_{1\to\ell} + \eta^2(1-\upsilon_{1\to\ell})}{1-\eta^2}$$

This property follows from the relation (2c) of Theorem 7.1 (p.292).

Property 7.3 (Principal subspaces). *The greatest canonical angle θ between the principal subspaces associated with the ℓ_0-th, ... , $(\ell_0+\ell-1)$-th eigenvalues is such that, letting $\epsilon = \inf\{(\lambda_{\ell_0-1} - \lambda_{\ell_0})\ ;\ (\lambda_{\ell_0+\ell-1} - \lambda_{\ell_0+\ell})\}$ for $\ell_0 > 1$ and $\epsilon = (\lambda_\ell - \lambda_{\ell+1})$ for $\ell_0 = 1$ (subspace of the first ℓ axes):*

$$\text{If} \quad \mu_1 - \mu_L < \epsilon \quad \text{then} \quad \theta < \pi/4 \quad \text{and} \quad \sin 2\theta \leq (\mu_1 - \mu_L)/\epsilon$$

This property follows from Theorem 7.4, p.294.

Comments

The *eigenvalues* decrease when points are grouped, but nothing can be said about the increasing or decreasing of inertia rates. The upper bounds of the *angle between principal subspaces* are increasing functions of the deviations between the largest and the smallest eigenvalues of the perturbation cloud (between or within), and they are decreasing functions of the minimal deviation between the eigenvalues that "border" the subspace under study.

The preceding properties are directly applicable to clouds stemming from CA or PCA (see Le Roux, 1977). For an example of application to CA, see Escofier & Le Roux (1972).

7.1.3 Stability and Partition into Two Classes

For a partition into two classes $C = \{c, c'\}$, the rank of the endomorphism Som_C is equal to one and its nonnull eigenvalue is equal to $\operatorname{Var} \mathrm{M}^C$. The previous intervals and upper bounds can be refined by taking into account the position of the line going through the points M^c and $\mathrm{M}^{c'}$ (and consequently through G) with respect to the principal axes of the cloud.

Let $\mathrm{Cta}_\ell(c,c')$ denote the absolute within–contribution of the dipole $(\mathrm{M}^c, \mathrm{M}^{c'})$ to axis ℓ, and $\mathrm{Cta}_{1\to\ell}(c,c') = \sum\limits_{r=1}^{\ell} \mathrm{Cta}_r(c,c')$ the one of the dipole to the first ℓ principal axes.

Let Ψ_ℓ be the angle between the subspace of the first ℓ axes and the line going through M^c and $\mathrm{M}^{c'}$; one has: $\cos^2\Psi_\ell = \mathrm{Cta}_{1\to\ell}(c,c')/\operatorname{Var}\mathrm{M}^C$.

Property 7.4 (Eigenvalues). *Between the eigenvalue λ_ℓ of the reference cloud and the eigenvalue λ'_ℓ of the within–cloud, one has:*

$$\max\{\lambda_{\ell+1};\lambda_\ell - \mathrm{Cta}_{1\to\ell}(c,c')\} \le \lambda'_\ell \le \lambda_\ell \quad (1 \le \ell < L)$$

This property follows from Theorem 7.2 (2), p.293.

Property 7.5 (Inertia rates). *Between the inertia rate $\tau_{1\to\ell}$ of the reference cloud and the inertia rate $\tau'_{1\to\ell}$ of the within–cloud pertaining to the first ℓ principal axes, one has the following inequalities:*

$$\frac{\tau_{1\to\ell} - \eta^2\cos^2\Psi_\ell}{1-\eta^2} \le \tau'_{1\to\ell} \le \frac{\tau_{1\to\ell}}{1-\eta^2}$$

Property 7.6 (Principal subspace $1\to\ell$). *The greatest canonical angle θ between the principal subspace of the cloud M^J and the principal subspace of the within–cloud $\mathrm{M}^{J(C)}$ (generated by the first ℓ principal axes) is such that:*

$$\text{If } m' = \frac{\lambda_\ell - \lambda_{\ell+1}}{\operatorname{Var}\mathrm{M}^C} > 1 \quad \text{then } \theta < \pi/4 \text{ and } \tan 2\theta \le \frac{\sin 2\Psi_\ell}{m' - \cos 2\Psi_\ell}$$

$$\text{If } \cos^2\Psi_\ell < m' < 1 \quad \text{then } \theta < \pi/4 \text{ and } \tan 2\theta \le \frac{\sin 2\Psi_\ell}{m' - \cos^2\Psi_\ell}$$

This property follows from Theorem 7.5 (2), p.294.

Applications

Culture Example (cf. chapter 5, §5.4, p.221). Consider the partition of the cloud of individuals into two classes $\{c,c'\}$ induced by Gender (cf. p.234). The variance of the between–cloud[2] is found to be 0.0113526.

[2]This variance is equal to the Φ^2 of the table crossing Gender with the 30 active modalities.

We will compare the initial cloud to the within-cloud. For this purpose, we have to compare the eigenvalues and the eigenvectors of the endomorphism *Som* to those of the endomorphism obtained by subtracting the endomorphism Som_C of rank unity.

From Table 5.12 (p.234), one calculates the absolute within-contributions of the dipole (c, c') to the first three axes: $\text{Cta}_1(c, c') = .000007$, $\text{Cta}_2(c, c') = .001295$ and $\text{Cta}_3(c, c') = .000107$.

Hence $\cos^2 \Psi_1 = \frac{.000007}{.0113526} = .000617$; $\cos^2 \Psi_2 = \frac{.000007+.001295}{.0113526} = .114687$; and $\cos^2 \Psi_3 = \frac{.000007+.001295+.000107}{.0113526} = .124113$.

• *Eigenvalues.* Recall (see p.224) the values of the first three eigenvalues of the reference cloud: $\lambda_1 = .29252$, $\lambda_2 = .24149$, $\lambda_3 = .22484$.

From Property 7.4 (p.273), we deduce the *interval of variation of the eigenvalues* of the within-cloud:

$$.29251 \le \lambda'_1 \le .29252 \ ; \quad .24019 \le \lambda'_2 \le .24149 \ ; \quad .22213 \le \lambda'_3 \le .22484$$

• *Principal subspaces.* From property 7.6 (p.273), we deduce that the angle θ_1 between the *first principal axis* of the reference cloud and the one of the within-cloud is such that $\tan 2\theta_1 \le .04965/(4.4950 + .99877) = 0.0090$, hence $\theta_1 \le 0°26$. Similarly, for the *first principal planes*, the greatest canonical angle $\theta_{1\to 2}$ is such that $\theta_{1\to 2} \le 7°95$ ($\tan 2\theta_{1\to 2} \le 0.28485$); for the subspaces of the *first three principal axes*[3], one has $\theta_{1\to 3} \le 8°$.

> These results indicate a *quasi-stability of the subspace of the first three principal axes*: in short, for the first three axes, there is practically no difference between genders.

Political space study (cf. chapter 9, §9.2, p.365). Consider the following partition of the 2980 individuals into two classes: class c with the 193 individuals of the NF electorate and class c' with the 2787 others.

Taking again the notation of the specific MCA of a subcloud (*MULTIPLE*, §5.2.2, p.210), let N denote the number of individuals ($N = 2980$), Q the number of questions ($Q = 20$) and K' the number of active modalities ($K' = 74$). We denote N_k the total number of individuals (among $N = 2980$) who have given response k, and n_k the number of those of class c (among $n = 193$) who have given response k.

The between variance is such that:

$$\text{Var}\,\text{M}^C = \frac{n(N-n)}{N^2}(\text{M}^c\text{M}^{c'})^2 = \frac{1}{Q} \times \frac{1}{n\,N(N-n)} \sum_{k\in K'} \frac{(N\,n_k - n\,N_k)^2}{N_k}$$

We thus find $\text{Var}\,\text{M}^C = 0.0193456$.

[3]One has $\lambda_4 = .20726$, hence $(\lambda_3 - \lambda_4)/\text{Var}\,\text{M}^C = 1.54854$ and $\tan 2\theta_{1\to 3} \le 0.28666$.

Let ψ_1 be the angle between the first axis and the line going through M^c and $M^{c'}$ (hence through G); $\cos^2\psi_1$ is the quality of representation of the mean-point modality NF (cf. Table 9.2.8, p.380). We define similarly ψ_2 (axis 2) and ψ_3 (axis 3).

More precisely, $\cos^2\psi_1 = .88832$, $\cos^2\psi_2 = .04330$, and $\cos^2\psi_3 = .008738$; then for the principal plane 1-2, $\cos^2\Psi_2 = .93162$ and for the principal subspace $1\to3$, $\cos^2\Psi_3 = .94036$. From properties 7.4 and 7.5 (p.273), we deduce the intervals of variation for eigenvalues and inertia rates (see Table 7.1).

Axis ℓ	λ_ℓ	Eigenvalues	λ'_ℓ
Axis 1	.21325	$.19606 \le \cdot \le .21325$	*.19662*
Axis 2	.16063	$.14261 \le \cdot \le .16063$	*.15942*
Axis 3	.11921	$.10102 \le \cdot \le .11921$	*.11900*

	$\tau_{1\to\ell}$	Inertia rates	$\tau'_{1\to\ell}$
1	.0785	$.0727 \le \cdot \le .0790$	*.0729*
$1\to2$	.1376	$.1319 \le \cdot \le .1386$	*.1319*
$1\to3$	.1814	$.1759 \le \cdot \le .1827$	*.1760*

Table 7.1. Eigenvalues (λ_ℓ) of initial cloud, interval of variation of the eigenvalues (λ'_ℓ) of within–cloud and its eigenvalues (in slanted characters); similarly for inertia rates.

- *First principal axis.* One has $(\lambda_1 - \lambda_2)/\operatorname{Var} M^C = 2.72$; from Property 7.6, we deduce that the angle θ_1 of the rotation of Axis 1 is such that $\tan 2\theta_1 \le 0.32415$, that is, $\theta_1 \le 9°$ (the value θ_1 is found to be 6°18).

- *First principal plane.* $\tan 2\theta_{1\to2} \le .050479/(2.141-.86324) = 0.39505$, that is, $\theta_{1\to2} \le 10°78$ (the greatest canonical angle between the two planes is found to be 2°33).

- *First three-dimensional principal subspace.* The greatest canonical angle between the two subspaces is such that $\tan 2\theta_{1\to3} = .047364/(1.995 - .88972) = 0.42515$, that is, $\theta_{1\to3} \le 11°52$ (it is found to be 1°91).

> The between–group effect on the rotation of the first three–dimensional principal subspace is very weak .

Remark. In the *Political Space* data, the Ψ_ℓ angles are small, $\cos 2\Psi_\ell$ is positive, therefore the upper bound of the canonical angles between the principal subspaces is larger, and the principal subspaces are less stable than in the *Culture Example*, for which the angle between the line supporting the between–cloud and the principal subspaces is near 90°.

7.1.4 Stability and Grouping of Two Points

If only two points M^j and $M^{j'}$ are grouped the perturbation of *Som* is of rank unity.

Consider the partition into $C = J - 1$ classes consisting in the class J<c>$= \{j, j'\}$ and the $J - 2$ singletons $\{j''\}$ ($j'' \notin J$<c>). One has (cf. formula (7.2), p.271): $Som = Som_C + Som_{J(C)}$.

For the class J<c'>$= \{j''\}$, one has $Som_{c'}(\overrightarrow{\alpha}) = f_{j''}\langle\overrightarrow{\mathrm{GM}^{j''}}|\overrightarrow{\alpha}\rangle\overrightarrow{\mathrm{GM}^{j''}}$ and $Som_{J(c')} = 0$.

For the class J<c>$= \{j, j'\}$, one has $Som_c(\overrightarrow{\alpha}) = f_c\langle\overrightarrow{\mathrm{GM}^c}|\overrightarrow{\alpha}\rangle\overrightarrow{\mathrm{GM}^c}$, and $f_c Som_{J(c)} = (f_j\, f_{j'}/f_c)\langle\overrightarrow{\mathrm{M}^j\mathrm{M}^{j''}}\,|\overrightarrow{\alpha}\rangle\overrightarrow{\mathrm{M}^j\mathrm{M}^{j''}}$.

Therefore the perturbation $Som_{J(C)}$ is of rank unity; its nonnull eigenvalue is equal to $\frac{n_j\, n_{j'}}{n\,(n_j+n_{j'})}(\mathrm{M}^j\mathrm{M}^{j'})^2$ (absolute within contribution of dipole) and the associated eigenvector is $\overrightarrow{\mathrm{M}^j\mathrm{M}^{j'}}$. Taking into account the position of the vector $\overrightarrow{\mathrm{M}^j\mathrm{M}^{j'}}$ with respect to the principal subspaces of the cloud M^J, upper bounds can be refined by applying Theorem 7.5 (p.294), as illustrated by the example hereafter.

Application

"Hair & Eye Color" (chapter 2, §2.3.5, p.51). The profiles of *Blue* and *Light Eyes* are quite similar (cf. Table 2.4, p.52), i.e. there is a quasi–distributional equivalence. Let us now study the effect of grouping columns $k1$ (*Blue Eyes*) and $k2$ (*Light Eyes*) on the results of CA.

One has $(\mathrm{M}^{k1}\mathrm{M}^{k2})^2 = 0.01676$, hence the absolute contribution of the dipole $(\mathrm{M}^{k1}, \mathrm{M}^{k2})$ is $\mathrm{Cta}_{(k1,k2)} = .00154$. The angle ψ_1 between the first principal line and the line going through points M^{k1} and M^{k2} is such that $\cos^2\psi_1 = (y_1^{k1} - y_1^{k2})^2/(\mathrm{M}^{k1}\mathrm{M}^{k2})^2$ (for y_1^{k1} and y_1^{k2}, coordinates of points M^{k1} and M^{k2} on Axis 1, see Table 2.7, p.53); one has $\cos^2\psi_1 = .0974$, i.e. $\psi_1 = 71°81$. Similarly, the angle ψ_2 between this line and the second principal line is such that $\cos^2\psi_2 = .3506$, hence the angle Ψ_2 between the line and the plane 1-2 is such that $\cos^2\Psi_2 = .0974 + .3506 = .4480$.

- *Interval of variation of eigenvalues.*

$.19909 = .19924 - .00015 \le \lambda'_1 \le .19924 \quad ; \quad .02940 \le \lambda'_2 \le .03009$

- *First principal line.* The rotation angle θ_1 of the first principal line is such that $\tan 2\theta_1 \le \sin 2\psi_1/(\frac{.19924-.03009}{.00154} - \cos 2\psi_1) = .0053$ ($\theta_1 < 0°15$).

- *First principal plane.* For the plane of the first two axes, knowing that $\lambda_2 - \lambda_3 = .02923$, one gets: $\tan 2\theta_{1\to2} \le .0521$ ($\theta_{1\to2} < 1°49$).

Remark. After grouping, the two eigenvalues of the contingency table (5×3) are found to be 0.19910 and .02956, and the first two principal axes are virtually unchanged. There is quasi–stability.

7.2 Influence of a Group of Points (Individuals)

In GDA, questions may be raised about the influence of a group of individuals, for example whether they have characteristics that differ from other individuals, or whether the contribution of the group to one or several axes is strong; in short: What would the outcome of analysis become without this group?

Methodologically, the following cases can occur:

— The axis is about the same, i.e. there is stability. The group of individuals *is not characteristic* of the axis.

— The axis vanishes without modification of the other interpretable axes: Then the axis is *specific* to the group of individuals. This case often occurs in opinion questionnaires, when a "nonrespondent" axis exists: Rather than deleting the "nonrespondent" individuals, one will prefer to put the nonresponse modalities as passive ones (cf. specific analysis, *MULTIPLE*, §5.2.2, p.210).

— A rotation of one of the first axes (or first principal plane or subspaces) occurs; then it is advisable to inquire about the *homogeneity* of data; discarding the group may be in order, especially if it appears isolated from the others.

If the aim is to compare the analysis of a table with that of a subtable, one faces the problem of comparing clouds that lie in different Euclidean spaces; deleting data entails a change of metric in CA, in standard PCA as well as in MCA. Most often, it seems methodologically preferable to fix the metric with respect to a reference cloud; this is the viewpoint of *specific analysis*, that we will adopt in this section. The problem of the change of metric will be dealt with in the next section (p.281).

We will firstly present the study of the deletion of a group of several points, with application to the *Political Space* study (§7.2.1), then we will outline the case of a single point (§7.2.2).

7.2.1 Influence of a Group of several Points (Individuals)

Let $J<c>$ denote the subset of J indexing the deleted points and $J<c'>$ the complementary subset; hence a partition of J into two classes c and c' ($C = \{c, c'\}$). We take up the notations of §7.1.3 (p.272).

Studying the influence of a group of points consists in comparing the analysis of the L–dimensional cloud M^J to that of the subcloud $\mathrm{M}^{J<c'>}$.

From the between–within breakdown of *Som*, one has (cf. p.271):

$Som = Som_C + Som_{J(C)}$, with $Som_{J(C)} = f_c Som_{J(c)} + f_{c'} Som_{J(c')}$

Then we have to compare the spectral analysis of *Som* to that of $Som_{J(c')} = \big((Som - Som_C) - f_c Som_{J(c)}\big)/f_{c'}$.

For this purpose, one considers that the initial cloud is submitted to two perturbations: 1) $-Som_C$ on *Som* (comparison of $Som_{J(C)}$ to *Som*); 2) $-f_c Som_{J(c)}/f_{c'}$ on $Som_{J(C)}/f_{c'}$ (comparison of $Som_{J(c')}$ to $Som_{J(C)}/f_{c'}$). The second perturbation will be studied hereafter, the first one was studied earlier (see §7.1.3, p.272-273).

We denote λ_ℓ the eigenvalues of *Som* (initial cloud M^J), λ'_ℓ those of $Som_{J(c')}$, μ_ℓ those of $Som_{J(c)}$ (subcloud of deleted points) and $\widetilde{\mu}_\ell$ those of $Som_{J(C)}$ (within–cloud).

Property 7.7 (Eigenvalues). *Between the eigenvalue λ_ℓ of the cloud M^J and the eigenvalue λ'_ℓ of the subcloud $\mathrm{M}^{J<c'>}$, one has the inequalities:*

$$\frac{1}{1-f_c}\Big(\lambda_\ell - f_c\mu_1 - \mathrm{Cta}_{1\to\ell}(c,c')\Big) \le \lambda'_\ell \le \frac{1}{1-f_c}\Big(\lambda_\ell - f_c\mu_L\Big)$$

Proof. From Property 7.4 (p.273), one deduces for the eigenvalues of $Som_{J(C)}$ that $\max\{\lambda_{\ell+1}; \lambda_\ell - \mathrm{Cta}_{1\to\ell}(c,c')\} \le \widetilde{\mu}_\ell \le \lambda_\ell$. Applying Weyl's inequalities (Theorem 7.1 *(2b)*, p.292) to $Som_{J(C)} - f_c Som_{J(c)}$, one gets the interval $[\widetilde{\mu}_\ell - f_c\mu_1 \,;\, \widetilde{\mu}_\ell - f_c\mu_L]$, with $\widetilde{\mu}_\ell - f_c\mu_1 \ge \lambda_\ell - \mathrm{Cta}_{1\to\ell}(c,c')\} - f_c\mu_1$ and $\widetilde{\mu}_\ell - f_c\mu_L \le \lambda_\ell - f_c\mu_L$; hence dividing by $f_{c'} = 1 - f_c$, the interval of variation of the eigenvalues λ'_ℓ of $Som_{J(c')}$. ◁

Property 7.8 (Principal subspaces $1\to\ell$). *The greatest canonical angle θ between the principal subspaces generated by the first ℓ principal axes is such that $\theta \le \theta' + \theta''$, with:*

$$\text{If } m' = \frac{\lambda_\ell - \lambda_{\ell+1}}{\mathrm{Var}\,\mathrm{M}^C} > 1 \quad \text{then } \theta' < \pi/4 \text{ and } \tan 2\theta' \le \frac{\sin 2\Psi_\ell}{m' - \cos 2\Psi_\ell}$$

$$\text{If } \cos^2\Psi_\ell < m' < 1 \quad \text{then } \theta' < \pi/4 \text{ and } \tan 2\theta' \le \frac{\sin 2\Psi_\ell}{m' - \cos^2\Psi_\ell}$$

and letting $\epsilon = \lambda_\ell - \lambda_{\ell+1} - \mathrm{Cta}_{1\to\ell}(c,c')$

$$\text{If } f_c(\mu_1 - \mu_L) < \epsilon \quad \text{then } \theta'' < \pi/4 \text{ and } \sin 2\theta'' \le \frac{f_c(\mu_1 - \mu_L)}{\epsilon}$$

Proof. The upper bound for θ' follows from Property 7.6 (p.273) applied to the pertubation of rank unity $-Som_C$ on the endomorphism Som.

The upper bound for angle θ'' follows from Theorem 7.4 (p.294) applied to the perturbation $-f_c Som_{J(c)}$ on $Som_{J(C)}$, since the coefficient $1/f_{c'}$ does not modify the eigenvectors. For this purpose a lower bound for $\widetilde{\mu}_\ell - \widetilde{\mu}_{\ell+1}$ is provided by $(\lambda_\ell - \mathrm{Cta}_{1\to\ell}(c,c')) - \lambda_{\ell+1}$.

Lastly, an upper bound of the angle θ is obtained by applying Lemma 7.2 (p.295). ◁

Application

Political space study. We study hereafter the influence of the subgroup of the 193 individuals of the NF electorate. As already seen, the principal axes of the within–cloud are very close to those of the initial cloud, so in the following study we neglect this perturbation and study the second one.

Let us recall that $\mathrm{Cta}_{1\to\ell}(c,c') = \mathrm{Var}\,\mathrm{M}^C \cos^2\Psi_\ell$, hence $\mathrm{Cta}_1(c,c') = .017185$, $\mathrm{Cta}_{1\to2}(c,c') = .018023$ and $\mathrm{Cta}_{1\to3}(c,c') = .018192$. The greatest eigenvalue μ_1 of the specific analysis of the subcloud of the NF electorate is equal to .297185, hence $f_c\mu_1 = .019247$ ($\mu_L \simeq 0$).

Applying Property 7.7 (p.278), one obtains the following intervals of variation for the eigenvalues of $Som_{J(c')}$.

Axis ℓ	λ_ℓ	Eigenvalues	λ'_ℓ
Axis 1	.21325	$.18906 \leq \cdot \leq .22802$	*.19668*
Axis 2	.16063	$.13190 \leq \cdot \leq .17175$	*.16427*
Axis 3	.11921	$.08743 \leq \cdot \leq .12747$	*.11958*

Table 7.2. Eigenvalues (λ_ℓ) of the initial cloud; interval of variation of the eigenvalues (λ'_ℓ) of the within–cloud and its eigenvalues (in slanted characters).

Let us now compare the first principal axes (cf. Property 7.8, p.278). One has $\epsilon = .035435$, hence $\sin 2\theta''_1 \leq .019247/.035435 = .5432$, that is, $\theta''_1 \leq 16°45$ ($\cos\theta''_1 \geq .959$). One finds $\cos\theta''_1 = .989$ ($\theta''_1 = 8°32$).

For plane 1-2 and the subspace 1→3, since the deviations $\lambda_2 - \lambda_3$ and $\lambda_3 - \lambda_4$ are inferior to $\lambda_1 - \lambda_2$, the inequalities cannot lead to smaller bounds for angles; one finds $\cos\theta''_{1\to2} \geq .886$ ($\theta''_{1\to2} \leq 27°67$), and $\cos\theta''_{1\to3} > .816$ ($\theta''_{1\to3} \leq 35°33$). If the analyses are performed, one finds $\cos\theta''_{1\to2} = .9974$ ($\theta''_{1\to2} = 4°14$) and $\cos\theta''_{1\to3} = .9975$ ($\theta''_{1\to3} = 4°09$).

The examination of the preceding results shows that the order of axes is not modified even though the variance of the first axis is much smaller. The rotation of the first axis remains weak.

7.2.2 Deleting a single point

Let us delete the point M^{j_0} of absolute frequency n_{j_0} and relative frequency $f_{j_0} = n_{j_0}/n$ and denote $J\langle c'\rangle = J - \{j_0\}$.

One has $Som = (1 - f_{j_0})Som_{J(c')} + 0 + Som_{c'} + Som_{j_0}$ (cf. equation (7.2), p.271). Now, one has $Som_{c'} + Som_{j_0} = Som_{j_0}/(1 - f_{j_0})$, hence:

$$Som_{J(c')} = \frac{1}{1 - f_{j_0}}\left(Som - \frac{1}{1 - f_{j_0}} Som_{j_0}\right)$$

$\overrightarrow{\mathrm{GM}}^{j_0}$ is an eigenvector of Som_{j_0}; the associated eigenvalue is equal to the absolute contribution of M^{j_0}, denoted Cta_{j_0}. Let Ψ_ℓ be the angle between the vector $\overrightarrow{\mathrm{GM}}^{j_0}$ and the subspace of the first ℓ principal axes, and $\mathrm{Cta}_{j_0(1\to\ell)} = \mathrm{Cta}_{j_0}\cos^2\Psi_\ell$ be the absolute contribution of M^{j_0} to this subspace, between the eigenvalue λ_ℓ of Som and the eigenvalue μ_ℓ of $Som_{J(c')}$, one has (Theorems 7.2 and 7.5, p.293-294):

$$\frac{1}{1 - f_{j_0}}\left(\lambda_\ell - \frac{1}{1 - f_{j_0}}\,\mathrm{Cta}_{j_0(1\to\ell)}\right) \leq \mu_\ell \leq \frac{1}{1 - f_{j_0}}\lambda_\ell$$

The greatest canonical angle between the subspaces $1\to\ell$ is such that:

If $m' = \dfrac{\lambda_\ell - \lambda_{\ell+1}}{\frac{1}{1-f_{j_0}}\,\mathrm{Cta}_{j_0}} > 1$, then: $\theta < \pi/4$ and $\tan 2\theta \leq \dfrac{\sin 2\Psi_\ell}{m' - \cos 2\Psi_\ell}$

If $\cos^2\Psi_\ell < m' < 1$, then: $\theta < \pi/4$ and $\tan 2\theta \leq \dfrac{\sin 2\Psi_\ell}{m' - \cos^2\Psi_\ell}$

Remarks (1) The problem of adding points is analogous to the one of deleting points.

(2) The above formulas, in terms of eigenvalues of initial cloud, contributions and qualities of representation of deleted points, are general, they apply whether the clouds have been constructed by PCA or MCA.

(3) Modifying the weights of some points of a cloud may be viewed as a particular case of deleting or adding points. Indeed, a point M^j of weight n_j is equivalent to n_j elementary points at the same location, and modifying its weight by putting it to n'_j amounts to deleting $n_j - n'_j$ elementary points if $n_j > n'_j$, or adding $n'_j - n_j$ points if $n'_j > n_j$.

Comment

The influence of a point depends both on its contribution to the subspace under study and on its quality of representation for this subspace. If the point lies in the subspace orthogonal to the subspace under study, the stability is complete; if the point lies in the subspace itself, there is stability if its contribution is not too large.

7.3 Effect of a Change of Metric

The stability of principal subspaces under a change of metric appears in biweighted PCA with the choice of weights of variables and/or individuals. It appears in CA when the influence of modalities is studied by comparing a table to a subtable (cf. Escofier & Le Roux, 1976); when such a study is done in the line of specific GDA, the metric remains that of the reference cloud (cf. below §7.4, p.283).

The problem of the change of metric can be formulated in the following terms. If the space $\mathcal{U}$ is equipped with two Euclidean metrics: one defined by the scalar product denoted $\langle\cdot|\cdot\rangle$ (with the associated norm $\|\cdot\|$), the other by the scalar product denoted $[\cdot|\cdot]$ (with the associated norm $[[\cdot|\cdot]]$).

It is known (e.g. Deheuvels, 1981, p.191-194) that there exists an endomorphism $\langle\cdot|\cdot\rangle$–symmetric and positive $Ech : \mathcal{V} \to \mathcal{V}$, such that:

$$\forall \overrightarrow{\alpha} \in \mathcal{V} \quad \forall \overrightarrow{\alpha}' \in \mathcal{V} : \langle Ech(\overrightarrow{\alpha})|\overrightarrow{\alpha}'\rangle = \langle \overrightarrow{\alpha}|Ech(\overrightarrow{\alpha}')\rangle = [\overrightarrow{\alpha}\,|\,\overrightarrow{\alpha}'] \tag{7.3}$$

Let $(\overrightarrow{\delta_k})_{k\in K}$ be a $\langle\cdot|\cdot\rangle$–orthogonal basis of $\mathcal{V}$ constituted of eigenvectors of Ech, i.e. such that $\forall k \in K : Ech(\overrightarrow{\delta_k}) = e^k\overrightarrow{\delta_k}$ $(e^k > 0)$.

In the basis $(\overrightarrow{\delta_k})_{k\in K}$, the change of metric is a mere change of scale. If $\|\overrightarrow{\delta_k}\| = \sqrt{\varpi_k}$ and $[[\overrightarrow{\delta_k}]] = \sqrt{\varpi'_k}$, then the eigenvalue e^k of Ech is equal to ϖ'_k/ϖ_k. In practice, this is the form under which the change of metric is made explicit.

For the Euclidean structure $\langle\cdot|\cdot\rangle$, the eigenvalue λ_ℓ and the principal axis $\overrightarrow{\alpha}_\ell$ of the cloud (M^J, f_J) verify the equation $Som(\overrightarrow{\alpha_\ell}) = \lambda_\ell\overrightarrow{\alpha_\ell}$ (cf. Theorem 3.3, p.88), where Som is the $\langle\cdot|\cdot\rangle$–symmetric endomorphism defined by $Som(\overrightarrow{\alpha}) = \sum\limits_{j\in J} f_j\langle\overrightarrow{\mathrm{GM}^j}|\overrightarrow{\alpha}\rangle\overrightarrow{\mathrm{GM}^j}$.

For the Euclidean structure $[\cdot|\cdot]$, the eigenvalue λ'_ℓ and the eigenvector $\overrightarrow{\alpha}'_\ell$ verify the equation $Som'(\overrightarrow{\alpha'_\ell}) = \lambda'_\ell\overrightarrow{\alpha}'_\ell$, where Som' is the $[\cdot|\cdot]$–symmetric endomorphism defined by $Som'(\overrightarrow{\alpha}) = \sum\limits_{j\in J} f_j[\overrightarrow{\mathrm{GM}^j}|\overrightarrow{\alpha}]\overrightarrow{\mathrm{GM}^j}$.

From Equation 7.3, one has: $[\overrightarrow{\mathrm{GM}^j}|\overrightarrow{\alpha}] = \langle\overrightarrow{\mathrm{GM}^j}|Ech(\overrightarrow{\alpha})\rangle$; hence one has $Som' = Som \circ Ech$.

Property 7.9 (Eigenvalues). *Let* $e_{max} = \max\limits_{k\in K} e^k$ *and* $e_{min} = \min\limits_{k\in K} e^k$. *One has:* $e_{min}\lambda_\ell \leq \lambda'_\ell \leq e_{max}\lambda_\ell$.

This property follows from Theorem 7.3 (p.293).

Property 7.10 (Principal Variables). *If* $\epsilon = \min\{\frac{\lambda_{\ell-1}}{\lambda_\ell} - 1), (1 - \frac{\lambda_{\ell+1}}{\lambda_\ell)}\}$, *the correlation* r_ℓ *between the* ℓ*–th principal variables, is such that:*

If $(e_{max}/e_{min} - 1)/\epsilon < 1$ *then* $\theta' < \pi/4$ *and* $|r_\ell| \geq \sqrt{\frac{e_{min}}{e_{max}}} \cos\theta'$ *with* θ' *such that* $\sin 2\theta' \leq (e_{max}/e_{min} - 1)/\epsilon$.

Proof. The ℓ–th principal variable y^J_ℓ of the cloud M^J with $\langle\cdot|\cdot\rangle$ metric is defined by $\left(\langle\overrightarrow{\mathrm{GM}}^j|\overrightarrow{\alpha_\ell}\rangle/\|\overrightarrow{\alpha_\ell}\|\right)_{j\in J} = Vac(\overrightarrow{\alpha_\ell}/\|\overrightarrow{\alpha_\ell}\|)$ and its variance is equal to λ_ℓ (cf. CLOUD, passage formulas (3.7), p.92). Similarly, the ℓ–th principal variable y'^J_ℓ of the cloud M^J with $[\cdot|\cdot]$ metric is defined by $\left([\overrightarrow{\mathrm{GM}}^j|\overrightarrow{\alpha}'_\ell]/[[\overrightarrow{\alpha}'_\ell]]\right)_{j\in J} = Vac'(\overrightarrow{\alpha}'_\ell/[[\overrightarrow{\alpha}'_\ell]]) = Vac\circ Ech(\overrightarrow{\alpha}'_\ell/[[\overrightarrow{\alpha}'_\ell]])$ and its variance is equal to λ'_ℓ. The correlation r_ℓ between the principal variables y^J_ℓ and y'^J_ℓ is such that:

$$r_\ell = \frac{1}{\sqrt{\lambda_\ell \lambda'_\ell}} \times \frac{\langle Vac(\overrightarrow{\alpha_\ell}) \mid Vac\circ Ech(\overrightarrow{\alpha}'_\ell)\rangle}{\|\overrightarrow{\alpha_\ell}\| \times [[\overrightarrow{\alpha}'_\ell]]} = \frac{1}{\sqrt{\lambda_\ell \lambda'_\ell}} \times \frac{\langle Som(\overrightarrow{\alpha_\ell}) \,|Ech(\overrightarrow{\alpha}'_\ell)\rangle}{\|\overrightarrow{\alpha_\ell}\| \times [[\overrightarrow{\alpha}'_\ell]]}$$

If $\theta^\star$ denotes the angle (for the $\langle\cdot|\cdot\rangle$ metric) between the principal vectors $\overrightarrow{\alpha}_\ell$ and $\overrightarrow{\alpha}'_\ell$, one has $|r_\ell| = \sqrt{\frac{\lambda_\ell}{\lambda'_\ell}} \times \frac{[[\overrightarrow{\alpha}'_\ell]]}{\|Ech(\overrightarrow{\alpha}'_\ell)\|}|\cos\theta^*| \geq \sqrt{\frac{e_{min}}{e_{max}}}|\cos\theta^\star|$.

From Theorem 7.8 (p.295) (applied to the transposes of the endomorphisms Som and Som'), the angle $\theta' = \min\{\theta^\star, \pi - \theta^\star\}$ is such that if $(e_{max}/e_{min}) - 1 < \epsilon$ then $\theta' < \pi/4$ and $\sin 2\theta' \leq (e_{max}/e_{min} - 1)/\epsilon$. Therefore, one deduces: $|r_\ell| \geq \sqrt{\frac{e_{min}}{e_{max}}} \cos\theta'$. ◁

Particular cases

• *Biweighted* PCA. Suppose that the reference cloud is equipped with the metric defined by ϖ_K, and that the modified metric is defined by ϖ'_K; for a score table (protocol of scores, *PRINCIPAL*, §4.1, p.132), one has $e^k = \varpi'_k/\varpi_k$, and for a table of measures (*PRINCIPAL*, §4.5, p.155), one has $e^k = \varpi_k/\varpi'_k$. These properties also show that if the variances of the initial variables are near one another, then the principal subspaces will also be near one another and the eigenvalues will be almost proportional.

• *Standard* PCA. If, as done in the preceding §, one deletes a group of individuals $J{<}c{>}$, the data can be calibrated by $\varpi'_k = \mathrm{Var}\, x^{J<c'>k}$ instead of $\mathrm{Var}\, x^{Jk}$. The study done with this new calibration (contrary to specific analysis) induces a transformation of data that may be viewed as a change of metric, with $e^k = \mathrm{Var}\, x^{Jk} / \mathrm{Var}\, x^{J<c>k}$.

• *Correspondence Analysis.* Let n_{JK} be the frequency–measure on $J \times K$ with $n_k = \sum_{j \in J} n_{jk}$ and $n = \sum_{k \in K} n_k$. If one deletes a subset $J{<}c'{>}$ of modalities and compares the CA of the initial table to the CA of the modified table (cf. Escofier & Le Roux, 1976), there is a change of metric.

Letting $n_{c'k} = \sum_{j \in J<c'>} n_{jk}$ and $n_{c'} = \sum_{k \in K} n_{c'k}$, the change of metric in $\mathbb{R}_K$ is defined by

$$e^k = \frac{n_k/n}{(n_k - n_{c'k})/(n - n_{c'})} = \frac{n - n_{c'}}{n} \times \frac{n_k}{n_k - n_{c'k}}$$

Comment

If e_{max}/e_{min} is near 1, the perturbation due to the change of metric is small. More precisely, one can say that there will exist some subspaces of principal variables that are stable under the change of metric if there exists, in the ranking of eigenvalues, a cutoff corresponding to a ratio markedly greater than e_{max}/e_{min}.

7.4 Influence of a Variable and/or a Modality

In PCA, one may discard one or several variables from the analysis; in CA or MCA, one may discard one or several modalities while preserving the same Euclidean structure. If after discarding, one is interested in the stability of an axis (or a principal subspace), one of the following cases may occur.

— The axis of interest remains about the same, i.e. one has stability of axis. The discarded elements do not have a great influence on the determination of axis; which does not mean that they can be neglected in the interpretation of the axis, especially if their quality of representation is good. The axis expresses an *overall trend of data*, such as a general axis as opposed to a specific one.

— The new axis is different but the plane spanned by this axis and the one of next rank is stable (e.g. if the two corresponding eigenvalues are close to each other). If there is only a rotation in the plane, one has *stability of the plane*, to be considered globally.

— The axis under study is found again among interpretable axes but at a more remote rank. In this case the interpretation should take into account the *hierarchy of axes*.

— The axis vanishes without modification of the other interpretable axes: one has a *specific axis* of the discarded variables or modalities.

— There is a perturbation of interpretable axes; this means that the discarded elements play an important role; the question of their *homogeneity* with the other elements should be raised.

All the foregoing cases refer to the stability of a Euclidean cloud of $\mathcal{U}$ by orthogonal projection onto a subspace $\mathcal{A} \subset \mathcal{U}$ ($\mathcal{A}$ strictly included in $\mathcal{U}$) going through the mean point G (cf. Specific analysis, CLOUD, §3.3.4, p.94).

We firstly recall some results and properties relating to the projection of a cloud (§7.4.1). We secondly deal with a perturbation cloud of any dimensionality (§7.4.2), then one–dimensional (§7.4.3); in each case we give formulas for PCA, CA and MCA. Lastly as an application, we study the influence of variables in the standard PCA of the *Parkinson study*.

7.4.1 Projection of a cloud

Let A^j be the orthogonal projection of the point M^j on $\mathcal{A}$, and B^j its orthogonal projection on $\mathcal{A}^\perp$ (supplementary orthogonal subspace of $\mathcal{A}$ going through G), one has: $\overrightarrow{\mathrm{GM}}^j = \overrightarrow{\mathrm{GA}}^j + \overrightarrow{\mathrm{GB}}^j$ with $\overrightarrow{\mathrm{GA}}^j \perp \overrightarrow{\mathrm{GB}}^j$.

Between the variances of clouds one has the following additive relation:

$$\mathrm{Var}\,\mathrm{M}^J = \mathrm{Var}\,\mathrm{A}^J + \mathrm{Var}\,\mathrm{B}^J \tag{7.4}$$

The endomorphism $Tom : \mathbb{R}^J \to \mathbb{R}^J$ associated with the cloud M^J is such that $x^J \mapsto \Big(\sum_{j' \in J} f_{j'} x^{j'} \langle \overrightarrow{\mathrm{GM}}^j | \overrightarrow{\mathrm{GM}}^{j'} \rangle\Big)_{j \in J}$ (cf. CLOUD, Definition 3.14, p.86); the principal variables of the cloud M^J are eigenvectors of *Tom*.

Let $\mathcal{X}_0$ be the subspace of $\mathbb{R}^J$ generated by the L nonnull principal variables of the cloud M^J; the endomorphism *Tom* restricted to $\mathcal{X}_0$ will also be denoted *Tom*. Similarly, with the cloud A^J there is associated the endomorphism $Tom_A : \mathcal{X}_0 \to \mathcal{X}_0$; and with the perturbation (or residual) cloud B^J there is associated the endomorphism $Tom_B : \mathcal{X}_0 \to \mathcal{X}_0$. Owing to the orthogonality of the supports of the two clouds A^J and B^J, one has:

$$Tom = Tom_A + Tom_B \tag{7.5}$$

To evaluate the influence of the perturbation, one will study the interval of variation of eigenvalues and the rotation of invariant subspaces of the endomorphism *Tom* (reference cloud) by subtracting endomorphism Tom_B.

7.4.2 Perturbation cloud of any dimensionality

Let L' $(1 \leq L' < L)$ denote the dimensionality of the perturbation cloud B^J; and let $(\beta_\ell)_{\ell=1\ldots L}$ denote the eigenvalues of Tom_B (ranked in decreasing order), with $\beta_\ell = 0$ for $\ell > L'$. The variance of the cloud B^J is equal to $\sum_{\ell=1}^{L'} \beta_\ell$; let us denote $\tau_B = \mathrm{Var}\,\mathrm{B}^J / \mathrm{Var}\,\mathrm{M}^J$ the relative contribution of the cloud B^J to the cloud M^J, and $v_{1\to\ell} = \sum_{\ell'=1}^{\ell} \beta_{\ell'} / \operatorname{Var} \mathrm{B}^J$ the inertia rate of the cloud B^J pertaining to its first ℓ axes.

Property 7.11 (Eigenvalues). *Between the eigenvalue λ_ℓ of the cloud M^J and the eigenvalue λ'_ℓ of the cloud A^J, one has the inequalities:*

$$\lambda_\ell - \beta_1 \leq \lambda'_\ell \leq \lambda_\ell \qquad (1 \leq \ell \leq L)$$

This property follows from Weyl's inequalities (cf. Theorem 7.1, p.292).

Property 7.12 (Inertia rates). *Between the inertia rates $\tau_{1\to\ell}$ of the cloud M^J and $\tau'_{1\to\ell}$ of the cloud A^J pertaining to the first ℓ principal axes, one has the following inequalities:*

$$\frac{\tau_{1\to\ell} - \tau_B\, v_{1\to\ell}}{1-\tau_B} \leq \tau'_{1\to\ell} \leq \frac{\tau_{1\to\ell} - \tau_B\, v_{L-\ell'+1\to L}}{1-\tau_B} \left(\leq \frac{\tau_{1\to\ell}}{1-\tau_B} \right)$$

One obtains these inequalities by dividing the relation *(2c)* of Theorem 7.1 (p.292) by $\operatorname{Var} \mathrm{A}^J = \operatorname{Var} \mathrm{M}^J (1-\tau_B)$.

Property 7.13 (Principal subspace $1\to\ell$). *The greatest canonical angle θ between the subspaces generated by the first ℓ principal variables of the two clouds M^J and A^J is such that:*

$$\textit{If } \epsilon = \lambda_\ell - \lambda_{\ell+1} > \beta_1 \textit{ then } \theta < \pi/4 \textit{ and } \sin 2\theta \leq \beta_1/\epsilon$$

This property follows from Theorem 7.4 (p.294).

Particular cases

Biweighted PCA. When in the PCA of a table of scores or of measures one discards a subset K_s of variables, the variance of the cloud M^B is equal to $\sum_{k\in K_s} \mathrm{Cta}_k$, where Cta_k is the absolute contribution of variable k to the variance of the cloud (cf. p.134 and p.155), and β_1 is the greatest eigenvalue of the PCA of the subtable of K_s discarded variables; as a first approximation, one can always majorize β_1 by $\sum_{k\in K_s} \mathrm{Cta}_k$.

Specific MCA. When in specific MCA, one discards a set K_s of modalities of active questions, one has $\operatorname{Var} \mathrm{B}^J = \sum_{k\in K_s} \mathrm{Cta}_k = (K_s - \sum_{k\in K_s} f_k)/Q$, where Q denotes the number of active questions (cf. *MULTIPLE*, property 5.25, p.206). If the K_s discarded modalities are all those of a same question (i.e. the question itself is discarded), one has $\beta_1 = 1/Q$, and the bounds only involve the number Q of questions: This result is uninteresting, it will be improved upon for a question with two modalities.

Correspondence Analysis. When modalities are discarded in CA, the preceding results apply in the framework of specific analysis, that is, assuming the metric is unchanged; in fact, the metric is usually modified (since the margins change), hence a further perturbation (for a detailed study, see Escofier & Le Roux, 1976).

Comments

The eigenvalues of the specific cloud are less than or equal to those of the reference cloud, but one cannot say anything about the increasing or decreasing of inertia rates.

The rotation of the principal subspaces $1 \to \ell$ is all the weaker as the ratio of the largest eigenvalue to the deviation between the eigenvalues ℓ and $\ell+1$ of the reference cloud is smaller. More precisely, one can state that there will exist subspaces of principal variables that are stable by deletion of variables or modalities only if there exists, in the ordering of eigenvalues, a deviation that is markedly superior to the variance of the discarded variables or modalities.

The above bounds, calculated only as functions of the eigenvalues of the residual cloud, are optimal. Still, they can be improved upon in the case of a one–dimensional residual cloud, when the correlation of its unique principal variable with the principal variables of the reference cloud is known.

7.4.3 One–dimensional Perturbation Cloud

The one–dimensional cloud B^J has a single nonnull principal variable, denoted b^J, which is eigenvector of the endomorphism Tom_B associated with the eigenvalue $\beta = \operatorname{Var} \mathrm{B}^J$; one then has $\tau_B = \beta / \operatorname{Var} \mathrm{M}^J$. Let r_ℓ denote the correlation coefficient of b^J with the ℓ–th principal variable of the cloud M^J. Let $r_{1\to\ell}$ be the multiple correlation of b^J with the first ℓ principal variables of cloud M^J, that is, $r^2_{1\to\ell} = \sum_{\ell'=1}^{\ell} r^2_{\ell'}$, and $r_{1\to 1} = |r_1|$.

Property 7.14 (Eigenvalues). *Between the eigenvalue λ_ℓ of the cloud M^J and the eigenvalue λ'_ℓ of the cloud A^J, one has the inequalities:*

$$\max\{\lambda_{\ell+1}\,;\,\lambda_\ell - \beta\, r^2_{1\to\ell}\} \le \lambda'_\ell \le \lambda_\ell$$

This property follows from the first inequality *(2)* of Theorem 7.2 (p.293).

Property 7.15 (Inertia rates). *Between the inertia rates $\tau_{1\to\ell}$ of cloud M^J and $\tau'_{1\to\ell}$ of cloud A^J relating to the subspaces of the first ℓ axes, one* *has the inequalities:*

$$\frac{\tau_{1\to\ell} - \tau_B r^2_{1\to\ell}}{1-\tau_B} \leq \tau'_{1\to\ell} \leq \min\{\frac{\tau_{1\to\ell}}{1-\tau_B}; 1\}$$

Property 7.16 (Principal subspaces $1\to\ell$). *The greatest canonical angle θ between the subspaces of the first ℓ principal variables of cloud M^J and of cloud A^J is such that, letting $\epsilon = \lambda_\ell - \lambda_{\ell+1}$:*

If $\epsilon > \beta$, *then* $\theta < \pi/4$ *and* $\tan 2\theta \leq \beta \dfrac{2r_{1\to\ell}(1-r^2_{1\to\ell})^{1/2}}{\epsilon + \beta - 2\beta r^2_{1\to\ell}}$

If $\beta r^2_{1\to\ell} \leq \epsilon < \beta$, *then* $\theta < \pi/4$ *and* $\tan 2\theta \leq \beta \dfrac{2r_{1\to\ell}(1-r^2_{1\to\ell})^{1/2}}{\epsilon - \beta r^2_{1\to\ell}}$

This property follows from Theorem 7.5 (p.294).

Similarly, one studies the rotation of the ℓ–th principal variable, by applying Theorem 7.6.

When the subspace of the ℓ–th and $(\ell+1)$–th principal variables is stable, one can also study the rotation of these two principal variables in this subspace, using Lemma 7.1 (p.294). The formulas for standard PCA are given hereafter and applied later on (§7.4.3).

Particular cases

Biweighted PCA. The foregoing results apply to the study of the influence of a variable x^{Jk} of the biweighted PCA of a table of scores x^{JK} (*PRINCIPAL*, §4.1). In this case, one has $\beta = \varpi_k \operatorname{Var} x^{Jk}$, and r_ℓ is the correlation of the variable x^{Jk} with the ℓ–th principal variable. For the biweighted PCA of a table of measures x^J_K (*PRINCIPAL*, §4.5), one has $\beta = \operatorname{Var} x^J_k / \varpi_k$.

Standard PCA. For the standard PCA of the table of variables $(x^{Jk})_{k\in K}$, one has $\beta = 1$ and the formulas simplify.

- *Eigenvalues.* $\max\{\lambda_{\ell+1}; \lambda_\ell - r^2_{1\to\ell}\} \leq \lambda'_\ell \leq \lambda_\ell$
- *Variance rules.* $\frac{K}{K-1}(\tau_{1\to\ell} - r^2_{1\to\ell}/K) \leq \tau'_{1\to\ell} \leq \min\{\frac{K}{K-1}\tau_{1\to\ell}; 1\}$
- *First principal subspace $1\to\ell$.*

  If $\epsilon > 1$, then $\theta < \pi/4$ and $\tan 2\theta \leq \dfrac{2r_{1\to\ell}(1-r^2_{1\to\ell})^{1/2}}{\epsilon + 1 - 2r^2_{1\to\ell}}$

  If $r^2_{1\to\ell} \leq \epsilon \leq 1$, then $\theta < \pi/4$ and $\tan 2\theta \leq \dfrac{2r_{1\to\ell}(1-r^2_{1\to\ell})^{1/2}}{\epsilon - r^2_{1\to\ell}}$

- *Principal variable ℓ.* If $\epsilon \leq \min\{\lambda_{\ell-1}, \lambda_\ell - r^2_{1\to\ell-1}\,;\, \lambda_\ell, \lambda_{\ell+1} - r^2_{1\to\ell}\}$, one has: $\sin\theta \leq |r_\ell|(1-r_\ell^2)^{1/2}/\epsilon$.

  This property follows from Theorem 7.6 (p.295).

- *Principal subspace $\ell\to\ell+1$.* If the subspace of the ℓ–th and $(\ell+1)$–th variables is stable, the rotation in the plane due to the deletion of an initial variable is such that:

$$\tan 2\theta = \frac{2|r'_\ell|(1-r'^2_\ell)^{1/2}}{\lambda_\ell - \lambda_{\ell+1} - 2r'^2_\ell + 1} \quad \text{with } r'_\ell = r_\ell/\sqrt{r_\ell^2 + r_{\ell+1}^2}$$

Specific MCA. To study the *influence of a modality* on the results of a specific MCA[4], one replaces in preceding formulas β by $\mathrm{Cta}_k = (1-f_k)/Q$, and r_ℓ^2 by the quality of representation of this modality on axis ℓ: $\cos^2\psi_\ell = f_k(y_\ell^k)^2/(1-f_k)$, where y_ℓ^k denotes the coordinate of modality k on axis ℓ; and one lets $\cos^2\Psi_\ell = \sum_{\ell'=1}^{\ell} \cos^2\psi_{\ell'}$.

For a *question with two modalities*, the perturbation cloud is one–dimensional ("lever principle"), and the endomorphism Tom_B is of rank one, with eigenvalue $\beta = 1/Q$, hence:

$$\lambda_\ell - \cos^2\Psi_\ell/Q \leq \lambda'_\ell \leq \lambda_\ell$$

$$\text{If } Q\epsilon > 1 \text{ then } \theta < \pi/4 \text{ and } \tan 2\theta \leq \frac{\sin 2\Psi_\ell}{Q\epsilon - \cos 2\Psi_\ell}$$

Comments

In MCA, the influence of a question with two modalities only depends on the quality of representation ($\cos^2\Psi_\ell$) of its modalities on the subspace under study, while the influence of a modality depends on the remoteness of this modality from the center of the cloud and on its weight; these facts provide response elements to the problems raised by infrequent modalities. If the modality lies at the periphery of the graph and if its quality of representation is very good (Ψ_ℓ close to 0), discarding the modality changes little the results of the analysis; the same property holds if the modality is poorly represented (Ψ_ℓ near $\pi/2$); if the quality of representation is intermediate, to assess its influence, it may be helpful to study the upper bounds of θ.

---

[4]For a detailed study, see Le Roux (1999).

Another interesting case in MCA is that of a specific axis, characterized by a set of well–represented modalities (quasi–one–dimensional cloud), which vanishes (among interpretable axes) after deleting these modalities. The study presented here allows to identify such axes, that is, if the elements are very well represented (qualities of representation near 1, or even one–dimensional cloud), one will apply the upper bounds of $\tan 2\theta$ to identify such an axis. Using Theorem 7.7 (p.295) will indicate whether this axis has been shifted to a more remote rank. This case is encountered in opinion questionnaires when among interpretable axes there is a non–response axis; as an example see Bonnet, Le Roux & Lemaine (1996).

Application to Parkinson Data

We now apply the foregoing results to the *Parkinson study* (cf. CASE STUDIES, §9.1, p.336, in particular to the results of §9.1.2) to study the ***influence of each initial variable*** on the first principal variable and on the subspace of the first two principal variables.

The deviation between the first two eigenvalues (cf. Table 9.1.6-a, p.343) is equal to $\epsilon_1 = 3.99269 - 1.82238 = 2.17031$; the deviation between the second and the third ones is $\epsilon_2 = 1.82238 - 0.17105 = 1.65133$. Since both deviations are greater than 1, we can assert that the rotation angle is less than $\pi/4$, which already expresses some conclusion of stability.

For each initial variable, one has $\beta_1 - \beta_L = 1$, therefore (Property 7.13, p.285), for the first principal variable, $\sin 2\theta_1 \leq 1/\epsilon_1 = 0.4608$, that is, $\cos\theta_1 \geq 0.971$ ($\theta_1 \leq 13°72$); and for the subspace of the first two principal variables, $\sin 2\theta_{1\to2} \leq 0.60557$, that is, $\cos\theta_{1\to2} \geq 0.9476$ ($\theta_{1\to2} \leq 18°64$)[5]. One is thus sure of a good stability of principal axes when some initial variable is deleted.

In order to study more precisely the ***influence*** of each initial variable, we take into account their correlation coefficients with the first two principal variables (cf. Table 9.1.6-b, p.343), the results are summarized in Table 7.3 (p.290). Figure 7.1 (p.290) gives the bound of the correlation ($\cos\theta_1$) between the first two variables (solid line) expressed as a function of the absolute value of correlation coefficient r_1 and the bound of the smallest canonical correlation ($\cos\theta_{1\to2}$) between the first principal planes expressed as a function of the multiple correlation $r_{1\to2}$ (dotted line).

Comments. There is a great stability of the first principal variable and mostly of the first principal subspace; the multiple correlations of the initial

[5]These values correspond to the minima of the curves of Figure 7.1, p.290.

	First princ. var. ($\tau_1 = .665$)			Plane 1-2 ($\tau_{1\to2} = .969$)		
	r_1	τ'_1	$\cos\theta_1$	$r_{1\to2}$	$\tau'_{1\to2}$	$\cos\theta_{1\to2}$
Velocity	.902	[.636; .800] *.653*	.973 *.976*	.983	[.970; 1] *.973*	.973 *.996*
Length	.665	[.710; .800] *.730*	.979 *.980*	.984	[.969; 1] *.974*	.974 *.990*
Swing	−.403	[.776; .800] *.775*	.992 *.992*	.976	[.972; 1] *.981*	.967 *.976*
Cycle	−.868	[.625; .800] *.669*	.972 *.972*	.999	[.963; 1] *.964*	.998 *1.*
Stance	−.958	[.615; .800] *.625*	.981 *.984*	.991	[.966; 1] *.968*	.984 *.999*
Db_Support	−.950	[.618; .800] *.627*	.979 *.987*	.973	[.974; 1] *.977*	.965 *.996*

First principal variable: correlation with each initial variable (r_1), interval of variation of the rates of inertia (τ'_1), lower bound of the correlation ($\cos\theta_1$) with the true value (in slanted characters).
Principal subspace 1-2: multiple correlation of each initial variable ($r_{1\to2}$), interval of variation of inertia rates ($\tau'_{1\to2}$), and lower bound of the minimal canonical correlation ($\cos\theta_{1\to2}$) with the true value (in slanted characters).

Table 7.3. ***Parkinson data***: Influence of initial variables.

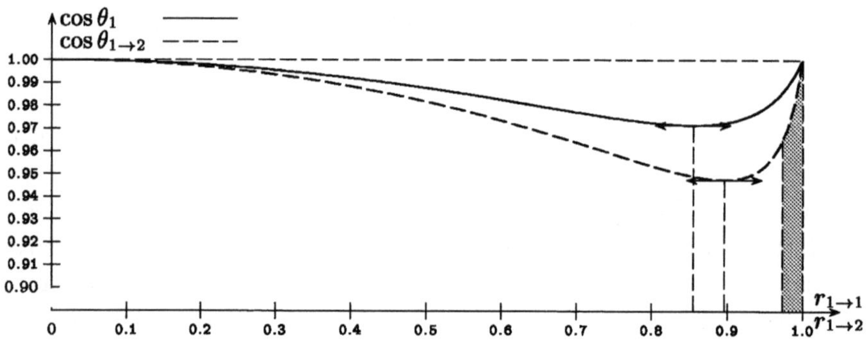

Figure 7.1. ***Parkinson data.*** Curves giving $\cos\theta$ as a function of $r_{1\to1} = |r_{\ell 1}|$ for the first principal variable (solid line) and as a function of $r_{1\to2}$ for the first principal plane (dotted line).

variables with the first two principal variables are all greater than .973; see gray zone in Figure 7.1.

For the *first principal variable*, deleting one variable among the four variables that are most correlated (in absolute value) with the first principal one is liable to entail a decrease of variance rates, whereas deleting the *Length* or the *Swing* variables may only increase the variance rate. The variables *Cycle* and *Velocity* entail the strongest (though moderate) rotation; they are the most influential on the first principal variable.

As for the *principal subspace 1-2*, deleting *Cycle* or *Stance* variables may entail a decrease of inertia rates, while leaving plane 1-2 stable. Yet, *Swing* remains the most influential variable; its deleting entails the weakest canonical correlation (.967), then comes variable *Length* (.974).

To complement the stability studies, the six corresponding PCA have been performed; their results (in slanted characters in Table 7.3, p.290) are in good agreement with the results of the stability studies.

Further complements. To show how any principal variable can be studied, we hereafter study the second principal variable. From the properties of p.288, we deduce the results of Table 7.4. One notices the influence of *Length* on axis 2.

	ϵ	$\sin\theta$	$\cos\theta$
Velocity	0.686	.523	.852 (*.975*)
Length	0.683	.731	.682 (*.972*)
Swing	0.698	.583	.813 (*.969*)
Cycle	0.653	.658	.753 (*.972*)
Stance	0.669	.370	.929 (*.984*)
Db_Support	0.704	.292	.957 (*.989*)

Table 7.4. *Parkinson data*: stability of the second principal variable

Then, assuming the stability of the subspace of the first two principal variables — in other words, studying the data reconstitution of order 2 — one gets the results of Table 7.5 (cf. formula p.288). One notices that in the subspace of the first two principal variables, *Cycle*, then *Velocity*, have the greatest influences.

	$\lambda'1$	$\frac{\lambda_1-\lambda_2}{r^2_{1\to 2}}$	$r'_1 = \frac{r_1}{r_{1\to 2}}$	$\cos\theta$
Velocity	3.256	2.247	+.918	.976
Length	3.649	2.241	+.676	.980
Swing	3.876	2.277	-.413	.992
Cycle	3.343	2.174	-.869	.972
Stance	3.120	2.209	-.966	.984
Db_Support	3.119	0.704	-.977	.989

Table 7.5. *Parkinson data*: rotation in the subspace of the first two principal variables.

7.5 Rotation of Invariant Subspaces: Basic Theorems

The study of the *eigenvalues* of the sum and product of symmetric endomorphisms has given rise to many publications such as Wielandt (1955), Anderson & Dasgupta (1963), Markus (1964), Wilkinson (1965), Thomson & Freede (1971). The classical results are recalled hereafter (Theorems 7.1 and 7.3).

For the addition of a perturbation B of rank unity, we will give more precise bounds expressed as functions of the position of the invariant subspace of B associated with the nonnull eigenvalue with respect to the invariant subspaces of C or A (Theorem 7.2, p.293).

On the other hand, there are few publications dealing with the study of the *rotation of invariant subspaces*, with the exception of the fundamental paper by Davis & Kahan (1970). These authors deal with the rotation of invariant subspaces of the sum of bounded operators ($C = A + B$) of a Hilbert space, and prove four theorems that give upper bounds for the rotation expressed as functions of the eigenvalues of A and/or C and of the norm of perturbation; the study being made for any norm of operators that is invariant by unitary transformations. We make use of two of these theorems, named by the authors the "$\sin\theta$ theorem" and "$\sin 2\theta$ theorem" (op. cit. p.10-11), to prove Theorems 7.4 (p.294) and 7.6 (p.295).

7.5.1 Eigenvalues

Let $\mathcal{V}$ be an n–dimensional Euclidean vector space. Let a_ℓ, b_ℓ and c_ℓ $(1 \leq \ell \leq n)$ be the eigenvalues (ranked in decreasing order, with their multiplicity order) of endomorphisms A, B and C. Let $(\alpha_\ell)_{\ell=1,\ldots n}$ (resp. $(\gamma_\ell)_{\ell=1,\ldots n}$) be an orthogonal basis of $\mathcal{V}$ constituted by eigenvectors of A (resp. C). If B is of *rank unity*, let b be its nonnull eigenvalue (assuming $b > 0$) associated with eigenvector β. Let us denote φ_ℓ (resp. ψ_ℓ) the angle between the eigenspace of B associated with b and the eigenspace of A (resp. C) generated by α_ℓ (resp. γ_ℓ); and Φ_ℓ (resp. Ψ_ℓ) the angle between the eigenspace of B associated with b and the invariant subspace of A (resp. C) generated by its ℓ first eigenvectors:

$$\cos^2 \Phi_\ell = \sum_{j=1}^{\ell} \cos^2 \varphi_j \qquad \cos^2 \Psi_\ell = \sum_{j=1}^{\ell} \cos^2 \psi_j$$

Theorem 7.1. *C=A+B, for any B* (e.g. Wilkinson (1965), p.100-101).
(1) Between the eigenvalues of A and C=A+B, one has the following relations:

(1a) For all integers j, k, ℓ, between 1 and n, such that $j + k \leq \ell + 1$,

$$c_\ell \leq a_j + b_k \quad \textit{and} \quad c_{n-\ell+1} \geq a_{n-j+1} + b_{n-k+1}$$

(1b) For $1 \leq \ell \leq n$: $a_\ell + b_n \leq c_\ell \leq a_\ell + b_1$ (Weyl's inequalities)

(1c) For $1 \leq \ell \leq n$: $\sum_{j=1}^{\ell} a_j + \sum_{j=n-\ell+1}^{n} b_j \leq \sum_{j=1}^{\ell} c_j \leq \sum_{j=1}^{\ell} a_j + \sum_{j=1}^{\ell} b_j$

(2) Between the eigenvalues of C and of $A = C - B$, one has the following relations:

(2a) For all integers j, k, ℓ, between 1 and n, such that $j + k \leq \ell + 1$,

$$a_\ell \leq c_j - b_{n-k+1} \quad \text{and} \quad a_{n-\ell+1} \geq c_{n-j+1} + b_k$$

(2b) For $1 \leq \ell \leq n$, $\quad c_\ell - b_1 \leq a_\ell \leq c_\ell - b_n$ (Weyl's inequalities)

(2c) for $1 \leq \ell \leq n \; : \; \sum_{j=1}^{\ell} c_j - \sum_{j=1}^{\ell} b_j \leq \sum_{j=1}^{\ell} a_j \leq \sum_{j=1}^{\ell} c_j - \sum_{j=n-\ell+1}^{n} b_j$

Theorem 7.2. *$C = A + B$ with B positive and of rank 1* (see Escofier, Le Roux, 1977).

(1) For $1 < \ell \leq n \; : \; a_\ell \leq c_\ell \leq \min\{a_\ell + b\sin^2\Phi_{\ell-1} \, ; \, a_{\ell-1}\}$

For $1 \leq \ell \leq n \; : \; \sum_{j=1}^{\ell} a_j + b\cos^2\Phi_\ell < \sum_{j=1}^{\ell} c_j \leq \sum_{j=1}^{\ell} a_j + b$

(2) For $1 \leq \ell < n \; : \; \max\{c_{\ell+1} \, ; \, c_\ell - b\cos^2\Psi_\ell\} \leq a_\ell \leq c_\ell$

For $1 \leq \ell \leq n \; : \; \sum_{j=1}^{\ell} c_j - b\cos^2\Psi_\ell \leq \sum_{j=1}^{\ell} a_j \leq \sum_{j=1}^{\ell} c_j$

Theorem 7.3. *$C = B \circ A$ (A positive and B positive–definite)* (see e.g. Anderson & Dasgupta, 1963).

Between the eigenvalues of A and $C = B \circ A$, one has the relations:

(1) For all integers j, k, ℓ between 1 and n such that $j + k \leq \ell + 1$:

$$c_\ell \leq a_j \, b_k \quad \text{and} \quad c_{n-\ell+1} \geq a_{n-j+1} \, b_{n-k+1}$$

(2) For all ℓ with $1 \leq \ell \leq n \; : \; a_\ell \, b_n \leq c_\ell \leq a_\ell \, b_1$

7.5.2 Invariant Subspaces

Let us recall that the *relative position of two subspaces* $\mathcal{A}$ and $\mathcal{C}$ of respective dimensionalities r and r' (with $r' \geq r$) is defined by r angles $\pi/2 \geq \theta_1 \geq \ldots \geq \theta_r \geq 0$ (Dixmier, 1948; Benzécri & al. 1973, p. 179) called *canonical angles*. In what follows, we will take as a measure of the deviation between two subspaces the *greatest canonical angle*, denoted θ, with, if $\theta(a|c)$ (with $0 \leq \theta(a|c) \leq \pi$) denotes the angle between the (nonnull) vectors a and c: $\theta = \max_{a \in \mathcal{A}, a \neq 0} \{\min_{c \in \mathcal{C}, c \neq 0} \theta(a|c)\}$.

Theorem 7.4 (p.294) deals with the case of the invariant subspaces associated with the eigenvalues of ranks $\ell_0, \ldots \ell_0 + \ell$ (we will say "intermediate" if $\ell_0 > 1$) for a symmetric perturbation of any rank.

Theorems 7.5 (p.294) and 7.6 (p.295) deal with the case of the adjunction of a symmetrical perturbation B of rank 1; the upper bounds of rotation are expressed not only as functions of the eigenvalues of A or of

C but also taking into account the position of the eigenspace of B with respect to the subspaces under study.

Theorem 7.7 (p.295) allows the comparison of the subspaces associated with eigenvalues of different ranks; a case often encountered in GDA when the rank of a principal axis is shifted without modifying the others.

Theorem 7.8 (p.295) studies the case of the multiplication by a positive–definite endomorphism.

Lastly, to study complex perturbations obtained by summing and/or composing two perturbations, one will use the Lemma 7.2 (p.295).

Theorem 7.4. *$C = A + B$ with any B* (Escofier & Le Roux, 1975).

The angle θ between the invariant subspaces of A and $C = A + B$ associated with the ℓ eigenvalues of ranks $\ell_0, \ldots \ell_0 + \ell - 1$ is such that:

- *for $\ell_0 = 1$ and $1 \leq \ell \leq n$, letting $\epsilon_1 = a_\ell - a_{\ell+1}$, $\epsilon_2 = c_\ell - c_{\ell+1}$;*
- *for $1 < \ell_0 \leq n-1$ and $1 \leq \ell \leq n - \ell_0$, letting*
  $\epsilon_1 = \inf\{(a_{\ell_0-1} - a_{\ell_0}), (a_{\ell_0+\ell-1} - a_{\ell_0+\ell})\}$,
  $\epsilon_2 = \inf\{(c_{\ell_0-1} - c_{\ell_0}), (c_{\ell_0+\ell-1} - c_{\ell_0+\ell})\}$

If $b_1 - b_n < \epsilon_1$ *then:* $\theta < \pi/4$ *and* $\sin 2\theta \leq (b_1 - b_n)/\epsilon_1$

If $b_1 - b_n < \epsilon_2$ *then:* $\theta < \pi/4$ *and* $\sin 2\theta \leq (b_1 - b_n)/\epsilon_2$

Remark. The bound is optimal: It is reached if B is of rank 1, as well as for endomorphisms of a two–dimensional Euclidean space.

Lemma 7.1. *If A, B and C are endomorphisms of a two–dimensional Euclidean space, with $C = A + B$ and B positive, one has:*

$$\tan 2\theta = \frac{(b_1 - b_2) sin 2\phi_1}{(a_1 - a_2) + (b_1 - b_2) cos 2\phi_1} = \frac{(b_1 - b_2) sin 2\psi_1}{(c_1 - c_2) - (b_1 - b_2) cos 2\psi_1}$$

Theorem 7.5. *$C = A + B$ with B positive and of rank 1* (Escofier & Le Roux, 1977). *The angle θ between the invariant subspaces of A and of $C = A + B$ associated with the first ℓ eigenvalues is such that:*

(1) If $m = \dfrac{a_\ell - a_{\ell+1}}{b} > 1$ *(or for $\ell = 1$ if $m + \cos 2\varphi_1 > 0$) :*

$$\theta < \pi/4 \quad \text{and} \quad \tan 2\theta \leq \frac{\sin 2\Phi_\ell}{m + \cos 2\Phi_\ell}$$

$$\textit{If } \sin^2 \Phi_\ell < m < 1 \;:\; \theta < \pi/4 \quad \text{and} \quad \tan 2\theta \leq \frac{\sin 2\Phi_\ell}{m - \sin^2 \Phi_\ell}$$

$$\textit{(2) If } m' = \frac{c_\ell - c_{\ell+1}}{b} > 1 \;:\; \theta < \pi/4 \quad \text{and} \quad \tan 2\theta \leq \frac{\sin 2\Psi_\ell}{m' - \cos 2\Psi_\ell}$$

$$\textit{If } \cos^2 \Psi_\ell < m' < 1 \;:\; \theta < \pi/4 \quad \text{and} \quad \tan 2\theta \leq \frac{\sin 2\Psi_\ell}{m' - \cos^2 \Psi_\ell}$$

Remark. For $m > 1$ and $m' > 1$, *the bounds are optimal*: They are reached if the eigenvector of B associated with b lies in the plane of the eigenvectors of ranks ℓ and $\ell+1$. For $m > 1$, the bound is maximum for $\cos 2\Phi_\ell = -1/m$, and the corresponding angle θ_{max} is such that $\tan 2\theta_{max} = 1/\sqrt{m^2-1}$, or else $\sin 2\theta_{max} = 1/m$ (upper bound of Theorem 7.4). Similarly, for $m' > 1$, the angle θ_{max} is such that $\sin 2\theta_{max} = 1/m'$ for $\cos 2\Psi_\ell = 1/m'$.

Theorem 7.6. *$C = A + B$ with B positive and of rank 1* (Escofier & Le Roux, 1977).
The angle θ between the invariant subspaces of A and of C associated with the eigenvalues of ranks $\ell_0, \ldots \ell_0 + \ell$ ($\ell_0 \geq 1$, $0 \leq \ell \leq n - \ell_0$) is such that:
If $\epsilon = \inf\{(a_{\ell_0-1} - a_{\ell_0}) - b\sin^2\Phi_{\ell_0-1}\ ;\ (a_{\ell_0+\ell} - a_{\ell_0+\ell+1}) - b\sin^2\Phi_{\ell_0+\ell}\} > 0$,

then: $\sin\theta \leq \dfrac{b}{2}\dfrac{\sin 2\Phi}{\epsilon}$ *with Φ such that* $\cos^2\Phi = \sum\limits_{j=\ell_0}^{\ell_0+\ell} \cos^2\phi_j$.

If $\epsilon' = \inf\{(c_{\ell_0-1} - c_{\ell_0}) - b\cos^2\Psi_{\ell_0-1}\ ;\ (c_{\ell_0+\ell} - c_{\ell_0+\ell+1}) - b\cos^2\Psi_{\ell_0+\ell}\} > 0$,

then: $\sin\theta \leq \dfrac{b}{2}\dfrac{\sin 2\Psi}{\epsilon'}$ *with Ψ such that* $\cos^2\Psi = \sum\limits_{j=\ell_0}^{\ell_0+\ell} \cos^2\psi_j$.

Remarks. This theorem allows to study the eigenspace associated with the eigenvalue ℓ_0. For $\ell_0 = 1$ (and $\ell = 0$), one will use Theorem 7.5 that gives a better upper bound.

Theorem 7.7. *Shifted subspaces* (Le Roux, 1999, p.83).
Let P_ℓ be the orthogonal projection on the eigenspace of C associated with γ_ℓ, and let $C' = C - (b\cos 2\psi_\ell)P_\ell$. The angle θ between the invariant subspaces of C' and of A associated with the eigenvalues of ranks $s, s+1, \ldots s+r$ is such that, letting $\epsilon = \inf\{(c'_{s-1} - c'_s), (c'_{s+r} - c'_{s+r+1})\}$:

$$\textit{If}\quad b\sin 2\psi_\ell < \epsilon \quad \textit{then}\quad \theta < \pi/4 \textit{ and } \sin 2\theta \leq \frac{b\sin 2\psi_\ell}{\epsilon}$$

Theorem 7.8. *$C = BA$, with A positive and B definite–positive.*
Letting $\epsilon = \inf\{(a_{\ell-1}/a_\ell) - 1\,;\, 1 - (a_{\ell+1}/a_\ell)\}$, the angle θ between the one–dimensional eigenspaces of A and of $C = BA$ associated with the (simple) ℓ–th eigenvalue is such that:

$$\textit{If}\ \left(\frac{b_1}{b_n} - 1\right) < \epsilon \quad \textit{then}\quad \theta < \pi/4 \textit{ and } \sin 2\theta \leq \left(\frac{b_1}{b_n} - 1\right)/\epsilon$$

Lemma 7.2. *Let $\mathcal{A}$, $\mathcal{B}$ and $\mathcal{C}$ be three subspaces of the same dimensionality. If ϕ_1 is the angle between $\mathcal{A}$ and $\mathcal{B}$, ϕ_2 the angle between $\mathcal{B}$ and $\mathcal{C}$, then the angle θ between $\mathcal{A}$ and $\mathcal{C}$ is such that: $\theta \leq \phi_1 + \phi_2$.*

For example, to get an upper bound of the angle θ between invariant subspaces of the endomorphisms A and $C = B_2(A + B_1)$, one takes, as an upper bound between the invariant subspaces of A and $A+B_1$, the angle ϕ_1 (theorem 7.4 or theorem 7.5), then one takes, as an upper bound between the invariant subspaces of $A+B_1$ and of $B_2(A+B_1)$, the angle ϕ_2 (theorem 7.8). Hence for θ the upper bound $\phi_1 + \phi_2$.

Chapter 8

Inductive Data Analysis

(‘*INDUCTIVE*’)

Far better an approximate answer to the right question ...
than an exact answer to the wrong one.
J. Tukey

Introduction

Many research studies involve *inductive objectives*, that is, the researcher would like to draw conclusions that go beyond the particular data set under study. With small data sets, however, inductive conclusions may be precarious, and this is the reason why researchers, in order to assess inductive conclusions, often resort to formal procedures, traditionally called ***statistical inference procedures***, ranging from the elementary χ^2–test to the most sophisticated procedures. In scientific publications, the use of inference procedures is the accepted way to validate hypotheses.

Now, a casual look at research studies involving GDA shows that statistical inference is scarcely used beyond the routine χ^2–test. This is an unusual state of things[1]. Our position, already stated in chapter *OVERVIEW*, is that Inductive Data Analysis (IDA), that is, statistical inference in the line of the Data Analysis Philosophy, could and should be used more liberally in GDA. In fact, IDA has been a long–standing (even if non–dominant) part of statistical inference, from Fisher to Tukey, in strong opposition to the Neyman–Pearson school.

[1] As pointed out in *OVERVIEW*, research papers that put forward conclusions on the sole basis of descriptive procedures tend to be marginalized by the norms prevailing in scientific publications.

So far in this book, we have only dealt with descriptive procedures, for which no statistical inference prerequisites are necessary. In the present chapter, we will review the inductive procedures that will be put into practice in the case studies (chapter 9), thus making the book self–contained even for those readers whose background in statistics inference is limited[2].

The chapter is organized as follows. First of all, we discuss current statistical inference in multivariate statistics, especially significance testing, and we characterize descriptive vs inductive procedures (§8.1). Then we revisit elementary univariate procedures in the line of IDA (§8.2). As enlargements of traditional inference, we discuss Combinatorial Inference, permutation tests and modeling (§8.3), and Bayesian Data Analyis (§8.4). Then we deal with inductive geometric procedures (§8.5). Finally, we present guidelines for the inductive analysis of effects (§8.6).

8.1 Inference in Multivariate Statistics

A lot of statistical inference work has been done not only in multivariate statistics in general, but also in CA: For an extensive survey, see Lebart, Morineau & Piron (1995, chap. 4)[3]. To name a few references: Lebart (1975b, 1976), Lebart, Morineau & Warwick (1984), Lebart & al (2003), Gilula & Haberman (1986), Saporta & Hatabian (1986), Daudin, Duby & Trécourt (1988), Gifi (1990), Saporta (1990), Tenenhaus & al. (1993); not to speak of the work done in traditional Multivariate Analysis and directly applicable to GDA, such as Rao (1964, 1973), Anderson (1963), etc.

Significance testing. Most inference procedures developed in multivariate statistics so far are significance tests, especially for the significance of axes. About significance testing, two points can be made.

A *first point* is that significance tests possess an ambivalent property, namely the "size–glued property" (already alluded to in *OVERVIEW* and *CORRESPONDENCE*): The degree of significance depends on the sample size as well as on the descriptive importance of effect. This property has the undesirable consequence that significance tests detect trivial effects with large samples and fail to detect large effects with small samples. As an example, finding a significant χ^2 in a contingency table shows there is

[2] As a prerequisite for this chapter, a good introductory text may be recommended, such as the book by Freedman & al (1991).

[3] Benzécri has constantly insisted on the *inductive logic* embodied in CA; and in Benzécri & al (1973, Vol 2), there is an entire chapter (p.210-230) on inference in CA.

some association between the two variables, but if the sample size is large, the magnitude of association detected may be small. In the same way, for the significance of axes, with the large sample sizes commonly used in GDA, significance tests will yield significant results even for axes whose descriptive importance is negligible; therefore, checking for significance of axes is useless when not misleading[4]. Still, there are situations when finding a significant result can be interesting; for instance, when specific effects are investigated on a small subsample of data (see next section).

A *second point* is that significance tests are not the whole of statistical inference. Other inference methods, to begin with confidence methods, may be more useful to fulfill inductive objectives and investigate hypotheses.

Descriptive vs inductive procedures. At this point, the question may be raised: What exactly should be meant by "descriptive" procedures and statistics, as opposed to "inductive" or "inference" ones?

There is a clear–cut operational response to this question. *Descriptive procedures do not depend on sample size*, that is, they can be based only on relative frequencies (Rouanet & al, 1998, p. 3); accordingly, descriptive statistics are statistics produced by descriptive procedures. Examples are the mean, the variance, the correlation coefficient, etc. By contrast, *inductive procedures* (or inference procedures), such as the χ^2–test, do effectively depend on sample size n. For a contingency table, the formula $\chi^2 = n\,\Phi^2$ relates two indices of departure from independence, namely the mean square contingency coefficient Φ^2, which is a descriptive statistic, and the test statistic χ^2, which is not a descriptive statistic. If the absolute frequencies of the table are all multiplied by a same coefficient $K > 1$, Φ^2 is invariant, whereas χ^2 is multiplied by K, and consequently (for a non-zero association) becomes more significant. The paradigm $\chi^2 = n\,\Phi^2$, that is,

"Test statistic = sample size × descriptive statistic"

provides the general key for articulating description and inference.

All statistical methods produce basic statistics that are descriptive (e.g. eigenvalues and principal components for PCA or CA), which at a later stage can be combined with sample size to yield inductive procedures.

Inductive Data Analysis. In the Data Analysis Philosophy, statistical procedures should dig out "what the data have to say", and depend as little

[4]In conjunction with the deceptive symmetry "rejection/acceptance of the null hypothesis", the size–glued property all too often lends itself to the *significance fallacy*: "Significant = There is an effect" vs "Nonsignificant = There is no effect".

as possible on gratuitous hypotheses, unverifiable assumptions, etc. Inductive Data Analysis is data–driven. Descriptive data analysis comes first and descriptive conclusions are stated. Then as a follow–up of descriptive analysis, the main objective of inductive data analysis is to corroborate (whenever possible) descriptive conclusions, allowing for non–controlled sources of variation. Looking at the importance of effects is done in the first stage of descriptive analysis, not as an afterthought following significance testing. If an effect is descriptively negligible, there is no point in inquiring about its statistical significance. In this way, there cannot be any seeming "conflict" between descriptive and inference conclusions (e.g. finding an axis that is descriptively negligible and still statistically significant).

When considering a hypothesis to investigate, the viewpoint of Inductive Data Analyis invites one to scrutinize the inference procedures relevant to the situation with the following two concerns in mind.

Validity of assumptions: What are the assumptions that seriously matter for validity? Are they reasonably met in the situation under study?

Relevance of hypotheses: Do the conclusions brought up by the procedures bring answers to research hypotheses of interest?

As an example, suppose a χ^2–test in a contingency table is significant. If the randomness assumption is questionable, doubt may be cast on the conclusion that there is some association between the two variables. On the other hand, when investigating the structures of association is the main objective, asserting the mere existence of association is of limited interest. As far as multivariate statistics is concerned, the conventional "general model" of Multivariate Analysis involves formidable assumptions (multinormality, variance and covariance homogeneity, etc.), that are unverifiable for small samples and patently false for large ones. From the reading of serious methodological texts such as Cox (1977), one might presume that, in most applications, a fantastic amount of statistical expert–knowledge has to be mobilized to figure out whether the conclusions of interest stand up or not.

Actually, the situation is not so desperate, for three reasons at least.

- Firstly, the usual normal–theory procedures about means are *robust* with respect to departure from normality assumptions (§8.2.2, p.304).
- Secondly, *specific modeling* can relieve one from the most severe assumptions of the general model (§8.2.2, p.307).
- Thirdly, *permutation modeling* is entirely normal–free and can even be made free from random assumptions altogether (§8.3.5, p.314).

8.2 Univariate Effects

In this section, we revisit, in the line of Inductive Data Analysis, the elementary procedures on means and effects in the univariate (numerical) case. We successively discuss hypotheses of interest, statistical modeling, and traditional inference.

8.2.1 Hypotheses of Interest

Hypotheses of interest, that is, the hypotheses that are contemplated for validation (inductive corroboration), can be research hypotheses explicitely stated before gathering data; or they can emerge from the descriptive analysis, guided by the structures of data. In any case, the first and crucial phase is the descriptive appraisal of observed effects.

Observed vs true effects. As an example, we take Student's historical data (Student, 1908), with a matched–pair design $S \star T_2$ ("Subjects crossed with two Treatments"), and numerical performances ("amount of additional sleep") for two drugs, for $n = 10$ subjects. From the basic n performance pairs, the protocol of individual differences $(d^i)_{i=1\ldots n}$ is derived (Table 8.1).

	$i1$	$i2$	$i3$	$i4$	$i5$	$i6$	$i7$	$i8$	$i9$	$i10$
d^i	+1.2	+2.4	+1.3	+1.3	+0.0	+1.0	+1.8	+0.8	+4.6	+1.4

Table 8.1. Student's data: Individual effects (in hours).

The mean is $\sum_{i \in I} d^i/n = +1.58$ hours; it is called *observed effect* and denoted d_{obs}. The variance is denoted v with $v = \sum_{i \in I} (d^i)^2/n - (\sum_{i \in I} d^i/n)^2 = 1.3616$, and the corrected variance[5] is denoted s^2 with $s^2 = \frac{n}{n-1} v = (10/9) \times 1.3616 = 1.5129 = (1.230)^2$. The *scaled effect* d_{obs}/s is equal to $\frac{1.58}{1.230} = 1.285$, a very large ratio, which leads to the *descriptive conclusion*:

The observed effect d_{obs} is positive and large in magnitude.

The intuitive idea of IDA is that observed values arise from a set of possible outcomes that might have been observed, i.e. the observed effect d_{obs} is supposed to be the realization of an Effect variable d varying in a *sample*

[5]In this chapter we follow the dominant custom in statistical inference by expressing the procedures in terms of *corrected variance* s^2 (dividing the sum of squares by the number of degrees of freedom q, here $n - 1 = 9$) rather than in terms of the genuine variance v (dividing the sum of squares by n).

space specified by the statistical modeling. We adopt the convenient "trueness metaphor", that is, the Effect variable d is thought to vary around an idealized value, called a *true effect* δ, to which the inference pertains[6]. For Student's data, starting from the descriptive conclusion, we will investigate firstly that the true effect δ is positive in sign, and further (possibly) the hypothesis that it is large in magnitude.

Observed effects. Various descriptive conclusions may be stated about effects; for instance, conclusions may pertain to the existence of the observed effect, its orientation (direction)[7] and its magnitude, as reflected in the following list:

Non–existence vs existence of effect. i) "There is virtually no observed effect" (d_{obs} close to zero); ii) "There is definitely an effect" (d_{obs} not close to zero).

Direction of effect. i) "The observed effect is definitely positive" ($d_{obs} > 0$); or ii) "It is definitely negative" ($d_{obs} < 0$).

Largeness of effect. i) "The observed effect is positive and large" (we write $d_{obs} \gg 0$); or ii) " It is negative and large" ($d_{obs} \ll 0$).

Smallness of effect. "The observed effect is small" (regardless of sign: $|d_{obs}|$ small).

The foregoing list is not exhaustive nor exclusive. Largeness conclusions are typically directional and are strengthenings of directional conclusions. Descriptive conclusions may be informal; the loose statement "There is no effect" may mean "Effect is small", and similarly "There is an effect" may mean "Effect is large"[8]. Descriptive conclusions are highly context–bound; a 1% increase in a percentage may be deemed to be "definitely positive" in an epidemiological study, but "virtually null" in an opinion survey.

Hypotheses about true effects. To the foregoing conclusions about the observed effect d_{obs} there correspond the following hypotheses about the true effect δ.

[6] In survey set–ups, "true values" are naturally thought of as "population values".

[7] We use "oriented" as a synonym of directional, as in chapter CLOUD.

[8] Such formulations reflect the pragmatic conversational rules that are well–known in psycholinguistics.

Null hypothesis: "True effect is null ($\delta = 0$)". *Non–null hypothesis*: "True effect is non–null ($\delta \neq 0$)".

Directional hypotheses: i) "True effect is positive ($\delta > 0$)"; or ii) "True effect is negative ($\delta < 0$)"; or iii) "True effect is on the same side as d_{obs}".

Largeness hypotheses: i) "True effect δ is large on the positive side ($\delta \gg 0$)"; or ii) "True effect is large on the negative side ($\delta \ll 0$)"; or iii) "True effect is large on the side of d_{obs}".

Largeness hypotheses are naturally directional.

Smallness hypothesis: "True effect is small ($|\delta|$ small)."

Smallness hypotheses are naturally non–directional.

The hypothesis ($\delta = 0$), that is, the usual *point null hypothesis*, plays a central role in the technique of significance tests[9]. However, at least in the univariate case, it is often a mere *dividing hypothesis* (a demarcation between the "positive" and "negative" zones) rather than a *sharp hypothesis*[10]. In what follows, we will leave aside the validation of sharp hypotheses. Conversely, in the univariate case at least, validating the hypothesis of the mere existence of effect ($\delta \neq 0$) is usually less interesting than validating a directional hypothesis. In research case studies, *directional hypotheses* are of prime interest. Then come *largeness hypotheses* — e.g. the new drug is markedly more effective than the standard one — and *smallness hypotheses* — e.g. the undesirable side–effects of two drugs are about the same ("bioequivalent" in pharmacological jargon). In the sequel, we will study firstly directional hypotheses, secondly largeness and smallness hypotheses.

Descriptive data analysis and the validation of hypotheses. Descriptive data analysis points to hypotheses that have a chance of being validated (i.e. enforced, substantiated, corroborated) by inference procedures, as opposed to hypotheses that can certainly not be validated by the data

[9]In the phrase "point null hypothesis", the word "point" does not refer to geometry, but underlines that the hypothesis has a null measure in parameter space — as opposed e.g. to a directional hypothesis ($\delta > 0$).

[10]The sharp hypothesis ($\delta = 0$), from which any deviation (however small) would be of interest, is usually not a hypothesis that one seriously attempts to validate; when comparing two drugs, or teaching methods, etc., one is normally not interested in detecting any non–null difference, but a notable difference. In any case, only Bayesian inference appears to be seriously equipped to tackle sharp hypotheses.

at hand. Thus the descriptive conclusion ($d_{obs} > 0$) suggests examining the directional hypothesis ($\delta > 0$); and the conclusion ($d_{obs} \gg 0$) suggests the largeness hypothesis ($\delta \gg 0$), as candidates for validation; both descriptive conclusions discard examining the hypothesis ($\delta < 0$) (and also ($|\delta|$ small)). Similarly the descriptive conclusions (d_{obs} close to 0) and ($|d_{obs}|$ small) suggest examining the *smallness hypothesis* ($|\delta|$ small), etc.[11]

8.2.2 Statistical Modeling

Statistical Modeling is the formalization of the link between observed and true values, and is the building block of statistical inference. Statistical modeling is usually framed as a *probabilization*, that is, an observation is supposed to be the realization of a variable x distributed in a probability space indexed by a parameter θ which is the object of inference to which hypotheses pertain. Thus, for the inference on means, the observed mean m_{obs} is supposed to be the realization of a variable m (statistic) that is the sum of the parameter (true mean) μ and a probabilistic error term e, that is $m = \mu + e$ ("observed" = "true" + "error"). Accordingly, for inference on effects, the observed effect d_{obs} is supposed to be the realization of a variable d that is the sum of the parameter (true effect) δ and a probabilistic error term e, that is, $d = \delta + e$ (again "observed" = "true" + "error"). The explicitation of the probability distribution of the error term (e.g. normal distribution) constitutes the *model specification*. Normal–theory modeling is the classical model specification for numerical data.

Normal modeling (univariate). To present normal–theory modeling, in brief normal modeling, we take (as usually done) one group of n numerical observations, that we denote $(d_i)_{i=1...n}$. In normal modeling, the group of observations is supposed to be a sample — independently and identically distributed normal variables (see density formulas p.332) — with mean δ (true effect, main parameter) and variance σ^2 (secondary parameter).

The parameters of the model are estimated by the corresponding descriptive statistics. Parameter δ is estimated by the mean $\sum d_i/n$ (maximum likelihood estimator); and parameter σ^2 is estimated either by the

[11] A hypothesis discarded by descriptive analysis cannot be shown to be valid; the subsequent inferential analyses may show, or not, that it is indeed invalid. Thus, in Student's data, the discarded hypothesis ($\delta < 0$) can be shown to be invalid because the inferential analysis leads to the conclusion ($\delta > 0$). In a situation where the conclusion ($\delta > 0$) could not be reached, the hypothesis ($\delta < 0$) could not be invalidated. As will be seen, the probability of hypotheses can in all cases be assessed by Bayesian procedures.

variance v or alternatively by the corrected variance $s^2 = n\,v/(n-1)$ (unbiased estimator)[12]. Then, applying probability theory, the following sampling distributions are derived:

• The statistic d is normally distributed with mean δ and variance σ^2/n; which we write $d \sim \mathcal{N}(\delta, \sigma^2/n)$ (reading "$\sim$": "is distributed as").

• For any σ^2, the variable $\sqrt{n}\,\frac{d-\delta}{s}$ (Student's ratio) is distributed as Student's elementary t–distribution with $q = n-1$ d.f. (see density p.332).

For the inference on means, what matters is the distribution of Student's ratio, and this distribution really depends very little on the normality assumption about the basic individual observations; hence the well–known *robustness* of Student's t–test with respect to non–normality.

Elementary Student's t–test. Under normal modeling, the Student's t–test puts to test the point null hypothesis $\mathcal{H}_0 : \delta = 0$. Taking the t–ratio, $t = \sqrt{n}\frac{d}{s}$, as a test statistic, one has under $\mathcal{H}_0\ :\ t \sim t_q$, or equivalently, in terms of the *scaled effect* $\frac{d}{s}$, one has:

$$\text{Under } \mathcal{H}_0 : \qquad \frac{d}{s} \sim t_q(0, \frac{1}{n}) \quad \text{(scaled Student's } t\text{–distribution)} \qquad (8.1)$$

Under $\mathcal{H}_0$, the sampling probability $P(|d| > |d_{obs}|)$ (probability for the test statistic to be more extreme on either side than its observed value), is called the two-sided p–value, or *two-sided observed level.* The *one-sided observed level* $p_{one-sided} = \frac{1}{2}\,p_{two-sided}$ is the probability under $\mathcal{H}_0$ that the test–statistic is more extreme than its observed value on the side of data[13]. Observed levels are equal to proportions of area under the density curve (one-sided level on figure 8.1).

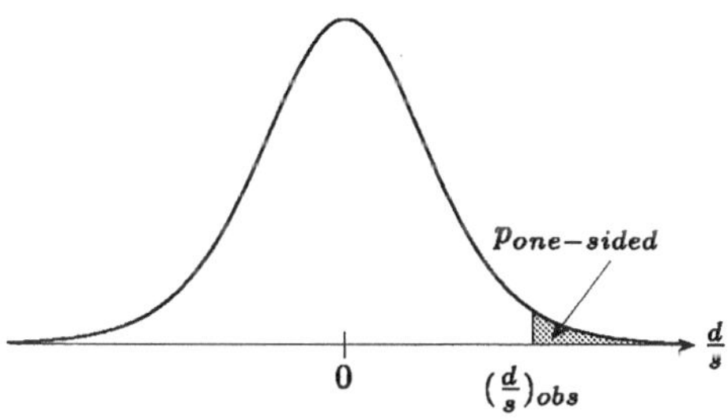

Figure 8.1. Sampling distribution of $\frac{d}{s}$ under $\mathcal{H}_0\ :\ \delta = 0$

---

[12] *Genuine vs corrected variance*: Whichever estimate is used is immaterial for the ensuing inference procedures. In terms of the genuine variance $v = (\sum(d_i - \bar{d})^2)/n$, the Student's ratio $t = \sqrt{n}\frac{d}{s}$ reads $t = \sqrt{n-1}\frac{d}{\sqrt{v}}$.

[13] "p–value" will refer to either $p_{two-sided}$ or $p_{one-sided}$ as context requires.

• *Student's data.* From $(d/s)_{obs} = +1.285$, we deduce $t = \sqrt{10} \times 1.285$, then $t = +4.06$. When referred to the t_9–distribution (9 d.f.), we find $p_{one-sided} = .0014$. The observed level is very small[14]: a finding that clearly means there is a strong evidence against the null hypothesis ($\delta = 0$).

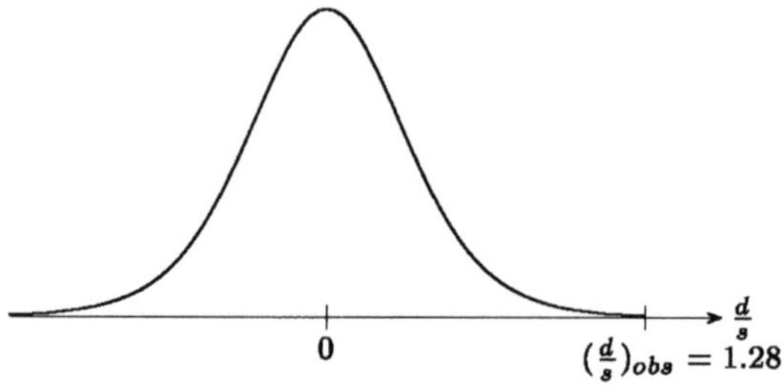

Figure 8.2. Student's data: $(\frac{d}{s})_{obs} = +1.28$, $P(\frac{d}{s} > +1.28/\delta = 0) = .0014$

Shifted hypotheses and confidence intervals. The t-test can be enlarged to cover shifted hypotheses of the kind $\delta = \delta_0$, where δ_0 is any given deviation from 0.

• *Student's data.* The hypothesis $\mathcal{H}'_0 : \delta_0 = 0.5$ (true difference of a half–hour) can be tested by the ratio $t' = \frac{d-0.5}{s/\sqrt{n}}$; hence $t'_{obs} = \frac{1.58-0.5}{1.230/\sqrt{10}} = +2.78$. At usual levels, the difference is significant, and we conclude that the true effect exceeds a half–hour.

The procedure of confidence intervals amounts to testing a set of shifted hypotheses at once. By definition, the confidence interval at level α (or confidence degree $1 - \alpha$) is the set of δ–values that do not differ significantly from d_{obs} at level α (two-sided), that is, that satisfy the following constitutive property (denoting $t_{n-1}[\alpha]$ the value $u > 0$ such that $P(|t_{n-1}| \geq u) = \alpha$).

$$|\delta - d_{obs}| < t_{n-1}[\alpha] \times \frac{s}{\sqrt{n}} \qquad (8.2)$$

• *Student's data.* Taking $\alpha = .05$, one has $t_9[.05] = 2.262$, the half–width of the interval is $2.262 \times 1.230/\sqrt{10} = 0.879$, hence the interval (0.70 ; 2.46), which is the set of the δ-values that do not differ significantly from d_{obs} at the two-sided .05 level. The value 0 lies outside the interval, which means that the hypothesis $\delta = 0$ is not compatible with the data at the two-sided .05 level, as already found with the test of no–effect. The value $\delta = 0.5$ also lies outside the interval, which means that the hypothesis $\delta = 0.5$ is not compatible with the data, as found with the test of the shifted hypothesis.

[14]In any case it is smaller than the traditional α–level landmarks (see p.308).

The lower limit of the confidence interval (0.70) is the highest difference not compatible with the data, at the one-sided .025 level. For increasing values of α, one gets a sequence of nested intervals: the larger α, the narrower the interval. For instance, taking $\alpha = .10$, we would find the interval with half-width 0.713 and lower limit 0.87.

Specific modeling and inference. In a matched-pair design $S \star T_2$, the basic data consist in a set of pairs $(x_i, y_i)_{i=1,\dots n}$. To compare the means of the *two matched groups*, we have only modeled the derived data set of the individual differences $(d_i = x_i - y_i)_{i=1,\dots n}$. In so doing, we have proceeded to a *specific modeling* on the relevant parameter δ (difference of means), involving the sole assumption of normality of individual differences. Instead, we could have proceeded to a *general modeling* on the bivariate data set, but at the expense of further assumptions (binormality and homogeneity of variances). In specific modeling, each specific inference, which only depends on the structure of the specific data set, is valid regardless of the complexity of the overall design.

The idea of basing an inference on the sole relevant data is highly intuitive, and in research practice, specific inferences abound, performed in an informal way. The Bayesian concept of partial sufficiency provides a formal justification of specific inference: See §8.4, p.321. Specific modeling and inference are in the line of the *specific analysis* approach mentioned in chapters 3 (CLOUD, §3.3.4, p.94) and 5 (MULTIPLE, §5.2, p.203). In the case studies (chapter 9), we will put specific modelings on the derived clouds (subclouds and/or projected clouds) corresponding to questions of interest.

Comparing the means of two independent groups. We now recall the $S < G_2 >$ structure ("Subjects nested within two groups"), with two independent groups $g1$ and $g2$ with n_1 and n_2 numerical observations respectively $(x^i)_{i \in I<g1>}$ and $(x^i)_{i \in I<g2>}$. The main parameter is the difference of true means $\delta = \mu^{g1} - \mu^{g2}$, estimated by the difference $d_{obs} = \overline{x}^{g1} - \overline{x}^{g2}$. In the most usual modeling (assuming homogeneity of variances), the $n_1 + n_2$ observations are supposed to be independently distributed, the observations in group g being distributed $\mathcal{N}(\mu^g, \sigma^2)$. The secondary parameter σ^2 (common within-group variance) is estimated by the corrected within mean-square $s^2 = \big((n_1 - 1)s^2_{g1} + (n_2 - 1)s^2_{g2}\big)/(n_1 + n_2 - 2)$, which is the weighted average of within-group corrected variances s^2_{g1} and s^2_{g2} (unbiased estimator) with $q = n_1 + n_2 - 2$ d.f. If we let $\widetilde{n} = 1/(\frac{1}{n_1} + \frac{1}{n_2})$, we have $d \sim \mathcal{N}(\delta, \sigma^2/\widetilde{n})$, hence $\sqrt{\widetilde{n}}\,\frac{d-\delta}{s} \sim t_q$. The usual Student's t-test for two independent groups follows, as well as the confidence interval procedure for the difference of means, replacing the sample size n of the one–group case by $\tilde{n}$ ("equivalent sample size").

8.2.3 Traditional Inference

Traditional inference, as reflected in research papers, has largely been influenced by Fisherian thinking. Its main tool is traditional significance testing, complemented by confidence intervals. *Traditional significance testing* consists in testing the point hypothesis of *no–effect* $\mathcal{H}_0 : \delta = 0$ (null hypothesis), and, using the familiar grid of landmarks[15], in declaring results as *significant* (small p–values) — S* (merely significant), S** (highly significant), the star system! — or else as nonsignificant (non–small p). Conclusions are: "There is evidence of effect, S*"; "There is strong evidence of effect, S**"; "There is no evidence of effect, NS".

The merits of traditional inference. There is much to be said for traditional inference. The p–value is an indicator of the degree of compatibility of the null hypothesis $\mathcal{H}_0$ with the data. The smaller the p-value, the stronger the evidence (in Fisher's words) against $\mathcal{H}_0$, thus bringing an answer to the legitimate query: Whether or not the descriptive conclusion "there is an effect" may be deemed to be "nonaccidental", and reflect a true effect that one is entitled to interpret. Thus, traditional significance testing maintains a safe dissymmetry significant vs nonsignificant. Only significant results (small p–values) entail positive conclusions; nonsignificant results are just statements of compatibility of the hypothesis ($\delta = 0$) with data (no evidence has been found against $\mathcal{H}_0$).

Furthermore, in traditional significance testing, a significant effect legitimately corroborates not only the *existence hypothesis* ($\delta \neq 0$), but also (at least in the one–dimensional case) the *directional hypothesis*: *δ on the side of d_{obs}*. As Tukey (1991) emphasizes, the more significant the effect, the more certain we are about its *direction*. Conversely, a nonsignificant result essentially means that we are *uncertain about the direction* of the effect. To conclude, for existence and/or directional hypotheses, traditional significance testing provides a safeguard against reckless inductive conclusions,

[15]The landmarks $\alpha = .05, .01$ are usually understood as two-sided; thus, one has S** if $p_{two-sided} < .01$, S* if $.01 < p_{two-sided} < .05$, NS if $p_{two-sided} > .05$; or equivalently S** if $p_{one-sided} < .005$, S* if $.005 < p_{one-sided} < .025$, NS if $p_{one-sided} > .025$. Additional landmarks are often considered, e.g. $p_{two-sided}$ significant at only .10 is interpreted as a presumption of effect.

if only the *safety rule* is respected: "Always use significance testing as a follow–up of the descriptive conclusion: There is an effect."

The limitations of traditional inference. To begin with, traditional significance testing does not handle largeness nor smallness hypotheses, as reflected in the two *Ritual Warnings*. *Warning # 1*: Statistical significance is not substantial (alias practical, psychological, sociological) significance! *Warning # 2*: No proof of effect is not proof of no effect! At this point, there is again a dissymmetry. Finding a significant effect does provide, in a sense, a first support toward a largeness conclusion, in so far as it establishes, at least, the *existence* of a true effect. On the contrary, a nonsignificant effect does not provide any support at all to the smallness hypothesis; for the smallness hypothesis, testing for no–effect is *irrelevant*, and had better be avoided, even as a first step. Whenever largeness or smallness hypotheses are at stake, traditional significance testing, if correctly interpreted, answers the wrong question, and if incorrectly interpreted, is misguided[16]. Then, there are the conceptual limitations of traditional inference that are linked with its historical commitment to a particular framework of statistical inference, namely the *frequentist framework* (see next paragraph). Within this framework, the p–value of a significance test cannot be interpreted as the probability of a hypothesis about true effect. Confidence intervals, which tell us about the *precision of estimates*, are often more interesting than testing the hypothesis of no–effect[17]. Yet, in practice, confidence procedures are seldom used beyond elementary cases, such as, in a design involving several groups, confidence intervals for the means of groups. Actually, there are conceptual difficulties with confidence intervals, again linked with the frequentist framework. In this framework, the natural interpretation of a confidence interval in probabilistic terms — in Student's example, "the probability is .95 that the true effect lies within the limits 0.70 and 2.46" is not possible.

Statistical frameworks. In any inference procedure, there is an algorithm, that tells how the procedure is to be performed, and a statistical framework, that *gives sense* to the outcome of the algorithm. Several statistical frameworks can be considered in statistical inference, which differ about the role and meaning assigned to probability in statistical procedures.

[16] In spite of warnings, "significance fallacy" is all too often committed: Significant effects are unduly interpreted as large effects, and nonsignificant effects as small ones.

[17] In econometrics, assessing the precision of estimates is of prime importance, and this is surely the main motivation for probabilistic modeling, as is clear in Malinvaud (1980).

In the *frequentist framework*[18], probabilities are assimilated to long–run relative frequencies, and statistical modeling is conceived as expressing the *data generating process* as a random process. Accordingly, parameters are regarded as "fixed though unknown" entities, about which no probabilistic statements are possible. Thus, within the frequentist framework, there is no room for probabilities of hypotheses. Besides, to overcome the limitations of traditional inference, it becomes exceedingly arduous to stick to the frequentist framework beyond the most elementary situations. Actually, statistical modeling, when stated in abstract probabilistic terms (as we have done), is not committed to the frequentist interpretation[19], and can fit alternative frameworks of statistical inference. In this chapter, we will outline a *two-fold enlargement* of statistical framework and procedures:

— *Combinatorial inference*, which *relaxes* the frequentist framework and bases inference on *proportions of samples*;

— *Bayesian inference*, which *enriches* the frequentist framework and allows *probabilities of hypotheses.*

8.3 Combinatorial Inference

8.3.1 Nonrandom Inductive Situations

Combinatorial procedures. In any inference procedure, the algorithm can be dissociated from the framework, and the procedure stripped down to the algorithm. Thus, if in Student's test we ignore the frequentist framework, the sampling distributions will still make sense in terms of the distributions over the n–size subsets — that is, "samples" in a plain set–theoretic sense — in a normal population; the p–value of the test will simply be the proportion of samples whose t–ratio is more extreme than the observed t–ratio. In other words, instead of assuming that the data set is a random sample from a normal distribution, we assess the typicality of the data set with respect to a normal distribution of a given mean. Given a property of interest, the *probability* for a *random* sample to have this property is replaced by the *proportion* of (set–theoretic) samples that have this property. We call the procedure so defined the *combinatorial* (or set–theoretic) t–test. The approach is clearly general, that is, any frequentist procedure

[18]The frequentist framework has been dominant in the developments of statistics in the twentieth century.

[19]Probability is about uncertainty, not about randomness. Suppes (2002) discusses at length no less than six different representations of probability, definitely breaking the sterile "objective/subjective" forced choice.

can give rise to a combinatorial procedure. By construction, combinatorial procedures effectively depend on sample size, therefore combinatorial inference is the first stage of statistical inference, as the straight extension of descriptive analysis. See Rouanet & al (1986), Rouanet & al (1990) and Rouanet & al (1998).

Nonrandom problems. There are inductive problems where randomness is basically irrelevant and for which combinatorial procedures bring appropriate answers. For instance, in a survey one may question whether the data set (random sample or not) is typical of a known population with respect to some variable (age, educational level, etc.); a small p–value of the combinatorial significance test will point that the data set is *atypical* of the population with respect to the variable under consideration. This combinatorial conception of significance level exactly corresponds to the "nonstochastic interpretation" presented by Freedman & Lane (1983).

As another example, suppose a trial jury of 9 members comprises only one woman; such a jury may be objectionable as atypical with respect to the sex ratio, even if it has been lawfully constituted by random sorting. This example shows that a nonrandom problem may arise even when randomness is assumed in the data generating process. Other nonrandom problems are *homogeneity*, for comparing independent or matched groups; and *nonassociation* between two variables. At this point, we come into *permutation tests*, that will be discussed later on (§8.3.5, p. 314).

8.3.2 Combinatorial Inference in GDA

Combinatorial inference is congenial to the descriptive–oriented key idea of GDA. In this subsection, we summarize the paper by Le Roux (1998), where typicality, homogeneity and nonassociation tests are presented in detail.

Typicality test for a mean and test–values. The typicality test for the mean is a combinatorial procedure for comparing the mean of a group of observations to a reference value. Given a numerical reference population of size N, mean μ and variance σ^2, and a *group of observations* of size n and observed mean m, the typicality of the group for the mean, with respect to the population, is tested as follows. The *sample space* is defined as the set $\mathcal{X}$ of the $\binom{N}{n}$ n–element subsets of the population. With each sample $x \in \mathcal{X}$ there is associated its mean $M(x)$, hence the *Mean statistic* defined as the mapping $M : x \longmapsto M(x)$. The sampling distribution of $M(x)$ is then derived, and the observed mean m is located in this distribution.

Approximate typicality test. The test–statistic $T = (M - \mu)/\sqrt{\text{Var}\, M}$ with $\text{Var}\, M = \frac{\sigma^2}{n}\frac{N-n}{N-1}$ is approximately a standard normal variable.

If t denotes the observed value of T, the observed level is approximately $p(T \geq t)$ (when $m > \mu$), or $p(T \leq t)$ (when $m < \mu$).

Application in GDA. Let us take as a "reference population" the projected cloud of the N individuals onto a principal axis (with mean $\mu = 0$ and variance $\sigma^2 = \lambda$), and as a group of observations a subcloud of interest of size n. The typicality test is thus seen to extend the descriptive concept of *contribution of a modality.*

The technique of *test–values* has been proposed by Morineau (1984), Lebart & al. (1984, p. 23-26, 95), and applied in Lebart & al (1995, p. 181), Alevizos and Morineau (1992, 1993). The inductive status of this technique is perfectly clear within the combinatorial framework.

Homogeneity tests: Comparisons of several means. In the line of typicality tests, combinatorial inference can be extended to various *homogeneity tests.* The rationale of homogeneity tests is apparent for structured data, where a space of possible protocols — the natural generalization of sample space — is generated by a permutation group associated with the data structure. Thus for comparing the means of *two independent groups* (structure $S < G_2 >$), a set of $\binom{n_1+n_2}{n_1}$ possible protocols is generated by exchanging the observations between the two groups; similarly for *two matched groups* (structure $S \star T_2$), a set of 2^n protocols is generated by permuting the observations within each pair. With each protocol there is attached a mean difference d, hence the *permutation distribution* of d over the protocol space, which is the analog of a sampling distribution. Taking d as a test–statistic, the proportion of d–values that are more extreme than the observed d_{obs} defines the observed p–level of the permutation test; when the p–level is small, the conclusion is that the two groups are heterogeneous. Here again, the test–statistic $T = D/\sqrt{\text{Var}\, D}$ can be approximated by a standard normal variable.

For $S \star T_2$ (two matched groups), one has $\text{Var}\, D = \sum_{i=1}^{n} (d^i)^2/n^2$.

For $S{<}G_2{>}$ (two independent groups), if σ^2_{obs} is the variance of the grouped data, one has $\text{Var}\, D = \frac{n_1+n_2}{n_1+n_2-1}(\frac{1}{n_1} + \frac{1}{n_2})\sigma^2_{obs}$.

For a partition of a population into K classes of means $(m^k)_{k \in K}$, the homogeneity tests consist in comparing some (or all) of these means $(m^k)_{k \in K'}$ (with $2 \leq K' \leq K$). In GDA, the homogeneity test for two classes extends the descriptive concept of *contribution of a deviation.* For more than two classes, its approximation by the χ^2 distribution generalizes the test–values.

8.3.3 Combinatorial Inference in MCA

We now specialize the general formulas to MCA, still taking as a "population" the cloud of N individuals $(\mathrm{M}^j)_{j\in J}$, and considering a principal variable y^J_ℓ with mean $\mu_\ell = 0$ and variance σ^2_ℓ ($= \lambda_\ell$ eigenvalue).

Typicality of a class. To study the typicality of class k (for the mean), we take $T = (\overline{y}^k_\ell/\sigma_\ell)\sqrt{n_k \times \frac{N-1}{N-n_k}} = \sqrt{N-1}\, y^k_\ell \sqrt{\frac{f_k}{1-f_k}} = \sqrt{N-1} \cos\theta_{k\ell}$ as a test statistic. Hence, denoting $\widetilde{k}$ the grouping of the modalities of question q other than k, one has: $t^2_k = (N-1)\cos^2\theta_{k\ell} = (N-1)v_{k\widetilde{k}} = (N-1)\times Q \times \mathrm{Cta}_{(k,\widetilde{k})\ell}$, that is, $t^2_k = \frac{N-1}{1-f_k} \times Q \times \mathrm{Cta}_{k\ell}$, where $v_{(k,\widetilde{k})\ell}$ denotes the variance of dipole $(k,\widetilde{k})$, and $\mathrm{Cta}_{(k,\widetilde{k})\ell}$ its within–contribution (cf. MULTIPLE, formula (5.14), p.196).

Every question q with K_q modalities induces a partition of the N individuals into K_q classes, hence the modality mean–point $\overline{\mathrm{M}}^k$, with principal coordinate $\overline{y}^k_\ell = \sigma_\ell y^k_\ell$. If $K' \subseteq K_q$, the variance of K' modality mean points $(\overline{\mathrm{M}}^k)_{k\in K'}$ is equal to $\sigma^2_\ell v_{K'\ell}$, and the within contribution is equal to $Q^2\sigma^2_\ell \,\mathrm{Cta}_{(K')\ell}$. With the partition of the cloud of individuals by a question q, there is associated the *correlation ratio* $\eta^2_{q\ell}$ with, for each axis, the relations $\eta^2_{q\ell} = v_{q\ell} = Q\,\mathrm{Cta}_{q\ell}$ (cf. MULTIPLE, Property 5.21, p.198).

Homogeneity of several classes. To study the homogeneity of K' classes of individuals associated with a question q ($K' \subseteq K_q$), we can take: $T^2 = (N-1) \times (N'/N) \times v_{K'\ell} = (N-1) \times Q \times \mathrm{Cta}_{(K')\ell}$ as a test statistic. As a particular case we have the overall homogeneity test of question q: $T^2 = (N-1) \times Q \times \mathrm{Cta}_{q\ell} = (N-1)\eta^2_{q\ell} = (N-1)v_{q\ell}$.

In MCA, the test statistics are functions of the descriptive statistics involved in the contributions of points and deviations (MULTIPLE, §5.1.2, p.196), in the line of the paradigm $\chi^2 = n\,\Phi^2$. The test value (typicality statistic) is the product of $\cos\theta$ by $\sqrt{N-1}$; the homogeneity statistics are functions of the within contributions and η^2 ratio, etc. A test statistic of the "deviation" sort (Z, T...) is the product by a term of $\sqrt{n}$ order of the corresponding descriptive statistic; a test statistic of the "quadratic" sort (T^2, etc.) is the product by a term of order n of a descriptive statistic of the square norm sort.

As direct inductive extensions of the descriptive aids to interpretation, combinatorial procedures deserve to be used in GDA much more widely than they are today.

8.3.4 Chance Formulations and Tests for Randomness

Chance formulations in statistical inference have a long–standing history, significance tests being interpreted as *tests for randomness.* A significant effect is regarded as an event that "cannot be attributed to chance", and for which one is authorized to search an interpretation[20]. This conception is basically sound — if it is well kept in mind that the p–level is not "the probability that the effect is due to chance"! — if only as a rhetorical device, and it illuminates the link between combinatorial inference and genuine probabilistic inference. Thus for the *typicality* test, a significant result means that the group of observations *cannot be assimilated to a random sample* of the population of interest; for the *homogeneity* test, a significant result means that the data set is not assimilable to a pair of random samples from a same population; etc.

8.3.5 Permutation Tests and Modeling

Permutation tests. Permutation tests, initiated by Fisher and by Pitman (1937), are classical nonparametric tests (Kendall & Stuart, 1973, Volume 2, p.506). Technically, they amount to the homogeneity tests presented above — though they are usually presented under probabilisitc formulations. Permutation modeling, which leads to permutation tests, can be regarded as the main alternative to normal modeling.

Permutation modeling for one group (univariate). To present permutation modeling, we take again the protocol of differences stemming from two matched groups. We consider a set of n numerical observations $(d^i)_{i\in I}$, with mean $d_{obs} = \sum_{i\in I} d^i/n$ and variance $v = \sum_{i\in I}(d^i)^2/n - d^2_{obs}$.

For testing $\mathcal{H}_0 : \delta = 0$, the protocol space is generated by changing each individual observation d_i into $-d_i$ one at a time, hence 2^n protocols, and the statistic d over this space. The permutation distribution of d has the properties $\text{Mean}\, d = 0$, and $\text{Var}\, d = \sum_{i\in I}(d^i)^2/n^2 = (v + d^2_{obs})/n$. Observed levels are defined by proportions of d–values more extreme than d_{obs}, namely $P(|d| > |d_{obs}|) = p_{two-sided}$ and $\frac{1}{2}p_{two-sided} = p_{one-sided}$.

The statistic d can also serve to test the *shifted hypothesis* $\mathcal{H}'_0 : \delta = \delta_0$. In this case, the permutation distribution of d has the two properties:

[20]The word "authorized" is in Laplace; remember also the title of Pearson (1900): "On the criterion that a given system of deviations ... is such that it can be reasonably supposed to have arisen in random sampling."

Mean $d = \delta_0$ and Var $d = \sum_{i \in I} \frac{(d^i - \delta_0)^2}{n^2}$, or, since $\sum_{i \in I} \frac{(d^i - \delta_0)^2}{n} = \sum_{i \in I} \frac{(d^i - d_{obs})^2}{n} + (\delta_0 - d_{obs})^2$, Var $d = \left(v + (\delta_0 - d_{obs})^2\right)/n$. Equivalently, one can therefore take as test statistic $z' = \sqrt{n}\,(d - \delta_0)/\sqrt{v + (\delta_0 - d)^2}$.

For moderate n, the distribution of z' can be approximated by a standard normal distribution, therefore d_{obs} differs significantly from δ_0 at the two-sided level α if and only if $\sqrt{n}\,|d_{obs} - \delta_0|/\sqrt{v + (\delta_0 - d_{obs})^2} > z_\alpha$. Confidence interval at level α follows; its half–width is the set of δ–values such that $\sqrt{n}\,|d_{obs} - \delta|/\sqrt{v + (\delta - d_{obs})^2} < z_\alpha$, or equivalently $|\delta - d_{obs}| < z_\alpha\sqrt{v/n}/\sqrt{1 - z_\alpha^2/n}$. This half–width is to be compared with that of normal theory (based on Student's distribution) $t_{n-1}[\alpha]\sqrt{v/(n-1)}$. Permutation results are more conservative than normal–theory ones, that is, they yield less significant results and wider confidence intervals. Still, for moderate n, and in the "useful zone" of α (roughly from .20 to .01), the values $z_\alpha/\sqrt{1 - z_\alpha^2/n}$ are numerically close to $t_{n-1}[\alpha]$, consequently the results of normal theory also provide an approximation to permutation results.

• *Student's data.* There are $2^{10} = 1024$ protocols; 9 observations are strictly positive and one is null, therefore $P(d > d_{obs}) = 1/2^9 = .00195$. The result (though highly significant) is less significant than that of normal theory[21] ($p = .0014$).

The approximate confidence interval at level $\alpha = .05$ has half–width $|\delta - d_{obs}| = 0.922$, hence the lower confidence limit $1.58 - 0.92 = 0.66$. The results of normal theory, half–width 0.879 and lower limit 0.70 respectively (cf. p.306), compare fairly well with these results, especially bearing in mind the really small size $n = 10$. For $\alpha = .10$, the half–width is equal to 0.71 in both modelings.

The range of permutation modeling. In the frequentist framework, permutation tests are confined to the situations where there is a random allocation of units to treatments: the so–called "physical act" of *randomization*[22]. Since in many situations — for instance in a "before–after" design, or when comparing two "natural groups" (e.g. boys vs girls etc.) — randomization is either inexistent or impossible, this conception severely restricts the range of application of permutation tests. In fact, permutation tests as combinatorial procedures have an unrestricted validity; furthermore, they lend themselves naturally to probabilistic interpretations[23].

[21] In permutation theory, $p_{one-sided}$ cannot be less than $1/2^n$, whereas in normal theory, there is no lower bound.

[22] See Edgington (1987, esp. p.17-21), Cox & Hinkley (1974, p.179-204).

[23] See Bayesian permutation modeling and the "direct road" (p.321).

The practice of permutation tests has also been hindered for long by computational obstacles. Nowadays, for small samples, exact computations can be performed[24]; for intermediate sizes, Monte Carlo techniques can be used; and for large samples, asymptotic approximations can be used. Also, as already seen, in the "useful zone", the permutation results may be approximated pretty well by those of normal–theory.

To sum up: Permutation modeling provides an alternative to normal modeling, that may be preferred especially in experimental and quasi–experimental setups: See e.g. the Parkinson case study (chapter 9, §9.1, p.336). It goes without saying that in each situation, the modeling should be expressed explicitely, even when the choice is unconsequential.

8.4 Bayesian Data Analysis

Probabilities of hypotheses. The users of statistical inference spontaneously interpret the results of procedures in terms of probabilities of hypotheses. Finding a "significant t" triggers the statement that "the treatment probably has an effect"; if the t is "highly significant", this probability is felt to be "beyond any reasonable doubt", etc. This leads one to the question: Is there a way to effectively *calculate the probabilities* of hypotheses of interest given data? The answer is *Yes*, provided one goes beyond the frequentist framework and turns to an approach that restores probability its natural interpretation, that is, going from the known (the data) to the unknown (the hypotheses of interest). Such an approach is provided by the Bayesian framework of statistical inference.

Bayesian modeling. Let us resume *statistical modeling* with the notation of §8.2.2 (p.304): parameter θ and sampling distribution $p(x|\theta)$ (written as a density). In Bayesian inference, a *prior distribution* $p(\theta)$ (again written as a density) is postulated, which expresses the uncertainty about θ independently from the data; this prior is combined with $p(x|\theta)$ and data x to yield the *posterior distribution* $p(\theta|x)$, via Bayes' theorem:

$$p(\theta|x) = \frac{p(\theta)\,p(x|\theta)}{\int_\theta p(\theta)\,p(x|\theta)\,d(\theta)} \tag{8.3}$$

The posterior distribution expresses the uncertainty about parameter θ, conditionally on data x: Bayesian inference not only gives sense to the probabilities of hypotheses about θ, but it renders their calculation feasible.

[24]See e.g. the Exact Test package by Metha & al.

Which distribution should be taken as a prior? This is the crucial question in Bayesian inference. In the Data Analysis Philosophy, the prior distribution should not reflect the researcher's beliefs, but tend to be *neutral* with respect to them. Neutral prior distributions (also called "ignorance" or else "reference" prior distributions) should be *diffuse*, at least locally, i.e. in the vicinity of parameter values favored by the data. The ensuing posterior distributions can then be regarded as expressing the information brought by data, and providing the Bayesian form of inductive data analysis, namely *Bayesian Data Analysis*[25]. In brief, Bayesian Data Analysis is Bayesian inference done with neutral priors. Technically, it is often convenient to take as neutral priors *noninformative distributions* in Jeffreys' sense, that is, *uniform priors* for centrality parameters (e.g. mean), and *loguniform priors* for dispersion parameters (e.g. variance).

Normal–modeling. For normal–modeling (§8.2.2, p.304), assuming a noninformative prior, the basic result is the following (Jeffreys, 1961). The distribution of the true effect δ is such that $\sqrt{n}\frac{\delta-d_{obs}}{s_{obs}} \sim t_q$, that is, the posterior distribution of the true effect δ is a *scaled t–distribution* :

$$\frac{\delta}{s_{obs}} \sim t_q\Big((\frac{d}{s})_{obs}, \frac{1}{n}\Big) \quad \text{or} \quad \sqrt{n}\,\frac{\delta - d_{obs}}{s_{obs}} \sim t_q \tag{8.4}$$

That is, if the observed effect and the true effect are both scaled by the observed value s_{obs}, the posterior distribution is obtained by shifting the sampling distribution under $\mathcal{H}_0$ by the amount $(\frac{d}{s})_{obs}$ (Figure 8.3, p.318).

Credibility intervals. In Bayesian Data Analysis, the posterior distribution of the true effect δ is centered on the observed effect d_{obs}, and the confidence interval at level α becomes the *γ–credibility interval* $\mathcal{I}_\gamma$, letting $\gamma = 1 - \alpha$, that is, in terms of the posterior probability, one has $P(\delta \in \mathcal{I}_\gamma) = 1 - \alpha = \gamma$.

Significance level and directional hypothesis. For $\alpha/2 = p_{one-sided}$, one obtains the *reinterpretation of the observed significance level* of the test of $\mathcal{H}_0 : \delta = 0$. In Bayesian Data Analysis, $1 - p_{one-sided}$ is the posterior probability that δ has the same sign as the observed effect d_{obs}; e.g. if $d_{obs} > 0$, one has $P(\delta > 0) = 1 - p_{one-sided}$ (Figure 8.3, p.318).

[25]Since the early days of Laplace (1825), Bayesian Data Analysis has had a very long history, even though in the twentieth century it was superseded by the frequentist school. Bayesian Data Analysis is attaining full recognition today (a paper in the journal *Science* in 2000!). For our long–standing commitment to Bayesian Data Analysis, see Rouanet & al (1976), Rouanet (1996), Rouanet & al (1998).

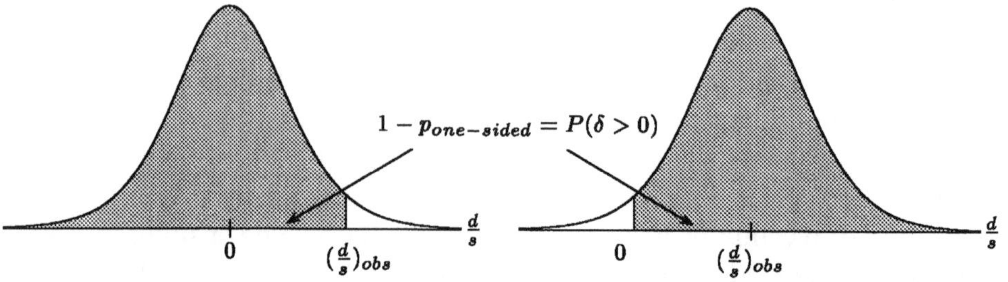

Figure 8.3. Sampling distribution under $\mathcal{H}_0 : \delta = 0$ (left); posterior distribution (right).

• *Student's data.* From the distribution $\frac{\delta}{s_{obs}} \sim t_9(\frac{1.58}{s_{obs}}, \frac{1}{10})$, with $s_{obs} = 1.230$, we get: $P(\delta > 0) = P(t_9 > \sqrt{10}\,\frac{0-1.58}{1.230}) = 1 - p_{one-sided} = .9986$ (cf. p.306: $p_{one-sided} = .0014$), hence the conclusion:

There is a very high posterior probability that the true effect is positive.

The p–value of test of $\mathcal{H}_0 : \delta = 0$ can thus be interpreted in three different ways. In the Combinatorial framework, p is the proportion of samples from a population that are more extreme than data. In the frequentist framework (normal–theory), p is the frequentist probability of a more extreme sample when the null hypothesis $\delta = 0$ is true. In the Bayesian framework (diffuse prior), p (one-sided) is the posterior probability that the true effect is in the direction opposite to the observed effect. The last interpretation, which was that of Laplace (for large samples at least), deserves the name of *Laplacian interpretation* of significance level[26].

Largeness and smallness hypotheses. From the posterior distribution of δ, the probability of *any hypothesis of interest* about this parameter can be calculated. In the Data Analysis Philosophy, all hypotheses that extend descriptive conclusions should be examined with the prospect of their corroboration, to begin with *directional hypotheses*, especially *largeness* and/or *smallness* hypotheses. If the bulk of the posterior distribution lies in the region of large effects (as for Student's data), the probability is high that the true effect is large; if it lies in the region of small effects, the probability is high that the true effect is small. More specifically, if a *credibility criterion* γ not far from 1 is set, together with a largeness limit d_{lar}, largeness of effect can be asserted if:

$$P(\delta > d_{lar}) = P(t_q > \sqrt{n}\,\frac{d_{lar}-d_{obs}}{s_{obs}}) > \gamma \quad (d_{obs} > d_{lar} > 0) \tag{8.5}$$

[26]Observe that with a diffuse prior, the p–level is in no case the probability of the point null hypothesis (a hypothesis that usually is not of interest, as already said); the prior probability of $\delta = 0$ is zero, hence the ensuing posterior probability is also zero.

Similarly taking a smallness limit $d_{sma} > 0$, smallness of effect can be asserted if $P(|\delta| < d_{sma}) > \gamma$, that is, for $|d_{obs}| < d_{sma}$, if

$$P(|\delta| < d_{sma}) = P(t_q < \sqrt{n}\tfrac{d_{sma}-d_{obs}}{s_{obs}}) - P(t_q > \sqrt{n}\tfrac{d_{sma}+d_{obs}}{s_{obs}}) \qquad (8.6)$$

Figure 8.4 summarizes the discussion.

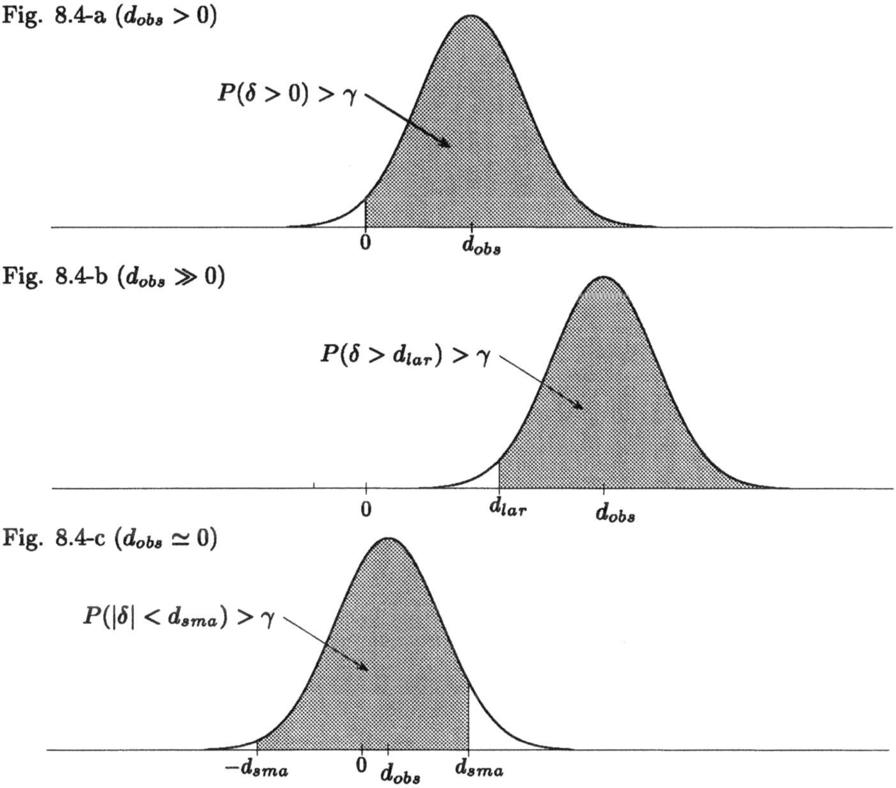

Figure 8.4. *Asserting inductive conclusions.* Fig. 8.4-a: Descriptive directional conclusion ($d_{obs} > 0$) and Bayesian directional conclusion $P(\delta > 0) > \gamma$ (taking e.g. $\gamma = .90$). Fig. 8.4-b: Descriptive largeness conclusion ($d_{obs} \gg 0$) and Bayesian largeness conclusion $P(\delta > d_{lar}) > \gamma$. Fig. 8.4-c: Descriptive smallness conclusion ($|d_{obs}| \simeq 0$) and Bayesian smallness conclusion $P(|\delta| < d_{sma}) > \gamma$.

• ***Student's data.*** Taking as a large effect an "amount of additional sleep" of 1 hour, we compute $P(\delta > 1) = P(t_9 > \sqrt{10}\frac{1-1.58}{1.230}) - P(t_9 > -1.49) = .92$.

The probability that the true effect is large on the positive side is .92.

At this point, it should be realized that establishing the largeness of an effect is a more demanding task than merely establishing its direction, so taking a less stringent criterion than $\gamma = .95$ (the Bayesian counterpart of the $\alpha = .05$) may be in order. In applications, we take $\gamma = .90$ for largeness as well as smallness. Here since $P(\delta > 1) > .90$, we can assert that for Student's data not only the mere directional hypothesis, but also the stronger largeness hypothesis is validated (Figure 8.5).

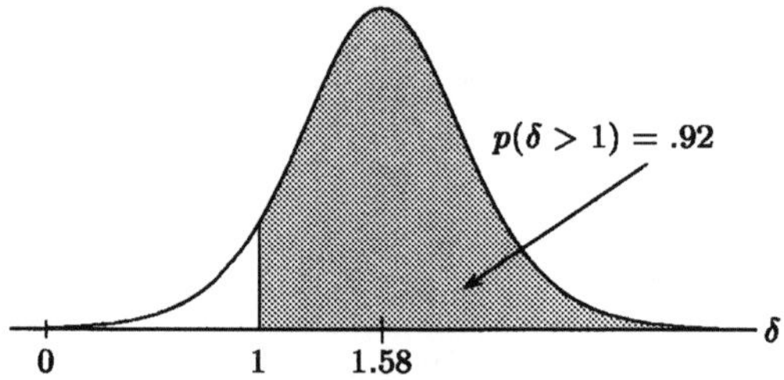

Figure 8.5. *Student's data.* Asserting direction of effect $P(\delta > 0) = .9986 = 1 - .0014$. Asserting largeness $P(\delta > 1) = .92$.

Naturally, when the posterior distribution has a wide dispersion — owing to small sample size and/or to large variance — the descriptive conclusions cannot be extended to inductive ones: see Figure 8.6.

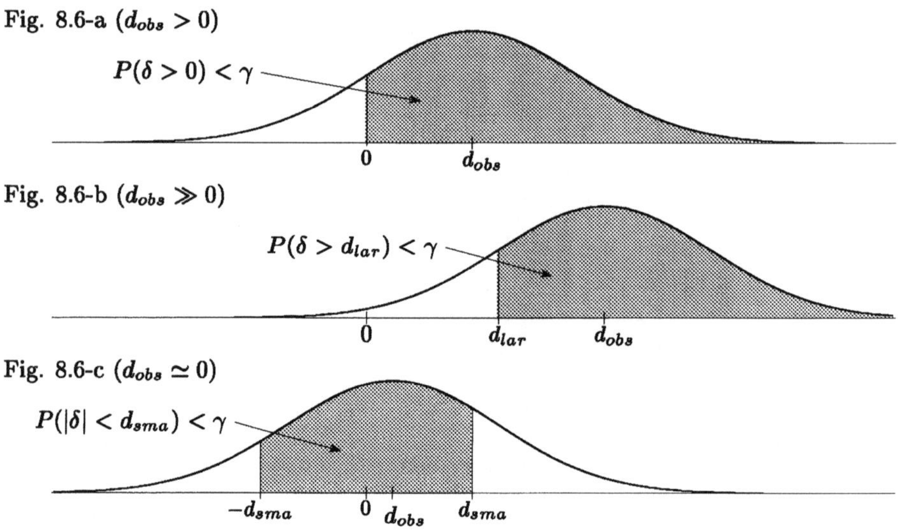

Figure 8.6. *No inductive conclusions.* Fig. 8.6-a: Descriptive directional conclusion ($d_{obs} > 0$), no Bayesian directional conclusion $P(\delta > 0) < \gamma$ (taking e.g. $\gamma = .90$). Fig. 8.6-b: Descriptive largeness conclusion ($d_{obs} \gg 0$), no Bayesian largeness conclusion $P(\delta > d_{lar}) < \gamma$. Fig. 8.6-c: Descriptive smallness conclusion ($|d_{obs}| \simeq 0$), no Bayesian smallness conclusion $P(|\delta| < d_{sma}) < \gamma$.

At this point, there is again an essential dissymetry. A directional or even a largeness hypothesis may be reached provided the observed effect d_{obs} is elevated enough, even if the sample size is comparatively small. Whereas for the smallness hypothesis, even with the most favorable data, namely

$d_{obs} = 0$, a sufficiently large sample size is indispensable; for instance, taking $d_{sma} = 0.4\,s$, at least $n = 19$ observations are needed for the conclusion $P(|\delta| < d_{sma}) > .90$ to be reached.

Bayesian permutation modeling. In the permutation test of $\mathcal{H}'_0 : \delta = \delta_0$ (§8.3.5, p.314), the proportions of protocols can be converted to probabilities, providing an operational characterization of null hypotheses, with no randomness assumption. Then, setting a noninformative prior, a posterior distribution on δ is derived that differs from that of the normal modeling; however, in the "useful zone" for credibility intervals, the latter provides an approximation to the former.

The direct road. Bayesian procedures are often derived by embedding a frequentist model into the Bayesian framework. Bayesian permutation modeling suggests an alternative "direct road" from combinatorial to Bayesian, based on exchangeability (de Finetti, 1974, Suppes, 2002), that bypasses the frequentist framework. This approach finds its place in the current objectively–minded Bayesian developments that aim at modeling situations where replication is either absent or impossible.

Specific inference. Specific inference, that is, "inference in context" based on the sole relevant data, is adequately captured by the Bayesian concept of *partial sufficiency*. This concept enables one to state the condition under which specific inference leads to the same conclusions as the general model, that is, whenever in the prior distribution of the general model the parameter subspace of the specific model is independent of the remaining parameter space[27]. This condition delineates the judicious utilization of specific modeling and inference (Rouanet & Lecoutre, 1983).

Comparison of two means. The Bayesian analysis directly extends Students's t–test : $\delta/s_{obs} \sim t_q\big((d/s)_{obs}, 1/n\big)$.

For *two matched groups*, one takes the protocol of n differences, with $q = n - 1$ and the corrected variance s^2 defined p.301.

For *two independent groups*, one takes $q = n_1 + n_2 - 2$ and the within mean square s^2 defined p.307, and one replaces n by $\tilde{n} = 1/\big(\frac{1}{n_1} + \frac{1}{n_2}\big)$ (equivalent sample size).

[27]Lindley (1965, p. 85) puts the matter this way, in connection with two matched groups: "We can write ... $p(x_1, x_2) = p(z, w) = p(z)\,p(w|z)$. Whilst $p(z)$ depends on δ, $p(w|z)$ *typically* will not, and, whatever parameters the latter does depend on, they *usually* do not occur in $p(z)$ and will, prior to the observations, be independent of those in $p(z)$. Consequently as far as inferences about δ are concerned, one may confine attention to $p(z)$ ... and the inference using z only will not lose any information." (italics ours).

8.5 Inductive Geometric Data Analysis

For multidimensional data conceptualized as clouds, numerical means become mean–points, and numerical effects become geometric vector–effects. In the case studies (chapter 9), the *specific approach* will be adopted, that is, starting from an overall cloud, statistical modeling will be put on specific clouds (subclouds and/or projected clouds) and hypotheses will be investigated pertaining to the mean–points of specific clouds. Accordingly, the present introduction to multidimensional procedures will focus on the inference on the mean–point of a cloud. We first recall the classical multinormal modeling, traditional significance testing and confidence ellipsoids; then we outline permutation modeling and Bayesian Data Analysis.

As everywhere in the book, the notation will be basically geometric, preserving the distinction between points and geometric vectors, whereas matrix notation (bound to a Cartesian frame) will be used for covariance matrices, in accordance with the current usage in Multivariate Analysis.

8.5.1 Geometric Statistical Modeling

Basic descriptive notions. Let us consider a cloud of n points in an r–dimensional Euclidean space[28] with an *observed mean–point* denoted D_{obs} or simply D. Referring the cloud to an orthonormal Cartesian frame, we denote $\mathbf{V}_{obs}$ or simply $\mathbf{V}$ the associated observed variance–covariance matrix (sums of squares divided by n), and $\mathbf{S}_{obs}$ or simply $\mathbf{S}$ the corrected matrix with q d.f. (sums of squares divided by q, with $q = n - 1$ in the elementary case). Taking origin O of the Cartesian frame as a reference point, we call *observed effect* the vector–deviation of the mean–point D_{obs} from the point O: $\mathrm{D}_{obs} - \mathrm{O} = \overrightarrow{\mathrm{OD}}_{obs}$, denoted[29] $\overrightarrow{d}_{obs}$ or simply $\overrightarrow{\mathrm{d}}$.

The coordinates of point D_{obs} (or of vector $\overrightarrow{d}_{obs}$) is represented by a column–matrix denoted $\mathbf{d}_{obs}$ or simply $\mathbf{d}$. With matrix $\mathbf{V}$ there is associated the κ–norm defined in chapter CLOUD (p.100), and denoted $|\cdot|_\kappa$, that is, $|\overrightarrow{\mathrm{OD}}|^2_\kappa = |\overrightarrow{\mathrm{d}}|^2_\kappa = \mathbf{d}'\,\mathbf{V}^{-1}\mathbf{d}$. In Multivariate analysis, it is customary to also consider the S–norm, associated with $\mathbf{S}$, that we denote $|\cdot|_{\mathrm{S}}$, such that

$$|\overrightarrow{\mathrm{OD}}|^2_{\mathrm{S}} = |\overrightarrow{\mathrm{d}}|^2_{\mathrm{S}} = \mathbf{d}'\,\mathbf{S}^{-1}\mathbf{d} \tag{8.7}$$

[28] In the notation of preceding chapters, r is equal to K, or L (dimensionality of support), or L' (projected cloud).

[29] As a general rule, variable entities (points, vectors or coordinates) will be denoted by italic letters, as opposed to the corresponding observed entities, written in roman letters or indexed by "obs", e.g. $\overrightarrow{\mathrm{d}}$ alias $\overrightarrow{d}_{obs}$.

The S–norm is proportional to the κ–norm; for $q = n-1$, one has $|\overrightarrow{\mathrm{d}}|^2_{\mathrm{S}} = \frac{n-1}{n}|\overrightarrow{\mathrm{d}}|^2_\kappa$ (one thus has $|\cdot|_{\mathrm{S}} < |\cdot|_\kappa$). These two norms will be distinguished from the geometric norm denoted $\|\cdot\|$e.g. $\mathrm{OD} = \|\overrightarrow{\mathrm{OD}}\| = \|\overrightarrow{\mathrm{d}}\| = (\mathbf{d'd})^{1/2}$.

• *Target Example* (CLOUD, p.125) with $r=2$, $n=10$, $q=9$, $\mathbf{V} = \boxed{\begin{matrix}40 & 8\\ 8 & 52\end{matrix}}$.
With respect to the Cartesian frame with origin O (center of the target), the mean–point D_{obs} has coordinates $\mathbf{d}' = \boxed{6\ 0}$, one has $|\overrightarrow{\mathrm{OD}}_{obs}|^2_\kappa = \mathbf{d}'\,\mathbf{V}^{-1}\mathbf{d} = 0.92857$, hence $|\overrightarrow{\mathrm{OD}}_{obs}|_\kappa = |\overrightarrow{d}_{obs}|_\kappa = 0.964$ (Figure 8.7-a).

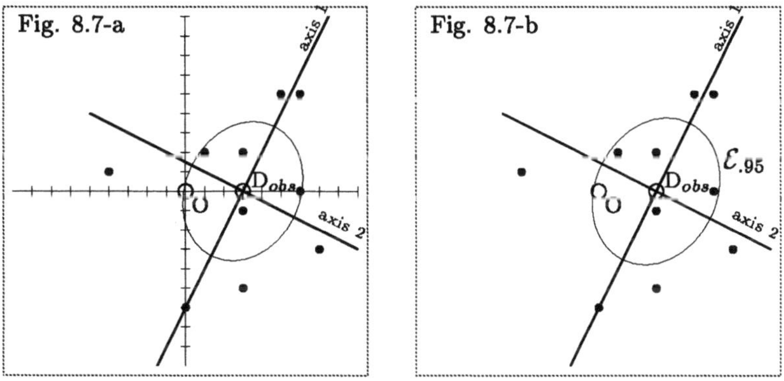

Figure 8.7. *Target Example* ($v = 40$, $v' = 52$, $c = 8$; $\lambda_1 = 56$, $\lambda_2 = 36$). Mean point D_{obs}, and reference point O (center of target). Fig. 8.7-a: indicator ellipse ($\kappa = 1$). Fig. 8.7-b: Confidence ellipse $\mathcal{E}_{.95}$ ($\alpha = .05$) under normal modeling ($F_{2,8}[.05] = 4.46$, $\kappa = 1.056$).

Statistical modeling. The intuitive rationale for statistical inference is that a point that has been observed is just one outcome among a set of possible outcomes, e.g. mean–point D_{obs} is supposed to be the realization of a *Mean–point* variable D (written in italics), varying in some *sample space* specified by the statistical modeling. Like in the univariate case, we will take up the convenient "trueness metaphor", that is, the Mean–point variable D will be thought to vary around a *true mean point* Δ on which the inference pertains, such that $D = \Delta + \overrightarrow{e}$, where $\overrightarrow{e}$ is an *error–vector* term. Accordingly, in terms of vector deviations from the point O, the observed effect is supposed to be the realization of a vector–variable $\overrightarrow{d}$ such that $\overrightarrow{d} = \overrightarrow{\delta} + \overrightarrow{e}$, where $\overrightarrow{\delta}$ denotes the *true vector–effect*.

In the inference procedure, the true mean–point Δ (parameter) is estimated by the observed mean–point D_{obs}, and the true effect $\overrightarrow{\delta}$ by the observed effect $\overrightarrow{d}_{obs}$ (See opposite Figure).

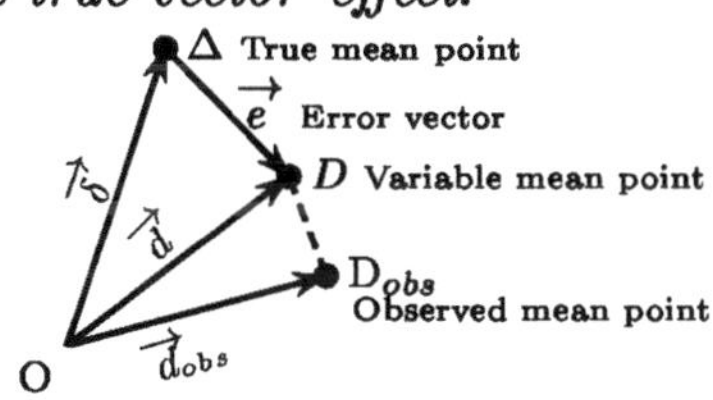

In *normal modeling*, the cloud of n points is a sample from an r–dimensional normal distribution centered on the true mean point Δ (main parameter). There is a secondary parameter, namely the covariance matrix $\mathbf{\Sigma}$, estimated by $\mathbf{V}$ (maximum likelihood estimate), or by $\mathbf{S}$ (unbiased estimate). The basic sampling properties involve the *Wishart*, χ^2 and *Snedecor* F distributions (see, e.g. Anderson, 1958). In matrix terms, denoting $\boldsymbol{d}$ (italics) the vector– variable whose realization is $\mathbf{d}$, and $\boldsymbol{S}$ (italics) the matrix–variable whose realization is $\mathbf{S}$, and reading $\sim$ "is distributed as", one has for fixed $\boldsymbol{\delta}$ and $\mathbf{\Sigma}$ (see e.g. Rouanet, Lecoutre, 1983):

• $\boldsymbol{d}$ and $\boldsymbol{S}$ are statistically independent.

• $\boldsymbol{d} \sim \mathcal{N}(\boldsymbol{\delta}, \mathbf{\Sigma}/n)$ (r-dimensional normal distribution of mean $\boldsymbol{\delta}$ and covariance matrix $\mathbf{\Sigma}/n$).

• $q\,\boldsymbol{S} \sim \mathcal{W}_{r,q}(\mathbf{\Sigma})$ ($\mathbf{\Sigma}$–scaled r- dimensional Wishart with r and q d.f.).

• $n\,(\boldsymbol{d} - \boldsymbol{\delta})'\,\boldsymbol{S}^{-1}(\boldsymbol{d} - \boldsymbol{\delta}) \sim \frac{r}{k}\,F_{r,q-r+1}$ (central Snedecor F distribution with r and $q-r+1$ d.f.), letting $k = (q-r+1)/q$.

Hotelling's test of no–effect. The null hypothesis of Hotelling's test is the point hypothesis that the true mean–point is O, $\mathcal{H}_0 : \Delta = \text{O}$, or equivalently that the true effect is null. Let $T^2 = n\,|\overrightarrow{d}\,|^2_S$ (Hotelling's test statistic). Under $\mathcal{H}_0$, one has $T^2 \sim \frac{r}{k}\,F_{r,q-r+1}$. Let $T^2_{obs} = n\,|\overrightarrow{d_{obs}}|^2_S$ (observed value of Hotelling's statistic). If $F[\alpha]$ denotes the critical value of the $F_{r,q-r+1}$ distribution at the upper level α, then if $\frac{k}{r}\,T^2_{obs} > F[\alpha]$, Hotelling's test is significant at level α. The observed level $p_{obs} = P(F_{r,q-r+1} > \frac{k}{r}\,T^2_{obs}) = P(T^2 > T^2_{obs})$ under $\mathcal{H}_0$ is the sampling probability that under $\mathcal{H}_0$ the S–norm of effect $\overrightarrow{d}$ exceeds its observed value: $p_{obs} = P(|\,\overrightarrow{d}\,|_S > |\,\overrightarrow{d_{obs}}|_S)$ under $\mathcal{H}_0$.

In matrix notation, one has: $\mathcal{H}_0 : \boldsymbol{\delta} = \mathbf{0}$; $T^2 = n\,\boldsymbol{d}'\,\boldsymbol{S}^{-1}\boldsymbol{d}$; $T^2_{obs} = n\,\mathbf{d}'\,\mathbf{S}^{-1}\mathbf{d}$; $p_{obs} = P(\boldsymbol{d}'\,\boldsymbol{S}^{-1}\boldsymbol{d} > \mathbf{d}'\,\mathbf{S}^{-1}\mathbf{d})$ under $\mathcal{H}_0$.

$$T^2 = n\,|\,\overrightarrow{d}\,|^2_S = n\,\boldsymbol{d}'\boldsymbol{S}^{-1}\boldsymbol{d}; \qquad \text{under } \mathcal{H}_0 : T^2 \sim \frac{r}{k}\,F_{r,q-r+1} \tag{8.8}$$

Particular cases

• $r = 1$. Hotelling's T^2 reduces to Student's t^2 and the p–value is thus the extension of the two-sided p–value of the univariate test $\delta = 0$.

• $q = n-1$ (*elementary case*). One has $q - r + 1 = n - r$; $k = \frac{n-r}{n-1}$,

$$T^2 \sim \frac{n-1}{n-r}\,r\,F_{r,n-r} \quad \text{under } \mathcal{H}_0; \quad T^2_{obs} = (n-1)\,\mathbf{d}'\,\mathbf{V}^{-1}\mathbf{d} \tag{8.9}$$

• *Large q*. One has $k \simeq 1$, and $r\,F_{r,q-r+1} \simeq F_{r,\infty} = \chi^2_r$, therefore $T^2 \sim \chi^2_r$ under $\mathcal{H}_0$.

Shifted Hotelling's test and confidence ellipsoids. From a pure geometry standpoint, there is nothing special about the point O, so Hotelling's test readily extends to the hypothesis that the true mean point coincides with any given point Δ_0, that is, to the shifted point hypothesis $\mathcal{H}'_0 : \Delta = \Delta_0$. Let $T'^2 = n\,|\overrightarrow{\Delta_0 D}|^2_S = n\,(\boldsymbol{d}-\boldsymbol{\delta}_0)'\,\boldsymbol{S}^{-1}(\boldsymbol{d}-\boldsymbol{\delta}_0)$, $n\,T'^2 \sim \frac{r}{k}\,F_{r,q-r+1}$ under $\mathcal{H}'_0$. The observed level of the shifted test is $p'_{obs} = P(|\overrightarrow{\Delta_0 D}|_S > |\overrightarrow{\Delta_0 \mathrm{D}}_{obs}|\mathrm{s})$ under $\mathcal{H}'_0$. The result is significant at level α if and only if $n\,T'^2_{obs} > \frac{r}{k}F[\alpha]$, that is, if $|\overrightarrow{\Delta_0 \mathrm{D}}_{obs}|^2 = (\mathbf{d}-\boldsymbol{\delta}_0)'\,\mathbf{S}^{-1}(\mathbf{d}-\boldsymbol{\delta}_0) > \frac{r}{kn}\,F[\alpha]$.

Consider now the points Δ such that $|\overrightarrow{\mathrm{D}\Delta}|^2_S = \frac{r}{kn}\,F[\alpha]$; the locus of these points, that are at a same S–distance from the mean point D_{obs} (in brief D), is an ellipsoid belonging to the inertia ellipsoids of the cloud (cf. chapter CLOUD, §3.4, p.95). The *observed confidence ellipsoid* at level α, denoted $\mathcal{E}_{1-\alpha}$, is defined by the following constitutive confidence property.

$$|\Delta - \mathrm{D}|^2_{\mathrm{S}} = |\overrightarrow{\mathrm{D}\Delta}|^2_{\mathrm{S}} < \frac{r}{k\,n}\,F[\alpha] \tag{8.10}$$

For any point Δ outside $\mathcal{E}_{1-\alpha}$, the hypothesis $\mathcal{H}'_0 : \Delta = \Delta_0$ is not compatible with data at level α; for any point $\Delta_0 \in \mathcal{E}_{1-\alpha}$, it is compatible with data.

• *Particular cases.* i) For $r = 1$, the confidence ellipsoid at level α reduces to the confidence interval at level α defined by $|\delta - d_{obs}| < t_\alpha \times s\sqrt{n}$.
ii) $q = n-1$ (*elementary case*). The constitutive property becomes

$$(|\overrightarrow{\mathrm{D}\Delta}|^2_{\mathrm{S}} < \frac{n-1}{n}\,\frac{r}{n-r}\,F_{r,n-r}[\alpha]), \quad \text{or } (|\overrightarrow{\mathrm{D}\Delta}|^2_{\kappa} < \frac{r}{n-r}\,F[\alpha]) \tag{8.11}$$

iii) For *large* q, the constitutive property becomes $(|\overrightarrow{\mathrm{D}\Delta}|^2_S < \frac{\chi_r{}^2[\alpha]}{n})$.

• *Target Example. Significance test*: Let us test the hypothesis that the true mean point Δ coincides with the center of the target: $\mathcal{H}_0 : \Delta = \mathrm{O}$. One has: $T^2_{obs} = 9\times|\overrightarrow{\mathrm{OD}}_{obs}|^2_\kappa = 9\times 0.92857 = 8.357$, $k = 8/9$ hence $\frac{k}{r}\,T^2_{obs} = \frac{8}{9}\times\frac{1}{2}\times 8.357 = 3.714$, a value between $F[.10] = 3.11$ and $F[.05] = 4.46$; the observed level is $p_{obs} = .072$. As a *conclusion*, the hypothesis that the true mean point Δ coincides with the center O of the target is compatible with the data at level $\alpha = .05$ (but not at level $\alpha = .10$). The *confidence ellipse* $\mathcal{E}_{1-\alpha}$ is the locus of the points Δ such that $(|\overrightarrow{\mathrm{D}\Delta}|^2_\kappa < \frac{2}{8}\,F[\alpha])$. For instance, taking $\alpha = .05$, hence $F[.05] = 4.46$, one has $|\overrightarrow{\mathrm{D}\Delta}|_\kappa < 1.056$ and the ellipse $\mathcal{E}_{.95}$ (Fig. 8.7-b, p. 323). As expected from the nonsignificant result at level .05, the point O lies inside $\mathcal{E}_{.95}$.

Remark. Like in univariate case, neither the levels of tests nor confidence ellipsoids can be interpreted in terms of probabilities about parameters, as long as one sticks to the frequentist framework.

Comparing the mean points of two independent groups. Like the t–test (p.307), Hotelling's test extends to the comparison of the mean points of two independent groups $S\langle G_2 \rangle$. The difference $\overrightarrow{\delta}$ of the true mean points is estimated by the difference $\overrightarrow{d_{obs}}$ of observed mean points. Assuming the multinormal model with homogeneity of within–group covariances, the common within–group covariance matrix $\mathbf{\Sigma}$ is estimated by $\mathbf{S} = \big((n_1-1)\mathbf{S}_{g1} + (n_2-1)\mathbf{S}_{g2}\big)/(n_1+n_2-2)$, weighted average of the two within–group corrected covariance matrices (letting $q = n_1+n_2-2$).

Letting $\widetilde{n} = 1/(\frac{1}{n_1}+\frac{1}{n_2})$, one has $\mathbf{d} \sim \mathcal{N}(\boldsymbol{\delta}, \mathbf{\Sigma}/\widetilde{n})$. Then we deduce, letting $k = (q-r+1)/q$, that $\widetilde{n}\,(\boldsymbol{d}-\boldsymbol{\delta})'\,\boldsymbol{S}^{-1}(\boldsymbol{d}-\boldsymbol{\delta}) \sim \frac{r}{k}\,F_{r,q-r+1}$,

Permutation modeling. Returning to the case of a single group of observations, we now sketch multidimensional permutation modeling, as an alternative to multinormal modeling, like in univariate case (§8.3.5, p. 314). We start with the test of $\mathcal{H}_0 : \Delta = \mathrm{O}$; the permutation space is the set of 2^n clouds generated by changing each individual point M^i into the point N^i symmetric with respect to O, i.e. such that $\overrightarrow{\mathrm{ON}^i} = -\overrightarrow{\mathrm{OM}^i}$. Under $\mathcal{H}_0$ the permutation distribution of variable point D is centered on point O; its covariance matrix is $\mathbf{B}/n$, where $\mathbf{B} = \mathbf{V} + \mathbf{dd}'$ (matrix of raw mean squares and products). Taking as a test statistic the B–norm of $\overrightarrow{\mathrm{O}D}$, i.e. $|\overrightarrow{\mathrm{O}D}|_{\mathrm{B}} = (\boldsymbol{d}'\mathbf{B}^{-1}\boldsymbol{d})^{1/2}$, the observed level p_{obs} is the proportion of D mean points such that $P(|\overrightarrow{\mathrm{O}D}|_{\mathrm{B}} \geq |\overrightarrow{\mathrm{OD}}_{obs}|_{\mathrm{B}})$ or equivalently, letting $X^2_{obs} = n\,\mathbf{d}'\mathbf{B}^{-1}\mathbf{d}$: $p_{obs} = P(n\,\boldsymbol{d}'\mathbf{B}^{-1}\boldsymbol{d} \geq X^2_{obs})$.

In the same way, for testing the shifted hypothesis $\mathcal{H}'_0 : \Delta = \Delta_0$, we construct the 2^n clouds generated by changing each individual point M^i into the point N'^i such that $\overrightarrow{\Delta_0\mathrm{N}'^i} = -\overrightarrow{\Delta_0\mathrm{M}^i}$. Under $\mathcal{H}'_0$, the distribution of variable–point D' is centered on point Δ_0, and its covariance matrix is $\mathbf{B}_0/n$, where $\mathbf{B}_0 = \mathbf{V} + (\mathbf{d}-\boldsymbol{\delta}_0)(\mathbf{d}-\boldsymbol{\delta}_0)'$. The point D_{obs} differs significantly from Δ_0 at level α if and only if $P\Big(|\overrightarrow{\Delta_0 D}|_{\mathrm{B}_0}) \geq |\overrightarrow{\Delta_0\mathrm{D}}_{obs}|_{\mathrm{B}_0}\Big) \leq \alpha$, or equivalently, letting $X'^2_{obs} = n\,(\mathbf{d}-\boldsymbol{\delta}_0)'\mathbf{B}_0^{-1}(\mathbf{d}-\boldsymbol{\delta}_0)$, if $P\Big(n\,(\boldsymbol{d}-\boldsymbol{\delta}_0)'\mathbf{B}_0^{-1}(\boldsymbol{d}-\boldsymbol{\delta}_0)\Big) \geq X'^2_{obs}) \leq \alpha$. For moderate n, the distribution of $n\,(\boldsymbol{d}-\boldsymbol{\delta}_0)'\mathbf{B}_0^{-1}(\boldsymbol{d}-\boldsymbol{\delta}_0)$ can be approximated by the χ^2_r distribution, therefore the point D_{obs} differs significantly from Δ_0 at level α if and only if $n\,(\mathbf{d}-\boldsymbol{\delta}_0)'\big(\mathbf{V}+(\mathbf{d}-\boldsymbol{\delta}_0)(\mathbf{d}-\boldsymbol{\delta}_0)'\big)^{-1}(\mathbf{d}-\boldsymbol{\delta}_0) \geq \chi^2_r[\alpha]$.

The approximate *confidence ellipsoid* at level α follows, as the locus of Δ points such that $(|\overrightarrow{\mathrm{D}\Delta}|^2_{\mathrm{B}_0} < \frac{1}{n}\chi^2_r[\alpha])$. Since the B_0–norm and the κ–norms are related by the formula $|\overrightarrow{\mathrm{D}\Delta}|^2_{\mathrm{B}_0} = |\overrightarrow{\mathrm{D}\Delta}|^2_\kappa/(1+|\overrightarrow{\mathrm{D}\Delta}|^2_\kappa)$, the approximate confidence property can equivalently be written: $|\overrightarrow{\mathrm{D}\Delta}|^2_\kappa = (\chi^2_r[\alpha]/n)/(1-\chi^2_r[\alpha]/n)$.

As a result, confidence ellipsoids are inertia ellipsoids, like in normal theory; the preceding square κ–norm is to be compared to that of normal theory (based on F–distribution), namely $\frac{r}{n-r}F_{r,n-r}[\alpha]$. Like in the univariate case, permutation results are more conservative, but here again, for moderate sample sizes they produce numerically close results in the "useful zone" of α–levels for significance and confidence.

• *Target example.* For the permutation test of $\mathcal{H}_0 : \Delta = 0$, we have the matrix $\mathbf{B} = \boxed{\begin{matrix}76 & 8\\ 8 & 52\end{matrix}}$, $|\overrightarrow{\mathrm{OD}}|^2_{\mathrm{B}} = \mathbf{d}'\mathbf{B}^{-1}\mathbf{d} = 0.48148$, and $X^2_{obs} = n\,|\overrightarrow{\mathrm{OD}}|^2_{\mathrm{B}} = 4.81$. This value lies between $\chi^2_2[.10] = 4.61$ and $\chi^2_2[.05] = 5.99$, therefore the level of the approximate test is between .10 and .05, like for Hotelling's test; more precisely we have $P(\chi^2_2 > 4.81) = .090$, to be compared with the p–value .072 of Hotelling's test. As expected, the permutation test is more conservative. To get the approximate confidence ellipse $\mathcal{E}'_{.95}$ at level $\alpha = .05$, we compute $\frac{5.991/10}{1-5.991/10} = 1.494$; hence $\kappa = 1.222$, a value larger than the value 1.056 of normal–theory. Figure 8.8 shows the observed significance ellipse $\mathcal{S}_0$ (centered at O and going through point D_{obs}), and ellipse $\mathcal{E}'_{.95}$.

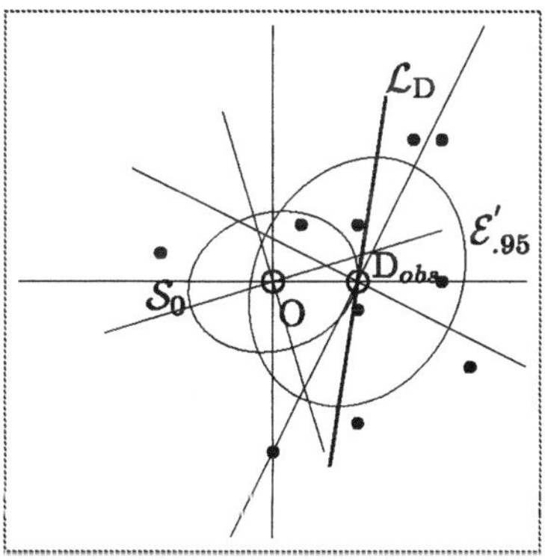

Figure 8.8. *Target example,* permutation modeling. Observed significance ellipse $\mathcal{S}_0$ and its tangent $\mathcal{L}_\mathrm{D}$ at point D_{obs}. Approximate confidence ellipse $\mathcal{E}'_{.95}$ ($\alpha = .05$) under permutation modeling ($\chi^2[.05] = 5.99$, $\kappa = 1.222$).

8.5.2 Bayesian Data Analysis

We will simply present Bayesian normal modeling; permutation modeling would follow a similar line (cf. the univariate case §8.3.5, p.314).

Multinormal Bayesian Data Analysis. In multinormal modeling, the noninformative prior on parameters $(\boldsymbol{\delta}, \boldsymbol{\Sigma})$ derived by Jeffreys' rule has a density $p(\boldsymbol{\delta}, \boldsymbol{\Sigma})$ proportional to $\det \boldsymbol{\Sigma}^{-(r+1)/2}$. Multinormal Bayesian inference involves distributions, such as the multivariate $\boldsymbol{t}$-distribution, not encountered in sampling inference; see Box & Tiao (1973), Press (1989) and Lecoutre (1999). Given D_{obs} or D (**d** in matrix notation), and **S**, one has (see Rouanet & Lecoutre, 1983) the following posterior distributions on the true point Δ ($\boldsymbol{\delta}$ in matrix notation) and on the S-norm of $\Delta - \mathrm{D}_{obs} = \overrightarrow{\mathrm{D}\Delta}$ ($\boldsymbol{\delta} - \mathbf{d}$ in matrix notation):

- $\boldsymbol{\delta} \sim \boldsymbol{t}_r(\mathbf{d}, \frac{1}{k}\frac{\mathbf{S}}{n})$ (multivariate $\boldsymbol{t}$-distribution centered on observed effect **d**, with a dispersion proportional to **S**).
- $n\,|\overrightarrow{\mathrm{D}\Delta}|_{\mathrm{S}}{}^2 = n\,(\boldsymbol{\delta} - \mathbf{d})'\,\mathbf{S}^{-1}(\boldsymbol{\delta} - \mathbf{d}) \sim \frac{r}{k}\,F_{r,q-r+1}$ (recall that $k = \frac{q-r+1}{q}$).

Credibility ellipsoids. In Bayesian Data Analysis, the posterior distribution of the true mean point Δ is centered on the observed mean point D_{obs}; the observed confidence ellipsoid $\mathcal{E}_{1-\alpha}$ becomes the *γ–credibility ellipsoid* $\mathcal{E}_\gamma$, letting $\gamma = 1 - \alpha$, that is, in terms of the posterior probability, we have: $P(\Delta \in \mathcal{E}_\gamma) = P(|\overrightarrow{\mathrm{D}\Delta}|^2_{\mathrm{S}} < \frac{r}{kn}\,F[\alpha])$, hence, since $|\overrightarrow{\mathrm{D}\Delta}|^2_{\mathrm{S}} \sim \frac{r}{k}\,F_{r,q-r+1}$, we have $P(\Delta \in \mathcal{E}) = 1-\alpha = \gamma$. Credibility ellipsoids are higher probability density zones, that is, every point inside an ellipsoid has a higher density than every point ouside it, and each ellipsoid is an isodensity contour.

Thus the family of inertia ellipsoids, which provides a descriptive summary of a cloud (chapter CLOUD, §3.4, p.95), provide in Bayesian Data Analysis a geometric summary of the posterior distribution of the true mean point of the cloud, interpretable in probabilistic terms.

Reinterpretation of the observed level of Hotelling's test. Let $\mathcal{E}_0$ be the inertia ellipsoid that goes through the point O; $\mathcal{E}_0$ is the locus of the Δ–points that are at the same S–distance, or equivalently at the same κ–distance, from the observed mean point D as the point O, i.e. such that $|\overrightarrow{\mathrm{D}\Delta}|_\kappa = |\overrightarrow{\mathrm{DO}}|_\kappa$. In terms of the posterior probability, one has: $P(|\overrightarrow{\mathrm{D}\Delta}|_\kappa > |\overrightarrow{\mathrm{DO}}|_\kappa) = p_{obs}$, that is, $\mathcal{E}_0$ is the ellipsoid with credibility $1 - p_{obs}$ (general notation for ellipsoids: $\mathcal{E}_0 = \mathcal{E}_{1-p_{obs}}$).

Therefore the observed level p_{obs} of $\mathcal{H}_0 : \Delta = \mathrm{O}$ is reinterpreted as the posterior probability that the true mean point Δ lies outside $\mathcal{E}_0$, that is, is more distant (in the sense of the κ–norm) from D_{obs} than the point O.

Distribution of magnitude of effect. Let us shift the γ–credibility ellipsoid by a translation vector equal to $-\overrightarrow{d_{obs}}$; the shifted ellipsoid is *centered at the origin* and is the locus of points Δ corresponding to effects of equal magnitudes in terms of the S–norm. The posterior distribution of the magnitude of effect, that is, of $|\overrightarrow{\mathrm{O}\Delta}|_{\mathrm{S}}$, has been studied under the name of "ψ-distribution" by Rouanet & Lecoutre (1983) and Lecoutre (1984), and implemented in software, see Lecoutre, Guigues & Poitevineau (1992).

One has: $n\,|\overrightarrow{\mathrm{O}\Delta}|_{\mathrm{S}}{}^2 = n\,\boldsymbol{\delta}'\,\mathbf{S}^{-1}\boldsymbol{\delta} \sim \frac{1}{k}\psi^2_{r,q-r+1}(k\,n\,|\overrightarrow{d_{obs}}|^2)$ (ψ^2 with r and $q-r+1$ d.f. and eccentricity $k\,n\,|\overrightarrow{d_{obs}}|^2 = rF_{obs}$). For $\overrightarrow{d_{obs}} = \overrightarrow{0}$, the ψ^2 distribution reduces to the F–distribution: $\psi^2_{r,q-r+1} = r\,F_{r,q-r+1}$. The ψ^2 distribution is especially useful in connection with the smallness hypothesis (see the *Parkinson case study* in chapter 9).

• *Target Example.* In Bayesian Data Analysis, the confidence ellipse at level $\alpha = .05$ (Figure 8.7, p.323) becomes the credibility ellipse $\mathcal{E}_{.95}$; that is, $\mathcal{E}_{.95}$ contains the true point Δ with a posterior probability equal to .95.

In Figure 8.9-a, ellipse $\mathcal{E}_0$ (going through the point O) is shown; $\mathcal{E}_0$ contains the true point Δ with a posterior probability equal to $1 - p_{obs} = 1 - .072 = .928$, hat is, $P(\Delta \in \mathcal{E}_0) = .928$.

In Figure 8.9-b ellipse $\mathcal{E}_{.90}$ is shown, defined from the critical value $F_{2,8}[.10] = 3.11$ by $|\overrightarrow{\mathrm{D}\Delta}|_\kappa{}^2 < \frac{2}{8} \times 3.11$, hence $|\overrightarrow{\mathrm{D}\Delta}|_\kappa < 0.882$; ellipse $\mathcal{E}_{.90}$ contains the true point Δ with a posterior probability equal to .90. (Observe that lines $\mathcal{L}_0$ and $\mathcal{L}'$ are parallel to line $\mathcal{L}_{\mathrm{D}}$ of Figure 8.8, p.327).

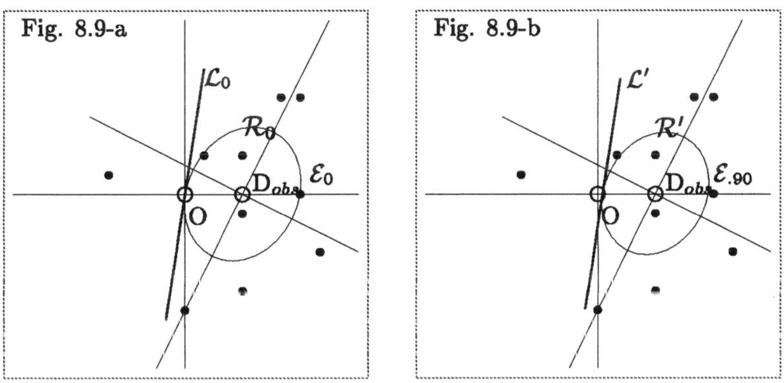

Figure 8.9. *Target Example.* Fig. 8.9-a: $\mathcal{E}_0$ ellipse ($p_{obs} = .072, 1-p_{obs} = .928$, $\kappa = 0.964$), tangent $\mathcal{L}_0$ and half–plane $\mathcal{R}_0$. Posterior probability $P(\Delta \in \mathcal{E}_0) = .928$. Fig. 8.9-b: $\mathcal{E}_{.90}$ ellipse ($F_{2,8}[.10] = 3.11$, $\kappa = 0.882$), tangent $\mathcal{L}'$ and half–plane $\mathcal{R}'$.

Probabilities of hypotheses. In Bayesian Data Analysis, the probability of *any hypothesis* about a parameter can be calculated from its posterior distribution — at least in principle — whether or not the hypothesis is backed by descriptive findings[30]. In practice, credibility ellipsoids may often provide bounds for the probabilities of hypotheses of interest. Hereafter, we investigate on the *Target example* multidimensional directional and largeness hypotheses.

Directional hypothesis. The observed mean point D_{obs} lies "on the right of the center 0 of the target". To operationalize the directional hypothesis, let us construct the tangent $\mathcal{L}_0$ to the ellipse $\mathcal{E}_0$ at the point 0 (Figure 8.9-a, p.329); line $\mathcal{L}_0$ divides the plane into two regions (half–planes), one of them $\mathcal{R}_0$ contains D_{obs}. We then operationalize the directional hypothesis as $\Delta \in \mathcal{R}_0$. One has $P(\Delta \notin \mathcal{E}_0) = .072$, therefore, since $\mathcal{E}_0 \subset \mathcal{R}_0$, one has $P(\Delta \notin \mathcal{R}_0) < .072/2 = .036$, therefore $P(\Delta \in \mathcal{R}_0) > .928 + .036 = .964$, so we can state: $P(\Delta \in \mathcal{R}_0) > .964$. For credibility $\gamma = .95$ (or less), we can assert the directional conclusion "The true point Δ lies on the right of the center 0 of the target".

*Largeness hypothesis.*Since $|\overrightarrow{d_{obs}}|_{\mathcal{K}} = .964$, we may consider that the deviation of mean point D_{obs} from the center is large. To operationalize the largeness hypothesis we consider the tangents to credibility ellipses at the points they intersect the horizontal axis.

For instance, for $\gamma = .90$ line $\mathcal{L}'$ divides the plane into two regions (half–planes), one of them $\mathcal{R}'$ contains mean point D_{obs}; see Figure 8.9-b, p.329. Since $\mathcal{E}_{.90} \subset \mathcal{R}'$, we can state: $P(\Delta \in \mathcal{R}') > .95$. This statement, however, does not establish the largeness of the true deviation, because the intersection point of $\mathcal{L}'$ and the horizontal axis is not far from point O. To be able to establish largeness of effect, we ought to find a line parallel to $\mathcal{L}$ both sufficiently far from point O and still such that the probability of the region on the right side of $\mathcal{L}$ would be high enough. This is not possible here, owing to the small sample size, so no largeness conclusion can be reached[31].

Comparing the mean points of two groups. The procedures apply directly to the comparison of the means of two matched groups, and they readily extend to the comparison of the means of two independent groups, like in the univariate case (cf. p.321 and p.326).

---

[30]A significant effect does not entail a large posterior probability for a largeness hypothesis, and a nonsignificant effect does not entail a large probability for a smallness hypothesis.

[31]Largeness conclusions will be reached, e.g. in the *Political space* study in chapter *CASE STUDIES*, p.392.

8.6 Guidelines for Inductive Analysis

1. *Description comes first!* Descriptive analyses guide subsequent inductive analyses. Only hypotheses that extend observed effects have a chance to get corroborated by inference procedures.

2. *Specific inference* (cf. §8.2.2, p.307 and §8.4, p.321). For each hypothesis of interest, derive a *specific relevant data set* and an indicator of the magnitude of effect.

For each specific analysis, the full strategy comprises three phases.

- *First phase: traditional inference*

  — *Traditional significance testing.* The hypothesis of no–effect can be investigated by procedures such as Student's t–test, or typicality test, or Hotelling's test (in the multivariate case). A significant result leads to asserting the existence of effect, and (at least in the one–dimensional case) its direction. A nonsignificant result means that the existence (and therefore the direction) cannot be asserted.

  — *Confidence regions*: intervals in the univariate case, ellipsoids in the multivariate case.

- *Second phase (optional): Bayesian reinterpretations*

  The results of traditional procedures can be expressed as probabilities of hypotheses: observed significance level as the probability of a directional hypothesis (at least in the univariate case); confidence interpreted as a probability (credibility).

- *Third phase (optional): largeness and smallness hypotheses*

  For largeness hypotheses, testing for no–effect is insufficient; for smallness hypotheses, testing for no–effect is irrelevant. Such hypotheses will be investigated by Bayesian analyses. Asserting largeness or smallness may involve choices. To begin with, credibility $\gamma = .90$ may be preferred to .95 (the complementary value of the familiar .05 α–level). Then, there is the choice of largeness (d_{lar}) and smallness (d_{sma}) limits. What is a "large effect?" ("How much of a difference makes a difference?") What is a "small effect?" Appropriate limits depend on the context of the study. Scaling effects by some reference variability index, such as the dispersion index of a known population ("psychometric scaling"), may be helpful[32].

[32] In applications, we have taken limits ranging from 0.20 to 0.60 SD for univariate

Appendix: Normal and Student's densities

The normal density function with mean μ and standard deviation σ is defined by (see density curve on Figure 8.10) :

$$p(z) = \frac{1}{\sqrt{2\pi}\,\sigma} \exp\Big[-\frac{1}{2}\Big(\frac{z-\mu}{\sigma}\Big)^2\Big] \qquad (z \in \mathbb{R}) \qquad (8.12)$$

We denote $\mathcal{N}(\mu, \sigma^2)$ the distribution of a normal variable and $\mathcal{N}(0,1)$ the distribution of the standard normal variable. Given a value u, we denote $P(z > u) = \int_u^\infty p(z)dz$, and the value $u > 0$ such that $P(|z| > u) = \alpha$ $(0 < \alpha \leq 1)$ is denoted z_α or $z[\alpha]$, that is, $P(|z| > z_\alpha) = \alpha$.

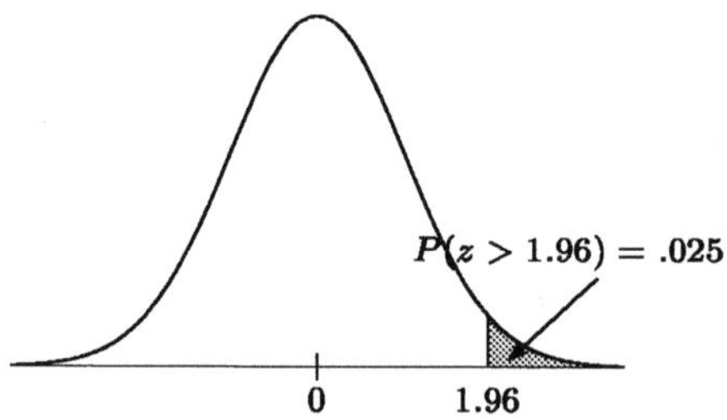

Figure 8.10. Density curves of Standard Normal distribution

The density function of Student's variable with q d.f. is defined:

$$p(t) = \frac{\Gamma[(q+1)/2]}{\sqrt{\pi q}\,\Gamma[q/2]} \times \frac{1}{(1+t^2/q)^{(q+1)/2}} \qquad (t \in \mathbb{R}) \qquad (8.13)$$

Given u, we denote $P(t_q > u) = \int_u^\infty p(t)dt$, and the value $u > 0$ such that $P(|t_q| > u) = \alpha$ $(0 < \alpha \leq 1)$ is denoted $t_q[\alpha]$: $P(|t_q| > t_q[\alpha]) = \alpha$. For

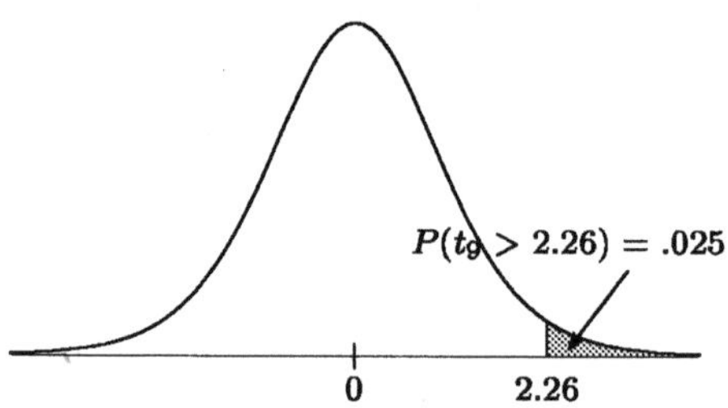

Figure 8.11. Density of Student's distribution with $q = 9$ d.f.

χ^2 and F distributions, $\chi^2[\alpha]$ and $F[\alpha]$ denote upper limit values, that is, $P(\chi^2 > \chi^2[\alpha]) = \alpha$ and $P(F > F[\alpha]) = \alpha$.

effects; for multidimensional ones, taking larger limits may be in order, owing to the wider dispersion of distributions. See the *Parkinson* and *Political Space* studies in chapter 9.

Chapter 9

Research Case Studies

(*'CASE STUDIES'*)

What we need is a rigorous method to extract structures from data.
J.P. Benzécri

Introduction

In this chapter, we apply Geometric Data Analysis to three research case studies chosen to exemplify the variety of real situations that can be handled in the present state of the art. In each study, the data analyzed are an Individuals×Variables table[1], and there are structuring factors on the individuals. The same strategy is used for all studies, with the following steps:

- Elementary statistical analyses and first codings of data.
- Choose a distance between individuals, and decide about active individuals, active variables, and structuring factors[2].
- Determine principal axes and calculate coordinates, then decide about how many axes to interpret, and interpret the axes retained.
- Investigate the cloud of individuals, in connection with the questions of interest and/or proceed to a Euclidean classification of individuals. The subclouds of individuals arising from analyses, especially in connection with structuring factors, are summarized by concentration ellipses, and breaking down of variance is performed in ANOVA line (typically between vs within).

[1]In the Parkinson data, some individuals (healthy people) are persons, other ones are defined as pairs (person, treatment). Cf. chapter *STRUCTURED*. p.255.

[2]For MCA: decide which structuring factors will be merely put as supplementary variables, and for specific MCA (cf. *MULTIPLE*, §5.2, p.203), decide which modalities of active questions will be put as passive ones.

- Attempt to extend the main conclusions by inductive analyses.

The case studies differ according to the kinds of questions of interest, and each of the foregoing steps may be more or less elaborate.

Parkinson study (§9.1). The data are part of a medical research on Parkinson's disease. They involve genuine experimental factors: Treatment factor (before and after drug intake) and Group factor (patients vs healthy persons). The dependent variables are measurements of gait performance.

After analyses variable by variable we proceed to GDA. The initial variables are on different scales; as a common reference scale, we take the SD of the healthy persons. Accordingly, a standard Principal Component Analysis (PCA) is performed on the reference group data, putting the patient data as supplementary elements. The PCA leads to a two–dimensional geometric model of data, with two axes interpreted as performance and style.

Taking the geometric space as a "geometric dependent variable" we proceed to two specific analyses, first studying the main effect of drug intake, then comparing patients after drug intake with healthy persons. The main descriptive conclusions are extended inductively by specific MANOVA procedures embedded in the geometric construction: significance testing, confidence intervals and ellipses, Bayesian methods.

Political Space study (§9.2). The data are part of a political science survey about the attitudes and values of French electors in 1997, together with their vote at the first round of parliamentary 1997 elections taken as a structuring factor of individuals. One question of interest was to identify the traditional political left vs right dimension in this space. Another one was to characterize the locations of electorates, especially National Front.

The questionnaire is analyzed by specific Multiple Correspondence Analysis to dispose of the problem of non–responses. Specific MCA leads to a three– dimensional model of data, with a first axis interpreted as an 'open vs closed' attitude toward "the outside"; this cleavage comes out prior to the traditional 'left vs right' oppositions. The examination of the electorates as subclouds of individuals highlights their wide dispersions and shows how largely they intermingle. The National Front electorate, mostly located on the 'closed' side of the first axis, appears to be in strong opposition to all other major (right as well as left) electorates.

As in the preceding case study, inductive data analyses have been performed to extend the main descriptive findings.

EPGY study (§9.3). The data are part of the large data base gathered over the years at Stanford University in the **E**ducation **P**rogram for **G**ifted **Y**outh (EPGY): see Tock & Suppes (2002). The data analyzed in collaboration with P. Suppes are the performances in mathematics of students in the third grade.

The overall aim of the study was to find out how the individual differences among gifted students are organized. In this respect, the body of existing knowledge about "ordinary students" appears to be of little relevance, so the present case study is really exploratory. Accordingly, only descriptive analyses have been performed. Basically, for each of five strands (Integers, Fractions, Geometry, Logic and Measurement), three indicators were available, namely error rates, latencies, and numbers of exercises to master the concepts of a strand. A question of interest was to investigate the trade–off between errors and latencies, and to figure out whether there might be a group of gifted students showing small error rates together with (comparatively) long latencies (a profile little encouraged by the current standard of educational testing!). The elementary analyses revealed the great heterogeneity of the distributions of indicators across strands. As an efficient way to ensure homogeneity (which is required to define a distance between individuals), an elaborate coding into categories has been performed, rendering the data amenable to Multiple Correspondence Analysis (MCA). Then an MCA has been performed on the 3×5 variables so constructed. The ensuing geometric space is bidimensional, at least, enabling to scrutinize the interplay between error rates and latencies at the individual level. An elaborate classification of students leads to a partition into five classes, one of them consisting in well–performing students exhibiting the particular profile mentioned above.

Remarks

Graphical units. For each study, two graphical scales are used, one being twice the other; distance units are always drawn on coordinate axes.

Stability studies. In chapter *STABILITY*, we have already used some data of the present chapter for stability studies, namely the Parkinson data (§7.4.3, p.289) and the Political data (§7.1.3, p.274 and §7.2, p. 279).

Data files. The data files of the case studies are available from the following anonymous ftp: `ftp.math-info.univ-paris5.fr`, they are under the repertory `pub/MathPsy/Kluwer/Data`, or they can directly be downloaded from `http://www.math-info.univ-paris5.fr/~lerb` (Brigitte Le ROUX's homepage) under the 'Livres' heading.

About software. See §9.4 (p.417).

9.1 The Parkinson Study

In an extensive research on parkinsonian patients' gait, Ferrandez & Blin studied an experimental group of patients together with a reference group of healthy elderly persons. Normal and fast gaits were investigated, and in each situation, variables pertaining to gait were recorded. There were 45 *Healthy persons* (women and men) between ages 60 and 92 (mean age 74) and heights between 1.48 and 1.80m (mean height 1.61). There were 15 *Patients* between ages 50 and 90 (mean age 72) and heights between 1.50 and 1.85m (mean height 1.62)[3].

In the present case study, we analyze the data set pertaining to normal gait. An observation consists in the recording of six numerical variables, namely *Velocity*, *Stride length*, and *Swing, Cycle, Stance and Double–Support durations* (see Figure 9.1.1).

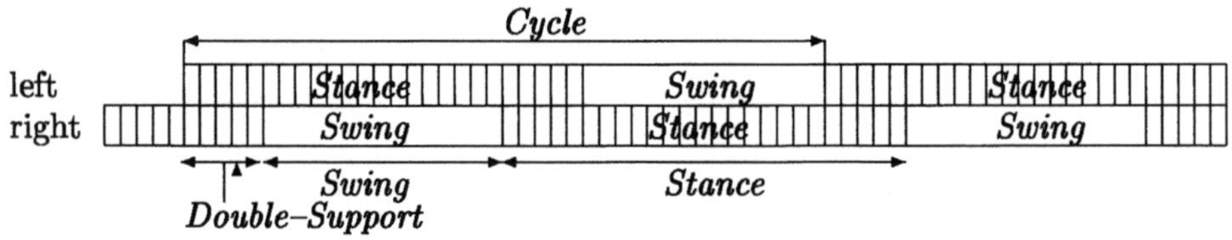

Figure 9.1.1. Description of the 4 duration variables for two strides.

There are three groups of 6–numerical observations, that is, a *reference group* of 45 healthy persons, observed once, and *two matched groups* of 15 parkinsonian patients, observed twice, before and after drug intake (L–Dopa); hence $45+(2\times 15)=75$ observations. In fact, owing to the protocol of data gathering, the 6 variables verify (at least approximately) the following three relations: Velocity = Length/Cycle, Cycle = Stance+Swing, Double–support = (Stance − Swing)/2.

The Parkinson study thus involves multivariate numerical data, with a genuine design and a treatment factor. To analyze such data, we will combine on the one hand, the approach of Geometric Data Analysis (GDA), by representing data as *clouds of points* and proceeding to the exploration of these clouds, on the other hand, the approach of MANOVA, with inductive analyses about the treatment effects and the differences between groups.

The main *questions* that will provide a guide throughout the study can be informally stated as follows:

---

[3]The detailed description of the tasks can be found in Ferrandez & Blin (1991), together with conventional statistical analyses. Our thanks go to the authors who communicated their data to us and commented upon an earlier version of the present text.

Question A. Does the drug have an effect?

Question B. Do patients after drug intake become more similar to healthy persons?

The statistical analysis will comprise three steps: firstly the analysis of initial variables (§9.1.1), secondly Geometric Data Analysis — PCA and Euclidean Classification (§9.1.2) — thirdly Inductive Data Analysis (§9.1.3).

9.1.1 Analysis of initial variables

Study of healthy persons

Table 9.1.1 gives the means and standard deviations (SD) of the 6 initial variables for the reference group and Table 9.1.2 their correlations.

	Mean	SD
Veloc.	0.777	0.228
Length	0.913	0.214
Swing	0.452	0.052
Cycle	1.202	0.155
Stance	0.751	0.118
Db_Supp.	0.145	0.050

Table 9.1.1. Means and standard deviations (SD)

	Veloc.	*Length*	*Swing*	*Cycle*	*Stance*	*Db_Supp.*
Velocity	1					
Length	.909	1				
Swing	−.054	.340	1			
Cycle	−.586	−.215	.782	1		
Stance	−.743	−.430	.586	.962	1	
Db_Supp.	−.899	−.746	.150	.727	.881	1

Table 9.1.2. Correlations between the six variables for $n = 45$ healthy persons (reference group).

Length and *Velocity* variables are positively correlated; the four duration variables are negatively correlated with *Velocity* and positively correlated among themselves, which suggests that they are inverse indicators of *Velocity* and *Length*. Now *Cycle*, *Stance* and *Double–Support* are negatively correlated with *Length*, whereas *Swing* is positively correlated with *Length* and practically uncorrelated with *Velocity*; thus the examination of correlations leaves us uncertain as to the status of *Swing* (a point that will be clarified by PCA).

From now on, we will *scale variables with respect to the reference group*, that is, for each variable we subtract its mean and we divide by its SD. To study in detail the structure of gait for the 45 healthy persons[4], bivariate *scatter diagrams* can be devised. We will comment on two of them, namely *Velocity* and *Length* which have the strongest correlation (.909), and *Velocity* and *Swing* which have the weakest correlation (−.054). Figure 9.1.2-a (next page) shows the strong positive correlation between *Velocity* and

[4]Healthy persons are labeled from $s17$ through $s61$. Only the labels of those mentioned in the text will appear in the figures.

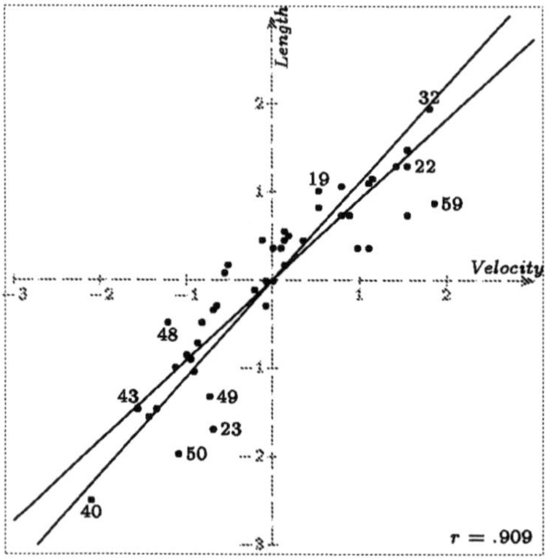

Figure 9.1.2-a. *Velocity* and *Length*

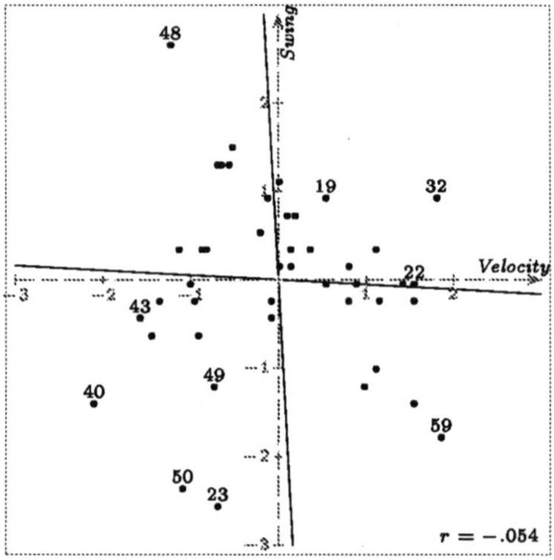

Figure 9.1.2-b. *Velocity* and *Swing*

Figure 9.1.2. Bivariate scatter diagrams and regression lines. The origin of axes corresponds to the mean of healthy persons and units of axes to their SD.

Stride length. In the SW quadrant, we notice persons $s40$ and $s50$, who walk slowly and with small steps; at the opposite, in the NE quadrant there are persons like $s32$ who walk quickly and stride along. Figure 9.1.2-b (this page) illustrates the quasi–null correlation between *Velocity* and *Swing*. On the left side, in SW there are $s50$ and $s23$ (low *Swing duration* and small steps), also $s40$ (already noticed on Figure 9.1.2-a), and in NW there is $s48$ (slow with a large *Swing duration*). On the right side, there is an opposition between $s59$ (fast with low *Swing duration*) and $s32$ (fast with large *Swing duration*).

> *To sum up: Cycle, Stance* and *Double–Support* durations are inverse indicators of *Velocity* and *Length*, the status of the *Swing* duration remaining to be clarified.

Study of patients

Let us now consider the 15 patients[5]. Their scaled deviations with respect to healthy persons are given in Table 9.1.3 (p.339). The observations of the 15 patients are reported on the correlation figures 9.1.3 (next page), with arrows linking the points "before" → "after". On Figure 9.1.3-a all patients, except $s2$ and $s13$, are located in the SW quadrant (low speed and

[5]Patients are labeled from $s2$ to $s16$ (the data of patient $s1$ were discarded of analyses).

	Velocity		*Length*		*Swing*		*Cycle*		*Stance*		*Db – Supp.*	
	*m*1	*m*2	*m*1	*m*2	*m*1	*m*2	*m*1	*m*2	*m*1	*m*2	*m*1	*m*2
*s*2	−0.38	0.01	0.27	0.78	0.53	1.30	1.34	1.21	1.51	1.09	1.50	0.50
*s*3	−1.17	−0.73	−1.32	−1.04	−0.43	−0.82	0.25	−0.46	0.50	−0.35	0.90	0.10
*s*4	−2.14	−1.96	−2.95	−2.63	−4.29	−3.13	−1.69	−0.98	−0.43	0.07	1.70	1.70
*s*5	−1.66	−0.34	−2.21	−0.57	−1.59	−0.43	−0.79	−0.46	−0.35	−0.35	0.70	−0.10
*s*6	−2.18	−0.78	−3.00	−1.32	−3.33	−1.40	−1.50	−0.92	−0.43	−0.52	1.30	0.30
*s*7	−2.10	−1.13	−2.86	−2.25	−2.94	−3.13	−1.24	−2.46	−0.35	−1.79	1.30	−0.50
*s*8	−0.95	−1.48	−0.99	−1.46	−0.63	−0.82	0.31	1.02	0.75	1.76	1.30	2.50
*s*9	−1.83	−0.60	−1.97	−0.99	0.15	−1.01	1.15	−0.72	1.51	−0.43	1.70	0.10
*s*10	−2.05	−0.60	−2.07	−0.95	−0.05	−1.21	1.92	−0.66	2.52	−0.26	3.29	0.30
*s*11	−2.27	−1.26	−2.53	−1.37	−1.40	−0.24	1.47	0.31	2.52	0.50	3.49	0.90
*s*12	−2.05	−1.53	−2.39	−1.97	−2.36	−2.17	0.50	−0.46	1.76	0.33	3.29	1.50
*s*13	−0.52	0.14	−0.25	0.59	−0.24	0.53	0.57	0.57	0.83	0.58	0.50	0.10
*s*14	−1.70	−1.70	−1.65	−1.23	−0.24	1.30	1.41	3.09	1.93	3.45	2.50	3.09
*s*15	−1.70	−1.17	−2.39	−2.02	−2.56	−2.75	−1.17	−1.69	−0.43	−1.03	0.90	0.30
*s*16	−2.14	−1.31	−2.16	−1.46	0.15	−0.05	2.38	0.38	2.95	0.50	3.49	0.70
mean	−1.66	−0.96	−1.90	−1.19	−1.28	−0.93	+0.33	−0.15	+0.99	+0.24	+1.86	+0.77
$\bar{d}$	+0.69		+0.71		+0.35		−0.48		−0.75		−1.09	

Table 9.1.3. *Scaled deviations* of patients before ($m1$) and after ($m2$) drug intake, means and differences of means ($\bar{d}$ = "after" − "before").

small steps). On Figure 9.1.3-b, 5 patients ($s4$, $s6$, $s7$, $s12$ and $s15$) have a very short *Swing duration*, the 10 others are roughly spread around the mean (0) of healthy persons, all speeds being much less than the mean of healthy persons.

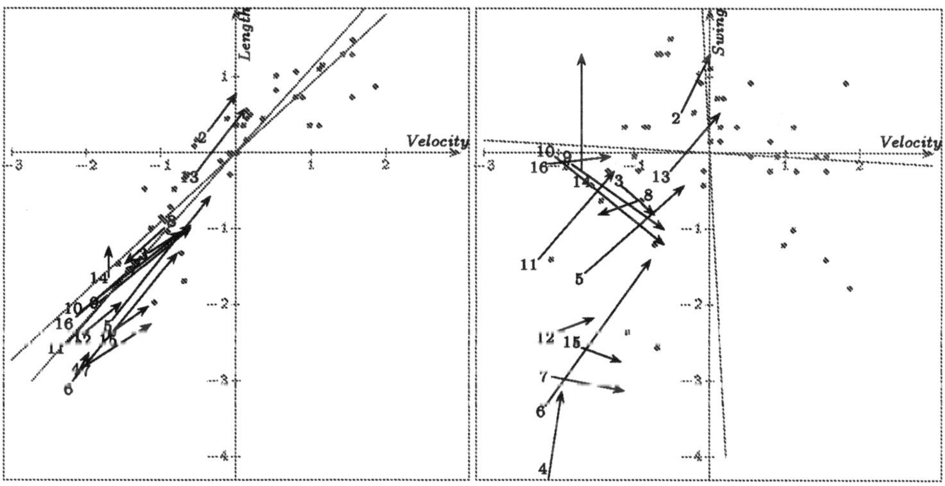

Figure 9.1.3-a: *Velocity* and *Length*   Figure 9.1.3-b: *Velocity* and *Swing*

Figure 9.1.3. Scatter diagrams with patients "before" and "after" (each patient is represented by an arrow)

Question A: Effect of Drug. The data of Table 9.1.3 (this page) are graphically represented on Figure 9.1.4 (p.341) variable by variable; each patient is represented by a vertical arrow (individual effect). For assessing individual effects, we will distinguish three zones: decrease ($\cdot < -0.5$), steadiness ($-0.5 \leq \cdot \leq +0.5$), increase ($\cdot > 0.5$). For each variable, a descriptive conclusion comes first, then the *homogeneity test* (permutation test) of the matched groups "before" and "after" is performed, and p–values (one- sided) are reported (cf. INDUCTIVE, §8.3.5, p.314). We follow the conventional α–landmarks (in terms of one-sided p–values) and star abbreviations[6]: S** (highly significant) if $p_{one-sided} < .005$; S* (significant) if $.005 \leq p_{one-sided} \leq .025$; NS (nonsignificant) if $p_{one-sided} > .025$.

— *Velocity* (Figure 9.1.4-a (next page)). After drug intake, 12 patients out of 15 increase their speeds; the speeds of $s14$ and $s4$ are unchanged, that of $s8$ decreases. On average, speed increases by 0.69 reference SD, and the increase is statistically significant ($\overline{d} = +0.69$, $p = .000$, S**).

— *Stride length* (Figure 9.1.4-b). All patients, except $s8$, take larger steps after drug intake. On average, *Stride length* increases by 0.71 SD and the increase is significant ($\overline{d} = +0.71$, $p = .000$, S**).

— *Swing duration* (Figure 9.1.4-e). The *Swing durations* of 6 patients are hardly changed by drug intake, they increase for 6 patients, they decrease for $s9$ and $s10$. On average, the increase of *Swing duration* is descriptively weak (0.35 SD) and nonsignificant ($\overline{d} = +0.35$, $p = .096$, NS).

— *Cycle duration* (Figure 9.1.4-c). For 8 patients *Cycles* decrease, for $s4$, $s6$, $s8$ and $s14$ they increase, for $s2$, $s5$ and $s13$ they remain steady. On average, *Cycle duration* decreases by 0.48 SD, but the decrease is nonsignificant ($\overline{d} = -0.48$, $p = .072$, NS).

— *Stance duration* (Figure 9.1.4-d). *Stance durations* decrease for the 8 patients for whom *Cycle durations* decrease; they increase for $s8$ and $s14$, also $s4$, and they remain steady for the 4 others. On average, *Stance duration* decreases by 0.75 SD ($\overline{d} = -0.75$, $p = .019$, S*).

— *Double–Support durations* (Figure 9.1.4-f). *Double–Support durations* decrease for 12 patients, they increase for $s8$ and $s14$, they remain steady for $s4$ and $s13$. On average, *Double–Support duration* decreases by 1.09 SD ($\overline{d} = -1.09$, $p = .002$, S**).

Question B: Comparing patients "after" drug intake and healthy persons. (Table 9.1.3, p.339, column $m2$ and Figures 9.1.4, p.341).

For *Velocity*, only 3 patients ($s2$, $s5$ and $s13$) are within 0.35 SD from

[6]Let us recall (§8.2.3, *INDUCTIVE*, p.308) that a significant effect allows us to draw an inductive conclusion on the direction of the true effect, while a nonsignificant effect leaves us uncertain about its direction.

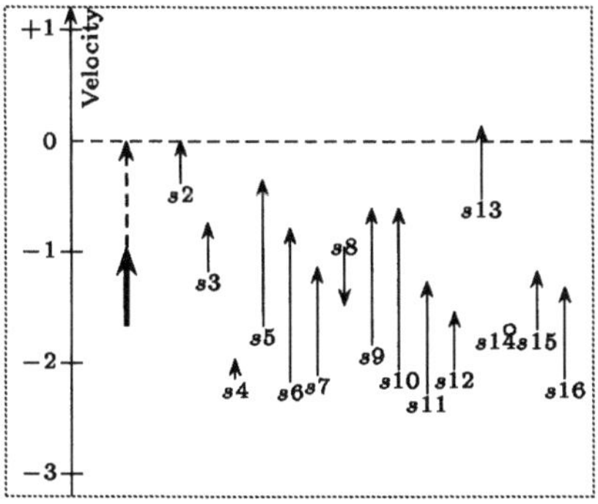

Figure 9.1.4-a: Velocity

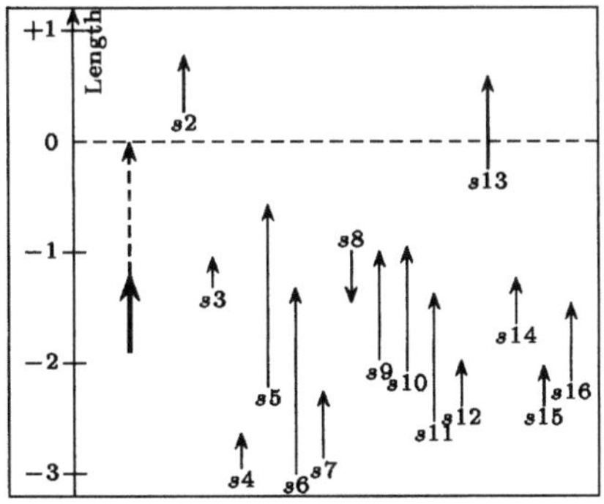

Figure 9.1.4-b: Length

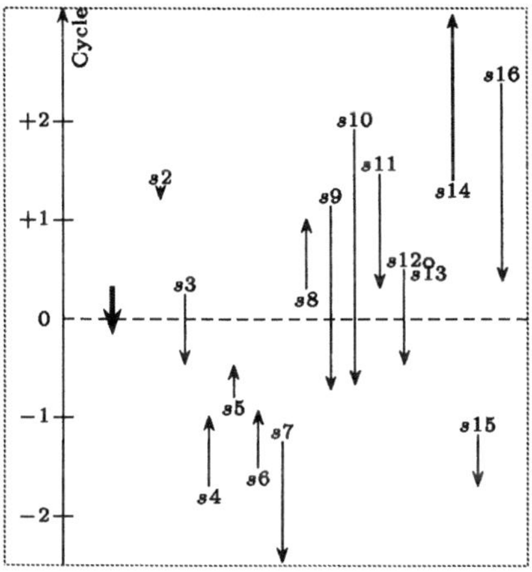

Figure 9.1.4-c: Cycle

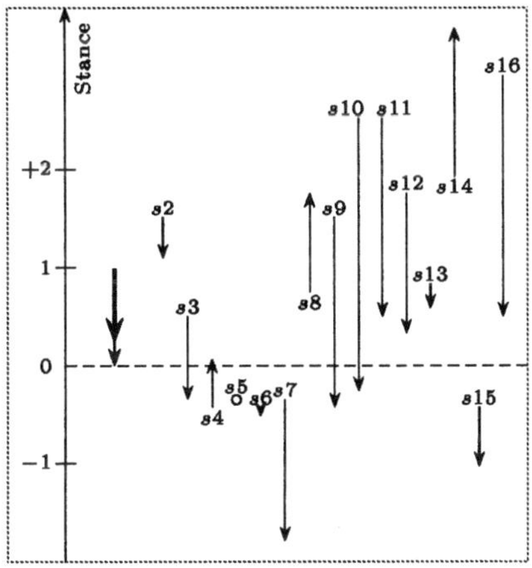

Figure 9.1.4-d: Stance

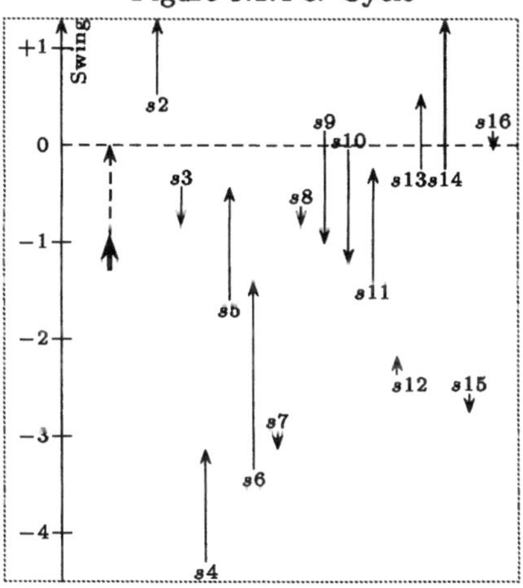

Figure 9.1.4-e: Swing

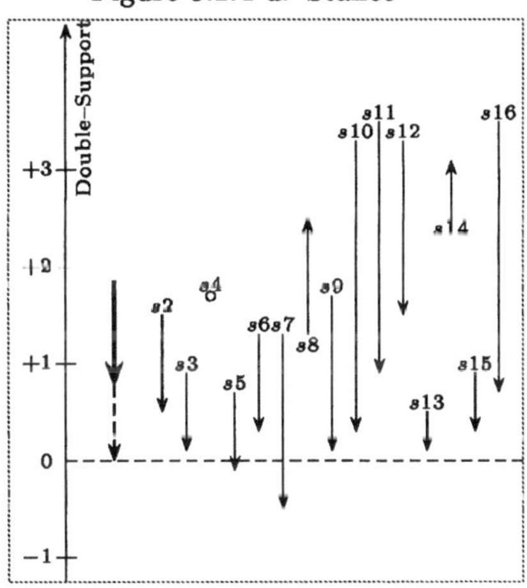

Figure 9.1.4-f: Double–Support

Figure 9.1.4. Individual effects of 15 patients and mean effects (in bold) for the 6 variables (units of vertical axes are reference SDs)

the reference mean; on average, the deviation is almost equal to 1 SD. For *Stride length*, all patients, except $s2$, $s5$ and $s13$, are beyond 1 SD below the reference mean. For *Cycle*, there are only 4 patients ($s3$, $s5$, $s11$ and $s16$) whose Cycle durations are between −0.5 and +0.5 SD. For *Stance*, there are 6, and for *Double–Support* there are 7. Table 9.1.5 and Figure 9.1.5 (this page) summarize the preceding findings about mean effects and differences[7].

	Question A		Question B	
	$\bar{d}$	p	$\bar{d}$	p
Velocity	+0.69	.000 **	−0.94	.000 **
Length	+0.71	.000 **	−1.19	.000 **
Swing	+0.35	.096 NS	−0.93	.004 **
Cycle	−0.48	.072 NS	+0.15	.325 NS
Stance	−0.75	.019 *	+0.24	.230 NS
Db-Supp	−1.09	.002 **	+0.77	.007 *

Table 9.1.5. Mean effects and differences, one-sided p–values (permutation tests)

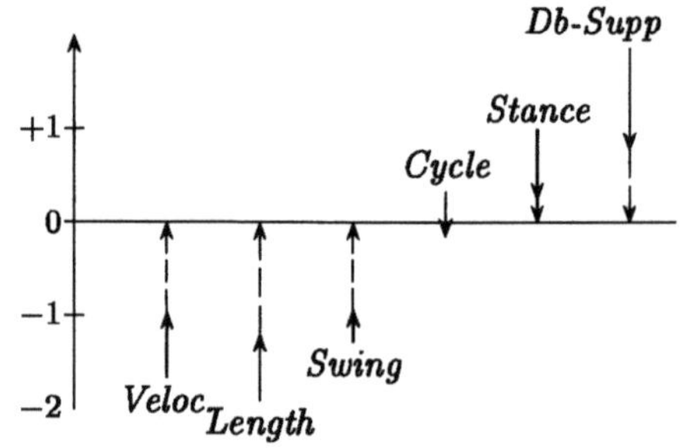

Figure 9.1.5. Mean effects and differences.

> *To sum up:* L–Dopa intake provokes an increase of *Velocity* and *Stride length*. Yet, for these two variables, the means of patients after drug intake are markedly inferior to the means of healthy persons. For *Cycle*, *Stance* and *Double–Support*, mean durations decrease: weakly for *Cycle*, a little more for the other two variables; for *Double–Support*, patients remain remote from the reference group. The mean *Swing* duration increases little after drug intake, it is still very inferior to the reference one.

9.1.2 Geometric Data Analysis

Principal Component Analysis

Principal Component Analysis (PCA, cf. *PRINCIPAL*, §4.3, p.150) has been performed on the six variables for the reference group (45 healthy persons), with 15 patients put as supplementary elements (cf. §4.1.5, p.146) . Since the 6 variables are not on an common scale, *standard* PCA (PCA of correlations) has been performed, i.e. the cloud of 45 individuals has been constructed from the 6 standardized initial variables. The *results* of PCA are shown in Tables 9.1.6 (p.343) and Figures 9.1.6 and 9.1.7 (p.345): circle

[7] p–values are exact for Question A and are Monte–Carlo approximations for Question B. Student's t–tests lead to p–values quite close (up to three decimal places) to those of permutation tests and to the same conclusions in significance terms.

$\lambda_1 = 3.9927$	$\lambda_2 = 1.8224$	$\lambda_3 = 0.1711$	$\lambda_4 = 0.0073$	$\lambda_5 = 0.0059$	$\lambda_6 = 0.0007$
$\tau_1 = .665$	$\tau_2 = .304$	$\tau_3 = .029$	$\tau_4 = .001$	$\tau_5 = .000$	$\tau_6 = .000$

Table 9.1.6-a: Eigenvalues (variances of axes) and variance rates.

	Veloc.	*Length*	*Swing*	*Cycle*	*Stance*	*Db_Supp.*
$r_{\ell 1}$	0.902	0.665	−0.403	−0.868	−0.958	−0.950
$r_{\ell 2}$	0.389	0.725	0.889	0.494	0.256	−0.210
R^2_{1-2}	0.966	0.968	0.953	0.998	0.983	0.947

Table 9.1.6-b: Correlations of initial variables with the first 2 principal variables ($r_{\ell 1}$ and $r_{\ell 2}$), and squares of multiple correlations (R^2_{1-2}).

	$\cos^2$	y_1	y_2		$\cos^2$	y_1	y_2		$\cos^2$	y_1	y_2
*s*17	.959	+3.091	−0.828	*s*32	.993	+2.887	2.015	*s*47	.499	+0.464	0.027
*s*18	.992	−2.420	1.632	*s*33	.486	−0.233	−0.185	*s*48	.987	−4.165	2.373
*s*19	.992	+0.501	1.492	*s*34	.992	+2.437	−0.923	*s*49	.905	+0.546	−2.253
*s*20	.886	−0.153	0.651	*s*35	.994	+1.961	0.951	*s*50	.976	+0.751	−3.800
*s*21	.924	+1.040	0.911	*s*36	.619	+0.475	0.632	*s*51	.999	−1.144	−1.261
*s*22	.999	+2.847	0.756	*s*37	.991	−2.080	−1.215	*s*52	.892	+0.198	0.981
*s*23	.990	+1.350	−3.758	*s*38	.909	−1.703	0.985	*s*53	.965	−2.319	0.186
*s*24	.989	+1.899	0.362	*s*39	.937	+1.802	0.593	*s*54	.902	+0.643	−0.751
*s*25	.783	+0.180	0.840	*s*40	.996	−3.105	−2.882	*s*55	.319	−0.169	0.343
*s*26	.964	−2.319	−0.501	*s*41	.995	−2.412	−1.535	*s*56	.843	−0.225	1.165
*s*27	.776	−0.354	0.750	*s*42	.972	+2.553	0.786	*s*57	.767	+0.545	0.182
*s*28	.948	+0.575	0.515	*s*43	.930	−3.967	−1.042	*s*58	.976	−1.142	1.243
*s*29	.999	+2.568	0.774	*s*44	.904	−1.599	0.983	*s*59	.958	+3.669	−0.943
*s*30	.998	−1.828	−0.136	*s*45	.988	−1.757	−0.690	*s*60	.998	−2.300	−0.290
*s*31	.998	−2.105	1.425	*s*46	.994	+1.667	0.258	*s*61	.995	+2.853	−0.815

Table 9.1.6-c: Coordinates and contributions of the 45 healthy persons.

	before			after		
	$\cos^2$	y_1	y_2	$\cos^2$	y_1	y_2
*s*2	.843	−2.212	0.928	.976	−1.286	1.854
*s*3	.994	−1.655	−1.288	.938	−0.193	−1.562
*s*4	.979	−0.948	−5.996	.983	−1.546	−4.651
*s*5	.971	−0.986	−3.174	.876	+0.158	−0.912
*s*6	.997	−1.073	−5.265	.991	−0.006	−2.335
*s*7	.998	−1.217	−4.800	.976	+1.535	−4.765
*s*8	.953	−1.747	−1.167	.914	−3.465	−1.430
*s*9	.993	−3.544	−1.048	.977	+0.074	−1.737
*s*10	.964	−5.217	−1.061	.997	−0.076	−1.814
*s*11	.935	−5.098	−2.401	.978	−1.777	−1.188
*s*12	.900	−3.877	−3.426	.976	−1.576	−3.270
*s*13	.816	−1.152	−0.151	.907	−0.420	1.012
*s*14	.967	−3.993	−1.038	.968	−5.907	1.014
*s*15	.997	−0.759	−4.111	.997	+0.434	−4.094
*s*16	.964	−5.820	−0.794	.929	−1.800	−1.069
Mean		−2.620	−2.323		−1.057	−1.663

Coordinates of vector effect: $\overline{d} = (1.563, 0.660)$

Table 9.1.6-d: Coordinates and contributions of the 15 patients (supplementary elements).

$\cos^2$ = quality of representation on plane 1-2; y_1 coordinate on Axis 1 (performance); y_2 coordinate on Axis 2 (style).

Table 9.1.6. Results of PCA (see *PRINCIPAL*, §4.6, p.153)

of correlations and cloud of individuals with its concentration ellipse (see CLOUD, §3.4.3, p.97). The 6 eigenvalues are shown in Table 9.1.6-a, the correlation coefficients of the 6 initial variables with the first two principal variables in Table 9.1.6-b. The first two principal coordinates (y_1, y_2), and qualities of representation in the first principal plane ($\cos^2$) are given for the 45 healthy persons in Table 9.1.6-c, and for the two observations "before" and "after" of the 15 patients in Table 9.1.6-d.

The 6 variables are redundant, owing to the three relations previously mentioned; PCA shows that in fact, ***the cloud is two-dimensional***[8]: The first two principal variables account for 97% of the variance of the cloud, and all initial variables have very strong multiple correlations with the first two principal variables ($R^2 \geq 0.947$, cf. Table 9.1.6-b, p.343). In addition, the quality of representation of persons in plane 1–2 (Table 9.1.6- c, p.343, column $\cos^2$) is quite good; it exceeds .84 for 38 persons, and goes below .50 only for 3 persons close to the "average person". The "circle of correlations" (Figure 9.1.6, p.345) highlights the ***circular order*** of variables. From right to left, one finds *Velocity*, *Length*, *Swing*, *Cycle*, *Stance* and *Double–Support*; the intermediate status of *Swing* appears clearly.

Interpretation of the first axis. Variables *Velocity* and *Length* are positively correlated with the first principal variable, the other variables (which vary contrariwise) are negatively correlated (Table 9.1.6, p.343). If the orientations of variables *Cycle*, *Stance*, *Double–Support* and *Swing* are changed, the correlations of the recoded variables are all positive. Furthermore, if for each person one calculates the mean of the recoded standard variables, one defines the ***Average variable*** whose correlation with the first principal variable is +.997 (+.998 if *Length* and *Swing* are discarded). That is, the first axis is a "***size factor***", where "size" here means performance.

In the ***cloud of individuals*** (Figure 9.1.7, p.345), on the extreme right of Axis 1, one finds well–performing persons, such as $s22$ and $s59$ who walk fast with large steps and whose *Cycle*, *Stance* and *Double–Support* durations are very short (see table hereafter). The reverse holds for the persons on the extreme left of Axis 1, such as $s43$ and $s48$. The *Swing* variable, whose correlation is much lower with Axis 1 than with Axis 2, does not appear in the opposite Table.

	Veloc.	*Length*	*Cycle*	*Stance*	*Db-S.*
$s22$	+1.5	+1.3	−1.0	−1.2	−1.5
$s59$	+1.9	+0.9	−1.8	−1.6	−1.3
$s43$	−1.6	−1.5	+1.4	+2.0	+2.7
$s48$	−1.2	−0.5	+2.7	+2.4	+1.3

Interpretation of Axis 1. Axis 1 is a performance axis.

[8] In chapter *STABILITY* (p.289), we showed that the first principal plane is stable.

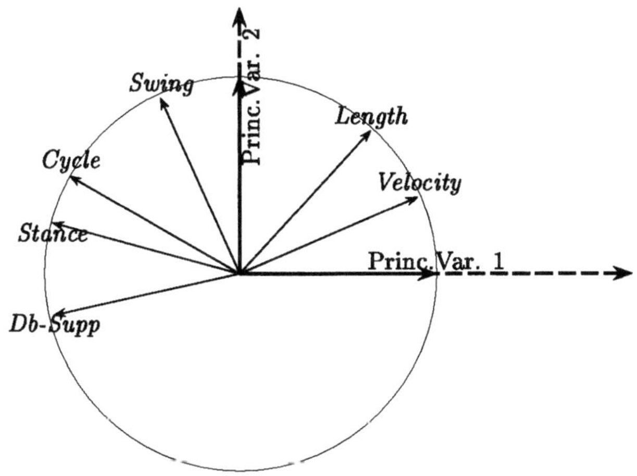

Figure 9.1.6. Space of variables: "circle of correlations".

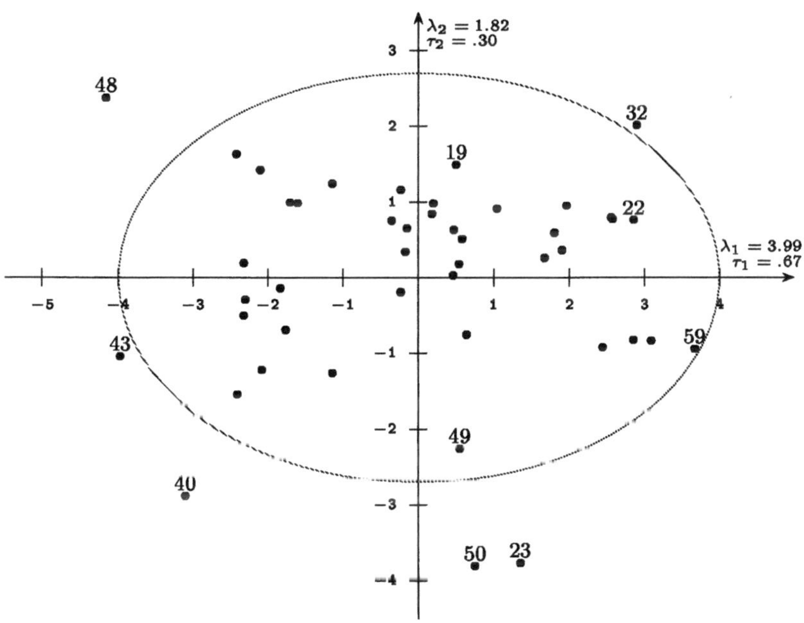

Figure 9.1.7. Cloud of individuals (45 healthy persons) in plane 1-2 with its concentration ellipse.

The apparent opposition on Axis 1 between *Velocity* and *Length* on the one hand, and *Cycle*, *Stance* and *Double–Support* durations on the other hand, reflects the status of *Velocity* and *Length* as *direct performance indicators*, and that of the other three variables as *inverse indicators*. The particular status of the *Swing* variable will now appear on Axis 2.

Interpretation of the second axis. The variables most correlated with the second principal variable are *Swing* ($r = .889$) and *Length* ($r = .725$) (see Table 9.1.6-b, p.343).

In the cloud of individuals, we examine six extreme points on Axis 2 (cf. Figure 9.1.7, p.345). On the side of good performances (right of figure), Axis 2 opposes persons such as $s59$ (below) with short *Swing duration* and medium *Stride length* to persons such as $s32$ (above) with large *Stride length* and medium *Swing duration* (see opposite Table). At the center of the cloud (medium performances), there is an opposition between persons such as $s50$ (below) with small steps and weak Swing durations, and persons such as $s19$ (above) with medium Swing duration and steps. On the side of low performances (left of figure), Axis 2 opposes persons such as $s40$ (below) with small steps and short Swing durations, to persons such as $s48$ (above) with long Swing durations and medium steps. In short, Axis 2 opposes persons who walk at a good pace with long Swing durations (above), to persons who take small steps with weak Swing durations (below).

	Swing	*Length*
$s32$	+0.9	+1.9
$s59$	−1.8	+0.9
$s19$	+0.9	+1.0
$s50$	−2.4	−2.0
$s48$	+2.7	−0.5
$s40$	−1.4	−2.5

Interpretation of Axis . Axis 2 is a style axis.

Supplementary Cloud of 15 patients

In order to situate patients with respect to healthy persons, the 2×15 profiles (scaled deviations) of patients ("before" and "after" drug intake) have been put as supplementary elements (cf. PRINCIPAL, §4.1.5, p.146). The 30 points are very well represented in plane 1-2 (Table 9.1.6-d, p.343, column $\cos^2$); one can therefore proceed to the comparative study from this plane. The effect of drug, for a patient, is two–dimensional; in Figure 9.1.9 (p.348), the individual effects are represented by arrows "before" → "after". Twelve individual effects are directed towards the East (performance improvement) or the NE (improvement in performance and stride length); whereas for patients $s4$, $s8$ and $s14$, the effects are oriented SW, that is, they regress after drug intake.

The 15 points of subcloud "before", represented in Figure 9.1.9 (■), are on the negative side of the performance axis (Axis 1) and of the style axis (Axis 2); the concentration ellipse of this subcloud is almost entirely contained in the SW quadrant (low performances, long Swing durations and small steps). As can be seen from this ellipse, the dispersion of the subcloud is as important as that of healthy persons, and the dispersion of patients is larger for style than for performance. The subcloud of 15 patients "after" (▲) is also located on the side of low performances. The concentration ellipse shows that the dispersion of this subcloud is large; the dispersion remains more marked for style than for performance. On the whole, the subcloud "after" appears to be shifted towards the right, getting closer to the healthy person subcloud. On Figure 9.1.9 (p.348), the two mean points "before" and "after" are represented. The angle between the *mean vector–effect* and Axis 1 is equal to +23°. The geometric distance from the "before" mean point to the origin–point (mean point of healthy persons) is equal to 3.5, that from the "after" mean point is equal to 2.0; thus we can say that on average the drug brings patients closer to healthy persons by about 44% (their getting closer is mainly due to performance).

One can situate precisely each patient with respect to healthy persons by means of the *extremality κ–index* (cf. *CLOUD*, §3.4.4, p.100), or else by the number of healthy persons that are more extreme (Table 9.1.7). After drug intake, extremality indices decrease for 9 patients out of 15; they remain steady for $s7$ and $s15$; they slightly increase for $s2$ and $s13$ and more markedly for $s8$ and $s14$ — for $s8$ the drug has negative effects for both style and performance, for $s14$ the drug has a paradoxical effect: negative on performance but "positive" on style (lengthening stride and increasing Swing duration). As for the differences between patients "after" and healthy persons, they are large: 13 κ–indices exceed 1.

		$s2$	$s3$	$s4$	$s5$	$s6$	$s7$	$s8$	$s9$	$s10$	$s11$	$s12$	$s13$	$s14$	$s15$	$s16$	$\bar{d}$
before	κ	1.30	1.26	4.47	2.40	3.84	3.61	1.23	1.94	2.73	3.14	3.20	0.59	2.14	3.07	2.97	1.34
	$n >$	21	18	0	4	0	0	21	7	3	0	0	37	4	0	0	18
after	κ	1.52	1.16	3.53	0.68	1.73	3.61	2.03	1.29	1.34	1.25	2.55	0.78	3.05	3.04	1.20	0.54
	$n >$	13	22	0	35	24	0	6	18	18	18	4	34	0	0	22	38

Table 9.1.7. Evaluating extremality of patients (κ and $n >$)

To sum up: Drug intake has an overall positive effect (Question A), but it is far from bringing patients to the level of healthy persons; a gap remains (Question B).

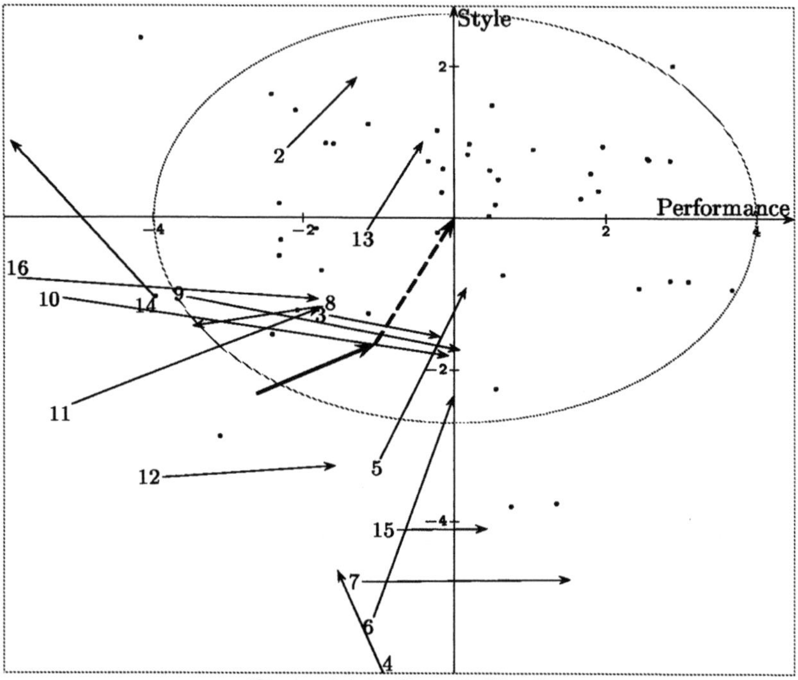

Figure 9.1.8. Overall Cloud

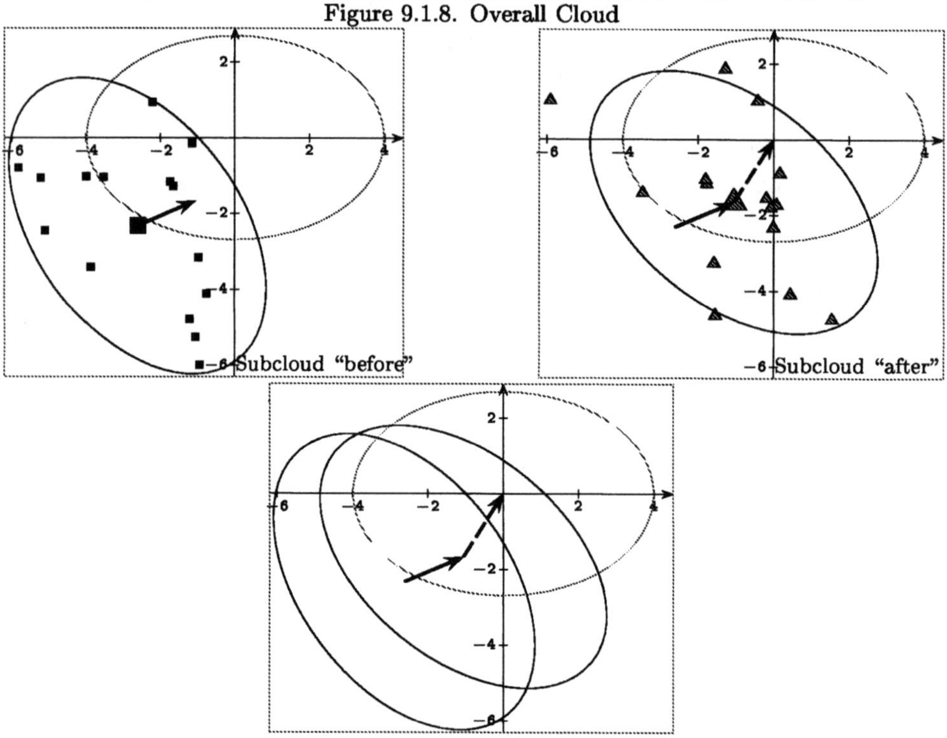

Figure 9.1.9. Subclouds of 15 patients ("before" and "after") with concentration ellipses.

Euclidean Classification

To refine PCA, a hierarchical ascending classification with aggregation according to variance (AHC) has been performed on healthy persons' data, taking the same distance as for PCA, i.e. the Euclidean distance from the 6 initial standard variables (cf. *CLOUD*, §3.6, p.106).

A partition into 6 classes has been retained, which accounts for 86% of the variance of the cloud; the 6 subclouds with their concentration ellipses are shown in Figure 9.1.10 (this page), and the superior hierarchical tree is shown in Figure 9.1.11 (next page). In order to interpret the successive dichotomies leading to the six–class partition, for each node of the subhierarchy, the mean scaled deviations of variables and the coordinates of the mean points on principal axes 1 and 2 have been reported in Table 9.1.9 (p.350).

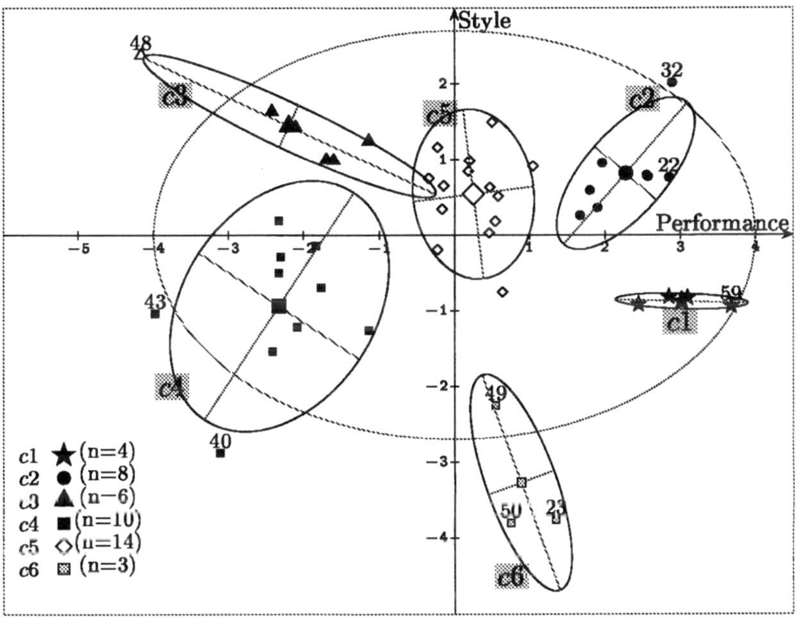

Figure 9.1.10. The 6 classes of healthy persons in plane 1-2.

The first dichotomy is by far the most important one ($\ell 89 - 2.85$) and involves performance; it separates classes $c3$ and $c4$ from the four others. Classes $c3$ and $c4$ bring together 16 very low–performing persons whose gaits, on average, are slow, with small steps and long Cycle and Stance durations; whereas the other classes gather 29 rather well–performing persons with overall inverse profiles. The second dichotomy, much less important

ℓ	δ_ℓ	*nodes*	η^2_ℓ
ℓ89	2.854	ℓ88 ℓ86	.476
ℓ88	0.840	ℓ87 ℓ77	.616
ℓ87	0.763	ℓ85 ℓ82	.743
ℓ86	0.488	ℓ84 ℓ83	.824
ℓ85	0.209	ℓ79 ℓ73	.859
ℓ84	0.153	ℓ81 ℓ80	.884
ℓ83	0.131	ℓ32 ℓ74	.906
ℓ82	0.055	ℓ78 ℓ71	.915
ℓ81	0.054	ℓ27 ℓ24	.924
ℓ80	0.050	ℓ72 ℓ67	.932

Table 9.1.8. Description of the superior nodes of hierarchy (cf. *CLOUD*, p.114).

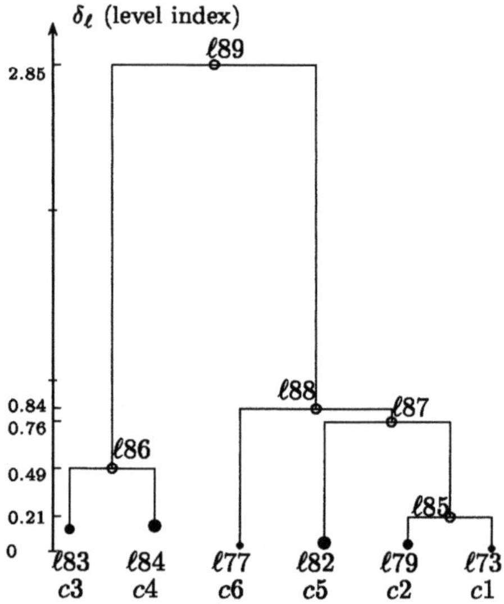

Figure 9.1.11: Superior hierarchical tree.

	Vel	*Len*	*Swi*	*Cyc*	*Sta*	*DbS*	*Perf*	*Sty*	*Vel*	*Len*	*Swi*	*Cyc*	*Sta*	*DbS*	*Perf*	*Sty*
ℓ88 : c1c2c5c6	+0.6	+0.4	−0.2	−0.5	−0.6	−0.6	1.3	+0.0	+	∘		+	+	+	+	∘
ℓ86 : c3c4	−1.0	−0.8	+0.4	+1.0	+1.1	+1.1	−2.3	−0.0	−	−		−	−	−	−−	∘
ℓ87 : c1c2c5	+0.7	+0.7	−0.0	−0.4	−0.5	−0.7	1.3	0.4	+	+	∘	∘		+	+	∘
ℓ77 : c6	−0.8	−1.7	−2.0	−1.6	−1.2	−0.2	0.9	−3.3	−	−−	∨∨	++		∘	+	−−
ℓ85 : c1c2	+1.3	+1.0	−0.4	−1.0	−1.1	−1.1	2.5	0.2	+	+		+	+	+	++	∘
ℓ82 : c5	+0.2	+0.4	+0.3	+0.1	−0.0	−0.3	0.2	0.5	∘	∘		∘	∘	∘	∘	+
ℓ84 : c4	−1.2	−1.2	−0.3	+0.7	+1.0	+1.4	−2.2	1.4		−	∘				−−	+
ℓ83 : c3	−0.7	−0.1	+1.5	+1.5	+1.3	+0.7	−2.3	−0.9		∘	∧∧				−−	−
ℓ79 : c2	+1.3	+1.2	+0.1	−0.7	−0.9	−1.2	3.0	−0.9			∘	+			++	−
ℓ73 : c1	+1.4	+0.6	−1.4	−1.6	−1.5	−1.1	2.3	0.8			∨	++			++	+

Table 9.1.9. Mean scaled deviations of the nodes of hierarchy. Mean scaled deviations have been encoded into 5 categories: −−, −, ∘, +, ++ as indicated on the following figure, according to whether indicators are direct (*Velocity* and *Length*) or inverse (*Cycle, Stance* and *Double–Support*); for *Swing*, we have put ∧ (long duration) and ∨ (short duration).

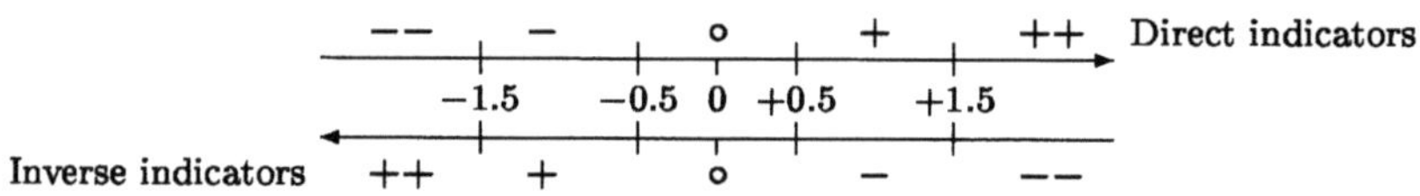

($\ell 88 = 0.84$), mainly concerns gait style; it separates a somewhat borderline group of 3 persons (class $c6$), with small steps and very short Swing and Cycle durations. At the next level, with an index level ($\ell 87 = 0.76$) very close to the preceding one, the separation concerns both performance and style; the 14 persons of class $c5$, with medium performances, are separated from the 12 persons of classes $c1$ and $c2$ who are fast, with short Cycles and large steps. Then there is a split firstly of classes $c3$ and $c4$ ($\delta_{\ell 86} = 0.49$), then of classes $c1$ and $c2$ ($\delta_{\ell 85} = 0.21$) owing to gait style.

If we seek a partition into fewer than 6 classes, the location of mean points of classes on Axis 1 suggests us to group together classes $c1$ and $c2$ of 12 well–performing persons, then classes $c3$ and $c4$ of 16 poorly–performing persons, lastly classes $c5$ and $c6$ of 17 moderately–performing persons. The between–variance of these three subclouds in plane 1-2 is equal to 3.603 (hence $\eta^2 = .60$): a value close to (though slightly less than) 3.694, the variance of the three–class partition produced by the hierarchy (with $\eta^2 = .62$). The partition suggested by the geometric cloud, whose interpretation is very clear, will surely be preferred as a classification summary:

Interpretation comes first!

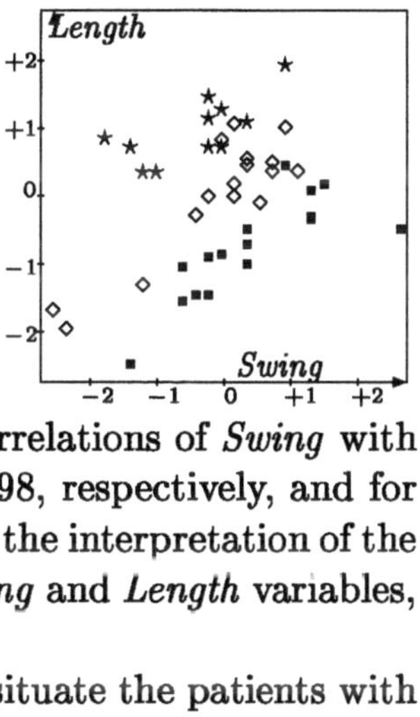

The correlations between variables *Length* and *Swing*, within each of the three classes, are equal to +0.71 for the medium–performance class (◇), +0.890 for the good–performance class (⋆) and +0.795 for the poor–performance class (▪) (opposite figure); whereas the overall correlation is only 0.34.

Furthermore, within those three classes, the correlations of *Swing* with the second principal variable are 0.96, 0.98 and 0.98, respectively, and for *Length* of stride 0.88, 0.97 and 0.90; which confirms the interpretation of the second axis as a style axis characterized by the *Swing* and *Length* variables, as well as the special status of *Swing*.

The two figures 9.1.12 (next page) allow us to situate the patients with respect to healthy persons; they show that before drug intake patients are close only to the poor–performance class $c4$, and that even after drug intake a gap remains: Patients never reach good–performance classes ($c1$ and $c2$).

Conclusions of descriptive geometric analysis. Two axes are sufficient to describe data. Axis 1 is a *performance* axis. Axis 2 is a *style* axis (*Stride Length* and *Swing duration*). Classification of persons yields three performance classes: good, medium and poor, each class being subdivided in two according to style.

On the whole, patients have poorer performances than healthy persons. They differ more as regards style than performance. Drug intake appears to bring patients closer to healthy persons, especially as regards performance (a welcome finding), although it does not bring any patient into a good–performance class (a less welcome finding).

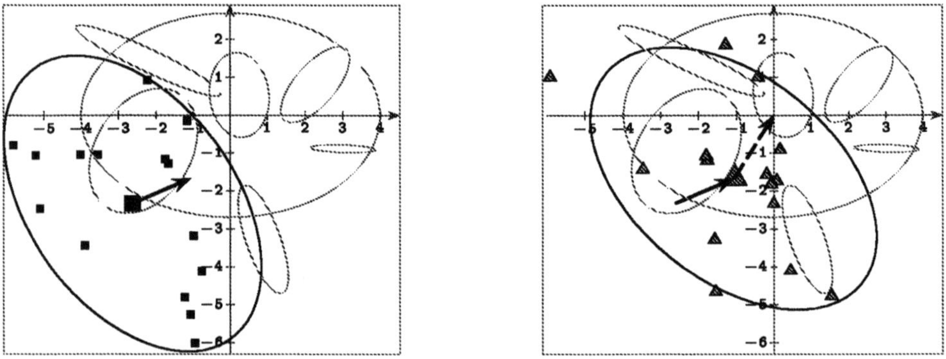

Figure 9.1.12. Ellipses of patients before and after drug intake.

9.1.3 Inductive Data Analysis

Some inductive data analysis was already presented with the preceding descriptive analysis of the initial variables, when we provided, together with observed effects, the attached p–values (significance levels). In the present section, we will proceed to deeper inductive analyses — including Bayesian ones — on the means of principal variables stemming form PCA, taken as an overall summary of data (Table 9.1.6, p.343).

Accordingly, we now reformulate Questions A and B as *hypotheses about true values and effects*. Taking into account the conclusions of descriptive analyses, Question B will be investigated, as well as Question A, in terms of *largeness hypothesis*, that is, the hypothesis that the true effect is large.

Question A. Is there a true effect of drug, and if yes, does its direction point to the reference true value (healthy persons)?

Question B. Is there a difference (gap) between the true value of patients after drug intake and the reference true value, and if yes, is this gap important?

The true value of a variable will be intuitively thought of as an average value of this variable on a population of parkinsonian patients (potential users of the drug), and a reference population of healthy persons.

We will successively perform *univariate analyses* on the two variables Performance and Style taken separately, then a *geometric inference* for the two variables taken jointly. For each question, a *specific inference* will be made on the derived relevant data set, with the following three phases in accordance with the strategy presented in chapter INDUCTIVE, §8.6, p.331: descriptive appraisal, then traditional inference and Bayesian data analysis. For drawing Bayesian conclusions, we will take a credibility $\gamma = .90$.

Univariate analyses

For asserting largeness of mean effects, we will take as a largeness limit the value $d_{lar} = 0.40\,s$, that is, we take the *scaled effect* $l_{lar} = 0.40$ for all univariate analyses[9].

About the largeness limit. The value $l_{lar} = 0.40$ is lower than the value 0.50 taken previously in our descriptive analyses to qualify the medium zone of individual effects. As pointed out in §8.6 (p.331), the value of largeness limit should be reconsidered in each concrete situation. The problem here is to assert that there is "some improvement" brought by the drug in terms of the average in a target population; for that purpose, a limit that is not too high seems reasonable.

Question A: Effect of Drug. Question A leads to comparing the two matched groups of patients "before" and "after" for each variable.

Performance. For this first study (Question A, Performance), we will work out the steps in detail.

— *Descriptive appraisal.* The mean performances of patients "before" and "after" are -2.62 and -1.06, respectively; hence the *observed effect* (mean of differences "after" – "before") is $d_{obs} = +1.56$; thereferore, we descriptively conclude to *improvement* of performance: $d_{obs} > 0$. To appraise largeness, we calculate the corrected standard deviation of the protocol of differencesand find $s = 2.02$; hence the *scaled effect* $d_{obs}/s = +0.77$. Since $0.77 > l_{lar} = 0.40$, we deem this scaled effect to be *large*: $d_{obs}/s > 0.40$.

[9]In inference procedures (cf. INDUCTIVE, footnote 5, p.301) we work with the *corrected variances* s^2, dividing the Sum of Squares by the number of d.f., and similarly for covariances below.

> *Descriptive Conclusion.* The mean observed effect shows a *large improvement* of performance.

— *Significance Testing.* The descriptive conclusion of improvement triggers the phase of significance testing, in which we attempt to extend this directional conclusion to the true effect. For this purpose, we will perform both the Fisher–Pitman permutation test for two matched groups and Student's t–test (cf. INDUCTIVE, §8.3.5, p.314 and §8.2.2, p.305)

The *permutation test* gives $p_{one-sided} = .0054$. *Student's t–test* leads to $t_{obs} = \sqrt{15} \times d_{obs}/s = +2.99$, hence (from Student's t distribution with 14 d.f.) $P(t_{14} > t_{obs}) = .0049$ (one-sided). Both tests lead to extend the conclusion of improvement to the true effect, that is, to state the following inductive directional conclusion.

> *Asserting direction of effect* (significance testing). The two groups "before" and "after" can be asserted to be heterogeneous, pointing to improvement of performance, at the one-sided level .025 (S*, at least).

— *Bayesian data analysis.* In turn, the significant conclusion invites to proceed to a Bayesian phase, in which we take as a posterior distribution of the true effect δ the scaled t–distribution centered at the observed effect d_{obs}. To begin with, we reinterpret the observed significance level (cf. §8.4, p.317). Since $P(t_{14} > t_{obs}) = .005$, we have $P(\delta > 0) = 1 - .005 = .995$: a high probability (greater than $\gamma = .90$).

> *Asserting direction of effect* (Bayesian). There is a high posterior probability that the true effect is in the direction of improvement: $P(\delta > 0) = .995$.

Then Bayesian data analysis permits us to attempt to extend the descriptive largeness conclusion ($d_{obs}/s > 0.40$), by calculating the posterior probability of the corresponding hypothesis $P(\delta/s > 0.40)$ (cf. formula (8.5), p.318)). We have $\sqrt{n} \times (0.40 - d_{obs}/s) = -1.43$ (with $n = 15$); therefore $P(\delta/s > 0.40) = P(t_{14} > -1.43) = .915$: a high probability (greater than $\gamma = .90$). The following Table summarizes the inductive analysis of Question A for Performance.

$d_{obs} = +1.56$ and $s = 2.02$ ($n = 15$)		Conclusion ($l_{lar} = 0.40$)
Description	$d_{obs}/s = 0.77 > 0.40$	Observed improvement is large
Permutation	$p_{one-sided} = .0054 < .025$ (S*)	Heterogeneity: improvement
Student	$P(t_{14} > 2.99) = .0049 < .005$ (S**)	$\delta > 0$: improvement
Bayes	$P(\delta > 0) = .995$	$P(\delta > 0)$ is high: improvement
	$P(\delta/s > 0.40) = .915$	*large improvement*

Asserting directional largeness. There is a high posterior probability that the true improvement in performance is large: $P(\delta/s > 0.40) = .915$.

It is seen that the last Bayesian conclusion encompasses the others and brings a more elaborate answer to the question asked.

Style. We proceed for *Style* as for *Performance*, starting with a descriptive conclusion that the effect is positive (increased *Stride Length* and *Swing duration*) and large. The results are gathered in the following Table.

$d_{obs} = +0.66$ and $s = 1.14$ $(n = 15)$		Conclusion ($l_{lar} = 0.40$)
Description	$d_{obs}/s = +0.58 > 0.40$	Large positive observed effect
Permutation	$p_{one-sided} = .021 < .025$ (S*)	Heterogeneity: positive effect
Student	$P(t_{14} > 2.24) = .021 < .025$ (S*)	$\delta > 0$: Positive effect
Bayes	$P(\delta > 0) = .979$	*Positive effect*
	$P(\delta/s > 0.40) = .751 < .90$	*No largeness conclusion*

The conclusions are that positiveness of true effect can be asserted, but not largeness; the probability of the hypothesis $(\delta/s > 0.40)$ is only .751, falling much below $\gamma = .90$.

Question B: Comparing patients "after" and healthy persons. Question B leads to comparing the two independent groups of the 15 patients "after" and of the 45 healthy persons.

Performance. The mean performance of healthy persons is 0, the one of patients "after" is -1.06 (Table 9.1.6, p.343). The observed difference "healthy" $-$ "after" is *positive* in sign, that is, the observed mean performance for patients "after" remains inferior to that of healthy persons; and this *difference is large*, since $d_{obs}/s = +0.54$.

One has (cf. §8.2.2, p.307) $\tilde{n} = 1/(\frac{1}{15}+\frac{1}{45}) = 11.25$ and $q = 15+45-2 = 58$, $t_{obs} = \sqrt{11.25} \times d_{obs}/s = 1.81$, and $\sqrt{11.25} \times (0.40 - d_{obs}/s) = -0.47$, hence $P(\delta/s > 0) = P(t_{58} > 1.81) = .038$ and $P(\delta/s > 0.40) = P(t_{58} > -0.47) = .68$. The results are summarized in the following Table.

$d_{obs} = +1.06$ and $s = 1.97$ $(n_1 = 15, n_2 = 45)$		Conclusion ($l_{lar} = 0.40$)
Description	$d_{obs}/s = +0.54 > 0.40$	Large diff. "healthy" $-$ "after"
Permutation	$p_{one-sided} = .040 > .025$ NS	Heterogeneity cannot be asserted
Student	$P(t_{58} > 1.81) = .038$, NS	$\delta > 0$?
Bayes	$P(\delta > 0) = .961$	*Difference is presumably positive*
	$P(\delta/s > 0.40) = .68 < .90$	*No largeness conclusion*

The "gap in performance" is descriptively large, but the Bayesian conclusion of largeness cannot be reached (at $\gamma = .90$)[10].

Style.

$d_{obs} = +1.66$ and $s = 1.54$ ($n_1 = 15$, $n_2 = 45$)		Conclusion ($l_{lar} = 0.40$)
Description	$d_{obs}/s = 1.08 > 0.40$	Large positive obs. difference
Permutation	$p_{one-sided} = .0009$ (S**)	Heterogeneity: positive difference
Student	$P(t_{58} > 3.62) \simeq .000$	S** : $\delta > 0$: positive difference
Bayesian	$P(\delta > 0) \simeq 1.000$	positive difference
	$P(\delta/s > 0.40) = .987$	*large positive difference*

For Style, the mean point of patients definitely remains remote from the mean point of healthy persons.

Inductive GDA (cf. INDUCTIVE, §8.5, p.322)

When Performance and Style are considered jointly, effects become geometric vectors, that we call *vector–effects*; we will express the procedures in the geometric mode; the figure 9.1.8 of the overall cloud (p.348) will provide a visual guide throughout all analyses. Geometric intuition, with the interplay between points and vectors (differences of points), will help to grasp the meaning of the procedures without being bothered by calculation details. For each question, we will derive a *specific cloud*, and perform a *specific inference* with the same three phases as above[11].

Question A: Effect of drug. The *two matched groups* are here the two subclouds of 15 patients "before" and "after" (Figure 9.1.8, p.348).

Constructing the specific cloud. To each patient there is associated a vector–effect (geometric deviation "after" − "before"). The specific cloud is constructed in the following three steps: i) Restrict the overall cloud (Figure 9.1.8, p.348) to the 15 patient vector–effects, with the mean vector-effect; ii) Shift bipoints by translation so as their initial points coincide with the origin of axes; iii) Take the endpoints of the vector–effects to represent effects, now conceptualized as points, that we call *point–effects.*

The *specific cloud* is the cloud of the individual point–effects, together with their mean point, which is the endpoint of the mean vector–effect:

[10]The nonsignificant results, which are mainly due to the small sample size, should evidently not be interpreted as corroborating the smallness hypothesis.

[11]The inductive analyses borrow from Rouanet & al (1998), chapter 7.

Figure 9.1.13 (this page). The specific inference for Question A will be based on this specific cloud.

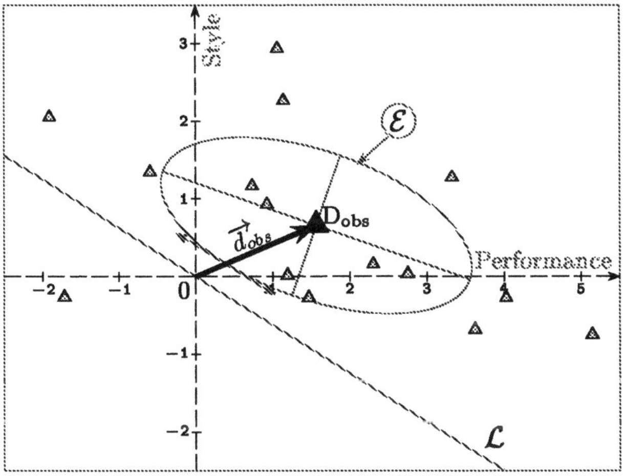

$n = 15$; $r = 2$; $q = 14\,d.f.$
$\overrightarrow{d_{obs}}$: $(1.563, 0.660)$,
$|\overrightarrow{d_{obs}}|_S = 1.33$.

$$\mathbf{S} = \begin{bmatrix} 4.093 & -1.108 \\ -1.108 & 1.307 \end{bmatrix}$$

Equation of $\mathcal{E}$:
$(\mathbf{d} - \mathbf{d}_{obs})'\mathbf{S}^{-1}(\mathbf{d} - \mathbf{d}_{obs}) = 1$
$\lambda'_1 = 4.48$ and $\lambda'_2 = 0.92$
$\alpha_1 = -19.25$ degrees.
Equation of $\mathcal{L}$:
$\mathbf{d}'_{obs}\mathbf{S}^{-1}\mathbf{d} = 0$,
letting $\mathbf{d}' = (d, d')$,
$0.673\,d + 1.076\,d' = 0$.

Figure 9.1.13. Question A. Specific cloud of individual point–effects with mean vector–effect (in bold) and indicator ellipse $\mathcal{E}$ (S–norm)

The mean observed effect, in brief *observed effect*, will be denoted $\overrightarrow{d_{obs}}$ as a vector, and $\mathrm{D_{obs}}$ as a point. Likewise, the corresponding *true effect* (inference parameter) will be denoted $\overrightarrow{\delta}$ as a vector, and Δ as a point.

Descriptive appraisal. Starting with the corrected (for $q = n - 1 = 14$ d.f.) covariance matrix $\mathbf{S}$ of the specific cloud, the diagonalization of this matrix (cf. CLOUD, p.124) yields the corrected eigenvalues $\lambda'_1 = 4.48$ and $\lambda'_2 = 0.92$ — that is, the corrected variances of the specific cloud along its principal axes — and the angle of the first principal axis with performance axis $\alpha_1 = -19.25$ degrees. As a *geometric summary* of the cloud, in the context of inference, we take the indicator ellipse $\mathcal{E}$ (for the S–norm), that is, the one whose semi–axis lengths are equal to the square roots of corrected variances, as shown in Figure 9.1.13 (this page).

We will operationalize the concepts of *direction and magnitude of observed effect* by means of the following geometric construction. Let $\mathcal{L}$ be the line going through the origin 0 whose direction is parallel to the tangent to $\mathcal{E}$ at the point where $\mathcal{E}$ meets the segment $(0; \mathrm{D_{obs}})$. Line $\mathcal{L}$ divides the plane in two regions (half–planes), one of which, the right–upper half–plane, denoted $\mathcal{R}_0$, contains the observed point–effect $\mathrm{D_{obs}}$. We will take property $(\mathrm{D_{obs}} \in \mathcal{R}_0)$ as our directional descriptive appraisal.

As an *index of magnitude* of an effect $\overrightarrow{d}$, we take its S-norm $|\overrightarrow{d}|_S$

(writing $|\cdot|_S$, as opposed to the geometric norm $\|\cdot\|$). In matrix notation we have $|\overrightarrow{d}|_S = (\mathbf{d}'\mathbf{S}^{-1}\mathbf{d})^{1/2}$ (cf. p.322). With respect to the axes of the reference cloud, the coordinates of $\overrightarrow{d_{obs}}$ are $(1.563, 0.660)$, hence $|\overrightarrow{d_{obs}}|_S = 1.33$. The magnitude 1.33 is undoubtly large[12].

> *Descriptive Conclusion.* The observed effect lies in the right–upper half–plane and is of large magnitude.

Significance testing. We now attempt to establish the *existence* of effect, by performing Hotelling's test for two matched groups (INDUCTIVE, §8.8, p.324). We have here $n = 15$ (sample size), $r = 2$ (cloud dimension), $q = 14$ (d.f. for covariance), hence $k = (q - r + 1)/q = 13/14$. The value of Hotelling's statistic is $T^2_{obs} = 15 \times (1.33)^2 = 26.49$, hence $(k/r)\,T^2_{obs} = 12.3$. From the $F_{2,13}$ distribution (central F with $r = 2$ and $q - r + 1 = 13$ d.f.), we find $p_{obs} = P(F_{2,13} > 12.3) = .001$, a highly significant result (S**).

> *Asserting existence of effect.* Data are in favor of the heterogeneity of the "before" and "after" groups, that is, of a non–null true effect: $p_{obs} = .001$ (S**).

This significant result does provide some support to the descriptive conclusion, in so far as it tells us that point null hypothesis $\mathcal{H}_0 : \Delta = \mathrm{O}$ is not compatible with data and therefore establishes, at least, the existence of a true effect. Yet it leaves unsettled the issues of direction and largeness of effect, thus falling short of the descriptive appraisal.

Bayesian data analysis. We will proceed to the inductive extension of the directional and largeness conclusions by means of Bayesian inference, taking as a *posterior distribution* of the true effect (parameter δ) the multidimensional scaled t-distribution, centered on the observed point–effect D_{obs}, hence for a γ–value between 0 and 1 the corresponding γ*–credibility ellipse*, which is the inertia ellipse corresponding to $|\Delta - \mathrm{D}_{obs}|^2_S = (r/kn)\ F_\alpha$, with $\alpha = 1 - \gamma$ (cf. INDUCTIVE, §8.5.1, p.325).

For $\gamma = 1 - p_{obs} = .999$, the credibility ellipse $\mathcal{C}_0$ going through the origin–point 0 provides the *Bayesian reinterpretation of the significance level*: $P(\Delta \text{ inside } \mathcal{C}_0) = .999$, i.e. the posterior probability that the true point–effect Δ is less distant (for the S–norm) from the observed point–effect D_{obs} than $|\overrightarrow{d_{obs}}|_S$, is $P(|\Delta - \mathrm{D}_{obs}|_S < 1.33) = 0.999$ (Figure 9.1.14).

[12]For univariate effects we took 0.40 as a limit for largeness. For two–dimensional effects a higher value seems appropriate to account for the wider dispersion of distributions (cf. CLOUD, §3.4.4, p.99); we take here the value 0.60 as in Rouanet & al (1998).

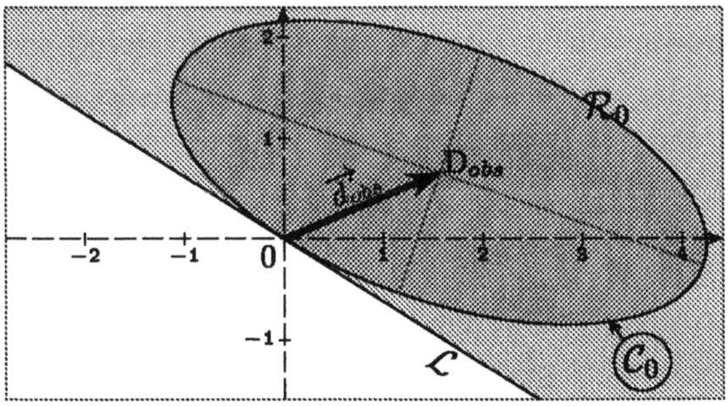

Figure 9.1.14. Question A. Reinterpretation of significance level $P(\Delta \text{ inside } \mathcal{C}_0) = .999$ and directional conclusion $P(\Delta \in \mathcal{R}_0) > .9995$ ($\mathcal{R}_0$ upper–right half–plane)

In accordance with the descriptive conclusion ($\mathrm{D}_{obs} \in \mathcal{R}_0$), the hypothesis "The true effect is on the side of the observed effect" will be operationalized as ($\Delta \in \mathcal{R}_0$). The posterior probability $P(\Delta \in \mathcal{R}_0)$ could be calculated from the t–distribution, but the consideration of the $\mathcal{C}_0$ ellipse provides a lower bound easily calculated from the standard F-distribution. Owing to the symmetry of the posterior distribution, one has $P(\Delta \notin \mathcal{R}_0) < p_{obs}/2 = .005$, therefore one has $P(\Delta \in \mathcal{R}_0) > .9995$: a very high credibility.

> *Asserting direction of effect.* There is a high posterior probability that the true effect lies "on the side of the observed effect": $P(\Delta \in \mathcal{R}_0) > .9995$.

We attempt to assert *directional largeness* in the line of the previous directional conclusion. Let $\overrightarrow{d_{lar}}$ be the vector proportional to $\overrightarrow{d_{obs}}$ such that $|\overrightarrow{d_{lar}}|_S = l_{lar}$, and $\mathrm{D_{lar}}$ the endpoint of $\overrightarrow{d_{lar}}$ (Figure 9.1.15 with $l_{lar} = 0.60$).

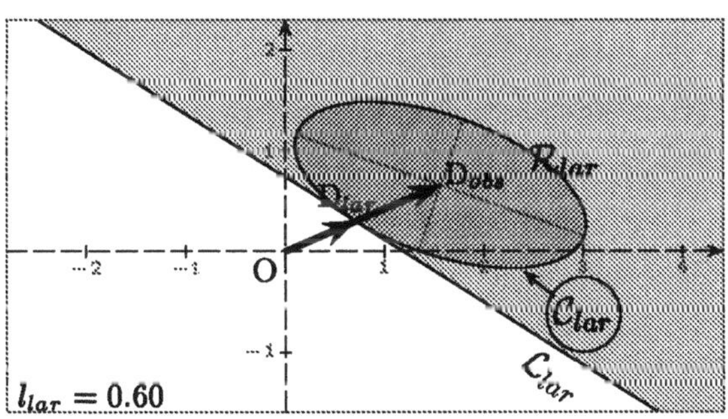

Figure 9.1.15. Question A. Directional largeness of effect: $P(\Delta \in \mathcal{R}_{lar}) > .973$.

Consider line $\mathcal{L}_{lar}$ going through $\mathrm{D_{lar}}$ parallel to $\mathcal{L}$, and let $\mathcal{R}_{lar}$ be the half–plane delimited by $\mathcal{L}_{lar}$ containing $\mathrm{D_{obs}}$; then we take ($\Delta \in \mathcal{R}_{lar}$) to define the hypothesis "$\overrightarrow{\delta}$ is large on the side of the observed effect". We then have $|\mathrm{D}_{obs} - \mathrm{D}_{lar}|_S = 1.33 - 0.60 = 0.73$; looking for the credibility ellipse $\mathcal{C}_{lar}$ such that ($|\Delta - \mathrm{D}_{obs}|_S < 0.73$), we find ($F_{2,13}$–distribution): $P(|\Delta - \mathrm{D}_{obs}|_S > 0.73) = .054$, hence $P(\Delta \notin \mathcal{R}_{lar}) < .054/2 = .027$, and the lower bound $P(\Delta \in \mathcal{R}_{lar}) > .973$: a high credibility (greater than .90).

> *Asserting directional largeness.* There is a high posterior probability that the true effect is large on the side of observed effect: $P(\Delta \in \mathcal{R}_{lar}) > .973$.

Question B: Comparing Patients after Drug Intake with Healthy Subjects. The two independent groups of Question B are the subcloud of 15 patients "after" and the one of the 45 healthy persons. The "effect of interest" is now a *difference*, namely the vector–difference of the mean point of the two subclouds. The steps are the same as for Question A.

Constructing the Specific Cloud. i) Restrict the overall cloud to the subcloud of 15 patients after drug and the subcloud of 45 healthy persons (see Figure 9.1.8, p.348); ii) Shift the two subclouds by translation so

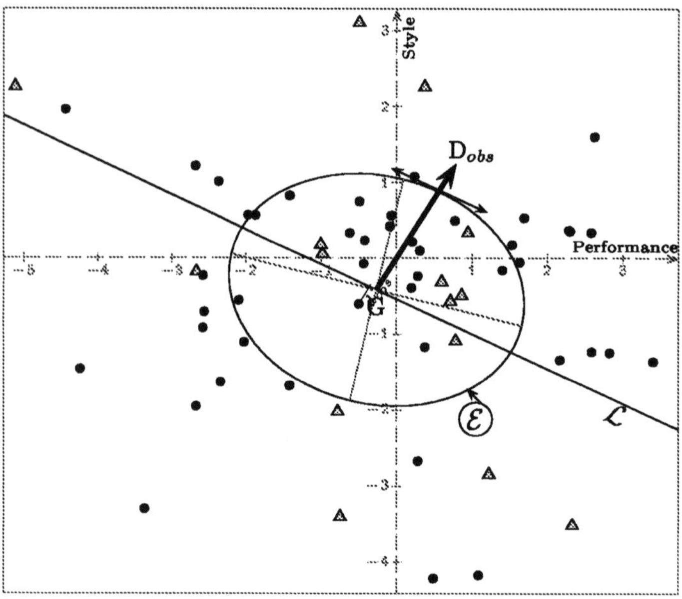

$\tilde{n} = 11.25$; $r = 2$;
$q = 58$ d.f.
$\overrightarrow{d_{obs}}$: (1.06,1.66);
$|\overrightarrow{d_{obs}}|_S = 1.28$

$$\mathbf{S} = \begin{bmatrix} 3.889 & -0.392 \\ -0.392 & 2.369 \end{bmatrix}$$

Equation of $\mathcal{E}$: $(\mathbf{d} - \mathbf{d}_{obs})'\mathbf{S}^{-1}(\mathbf{d} - \mathbf{d}_{obs}) = 1$
$\lambda'_1 = 3.98$; $\lambda'_2 = 2.27$,
$\alpha_1 = -13.7$ degrees.

Equation of $\mathcal{L}$:
$\mathbf{d}'_{obs}\mathbf{S}^{-1}\mathbf{d} = 0$
letting $\mathbf{d}' = (d, d')$,
$0.348\, d + 0.760\, d' = 0$.

Figure 9.1.16. Question B. Specific cloud: Within–cloud of 15 patients after drug intake (triangles) and 45 healthy persons (bullets), with vector-difference (in bold); and indicator ellipse $\mathcal{E}$.

that they have the same mean point G which is the barycenter of the union of the two subclouds; hence the *within–cloud* (cf. CLOUD §3.5.2, p.103), which defines the specific cloud, together with the vector–difference, as shown in Figure 9.1.16 (preceding page).

Descriptive Appraisal. Starting with the corrected (for $q = 45 + 15 - 2 = 58$ d.f.) covariance matrix $\mathbf{S}$ of the within–cloud, the corrected variances along its principal axes are $\lambda'_1 = 3.98$ and $\lambda'_2 = 2.27$, and the angle of the first principal axis with the performance axis is $\alpha_1 = -13.7$ degrees. The corresponding indicator ellipse $\mathcal{E}$ is shown in Figure 9.1.16.

Let $\mathcal{L}$ be the line going through point G and parallel to the tangent to $\mathcal{E}$ at the point where $\mathcal{E}$ cuts the segment $(\mathrm{G}; \mathrm{D_{obs}})$; hence the right–upper half–plane $\mathcal{R}_0$. We take property $(\mathrm{D_{obs}} \in \mathcal{R}_0)$ as a directional descriptive appraisal. Taking the S–norm (with respect to the within–cloud) of the difference–vector $\overrightarrow{d_{obs}}$, we find $|\overrightarrow{d_{obs}}|_S = 1.28$: a large value.

> *Descriptive conclusion.* The observed point–difference lies in the right–upper half–plane and is of large magnitude.

Significance Testing. We now establish the existence of a difference between mean points by Hotelling's test for two independent groups (cf. §8.5.1, p.326). We now have $\widetilde{n} = 1/(1/45+1/15) = 11.25$, $q = 45+15-2 = 58$ and $k = (q - r + 1)/q = 57/58 = 0.983$, hence $T^2_{obs} = 11.25 \times (1.28)^2 = 18.43$ and $\frac{k}{r} T^2_{obs} = 9.02$, hence from the distribution of $F_{2,57}$: $p_{obs} = 0.0004$, a highly significant result.

> *Asserting existence of effect.* Data are in favor of the heterogeneity between patients after drug intake and healthy persons, that is, of a non–null true difference: $p_{obs} < 0.004$ (S**).

Bayesian Data Analysis. To begin with, we have the *Bayesian reinterpretation of the significance* level: The posterior probability that the true point Δ is less distant from the observed point D_{obs} than $|\overrightarrow{d_{obs}}|_S = 1.28$ is $1 - p_{obs} = .9996$: $P(|\Delta - \mathrm{D}_{obs}|_S < 1.28) = .9996$ (cf. Figure 9.1.17, p.362): The probability $P(\Delta \in \mathcal{R}_0)$ is higher than .9998: a very high credibility.

> *Asserting direction of effect.* There is a high posterior probability that the true difference lies "on the side of the observed difference": $P(\Delta \in \mathcal{R}_0) > .9998$.

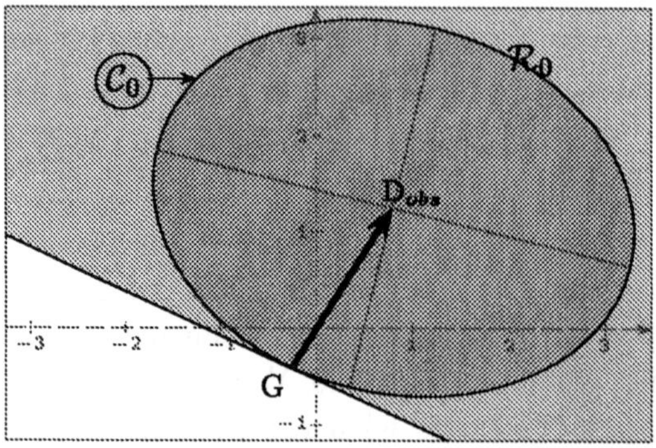

Figure 9.1.17. Question B. Reinterpretation of significance level $P(\Delta \text{ inside } \mathcal{C}_0) = .9996$ and directional conclusion $P(\Delta \in \mathcal{R}_0) > .9998$ ($\mathcal{R}_0$ upper–right half–plane).

To attempt asserting of *directional largeness*, we compute: $|\overrightarrow{d_{obs}}|_S - l_{lar} = 1.28 - 0.60 = 0.68$. From the $F_{2,57}$–distribution, we find: $P(\delta \in \mathcal{C}_{lar}) = P(|\Delta - \mathrm{D}_{obs}|_S > 0.68) = .090$. Therefore $P(\delta \notin \mathcal{R}_{lar}) < .045$, and one deduces $P(\delta \in \mathcal{R}_{lar}) > .955$: a high credibility (Figure 9.1.18).

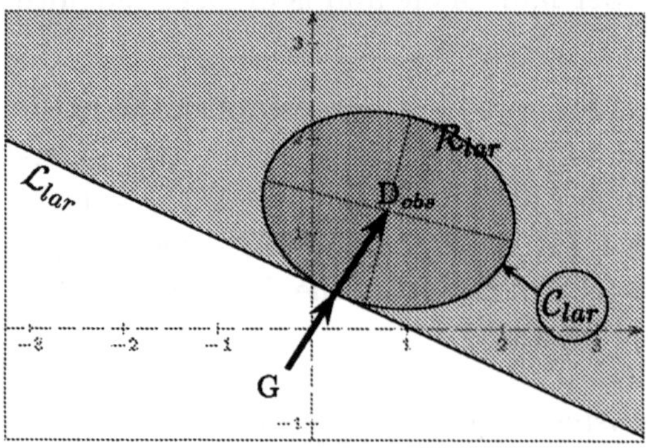

Figure 9.1.18. Question B. Directional largeness of difference: $P(\Delta \in \mathcal{R}_{lar}) > .955$.

Asserting directional largeness. There is a high posterior probability that the true difference is large on the side of observed effect: $P(\Delta \in \mathcal{R}_{lar}) > .955$ ($l_{lar} = 0.60$)

Probabilities of hypotheses of interest

In the foregoing inductive analyses, we have investigated hypotheses that were extensions of descriptive conclusions, and therefore had a chance of being substantiated by data. Bayesian data analysis also permits us to calculate the probabilities of *any hypothesis of interest*, in the first place, research hypotheses that have motivated the experiment but have not been substantiated by data.

Smallness hypothesis for Question B. As an example, consider the smallness hypothesis, stating that the true difference "healthy" – "after" is small; we will investigate this hypothesis first for Performance then for Performance and Style taken jointly.

— *Performance* (univariate analysis). The smallness hypothesis can be operationalized by stating that δ is in absolute value less than some upper limit for smallness. Taking for convenience the value 0.40 (taken earlier for largeness)[13], the smallness hypothesis is $(|\delta|_S/s < 0.40)$. Now we have $P(\delta/s > 0.40) = .675$ (as already found), hence $P(\delta/s < 0.40) = .325$; similarly we find $P(\delta/s < -0.40) = .001$; hence $P(|\delta|/s < 0.40) = .324$: a rather low credibility. This result shows how misleading it would be to interpret the nonsignificant result of $\mathcal{H}_0 : \delta = 0$ as providing evidence in favor of the smallness hypothesis.

— *Performance and Style* (geometric inference). The smallness hypothesis can be operationalized by stating that the S–norm of $\overrightarrow{\delta}$ is less than some specified smallness limit.

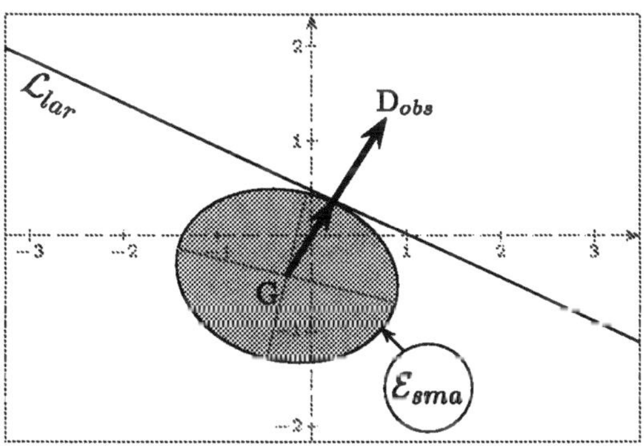

Figure 9.1.19. Question B. Smallness hypothesis: $P(\Delta \text{ inside } \mathcal{E}_{sma}) < .045$ ($l_{sma} = 0.60$).

Geometrically, the smallness zone is the inside of an ellipse centered at the origin–point shifted by translation from a credibility ellipse. Taking for convenience the smallness limit 0.60 (taken earlier for two–dimensional

[13]The value $l_{sma} = 0.40$ is admittedly not really small; but taking a smaller l_{sma} value would evidently lead to an even lower credibility for the smallness hypothesis.

largeness), the smallness hypothesis is $(|\overrightarrow{\delta}| < 0.60)$. The corresponding ellipse $\mathcal{E}_{sma}$ is shown in Figure 9.1.19 (p.363). From the result on directional largeness (p.362), we have $P(\Delta \text{ inside } \mathcal{E}_{sma}) < .045$: a very low credibility.

Miracle Drug. In the Parkinson Study, we have not met the situation where the descriptive conclusion is "the difference is small" and what is sought is to assert smallness. To illustrate such a situation, let us figure out a "Miracle Drug" leading to the following fictitious data set. Suppose the points "after" are obtained by shifting, for each patient, each real point by a translation vector equal to the mean vector-difference $\overrightarrow{d_{obs}}$; then for the Miracle Drug the mean point "after" coincides with the mean point of healthy persons, i.e. $\mathrm{D}_{obs} = \mathrm{O}$, which is descriptively the situation most in favor of smallness of difference[14]. Consequently, we will attempt to assert smallness of true difference, by means of Bayesian data analysis.

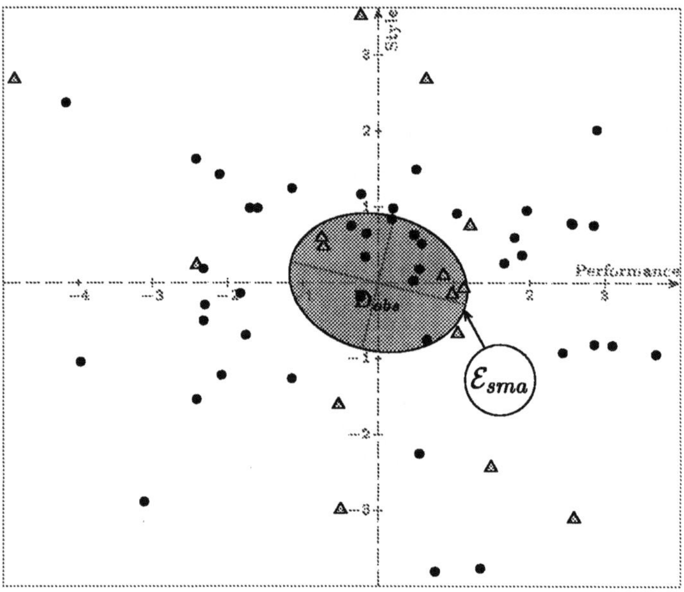

Figure 9.1.20. Miracle Drug ($\mathrm{D}_{obs} = \mathrm{O}$): patients "after" and healthy persons and Credibility ellipse for the smallness hypothesis ($l_{sma} = 0.60$).

The posterior distribution is centered around the origin–point, with the same credibility ellipses as in Question B. We have therefore $P(|\delta| < 0.60) = 0.85$: a credibility certainly more encouraging than the one found

[14]With significance testing, we have here $p_{obs} = 1$ (a "perfectly nonsignificant" result), which simply reflects that the null hypothesis is perfectly compatible with data, with no implication about the probability of a small difference.

with actual data — yet still falling short of $\gamma = .90$. Thus, even with the Miracle Drug (observed difference "as small as small can be"), we are not in a position to assert smallness at credibility $\gamma = .90$. The reason is that the sample size is not big enough; more data are needed[15]!

9.2 French Political Space

This case study is devoted to the construction and analysis of the political space of values and attitudes of the French electors in the late nineties, built from the data of a 1997 post–electoral survey[16].

What place do individuals occupy in quantitative studies of political science? In the most sophisticated studies, the modeling is focused on variables (e.g. attitudes and opinions), as if individuals were just there to supply a random residual term. In the present study, we reinstate the individuals at the heart of the analysis. By means of Multiple Correspondence Analysis, we construct a geometric model of data, with a cloud of modalities and a cloud of individuals. By introducing the vote as a structuring factor of the set of individuals, we will conduct the study of electorates directly on the subclouds associated with electorates. The cloud of individuals will thus become a genuine "political map", exhibiting the cleavages between electorates and how voters are spread over these cleavages.

The data set stems from a survey carried out between the two rounds of the 1997 parliamentary elections ("élections législatives")[17], on a national sample of 3010 individuals representative of registered voters. The survey was conducted by telephone in 549 precincts, with stratification by region and urban size, using quotas: gender, age, occupation of the head of household, level of education. The survey comprised questions about life conditions and economy prospects, about campaign issues and politics (*stricto sensu*), about ideological attitudes such as confidence toward left and right wings, about the functioning of democracy and attitudes toward the European Union (EU); finally, the question was asked about the choice at the first round of elections.

[15]Equivalently, the smallness conclusion might be reached if a prior distribution incorporating external evidence of smallness were taken instead of neutral prior.

[16]The descriptive part of the study follows Chiche, Le Roux, Perrineau, Rouanet (2000). The party competition in France is discussed by Adams (2001), the 1997 parliamentary elections by Perrineau, Ysmal & al (1998).

[17]The survey was sponsored by the newspaper *Libération* and the political experts of CEVIPOF, with the help of FNSP, CIDSP-Grenoble and CRAPS-Lille, and carried out by SOFRES poll institute.

The statistical analysis will comprise four parts: i)*Elementary descriptive statistics* (§9.2.1); ii) *Geometric Analysis* (MCA), with interpretation of axes (§9.2.2); iii) in the cloud of individuals, *study of the electorates* of the major political parties and of Non Respondents (§9.2.3). iiii) *Inductive analyses* extending the major descriptive conclusions of part iii) (§9.2.4).

9.2.1 Elementary Descriptive Statistics

Construction of the political space

To construct the political space of values and attitudes, we have selected 20 questions referring to the following six themes[18] (with the purpose of balancing the six themes as equally as possible): *Ethnocentrism* (4 questions); *Authoritarianism* (2 questions); *Social* (4 questions); *Economy* (4 questions); *Politics* (*stricto sensu*, 2 questions); *Supranationality* (4 questions). For each question, a sentence was enunciated, and the usual response format was the 4–modality scale: 1 strong agreement, 2 weak agreement, 3 weak disagreement, 4 strong disagreement with the sentence. Modalities 1 and 4 thus correspond to extreme attitudes, and modalities 2 and 3 to moderate ones,[19] and only one answer is required per question. Question School had only 2 modalities, and the questions EU–power and End–EU only 3 modalities. For all questions, there is a "No Answer" modality. All in all, we have 96 modalities (including 20 "No Answer" modalities). The detailed formulations of the 20 selected questions and of the question about the Vote are reproduced hereafter, with the response marginal frequencies (one–way tables). Percentages are calculated on the 2980 individuals who answered more than 5 questions.

Response frequencies of the 20 questions retained for analysis

- **Ethnocentrism**

[Immigrants] There are too many immigrants in France.

1) Strongly agree	914	30.8%		3) Somewhat disagree	575	19.3%
2) Somewhat agree	842	28.3%		4) Strongly disagree	597	20%
				0) No Answer	52	1.6%

[18]The allocation of questions to themes is crucial for the construction of the space; it does not affect the geometric analysis, but will be again crucial as an aid for interpretation (cf. *MULTIPLE*, p.215).

[19]The numbers 1 through 4 refer to the sentence enunciated in the survey. For example, Immigrants 1 means "strongly agrees" with the sentence "There are too many immigrants in France"; North-Africans 3 means "somewhat disagrees" with the sentence "North-Africans who live in France will be one day French people like others"; etc.

[North-Africans] North-Africans who live in France will be one day French people like others.

1) Strongly agree	855	28.7%		3) Somewhat disagree	545	18.3%
2) Somewhat agree	1004	33.8%		4) Strongly disagree	519	17.6%
				0) No Answer	50	1.6%

[Races] There are less gifted races than others.

1) Strongly agree	245	8.2%		3) Somewhat disagree	371	12.5%
2) Somewhat agree	298	10%		4) Strongly disagree	2016	67.7%
				0) No Answer	50	1.5%

[At home] One no longer feels at home like one once did.

1) Strongly agree	779	26.1%		3) Somewhat disagree	623	21%
2) Somewhat agree	547	18.4%		4) Strongly disagree	1005	33.7%
				0) No Answer	26	0.8%

• Authoritarianism

[Death Penalty] The Death penalty should be reinstated.

1) Strongly agree	857	28.8%		3) Somewhat disagree	372	12.5%
2) Somewhat agree	616	20.8%		4) Strongly disagree	1104	37.1%
				0) No Answer	31	0.9%

[School] In regard to education, could you tell me which statement you agree more with?

1) School should primarily develop the sense of **discipline** and effort.	1506	50.5%
2) School should primarily develop **critical mind.**	1440	48.4%
0) No Answer	34	1.1%

• Social

[Strike Effectiveness] Do you think that today strike is a means of action ...

1) Very effective	304	10.3%		3) Little effective	1157	38.8%
2) Rather effective	1120	37.6%		4) Not at all effective	389	13.1%
				0) No Answer	10	0.3%

[Strike 95] At the time of the strike movement in November/December 1995, how did you feel with respect to this movement ...

1) Great solidarity	561	18.8%		3) Little solidarity	749	25.2%
2) Some solidarity	1020	34.3%		4) No solidarity	598	20.1%
				0) No Answer	52	1.7%

Can you tell me if he following word evokes for you something very positive ...

[Unions]

1) Very positive	407	13.7%		3) Fairly negative	894	29.9%
2) Fairly positive	1238	41.5%		4) Very negative	415	14%
				0) No Answer	26	0.9%

[Public services]

1) Very positive	749	25%		3) Fairly negative	606	20.4%
2) Fairly positive	1439	48.3%		4) Very negative	173	5.8%
				0) No Answer	13	0.4%

• Economy

[Liberalism][20]

1) Very positive	504	17%		3) Fairly negative	677	22.8%
2) Fairly positive	1477	49.5%		4) Very negative	261	8.8%
				0) No Answer	61	2.0%

[20]In the French context, "liberalism" always refers to "economic liberalism", not to "cultural liberalism".

[**Profit**]

1) Very positive	283	9.5%	3) Fairly negative	988	33.2%
2) Fairly positive	1264	42.6%	4) Very negative	406	13.7%
			0) No Answer	39	1%

[**Privatization**]

1) Very positive	418	14.1%	3) Fairly negative	898	30.2%
2) Fairly positive	1164	39.1%	4) Very negative	421	14.2%
			0) No Answer	79	2.4%

[**Globalization**] Do you think that globalization still leaves to the French government a working margin in the economy area?

1) Very large	144	4.8%	3) Fairly small	1270	42.6%
2) Fairly large	1102	37.0%	4) Very small	338	11.3%
			0) No Answer	126	4.2%

- **Politics**

[**Democracy**] Do you think at the moment democracy in France works ...

1) Very well	109	3.7%	3) Not very well	1241	41.6%
2) Fairly well	1106	37.2%	4) Not well at all	517	17.3%
			0) No Answer	7	0.2%

[**Politicians**] In your opinion, are politicians concerned with what people like yourself think?

1) Very much	90	3%	3) Little	1436	48.2%
2) Rather	480	16.1%	4) Not at all	962	32.4%
			0) No Answer	12	0.3%

- **Supranationality**

[**Euro**] Are you for or against the prospect of the EU having a unique currency, which means that the Franc will be replaced by the Euro?

1) Definitely favorable	581	19.5%	3) Somewhat unfavorable	622	20.9%
2) Somewhat favorable	1248	41.9%	4) Definitely unfavorable	446	15.0%
			0) No Answer	83	2.8%

[**EU Power**] In your opinion should in the future the political power of the EU (European Union)— as opposed to those of the member states — be

1) **reinforced**	1118	37.5%	3) kept at present level	1233	41.4%
2) **diminished**	473	15.9%	0) No Answer	156	5.2%

[**End EU**] If you learned, tomorrow, that the EU is abandoned, would you feel

1) **regrets**	1410	47.3%	3) **relief**	536	18.0%
2) **indifference**	980	32.9%	0) No Answer	54	1.8%

[**EU protection**] With European Union, France will be better protected against the risks linked with the globalization of economy.

1) Strongly agree	587	19.7%	3) Somewhat disagree	611	20.5%
2) Somewhat agree	1241	41.6%	4) Strongly disagree	448	15%
			0) No Answer	93	3.1%

Question about Vote[21]

[**Vote**] How did you vote on the first round of the legislative elections?

1) Communist	225	7.6%	*6.4%*	
2) Extreme left	30	1.0%	*1.7%*	
3) Socialist	638	21.4%	*18.1%*	(Socialist Party & diverse Left)
4) Greens	165	5.6%	*5.4%*	(Ecologist Party & others ecologists)

[21]The last column, in italics, indicates the percentages of results for metropolitan France (Source: Ministry of the Interior).

5) Traditional Right	690	23.1%	*23.5%*	(UDF, RPR Parties & diverse Right)
6) Independent Right	58	1.9%		
7) National Front	193	6.5%	*9.9%*	
8) Others	14	0.5%	*0.1%*	
9) NonRespondents	967	32.4%	*34.9%*	(Blank, Null, No vote & No Answer)

Comparing these results with national results, one notices that the rates are of the same order of magnitude, except for Socialist (overestimated in the sample) and National Front Vote (underestimated in the sample).

9.2.2 Geometric Analysis and Interpretation of Axes

Discarding 30 individuals (out of 3010), who answered no more than 5 questions (out of the 20 selected questions), we are left with $3010 - 30 = 2980$ individuals. In addition, to avoid infrequent modalities, and according to the "five percent rule" (cf. *MULTIPLE*, p.216), for the question *Democracy* we have pooled the modality "works very well" (3.7%, Democracy 1) with "works rather well" (Democracy 2), hence the pooled modality denoted Democracy 1_2). Similarly for the question *Politicians*, we have pooled "very much concerned" (3%, Politicians 1) with "somewhat concerned" (Politicians 2), hence the pooled modality denoted Politicians 1_2; hence $96 - 2 = 94$ basic modalities for the construction of the space.

Results of specific Multiple Correspondence Analysis

For all 20 questions, the rates of "No Answer" modalities are quite low, all inferior to 3.1%, except EU *Power* (5.2%) and *Globalization* (4.2%), therefore we have proceeded to a *specific* MCA (cf. *MULTIPLE*, §5.2.1, p.204), from the data of 2980 individuals, putting the 20 no–answer modalities as *passive modalities*, and the $94 - 20 = 74$ others as *active modalities*.

Eigenvalues and inertia rates. Applying the formula for the variance of cloud in specific MCA (cf. *MULTIPLE*, property 5.25, p.206) we have: $\frac{74}{20} - 1 + \frac{1061}{20\times 2980} = 2.7178$. The variances (eigenvalues) of the first six axes and their inertia rates are given in Table 9.2.1 (for modified rates cf. *MULTIPLE* p.209, and formula (5.16), p.201). The standard deviations of

Axes	1	2	3	4	5	6
Variances (λ)	.2133	.1606	.1192	.0806	.0758	.0689
Inertia Rates	7.8%	5.9%	4.4%	3.0%	2.8%	2.5%
Modified Rates	49%	24%	11%	3%	2%	2%

Table 9.2.1. Variances (eigenvalues), raw and modified rates of inertia

principal variables, denoted ξ, are the square roots of eigenvalues. For the first three axes, one has: $\xi_1 = .462$, $\xi_2 = .401$ and $\xi_3 = .345$.

Principal coordinates. The principal coordinates of the 74 active modalities on the first three axes are given in Table 9.2.2. From this table, one could represent the whole set of 74 modalities in the first planes. In the sequel, we will confine the representation to the modalities relevant to the interpretation of the successive axes 1, 2, 3; cf. Figures 9.2.1 (p.373), 9.2.2 (p.375) and 9.2.3 (p.377). The cloud of individuals will be depicted in plane 1-2 (Figure 9.2.4, p.378) and in plane 2-3 (Figure 9.2.5, p.379).

		Weights	Coordinates		
		f_k/Q	Axis 1	Axis 2	Axis 3
Immigrants	1	.0156	0.849	0.255	0.274
	2	.0144	−0.115	0.495	−0.129
	3	.0098	−0.647	−0.034	−0.204
	4	.0102	−0.517	−1.028	−0.035
North Africans	1	.0146	−0.303	−0.733	0.158
	2	.0172	−0.366	0.242	−0.117
	3	.0093	0.116	0.703	−0.213
	4	.0089	1.091	0.005	0.205
Races	1	.0042	1.173	−0.051	0.639
	2	.0051	0.590	0.434	0.033
	3	.0063	0.216	0.707	−0.118
	4	.0344	−0.278	−0.187	−0.065
At home	1	.0133	0.978	0.092	0.225
	2	.0093	0.137	0.646	−0.226
	3	.0106	−0.467	0.399	−0.147
	4	.0172	−0.544	−0.666	0.028
Death Penalty	1	.0146	0.875	0.099	0.246
	2	.0105	0.085	0.634	−0.161
	3	.0064	−0.409	0.512	−0.098
	4	.0189	−0.593	−0.606	−0.074
School	1	.0257	0.398	0.262	0.104
	2	.0246	−0.410	−0.269	−0.107
Strike Effectiveness	1	.0052	0.347	−0.942	−0.152
	2	.0191	−0.283	−0.121	−0.280
	3	.0198	−0.077	0.367	0.023
	4	.0066	0.760	0.001	0.843
Strike95	1	.0096	0.167	−1.063	−0.504
	2	.0174	−0.129	0.074	−0.398
	3	.0128	−0.220	0.503	0.143
	4	.0102	0.329	0.224	0.971
Unions	1	.0070	−0.022	−1.201	−0.307
	2	.0211	−0.331	−0.038	−0.193
	3	.0153	0.059	0.596	−0.025
	4	.0071	0.886	−0.016	0.937
Public services	1	.0128	−0.045	−0.758	−0.054
	2	.0246	−0.173	0.256	−0.135
	3	.0104	0.154	0.436	0.154
	4	.0030	1.058	−0.367	0.788

		Weights	Coordinates		
		f_k/Q	Axis 1	Axis 2	Axis 3
Liberalism	1	.0086	0.171	−0.255	0.692
	2	.0252	−0.163	0.344	0.078
	3	.0116	0.006	−0.035	−0.505
	4	.0045	0.487	−1.370	−0.415
Profit	1	.0048	0.242	−0.224	0.950
	2	.0216	−0.214	0.294	0.174
	3	.0169	−0.009	0.049	−0.357
	4	.0069	0.507	−0.883	−0.334
Privatization	1	.0071	0.348	0.051	1.318
	2	.0199	−0.272	0.362	0.113
	3	.0153	−0.122	−0.007	−0.559
	4	.0072	0.666	−1.049	−0.438
Globalization	1	.0025	0.097	−0.680	0.957
	2	.0188	−0.335	0.125	0.223
	3	.0217	0.024	0.088	−0.271
	4	.0058	0.876	−0.503	−0.129
Democracy	1_2	.0208	−0.444	0.019	0.300
	3	.0212	0.035	0.181	−0.270
	4	.0088	0.955	−0.475	−0.062
Politicians	1_2	.0097	−0.466	0.038	0.584
	3	.0245	−0.258	0.120	−0.075
	4	.0164	0.657	−0.198	−0.241
Euro	1	.0099	−0.519	−0.390	0.946
	2	.0213	−0.413	0.152	−0.050
	3	.0106	0.380	0.339	−0.565
	4	.0076	1.265	−0.362	−0.267
EU power	1	.0191	−0.355	−0.171	0.246
	2	.0081	0.830	−0.116	−0.388
	3	.0211	−0.038	0.205	−0.049
End EU	1	.0241	−0.594	−0.183	0.406
	2	.0167	0.198	0.327	−0.404
	3	.0092	1.175	−0.117	−0.318
EU protection	1	.0100	−0.333	−0.442	0.813
	2	.0212	−0.385	0.206	−0.016
	3	.0104	0.222	0.305	−0.619
	4	.0077	1.193	−0.434	−0.148

Table 9.2.2. Coordinates of 74 active modalities for the first three axes.

Contributions of modalities. Table 9.2.3 (facing page) gives the relative contributions of the 74 active modalities to the cloud (column Ctr)

		Ctr	Axis 1	Axis 2	Axis 3
Immigrants	1	.013	.0519	.0062	.0096
	2	.013	.0009	.0216	.0020
	3	.015	.0189	.0001	.0034
	4	.015	.0125	.0659	.0001
North Africans	1	.013	.0062	.0479	.0030
	2	.012	.0106	.0061	.0019
	3	.015	.0006	.0282	.0035
	4	.015	.0486	.0000	.0031
Races	1	.017	.0265	.0001	.0141
	2	.017	.0082	.0059	.0000
	3	.016	.0014	.0194	.0007
	4	.006	.0122	.0074	.0012
At Home	1	.014	.0586	.0007	.0055
	2	.015	.0008	.0238	.0039
	3	.015	.0107	.0104	.0019
	4	.012	.0234	.0466	.0001
Death penalty	1	.013	.0516	.0009	.0073
	2	.015	.0004	.0258	.0022
	3	.016	.0049	.0102	.0005
	4	.012	.0305	.0424	.0008
School	1	.009	.0188	.0108	.0023
	2	.010	.0191	.0109	.0023
Strike efficacy	1	.017	.0029	.0282	.0010
	2	.011	.0070	.0017	.0124
	3	.011	.0005	.0163	.0001
	4	.016	.0177	.0000	.0389
Strike95	1	.015	.0021	.0662	.0201
	2	.012	.0013	.0006	.0228
	3	.014	.0029	.0198	.0021
	4	.015	.0051	.0031	.0793
Unions	1	.016	.0000	.0614	.0054
	2	.011	.0107	.0002	.0065
	3	.013	.0002	.0332	.0001
	4	.016	.0256	.0000	.0513
Public services	1	.014	.0001	.0449	.0003
	2	.010	.0034	.0098	.0037
	3	.015	.0011	.0120	.0020
	4	.017	.0152	.0024	.0151

		Ctr	Axis 1	Axis 2	Axis 3
Liberalism	1	.015	.0012	.0034	.0340
	2	.009	.0031	.0183	.0012
	3	.014	.0000	.0001	.0243
	4	.017	.0049	.0512	.0063
Profit	1	.017	.0013	.0015	.0360
	2	.011	.0045	.0114	.0054
	3	.012	.0000	.0002	.0177
	4	.016	.0082	.0331	.0064
Privatization	1	.016	.0040	.0001	.1022
	2	.011	.0068	.0159	.0021
	3	.013	.0011	.0000	.0395
	4	.016	.0147	.0484	.0114
Globalization	1	.018	.0001	.0070	.0186
	2	.012	.0097	.0018	.0077
	3	.011	.0001	.0010	.0132
	4	.016	.0204	.0089	.0008
Democracy	1_2	.011	.0188	.0000	.0154
	3	.011	.0001	.0043	.0127
	4	.015	.0371	.0122	.0003
Politicians	1_2	.015	.0197	.0001	.0274
	3	.010	.0075	.0022	.0011
	4	.012	.0326	.0039	.0079
Euro	1	.015	.0123	.0092	.0731
	2	.011	.0168	.0030	.0004
	3	.015	.0071	.0075	.0280
	4	.016	.0562	.0061	.0045
EU power	1	.011	.0111	.0034	.0195
	2	.015	.0256	.0007	.0100
	3	.011	.0001	.0054	.0004
End EU	1	.010	.0392	.0049	.0327
	2	.012	.0030	.0109	.0225
	3	.015	.0583	.0008	.0076
EU protection	1	.015	.0051	.0120	.0547
	2	.011	.0144	.0055	.000
	3	.015	.0024	.0059	.0330
	4	.016	.0501	.0088	.0014

Table 9.2.3. Relative contributions of 74 active modalities to the cloud (Ctr) and to the first three axes.

and to the first three axes.

Interpretation of axes

The "interpretation" at this stage is the *statistical interpretation* that refers to the analyzed data and should always be done before proceeding to comments that make use of external information.

How many axes to be interpreted? The sum of the first three modified rates reaches 84%, which suggests retaining three axes. Another argument in favor of the three–dimensional solution is the following. Since for the Authoritarianism and Politics themes there are only two questions, we performed a control analysis giving double weights to the 4 questions, and we found that the first three axes are well correlated (0.962, 0.950 and 0.971 respectively) with those of the analysis presented here.

First interpretation. Since the first three eigenvalues are well separated, we will proceed to the interpretation of each of the first three axes.

Adding the contributions pertaining to the questions of each of the six themes (cf. Table 9.2.3), we obtain Table 9.2.4, which gives the contributions of each of the six themes to the first three axes. This table provides a first interpretation of axes based on the predominant contributions of themes.

	Axis 1	Axis 2	Axis 3
Ethnocentrism	**.29**	**.29**	.05
Authoritarianism	**.13**	.10	.02
Social	.10	**.30**	**.26**
Economy	.08	**.20**	**.33**
Politics	.11	.02	.07
Supranationality	**.30**	.08	**.28**
	1.	1.	1.

Table 9.2.4. Relative contributions (Ctr) of themes to the first three axes.

Axis 1. The Supranationality and Ethnocentrism themes together account for 59% of the variance of Axis 1. Adding the Authoritarianism theme, we reach 72%; adding the Politics theme, we reach 82%.

Axis 2. The Social and Economy themes contribute to 50% of the variance; adding the Ethnocentrism theme, we reach 79% of the variance.

Axis 3. The Economy and Social themes account for 59% of the variance; adding the Supranationality theme, we reach 87%; contrary to the first two axes, Axis 3 is virtually exempt from the Ethnocentrism theme (only 5% of variance).

> *To sum up: Axis 1* is in the first place a supranationality and ethnocentrism axis; *Axis 2* is roughly a social axis; *Axis 3* is in the first place an economy axis.

The detailed interpretation of each axis will be based on the contributions of modalities (cf. Table 9.2.3.); we will in any case retain the modalities — or barycenters of modalities close to one another — whose contributions verify the *basic criterion*, that is, exceed the average contribution which is is here $1/74 = .0135 (= 1.35\%)$.

Interpretation of Axis 1 ($\lambda_1 = 0.2133$). From the Table 9.2.3 (p.371), Table 9.2.5 can be constructed, showing the relative contributions of the 22 basic modalities and the 5 barycenters of close modalities[22] which contribute most to Axis 1, divided according to the signs of their coordinates:

[22]For instance, the AtHome3 and AtHome4 modalities are close on Axis 1 (coordinates: -0.467 and -0.544); the absolute contribution of their barycenter is equal to $\frac{1}{20}\frac{1628}{2980}\left(\frac{(-0.457\times 623-0.544\times 1005)^2}{1628}\right) = .007245$, hence the relative contribution $.007245/.2133 = 3.4\%$.

negative (left side) vs positive (right side), and ranked according to decreasing contributions to Axis 1, until criterion (.0135) is reached.

Relative contributions (Ctr)		left side	right side
At home	1		.0586
End EU relief	3		.0583
Euro	4		.0562
Immigrants	1		.0519
Death Penalty	1		.0516
EU protection	4		.0501
North-Africans	4		.0486
End EU regrets	1	.0392	
Democracy	4		.0371
At home	3&4	.0340	
Politicians	4		.0326
Immigrants	3&4	.0315	
Death penalty	4	.0305	
Euro	1&2	.0291	

Relative contributions (Ctr)		left side	right side
Races	1		.0265
EU diminished	2		.0256
Unions	4		.0256
Globalization	4		.0204
Politicians	1_2	.0197	
EU protection	1&2	.0196	
School critical mind	2	.0191	
School discipline	1		.0188
Democracy	1_2	.0188	
StrikeEffectiveness	4		.0177
North-Africans	1&2	.0168	
Public services	4		.0152
Privatization	4		.0147
Total		.2581	.6095

Table 9.2.5. 22 modalities and 5 barycenters most contributing to Axis 1 (87% of variance), ranked in decreasing order of contributions.

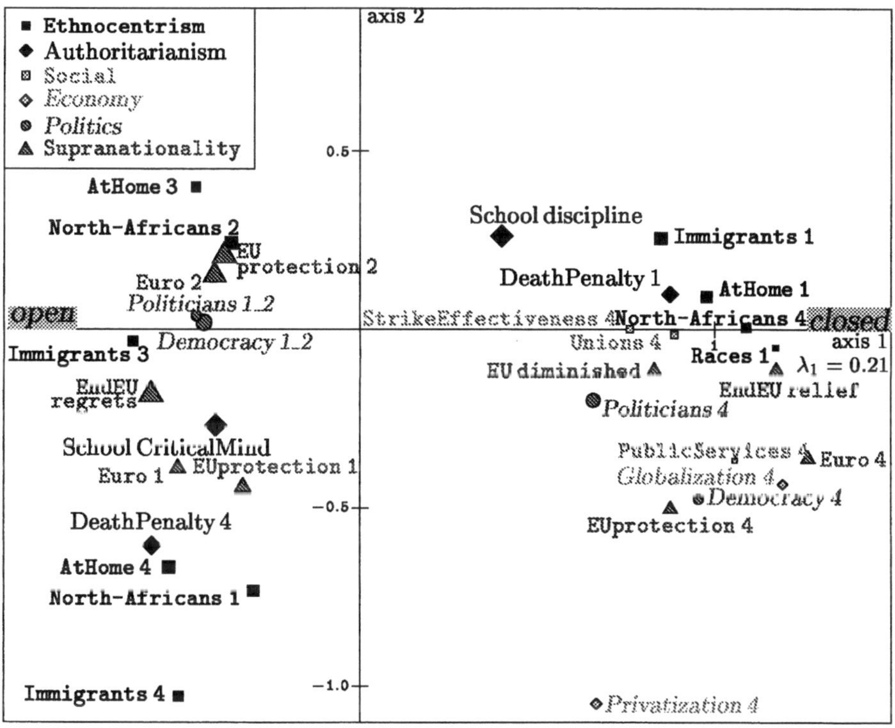

Figure 9.2.1. Interpretation of Axis 1 (32 modalities in plane 1-2).

The 32 modalities (out of the 74 ones) are reported on Figure 9.2.1 (p.373). Together they contribute to 87% of the variance of Axis 1. A striking feature is that the 17 modalities on the right side are on the whole farther from the origin point than the 15 modalities on the left, that they contribute much more (61%) than those on the left (26%), and that they all reflect extreme opinions. Let us now examine the modalities in detail.

• On one side of Axis 1 (right side on the graph) there are modalities reflecting xenophobia (AtHome 1, Immigrants 1, North-Africans 4, Races 1), hostility to European Union (EndEU relief 3, EU–protection 4, Globalization 4, EU diminished 2, Euro 4), in favor of reinstating death penalty and for school developing discipline (School 1, DeathPenalty 1).

• On the other side (left side on graph), we find the opposite modalities, namely more favorable to European Union (EndEU regrets 1, Euro 1&2), lack of xenophobia (AtHome 3&4, Immigrants 3&4, North-Africans 1&2), against death penalty (DeathPenalty 4), and for school developing critical mind (School 2).

These oppositions go with further ones in the political theme, namely, on the right side, pessimism about how democracy works (Democracy 4), and how politicians are concerned about people (Politicians 4); versus more optimistic views (Politicians 1_2, Democracy 1_2). The remaining modalities contributing to Axis 1 are all on the right side and reflect a definite distrust of unions (Unions 4), strikes (StrikeEffectiveness 4), public services (Public Services 4), and even privatization (Privatization 4).

As a conclusion, Axis 1 basically opposes one cluster of opinions sharply hostile to “the outside”, to a cluster of more receptive ones, that with P. Perrineau we will designate by a “closed attitude” and an “open attitude”, respectively.

Conclusion: Axis 1 opposes a closed attitude to an open one.

Interpretation of Axis 2 ($\lambda_2 = 0.1606$). We apply the same line of analysis as for Axis 1. From Table 9.2.2 (p.370) and 9.2.3 (p.371), we construct Table 9.2.6, showing the 20 basic modalities and 1 barycenter of modalities close on Axis 2 that contribute most to Axis 2, divided according to the signs of coordinates: negative (left) vs positive (right), and ranked according to decreasing contributions to Axis 2 until criterion (again .0135).

These 22 modalities are represented on Figure 9.2.2 (p.375) in plane 2-3; together they contribute to 77% of the variance of Axis 2.

Like for Axis 1, one finds a marked unbalance between the contributions of the two sides of axis (54% on left side, 23% on right side). Out of the 22

Relative contributions (Ctr)		left side	right side
Strike95	1	.0662	
Immigrants	4	.0659	
Unions	1	.0614	
Liberalism	4	.0512	
Privatization	4	.0484	
North-Africans	1	.0479	
At Home	4	.0466	
Public Services	1	.0449	
Death Penalty	4	.0424	
Death Penalty	2&3		.0360
Unions	3		.0332

Relative contributions (Ctr)		left side	right side
Profit	4	.0331	
StrikeEffectiveness	1	.0282	
North-Africans	3		.0282
At home	2		.0238
Immigrants	2		.0216
Strike95	3		.0198
Races	3		.0194
Liberalism	2		.0183
StrikeEffectiveness	3		.0163
Privatization	2		.0159
Total		.5362	.2325

Table 9.2.6. 20 modalities and 1 barycenter contributing to more than .0135 of the variance of Axis 2 (77% of variance) ranked in decreasing order of contributions.

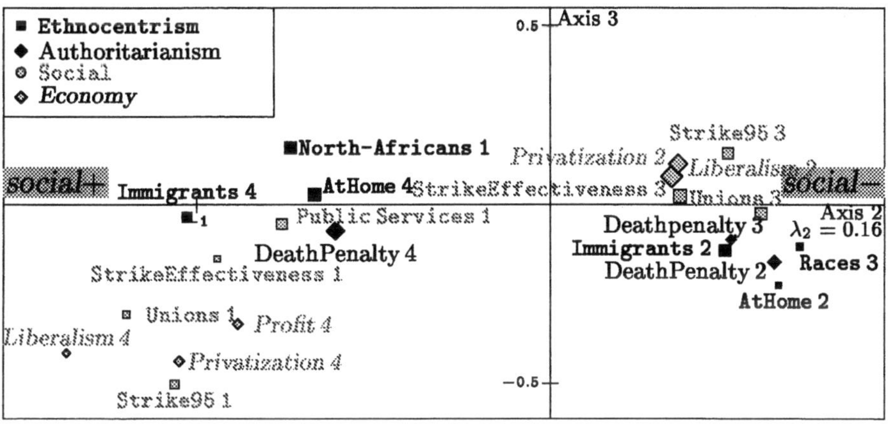

Figure 9.2.2. Interpretation of axis 2 (22 modalities in plane 2-3)

modalities, 12 pertain to the Economy and Social themes and contribute to 42% of the variance of axis. Examining the modalities in detail.

• On one side (left of graph), one finds the opinions strongly favorable to unions (Unions 1), to strike (Strike95 1, StrikeEffectiveness 1) and public services (PublicServices 1), and unfavorable to economic liberalism (Liberalism 4), privatization (Privatization 4) and profit (Profit 4); on this same side, one also finds the opinions against death penalty (Death penalty 4) and against xenophobia (Immigrants 4, North-Africans 1; AtHome 4). Observe that all these modalities are extreme ones (1 or 4).

• On the other side (right of graph), one finds the opposite opinions: rather unfavorable to unions (Union 3) and strike (Strike95 3, StrikeEffectiveness 3); rather favorable to liberalism (Liberalism 2) and privatization

(Privatization 2); one also finds the two moderate opinions about death penalty (Death Penalty 2&3), and the moderately xenophobic opinions (North-Africans 3, AtHome 2, Immigrant 2, Races 3). Observe that all these modalities are moderate ones (2 or 3).

All in all, these oppositions reflect the traditional left/right wings oppositions.

> *Conclusion*: Axis 2 opposes a 'Social +' attitude to a 'Social −' attitude.

Interpretation of Axis 3 ($\lambda_3 = 0.1192$). Again in the same line of analysis, we construct Table 9.2.7, showing the 18 basic modalities and the 6 barycenters of modalities close on Axis 3, that contribute most to Axis 3, among which 15 with positive coordinates (top), and 15 with negative ones (bottom).

Ctr		top	bottom
Privatization	1	.1022	
Strike95	4	.0793	
Euro	1	.0731	
EU Protection	1	.0547	
Unions	4	.0513	
Privatization	3&4		.0508
Strike95	1&2		.0429
StrikeEffectiveness	4	.0389	
Profit	1	.0360	
Liberalism	1	.0340	
EU protection	3		.0330
End EU regrets	1	.0327	

Ctr		top	bottom
Liberalism	3&4		.0306
Euro	3		.0280
Politicians	1_2	.0274	
Profit	3&4		.0241
End EU indifference	2		.0225
EU reinforced	1	.0195	
Globalization	1	.0186	
Democracy	1_2	.0154	
Public Services	4	.0151	
Races	1	.0141	
Globalization	3&4		.0139
StrikeEffectiveness	1&2		.0134
Total		.6123	.2592

Table 9.2.7. 18 basic modalities and 6 barycenters contributing to more than .0135 of the variance of Axis 3 (87% of variance) ranked in decreasing order of contributions.

On Figure 9.2.3 (next page), the 30 modalities contributing to Axis 3 are represented in plane 2-3. Once again, the picture is highly unbalanced.

• On one side (top), one finds the opinions strongly favorable to liberalism (Privatization 1, Profit 1, Liberalism 1), and the opinions strongly unfavorable to unions and strikes (Strike95 4, Unions 4, StrikeEffectiveness 4) and to public services (PublicServices 4); one also finds the opinions very favorable to EU and globalization (Euro 1, EU-protection 1, End EU regrets 1, EU reinforced 1, Globalization 1); again, one finds the modalities optimistic about democracy (Politicians 1_2, Democracy 1_2) ... and the most xenophobic modality (Races 1).

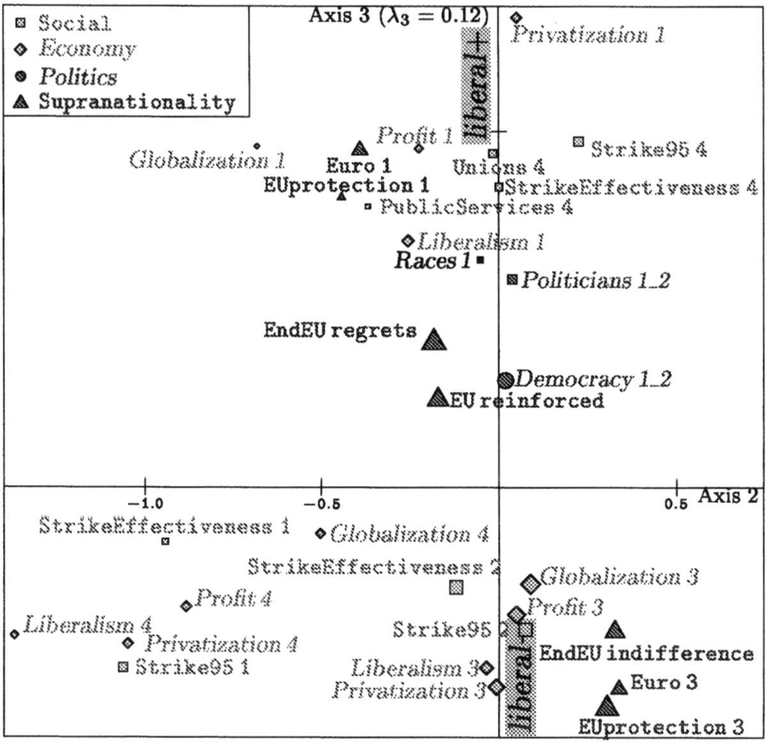

Figure 9.2.3. Interpretation of axis 3 (30 modalities in plane 2-3)

• On the other side (bottom), one finds the opposite (not as sharp) opinions: unfavorable to liberalism (Privatization 3&4, Liberalism 3&4, Profit 3&4), favorable to strike (Strike95 1&2), reluctance toward EU (EU-protection 3, Euro 3, End EU indifference 2, Globalization 3&4).

Conclusion: Axis 3 opposes a 'liberal +' to a 'liberal −' attitude.

First conclusions. For the political scientist, the order of axes is revealing. In 1997, the first cleavage among French voters concerns the perception of "the outside": immigrants, foreigners, Europe, globalization. Only afterwards do cleavages appear about social and economy values, that shaped the French political space for decades, around the opposition left/right wings. Furthermore, to make sense of the oppositions concerning social and economic attitudes, one single dimension is not sufficient, the plane 2-3 is necessary[23]. These findings will be basic for the study of electorates.

[23]The stability study done in chapter 7 (p.279) shows that if the NF electorate is taken out, the variance of Axis 1n decreases markedly, yet leaving the other axes unchanged.

9.2.3 Cloud of Individuals

We now arrive to the central object of this study, namely the analysis of electorates in the space of values and attitudes constructed by the geometric analysis. Introducing the electoral choice as a structuring factor, we distinguish the individuals according to their votes (at the first round of the parliamentary elections). We will concentrate the analyses on the four major political parties (or groupings of parties), namely Communist, Socialist, Traditional Right, and National Front, as well as the Non Respondents to the question about Vote. The two figures 9.2.4 (this page) and 9.2.5 (next page) show the cloud of individuals in planes 1-2 and 2-3 respectively, together with the mean points of the subclouds of interest; for instance, with the subcloud of communist voters is shown its mean point.

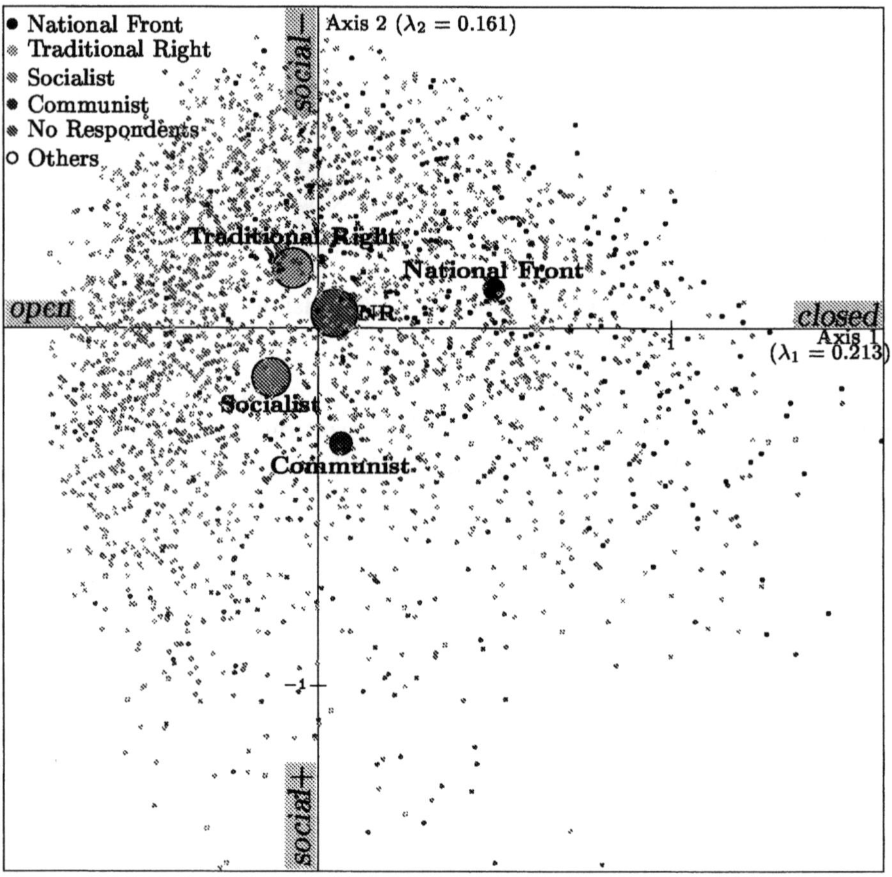

Figure 9.2.4. Plane 1-2: Cloud of 2980 individuals, and mean points of the four main electorates and "Non Respondents" (NR) subclouds.

In plane 1-2 (Figure 9.2.4, p.378), we see, at first sight, the large dispersion of the major 4 electorates, with a high density of Traditional Right points in the half plane 'social –' ; of Socialist points in the half- plane 'open'; of Communist points in the half-plane 'social+'; most of the National Front points are clearly situated on the 'closed' side.

In plane 2-3 (Figure 9.2.5), the Traditional Right points (blue) are mostly located in the quadrant 'social –', liberal +'; socialist and communist points are mostly in the quadrant 'social+, liberal-'; most National Front points (black) lie in the central part of the cloud, more or less shifted toward the 'social –' pole.

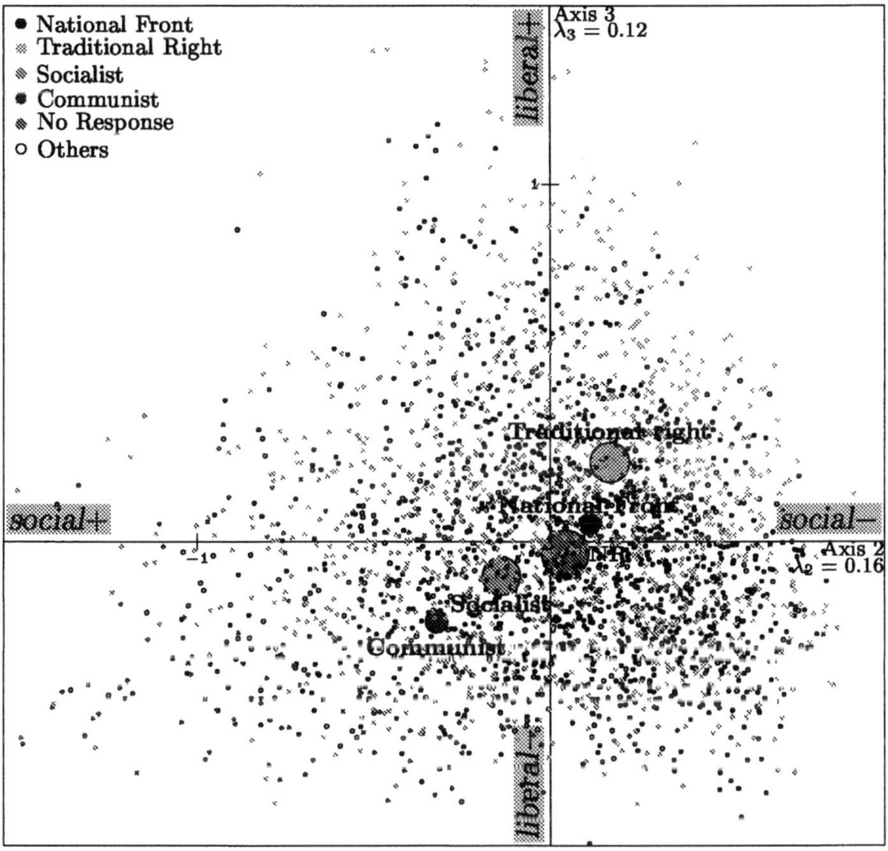

Figure 9.2.5. Plane 2-3: Cloud of 2980 individuals, and mean points of the four main electorates, and "non respondents" (NR) subclouds.

Mean points of electorates. Table 9.2.8 (p.380) shows the coordinates of the mean points of the 5 subclouds (Communist, Socialist, Traditional Right, National Front and NR) on the first three principal axes in the space of individuals, together with their qualities of representation; all qualities of representation in this subspace are at least .84. The difference of coordinates between two modality mean–points can be calibrated by the standard deviation (SD) of axis. Thus, on Axis 1, the calibrated deviation between National Front and Traditional Right is $(0.498 - (-0.0072))/.462)) = 1.23$ (greater than 1 standard deviation): Descriptively this deviation is large. The smallest deviation from NF to the three other major electorates is to Communist; it amounts to 0.94 standard deviation ($\xi_1 = .462$), again quite large. As for Non Respondents, it is clear that their mean point is quite close to the mean point G of the cloud: for the first three axes, one has respectively $0.1\xi_1$, $0.1\xi_2$ and $0.1\xi_3$ respectively. For plane 2-3, we can take as index of magnitude of deviation the κ–index (cf. chapter CLOUD, p.100).

For instance, the deviations between the coordinates of Communist and Traditional Right mean–points are given by $\mathbf{d}' = \boxed{0.49106 \quad 0.44644}$; one has $|\overrightarrow{d_{obs}}|^2_\kappa = \mathbf{d}'\mathbf{V}^{-1}\mathbf{d} = \frac{0.49106^2}{.1606} + \frac{0.44644^2}{.1192} = (1.78)^2$ with $\mathbf{V} = \begin{array}{|cc|} .1606 & 0 \\ 0 & .1192 \end{array}$.

Descriptively this deviation is large.

	Coordinates		
	Axis 1	Axis 2	Axis 3
NR	+0.046	+0.042	−0.034
Comm.	+0.065	−0.323	−0.224
Soc.	−0.133	−0.138	−0.097
Trad. Right	−0.072	+0.168	+0.222
Nat. Front	+0.498	+0.110	+0.049

	Qualities of representation			
	Axis 1	Axis 2	Axis 3	Sum
Comm.	.023	.567	.273	.863
Soc.	.323	.347	.170	.840
Trad. Right	.056	.304	.532	.892
Nat. Front	.888	.043	.009	.940

	Comm.	Soc.	Tr.R.	Nat.Fr.
Comm.	—			
Soc.	0.30	—		
Tr.Right	0.68	0.45	—	
Nat.Front	0.67	0.69	0.60	—

Table 9.2.8. Mean points Coordinates, qualities of representation and distances between mean points in the three–dimensional principal subspace of individuals.

Table 9.2.8 also shows the 6 distances between the 4 points in the three–dimensional subspace. This structure of distances cannot be approximated by a plane picture, it is adequately represented by a tetrahedron (Figure 9.2.6, p.381). The large distance that separates the National Front mean point from the other three ones — especially Traditional Right — is striking. This fact puts radically into question the conventional unidimensional scheme — extreme–left/left/right/extreme–right — that would permit the alliances Communist-Socialist on the one hand and Right and Extreme Right on the other hand. The National Front mean point does not lie in the subspace of the other three parties; the National Front is "elsewhere".

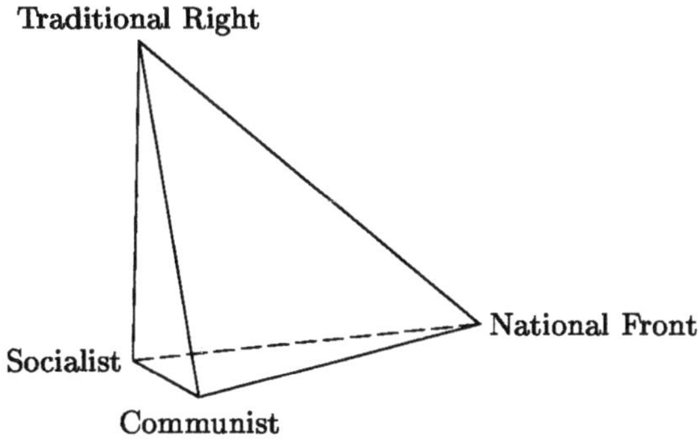

Figure 9.2.6. Four electorate mean points in principal three–dimensional space

Subclouds of electorates. For axes 1 through 3, Table 9.2.9 (this page) gives the variance of the subcloud of Non Respondents (NR), then the variances of the subclouds of the 4 electorates, the within–variance (weighted average of subcloud variances); and the between–variance (the variance of the subcloud of the 4 mean points). This table shows that the variances within electorates are much larger than the variance between electorates; for each axis, the within–variance is about or more than four times larger than the between–variance. In other words, the average quadratic distance between pairs of individuals within a same electorate is about twice as large as the average distance between the mean points of electorates.

	Frequencies	Axis 1	Axis 2	Axis 3
NR	965	0.212	0.128	0.103
Communists	225	0.196	0.221	0.083
Socialists	638	0.194	0.165	0.070
Traditional Right	690	0.156	0.098	0.132
National Front	193	0.198	0.097	0.111
Within (average)	1746	0.180	0.138	0.101
Between	1746	0.036	0.033	0.029

Table 9.2.9. Variances of 4 electorates and NR, and within and between variances.

In order to summarize geometrically the subclouds of electorates in the plane, we have constructed for each electorate its concentration ellipse (see Figures 9.2.7 and 9.2.8)[24]. One notices that in plane 1-2, as well as in plane 2-3, there is a large overlap among concentration ellipses, exhibiting

[24]As seen in the chapter *CLOUD* §3.4.4 (p.99), the concentration ellipse of a cloud is a geometric summary of the cloud. The area of ellipse indicates its overall spread. The eccentricity coefficient summarizes the variation of dispersions along the various directions. For a principal direction, the length of diameter corresponds to 4 SDs; for any other direction, the dispersion can be assessed by orthogonally projecting the ellipse onto this direction (cf. *CLOUD*, projection property 3.17, p.96).

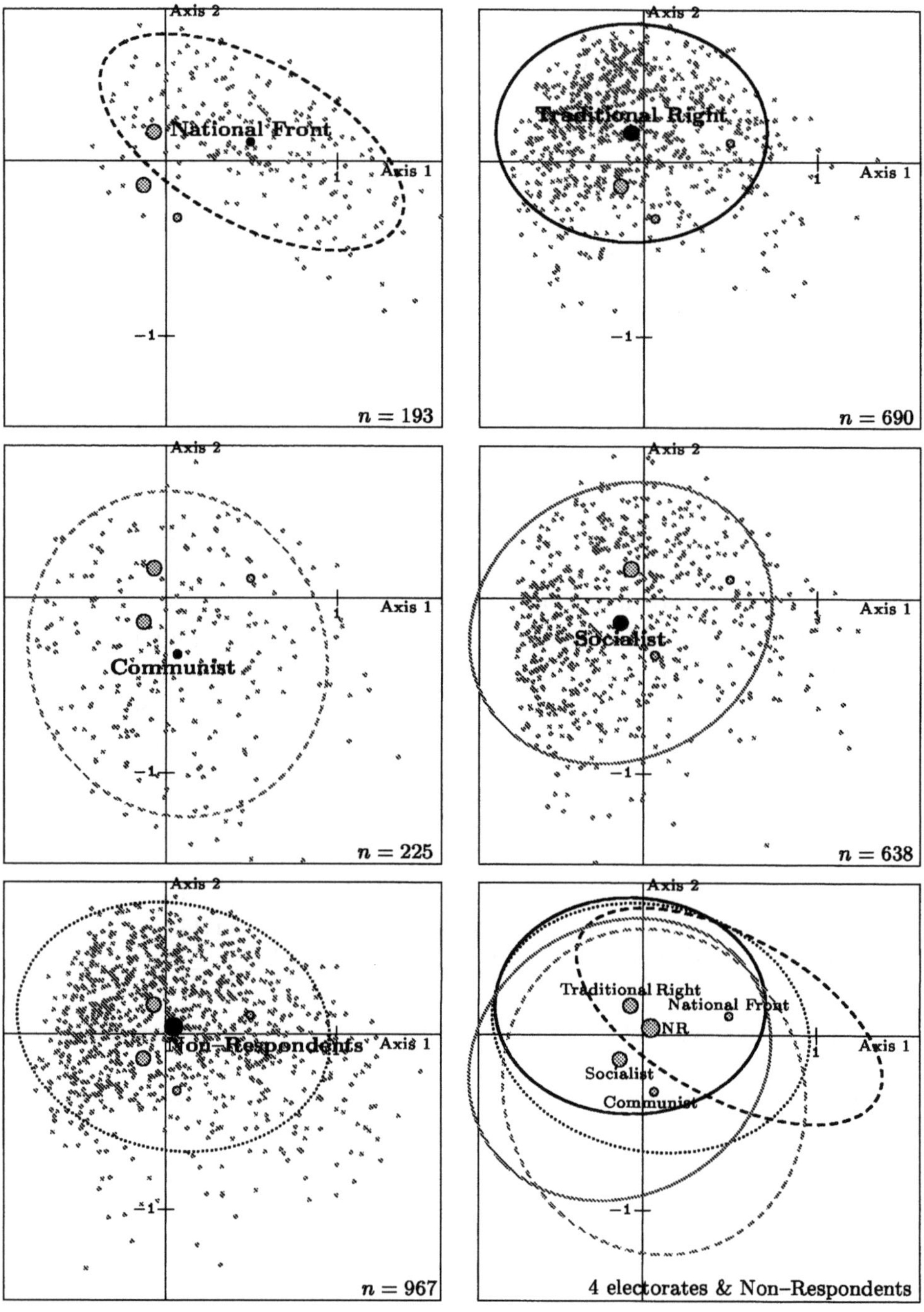

Figure 9.2.7. Plane 1-2: Concentration ellipses of the 4 electorates and Non–respondents with their centers.

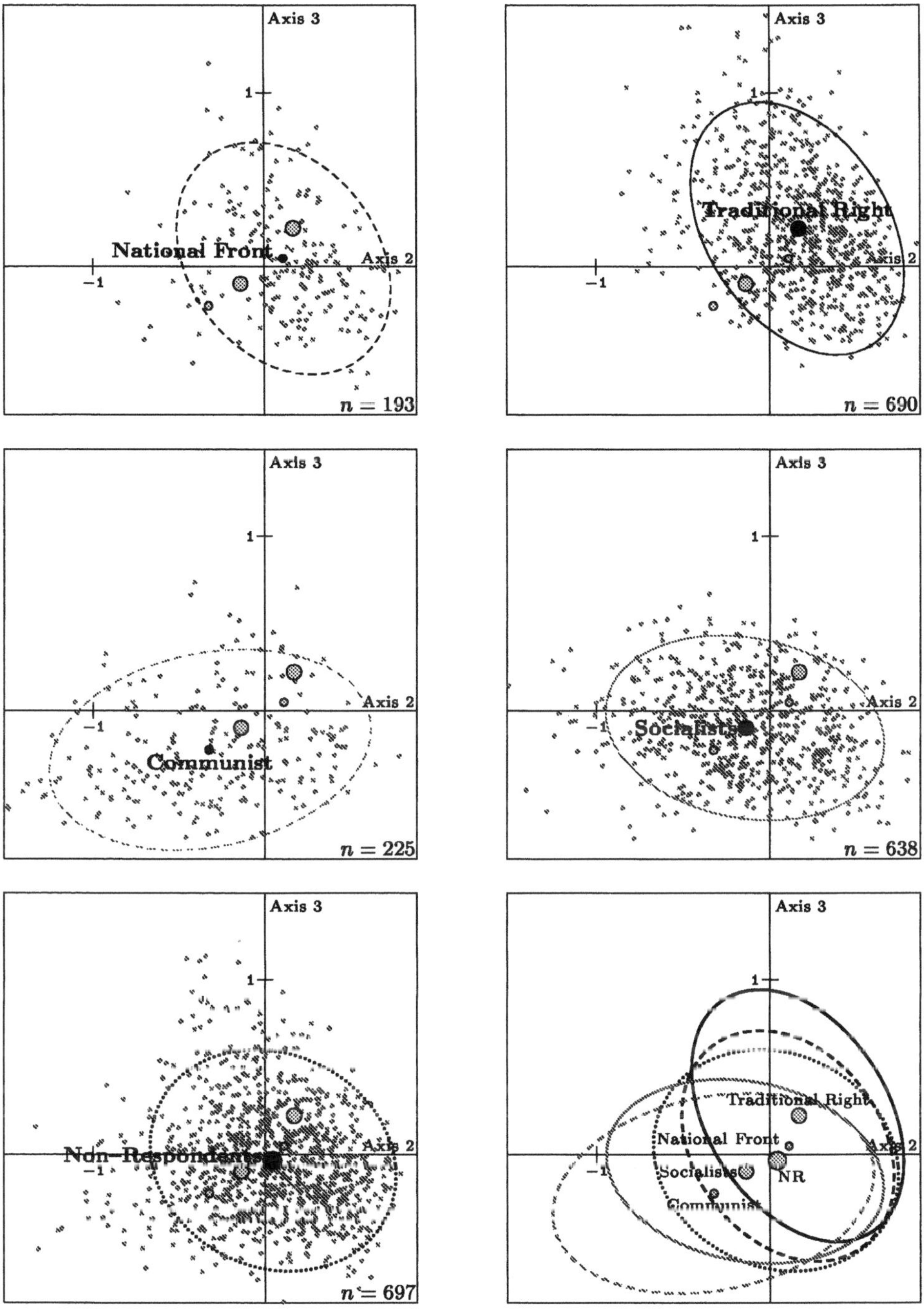

Figure 9.2.8. Plane 2-3 — Concentration ellipses of the 4 electorates and Non-Respondents with their centers.

the wide dispersion of electorates. In plane 1-2, the ellipse of National Front electorate has the smallest area, and also the largest eccentricity coefficient, which reflects the largest dispersion along the second bissector of the plane. In plane 2-3, the ellipses of Traditional Right and Communist electorates have the largest areas; the eccentricity indices are around .7 for the four electorates; the longest elongations are along axis 3 for FN and TR, and along axis 2 for Communist and Socialist (see also Table 9.2.9, p.381)[25].

Descriptive conclusions

The geometric analysis of the survey highlights two points that are crucial for understanding current voting behavior in France. On the one hand, the large variability within electorates points to the complexity of voting behavior, which aggregates, in a single ballot, quite diversified values; this variability shows how the analyses merely framed in terms of explanatory variables of voting behavior may overlook this complexity. On the other hand, the three–dimensional political space that emerges from the geometric analysis departs markedly from the unidimensional scheme Communist/Socialist/Traditional Right/National Front.

The left/right wings cleavage has not been found on any of principal axes, even on Axis 1. It seems clear that the structure of party preferences cannot be characterized as unidimensional. At this point, we meet the discrepancy between the "voter demand" in terms of values and attitudes and the "political supply" offered by the political system; a discrepancy commented upon by political scientists as one of the major factors responsible of the current crisis in political representation in France. These conclusions back the analyses of the 1997 elections by Perrineau, Ysmal & al (1998).

9.2.4 Inductive Analyses

Starting with the results of the geometric analysis, we now proceed to inductive analyses aiming at extending the major descriptive conclusions of §9.2.3. We regard the data set under study as a sample from the population of electors, and each observed electorate mean point as an estimate of the corresponding population mean point, or "true mean point" of the electorate. We firstly present *inferences on the mean points of electorates*

[25]Looking at the ellipses shows that the axes of the specific analysis of the subcloud (see *MULTIPLE*, §5.2.2, p.210) of the NF electorate differ from those of the overall cloud. Yet, if the NF electorate is taken out, the axes remain (cf. chapter *STABILITY*, p.279).

(significance tests and confidence methods), then *specific inferences* for comparisons among mean points (significance tests and Bayesian analyses).

In accordance with the descriptive findings, the inferences will pertain to axis 1 on the one hand, to plane 2-3 on the other hand. For each inference, we adopt the specific approach (§8.2.2, p.307), and we put the normal modeling on the relevant derived data. In this study, the calculation formulas will be expressed in terms of descriptive statistics (variance v rather than corrected variance s^2, κ–norm $|\cdot|_\kappa$ rather than S–norm $|\cdot|_S$).

Inferences on the mean points of electorates

Electorates on Axis 1. Putting the normal specific modeling on Axis 1, for each of the 7 electorates[26] and the group of Non–Respondents (NR) to Vote, we take the observed mean $\widehat{y}^k$ as an estimate of the true mean μ^k (parameter), and we proceed to significance tests and confidence intervals.

Significance tests and directional conclusions

For electorate k, we put to test the *null hypothesis* $\mathcal{H}_0 : \mu^k = 0$, i.e. we ask: "Is the hypothesis that on Axis 1 the mean μ^k coincides with the central value 0 of the cloud compatible with data?" We use two test procedures, denoting n_k the size of electorate k and v_k its variance:
i) *typicality test*, taking the statistic $T = (\widehat{y}^k/\xi_1)\sqrt{n_k(N-1)/(N-n_k)}$ (*INDUCTIVE*, p.313), with $\xi_1 = 0.462$ and $N = 2980$, and comparing T_{obs} to the standard normal distribution;
ii) *Student's t–test* with $t = \frac{\widehat{y}^k}{\sqrt{v_k/(n_k-1)}}$ (formula written here in terms of the variance v_k)[27], and comparing t_{obs} to the Student's distribution with $n_k - 1$ df., for $\alpha = .05$ ($S\star$) and .01 ($S\star\star$). Table 9.2.10 gathers the results.

Both tests lead to concordant conclusions. For the major 4 electorates and NR, the results are significant, that is, the hypothesis that on Axis 1 the true mean coincides with the central value of the cloud is not compatible with the data. Directional conclusions follow for Socialist and Traditional Right, whose true points can be asserted to lie on the left side of Axis 1, whereas those of Communist, Non–Respondents and National Front can be asserted to lie on the right side. The non–significant result for Extreme Left means that no directional conclusion can be reached for Extreme Left.

The directional conclusions do not have implications concerning the largeness of the deviations from the mean point G of the cloud: The true

[26]The group of "Others" does not define an electorate.

[27]In terms of the corrected variance s_k^2, the formula of the Student's ratio reads $t = \sqrt{n_k}\,\widehat{y}^k/s_k$ (cf. footnote 5, p.301).

k	$\widehat{y}^k$	n_k	v_k	T_{obs}	t_{obs}	Confidence limits
National Front	+0.498	193	0.197	+15.5 ⋆⋆	+15.5 ⋆⋆	[+0.435; +0.561]
Indep. Right	+0.188	58	0.173	+3.1 ⋆	+3.4 ⋆⋆	[+0.078; +0.298]
Trad. Right	−0.072	690	0.156	−4.7 ⋆⋆	−4.8 ⋆⋆	[−0.102; −0.043]
NR	+0.046	967	0.212	+3.8 ⋆⋆	+3.1 ⋆⋆	[+0.017; +0.075]
Greens	−0.163	165	0.160	−4.7 ⋆⋆	−5.2 ⋆⋆	[−0.225; −0.101]
Socialist	−0.133	638	0.194	−8.2 ⋆⋆	−7.6 ⋆⋆	[−0.167; −0.099]
Extreme Left	−0.032	30	0.195	−0.4 *NS*	−0.4 *NS*	[−0.199; +0.136]
Communist	+0.065	225	0.196	+2.2 ⋆	+2.2 ⋆	[+0.007; +0.123]

Table 9.2.10. Axis 1 ($\xi_1 = 0.462$). For each group k: mean $\widehat{y}^k$, sample size n_k, variance v_k, typicality statistic T_{obs}; Student's ratio (t_{obs}). ⋆ S.05 (two–sided), ⋆⋆ S.01 (two–sided) and beyond, NS nonsignificant at .05; confidence limits at level $\alpha = .05$.

mean points of NF and NR can both be asserted to lie at the right side of point G, but for National Front (NF), the observed deviation from point G is considerable whereas for Non–Respondents (NR) it is negligible[28].

Confidence intervals

For each electorate, we calculate the interval at level $\alpha = .05$ (normal modeling, formula (8.2), p.306: $\widehat{y}^k \pm t_{n_k-1}[.05]\sqrt{v_k/(n_k-1)}$, where $t_{n_k-1}[.05]$ denotes the critical value at level .05 of Student's distribution with $n_k - 1$ d.f). Results are gathered in Table 9.2.10 (last column) and Figure 9.2.9. Confidence intervals bring an essential complement to significance tests, because they evaluate the precision of the estimations of the true means of electorates. The width of interval k (which does not depend on $\widehat{y}^k$) is such that the larger the sample size, the narrower the interval, that is, the better the precision with which μ^k is estimated.

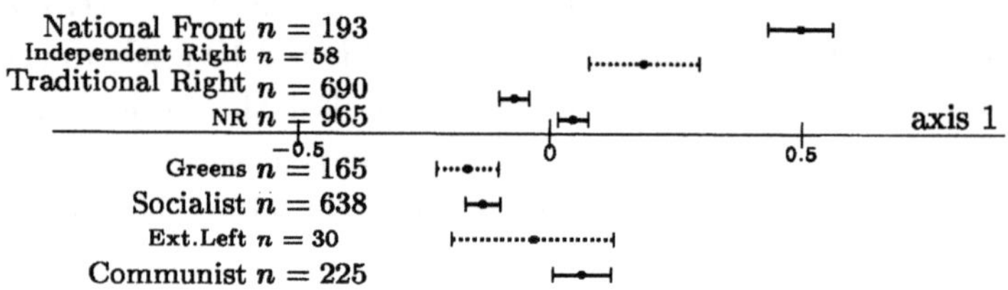

Figure 9.2.9. Axis 1: Confidence intervals at level $\alpha = .05$ for the mean points of the 7 electorates and the NR group.

For Socialist and Traditional Right, the intervals lie entirely on the left side of point O, whereas for NR and Communist they lie on the right side;

[28]The observed significance level depends both on the mean $\widehat{y}^k$ (deviation from 0) and on the sample size n_k (the "size–glued" property, cf. *INDUCTIVE*, p.298). The significant result for NR is primarily due to the large sample size. The nonsignificant result at .05 for Extreme Left is due to $\widehat{y}^k$ not large and sample size n_k very small.

and as far as NF is concerned, its interval is far away on the right side.

The reinterpretation of confidence intervals as Bayesian credibility intervals, i.e. in posterior probability terms, is highly meaningful. For instance, for NF, the probability that the true mean μ^k lies between the limits $0.435 = 0.94\,\xi_1$ and $0.561 = 1.21\,\xi_1$ (very large deviations from 0) is 95%. As a consequence, the probability that the true mean μ^k is higher than 0.435 (lower limit) exceeds 97.5%: $P(\mu^k > 0.94\,\xi_1) = .975$. Alternatively, taking a credibility $\gamma = .90$, we have: $P(\mu^k > 0.99\,\xi_1) = .90$. We therefore can assert the *largeness conclusion*: On Axis 1, there is a high probability that the NF mean point lies on the right side far away from the mean point G of the cloud: $P(\mu^k > 0.99\,\xi_1) = .90$.

Electorates on plane 2-3. Putting now the binormal specific modeling on plane 2-3, we take the observed mean point C^k of electorate k as an estimate of the "true mean point" Γ^k (parameter) (cf. figure in chapter *INDUCTIVE*, p.323). For each electorate, we proceed to Hotelling's test and construct confidence ellipses.

- *Hotelling's test*

For electorate k, we put to test the null hypothesis: $\mathcal{H}_0 : \Gamma^k = G$, i.e. we ask: "Is the hypothesis that in plane 2-3 the mean point Γ^k coincides with the mean point G of the cloud compatible with data?"

For each electorate k, we denote $|\overrightarrow{GC^k}|_\kappa$ the κ–norm of $\overrightarrow{GC^k}$, that is, $|\overrightarrow{GC^k}|^2_\kappa = \mathbf{y}'\mathbf{V}^{-1}\mathbf{y}$, where $\mathbf{y}$ denotes the column of coordinates of point C^k in plane 2-3, and $\mathbf{V}$ the covariance matrix of electorate k in plane 2-3. Then applying the formulas (8.9) (*INDUCTIVE*, p.324) the Hotelling's statistic is $T^2_k = (n_k - 1)|\overrightarrow{GC^k}|^2_\kappa$, then we compare T^2_{obs} to $\frac{n_k-1}{n_k-r}\,rF_{r,n_k-1}[\alpha]$, with $r = 2$ (two dimensional- space), or, for large n, T^2_{obs} to $\chi^2_r[\alpha]$). Table 9.2.11 gathers the results.

Electorates	n_k	$\lvert\overrightarrow{GC^k}\rvert_\kappa$	T^2_k	Confidence κ_α
National Front	193	0.431	35.9 **	0.178
Independent Right	58	0.866	43.5 **	0.336
Traditional Right	690	1.010	703.8 **	0.003
Non-Respondents	967	0.150	22.2 **	0.079
Greens	165	0.464	35.5 **	0.193
Socialist	638	0.548	192.0 **	0.097
Extreme Left	30	1.061	33.8 **	0.488
Communist	225	0.943	200.0 **	0.165

Table 9.2.11. Inference in plane 2-3. For each electorate k, sample size n_k, κ–norm of $C^k - G$, Hotelling's T^2, confidence κ_α at level $\alpha = .05$.

The 8 results are extremely significant, which means that for each of the 7 electorates and the NR group, one can assert that for plane 2-3 the true mean point differs from the mean point G of the cloud.

• *Confidence ellipses*

For electorate k, the confidence ellipse at level α, centered on point C^k, is the ellipse of inertia (*CLOUD*, §3.4, p.95 and p.100) with $\kappa_\alpha^2 = \frac{2}{n-2} F_{2,n-2}[\alpha]$ (For large q; $\kappa_\alpha^2 = \chi_\alpha^2/n_k$); see *INDUCTIVE*, p.325. Constructing the 8 confidence ellipses at level $\alpha = .05$, we obtain the results gathered in Table 9.2.11 (page before) (the last column gives κ_α) and Figure 9.2.10.

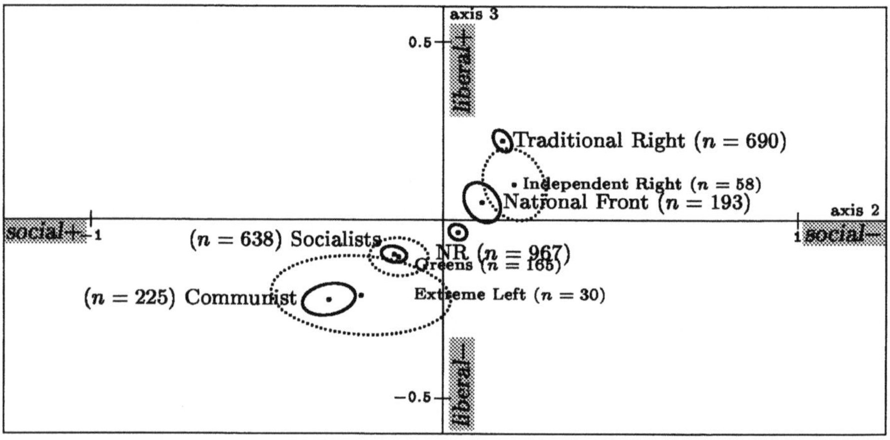

Figure 9.2.10. Plane 2-3: Confidence ellipses of the 7 electorates and Non–Respondents with their centers.

Here again, confidence methods bring an essential information about the precision of the estimation of the Γ^k points by the C^k points. The examination of Figure 9.2.10 suggests that the mean points of Independent and Traditional Right lie on the side of 'social−, liberal+', and that those of Communist, Socialist and Greens lie in the quadrant 'social+, liberal−'. Here again, the Bayesian reinterpretation is illuminating. The ellipse at level $\alpha = .05$, therefore at credibility $\gamma = .95$, can be asserted to contain the true mean point with a posterior probability equal to 95%. As a consequence, when an ellipse is itself contained in a region of the plane (such a quadrant, etc.), one can assert that the true mean point belongs to this region with a probability higher than 95%.

— *Remark*: As already pointed out in chapter *INDUCTIVE*, the families of ellipses of inertia, which provide descriptive summaries of clouds, also provide in Bayesian Data Analysis geometric summaries of the posterior distributions of the mean points of clouds.

Specific comparisons among mean points

In the following three analyses, we illustrate the full strategy of Inductive Data Analysis (cf. INDUCTIVE, §8.6, p.331) including Bayesian analyses, to investigate hypotheses of special interest such as largeness and smallness hypotheses (see INDUCTIVE, §8.4, p.318). The first two analyses refer to Axis 1, the third one to plane 2-3.

Axis 1, National Front vs the other three major electorates: largeness hypothesis. On Axis 1, the mean point F of National Front, with an abscissa of +0.4982, is in sharp opposition to the mean points of the three electorates Communist, Socialist and Traditional Right; the latter three points are rather close to one another, and their average point E is not far from G, with the abscissa $\frac{+0.065\times 225-0.133\times 638-0.072\times 690}{225+638+690} = -0.0774$. Let d_{obs} denote the difference in abscissas between points F and E: $d_{obs} = 0.4982 - (-0.0774) = +0.5756 = +1.25\,\xi_1$, a very large deviation. That is, as descriptive conclusions we not only have $(d_{obs} > 0)$ (directional), but also $(d_{obs} \gg 0)$ (largeness). We will now attempt to extend these two conclusions inductively.

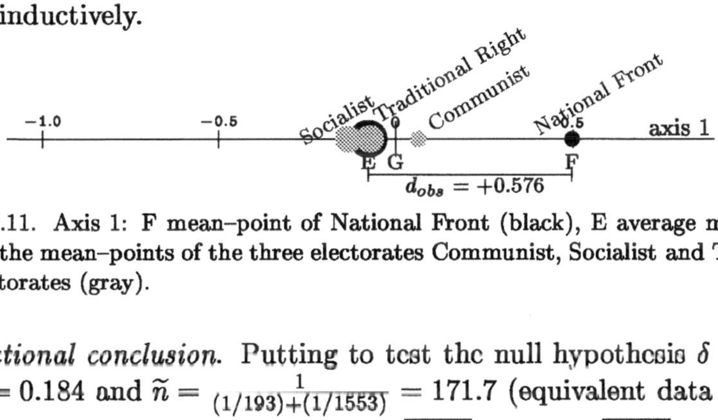

Figure 9.2.11. Axis 1: F mean-point of National Front (black), E average mean-point (black) of the mean-points of the three electorates Communist, Socialist and Traditional Right electorates (gray).

Directional conclusion. Putting to test the null hypothesis $\delta = 0$, one has: $s^2 = 0.184$ and $\widetilde{n} = \frac{1}{(1/193)+(1/1553)} = 171.7$ (equivalent data size) (cf. INDUCTIVE, §8.2.2, p.307), hence $t_{obs} = \sqrt{171.7}\times 0.576/\sqrt{0.184} = 17.6$. The result is extremely significant, therefore one can assert that on Axis 1 the true mean of National Front is greater than the average true mean of the other three major parties. In the Bayesian reinterpretation, the probability is extremely high that the true deviation δ is positive: $P(\delta > 0) \simeq 1$.

Largeness conclusion. The posterior distribution of δ is centered on $d_{obs} = +0.5756$. For $\gamma = .90$ the lower credibility limit is $d_{obs} - z_{.20}\, s/\sqrt{\widetilde{n}} = 0.5756 - 1.282\sqrt{0.184/171.7} = +0.53 = +1.14\,\xi_1$, therefore one has $P(\delta > 1.14\,\xi_1) = .90$. (cf. Figure 9.2.12) Alternatively, one has $P(\delta > +\xi_1) = .9997$. Both statements back the following largeness conclusion.

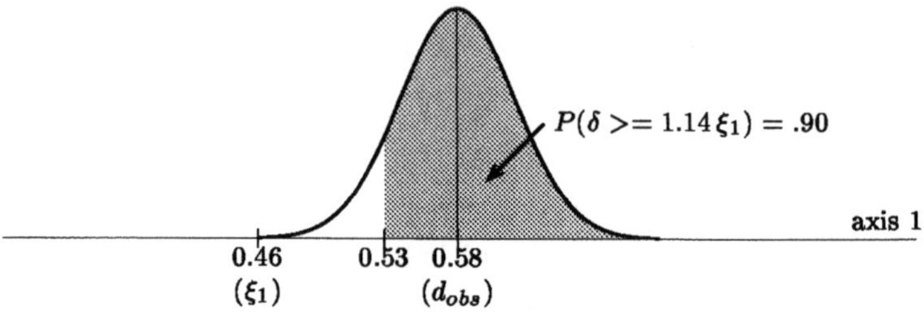

Figure 9.2.12. Axis 1: National Front vs Communist, Socialist and Trad. Right; asserting importance of deviation. $P(\delta > +1.14\,\xi_1) = .90$, $P(\delta > +\xi_1) = .9997$.

Asserting largeness of deviation: On Axis 1, there is a high probability that the true deviation of the mean point of National Front from the average mean point of the three major parties exceeds 1 SD: $P(\delta > +\xi_1) = .9997$.

Axis 1 — Comparing NR to the mean point of the cloud: smallness hypothesis. On Axis 1, the observed mean point N of NR to the question Vote has a small positive abscissa $d_{obs} = +0.046$, so two descriptive conclusions can be stated. The first one is the directional conclusion ($d_{obs} > 0$), that has been already extended inductively (see Table 9.2.10 and Figure 9.2.10), that is, one can assert that on axis 1 the true mean point δ of NR lies on the right side of axis. The second one — certainly more interesting — is that the deviation of NR mean point from the mean point G of the cloud is negligible: $d_{obs} = 0.02\,\xi_1$.

We now attempt to extend the latter conclusion inductively, that is, to establish that the true deviation from 0 is small (smallness hypothesis, INDUCTIVE, §8.4, p.318). The posterior distribution of δ is centered on $d_{obs} = +0.046$, and its upper .90–credibility limit is $0.046 + 1.282\sqrt{0.212/965} = +0.065$, hence (since $P(\delta < -0.065) \simeq 0$) we can state $P(|\delta| < 0.065) = P(\delta < 0.065) - P(\delta < -0.065) = .90$[a].

[a]If the term $P(\delta < -0.065)$ could not be neglected, an iterative process would be needed to arrive to the smallness assertion.

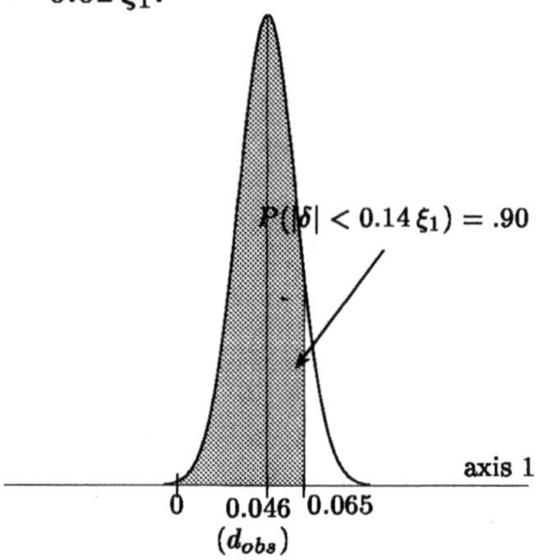

Figure 9.2.13. Axis 1: Asserting smallness of the deviation of the mean–point "Non respondents" from the mean point G of the cloud: $P(|\delta| < 0.14\,\xi_1) = .90$.

In so far as the value $0.065 = 0.14\,\xi_1$ can be deemed to be a small deviation from 0, we can assert the following smallness conclusion.

> *Asserting smallness of deviation*: On Axis 1, there is a high probability that the deviation of the true mean point of Non Respondents from the mean point of the cloud is small: $P(|\delta| < 0.14\,\xi_1) = .90$.

Plane 2-3 — Communist vs Traditional Right: largeness hypothesis. In plane 2-3, the mean points C (Communist) and R (Traditional Right) are such that the vector–deviation $\mathrm{R} - \mathrm{C} = \overrightarrow{d_{obs}}$ has as principal coordinates $(+0.491, +0.446) = (+1.224\,\xi_2, +1.292\xi_3)$. Descriptively speaking, when going from C to R, one moves toward the direction 'Social–' 'Liberal+', and the deviation is large.

We will now attempt to extend this conclusion: i) specific cloud and descriptive conclusions, ii) significance test, iii) Bayesian analyses.

i) *Specific cloud and descriptive conclusions*

• The overall cloud of individuals is restricted to the two subclouds of Communist ($n = 225$) and Traditional Right ($n = 690$) electorates, with their respective mean points C and R; then point D, barycenter of C and R, is constructed: $\mathrm{D} = (225\mathrm{C} + 690\mathrm{R})/915$ (cf. Figure 9.2.14, p.392).

• Each of the two clouds is shifted so that both have the same mean point D, hence the *within–cloud* (cf. CLOUD, definition 3.25, p.103), which is the specific cloud relevant to the comparison under study, together with the vector–deviation $\overrightarrow{d_{obs}}$, or equivalently the point O defined by $\mathrm{O} = \mathrm{D} - \overrightarrow{d_{obs}}$. Then, the covariance matrix $\mathbf{W}$ of the within–cloud is calculated, as the weighted average of the two covariance matrices of the two clouds[29]:

$$\mathbf{W} = \tfrac{225}{915}\begin{array}{|cc|}\hline 0.221 & +0.028 \\ & 0.083 \\ \hline\end{array} + \tfrac{690}{915}\begin{array}{|cc|}\hline 0.098 & -0.040 \\ & 0.132 \\ \hline\end{array} = \begin{array}{|cc|}\hline 0.128 & -0.023 \\ & 0.120 \\ \hline\end{array}$$

The concentration ellipse of the within–cloud is a weighted average of the two concentration ellipses; the half–lengths of its principal axes are 0.768 and 0.635, and the angle of its first axis with the horizontal axis is $-40°$. Figure 9.2.15 (next page) depicts the concentration ellipses of the two clouds, centered at point D, together with the concentration ellipse of the within– cloud. We have $|\overrightarrow{d_{obs}}|^2_{\mathrm{W}} = \mathbf{d}'\mathbf{W}^{-1}\mathbf{d} = (2.08)^2$, that is, point O is slightly outside the within concentration ellipse[30].

[29] One has $\mathbf{S} = \frac{n_1+n_2}{n_1+n_2-2}\,\mathbf{W}$.

[30] The index $|\overrightarrow{d_{obs}}|_{\mathrm{W}}$ calculated from the within covariance $\mathbf{W}$ in the specific analysis is to be compared with $|\overrightarrow{d_{obs}}|_{\kappa}$ (such that $|\overrightarrow{d_{obs}}|^2_{\kappa} = \mathbf{d}'\mathbf{V}^{-1}\mathbf{d}$) calculated from the overall covariance matrix $\mathbf{V}$ (cf. p.380); both indices point to the largeness of $\overrightarrow{d_{obs}}$.

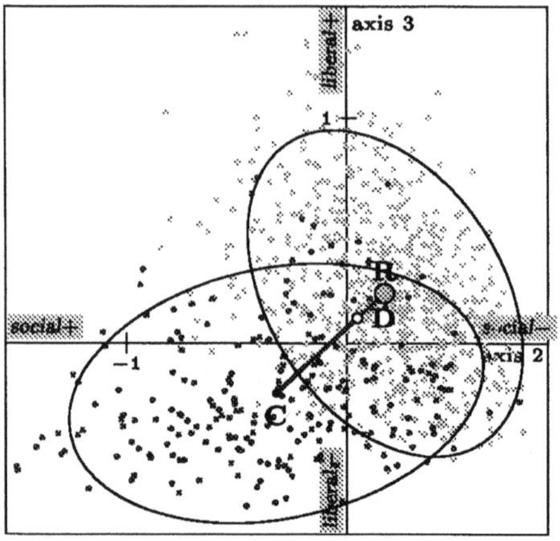

Figure 9.2.14. Subclouds of Communist and Traditional Right electorates, with their mean points C and R and concentration ellipses, vector–deviation R − C and the barycenter D of points C and R.

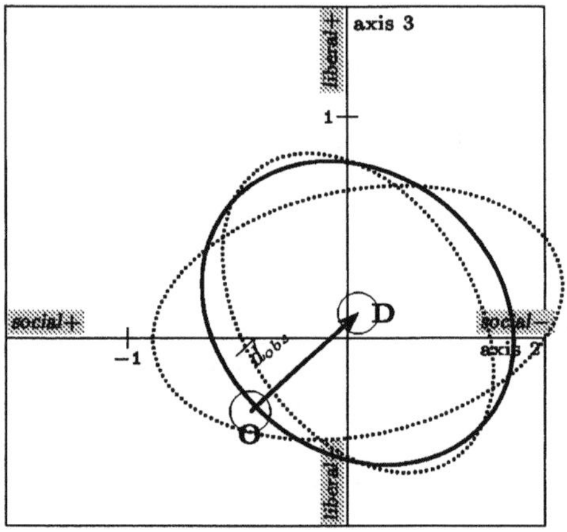

Figure 9.2.15. Concentration ellipses of the two clouds Communist and Traditional Right and of the within–cloud (specific cloud) centered at point D, and deviation–vector D − O = R − C.

• Let us now construct the inertia ellipse $\mathcal{E}_0$ that goes through point O, and the tangent $\mathcal{L}$ to $\mathcal{E}_0$ at point O, hence the upper half–plane $\mathcal{R}_0$ on the side of 'Social−,Liberal+': see Figure 9.2.16.

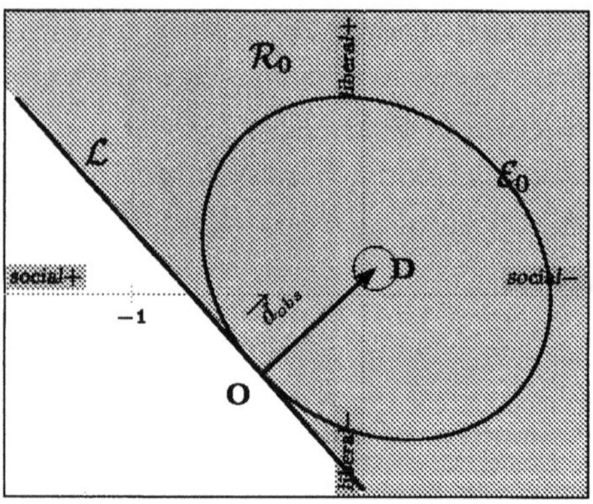

Figure 9.2.16. $\mathcal{E}_0$ ellipse, tangent $\mathcal{L}$ and half–plane $\mathcal{R}_0$

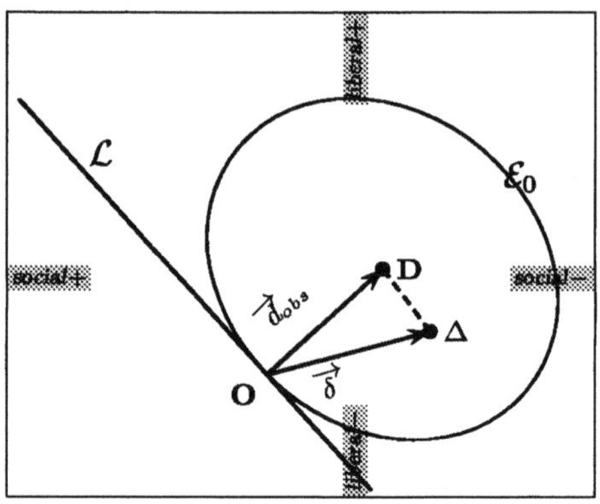

Figure 9.2.17. Observed mean point D, true mean point Δ and associated deviation–vectors.

We state the following descriptive conclusions:

Directional and largeness descriptive conclusions: The vector $\vec{d}_{obs} = \overrightarrow{OD} = D - O$ is on the side of the half–plane $\mathcal{R}_0$, and its magnitude is large: $(|\overrightarrow{OD}|_W > 2)$.

ii) *Significance test (Hotelling)*

Under the normal modeling on the specific cloud, the observed deviation $\overrightarrow{d_{obs}}$ is the estimate of the true deviation $\overrightarrow{\delta}$; we now put to test the null hypothesis $\mathcal{H}_0 : \overrightarrow{\delta} = \overrightarrow{0}$ by Hotelling's test. One has $T^2_{obs} = \widetilde{n} |\overrightarrow{d_{obs}}|^2_{\text{S}}$ with $|\overrightarrow{d_{obs}}|^2_{\text{S}} = (913/915)|\overrightarrow{d_{obs}}|^2_{\text{W}}$ and $\widetilde{n} = 1/(1/225 + 1/690) = 169.87$, hence $T^2_{obs} = 169.87 \times 4.3512 = 738$. The observed p–level is virtually 0, i.e the deviation $D - G = C - R$ is extremely significant.

> *Asserting existence of deviation.* Data are in favor of a non–null deviation between the true mean points of Communist and Traditional Right.

iii) *Bayesian analyses*

Bayesian analyses are based on the posterior distribution of the true mean point $\Delta = \text{O} + \overrightarrow{\delta}$ centered on point $\text{D} = \text{O} + \overrightarrow{d_{obs}}$ (Figure 9.2.17).

• *Credibility ellipses.* The posterior distribution of point Δ is centered on point D; the γ–credibility ellipse is the locus of the points such that $\kappa^2 = \chi^2[\alpha]/\widetilde{n}$, where $\chi^2[\alpha]$ is the critical value of χ^2 with 2 d.f. at upper level $\alpha = 1 - \gamma$. For example, taking $\gamma = .90$, one has $\chi^2[.10] = 4.605$, hence $\kappa^2 = 4.605/169.87 = 0.027 = (0.165)^2$, therefore $P(\Delta \in \mathcal{E}_{.90}) = P(|\overrightarrow{\Delta\text{D}}|_{\text{W}} < .165) = .90$. Ellipse $\mathcal{E}_{.90}$ is depicted in Figure 9.2.18.

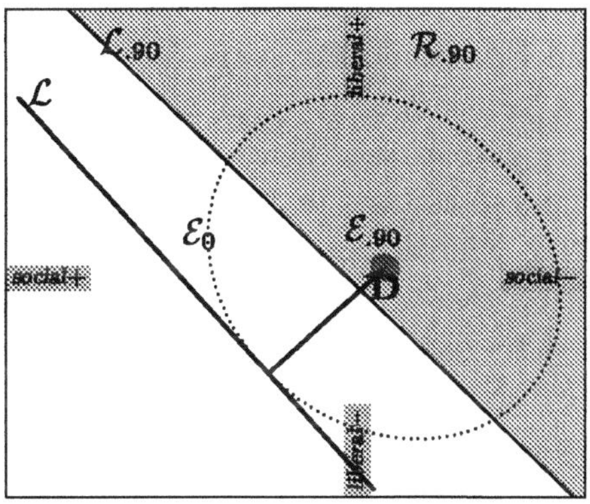

Figure 9.2.18. Ellipse $\mathcal{E}_0$, credibility ellipse $\mathcal{E}_{.90}$ with tangent $\mathcal{L}_{.90}$, and half–plane $\mathcal{R}_{.90}$.

• *Directional conclusion.* Let $\mathcal{E}_0$ denote the credibility ellipse going through point G (Figure 9.2.16, p. 392); one has $P(\Delta \in \mathcal{E}_0) = P(|\overrightarrow{\Delta\text{D}}|_{\text{W}} < |\overrightarrow{\text{OD}}|_{\text{W}}) = 1 - p_{obs}$, where p_{obs} is the observed p level of Hotelling's test. Here we have $p_{obs} \simeq 0$, hence $1 - p_{obs} \simeq 1$, therefore $P(\Delta \in \mathcal{E}_0) = P(|\overrightarrow{\Delta\text{D}}|_{\text{W}} < 2.086) \simeq 1$. Since $\mathcal{E}_0 \subset \mathcal{R}_0$, one has a fortiori $P(\Delta \in \mathcal{R}_0) \simeq 1$, that is, we can extend the descriptive directional conclusion and state:

> *Asserting direction of deviation.* There is an extremely high posterior probability that the deviation between the true mean points of Communist and Traditional Right lies "on the side of the observed deviation": $P(\Delta \in \mathcal{R}_0) \simeq 1$.

• *Largeness conclusion.* For any credibility ellipse, the tangent line parallel to $\mathcal{L}$ delineates a half–plane on the side of point D. For example, with $\mathcal{E}_{.90}$ there is associated the half–plane $\mathcal{R}_{.90}$. The conclusion $P(\Delta \in \mathcal{E}_{.90}) = .90$, entails that $P(\Delta \in \mathcal{R}_{.90}) > .90$: see Figure 9.2.18. In so far as the value $2.08 - 0.16 = 1.92$ is deemed to be a large deviation, we can extend the descriptive largeness conclusion and state:

> *Asserting largeness of deviation.* There is a high posterior probability that the deviation between the true mean points of Communist and Traditional Right is large on the side of the observed deviation. $P(\Delta \in \mathcal{R}_{.90}) > .90$.

9.3 Individual Differences among Gifted Students (EPGY Study)

The **E**ducation **P**rogram for **G**ifted **Y**outh (EPGY) at Stanford University is a continuing project dedicated to developing and offering multimedia computer–based distance–learning courses. Through EPGY, students have access to courses in a large variety of subjects; for instance in Mathematics, EPGY offers a complete sequence of courses at levels ranging from kindergarten through advanced–undergraduate; see Tock & Suppes (2002).

The data analyzed in this case study are those of 533 students in the third grade; they concern the EPGY course of Mathematics with its five strands: Integers, Fractions, Geometry, Logic and Measurement. The objective is to construct a geometric space of the individual differences[31].

In this section we describe the data set and its coding (§9.3.1), we proceed to MCA (§9.3.2) and interpret axes (§9.3.3); we study the cloud of individuals (§9.3.4); we proceed to a Euclidean classification (§9.3.5).

9.3.1 Data and Coding

Active variable and structuring factors

For each strand, we have taken the three following types of variables: *Error rates*, *Latencies* for correct answers, *Number of exercises* to master the concepts of strand. Crossing the three types of variables with the five strands, we get *fifteen active variables* in all. As *structuring factors* of individuals, we will take the *number of hours* spent on the computer, *Gender* and *Age*.

[31]This case study is part of a body of research, that can be found at the Web–site `http://epgy.stanford.edu/research/`.

Univariate analyses and coding of variables

• The distributions of *Error rates* differ among the strands (see Figures 9.3.1). The Integers, Fractions and Measurement distributions are $\mathcal{I}$–shaped whereas Geometry and Logic ones are more bell–shaped.

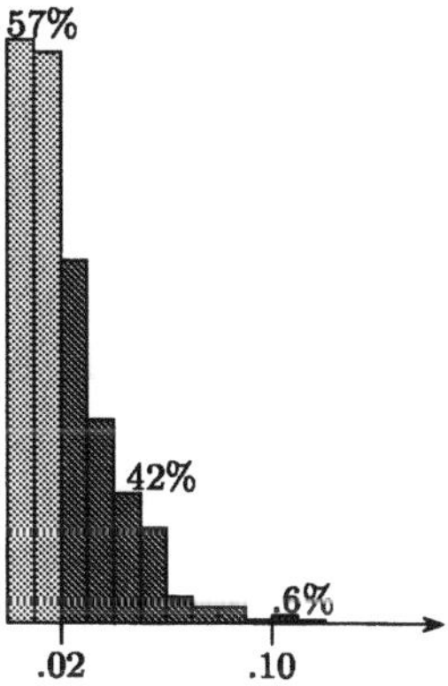

Figure 9.3.1-a. Integers

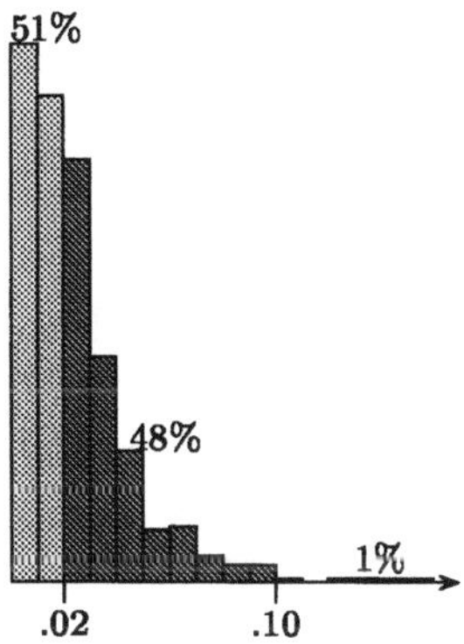

Figure 9.3.1-b. Fractions

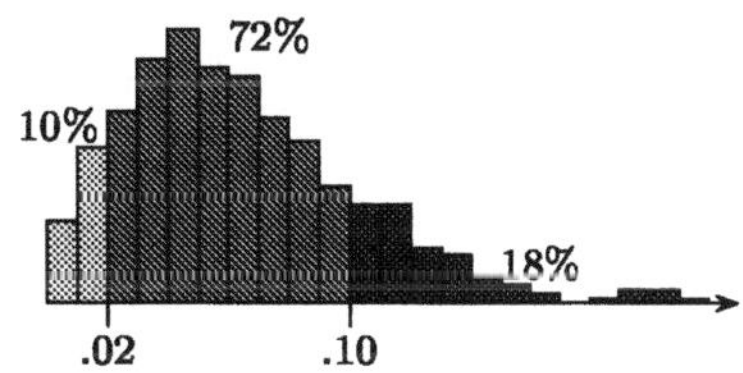

Figure 9.3.1-c. Geometry

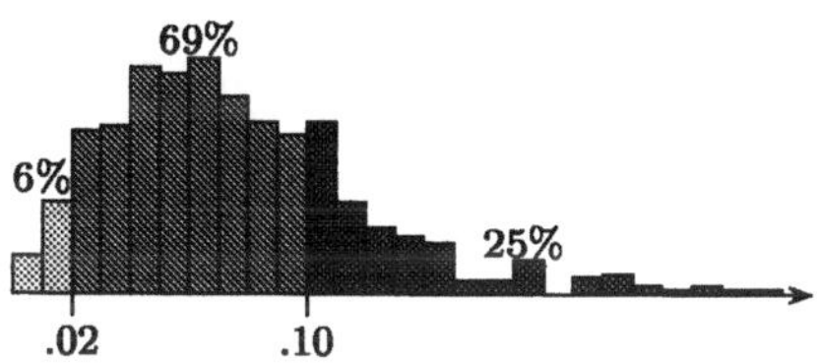

Figure 9.3.1-d. Logic

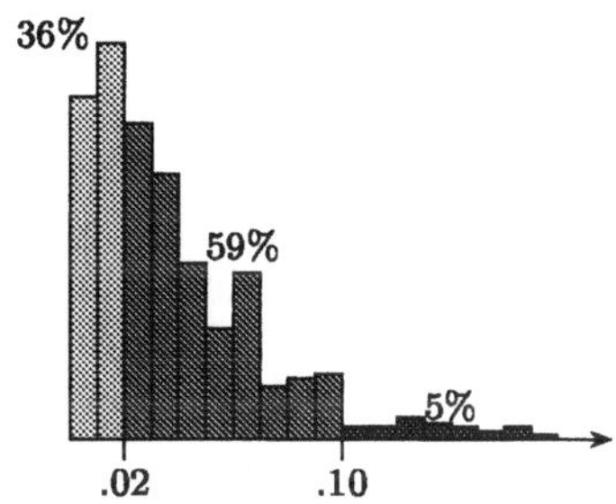

Figure 9.3.1-e. Measurement

Figure 9.3.1. Distributions of *Error rates* and percentages for the 3 categories defined in Table 9.3.2 (p.396).

— *Remark.* The numbers of students who make *no error* are 9 in Integers, 70 in Fractions, 4 in Geometry, 0 in Logic, 49 in Measurement.

• The five distributions of Latencies are more or less bell–shaped (Figures 9.3.2, p.397).

• The *Number of exercises* is a discrete variable. For each of five strands, Table 9.3.1 (next page) gives the distribution (absolute frequencies) of the number of exercises to master the concepts of the strand[32].

[32]The *Number of exercises* in Geometry is different from the one in the other strands because some concept classes in Geometry have no criterion to mastery. This is due to the fact that in those concept classes the student is given a geometrical construction consisting of a number of steps, none of which can be omitted.

Integers

4	*5*
480	53

Fractions

4	*5*	*6*	*7*
482	49	1	1

Geometry

6	*8*	*9*	*10*	*11*	*12*	*13*	*14*	*15*	*16*
1	1	19	61	118	203	100	20	7	3

Logic

4	*5*	*6*	*7*	*9*	*10*
165	301	55	9	2	1

Measurement

4	*5*	*6*
387	125	21

Table 9.3.1. For each strand, absolute frequencies of the *Number of exercises* (in italics) to mastery.

Given the heterogeneity of variables and of their distributions, the most appropriate Geometric Data Analysis is Multiple Correspondence Analysis (MCA) after coding of variables into categories[33] (2, 3 or 4) that we describe below, according to the specificities of variables, in order to attain as much homogeneity as possible, which is required to define a distance between individuals.

Coding of Error rates. We have taken a common coding defined by two cuts at 2% and 10%, generating 3 categories: 1 ($\cdot < 0.02$), 2 ($0.02 \leq \cdot < 0.10$) and 3 ($\cdot \geq 0.10$).

Categories		Integers	Fractions	Geometry	Logic	Measurement
1	$\cdot < 0.02$	305	274	55	31	192
2	$0.02 \leq \cdot < 0.10$	225	254	382	367	316
3	$\cdot \geq 0.10$	3	5	96	135	25

Table 9.3.2. Absolute frequencies of the 3 × 15 categories of *Error rates*.

Coding of Latencies. *Latencies* widely differ among strands[34]. Consequently we have, for each strand, taken a 4–category coding defined by the inferior quintile (20%), median (50%), and superior quintile (80%); see Table 9.3.3 and Figure 9.3.2 (p.397). Absolute frequencies are 106 for category 1, 160 for categories 2 and 3, 107 for category 4: For instance, 106 students have a latency less than 19.84 in Integers, less than 10.06 in Fractions etc. Hence we obtain $4 \times 5 = 20$ categories.

Quintiles have been taken in order to give more importance to the two groups of individuals with short and long latencies (cf. formula (5.3) of the distance definition in MCA, p.181).

[33]In this case study, we prefer to speak of categories (rather than modalities) to allude to the coding procedure (categorization) applied to numerical variables.

[34]This discrepancy may be attributable to the differing organizations of exercises among the strands.

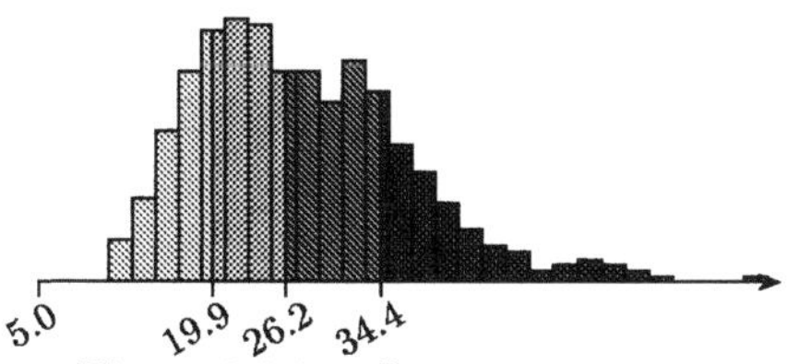

Figure 9.3.2-a. Integers

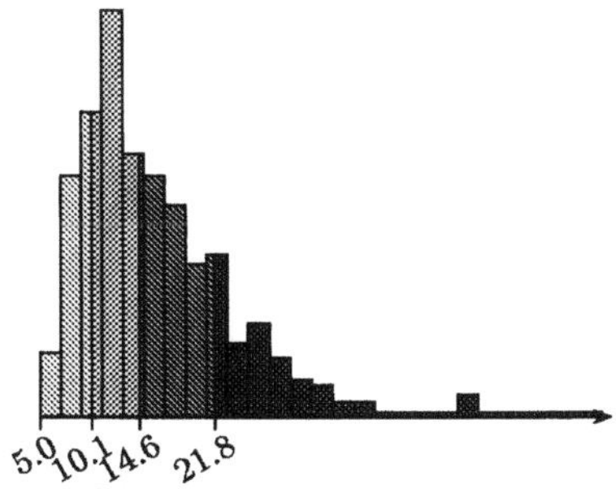

Figure 9.3.2-b. Fractions

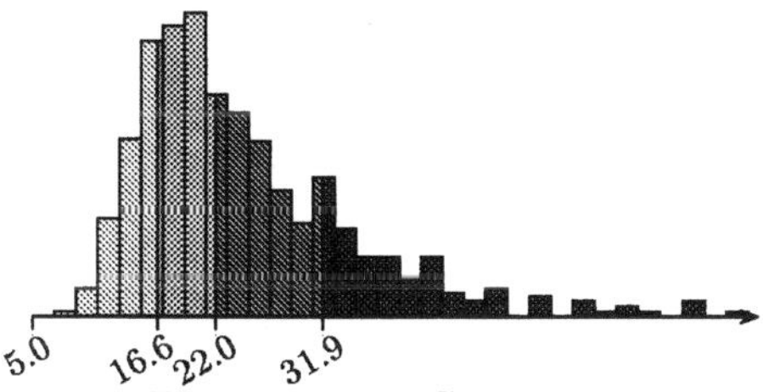

Figure 9.3.2-c. Geometry

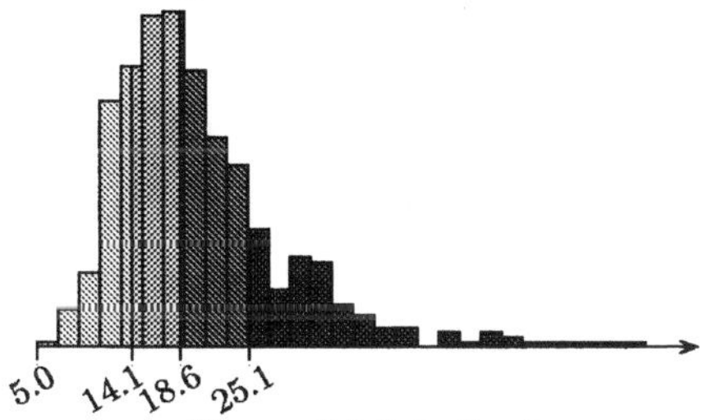

Figure 9.3.2-d. Logic

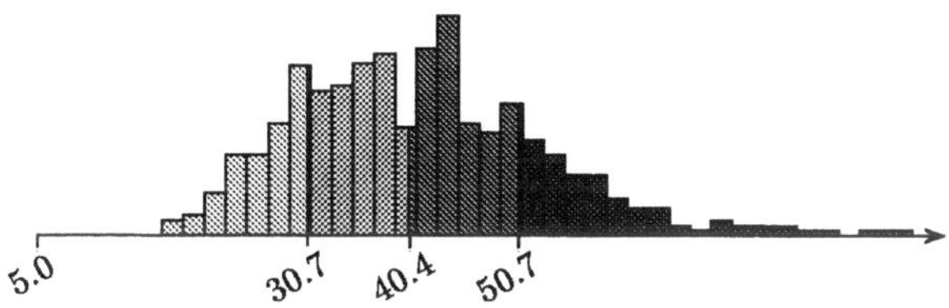

Figure 9.3.2-e. Measurement

Figure 9.3.2. Distributions of *Latencies*, with cut values (see Table 9.3.3).

Category	1	2	3	4
Integers	$\cdot < 19.84$	$19.84 \leq \cdot < 26.18$	$26.18 \leq \cdot < 34.35$	$\cdot > 34.35$
Fractions	$\cdot < 10.06$	$10.06 \leq \cdot < 14.64$	$16.64 \leq \cdot < 21.79$	$\cdot > 21.79$
Geometry	$\cdot < 16.55$	$16.55 \leq \cdot < 21.95$	$21.95 \leq \cdot < 31.92$	$\cdot > 31.92$
Logic	$\cdot < 14.05$	$14.05 \leq \cdot < 18.60$	$18.60 \leq \cdot < 25.13$	$\cdot > 25.13$
Measurement	$\cdot < 30.66$	$30.66 \leq \cdot < 40.35$	$40.35 \leq \cdot < 50.71$	$\cdot > 50.71$
Frequency	106	160	160	107

Table 9.3.3. Categories and Cut values (in seconds) for *Latencies* coding.

Coding of the Number of exercises to mastery. From Table 9.3.1 (p.396) we have proceeded to the coding of the *Number of exercises*. For Integers, Fractions, and Measurement, we code two categories: four exercises (1) and more than four exercises to mastery (2). For Geometry and Logic, we code three categories: In Geometry, the categories are less than eleven exercises (1), eleven or twelve exercises (2), and more than twelve exercises (3). In Logic, the categories are four exercises (1), five exercises (2), and more than five exercises (3). These categories have been devised so that at least 5% lie within each category, while attempting to balance the number of categories per strand. For example, for Fractions there is one individual that, on average, needs 6 exercises and there is one that needs 7 exercises (cf. Table 9.3.1, p.396), both individuals have been allocated to category 2. For each strand, Table 9.3.4 gives the absolute frequencies of categories.

Cate-gories	Integers	freq.	Fractions	freq.	Geometry	freq.	Logic	freq.	Measurement	freq.
1	4	480	4	482	≤ 10	82	4	165	4	387
2	≥ 5	53	≥ 5	51	11, 12	321	5	301	≥ 5	146
3					≥ 13	130	≥ 6	67		

Table 9.3.4. For each strand, categories of *Number of exercises* and their absolute frequencies.

In total one thus obtains $15 + 20 + 12 = 47$ categories (or modalities).

Response patterns

With each variable and each individual, there is associated a numerical value, hence a category of the corresponding coded variable. For an individual, the 15 categories corresponding to the 15 coded variables define the individual's *response pattern*. For example, for the first individual in the file, table below gives the 15 values and the 15 categories:

	Integers	Fractions	Geometry	Logic	Measurement
Raw → Coded *Error rates*	.014 → 1	.015 → 1	.034 → 2	.054 → 2	.069 → 2
Raw → Coded *Latencies*	27.2 → 3	13.7 → 2	26.5 → 3	17.5 → 2	37.5 → 2
Raw → Coded *Exercises*	4 → 1	4 → 1	12 → 2	5 → 2	6 → 2

The corresponding response pattern is therefore 11222 32322 11222.

The number of possible patterns is equal to

$$3^5 \times 4^5 \times (2 \times 2 \times 3 \times 3 \times 2) = 243 \times 1024 \times 72 = 17\,915\,904.$$

The number of observed patterns is 520, quite close to the number of individuals (533), which expresses almost the maximum of individual differences at the level of coded variables[35].

[35] As active variables, we could have taken the averages over strands of the three types of variables; however in doing so, the distance between individuals would only depend on the three types of variables, and we would lose an important source of variation between individuals. As a matter of comparison, if we had taken only the three types of variables as active variables, we would have much fewer possible patterns (say about $3 \times 4 \times 3 = 36$) than individuals.

9.3.2 Results of Multiple Correspondence Analysis

We have $Q = 5 \times 3 = 15$ active variables. For Integers and Fractions, Category 3 of *Error rates* has a frequency less than 1% (Table 9.3.2, p.396), therefore we have grouped this category with Category 2. As a result, the total number of modalities (categories) is equal to $K = 47 - 2 = 45$.

The basic results of MCA are the variances of axes (eigenvalues), the principal coordinates of the 45 categories and of the 533 individuals (i.e. 520 response patterns), together with their contributions to axes, and the geometric representations of the two clouds (categories and individuals).

Variances of axes and inertia rates

There are $K - Q = 45 - 15 = 30$ (nontrivially null) eigenvalues, and the variance of the cloud is equal to 2 (cf. *MULTIPLE*, Property 5.5, p.186). The first four eigenvalues, and the corresponding inertia rates (cf. *MULTIPLE*, p.201 and formula (5.16), p.201) are given in Table 9.3.5.

	Axis 1	Axis 2	Axis 3	Axis 4
Eigenvalues (λ)	.3061	.2184	.1460	.1199
Raw rates of inertia	15.3%	10.9%	7.3%	6.0%
Modified rates	63.1%	25.4%	6.9%	3.1%
Cumulated modified rates	63.1%	88.5%	95.4%	98.6%

Table 9.3.5. Eigenvalues; raw and modified rates of inertia.

Principal coordinates

The principal coordinates of the 45 active categories on the first four axes are given in Table 9.3.6 (p.400). We do not give the table of coordinates of individuals, but it is easy to deduce it from Table 9.3.6[36].

Contributions

Table 9.3.7 (p.401) gives the relative *contributions of modalities* to the first four axes; it can easily be obtained from absolute frequencies (Tables 9.3.2, 9.3.3 and 9.3.4, p.396-398) and coordinates (Table 9.3.6, p.400) of categories[37]. From Table 9.3.7 we deduce the *contributions of the 15 variables* (Table 9.3.8), the *contributions of the 5 Strands* (Table 9.3.9) and the *contributions of the 3 Types of variables* (Table 9.3.10, p.401).

[36]Cf. *MULTIPLE*, Property 5.7 (p.188); for an example of calculation, see *Culture example* (p.225).

[37]For definitions of contributions, see *MULTIPLE*, p.194-196; for examples of calculations, see *Culture example*, p.225-226.

Error rates					
		Axis 1	Axis 2	Axis 3	Axis 4
Integers	I1	−.535	.326	−.054	.071
	I2	.715	−.436	.073	−.095
Fractions	F1	−.469	.322	−.165	.045
	F2	.496	−.341	.174	−.048
Geometry	G1	−1.160	.194	−.191	1.826
	G2	−.167	.155	.049	−.384
	G3	1.331	−.727	−.084	.482
Logic	L1	−1.128	.356	.332	2.513
	L2	−.307	.234	−.063	−.309
	L3	1.094	−.717	.095	.264
Measure–ment	M1	−.780	.215	−.093	.458
	M2	.323	−.058	.084	−.388
	M3	1.905	−.926	−.349	1.389

Number of exercises					
		Axis 1	Axis 2	Axis 3	Axis 4
Integers	I1	−.181	.089	−.002	−.086
	I2	1.640	−.810	.022	.779
Fractions	F1	−.117	.099	−.009	−.032
	F2	1.104	−.934	.084	.304
Geometry	G1	−.456	−.457	−.723	−.068
	G2	−.095	−.051	.223	.078
	G3	.522	.414	−.095	−.151
Logic	L1	−.816	.263	.071	.736
	L2	.149	.074	−.071	−.561
	L3	1.340	−.979	.146	.710
Measure–ment	M1	−.382	.164	.004	−.013
	M2	1.012	−.436	−.011	.035

Latencies					
		Axis 1	Axis 2	Axis 3	Axis 4
Integers	I1	.728	1.168	−.794	.043
	I2	.199	.283	.660	.105
	I3	−.242	−.448	.374	−.105
	I4	−.657	−.912	−.759	−.042
Fractions	F1	.769	1.107	−.779	.163
	F2	.109	.342	.715	−.033
	F3	−.256	−.451	.325	.108
	F4	−.542	−.933	−.784	−.274
Geometry	G1	.905	.960	−.737	.211
	G2	.137	.292	.679	−.059
	G3	−.348	−.325	.380	.015
	G4	−.582	−.902	−.854	−.143
Logic	L1	.913	.925	−.695	.180
	L2	.218	.311	.630	−.047
	L3	−.484	−.237	.437	.118
	L4	−.506	−1.026	−.906	−.284
Measure–ment	M1	.797	.804	−.744	.218
	M2	.188	.395	.495	−.201
	M3	−.265	−.401	.375	.023
	M4	−.674	−.787	−.564	.050

Table 9.3.6. Coordinates of the 45 active modalities (categories).

9.3.3 Interpretation

How many axes to interpret?

Looking at modified rates, it is clear that one axis is not sufficient (63.1%), whereas taking two axes brings the rate up to 88.5%. The first two axes (cf. Table 9.3.5, p.399) are more important than the following ones ($\lambda_1 = .306$, $\lambda_2 = .218$); the following two axes have eigenvalues close to each other ($\lambda_3 = .146$, $\lambda_4 = .120$). Taking four axes brings the modified inertia rate to 99%. We will in any case interpret the first two axes, and attempt to interpret the next two ones.

First interpretation

• In Table 9.3.10 (p.401) are shown the relative contributions of the 3 *Types of variables*. For Axis 1, *Error rates* account for 43% of the variance, the *Number of exercises* for 30%, then the *Latencies* for 27%. For Axis 2, *Latencies* account for 68% of the variance; *Error rates* for only 18% and *Number of exercises* for 14%. Therefore the first axis is mainly the axis of *Error rates*, and the second axis is mainly the axis of *Latencies*. One notices that *Latencies* account for 92% of the variance of Axis 3, and *Error rates* for

67% of the variance of Axis 4. The third axis, which is specific of *Latencies*, is a refinement of the second axis, and the fourth axis a refinement of the first one; we will discuss these points later on.

• In Table 9.3.9 (this page) are shown the relative contributions of the 5 *strands* to the first four axes: For the first three axes, the contributions

		Axis 1	Axis 2	Axis 3	Axis 4
Error rates					
Integers	I1	**.036**	.019	.001	.002
	I2	**.048**	**.025**	.001	.002
Fractions	F1	**.025**	.016	.006	.001
	F2	**.026**	.017	.007	.001
Geometry	G1	**.030**	.001	.002	.191
	G2	.004	.005	.001	**.059**
	G3	**.069**	**.029**	.001	**.023**
Logic	L1	.016	.002	.003	**.204**
	L2	.014	.011	.001	**.037**
	L3	**.000**	**.040**	.001	.010
Measurement	M1	**.048**	.005	.001	**.042**
	M2	.013	.001	.002	**.050**
	M3	**.037**	.012	.003	**.050**
Latencies		Axis 1	Axis 2	Axis 3	Axis 4
Integers	I1	**.023**	**.083**	**.057**	.000
	I2	.003	.007	**.060**	.002
	I3	.004	.018	.019	.002
	I4	.019	**.051**	**.053**	.000
Fractions	F1	**.026**	**.074**	**.055**	.003
	F2	.001	.011	**.070**	.000
	F3	.004	.019	.015	.002
	F4	.013	**.053**	**.056**	.008
Geometry	G1	**.036**	**.056**	**.049**	.005
	G2	.001	.008	**.063**	.001
	G3	.008	.010	.020	.000
	G4	.015	**.050**	**.067**	.002
Logic	L1	**.036**	**.052**	**.044**	.004
	L2	.003	.009	**.054**	.000
	L3	.015	.005	**.026**	.002
	L4	.011	**.065**	**.075**	.009
Measurement	M1	**.028**	**.039**	**.050**	.005
	M2	.002	.014	**.034**	.007
	M3	.005	.015	.019	.000
	M4	.020	**.038**	**.029**	.000
Number of exercises		Axis 1	Axis 2	Axis 3	Axis 4
Integers	I1	.006	.002	.000	.004
	I2	**.058**	.020	.000	**.034**
Fractions	F1	.003	.003	.000	.001
	F2	**.025**	**.025**	.000	.005
Geometry	G1	.007	.010	**.037**	.000
	G2	.001	.000	.014	.002
	G3	.014	.013	.001	.003
Logic	L1	**.045**	.007	.001	**.093**
	L2	.003	.001	.001	**.099**
	L3	**.049**	**.037**	.001	**.035**
Measurement	M1	**.023**	.006	.000	.000
	M2	**.061**	.016	.000	.000

Table 9.3.7. Relative contributions (Ctr) of the 45 modalities (in bold Ctr > .022).

	Axis 1	Axis 2	Axis 3	Axis 4
Ctr of *Error rates*				
Integers	**.083**	.043	.002	.004
Fractions	.051	.034	.013	.001
Geometry	**.104**	.035	.003	**.273**
Logic	**.096**	.053	.005	**.251**
Measurement	**.098**	.018	.006	**.142**
Ctr of *Latencies*				
Integers	.048	**.159**	**.189**	.004
Fractions	.043	**.157**	**.190**	.013
Geometry	.059	**.123**	**.199**	.008
Logic	.066	**.131**	**.200**	.015
Measurement	.054	**.106**	**.132**	.012
Ctr of *Number of exercises*				
Integers	.065	.022	.000	.037
Fractions	.028	.028	.000	.005
Geometry	.023	.023	.051	.006
Logic	**.097**	.044	.003	**.227**
Measurement	**.084**	.022	.000	.000

Table 9.3.8. Relative contributions (Ctr) of the 15 variables (in bold, Ctr > 1/15 = .067).

Ctr of Strands	Axis 1	Axis 2	Axis 3	Axis 4
Integers	.196	**.225**	.191	.045
Fractions	.122	**.219**	**.209**	.020
Geometry	.186	.182	**.254**	**.287**
Logic	**.259**	**.228**	**.208**	**.493**
Measurement	**.237**	.146	.138	.155

Table 9.3.9. Relative contributions (Ctr) of the 5 Strands (in bold, Ctr > .2).

Ctr of Types of Variables	Axis 1	Axis 2	Axis 3	Axis 4
Error rates	**.433**	.184	.029	**.671**
Latencies	.271	**.677**	**.916**	.053
Exercises	.296	.139	.055	.276

Table 9.3.10. Relative contributions of the 3 Types of variables (in bold, Ctr > .333).

range from 12% to 26%, whereas the contribution of Logic predominates for Axis 4 (49%).

• If one looks at the contributions of the 15 *variables* in detail (Table 9.3.8, p.401), one sees that the *Number of exercises* in Fractions and in Geometry hardly contribute (3% and 2%) to Axis 1, that *Latencies* of the 5 strands contribute almost equally to Axis 2.

• If one now examines the three figures 9.3.3 (this page) corresponding to the 3 types of variables in *plane 1-2*, one notes that for each type variable there is a coherence between strands except for the *Number of exercises* in Geometry (Figure 9.3.3-c).

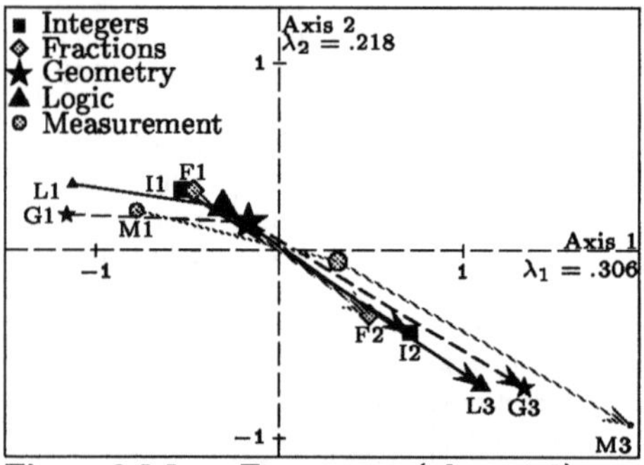

Figure 9.3.3-a. *Error rates* (plane 1-2).

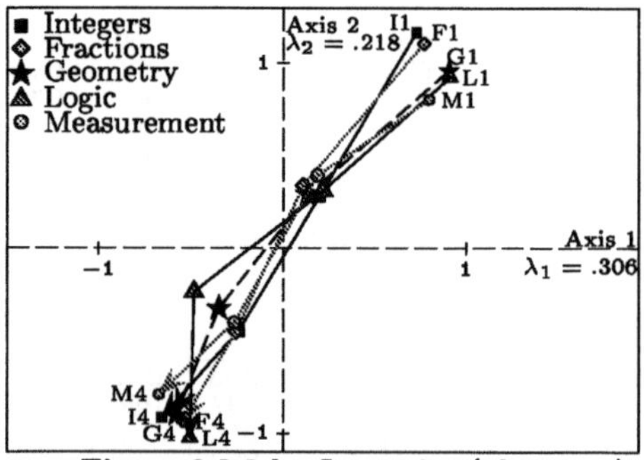

Figure 9.3.3-b. *Latencies* (plane 1-2).

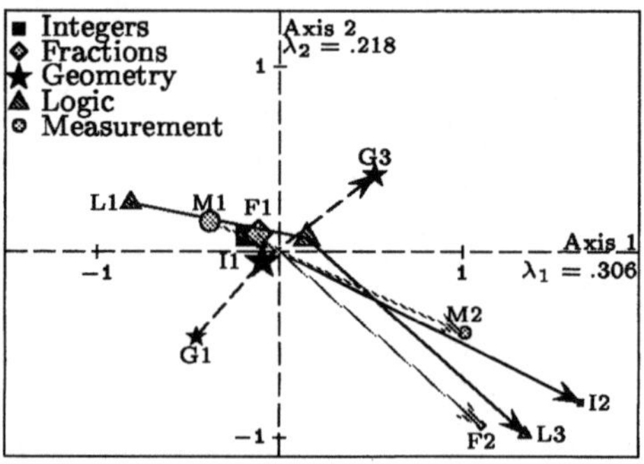

Figure 9.3.3-c. *Number of exercises* to mastery (plane 1-2).

Figure 9.3.3. Clouds of modalities (3 types of variables) in plane 1-2.

If one examines plane 2-3 (Figure 9.3.4 left, p.403), one observes a Guttman effect for *Latencies*, that is, the horizontal axis (Axis 2) orders categories from left to right according to decreasing latencies; and Axis 3 opposes medium categories (2 and 3) to extreme ones (1 and 4). Similarly, one observes a Guttman effect of *Error rates* in plane 1-4 (Figure 9.3.4, right). These Guttman effects observed in planes 2-3 (*Latencies*) and 1-4 (*Error rates*) show that the data are essentially articulated around 2 scales:

one of *Error rates* (Axis 1) and one of *Latencies* (Axis 2). We are therefore going to resume in detail the interpretation of axes 1 and 2.

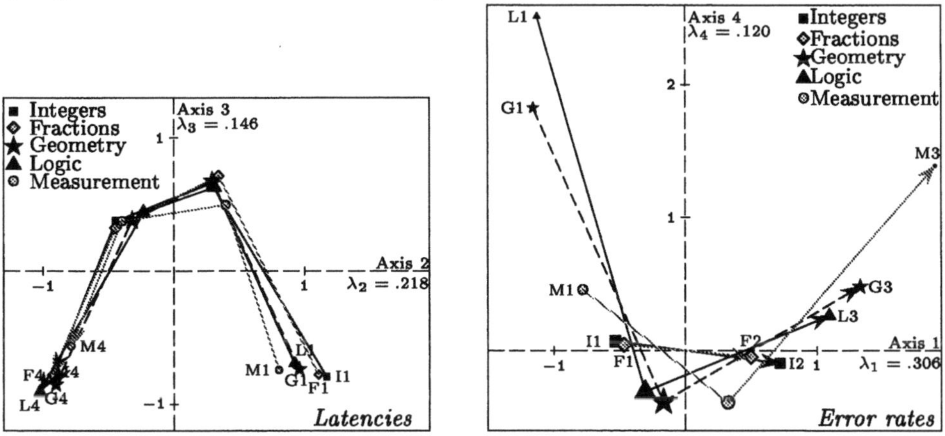

Figure 9.3.4. Guttman effects in cloud of categories in planes 2-3 (left) and 1-4 (right).

Interpretation of Axis 1 ($\lambda_1 = .3061$).

The number of categories whose contributions to Axis 1 are greater than average ($1/Q = 1/45 = .022$) is equal to 20 (see Table 9.3.7, p.401); to which we will add the low error rate category for Logic (L1, Ctr = .016), hence 21 categories, depicted on Figure 9.3.5 (p.404). On one side of axis (left on figure), one finds the 5 low *Error* rate categories (*Error* I1, F1, G1, L1, M1), and the small Number of *Exercises* categories in Logic and Measurement (*Exercises* L1, M1). On the other side (right), one finds the 5 high Error rate categories (*Error* I2, F2, G3, L3, M3), the large Number of Exercises (except in Geometry) (*Exercises* I2, F2, L3, M2), and the 5 short latency categories (*Latency* I1, F1, G1, L1, M1). The interpretation of Axis 1 will be based on these 21 categories, which account for 81% of the variance of axis. The opposition between high and low error rates is very important: the deviation between the two corresponding barycenters accounts for 35% of the variance of Axis 1 and 81% of the variance of the *Error rate* variable on Axis 1. The contributions of short latency categories for the 5 strands are greater than average contribution (see Table 9.3.7, p.401). These 5 categories are located on the right of the origin of axes (cf. Figure 9.3.5, p.404 and Table 9.3.6, p.400), which shows a link between high error rates and short latencies. The opposition between low error rates and short latencies accounts for 28% of the variance of Axis 1, and the one between small and large numbers of exercises for 24%. The opposition between the 7 categories on the left of origin and the 14 on the right of origin accounts for 67% of the variance of Axis 1.

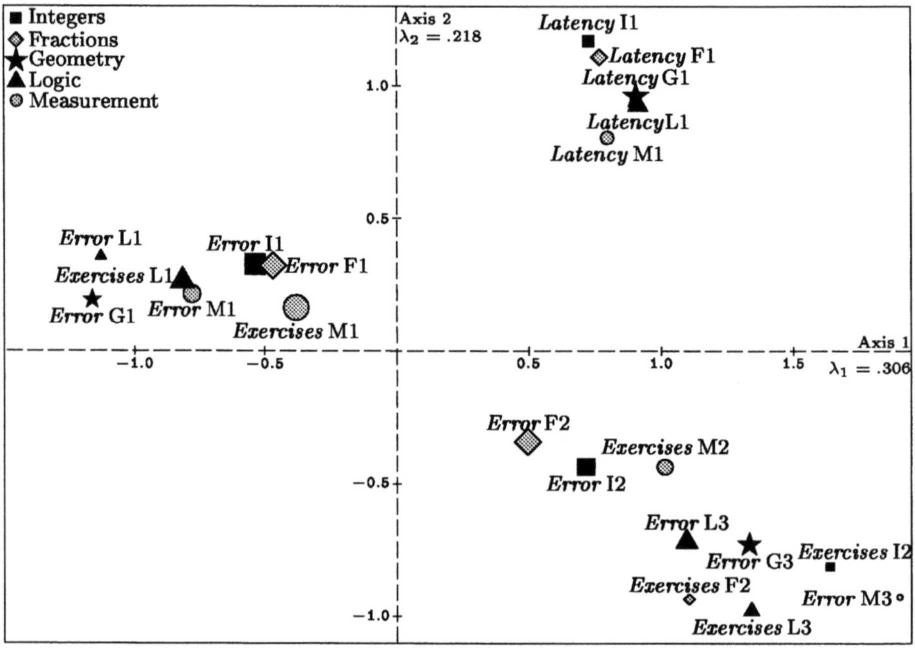

Figure 9.3.5. Interpretation of Axis 1 (21 categories).

> *The first axis is the axis of Error rates and Numbers of exercises.* It opposes on one side low error rates and small numbers of exercises and on the other side high error rates and large numbers of exercises, the latter being associated with short latencies.

Interpretation of Axis 2 ($\lambda_2 = .2184$)

On Figure 9.3.6 (p.405), the 15 categories whose contributions to Axis 2 are greater than average are depicted. At the top of the figure, one finds the 5 short *latency* categories (*Latency* I1, F1, G1, L1, M1). At the bottom, one finds the 5 long *latency* categories (*Latency* I4, F4, G4, L4, M4), the 3 high *error* rate categories in Integers, Logic and Geometry (*Error* I2, G3, L3) and the two large Number of *exercises* categories in Fractions and Logic (*Exercises* F2, L3). These 15 categories account for 72% of the variance of Axis 2. The interpretation of Axis 2 will be based on these 15 categories. There is an opposition between short and long latency categories, the deviation between the two corresponding mean points accounts for 55% of

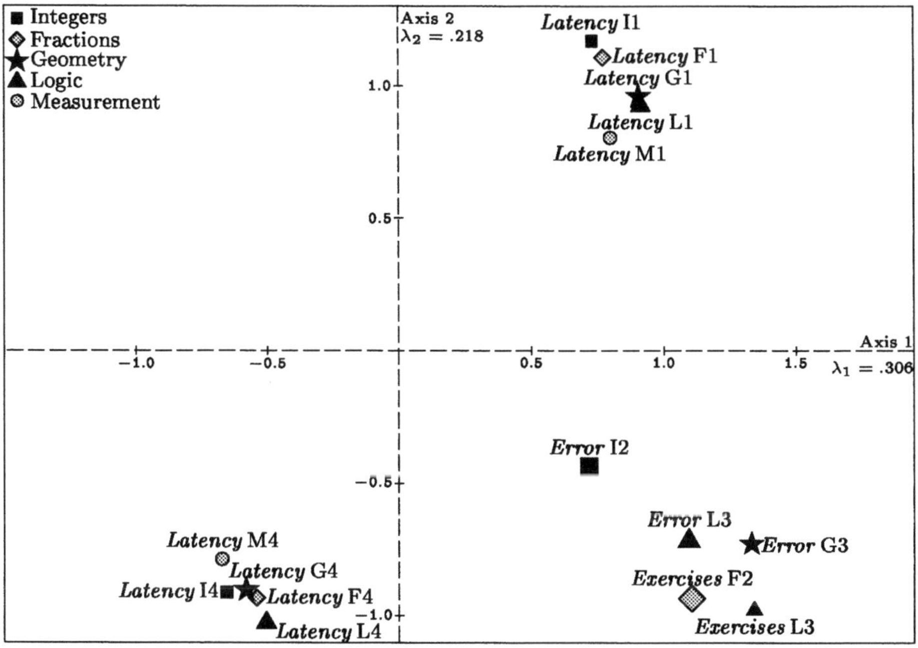

Figure 9.3.6. Interpretation of Axis 2 (15 categories).

the variance of Axis 2 and for 81% of the variance of *Latency* variable on Axis 2. The opposition between the 5 short latency categories and the 3 high error rate categories in Integers, Geometry and Logic accounts for 35% of Axis 2; the opposition between the 5 short latency categories and the 2 large Number of Exercises in Fractions and Logic accounts for 21% of the variance of Axis 2. More generally, the opposition between the 5 short latency categories and the 10 aforementioned categories accounts for 65% of the variance of Axis 2.

> *The second axis is the axis of latencies.* It opposes short latencies and long latencies, the latter being associated with high error rates and large numbers of exercises.

Typical Response Patterns

From the interpretations of axes and the distribution of categories in plane 1-2, the following response patterns emerge as typical patterns:
pattern 11111 11111 11111 (point A) (low error rates, short latencies, small number of exercises); Pattern 11111 44444 11111 (point B) (low error rates, long latencies, small number of exercises); pattern 22333 11111 22332 (point D) (high error rates, short latencies, large number of exercises); and pattern

22333 44444 22332 (point C) (high error rates, long latencies, large number of exercises). Notice that none of the 533 individuals matches any one of these typical response patterns.

9.3.4 Cloud of Individuals

Cloud in plane 1-2 (Figure 9.3.7). The cloud of individuals (533 students) consists in 520 observed response patterns, to which we add the 4 typical response patterns; it is represented on Figure 9.3.7. The individuals

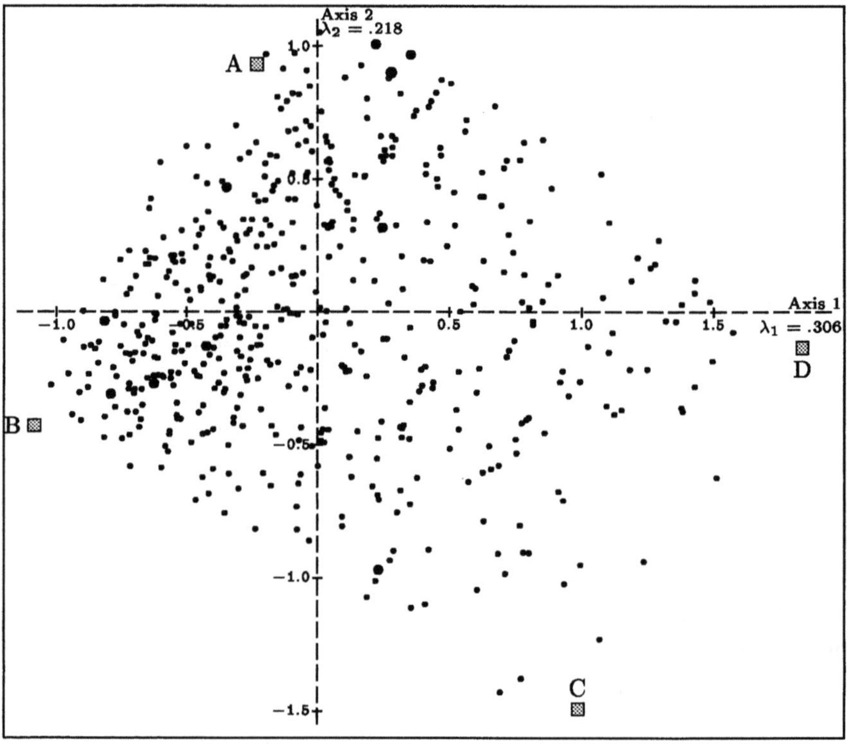

Figure 9.3.7. Cloud of individuals (patterns) with typical response patterns in plane 1-2.

are roughly scattered inside the quadrilateral ABCD defined by the 4 ideal patterns, with a high density of points along the side AB and a low density along the opposed side. This shows there are many students who make few errors whatever their latencies. On the other hand, students with high error rates are less numerous and very scattered. The Table 9.3.11 (p.407) gives the variances of subgroups of individuals with low (I1, F1, G1, L1, M1) and high (I2, F2, G3, L3, M3) error rates in plane 1-2.

Error rates	Integers	Fractions	Geometry	Logic	Measurement
low	0.307	0.362	0.211	0.204	0.257
high	0.469	0.500	0.387	0.399	0.360

Table 9.3.11. Plane 1-2: Variances of subclouds of individuals with low and high error rates.

Crossing of Error rates and latencies. If for each strand, we cross error rate and latency categories, with each composite category there is associated a subcloud of individuals with its mean–point.

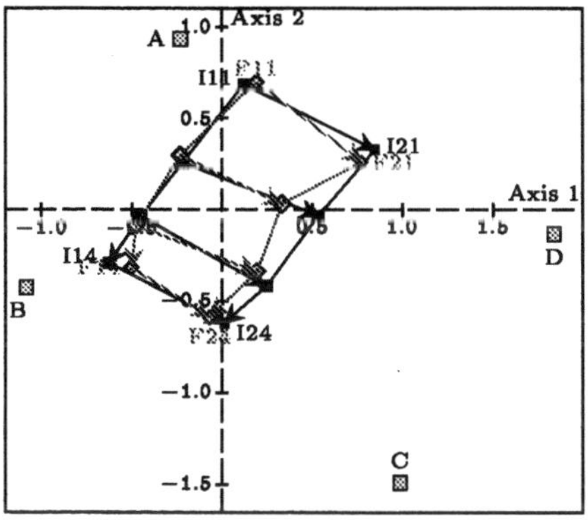

Figure 9.3.8-a: Error rates×Latencies for Integers (black) and Fractions (gray).

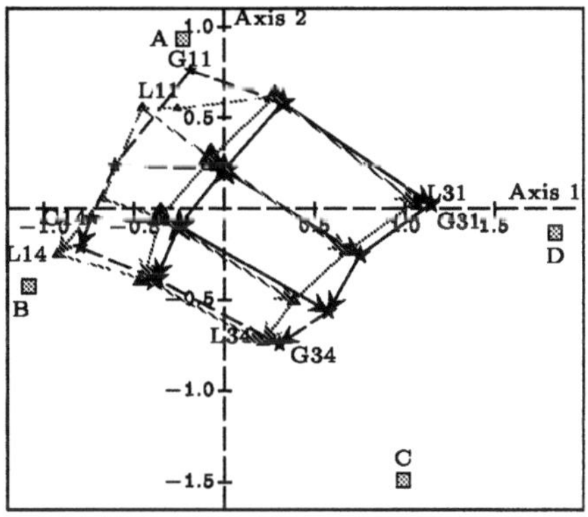

Figure 9.3.8-b: Error rates×Latencies for Geometry (black) and Logic (gray)

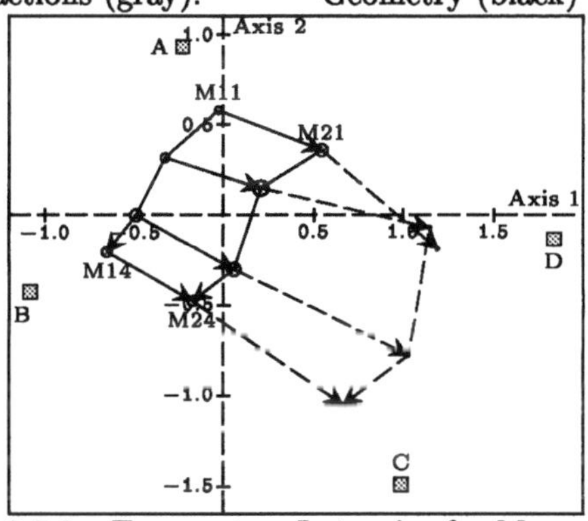

Figure 9.3.8-c Error rates×Latencies for Measurement

Figure 9.3.8. Error rates×Latencies in plane 1-2 (Cloud of individuals).

Figure 9.3.8-a shows the 2×4 mean–points for Integers and the 2×4 mean–points for Fractions (marker sizes are proportional to the frequencies of subgroups). One notices that the 8 "Integers" mean–points are very close to the 8 "Fractions" mean–points, and that these points are closer to side AB than to side CD of the quadrilateral; this means that error rates are globally low. Similarly, Figure 9.3.8-b shows the $3 \times 4 = 12$ "Geometry" mean–points and the 12 "Logic" mean–points. One notices the proximity of homologous points, with some mean–points of small groups quite close to side AB, that is, corresponding to students with globally low error rates. Similarly, Figure 9.3.8-c shows the 3×4 "Measurement" mean–points. There are very few individuals with high error rates whose mean–points are close to side CD, that is, corresponding to students with globally high error rates.

Figures 9.3.8 (page before) show well that quadrilateral ABCD is a frame that brings forth the following geometric model. When one goes down along the AB direction, latencies increase, while error rates remain constant; when one goes down along the AD direction, error rates increase, while latencies remain constant.

More on Guttman Effects. Guttman effects (*MULTIPLE*, p.220) can be investigated in the cloud of individuals. Figure 9.3.9 (left) shows the 45 observed category mean– points (among $2 \times 2 \times 3 \times 3 \times 3 = 108$ possible) for *Error rates* in plane 1-4. The error rate scale appears very distinctly, showing a strong homogeneity across strands. Similarly Figure 9.3.9 (right) shows, for *Latencies* in plane 2-3, the 214 (among $4^5 = 1024$) observed category mean–points (labels are written for patterns with frequency ≥ 9). The scale of latencies is not so sharply distinct, as some subjects have both short and long latencies across strands (for example *43441*).

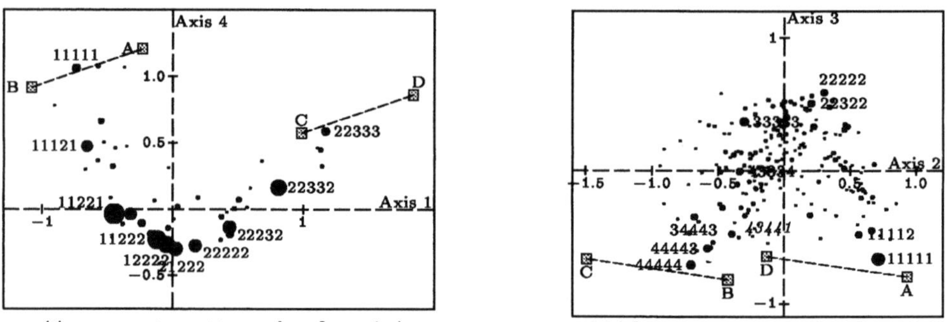

44 error rate patterns in plane 1-4 214 latency patterns in plane 2-3

Figure 9.3.9. Guttman effects in cloud of individuals (planes 1-4 and 2-3).

Comparing Integers and Geometry. Is there any difference in students' behavior in Integers vs Geometry?

Among the 55 students who have a low *error rate* in Geometry, 52 have also a low error rate in Integers. More than half of these 52 students have long latencies (categories 3 and 4) both in Integers and in Geometry. We have depicted in Figure 9.3.10 (plane 1-2) the category mean–points associated with the observed combinations of *latencies* in Integers and Geometry (15 observed out of $4 \times 4 = 16$ possible). If, for each latency category in Integers, one joins category points G1, G2, G3, G4 in Geometry (lines I1G1 through I1G4, I2G1 through I2G4, I3G1 through I3G4, I4G2 through I4G4, see graph on the left), the segments are roughly parallel to Axis 1, that is, when going from G1 to G4, error rates decrease and latencies weakly increase. If similarly, for each latency category in Geometry, one joins category points I1, I2, I3, I4 in Integers (see graph on the right), the segments are roughly parallel to Axis 2, that is, when going from I1 to I4, latencies increase and error rates remain about steady. Such findings suggest that to improve performance in Geometry, some increase in latency is needed.

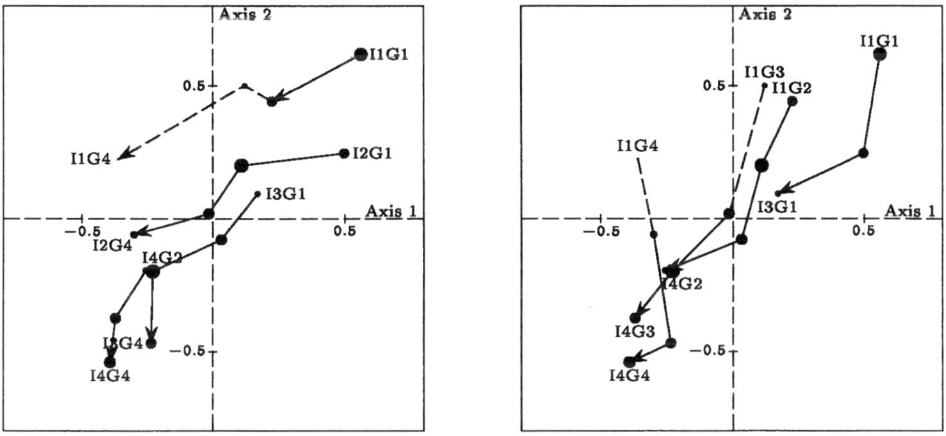

Figure 9.3.10. Comparing Integer and Geometry latencies (plane 1-2).

Structured Data Analysis

We now study the following structuring factors: Time spent on computer, Age and Gender, allowing for missing data (133, 57, 46 respectively).

Number of hours spent on computer. The number of hours spent on computer ranges from 6h30 to 248h, the median being equal to 21h20. For about 20% of students, time is less than 15h, and for about 20% it is greater than 34h. Eight students spent between 70h and 122h30, one student spent 248h. The number of hours is correlated with the second axis (−.731, Spearman); globally, the shorter the latencies, the smaller the number of hours, as would be expected. If we code the number of hours into 4 classes (cuts at inferior quintile, median and superior quintile, as we did above for latencies), one sees, on the table 9.3.12 (next page), that for the first axis, the within variance is much larger than the between one, and that for the second axis, the between variance is almost equal to the within variance.

variances	Axis 1	Axis 2
between	.0176	.1009
within	.2748	.1087

Table 9.3.12. Between and within Variances for Number of Hours.

Age and Gender. The ages of students, at the end of the course, range from 5 years to $11\frac{1}{2}$ years with a mode between 8 and 9 years. If one codes the age in 4 classes from the quartiles (7.76, 8.415 and 9.05), one observes an age effect with the 4 age categories ordered on axes 1 and 2. The deviation between the extreme age classes (85% of the variance of Age on Axis 1 and 78% on Axis 2) is equal to 0.89 SD of Axis 1 (an important deviation) but only to 0.46 SD of Axis 2 (a medium deviation). One can therefore say that, on average, when age increases, error rates increase and latencies slightly decrease, see Figure 9.3.11 (this page). Nevertheless, within each age class, students are very scattered, as shown in Table 9.3.13 (next page) of double breaking down of variance.

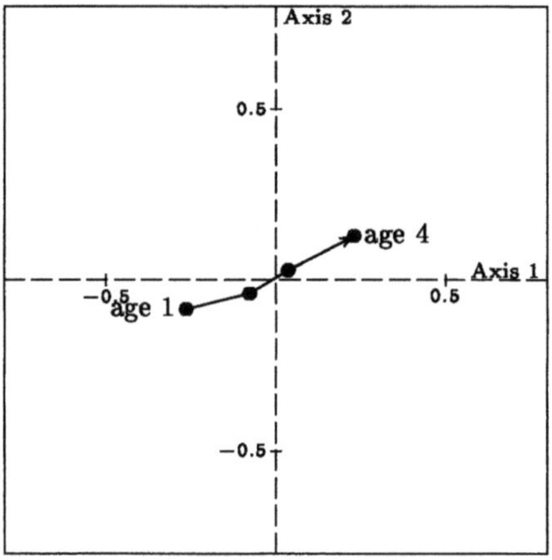

Figure 9.3.11. Mean–points of the 4 age categories (plane 1-2).

The students are composed of 283 boys and 204 girls. There is virtually no difference between boys and girls as shown in Table 9.3.14 (facing page).

If we cross Age and Gender, generating $4 \times 2 = 8$ classes, one notices the large dispersion within the 8 classes in plane 1-2 — the within variance is equal to $.2683 + .2055 = .4738$, with a between–variance only equal to

.5143 − .4738 = .0405. The interaction between Age and Gender is very low, as shown in Table 9.3.15. For each axis, the sum of the variances of Age and Gender factors and of their interaction is almost equal to the one of their crossing, which reflects that the crossing is nearly orthogonal (see Table 9.3.16).

variances	Axis 1	Axis 2
between	.0317	.0065
within	.2684	.2102

Table 9.3.13. Between and within Variances for Age.

variances	Axis 1	Axis 2
between	.0001	.0012
within	.2927	.2112

Table 9.3.14. Between and within Variances for Gender.

	Axis 1	Axis 2
Age×Gender	.0306	.0099
Age	.0301	.0067
Gender	.0000	.0012
Interaction	.0006	.0016
Within (Age×Gender)	.2683	.2055
Total variance ($n = 468$)	.2989	.2154

Table 9.3.15. Double breaking down of variances for the crossing Age×Gender.

	Boys	Girls	
$\cdot < 7.76$	68	49	117
$7.76 < \cdot < 8.415$	74	46	120
$8.415 < \cdot < 9.05$	66	51	117
$\cdot > 9.05$	60	54	114
Total	268	200	468

Table 9.3.16. Absolute frequencies for the crossing Age×Gender.

9.3.5 Euclidean Classification

We have proceeded to a Euclidean classification of the 533 individuals from the 15 coded variables. Figure 9.3.12 (next page) shows the superior tree resulting in the partition into 6 classes with the level indices. Clearly, two partitions emerge: a three–class partition generated by 2 successive dichotomies, and a six–class partition generated by 5 dichotomies. We will comment on the successive dichotomies leading to these two partitions, then, we amend the partition into 6 classes of the AHC to get a partition into 5 classes that we will retain as a final summary. The 5 successive partitions of AHC will be designated by CA (2 classes ca1 and ca2), CB (3 classes cb1, cb2 and cb3), CC (4 classes cc1, ... cc4), CD (5 classes) and CE (6 classes), the final partition into 5 classes will be designated by C. The succession of dichotomies is represented on Figure 9.3.13 (p.413). Each dichotomy corresponds to two subclouds (gray and black points); the ellipse colored in gray denotes the subcloud that will be subdivided at the following step.

Partition into 3 classes

First dichotomy. The first dichotomy (level index .176, partition CA) separates the students in 2 classes ca1 of 255 students and ca2 of 278 students (cf. Figure 9.3.13-a, p.413), and is characterized as follows.

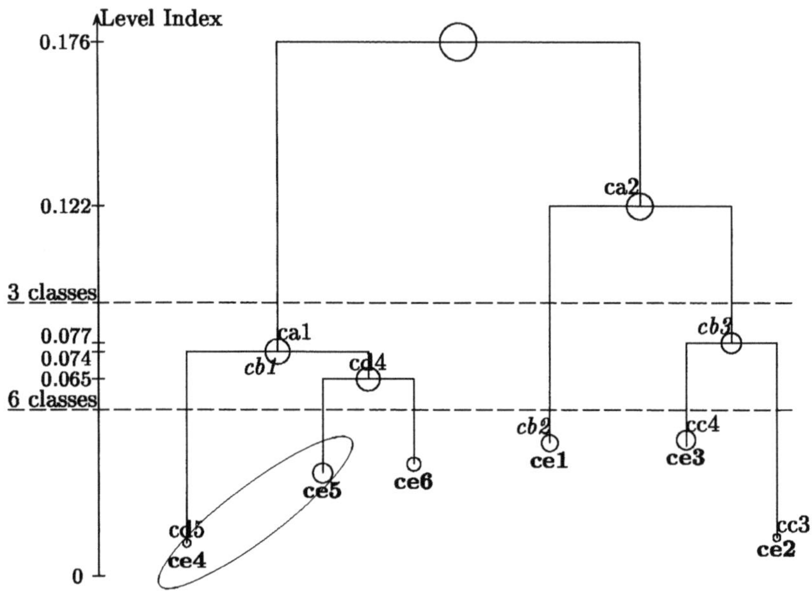

Figure 9.3.12. Superior hierarchical tree resulting in six–class partition.

Error rates. The students that have low error rates are a large majority in class ca1: they are 196/305 for Integers (I1), 169/274 for Fractions (F1), 48/55 for Geometry (G1), 28/31 for Logic (L1) and 160/192 for Measurement (M1).

Latencies. The students that have long latencies (category 4) are about 80% in class ca1. More than 70% of the students in class ca1 have latencies above median in all strands (categories 3 and 4).

Number of exercises. Among the students that need more than 4 exercises, in class ca1, there are 2/53 for Integer (I2), 2/51 for Fractions (F2), 18/146 in Logic (L2, L3); and 38/130 need more than 12 exercises in Geometry (G3)and 6/61 do more than 5 exercises in Logic (L2).

> Class ca1 is characterized by students with low error rate; it includes almost all students with long latencies and who need few exercises to mastery. An extreme pattern of this class is 11111 44444 11111 (point B).

Second dichotomy. The second dichotomy (level index .122) generates the three–class partition CB. It comprises class cb1 (alias ca1) and splits class ca2 into two classes cb2 of 111 students and cb3 of 167 students (Figure 9.3.13-b, p.413). It separates out class cb2. Class cb2 is rather compact; its first characteristic is that almost all students of this class (except 3

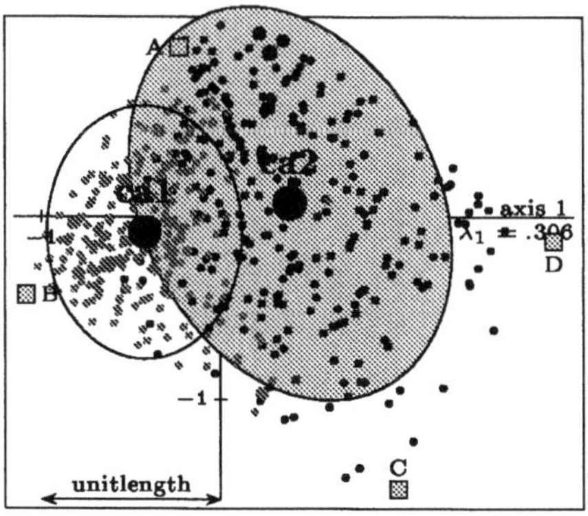

Figure 9.3.13-a. First dichotomy (two-class partition CA).

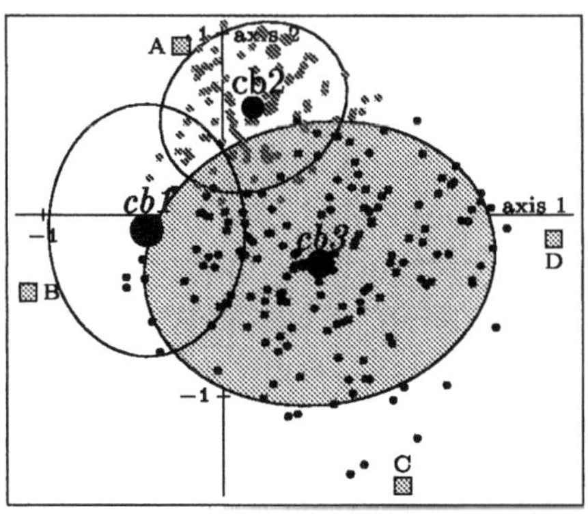

Figure 9.3.13-b. Second dichotomy (partition CB).

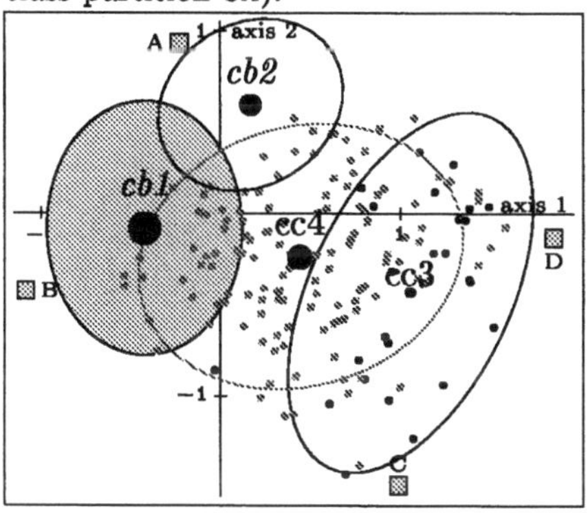

Figure 9.3.13-c. Third dichotomy (four-class partition CC).

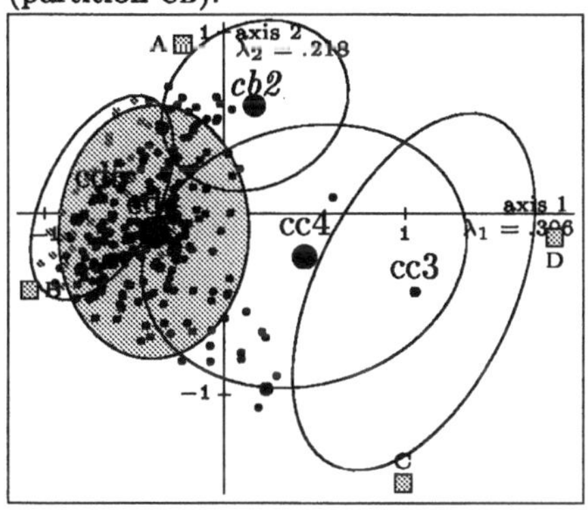

Figure 9.3.13-d. Fourth dichotomy (five-class partition CD).

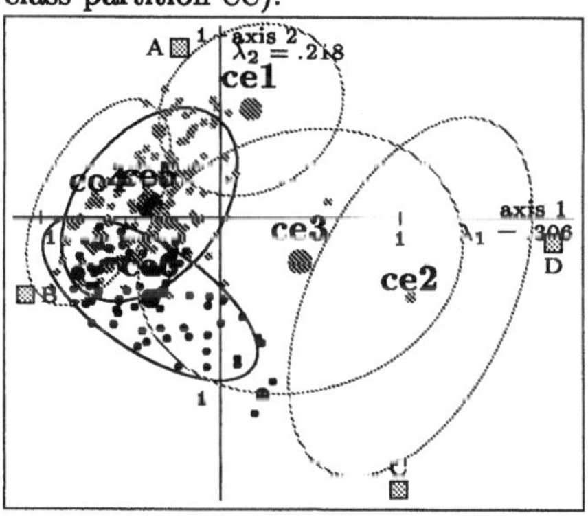

Figure 9.3.13-e. Fifth dichotomy (six-class partition (CE).

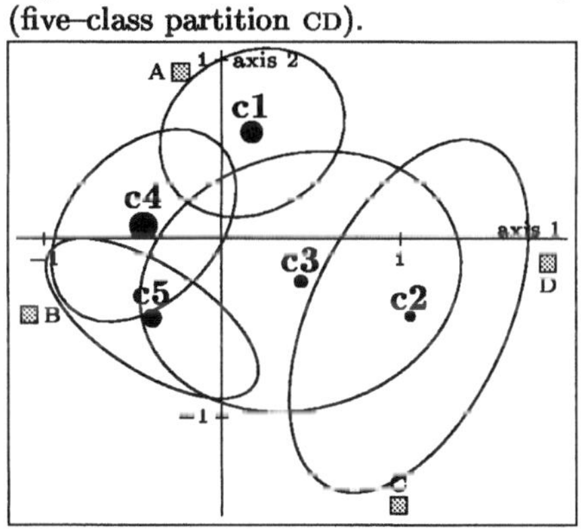

Figure 9.3.13-f. Concentration ellipses of the final five-class partition (C).

Figure 9.3.13. Ellipses of classes of partitions.

in Fractions and in Measurement, 9 in Geometry and 15 in Logic) have latencies below median. The average error rates are situated around the median; 2/3 of students have error rates inferior to .02 in Integers and Fractions, and are between .02 and .10 in the 3 other strands for almost all (respectively for 104, 100 and 85). Except in Geometry, the students of this group need few exercises for mastery: none does more than 4 in Integers and Fractions; 22 more than 4 in Measurement; in Logic 86 do 5 and 24 do less of 5; in Geometry, almost all (except 7) do more than 11 exercises. This class is therefore characterized by short latencies, medium error rates, and small numbers of exercises except in Geometry. For class cb3, the distribution of latencies is for each strand near the overall distribution, with under-representation of long latencies. The between and within variances of partition CB on the first 2 axes are given hereafter.

The between–variance is very superior to the within–variance for Axis 1 and slightly inferior for Axis 2.

	Axis 1	Axis 2
Between: CB	.1813	.0993
Within: $I(\text{CB})$	.1248	.1191

> The partition into three classes contains a class (cb1) with low error rates (and rather long latencies), a class (cb2) with short latencies and small numbers of exercises to mastery and a class (cb3) with high error rates.

	Frequencies	Error rates	latencies	# of exercises
cb1	255	low	rather long	rather small
cb2	111		short	small except in Geometry
cb3	167	high	medium	rather large

Partition into six classes

The subsequent dichotomies lead to refining the partition into 3 classes by subdividing the classes of low and high error rates and with large dispersion of latencies.

Third dichotomy. The third dichotomy (level index $.077 \ll .122$) generates the four–class partition CC. It splits class cb3 into two classes cc3 with 25 students and cc4 with 142 students; both classes are very scattered (cf. Figure 9.3.13-c, p.413). Class cc3 is characterized by students who have almost all high error rates and who do large numbers of exercises; latencies are scattered on all categories with a weak majority with latencies inferior to the median. Class cc4 is characterized by high error rates but on the average inferior to those of class cc3, the proportions of students with extreme latencies are comprised between 10% and 18% ($< 20\%$), except for short latencies in Logic and Measurement (23%). The students who need many exercises to master a notion are mostly in this class (34/53 in Integers, 43/51 in Fractions, 48/67 in Logic and 81/146 in Measurement).

Fourth dichotomy. The fourth dichotomy (level index .074) generates the five–class partition CD. It splits class cc1 (alias ca1, cb1) into class cd4 (227 students) and class cd5 (28 students) (cf. Figure 9.3.13-d, p.413). The class cd5 is composed of students who have, in large majority, low error rates (they are 21 in fractions, 28 in Logic, 17 in Geometry, and for no strand the error rate is greater than .10); with medium latencies (between the first quintile and the fourth quintile), and not very large numbers of exercises (27 or 28 do only 4 exercises, except in Geometry where 22 students do 11 or 12 exercises and only 5 students do less than 11).

Fifth dichotomy. The fifth dichotomy (level index .065) generates the six class partition CE. It divides class cd4 in two classes ce5 (150 students) and ce6 (77 students) (cf. Figure 9.3.13-e, p.413). For class ce5, one has low error rates, medium latencies (proportions of short latencies vary between 5% and 11% $\ll$ 20% and the long ones between 15% and 19%). This class is very close to class ce4, with nevertheless error rates slightly greater than average. Class ce6 is the one of the long latencies (above median for all the students), and with proportions of small numbers of exercises superior to the proportion over the 533 students except in Logic (cf. following table).

Strands		Class ce6 ($n = 77$)	All (n=533)
Integer	< 4 exercises	97%	90%
Fractions	< 4 exercises	99%	90%
Geometry	< 11 exercises	34%	14%
Logic	< 5 exercises	29%	31%
Measurement	< 4 exercises	83%	73%

Amendment: Final partition into 5 classes (Figure 9.3.13-f, p.413)

The concentration ellipses of classes ce4 and ce5 in the plane 1-2 appear to be quite near to each other, which invites grouping these two classes together. The partition into 5 classes (c1, c2, c3, c4, c5) thus obtained has a between-variance on the 4 first axes equal to 0.3863, which is greater than the variance of the partition into 5 classes of the hierarchical classification (= 0.3676). The between–variance and the within–variance of this partition into 5 classes on the two first axes are given hereafter.

The between–variance is greater than the within–variance. We will summarize the data with this partition into 5 classes characterized as follows.

	Axis 1	Axis2
Between–Variance	.1964	.1277
Within–Variance	.1097	.0907

	frequencies	*Error rates*	*Latencies*	*Exercises*
c1	111		short	small except in Geometry
c2	25	high		rather large
c3	142	high		
c4	178	low		rather small
c5	77		long	rather small

There are two compact classes of well–performing students; one class is close to point A (class *c*1) with short latencies and medium error rates, and the other one is close to point B (class c4) with rather low error rates and medium to long latencies. Class c4 includes students with low error rates, especially in Geometry (46/55) and in Logic (28/31).

Conclusions

Starting from the three types of variables (*Error rates*, *Latencies*, *Number of exercises*) and from the five strands (Integers, Fractions, Geometry, Logic, Measurement), we have used MCA to construct a geometric space of individual differences. The geometric analysis shows a good homogeneity of strands for each type of variable. It also shows that individual differences are articulated around two scales: one of *Error rates* and *Number of exercises* and one of *Latencies*. The *Error rate* scale is clear–cut, showing strong homogeneity; the one of *Latencies* is less sharp, as some subjects have both short and long latencies across strands.

The individual points are scattered within a quadrilateral ABCD: When going down along the AB direction, latencies increase, while error rates remain constant; when going down along the AD direction, error rates increase, while latencies remain constant. The scattering of points within the quadrilateral is not uniform, showing a low density along side CD and a high density along AB. A Euclidean classification of individuals has been performed leading to a five–class partition. There are two compact classes of well–performing students; one class is close to point A (class *c*1) with short latencies and medium error rates, and the other one is close to point B (class c4) with rather low error rates and medium to long latencies (a profile little encouraged by current standards of educational testing).

Once the geometric space is constructed, one can study in detail the cloud of individuals by means of structuring factors. In the present study, we have investigated Age and Gender. There is an effect of Age: When age increases, error rates increase and latencies slightly decrease. There is virtually no difference between boys and girls.

One could enrich these analyses by introducing other structuring factors, as for instance the scores on final tests.

9.4 About Software

To carry out the case studies, we have used SPSS for the descriptive elementary statistics and data codings, ADDAD for PCA and MCA. Starting from principal coordinates calculated with ADDAD, the exploration of clouds and the double breakings of variance have been made with EyeLID.

The EyeLID program, developed by Bernard, Baldy and Rouanet (1988), is a free DOS package that combines two original features: a *Language for Interrogating Data* ("LID"), which designates relevant data sets in terms of structuring factors and constitutes a command language for derivations, and the *Visualization* ("Eye") of the clouds designated by EyeLID requests[38].

In addition, concentration ellipses have been determined by the `ellipse` program, which prepares a request file for the drawings done by the "freeware" WGNUPLOT.

Univariate combinatorial inferences have been made with STATXACT package. Bayesian inferences have been made with "Le Bayésien" package developed by Lecoutre & Poitevineau[39].

How to perform GDA procedures

• To perform a GDA method — CA, PCA or MCA — some users may prefer to program the basic algorithms directly from the computation procedures of the book, resorting to a library program for diagonalizing a symmetric matrix or performing a SVD. The contributions of points to axes are obtained from the coordinates and weights of points, as well as the qualities of representations (which involve the distances of points from the mean point of cloud).

• For PCA and MCA, the packages developed in France — like ADDAD or SPAD — provide the principal coordinates of both variables or modalities and individuals; they give the relative contributions, that is, the contributions to axes (Ctr) and the qualities of representation; they also give the principal coordinates of supplementary elements. They give the appropriate output for Euclidean classification.

• Most standard statistical packages, in addition to conventional PCA, have now incorporated the basic procedures of CA, and even MCA, providing at

[38]For illustrations of the command language, see Bernard, Le Roux, Rouanet, Schiltz (1989), and Bonnet, Le Roux & Lemaine (1996).

[39]Information about the software used in this book will be found on Brigitte Le Roux's homepage: "http://math-info.univ-paris5.fr/~lerb" under the "Logiciels" heading.

least eigenvalues and relevant coordinates. Checks may be needed to recover the proper principal coordinates for genuine geometric representations.

For MCA, some packages give "object scores" that are *not* the coordinates of points of principal axes (i.e. calibrated principal variables) but standardized principal variables. To recover the proper coordinates and proceed to genuine geometric representations of the cloud of individuals, one has to multiply these scores by the square roots of eigenvalues.

Similarly, from the output of a conventional PCA package, one may have to recover the principal coordinates of individuals, so as to ensure the invariance of the distance between individuals defined by the formula (4.1), (p.131) or the reconstitution formula of distances (formula (4.7), p.136).

For *classification*, choosing Ward's criterion, one has to manage to construct the superior hierarchichal tree corresponding to the partition retained for interpretation.

Chapter 10

Mathematical Bases

(*'MATH.BASES'*)

The importance of invariance properties in geometric settings cannot be overemphasized.
P. Suppes

Introduction

In the present chapter, we review the background necessary to render the book self–contained mathematically. In accordance with the formal–geometric approach of GDA (see chapter *OVERVIEW*), the central place is granted to abstract linear algebra; let us briefly comment on this point.

The clouds of points produced by GDA procedures are not ready–made geometric objects; they are constructed from numerical tables[1]. The construction is an elaborate process in two phases: i) Formalization, to express the basic statistical objects in mathematical terms; ii) Application of mathematical theory, to characterize statistical procedures and conduct relevant proofs. To account for this process, abstract linear algebra is ideally suited, because it allows for clearly distinguishing *vectors* (elements of a vector space), *points* (elements of a geometric space) and *sets of numbers* (e.g. coordinates), and describing the interplay between these entities.

The current matrix approach to statistics, which exclusively relies on matrix manipulations, cannot account for this process, as discussed below (p.449). This being said, *matrix formulas* are useful as a *shorthand notation*, as done in the book (and in the present chapter in §10.2 and §10.3).

[1]This situation is seen to differ from that of geometry in mathematics, where the objects can be studied directly, taking an arbitrary orthonormal reference basis.

The chapter is organized as follows. We firstly recall (with the short-hand idea in mind) the basic matrix operations (§10.1). Then we review finite–dimensional vector spaces (§10.2), Euclidean properties (§10.3), multidimensional geometry (§10.4), and spectral decomposition (§10.5).

Remarks: The notation used in the present chapter is distinct from that of the remainder of the book. The review of this chapter, where most proofs are sketched or simply omitted, is no substitute for the study of textbooks (cf. references p.422-423).

10.1 Matrix Operations

Rows×Columns tables of real numbers are called *matrices* if they obey the rules of matrix calculus. A matrix of *terms* or *cells* $[a_{ij}]$ (where $a_{ij} \in \mathbb{R}$), with rows indexed by a set I and columns indexed by a set J, is called a $I \times J$ matrix, and written as a bold uppercase letter $\mathbf{A} = [a_{ij}]$.

10.1.1 Basic Operations

Transposition. Interchanging rows and colums is denoted by the prime symbol ($'$). The *transpose* of the $I \times J$ matrix $\mathbf{A} = [a_{ij}]$ is the $J \times I$ matrix $\mathbf{A}' = [a_{ji}]$.

Addition. The sum (denoted $+$) of two matrices $\mathbf{A}$ and $\mathbf{B}$ (both $I \times J$) is the matrix $\mathbf{A}+\mathbf{B} = [a_{ij} + b_{ij}]$. One has the *properties*:
$\mathbf{A}+\mathbf{B}=\mathbf{B}+\mathbf{A}$; $(\mathbf{A}+\mathbf{B})+\mathbf{C}=\mathbf{A}+(\mathbf{B}+\mathbf{C})$; $(\mathbf{A}+\mathbf{B})'=\mathbf{A}'+\mathbf{B}'$.
The *null matrix* (with all terms equal to 0), denoted $\mathbf{0}$, verifies the property
$\forall \mathbf{A} : \mathbf{A}+\mathbf{0}=\mathbf{0}+\mathbf{A}=\mathbf{A}$.

Product. The *product* of $\mathbf{A} = [a_{ij}]$ by $\mathbf{B} = [b_{jk}]$ (the same set J indexing the columns of $\mathbf{A}$ and the rows of $\mathbf{B}$) is defined as the $I \times K$ matrix $\mathbf{C} = \mathbf{AB}$, such that $\mathbf{C} = [c_{ik}]$ with $c_{ik} = \sum_j a_{ij} b_{jk}$. One has the *properties*:
$\mathbf{A}(\mathbf{B}+\mathbf{C}) = \mathbf{AB}+\mathbf{AC}$; $(\mathbf{B}+\mathbf{C})\mathbf{D} = \mathbf{BD}+\mathbf{CD}$; $(\mathbf{AB})' = \mathbf{B}'\mathbf{A}'$.

The product of $\mathbf{A} = [a_{ij}]$ by a scalar λ is $\lambda\mathbf{A} = [\lambda a_{ij}]$.

10.1.2 Row and Column Matrices

A $I \times 1$ matrix, or *I–column matrix*, is denoted by a bold lowercase letter $\mathbf{x} = [x_i]$. A $1 \times J$ matrix or J–row is also denoted by a bold lowercase letter, and often written as the transpose $\mathbf{y}'$ of a J–column $\mathbf{y}$. A 1×1 matrix (number) is written as a italic lowercase letter. The family $(1)_{i \in I}$ is denoted as the I–column $\mathbf{e}$.

Properties. If $\mathbf{A}$ is a $I \times J$ matrix and $\mathbf{B}$ a $J \times I$ matrix, both $\mathbf{AB}$ and $\mathbf{BA}$ exist, but in general $\mathbf{AB} \neq \mathbf{BA}$. For example, if $\mathbf{x}$ and $\mathbf{y}$ are I–columns, $\mathbf{x'y} = \mathbf{y'x} = \sum_i x_i y_i$ is a number, whereas $\mathbf{xy'} = [x_i y_{i'}]$ and $\mathbf{yx'} = [y_i x_{i'}]$ are $I \times I$ matrices.
Observe that one can have $\mathbf{AB} = \mathbf{0}$ with $\mathbf{A} \neq \mathbf{0}$ and $\mathbf{B} \neq \mathbf{0}$.
If $\mathbf{C} = \mathbf{AB}$, then $c_{ik} = \mathbf{a}'_i \mathbf{b}_k$ (product of row $\mathbf{a}'_i$ of $\mathbf{A}$ by column $\mathbf{b}_k$ of $\mathbf{B}$), and $\mathbf{C} = \sum_j \mathbf{a}_j \mathbf{b}'_j$ (product of column $\mathbf{a}_j$ of $\mathbf{A}$ by row $\mathbf{b}'_j$ of $\mathbf{B}$).

10.1.3 Square Matrices

A matrix is square if its rows and columns are indexed by a same set I. The cells $(c_{ii})_{i \in I}$ of the square matrix $\mathbf{C}$ are called its *diagonal terms.*

Symmetric matrix. A square matrice $\mathbf{S} = [s_{ii'}]$ is *symmetric* if $\mathbf{S}' = \mathbf{S}$ ($s_{ii'} = s_{i'i}$). *Property*: $\forall \mathbf{A} = [a_{ij}]$: $\mathbf{AA}'$ is $I \times I$ symmetric and $\mathbf{A}'\mathbf{A}$ is $J \times J$ symmetric.

Diagonal matrix. A square matrix $\mathbf{D}$ is *diagonal* if its off–diagonal terms are zero ($d_{ii'} = 0$ if $i \neq i'$); the diagonal terms of $\mathbf{D}$ will be written $(d_i)_{i \in I}$. A diagonal matrix with equal diagonal terms is called a *scalar matrix.* The scalar matrix with all diagonal terms equal to 1 is called the *identity matrix*, and denoted $\mathbf{I}$, or $\mathbf{I}_I$ ($I \times I$ identity matrix), with $\forall \mathbf{C} = [c_{ii'}]$: $\mathbf{IC} = \mathbf{CI} = \mathbf{C}$.
Premultiplication and postmultiplication by a diagonal matrix. Premultiplying the $I \times J$ matrix $\mathbf{A} = [a_{ij}]$ by the diagonal matrix $\mathbf{D} = [d_i]$ multiplies the ith row by d_i, hence the matrix $[d_i a_{ij}]$. Postmultiplying by the diagonal $\mathbf{D} = [d_j]$ multiplies the jth column by d_j, hence the matrix $[a_{ij} d_j]$.

Trace. The *trace* of a square matrix $\mathbf{C}$ is the number $\sum_i c_{ii}$ (sum of diagonal terms), and denoted $\operatorname{tr} \mathbf{C}$.
Property: If $\mathbf{AB}$ and $\mathbf{BA}$ are both square matrices, one has $\operatorname{tr} \mathbf{AB} = \operatorname{tr} \mathbf{BA}$. In particular, if $\mathbf{x}$ and $\mathbf{y}$ are I–columns, on has $\operatorname{tr} \mathbf{xy'} = \operatorname{tr} \mathbf{y'x} = \operatorname{tr} \mathbf{x'y} = \sum_i x_i y_i$. Observe that $\operatorname{tr} \mathbf{AB} \neq \operatorname{tr} \mathbf{A} \times \operatorname{tr} \mathbf{B}$.

Determinant. Associated with every square matrix is a unique scalar number called its *determinant.* We will denote the determinant of $\mathbf{C}$ by $\det \mathbf{C}$. The determinant of the $I \times I$ matrix $\mathbf{C} = [c_{ii'}]$ is the sum

$$\sum_{\sigma \in S_I} (-1)^s \prod_{i \in I} c_{\sigma(i)\, i}$$

The sum is taken over the set $S(I)$ of all $I!$ permutations of the elements of I; a permutation is denoted $\left(\sigma(i)\right)_{i \in I})$ and s denotes the number of inversions in the permutation that is the total number of times in which an element is followed by elements which precede it in the initial order.

By definition, $\mathbf{C}$ is *singular* if $\det \mathbf{C} = 0$.
Properties: $\det \mathbf{C}' = \det \mathbf{C}$; $\det(\mathbf{AB}) = \det \mathbf{A} \det \mathbf{B}$; $\mathbf{AB} = \mathbf{AC}$ and $\mathbf{A}$ nonsingular $\Rightarrow \mathbf{B} = \mathbf{C}$.

The determinant of a diagonal matrix is the product of its diagonal elements: $\det \mathbf{D} = \sum_i d_i$ (with $\mathbf{D} = [d_i]$).

Inverse. The *inverse* of a nonsingular square matrix $\mathbf{C}$ is the unique square matrix denoted $\mathbf{C}^{-1}$ such that $\mathbf{CC}^{-1} = \mathbf{C}^{-1}\mathbf{C} = \mathbf{I}$.
Properties: $(\mathbf{C}')^{-1} = (\mathbf{C}^{-1})'$; $\det \mathbf{C}^{-1} = (\det \mathbf{C})^{-1}$; $(\mathbf{AB})^{-1} = \mathbf{B}^{-1}\mathbf{A}^{-1}$.

Idempotent matrix. A square matrix $\mathbf{A}$ is called idempotent if $\mathbf{A}^2 = \mathbf{A}$ (whereas $\mathbf{A}^2$ denotes the product $\mathbf{AA}$), or equivalently $\mathbf{A}(\mathbf{I} - \mathbf{A}) = \mathbf{0}$.

10.1.4 Rank of a Matrix

A product of a I–column by a J–row is a $I \times J$ matrix of rank 1. In a matrix of rank 1, all rows are proportional and all columns are proportional.

Property & Definition 10.1. *Any $I \times J$ matrix can be expressed as a sum of matrices of rank 1; by definition the rank L of the matrix is the minimum number of matrices of rank 1 necessary to express the matrix.*

Thus if the rank of $\mathbf{M}$ is equal to L, $\mathbf{M}$ can be expressed as the sum of L matrices of rank 1: $\mathbf{M} = \sum_\ell \mathbf{a}_\ell \mathbf{b}'_\ell$. Equivalently $\mathbf{M}$ is a product of a $I \times L$ matrix $\mathbf{A}$ by a $L \times J$ matrix $\mathbf{B}$ with $\mathbf{M} = \mathbf{AB}$.

10.2 Finite–dimensional Vector Space

The concept of *vectors* (elements of a vector space) applies to entities which may not be reducible to numbers; for example, forces in mechanics are formalized as vectors, in so far as they add up vectorially ("the parallelogram of forces") and are multiplied by numbers ("scalars") to yield other forces. Abstract linear algebra, in short *linear algebra*, is the general theory of *vector spaces*[2]: in this book, *finite–dimensional* vector spaces over $\mathbb{R}$ (the field of real numbers).

Abstract linear algebra involves two sorts of notions and properties: *purely vectorial* ones, and *Euclidean* ones, that are well covered in mathematical textbooks, such as Halmos (1958), Queysanne (1964), Godement

[2] "Abstract" in this context means "general", and is not to be opposed to "concrete", but to "numerical".

(1966), MacLane & Birkhoff (1967). In this section, we recall the formal definition of a vector space and briefly review the purely vectorial notions, then we recall the basic Euclidean notions and properties.

10.2.1 Vector Space

Definition 10.1 (Vector space). *A triple* $(\mathcal{A}, +, .)$ *is a vector space over* $\mathbb{R}$ *(scalars) if* $\mathcal{A}$ *is a nonempty set , + (vector addition) is a binary operation on* $\mathcal{A}$*, and . (multiplication by scalars, here elements of* $\mathbb{R}$*) is a mapping* $\mathbb{R} \times \mathcal{A} \to \mathcal{A}$*, such that for the element* $0 \in \mathcal{A}$ *(null vector),* $\forall \alpha, \beta, \gamma \in \mathcal{A}$ *and* $\forall a, b \in \mathbb{R}$*, the following axioms hold:*

1. *Vector addition. (i)* $(\alpha + \beta) + \gamma = \alpha + (\beta + \gamma)$*; (ii)* $\alpha + \beta = \beta + \alpha$*; (iii)* $0 + \alpha = \alpha$*; (iv) There exists* $-\alpha \in \mathcal{A}$ *such that* $\alpha + (-\alpha) = 0$.

2. *Multiplication by scalars. (i)* $a.(\alpha + \beta) = a.\alpha + a.\beta$*; (ii)* $(a + b).\alpha = a.\alpha + b.\alpha$*; (iii)* $(ab).\alpha = a.(b.\alpha)$*; (iv)* $1.\alpha = \alpha$.

For vector spaces in general, we will denote vectors by Greek alphabet letters (α, β ...) and the null vector by 0; for the multiplication by scalars we omit the point.

10.2.2 Linear Mapping and Dual Space

First notions. Given a subset $\alpha, \beta, \gamma \ldots$ of vectors in a vector space $\mathcal{A}$, and a, b, $c \ldots$ scalars, the set of all *linear combinations* $a\alpha + b\beta + c\gamma \ldots$ is a vector subspace $\mathcal{L} \subseteq \mathcal{A}$ called the *vector subspace generated* (or spanned) by $\alpha, \beta, \gamma \ldots$. If none of the vectors $\alpha, \beta, \gamma \ldots$ is a linear combination of the other ones, these vectors are said to be *linearly independent.*

A family of vectors generating $\mathcal{L}$ is a generating subset of $\mathcal{L}$; a family of linearly independent vectors generating $\mathcal{L}$ is a *basis* of $\mathcal{L}$. All bases of a vector space have the same number of vectors. If this number is p ($p < \infty$), the vector space is said to be $p-$ *dimensional* (finite–dimensional vector space).

> *In the remainder of the chapter, "vector space" will always mean a finite–dimensional vector space over* $\mathbb{R}$.

Definition 10.2 (Linear mapping, or Homomorphism). *If* $\mathcal{A}$ *and* $\mathcal{B}$ *are vector spaces over* $\mathbb{R}$*, a mapping* $f : \mathcal{A} \to \mathcal{B}$ *is a linear mapping, or homomorphism, if it satifies the linearity property* $f(a\alpha + b\beta) = af(\alpha) + bf(\beta)$*. An endomorphism is a homomorphism from* $\mathcal{A}$ *to* $\mathcal{A}$.

Property 10.1. *A linear mapping $f : \mathcal{A} \to \mathcal{B}$ is*
(i) injective if and only if $\forall \alpha \in \mathcal{A} : f(\alpha) = 0 \Rightarrow \alpha = 0$;
(ii) surjective if and only if $f(\mathcal{A}) = \mathcal{B}$;
(iii) An isomorphism is a one–one (bijective, that is injective and surjective) homomorphism;
(iv) With a isomorphism $h : \mathcal{A} \to \mathcal{B}$ there is associated the inverse isomorphism $h^{-1} : \mathcal{B} \to \mathcal{A}$ such that $h^{-1}(h(\alpha)) = \alpha$ and $h(h^{-1}(\beta)) = \beta$.

Composition of mappings. The composed mapping of two mappings $f : \mathcal{A} \to \mathcal{B}$ and $g : \mathcal{B} \to \mathcal{C}$ is denoted by $g \circ f : \mathcal{A} \to \mathcal{C}$ and is such that $\forall \alpha \in \mathcal{A} : g \circ f(\alpha) = g\big(f(\alpha)\big) \in \mathcal{C}$.

Definition 10.3 (Projection). *If $\mathcal{A}$ is the direct sum of two subspaces $\mathcal{L}$ and $\mathcal{L}'$, so that every $\alpha \in \mathcal{A}$ can be written uniquely in the form $\alpha = \gamma + \gamma'$ with $\gamma \in \mathcal{L}$ and $\gamma' \in \mathcal{L}'$, the projection on $\mathcal{L}$ along $\mathcal{L}'$ is the linear mapping p defined by $p(\alpha) = \gamma$.*

Property 10.2. *An endomorphism $p : \mathcal{A} \to \mathcal{A}$ is a projection if and only if it is idempotent, that is, $p \circ p = p$.*

With every projection $p : \mathcal{A} \to \mathcal{A}$ onto $\mathcal{L}$ there is associated a surjective mapping $p_L : \mathcal{A} \to \mathcal{L}$ such that $p = p_L \circ i$, where i is the injection $\mathcal{L} \to \mathcal{A}$ $(\forall \gamma \in \mathcal{L} : i(\gamma) = \gamma \in \mathcal{A})$.

Definition 10.4 (Rank of a linear mapping). *The rank of a linear mapping f is the dimensionality of the image $f(\mathcal{A}) \subseteq \mathcal{B}$.*

Theorem 10.1 (Rank theorem). *Given a linear mapping $f : \mathcal{A} \to \mathcal{B}$, and $\mathcal{L} \subseteq \mathcal{A}$ a supplementary subspace of subspace $\mathcal{L}_0$ generated by the vectors of $\mathcal{A}$ verifying $\forall \alpha \in \mathcal{A} : f(\alpha) = 0$ (kernel of f), the mapping f induces an isomorphism $b : \mathcal{L} \to f(\mathcal{A}) \subseteq \mathcal{B}$.*

Property 10.3 (Fundamental decomposition of a mapping). *Given a linear mapping $f : \mathcal{A} \to \mathcal{B}$, let $\mathcal{L}$ be a supplementary subspace of the subspace $\mathcal{L}_0$ (kernel of f); if $p_L : \mathcal{A} \to \mathcal{L}$ denotes the projection onto $\mathcal{L}$ and b the bijection $\mathcal{L} \to f(\mathcal{A})$, then one has $f = i \circ b \circ p_L$.*

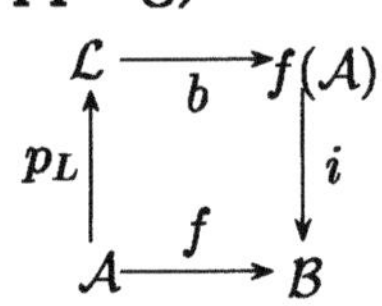

Definition 10.5 (Linear form). *A linear form on $\mathcal{A}$ is a linear mapping from $\mathcal{A}$ to $\mathbb{R}$.*

Definition 10.6 (Bilinear form). *A bilinear form on $\mathcal{A}\times\mathcal{B}$ is a function $\varphi : \mathcal{A}\times\mathcal{B} \to \mathrm{I\!R}$ with the property that for each fixed value of either argument they depend linearly of the other argument.*

More precisely, $\forall\alpha \in \mathcal{A}$ and $\forall\beta \in \mathcal{B}$, consider the mappings $\varphi_B : \mathcal{A} \to \mathrm{I\!R}$ defined by $\varphi_\beta(\alpha) = \varphi(\alpha,\beta)$ and $\varphi_A : \mathcal{B} \to \mathrm{I\!R}$ defined by $\varphi_\alpha(\beta) = \varphi(\alpha,\beta)$, then both φ_α and φ_β are linear forms.

Property 10.4 (Symmetric bilinear form). *A bilinear form σ on $\mathcal{A}\times\mathcal{A}$ is symmetric if $\forall(\alpha,\alpha') \in \mathcal{A}\times\mathcal{A} : \sigma(\alpha,\alpha') = \sigma(\alpha',\alpha)$.*

Definition 10.7 (Quadratic form). *With the symmetric bilinear form $\sigma : \mathcal{A}\times\mathcal{A} \to \mathrm{I\!R}$ there is associated the quadratic form $\widetilde{\sigma}$ on $\mathcal{A}$ defined by $\widetilde{\sigma}(\alpha) = \sigma(\alpha,\alpha)$.*

The symmetric bilinear form σ associated with the quadratic form $\widetilde{\sigma}$ on $\mathcal{A}$ is defined by $\forall(\alpha,\alpha') \in \mathcal{A}\times\mathcal{A} : \sigma(\alpha,\alpha') = \big(\widetilde{\sigma}(\alpha+\alpha') - \widetilde{\sigma}(\alpha) - \widetilde{\sigma}(\alpha')\big)/2$.

In the sequel, a quadratic form will be denoted in the same way as the corresponding bilinear form.

Definition 10.8 (Dual space). *The set of all linear forms on $\mathcal{A}$ generates a vector space $\mathcal{A}^*$ called the dual space of $\mathcal{A}$.*

If $(\alpha_i)_{i\in I}$ is a basis of $\mathcal{A}$, we denote α_i^* the linear form on $\mathcal{A}$ defined by $\alpha_i^*(\alpha_i) = 1$ and $\alpha_i^*(\alpha_{i'}) = 0$ if $i \neq i'$, then the family $(\alpha_i^*)_{i\in I}$ is a basis of $\mathcal{A}^*$ called the *dual basis* of $(\alpha_i)_{i\in I}$. Therefore we have $\dim\mathcal{A}^* = \dim\mathcal{A}$. However, between $\mathcal{A}$ and $\mathcal{A}^*$ there is no canonical (i.e. independent from bases) isomorphism.

Remark. The relations which define the linear forms α_i^* can also be written $\alpha_i^*(\alpha_{i'}) = \delta_{ii'}$ with $\delta_{ii'} = 1$ if $i = i'$ and $\delta_{ii'} = 0$ if $i \neq i'$ (*Kronecker symbol*).

Theorem 10.2 (Bidual space). *Let $\mathcal{A}$ be a finite–dimensional vector space, then, with every linear form α^{**} on $\mathcal{A}^*$ there is associated a vector $\alpha \in \mathcal{A}$ such that $\alpha^{**}(\alpha^*) = [\alpha|\alpha^*] = \alpha^*(\alpha)$ for every $\alpha^* \in \mathcal{A}^*$; the mapping $\alpha^{**} \to \alpha$ betwen $\mathcal{A}$ and $\mathcal{A}^{**}$ is an isomorphism.*

For finite–dimensional space we identify $\mathcal{A}^{**}$ with $\mathcal{A}$ by the natural isomorphism.

Definition 10.9 (Canonical bilinear form). *The bilinear form on $\mathcal{A}\times\mathcal{A}^*$ defined by $(\alpha,\gamma^*) \mapsto \gamma^*(\alpha) = [\alpha|\gamma^*]$ does not depend on bases, and therefore is called the canonical bilinear form on $\mathcal{A}\times\mathcal{A}^*$.*

Definition 10.10 (Transposed homomorphism). *If $f : \mathcal{A} \to \mathcal{B}$ is a homomorphism, the transposed (or dual) homomorphism ${}^t f : \mathcal{B}^* \to \mathcal{A}^*$ is defined by ${}^t f(\beta^*) = \beta^* \circ f$.*

Property 10.5. $\forall \alpha \in \mathcal{A} : [f(\alpha)|\beta^*]_B = [\alpha|{}^t f(\beta^*)]_A$

Remark. The concepts around duality, which are admittedly rather "abstract" in their general formulation, are easily handled for vector spaces equipped with intrinsic bases and a preferred isomorphism between dual spaces (see p.431); as is the case for CA and PCA (cf. chapters 2 and 4).

10.2.3 Coordinates and Change of Basis

Numerical vector spaces

$\mathbb{R}^n$ as a vector space. If n is an integer with $n \geq 1$, the set $\mathbb{R}^n$ of the numerical n–tuples $(x_i)_{i=1,\dots n}$ is naturally structured as a n–dimensional vector space with the canonical basis $(\delta_i)_{i=1,\dots n}$ where $\delta_i = (\delta_{i'i})_{i'=1\dots n}$ with $\delta_{ii} = 1$ and $\delta_{ii'} = 0$ for $i \neq i'$.

$\mathbb{R}^I$ as a vector space. If I is a nonempty finite set, the set of mappings $I \to \mathbb{R}$, denoted $\mathbb{R}^I$, is structured as a I–dimensional vector space, with the canonical basis $(\delta_i)_{i\in I}$, that is, if $(x_i)_{i\in I} = x_I \in \mathbb{R}^I$, one has[3] $x_I = \sum_i x_i \delta_i$.

Coordinates. Let $(\alpha_i)_{i\in I}$ be a basis of vector space $\mathcal{A}$; then any vector $\gamma \in \mathcal{A}$ can be uniquely expressed by its *coordinates* $(x_i)_{i\in I}$ with respect to this basis, as $\gamma = \sum_i x_i \alpha_i$. That is, with basis $(\alpha_i)_{i\in I}$ there are associated the two isomorphisms c (vector $\mapsto$ coordinates) and c^{-1} (coordinates $\mapsto$ vector).

$$\begin{array}{rccc} c: & \mathcal{A} & \to & \mathbb{R}^I \\ & \alpha_i & \mapsto & \delta_i \\ & \gamma & \mapsto & x_I = (x_i)_{i\in I} \end{array} \qquad \begin{array}{rccc} c^{-1}: & \mathbb{R}^I & \to & \mathcal{A} \\ & \delta_i & \mapsto & \alpha_i \\ & x_I & \mapsto & \gamma = \sum_i x_i \alpha_i \end{array}$$

Using coordinates in vector operations enables one to "think in terms of vectors" while "working with numbers". In what follows, coordinates will be denoted by Latin alphabet letters $(x, y \dots)$.

If $(\alpha_i^*)_{i\in I}$ is the dual basis of $(\alpha_i)_{i\in I}$, every linear form γ^* can be written $\gamma^* = \sum_i y_i \alpha_i^*$. If $\alpha = \sum_i x_i \alpha_i$ is a vector of $\mathcal{A}$, one has $\gamma^*(\alpha) = \sum_i y_i \alpha_i^*(\gamma) = \sum_i y_i x_i = [\alpha|\gamma^*]$.

Change of basis. *Coordinates of vectors are basis–bound*, that is, if in a vector space the basis changes, the vectors remain the same whereas their coordinates change. Let $(\alpha_i)_{i\in I}$ be a basis of $\mathcal{A}$, and $\mathcal{L}$ be a subspace

[3]In this chapter, all indices are written as subscripts, without reference to the measure vs variable duality as elsewhere in the book.

of $\mathcal{A}$ (possibly $\mathcal{A}$ itself), with a basis $(\epsilon_\ell)_{\ell\in L}$, such that $\epsilon_\ell = \sum_i e_{i\ell}\,\alpha_i$. Every vector $\gamma \in \mathcal{L}$ can be referred either to $(\alpha_i)_{i\in I}$ ("old basis") as $\gamma = \sum_i x_i\,\alpha_i$, or to $(\epsilon_\ell)_{\ell\in L}$ ("new basis") as $\gamma = \sum_\ell y_\ell\,\epsilon_\ell$. One has therefore: $\gamma = \sum_\ell y_\ell \sum_i e_{i\ell}\alpha_i = \sum_i(\sum_\ell e_{i\ell}y_\ell)\alpha_i$. Hence the *passage formulas* $\forall i : x_i = \sum_\ell e_{i\ell}\,y_\ell$. Observe that while the new basis $(\epsilon_\ell)_{\ell\in L}$ is given as a function of the old one $(\alpha_i)_{i\in I}$, the passage formulas give the old coordinates $(x_i)_{i\in I}$ as a function of the new ones $(y_\ell)_{\ell\in L}$.

Linear mapping. Let $\mathcal{A}$ and $\mathcal{B}$ be two vector spaces with respective bases $(\alpha_i)_{i\in I}$ and $(\beta_j)_{j\in J}$ and a *linear mapping* $f : \mathcal{A} \to \mathcal{B}$. Let $f(\alpha_i) = \sum_j m_{ji}\,\beta_j$. If $\gamma = \sum_i x_i\,\alpha_i$, one has $f(\gamma) = \sum_j(\sum_i m_{ji}x_i)\beta_j$. Thus letting $f(\gamma) = \sum_j y_j\,\beta_j$, one has $\forall j : y_j = \sum_i m_{ji}\,x_i$.

A particular mapping is the *canonical injection* $i : \mathcal{L} \to \mathcal{A}$ defined by $\gamma \mapsto i(\gamma) = \gamma$ (When $\mathcal{L} = \mathcal{A}$, the injection i is the identity mapping $id : \mathcal{A} \to \mathcal{A}$). Referring $\mathcal{L}$ to $(\epsilon_\ell)_{\ell\in L}$ and $\mathcal{A}$ to $(\alpha_i)_{i\in I}$, if $\gamma = \sum_i x_i\,\alpha_i = \sum_\ell y_\ell\,\epsilon_\ell$, one obtains $x_i = \sum_\ell e_{i\ell}\,y_\ell$, i.e. the passage formula, establishing the link between the change of basis and the injective mapping.

Matrix notation, I

Once bases are selected, linear algebra operations can be expressed in matrix notation, writing coordinates as columns or rows. In what follows, referring $\mathcal{A}$ to basis $(\alpha_i)_{i\in I}$, the coordinates $x_I = (x_i)_{i\in I}$ of vector $\gamma = \sum_i x_i\,\alpha_i$ will be represented by the I–column $\mathbf{x}$. We now give the matrix expressions for a change of basis and for a linear mapping, always writing coordinate families as *columns*.

Change of basis. Let $\mathbf{E}$ be the $I \times L$ matrix whose ℓ–th column $\mathbf{e}_\ell$ is the family $(e_{i\ell})_{i\in I}$ of the coordinates of ϵ_ℓ in basis $(\alpha_i)_{i\in I}$. Let $\mathbf{x}$ be the I–column of the old coordinates of a vector in basis $(\alpha_i)_{i\in I}$, $\mathbf{y}$ be the L–column of its new coordinates in basis $(\epsilon_\ell)_{\ell\in L}$. In matrix terms, the passage formula $\forall i : x_i = \sum_\ell e_{i\ell}y_\ell$ reads: $\mathbf{x} = \mathbf{E}\mathbf{y}$.

Linear mapping. Let $\mathbf{M}$ be the $J \times I$ matrix whose i–th column m_{Ji} is the family of the coordinates of $f(\alpha_i)$ in basis $(\beta_j)_{j\in J}$. Let $\mathbf{x}$ denote the coordinates of vector γ in basis $(\alpha_i)_{i\in I}$,and $\mathbf{y}$ the coordinates of $f(\gamma)$ in basis $(\beta_j)_{j\in J}$. In matrix terms, with respect to the two bases, the mapping reads $\mathbf{y} = \mathbf{M}\mathbf{x}$. As a particular case, for the *canonical injection* $i : \mathcal{L} \to \mathcal{A}$, $\mathbf{E}$ is the $I \times L$ matrix representing mapping i in the bases $(\epsilon_\ell)_{\ell\in L}$ and $(\alpha_i)_{i\in I}$, and we get the change of basis formula $\mathbf{x} = \mathbf{E}\mathbf{y}$.

Transposed homomorphism. If the dual spaces $\mathcal{A}^*$ and $\mathcal{B}^*$ are referred to the dual bases $(\alpha_i^*)_{i\in I}$ and $(\beta_j^*)_{j\in J}$, the transposed homomorphism ${}^tf$:

$\mathcal{B}^* \to \mathcal{A}^*$ is associated with $\mathbf{M}'$, the *transposed matrix* of $\mathbf{M}$.

Composition of mappings. The *composition of mappings* is expressed by *matrix multiplication*, that is, if f, g, h are three linear mappings and $\mathbf{F}$, $\mathbf{G}$, $\mathbf{H}$ their associated matrices, then if $h = g \circ f$, we have $\mathbf{H} = \mathbf{GF}$.

Bilinear and quadratic forms. With respect to bases $(\alpha_i)_{i \in I}$ of $\mathcal{A}$ and $(\beta_j)_{j \in J}$ of $\mathcal{B}$, the matrix of the bilinear form φ on $\mathcal{A} \times \mathcal{B}$ is a $I \times J$ matrix $\mathbf{\Phi} = [\varphi_{ij}]$ with $\varphi_{ij} = \varphi(\alpha_i, \beta_j)$; the matrix of a symmetric bilinear form σ on $\mathcal{A} \times \mathcal{A}$ and of the associated quadratic form is the symmetric matrix $\mathbf{\Sigma} = [\sigma_{ii'}]$ with $\sigma_{ii'} = \sigma(\alpha_i, \alpha_{i'})$.

10.3 Euclidean Vector Space

The basic Euclidean notions in a vector space are those of scalar product, orthogonality, Euclidean norm, orthogonal projection, angles. They are more intuitive in character than the purely vectorial notions, owing to the geometric representations they afford (see hereafter).

10.3.1 Basic Notions

Definition 10.11 (Scalar product). *Let $\mathcal{A}$ be a finite-dimensional vector space over $\mathbb{R}$; a bilinear form $\langle \cdot | \cdot \rangle$ from $\mathcal{A} \times \mathcal{A}$ to $\mathbb{R}$ is a scalar product on $\mathcal{A}$ if it satisfies the following two properties for α, α' in $\mathcal{A}$ and $a \in \mathbb{R}$:*
(1) Definite positiveness. $\langle \alpha | \alpha \rangle \geq 0$ *with* $\alpha \neq 0 \iff \langle \alpha | \alpha \rangle > 0$
(2) Symmetry. $\langle \alpha | \alpha' \rangle = \langle \alpha' | \alpha \rangle$

A vector space with a scalar product is a *Euclidean vector space.*

Definition 10.12 (Orthogonality). *Two (non-null) vectors α and α' are called orthogonal if $\langle \alpha | \alpha' \rangle = 0$ (one writes $\alpha \perp \alpha'$).*

Definition 10.13 (Euclidean norm). *The Euclidean norm $\| \cdot \|$ on $\mathcal{A}$ is the positive function defined from the scalar product $\langle \cdot | \cdot \rangle$, by $\|\alpha\|^2 = \langle \alpha | \alpha \rangle$.*

The Euclidean norm satisfies the three properties of a norm:
i) *Definite positiveness.* $\|\alpha\| \geq 0$ with $\|\alpha\| = 0 \iff \alpha = 0$ (null vector)
ii) *Semi-linearity.* $\forall a \in \mathbb{R} : \|a\, \alpha\| = |a|\, \|\alpha\|$
iii) *Triangle inequality.* $\|\alpha + \alpha'\| \leq \|\alpha\| + \|\alpha'\|$

Property 10.6. *Schwarz inequality:* $|\langle \alpha | \alpha' \rangle| \leq \|\alpha\|\, \|\alpha'\|$ *with* $|\langle \alpha | \alpha' \rangle| = \|\alpha\| . \|\alpha'\| \iff$ *α and α' are proportional (or: $\alpha = 0$ or $\alpha' = 0$)*
Classical identities:
$\|\alpha + \alpha'\|^2 = \|\alpha\|^2 + \|\alpha'\|^2 + 2\langle \alpha | \alpha' \rangle \qquad \|\alpha - \alpha'\|^2 = \|\alpha\|^2 + \|\alpha'\|^2 - 2\langle \alpha | \alpha' \rangle$
Pythagoras' theorem: $\alpha \perp \alpha' \iff \|\alpha + \alpha'\|^2 = \|\alpha\|^2 + \|\alpha'\|^2$

10.3.2 Geometric Representations

Whenever in an n–dimensional vector space ($n \geq 2$), several vectors span a two–dimensional subspace, these vectors can be represented by segments in a plane geometric figure[4]. Thus two linearly independent vectors α and α' can be represented by two sides of a triangle whose lengths are $\|\alpha\|$ and $\|\alpha'\|$ and whose angle φ is given by

$$\cos\varphi = \frac{\langle\alpha|\alpha'\rangle}{\|\alpha\| \times \|\alpha'\|} \qquad \text{with} \quad 0 \leq \varphi \leq \pi$$

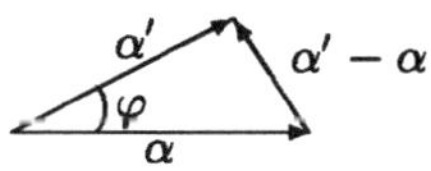

$\langle\alpha|\alpha'\rangle > 0 \quad 0 < \varphi < \pi/2$

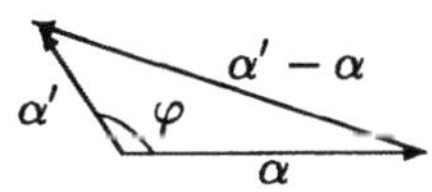

$\langle\alpha|\alpha'\rangle < 0 \quad pi/2 < \varphi < \pi$

In a similar way, vectors spanning a three–dimensional subspace can be represented by a three–dimensional figure. Thus three linearly independent vectors can be represented by three sides of a tetrahedron. A set of vectors spanning a p–dimensional subspace, with $p > 3$, will be represented by two or tree–dimensional figures.

10.3.3 Orthogonal Projection

Definition 10.14 (Orthogonal projection). *Let $\mathcal{L}$ be a L–dimensional subspace of $\mathcal{A}$ (with $L \geq 1$), the orthogonal projection of vector α onto $\mathcal{L}$ is the vector $\widetilde{\alpha} \in \mathcal{L}$ such that the vector $\alpha - \widetilde{\alpha}$ is orthogonal to every vector in $\mathcal{L}$.*

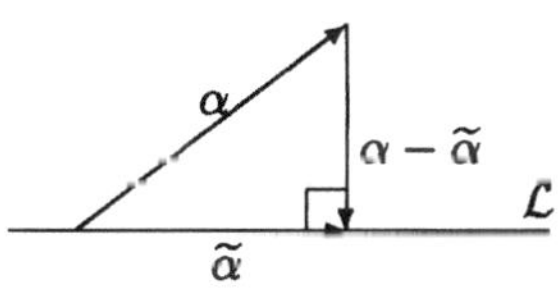

dim. $\mathcal{L} = 1$

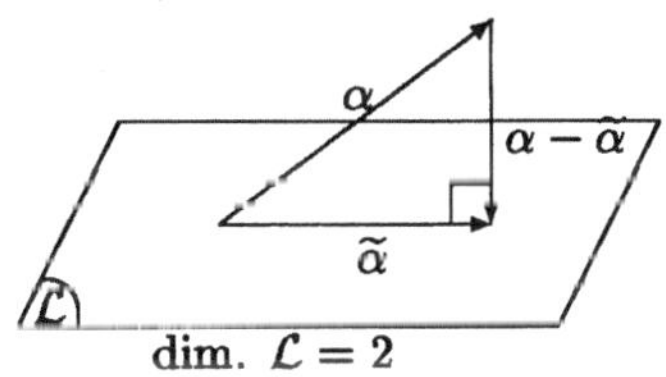

dim. $\mathcal{L} = 2$

If α is orthogonal to $\mathcal{L}$, projection $\widetilde{\alpha}$ is the null vector; if $\alpha \in \mathcal{L}$, $\widetilde{\alpha} = \alpha$.

Property 10.7 (Projection onto a vector). *The orthogonal projection of vector γ onto vector α ($\neq 0$) is the vector $\widetilde{\gamma}$ proportional to α such that $\gamma - \widetilde{\gamma}$ is orthogonal to α, that is $\widetilde{\gamma} = \left(\langle\alpha|\gamma\rangle/\|\alpha\|^2\right)\alpha$, with $\|\widetilde{\gamma}\| = |\langle\alpha|\gamma\rangle|/\|\alpha\|$.*

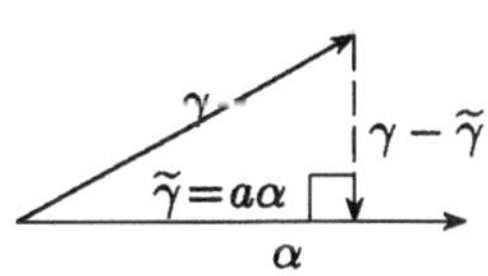

[4]Geometric representations follow from the canonical property of affine spaces stated below p.436.

Square of norm and scalar product: $\|\widetilde{\alpha}\|^2 = \langle\widetilde{\alpha}|\alpha\rangle$.
Proof. $\widetilde{\alpha} \perp \alpha - \widetilde{\alpha} \iff \langle\widetilde{\alpha}|\alpha - \widetilde{\alpha}\rangle = 0 \iff \langle\widetilde{\alpha}|\alpha\rangle = \langle\widetilde{\alpha}|\widetilde{\alpha}\rangle$. ◁

Positivity: the angle of a vector and its projection is acute (or right if $\alpha \perp \mathcal{A}$); $\cos(\alpha, \widetilde{\alpha}) = \dfrac{\|\widetilde{\alpha}\|^2}{\|\alpha\|.\|\widetilde{\alpha}\|} = \dfrac{\|\widetilde{\alpha}\|}{\|\alpha\|} \geq 0$, therefore $0 \leq \text{angle}(\alpha, \widetilde{\alpha}) \leq \dfrac{\pi}{2}$.

Contracting property. $\|\widetilde{\alpha}\| \leq \|\alpha\|$.
Proof. One has: $\langle\widetilde{\alpha}|\alpha\rangle \leq \|\widetilde{\alpha}\|.\|\alpha\|$, that is $\|\widetilde{\alpha}\|^2 \leq \|\widetilde{\alpha}\|.\|\alpha\|$; hence if $\widetilde{\alpha} \neq 0$, $\|\widetilde{\alpha}\| \leq \|\alpha\|$. If $\widetilde{\alpha} = 0$, we have $0 \leq \|\alpha\|$ (property of norm). The orthogonal projection is a contracting mapping. ◁

Metric characterization of orthogonal projection. Given $\alpha \in \mathcal{A}$ and a subspace $\mathcal{L} \subseteq \mathcal{A}$; among all vectors of subpace $\mathcal{L}$, the orthogonal projection $\widetilde{\alpha}$ of α onto $\mathcal{L}$, is the closest to α in the least square sense, that is, is such that $\|\alpha - \widetilde{\alpha}\|$ is minimal.

Theorem of three perpendiculars. Let $\mathcal{L}'$ be a subspace of $\mathcal{L}$; if $\widetilde{\alpha}$ is the orthogonal projection of α onto $\mathcal{L}$, and $\widetilde{\alpha}'$ the orthogonal projection of $\widetilde{\alpha}$ onto $\mathcal{L}'$, then $\widetilde{\alpha}'$ is the orthogonal projection of α onto $\mathcal{L}'$.

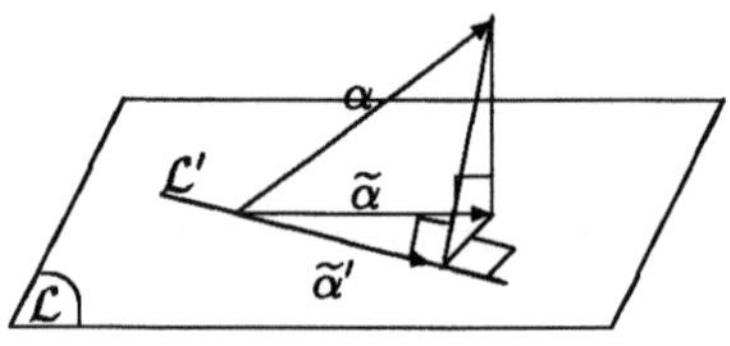

dim. $\mathcal{L} = 2$ and dim. $\mathcal{L}' = 1$

Invariance property of scalar product. Given α and γ two vectors of $\mathcal{A}$ and let $\mathcal{L}$ be a subspace containing α and let $\widetilde{\gamma}'$ be the orthogonal projection of γ onto α; then $\langle\gamma|\alpha\rangle = \langle\widetilde{\gamma}|\alpha\rangle$, that is the scalar product of two vectors is unchanged if one of them is replaced by its orthogonal projection onto a subspace containing the other one. In particular, the scalar product of two vectors is equal to the scalar product of one of the two vectors by the orthogonal projection onto it of the other one: $\langle\alpha|\gamma\rangle = \langle\widetilde{\alpha}|\gamma\rangle = \langle\alpha|\widetilde{\gamma}\rangle$.
Proof. $\langle\alpha|\gamma\rangle = \langle\alpha|\widetilde{\gamma} + (\gamma - \widetilde{\gamma})\rangle = \langle\alpha|\widetilde{\gamma}\rangle + \langle\alpha|\gamma - \widetilde{\gamma}\rangle$; if $\widetilde{\gamma}$ is the orthogonal projection of γ onto α, one has $\langle\alpha|\gamma - \widetilde{\gamma}\rangle = 0$, hence the property. ◁

Additive decomposition property. Let us project vector γ onto L supplementary subspaces $(\mathcal{A}_\ell)_{\ell \in L}$ (such that $\forall \ell \neq \ell' : \mathcal{A}_\ell \cap \mathcal{A}'_\ell = 0$ and $\cup_\ell \mathcal{A}_\ell = \mathcal{A}$), that is: $\gamma = \sum_\ell \gamma_\ell$. Let $\widetilde{\gamma}_\ell$ be the orthogonal projection of γ onto $\mathcal{A}_\ell$. Then one has $\|\gamma\|^2 = \sum_\ell \langle\gamma_\ell|\widetilde{\gamma}_\ell\rangle$.
Proof. $\langle\gamma|\gamma\rangle = \langle\sum_\ell \gamma_\ell|\gamma\rangle = \sum_\ell \langle\gamma_\ell|\gamma\rangle = \sum_\ell \langle\gamma_\ell|\widetilde{\gamma}_\ell\rangle$ (replacing γ by $\widetilde{\gamma}_\ell$ by invariance property). ◁

Observe that the additive decomposition of $\|\gamma\|^2$ may include negative components.

10.3.4 Euclidean Structure in Matrix Notation

Scalar product in matrix notation. Let $\mathbf{x}_1$ be the I–column of coordinates of γ_1 and $\mathbf{x}_2$ be the I–column of coordinates of γ_2 with respect to the basis $(\alpha_i)_{i\in I}$; letting $q_{ii'} = \langle\alpha_i|\alpha_{i'}\rangle$, one has $\langle\gamma_1|\gamma_2\rangle = \sum_i \sum_{i'} x_{i1}x_{i'2}q_{ii'}$. Letting $\mathbf{x}_1$ and $\mathbf{x}_2$ be the coordinates of γ_1 and γ_2 with respect to $(\delta_j)_{j\in J}$, one has $\langle\gamma_1|\gamma_2\rangle = \mathbf{x}_1'\mathbf{Q}\mathbf{x}_2 = \mathbf{x}_2'\mathbf{Q}\mathbf{x}_1$.

If a subspace $\mathcal{L} \subseteq \mathcal{A}$ is referred to a basis $(\epsilon_\ell)_{\ell\in L}$ with $\epsilon_\ell = \sum_i e_{i\ell}\alpha_i$, the scalar product is invariant whereas its matrix expression is not. Letting $\mathbf{y}_1$ and $\mathbf{y}_2$ be the coordinates of γ_1 and γ_2 with respect to the basis $(\epsilon_\ell)_{\ell\in L}$ of $\mathcal{L}$, one has $\langle\gamma_1|\gamma_2\rangle = \mathbf{y}_1'\mathbf{R}\mathbf{y}_2 = \mathbf{y}_2'\mathbf{R}\mathbf{y}_1$, letting $r_{\ell\ell'} = \langle\epsilon_\ell|\epsilon_{\ell'}\rangle$, with $\mathbf{R} = \mathbf{E}'\mathbf{Q}\mathbf{E}$.

Elementary Euclidean structure on $\mathbb{R}^n$. The vector space $\mathbb{R}^n$ can be structured with the *elementary scalar product*, for which the canonical basis is taken to be orthonormal, that is, $\|\delta_i\| = 1$ and $\langle\delta_i|\delta_{i'}\rangle = 0$ for $i \neq i'$.

Euclidean terminology in matrix calculus. The matrix product $\mathbf{x}'\mathbf{y} = \mathbf{y}'\mathbf{x}$ is often called the "scalar product" of column–vectors $\mathbf{x}$ and $\mathbf{y}$; the columns $\mathbf{x}$ and $\mathbf{y}$ are said to be "orthogonal" if $\mathbf{x}'\mathbf{y} = 0$; an $I \times L$ matrix $\mathbf{E}$ is called (column)–"orthogonal" if $\mathbf{E}'\mathbf{E} = \mathbf{I}_L$. Etc. This terminology makes sense in terms of the elementary scalar product on $\mathbb{R}^n$. It becomes confusing whenever non– orthonormal reference bases are involved (see p.449).

10.3.5 Dual Spaces and Scalar Products

Let $\langle\cdot|\cdot\rangle$ be a scalar product on $\mathcal{A}$. For each $\gamma_1 \in \mathcal{A}$, the mapping $\mathcal{A} \to \mathbb{R}$ such that $\gamma_1 \mapsto \langle\gamma_1|\cdot\rangle$ is a linear form on $\mathcal{A}$. Consequently, the scalar product $\langle\cdot|\cdot\rangle$ induces a unique isomorphism $q : \mathcal{A} \to \mathcal{A}^*$ such that $\gamma_1 \mapsto q(\gamma_1) = \langle\gamma_1|\cdot\rangle$, i.e. $\big(q(\gamma_1)\big)(\gamma_2) = \langle\gamma_1|\gamma_2\rangle = [q(\gamma_1)|\gamma_2]$. Isomorphism q inherits the positivity and symmetry properties of scalar product, that is

$$[q(\gamma_1)|\gamma_1] \geq 0 \quad (= 0 \iff \gamma_1 = 0)[q(\gamma_1)|\gamma_2] - [\gamma_1|q(\gamma_2)]$$

Conversely, let $q : \mathcal{A} \to \mathcal{A}^*$ $(\gamma_1 \mapsto q(\gamma_1))$ be a positive definite and symmetric isomorphism. Then:

• q induces a scalar product on $\mathcal{A}$ defined by

$$\langle\gamma_1|\gamma_2\rangle = [q(\gamma_1)|\gamma_2] = [\gamma_1|q(\gamma_2)]$$

• The inverse isomorphism $q^{-1} : \mathcal{A}^* \to \mathcal{A}$ induces a scalar product on $\mathcal{A}^*$ defined by

$$\langle\gamma_1^*|\gamma_2^*\rangle = [q^{-1}(\gamma_1^*)|\gamma_2^*] = [\gamma_1^*|q^{-1}(\gamma_2*)]$$

Matrix formulas

Property 10.8. *Given a basis $(\alpha_i)_{i\in I}$ of $\mathcal{A}$, let $\langle\alpha_i|\alpha_{i'}\rangle = q_{ii'}$ hence the positive definite matrix $\mathbf{Q} = [q_{ii'}]$. With respect to the bases $(\alpha_i)_{i\in I}$ and $(\alpha_i^*)_{i\in I}$, isomorphism $q : \mathcal{A} \to \mathcal{A}^*$ is expressed in matrix terms by $\mathbf{x}^* = \mathbf{Q}\mathbf{x}$.*

10.3.6 Adjoint Homomorphisms

Let two Euclidean vector spaces on $\mathbb{R}$, denoted $\mathcal{A}$ and $\mathcal{B}$, with scalar products $\langle\cdot|\cdot\rangle_A$ and $\langle\cdot|\cdot\rangle_B$, and Euclidean norms $||\cdot||_A$ and $||\cdot||_B$.

Theorem 10.3 (Adjoint of homomorphism). *If $f : \mathcal{A} \to \mathcal{B}$ is a homomorphism from $\mathcal{A}$ to $\mathcal{B}$, then there is a unique homomorphism $f^* : \mathcal{B} \to \mathcal{A}$ such that:*

$$\forall\alpha \in \mathcal{A} \quad \textit{and} \quad \forall\beta \in \mathcal{B} \ : \ \langle\alpha|f^*(\beta)\rangle_A = \langle f(\alpha)|\beta\rangle_B$$

The homomorphism f^* is called the *adjoint* of f. By the symmetry of scalar product, $f^{**} = f$. Summarizing:

$$\mathcal{A} \underset{f^*}{\overset{f}{\rightleftarrows}} \mathcal{B} \qquad \begin{array}{rcl} \alpha & \longmapsto & f(\alpha) \\ f^*(\beta) & \longleftarrow & \beta \\ \langle\alpha|f^*(\beta)\rangle_A & = & \langle f(\alpha)|\beta\rangle_B \end{array}$$

Let $q : \mathcal{A} \to \mathcal{A}^*$ be the isomorphism induced by $\langle\cdot|\cdot\rangle_A$ and $r : \mathcal{B} \to \mathcal{B}^*$ the isomorphism induced by $\langle\cdot|\cdot\rangle_B$. Then $f^* = q^{-1} \circ {}^t f \circ r$ is the adjoint of f (see opposite scheme).

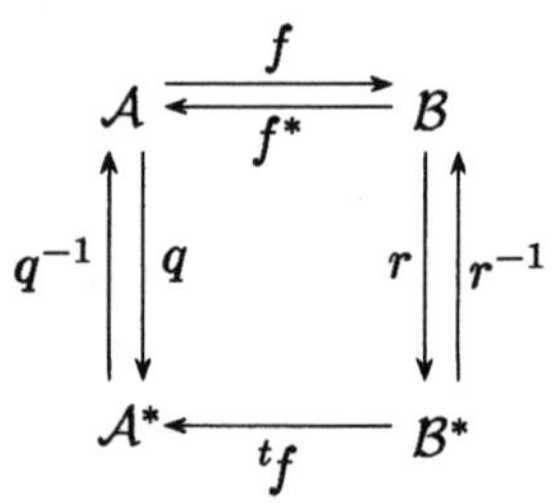

Adjoint of an endomorphism. As a particular case, let $h : \mathcal{A} \to \mathcal{A}$ (endomorphism on $\mathcal{A}$). Then the adjoint endomorphism is $h^* = q^{-1} \circ {}^t h \circ q$.

Property 10.9 (Bilinear form). *With the adjoint homomorphisms f and f^* there is associated a bilinear form $\varphi : \mathcal{A} \times \mathcal{B} \to \mathbb{R}$ such that*

$$\begin{array}{rccl} \varphi : & \mathcal{A} \times \mathcal{B} & \to & \mathbb{R} \\ & (\alpha, \beta) & \mapsto & \varphi(\alpha,\beta) = \langle\alpha|f^*(\beta)\rangle_A = \langle f(\alpha)|\beta\rangle_B \end{array}$$

The bilinear form $\varphi : \mathcal{A} \times \mathcal{B} \to \mathbb{R}$ associated with the adjoint homomorphisms f and f^* induces two composed linear mappings, that is, $r \circ f : \mathcal{A} \to \mathcal{B}^*$ and $q \circ f^* : \mathcal{B} \to \mathcal{A}^*$.

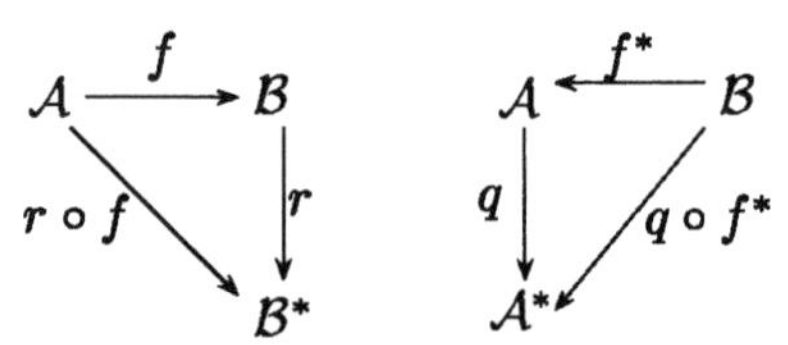

For $r \circ f$, we thus have: $\alpha \mapsto r(f(\alpha)) = \langle f(\alpha)|\cdot\rangle_B$.

Definition 10.15 (Symmetric endomorphism). *A linear mapping $h : \mathcal{A} \to \mathcal{A}$ (endomorphism on $\mathcal{A}$) is called self–adjoint, or symmetric, if $h^* = h$, that is if*

$$\forall \alpha \in \mathcal{A} \quad \forall \alpha' \in \mathcal{A} : \quad \langle h(\alpha)|\alpha'\rangle = \langle \alpha|h(\alpha')\rangle$$

Then by definition h is self–adjoint $\iff h^* = h \iff q^{-1} \circ {}^t h \circ q = h$ $\iff {}^t h \circ q = q \circ h$.

Definition 10.16 (Positive symmetric endomorphism). *A symmetric endomorphism h is called positive if $\forall \alpha \in \mathcal{A} : \langle h(\alpha)|\alpha\rangle \geq 0$, strictly positive if $\forall \alpha \in \mathcal{A}, \alpha \neq 0 : \langle h(\alpha)|\alpha\rangle > 0$.*

Property 10.10. *With two adjoint homomorphisms f and f^* there are associated the following two positive symmetric endomorphisms[5]:*

$$s = f^* \circ f : \mathcal{A} \to \mathcal{A} \quad \text{and} \quad t = f \circ f^* : \mathcal{B} \to \mathcal{B}$$

Property 10.11 (Bilinear and quadratic form). *With the symmetric endomorphism $s : \mathcal{A} \to \mathcal{A}$, there are associated (i) the symmetric bilinear form $\sigma : \mathcal{A} \times \mathcal{A} \to \mathbb{R}$ defined by $\forall \alpha \in \mathcal{A}$ and $\forall \alpha' \in \mathcal{A} : \sigma(\alpha, \alpha') = \langle \alpha|s(\alpha')\rangle_A = \langle s(\alpha)|\alpha'\rangle_A$; (ii) the quadratic form also denoted $\sigma : \mathcal{A} \to \mathbb{R}$ with $\forall \alpha : \sigma(\alpha) = \langle \alpha|s(\alpha)\rangle_A$.*

In particular, the quadratic form on $\mathcal{A}$ associated with $f^* \circ f$ induces the linear mapping ${}^t f \circ r \circ f : \mathcal{A} \to \mathcal{A}^*$.

Matrix formulas

Let $f : \mathcal{A} \to \mathcal{B}$; let $(\alpha)_{i \in I}$ be a basis of $\mathcal{A}$ and $(\beta)_{j \in J}$ a basis of $\mathcal{B}$. Let $\mathbf{Q} = [q_{ii'}]$ with $q_{ii'} = \langle \alpha_i|\alpha_{i'}\rangle_A$ and $\mathbf{R} = [r_{jj'}]$ with $r_{jj'} = \langle \beta_j|\beta_{j'}\rangle_B$.

Let $\mathbf{M} = [m_{ji}]$ be the matrix associated with f. Then the matrix associated with the *adjoint* f^* is $\mathbf{M}^* = \mathbf{Q}^{-1}\mathbf{M}'\mathbf{R}$ (using 10.3, p.432), hence $\mathbf{Q}\,\mathbf{M}^* = \mathbf{M}'\mathbf{R}$.

The matrix associated with the *bilinear form* $\varphi : \mathcal{A} \times \mathcal{B} \to \mathbb{R}$ is by definition the matrix associated with $r \circ f$, that is $\mathbf{\Phi} = \mathbf{RM}$ (the matrix associated with $q \circ f^*$ being $\mathbf{QM}^* = \mathbf{M}'\mathbf{R}$).

The matrix $\mathbf{H}$ associated with a *self–adjoint endomorphism* h verifies the property $\mathbf{Q}^{-1}\mathbf{H}'\mathbf{Q} = \mathbf{H}$, i.e. $\mathbf{H}'\mathbf{Q} = \mathbf{QH}$. The matrix of a self–adjoint ("symmetric") endomorphism is not symmetric (in general); it is symmetric when the basis of $\mathcal{A}$ is orthonormal, that is, one has $\mathbf{H}' = \mathbf{H}$ if $\mathbf{Q} = \mathbf{I}$ (or more generally if $\mathbf{Q}$ is proportional to $\mathbf{I}$).

[5]Conversely, if s is a positive symmetric endomorphism, there are two adjoint homomorphisms f and f^* such that $s = f^* \circ f$; and similarly for t.

The matrix associated with the *symmetric endomorphism* $s = f^* \circ f$ is $\mathbf{S} = \mathbf{Q}^{-1}\mathbf{M}'\mathbf{R}\mathbf{M}$ (not symmetric in general).

The matrix of the *quadratic form* associated with s is $\mathbf{QS} = \mathbf{M}'\mathbf{R}\mathbf{M}$ (symmetrical), which is also the matrix associated with $r \circ f : \mathcal{A} \to \mathcal{B}^*$.

10.3.7 More on Orthogonal Projection

Property 10.12. *Let $\mathcal{A}$ be a Euclidean vector space with the scalar product $\langle\cdot|\cdot\rangle$ inducing the isomorphism $q : \mathcal{A} \to \mathcal{A}^*$. Let $\mathcal{L}$ be a subspace of $\mathcal{A}$. Let $i : \mathcal{L} \to \mathcal{A}$ (canonical injection); then $i^* : \mathcal{A} \to \mathcal{L}$ (adjoint homomorphism of i) is the orthogonal projection onto $\mathcal{L}$.*

Proof. The restriction of the scalar product to $\mathcal{L}$ induces the isomorphism $r = {}^t i \circ q \circ i : \mathcal{L} \to \mathcal{L}^*$. One has $i^* = r^{-1} \circ {}^t i \circ q$. Let $\widetilde{\alpha} = i^*(\alpha)$, then $\forall \gamma \in \mathcal{L} :\ \langle\widetilde{\alpha}|\gamma\rangle = \langle i^*(\alpha)|\gamma\rangle = \langle\alpha|i(\gamma)\rangle = \langle\alpha|\gamma\rangle$, that is, $\alpha - \widetilde{\alpha} \perp \gamma$ ◁

Hereafter the mapping i^* will be denoted p_L with $p_L = ({}^t i \circ q \circ i)^{-1} \circ {}^t i \circ q$.
The endomorphism on $\mathcal{A}$ orthogonal projection onto $\mathcal{L}$ is denoted $p = i \circ p_L$, that is, $p = i \circ ({}^t i \circ q \circ i)^{-1} \circ {}^t i \circ q$.
The endomorphism on $\mathcal{L}$ associated with p_L is $p_L \circ i = Id_L$ (Id_L denoting the identity mapping of $\mathcal{L}$).
The bilinear (and quadratic) form on $\mathcal{A} \times \mathcal{A}$ defined by $(\alpha, \alpha') \mapsto \langle\widetilde{\alpha}|\alpha'\rangle$ induces the homomorphism ${}^t p \circ q : \mathcal{A} \to \mathcal{A}^*$, with

$$ {}^t p \circ q = q \circ i \circ ({}^t i \circ q \circ i)^{-1} \circ {}^t i \circ q : \mathcal{A} \to \mathcal{A}^* $$

Matrix formulas

Still taking bases $(\epsilon_\ell)_{\ell \in L}$ and $(\alpha_i)_{i \in I}$, and denoting $\mathbf{E}$ the $I \times L$ matrix of canonical injection $i : \mathcal{L} \to \mathcal{A}$, then:
The $L \times L$ matrix of isomorphism $r : \mathcal{L} \to \mathcal{L}^*$ is $\mathbf{R} = \mathbf{E}'\mathbf{Q}\mathbf{E}$.
The $L \times I$ matrix of $p_L : \mathcal{A} \to \mathcal{L}$ is $\mathbf{P}_L = (\mathbf{E}'\mathbf{Q}\mathbf{E})^{-1}\mathbf{E}'\mathbf{Q}$.
The $I \times I$ matrix of $p : \mathcal{A} \to \mathcal{A}$ is $\mathbf{P}$ with $\mathbf{P} = \mathbf{E}(\mathbf{E}'\mathbf{Q}\mathbf{E})^{-1}\mathbf{E}'\mathbf{Q}$. The matrix $\mathbf{P}$ is idempotent, but not symmetrical (except if $\mathbf{Q} = \mathbf{I}$).
The $I \times I$ matrix of the bilinear (and quadratic) form on $\mathcal{A} \times \mathcal{A}$ is $\mathbf{\Sigma}$ with $\mathbf{\Sigma} = \mathbf{Q}\mathbf{E}(\mathbf{E}'\mathbf{Q}\mathbf{E})^{-1}\mathbf{E}'\mathbf{Q}$ (symmetric).

Particular cases. The foregoing formulas simplify in the following three particular cases.

1) Basis $(\epsilon_\ell)_{\ell \in L}$ is orthonormal ($\mathbf{E}'\mathbf{Q}\mathbf{E} = \mathbf{I}_L$).—
$\mathbf{R} = \mathbf{I}_L$; $\mathbf{P}_L = \mathbf{E}'\mathbf{Q}$; $\mathbf{P} = \mathbf{E}\mathbf{E}'\mathbf{Q} (\neq \mathbf{Q}\mathbf{E}\mathbf{E}' = \mathbf{P}')$; $\mathbf{\Sigma} = \mathbf{Q}\mathbf{E}\mathbf{E}'\mathbf{Q}$.

2) Basis $(\alpha_i)_{i \in I}$ is orthonormal ($\mathbf{Q} = \mathbf{I}$).
$\mathbf{R} = \mathbf{E}'\mathbf{E}$; $\mathbf{P}_L = (\mathbf{E}'\mathbf{E})^{-1}\mathbf{E}'$; $\mathbf{P} = \mathbf{\Sigma} = \mathbf{E}(\mathbf{E}'\mathbf{E})^{-1}\mathbf{E}'$ (symmetric).

3) Both bases $(\alpha_i)_{i\in I}$ and $(\epsilon_\ell)_{\ell\in L}$ are orthonormal:
$\mathbf{Q} = \mathbf{I}$ and $\mathbf{E}'\mathbf{E} = \mathbf{I}_L$, then $\mathbf{R} = \mathbf{I}_L$; $\mathbf{P}_L = \mathbf{E}'$; $\mathbf{P} = \mathbf{\Sigma} = \mathbf{E}\mathbf{E}'$.

10.4 Multidimensional Geometry

Multidimensional Geometry is the extension of plane and three–dimensional geometry (taught in high school) to higher (finite) dimensional spaces. The primitive concepts of multidimensional geometry are basically those of elementary geometry: points, lines, planes, geometric vectors, etc. Multidimensional geometry enables one to carry over both the spatial intuition and the rigorous coordinate–free mode of reasoning proper to pure geometry. The familiar geometric notation can be preserved: capital letters (M, P, A, etc.) for points, and "arrowed" letters for *geometric vectors* ($\vec{v}$, $\overrightarrow{\mathrm{MN}}$, etc.). Multidimensional geometry involves *affine properties*, dealing with lines, parallelism, etc., without a distance function, and *Euclidean properties*, dealing with (Euclidean) distances, orthogonality, angles, etc.

On the one hand, from a mere mathematical standpoint, going from vector spaces to geometry is a small step, because the underlying structures of multidimensional geometry are basically those of linear algebra. This presumably explains the limited coverage of multidimensional geometry in textbooks; see, however, MacLane & Birkhoff (1967, chapter 12), Coxeter (1961), Lehman & Bkouche (1968), Suppes et al (1989, Vol 2 §12), and Benzécri (1992, chapters 9 and 10). On the other hand, from a conceptual standpoint, going from vector spaces to geometry is a big step, because it establishes the specificity of geometric concepts with respect to vectorial ones, to begin with the notions of points and distances between points[6].

In this section, we first characterize the notion of affine space and discuss barycenter in some detail. Then we outline the following fundamental affine notions: linear affine variety; Cartesian and affine frame; simplex, linear mapping. Finally, we review Euclidean geometric properties.

10.4.1 Affine Properties and Barycenter

Definition 10.17 (Affine space). *A nonempty set $\mathcal{U}$ is called an affine space and its elements are called points, if $\mathcal{U}$ is equipped with a finite–dimensional vector space $\mathcal{V}$ (on $\mathbb{R}$), together with a mapping $\mathcal{V}\times\mathcal{U}$ into $\mathcal{U}$, such that with each pair $(\vec{v}, \mathrm{M})$ there is associated the point $\mathrm{M}+\vec{v}$ satisfying*

[6] It is to emphasize the specificity of geometry that we have defined the Euclidean structure on a vector space not in terms of distance but merely in terms of norm.

the following properties:

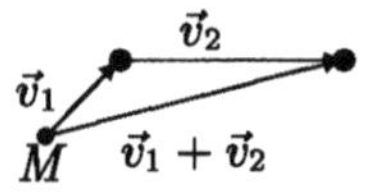

P1: $\forall \vec{v}_1\, \vec{v}_2\, \forall \mathrm{M} : (\mathrm{M} + \vec{v}_1) + \vec{v}_2 = \mathrm{M} + (\vec{v}_1 + \vec{v}_2)$;
P2: $\forall \mathrm{M} : \mathrm{M} + \overrightarrow{0} = \mathrm{M}$;
P3: for any bipoint (pair of points) (M,N), there exists a unique vector $\vec{v}$ *such that* $\mathrm{N} = \mathrm{M} + \vec{v}$.

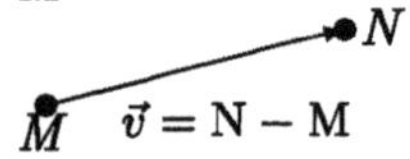

$\mathcal{V}$ is called the *vector space underlying the affine space* $\mathcal{U}$. By definition, the dimensionality of $\mathcal{U}$ is that of $\mathcal{V}$. The vector associated with the bipoint (M,N) is denoted $\overrightarrow{\mathrm{MN}}$ or N − M, as the deviation (vectorial difference) of point N from point M ("endpoint minus origin", Grassman's notation).

The back and forth process between affine and vector spaces is conveyed by the following two properties.

Property 10.13 (Canonical affine space). *With a vector space* $\mathcal{V}$ *there is associated a canonical affine space* $\mathcal{U} = \mathcal{V}$ *with* $\mathcal{V}$ *as underlying vector space.*

Property 10.14 (Vectorial extension of an affine space). *Any* n*–dimensional affine space* $\mathcal{U}$ *can be embedded in an* $n+1$*–dimensional vector space* $\mathcal{V}^{\#}$.

These properties enable one, starting with a numerical vector space, to represent its vectors as points in an affine space $\mathcal{U}$, then as geometric vectors in the underlying vector space $\mathcal{V}$, or alternatively in the extended vector space $\mathcal{V}^{\#}$ (as in projective geometry). This construction is standard in Geometric Data Analysis. For example, in CA, normalized measures over J, which are basically vectors in $\mathbb{R}_J$ (J–dimensional vector space), are represented by points in a hyperplane ($J-1$–dimensional affine space), then by vectors in the underlylying vector space, and also by vectors of the extended J–dimensional vector space (Chapter *CORRESPONDENCE*, §2.1.5). A similar representation is in chapter *PRINCIPAL* for score profiles.

Let us now consider a family of points $(\mathrm{M}^j)_{j\in J}$ in an affine space $\mathcal{U}$, with coefficients (weights) $(u_j)_{j\in J}$, together with a point P; and let us study the weighted vectorial sum $\sum\limits_{j\in J} u_j \overrightarrow{\mathrm{PM}^j}$.

Property 10.15 (Barycenter). *If* $\sum\limits_{j\in J} u_j \neq 0$, *the sum* $\sum\limits_{j\in J} u_j \overrightarrow{\mathrm{PM}^j}$ *defines a point independent of* P, *denoted* G *such that* $(\sum\limits_{j\in J} u_j)\overrightarrow{\mathrm{PG}} = \sum\limits_{j\in J} u_j \overrightarrow{\mathrm{PM}^j}$. *Point* G *is called the barycenter of the weighted points* $(\mathrm{M}^j, u_j)_{j\in J}$.

Proof. Starting with a point P', we get G' such that $(\sum_{j\in J} u_j)\overrightarrow{\mathrm{P'G'}} = \sum_{j\in J} u_j\overrightarrow{\mathrm{P'M}^j}$. Now $\sum_{j\in J} u_j\overrightarrow{\mathrm{P'M}^j} = (\sum_{j\in J} u_j)\overrightarrow{\mathrm{P'P}} + \sum_{j\in J} u_j\overrightarrow{\mathrm{PM}^j} = (\sum_{j\in J} u_j)\overrightarrow{\mathrm{P'P}} + \sum_{j\in J} u_j\overrightarrow{\mathrm{PG}} = (\sum_{j\in J} u_j)\overrightarrow{\mathrm{P'G}}$. Therefore $(\sum_{j\in J} u_j)\overrightarrow{\mathrm{P'G'}} = (\sum_{j\in J} u_j)\overrightarrow{\mathrm{P'G}}$, hence $\mathrm{G'} = \mathrm{G}$. ◁

Barycenter G can be written $\mathrm{G} = \mathrm{P} + (\sum_{j\in J} u_j\overrightarrow{\mathrm{PM}^j})/(\sum_{j\in J} u_j)$, or $\mathrm{G} = (\sum_{j\in J} u_j\mathrm{M}^j)/(\sum_{j\in J} u_j)$, as the weighted average of points $(\mathrm{M}^j)_{j\in J}$.

Corollary 10.1. *The weighted sum of the vector deviations* $\overrightarrow{\mathrm{GM}^j}$ *of points* $(\mathrm{M}^j)_{j\in J}$ *from their barycenter* G *is the null vector:* $\sum_{j\in J} u_j\overrightarrow{\mathrm{GM}^j} = \overrightarrow{0}$.

Property 10.16 (Contrast). *If* $\sum_{j\in J} u_j = 0$, *the vectorial sum* $\sum_{j\in J} u_j\overrightarrow{\mathrm{PM}^j}$ *defines a vector independent of* P.

Proof. Taking point P', we have $\sum_{j\in J} u_j\overrightarrow{\mathrm{P'M}^j} = (\sum_{j\in J} u_j)\overrightarrow{\mathrm{P'P}} + \sum_{j\in J} u_j\overrightarrow{\mathrm{PM}^j} = \overrightarrow{0} + \sum_{j\in J} u_j\overrightarrow{\mathrm{PM}^j} = \sum_{j\in J} u_j\overrightarrow{\mathrm{PM}^j}$. ◁

When $\sum_{j\in J} u_j = 0$, the vector $\overrightarrow{e} = \sum_{j\in J} u_j\overrightarrow{\mathrm{PM}^j}$, independent of P, will be denoted $\sum_{j\in J} u_j\mathrm{M}^j$, as the effect of contrast u_J among points $(\mathrm{M}^j)_{j\in J}$; this notation extends Grassman's notation for the difference between two points.

To sum up: An average of points is a point (barycenter); a contrast among points is a vector. These basic operations on the points of an affine space extend those of the points on a line or in a plane. They are more restrictive than the operations on vectors: vectors can be added, points cannot.

Particular case. $J = \{j, j'\}$ (two elements). For $u_j = u_{j'}$, G is the *midpoint* de $(\mathrm{M}^j, \mathrm{M}^{j'})$: $\mathrm{G} = (\mathrm{M}^j + \mathrm{M}^{j'})/2$. For $u_j = -1$ and $u_{j'} = 1$, one gets the *deviation-vector* of point $\mathrm{M}^{j'}$ from M^j : $\mathrm{M}^{j'} - \mathrm{M}^j = \overrightarrow{\mathrm{M}^j\mathrm{M}^{j'}}$.

The geometric construction of the barycenter of the two weighted points (M^j, u_j) and $(\mathrm{M}^{j'}, u_{j'})$, with $u_j = 3$ and $u_{j'} = 2$, is illustrated on the opposite figure.

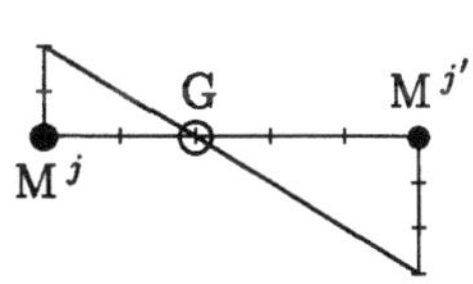

10.4.2 Other Basic Affine Notions

Variety, hyperplane, line, axis

A subset $\mathcal{U}'$ of $\mathcal{U}$ is called an *affine subspace*, or a linear affine variety (a *variety* for short), if every barycenter of points of $\mathcal{U}'$ is a point of $\mathcal{U}'$. A *line* is a one–dimensional variety, a *plane* a two–dimensional one. A *hyperplane* is a variety whose dimensionality is equal to that of the whole space minus 1; thus for a three–dimensional space, a hyperplane reduces to a plane.

If one chooses a point P_0 in the variety, the vectors $\overrightarrow{\mathrm{P}_0\mathrm{P}}$, with $\mathrm{P} \in \mathcal{U}$, generate a vector subspace independent of P_0, that characterizes the *direction* of the variety. In particular, the direction of a line is characterized by a one–dimensional vector space; any vector $\vec{\alpha}$ generating this vector space is called a *director vector* of the line, and the line will be said to be of direction $\vec{\alpha}$; a director vector also provides an *orientation* to the line, that is, it defines an *oriented line*, or *axis*.

Cartesian frame, affine frame, simplex

Let $\mathcal{U}'$ be a K'–dimensional[7] affine subspace of $\mathcal{U}$, a Cartesian frame of $\mathcal{U}'$ is a pair $\left(\mathrm{O}, (\vec{\delta}_k)_{k\in K'}\right)$, where O is a point of $\mathcal{U}'$, and $(\vec{\delta}_k)_{k\in K'}$ is a basis of the underlying vector space of $\mathcal{U}'$. By definition, the coordinates of a point $\mathrm{M} \in \mathcal{U}'$ in this Cartesian frame are those of vector $\overrightarrow{\mathrm{OM}}$ in the basis $(\vec{\delta}_k)_{k\in K'}$. A family of points $(\mathrm{P}^k)_{k\in K'}$ is an *affine frame* of the affine subspace $\mathcal{U}'$ if every point M in this subspace can be expressed in a unique way as a barycenter of points P^k. For a K'–multidimensional subspace, an affine frame consists in $K' + 1$ points; therefore $K + 1$ for the whole K–dimensional space, K for a hyperplane, 3 for a plane, 2 for a line.

If $\left(\mathrm{O}, (\vec{\delta}_k)_{k\in K}\right)$ is a Cartesian frame of $\mathcal{U}$, the family of K points $(\mathrm{P}^k)_{k\in K}$ defined by $\forall k : \mathrm{P}^k = \mathrm{O} + \vec{\delta}_k$ is an affine frame of a hyperplane of $\mathcal{U}$; the set of all barycenters *with positive weights* of these K points is a subset of this hyperplane, and is called the simplex of edges $(\mathrm{P}^k)_{k\in K}$.

Affine mapping

Let $\mathcal{U}$, $\mathcal{U}'$ be two affine spaces and $\mathcal{V}$, $\mathcal{V}'$ their underlying vector spaces; a mapping $f : \mathcal{U} \longrightarrow \mathcal{U}'$ is an *affine mapping* if there exists a linear mapping $\ell : \mathcal{V} \longrightarrow \mathcal{V}'$ such that

$$\forall M \in \mathcal{U}, \forall \vec{v} \in \mathcal{V} : f(M + \vec{v}) = f(M) + \ell(\vec{v})$$

[7]According to our general convention, K' denotes both a finite set (indexing the basis) and its cardinality (dimensionality of space).

An affine mapping is determined by the image of a point together with the associated linear mapping. A mapping is affine if and only if it preserves barycenters. Examples are projection, homothety and affinity.

10.4.3 Euclidean Geometric Properties

Basic Euclidean notions

The Euclidean notions and properties of a multidimensional geometric space follow from those of a vector space (p.428). Let $\mathcal{U}$ be a K–dimensional affine space and $\mathcal{V}$ the underlying vector space on $\mathbb{R}$, now equipped with the scalar product denoted $\langle\cdot|\cdot\rangle$, and the associated norm denoted $\|\cdot\|$. The space $\mathcal{U}$ becomes a *Euclidean space* if the *distance* between two points M and N in $\mathcal{U}$ is defined as the norm of geometric vector $\overrightarrow{\mathrm{MN}}$: $\mathrm{MN} = \|\overrightarrow{\mathrm{MN}}\|$.

Property 10.17 (Distance). *The Euclidean distance verifies the properties of a distance (metric space):*
(1) Positivity. $\forall \mathrm{A}, \mathrm{B} \in \mathcal{U} : \mathrm{AB} \geq 0$ *with* $\mathrm{AB} = 0 \iff \mathrm{A} = \mathrm{B}$
(2) Symmetry. $\forall \mathrm{A}, \mathrm{B} \in \mathcal{U} : \mathrm{AB} = \mathrm{BA}$
(3) Triangle inequality. $\forall \mathrm{A}, \mathrm{B}, \mathrm{C} \in \mathcal{U} : \mathrm{AC} \leq \mathrm{AB} + \mathrm{BC}$
with $\mathrm{AC} = \mathrm{AB} + \mathrm{BC} \iff$ *A, B, C aligned, with B between A and C.*

For dimensionality 1 (geometric line), the Euclidean structure is canonically induced by the affine structure; for a multidimensional space, the Euclidean distance must be specified. A Euclidean distance induces the notion of *angle* between geometric vectors as in elementary geometry.

Property 10.18 (Parallelogram). *If four points* (M, P, Q, R) *are such that* (M, Q) *and* (P, R) *have the same midpoint, then one has:*
$\mathrm{MQ}^2 + \mathrm{PR}^2 = \mathrm{MP}^2 + \mathrm{PQ}^2 + \mathrm{QR}^2 + \mathrm{RM}^2$

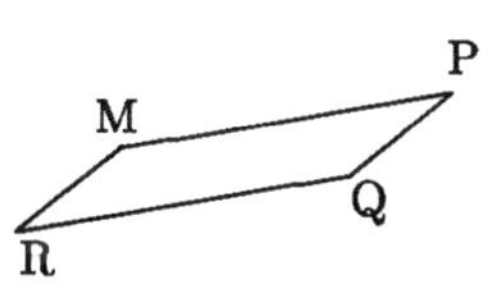

Theorem 10.4 (Pythagoras' theorem). *For three points* (M,P,Q), *one has the equivalence:* $\overrightarrow{\mathrm{MP}} \perp \overrightarrow{\mathrm{PQ}} \iff \mathrm{MP}^2 + \mathrm{PQ}^2 = \mathrm{MQ}^2$

Orthogonal projection. In elementary geometry, one defines the orthogonal projection of a point on a line or a plane; the extension of this notion is the orthogonal projection of a point on a subspace.

Definition 10.18 (Projected point on a subspace). *Let* $\mathcal{L}$ *be a subspace of a Euclidean space; for any point* $\mathrm{M} \notin \mathcal{L}$*; there exists a unique point* $\mathrm{M}' \in \mathcal{L}$ *such that* MM' *is orthogonal to* $\mathcal{L}$*: Point* M' *is called the orthogonal projection (or projected point) of point* M *on* $\mathcal{L}$*.*

By definition: $\forall \mathrm{M} \in \mathcal{L} : \mathrm{M}' = \mathrm{M}$.

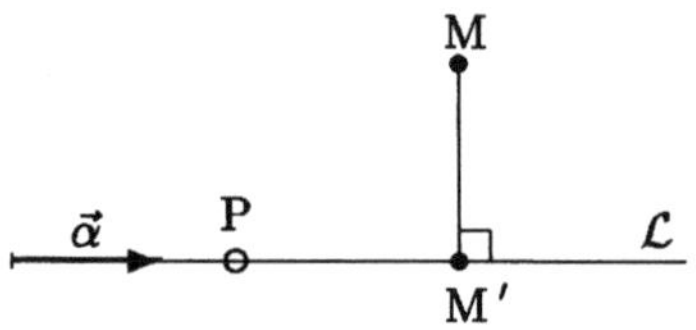

$dim\, \mathcal{L} = 1$ (line)

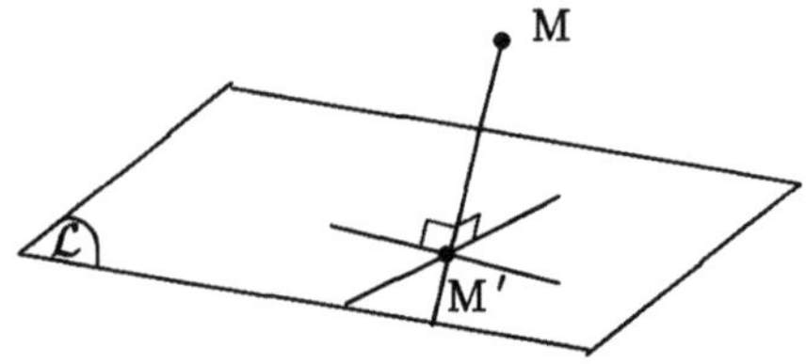

$dim\, \mathcal{L} = 2$ (plane)

Property 10.19 (Distance from a point to a subspace). *Among all points in $\mathcal{L}$, the projected point* M′ *is the point at the minimum distance from point* M. *The* MM′ *distance defines the distance from point* M *to subspace $\mathcal{L}$.*

Property 10.20 (Distance from a point to a hyperplane). *If in an orthonormal frame $(\mathrm{O}, (\overrightarrow{\delta_k})_{k\in K})$ of $\mathcal{U}$, a hyperplane has equation $\sum_{k\in K} u_k x^k + u_0 = 0$, the distance from point* M *with coordinates $(y^k)_{k\in K}$ to the hyperplane is given by:*

$$\mathrm{MH} = \Big| \sum_{k\in K} u_k y^k + u_0 \Big| \Big/ \sqrt{\sum_{k\in K} u_k^2}$$

Property 10.21 (Projection of a point on a line). *Given a line $\mathcal{D}$ determined by point* P *and vector $\overrightarrow{a}$, the orthogonal projection of* M *on $\mathcal{D}$ is the point* M′ *such that:*

$$\mathrm{M}' = \mathrm{P} + \frac{\langle \overrightarrow{\mathrm{PM}} | \overrightarrow{a} \rangle}{\|\overrightarrow{a}\|^2}\, \overrightarrow{a}$$

Proof. $\overrightarrow{\mathrm{MM}'} \perp \overrightarrow{a} \iff \langle \overrightarrow{\mathrm{MM}'} | \overrightarrow{a} \rangle = 0$, that is $\langle \overrightarrow{\mathrm{PM}'} - \overrightarrow{\mathrm{PM}} | \overrightarrow{a} \rangle = 0$. Letting $\overrightarrow{\mathrm{PM}'} = y\overrightarrow{a}$, one has: $y\langle \overrightarrow{a} | \overrightarrow{a} \rangle = \langle \overrightarrow{\mathrm{PM}} | \overrightarrow{a} \rangle$. ◁

Property 10.22 (Orthogonal projection of a bipoint). *Let $\mathcal{L}_1$ and $\mathcal{L}_2$ be parallel and same–dimensional subspaces; let $(\mathrm{M}_1,\mathrm{N}_1)$ and $(\mathrm{M}_2,\mathrm{N}_2)$ be the bipoints that are orthogonal projections of (M,N) onto these two subspaces respectively; then bipoints $(\mathrm{M}_1, \mathrm{N}_1)$ and $(\mathrm{M}_2, \mathrm{N}_2)$ are equipollent, that is $\overrightarrow{\mathrm{M}_1\mathrm{N}_1} = \overrightarrow{\mathrm{M}_2\mathrm{N}_2}$ or else $\overrightarrow{\mathrm{M}_1\mathrm{M}_2} = \overrightarrow{\mathrm{N}_1\mathrm{N}_2}$.*

Definition 10.19 (Orthogonal affinity). *Given a subspace $\mathcal{L}$, if* M′ *is the orthogonal projection of* M *onto $\mathcal{L}$, an orthogonal affinity of basis $\mathcal{L}$ and ratio λ (with $\lambda \neq 0$) is the mapping which with point* M *associates the point* M″ *such that:* $\mathrm{M}'' = \mathrm{M}' + \lambda\, \overrightarrow{\mathrm{M}'\mathrm{M}}$.

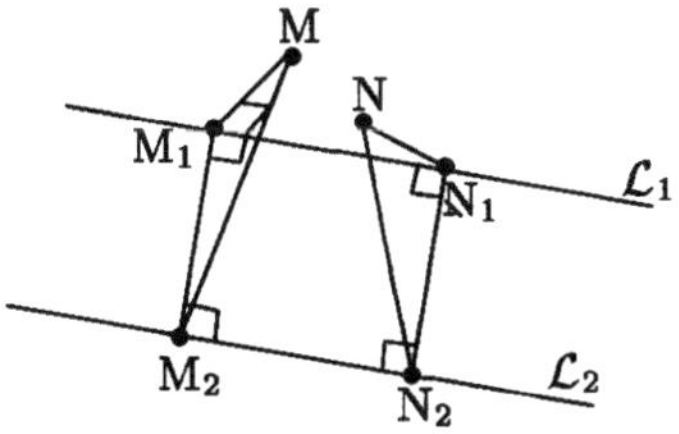

dim $\mathcal{U} = 3$; dim $\mathcal{L}_1 = \dim\mathcal{L}_2 = 1$

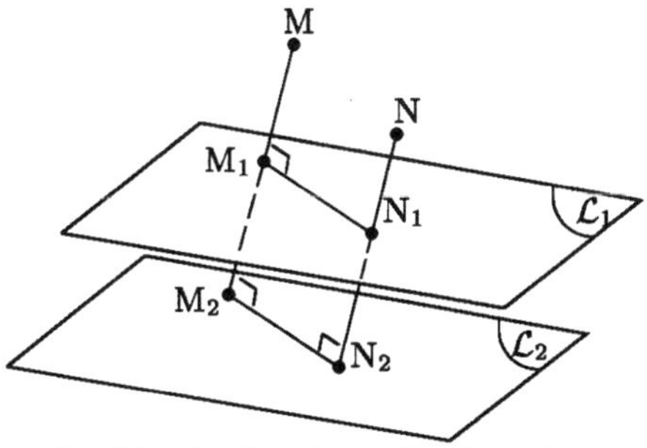

dim $\mathcal{U} = 3$, dim $\mathcal{L}_1 = \dim\mathcal{L}_2 = 2$

Hyperellipsoid

The hyperellipsoid — in short, ellipsoid — is the multidimensional extension of the two–dimensional ellipse and the three–dimensional ellipsoid.

Definition 10.20 (Hyperellipsoid). *Let $(\mathrm{O}, (\overrightarrow{\delta_k})_{k\in K})$ be an orthonormal frame of $\mathcal{U}$. In this frame the Cartesian equation $\sum\limits_{k\in K} x_k^2/a_k^2 = 1$ defines the hyperellipsoid $\mathcal{E}$ with center O and principal axes $(\overrightarrow{\delta_k})_{k\in K}$ whose half–lengths are equal to $(a_k)_{k\in K}$.*

Property 10.23 (Tangential equation of a hyperellipsoid). *The hyperplane with equation $\sum\limits_{k\in K} u_k x^k = u_0$ is a tangent hyperplane to the hyperellipsoid $\mathcal{E}$ if and only if coefficients $(u_k)_{k\in K}$ satisfy the following equation $\sum\limits_{k\in K} a_k^2 u_k^2 = u_0^2$, called the tangential equation of the hyperellipsoid.*

Ellipse

In a plane $\mathcal{U}$, referred to an orthonormal frame, the Cartesian equation $x^2/a^2 + y^2/b^2 = 1$ (where $a \geq b > 0$) defines an ellipse having center O, and half–lengths axes equal to a (major principal axis) and b (minor principal axis). Hereafter we recall some elementary definitions and properties.

Definition 10.21 (Eccentricity of ellipse). *The eccentricity of ellipse is the coefficient e defined by $e^2 = 1 - b^2/a^2$. If the ellipse is a circle $a = b$ and $e = 0$.*

Property 10.24 (Ellipse area).
The area of ellipse is equal to πab.

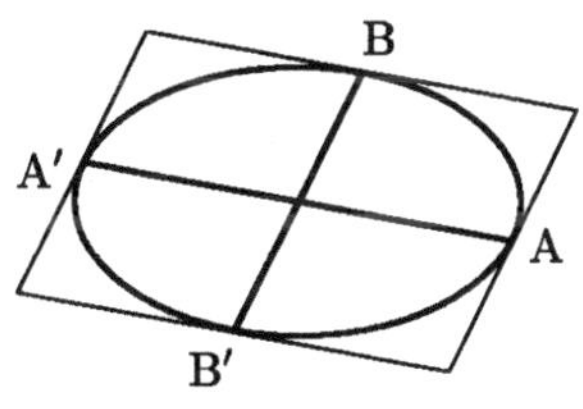

Definition 10.22 (Conjugate diameters).
The conjugate diameter of the diameter AA′ of an ellipse is the diameter BB′ parallel to the tangents to the ellipse at points A and A′.

If BB′ is conjugate to AA′, AA′ is conjugate to BB′: see figure p.441.

Property 10.25. *If θ and θ' denote the angles of two mutually conjugate diameters with the major principal axis, one has:* $\tan\theta\tan\theta' = -b^2/a^2$.

Dimensional considerations and homogeneity. In classical geometry, the distance between points is a *length* whose numerical expression depends on the length taken as a unit. Thus 3 cm = 30 mm: With a unit ten times smaller, the numerical expression of length is multiplied by ten: The length is a magnitude of *dimension 1*. The product of two lengths has a numerical expression multiplied by 100, and is as such of *dimension 2* (area). The ratio of two lengths is invariant, i.e. is "without dimension".

Pure geometric properties do not depend on unit length; the terms involved in such properties are of identical dimensions and are said to be *homogeneous*. For instance, in Pythagoras' theorem, both sides of the '=' sign are of dimension 2. In pure geometry, non–homogeneous relations (such as $AB^2 = AC$) are meaningless. The pure geometry concept extends to multidimensional geometry, assimilating the Euclidean distance with a length. A geometric vector will be called unit–norm vector if its length is equal to the distance unit. A Cartesian frame of a Euclidean space will be called orthonormal if its basis vectors are orthogonal and their norms are equal to the distance unit. Dividing a vector by its norm enables one to state pure geometric properties.

10.5 Spectral Theorem

The spectral theorem, alias *diagonalization theorem*, is the "mathematical heart" of GDA. It is motivated by extremality problems, that lead to *spectral equations*, whose solutions entail a *spectral decomposition*. In the first subsection, we recall the diagonalization of a symmetric endomorphism, which is the classical version of the spectral theorem, then we present the spectral theorem for two adjoint homomorphisms, which is the version directly applicable to GDA. The second subsection is devoted to matrix formulations.

10.5.1 General Formulations

Symmetric endomorphism

Definition 10.23 (Eigenvector and eigenvalue). *An eigenvector of the endomorphism $h : \mathcal{A} \to \mathcal{A}$ is a vector $\alpha \neq 0$ such that $h(\alpha)$ is a scalar* *multiple of α. A scalar $\lambda \in \mathbb{R}$ is an eigenvalue of h when there is a vector $\alpha \neq 0$ such that $h(\alpha) = \lambda\alpha$ (spectral equation)*

If λ is an eigenvalue of h, the set of all vectors $\alpha \in \mathcal{A}$ with $h(\alpha) = \lambda\alpha$ is a non– zero subspace of $\mathcal{A}$ called *eigenspace* of λ; it consists of the zero vector and all the eigenvectors associated to λ.

Definition 10.24. *A endomorphism $h : \mathcal{A} \to \mathcal{A}$ is diagonalizable if and only if the eigenvectors of h span $\mathcal{A}$.*

Theorem 10.5 (Spectral theorem). *Any symmetric endomorphism is diagonalizable.*

This theorem is the classical version of the spectral theorem and its demonstration figures in all abstract linear algebra textbooks (see p.423).

Property 10.26. *Eigenvectors of a symmetric endomorphism that are associated with two distinct eigenvalues are orthogonal.*

Proof. Let $h(\alpha_\ell) = \lambda_\ell \alpha_\ell$ and $h(\alpha_{\ell'}) = \lambda_{\ell'} \alpha_{\ell'}$.
From $\langle h(\alpha_\ell)|\alpha_{\ell'}\rangle = \langle \alpha_\ell | h(\alpha_{\ell'})\rangle$, one deduces: $\lambda_\ell \langle \alpha_\ell | \alpha_{\ell'}\rangle = \lambda_{\ell'} \langle \alpha_\ell | \alpha_{\ell'}\rangle$;
therefore $\lambda_\ell \neq \lambda_{\ell'} \Rightarrow \langle \alpha_\ell | \alpha_{\ell'}\rangle = 0$. ◁

The L non–null eigenvalues of a symmetric endomorphism are the roots of an algebraic equation of degree L (where L is the rank of the endomorphism). Eigenvalues may be simple or multiple. The L eigenvalues will be written from λ_1 (maximum) to λ_L (minimum), such that $\lambda_1 \geq \lambda_2 \geq \ldots \lambda_L$. For a value of multiplicity r, there exists an r–dimensional eigenspace, within which one can always find r orthogonal eigenvectors.

Property 10.27 (Spectral decomposition). *Let $(\alpha_\ell)_{\ell=1,\ldots L}$ be an orthonormal family of eigenvectors of the symmetric endomorphism h on $\mathcal{A}$ associated with eigenvalues $(\lambda_\ell)_{\ell=1,\ldots L}$; then h admits the spectral decomposition:*

$$h = \sum_{\ell=1}^{L} h_\ell \quad \text{with} \; \forall \alpha \in \mathcal{A} \; : \; h_\ell(\alpha) = \lambda_\ell \langle \alpha | \alpha_\ell \rangle \, \alpha_\ell$$

where the endomorphism h_ℓ on $\mathcal{A}$ defined by $x \mapsto \langle x | \alpha_\ell \rangle \, \alpha_\ell$ is the orthogonal projector on the subspace spanned by α_ℓ.

Property 10.28. *The eigenvectors of h are the eigenvectors of $h - h_\ell$ associated with the same eigenvalues, except α_ℓ which is associated with 0.*

Proof. $(h - h_\ell)(\alpha_\ell) = h(\alpha_\ell) - \lambda_\ell\langle\alpha_\ell|\alpha_\ell\rangle\,\alpha_\ell = \lambda_\ell\alpha_\ell - \lambda_\ell\alpha_\ell = 0$; for $\ell \neq \ell'$, $(h - h_\ell)(\alpha_{\ell'}) = h(\alpha_{\ell'}) - \lambda_\ell\langle\alpha_\ell|\alpha_{\ell'}\rangle\,\alpha_\ell = \lambda_{\ell'}\alpha_{\ell'} - 0$ (cf. property 10.26). ◁

Property 10.29 (Positive symmetric endomorphism). *All eigenvalues of a positive symmetric endomorphism are positive (or null).*

In GDA the spectral theorem for a symmetric endomorphism will be specialized to *positive* endomorphisms associated with the two adjoint homorphisms $s = f^* \circ f$ and $t = f \circ f^*$. (p.433).

Adjoint homomorphisms

Definition 10.25 (Eigenvectors and singular values). *Let two adjoint homomorphisms $f : \mathcal{A} \to \mathcal{B}$ and $f^* : \mathcal{B} \to \mathcal{A}$. A pair of nonnull vectors ($\alpha \in \mathcal{A}, \beta \in \mathcal{B}$) of equal norms ($||\alpha||_A = ||\beta||_B$) is called an eigenvector pair of (f, f^*) associated with the singular value $\xi \geq 0$ if the following spectral equations hold:*

$$\begin{cases} f(\alpha) & = \xi\,\beta \\ f^*(\beta) & = \xi\,\alpha \end{cases} \tag{10.1}$$

Property 10.30. *If (α, β) and (α', β') are eigenvector pairs associated with distinct singular values (ξ and $\xi' \neq \xi$), then vectors α and α' are orthogonal, and β and β' are orthogonal.*

Proof. $\langle f(\alpha)|f(\alpha')\rangle_B = \xi'\langle f(\alpha)|\beta'\rangle_B = \xi'\langle\alpha|f^*(\beta')\rangle = \xi'^2\langle\alpha|\alpha'\rangle$. Similarly one has: $\langle f(\alpha)|f(\alpha')\rangle_B = \xi^2\langle\alpha|\alpha'\rangle$. Therefore $\xi \neq \xi' \Rightarrow \langle\alpha|\alpha'\rangle = 0$. It is similarly shown that $\langle\beta|\beta'\rangle = 0$. ◁

Homomorphisms and endomorphisms. If $\alpha \in \mathcal{A}$ is eigenvector of f, the pair $(\alpha, f(\alpha))$ is an eigenvector pair of f and f^* associated with values 1 and $\lambda = \xi^2$. One has: $s(\alpha) = f^* \circ f(\alpha) = \lambda\,\alpha$, therefore α is an eigenvector of s associated with the eigenvalue λ. Similarly, $f \circ f^*(\beta) = \lambda\,\beta$, therefore β is an eigenvector of t associated with λ. Conversely, if α is an eigenvalue of s associated with λ, one has $s(\alpha) = \lambda\,\alpha = f^*(f(\alpha))$. For $\lambda > 0$, let $\beta = f(\alpha)/\xi$ with $\xi = \sqrt{\lambda}$; one has $f^*(\xi\,\beta) = \xi^2\alpha$, that is $f^*(\beta) = \xi\,\alpha$; therefore the pair (α, β) is an eigenvector pair of f and g associated with the singular value $\xi = \sqrt{\lambda}$.

Theorem 10.6 (Spectral theorem for adjoint homomorphisms). *Let f and f^* be two adjoint homomorphisms of rank L, there exists L terms $(\alpha_\ell, \beta_\ell, \xi_\ell)$, with $||\alpha_\ell||_A = ||\beta_\ell||_B$, such that:*

$$\begin{cases} f(\alpha_\ell) &= \xi_\ell\,\beta_\ell \\ f^*(\beta_\ell) &= \xi_\ell\,\alpha_\ell \end{cases} \quad \textit{with } \xi_1 \geq \xi_2 \ldots \geq \xi_L > 0$$

Singular values may be simple or multiple; if ξ_ℓ is simple, the pair $(\alpha_\ell, \beta_\ell)$ is unique *up to orientation.* If ξ_ℓ is of multiplicity r, one can always find r pairs of orthogonal eigenvectors.

Remark. If h is a positive symmetric endomorphism, the spectral theorem applied to h *as a homomorphism* amounts to its diagonalization, and one has: $\alpha_\ell = \beta_\ell$ and $h(\alpha_\ell) = \lambda_\ell\,\alpha_\ell$.

With the term $(\alpha_\ell, \beta_\ell, \xi_\ell)$ (with $||\alpha_\ell|| = ||\beta_\ell|| = 1$), there are associated the two adjoint homomorphisms of rank 1:

$$f_\ell : \begin{array}{ccc} \mathcal{A} & \to & \mathcal{B} \\ x & \mapsto & \xi_\ell \langle x|\alpha_\ell\rangle_A\,\beta_\ell \end{array} \qquad\qquad f_\ell^* : \begin{array}{ccc} \mathcal{B} & \to & \mathcal{A} \\ y & \mapsto & \xi_\ell \langle y|\beta_\ell\rangle_B\,\alpha_\ell \end{array}$$

Homomorphism f is fitted by a homomorphism of rank 1 by replacing f by f_1 associated with the singular value ξ_1. The homomorphism $r_1 = f - f_1$ has the same eigenvectors and eigenvalues as f, except for α_1, which belongs to the kernel of r_1 $(r_1(\alpha_1) = 0)$.

Indeed, if $(\alpha_\ell, \beta_\ell, \xi_\ell)$ is an eigen term of f, and $r_1 = f - f_1$, one has:
$r_1(\alpha_\ell) = \xi_\ell\,\beta_\ell - \xi_1 \langle \alpha_\ell|\alpha_1\rangle_A\,\beta_1$ and $r_1^*(\beta_\ell) = \xi_\ell\,\alpha_\ell - \xi_1 \langle \beta_\ell|\beta_1\rangle_B\,\alpha_1$

For $\ell \neq 1$ one has: $\langle \alpha_\ell|\alpha_1\rangle_A = \langle \beta_\ell|\beta_1\rangle_B = 0$, therefore $(\alpha_\ell, \beta_\ell)$ is an eigenvector pair of r_1 associated with ξ_ℓ.

For $\ell = 1$, one has: $r_1(\alpha_1) = r_1^*(\beta_1) = 0$.

Property 10.31 (Spectral decomposition). *A homomorphism f of rank L is the sum of L homomorphisms of rank 1 (where $||\alpha_\ell|| = ||\beta_\ell|| = 1$):*

$$f = \sum_{\ell=1}^{L} f_\ell \; \textit{with } f_\ell(x) = \xi_\ell \langle x|\alpha_\ell\rangle_A\,\beta_\ell \qquad \operatorname{tr}(f^* \circ f) = \sum_{\ell=1}^{L} \xi_\ell^2 = \sum_{\ell=1}^{L} \lambda_\ell$$

For the adjoint homomorphism f^*, for the bilinear form φ, and for the endomorphisms s and t and quadratic forms σ and τ, related decompositions readily follow:

– Adjoint homomorphism	f^* :	$f_\ell^*(y) = \xi_\ell \langle y|\beta_\ell\rangle_B\,\alpha_\ell$
– Bilinear form	φ :	$\varphi_\ell(x,y) = \xi_\ell \langle x|\alpha_\ell\rangle_A\,\langle y|\beta_\ell\rangle_B$
– Endomorphisms	s :	$s_\ell(x) = \lambda_\ell \langle x|\alpha_\ell\rangle_A\,\alpha_\ell$
	t :	$t_\ell(x) = \lambda_\ell \langle y|\beta_\ell\rangle_B\,\beta_\ell$
– Quadratic forms	σ :	$\sigma_\ell(x) = \lambda_\ell \langle x|\alpha_\ell\rangle_A^2$
	τ :	$\tau_\ell(x) = \lambda_\ell \langle y|\beta_\ell\rangle_B^2$

Extremality properties

Property 10.32 (Maximization). *Given a homomorphism $f : \mathcal{A} \to \mathcal{B}$, the ratio $\frac{||f(\alpha)||_B}{||\alpha||_A}$ ($\alpha \in \mathcal{A}$, $\alpha \neq 0$) is maximum for any eigenvector of f associated with the largest singular value ξ_1, and the value of maximum is ξ_1. More generally, if $\alpha \neq 0$ belongs to the subspace generated by the eigenvectors $(\alpha_\ell)_{\ell=p\ldots q}$, one has: $\xi_p \geq \|f(\alpha)\|/\|\alpha\| \geq \xi_q$ ($p \leq q$).*

For the *bilinear form* φ (definition 10.9, p.432):
the ratio $|\varphi(\alpha,\beta)|/(||\alpha||_A\ ||\beta||_B)$ is maximum for any pair of vectors proportional to (α_1, β_1), and the maximum is equal to ξ_1.
For the *endomorphism* s and the *quadratic form* σ: the ratio (*Rayleigh coefficient*) $\langle s(\alpha)|\alpha\rangle/||\alpha||^2 = \sigma(\alpha)/||\alpha||^2$ is maximum for any vector proportional to α_1, and the maximum is equal to λ_1.

The fitted endomorphism $f_1^* \circ f_1$ is, among the endomorphisms of rank 1, *the closest to* f, in the sense that it minimizes the sum of eigenvalues of the residual endomorphism $r_1^* \circ r_1$ (trace of endomorphism). One has: tr $(f_1^* \circ f_1) = \lambda_1$; hence the decomposition :

$$\text{tr } (f^* \circ f) = \lambda_1 + \text{tr } (r_1^* \circ r_1) \text{ with tr } (r_1^* \circ r_1) = \sum_{\ell=2}^{L} \lambda_\ell$$

Property 10.33 (Incomplete decomposition). *For $1 \leq L' \leq L$, the homomorphism of rank L' closest to f is the sum of the L' first homomorphisms of rank 1, hence the incomplete decomposition formula:*

$$\widetilde{f} = \sum_{\ell=1}^{L'} f_\ell\ , \textit{ with } \text{tr } (\widetilde{f}^* \circ \widetilde{f}) = \sum_{\ell=1}^{L'} \xi_\ell^2 = \sum_{\ell=1}^{L'} \lambda_\ell$$

This property entails the *heredity property* of spectral decomposition: from f_1 (best fit of rank 1), one gets the best fit of rank 2 by adding f_2, etc.

10.5.2 Matrix Expressions of Spectral Theorem

Symmetric endomorphism and diagonalization of a symmetric matrix

Restricting henceforth to the principal space $\mathcal{L}$ associated with non–null eigenvalues, let $(\alpha_\ell)_{\ell \in L}$ be a *principal orthonormal basis* of this space[8]. The restriction of the endomorphism h to $\mathcal{H}$ is represented by the diagonal matrix $\mathbf{\Lambda} = [\lambda_\ell]$ (hence the name "diagonalization" given to spectral theorem).

[8]The principal basis is unique (up to orientation of vectors) if all eigenvalues are simple.

If the vector space is referred to basis $(\alpha_i)_{i\in I}$ (initial basis), the spectral decomposition lead to matrix formulas. Let $\mathbf{E}$ the $I \times L$ matrix representing the L orthonormal eigenvectors in basis $(\alpha_i)_{i\in I}$. Applying the projection formula (p.434) and letting $\mathbf{H}$ the $I \times I$ matrix associated with the endomorphism h in the initial basis, we get:

$$\mathbf{E'QHE} = \mathbf{\Lambda} \text{ and } \mathbf{H} = \mathbf{E\Lambda E'Q}, \text{ or } \mathbf{H\,E} = \mathbf{E\,\Lambda} \quad \text{with } \mathbf{E'QE} = \mathbf{I}$$

If the initial basis is orthonormal, i.e. if $\mathbf{Q} = \mathbf{I}$, $\mathbf{H}$ is symmetrical, and the foregoing formulas reduce to the classical *diagonalization of a symmetric matrix*:

$$\mathbf{H} = \mathbf{E\Lambda E'} = \sum_{\ell=1}^{L} \lambda_\ell \mathbf{e}_\ell \mathbf{e'}_\ell, \text{ with } \mathbf{E'E} = \mathbf{I}$$

The columns $(\mathbf{e}_\ell)_{\ell\in L}$ are called the *eigencolumns of matrix* $\mathbf{H}$ associated with the eigenvalues $(\lambda_\ell)_{\ell\in L}$, with $\begin{cases} \mathbf{e}'_\ell \mathbf{e}_\ell = 1 \\ \mathbf{e}'_{\ell'} \mathbf{e}_\ell = 0 \text{ if } \ell \neq \ell' \end{cases}$

The diagonalization of symmetric matrices is the computational standard format when statistical procedures are framed in terms of symmetric endomorphisms.

Adjoint homomorphisms and SVD

For two adjoint homomorphisms $f : \mathcal{A} \to \mathcal{B}$ and $f^* : \mathcal{B} \to \mathcal{A}$, let $(\alpha_\ell)_{\ell\in L}$ and $(\beta_\ell)_{\ell\in L}$ be two orthonormal bases of the two respective principal spaces $\mathcal{L}$ and $\mathcal{M}$. The restrictions of f and f^* to $\mathcal{L}$ and $\mathcal{M}$ are represented by the same diagonal matrix of singular values $\mathbf{\Xi} = [\xi_\ell]$.

Taking as initial bases $(\alpha_k)_{k\in K}$ of $\mathcal{A}$ and $(\beta_j)_{j\in J}$ of $\mathcal{B}$, with scalar products represented by $\mathbf{Q}$ and $\mathbf{R}$ respectively, let $\mathbf{M} = [m_{jk}]$ denote the $(J \times K)$ matrix of homomorphism f.

Let $\mathbf{E}$ be the $K \times L$ matrix representing the L orthonormal eigenvectors $(\alpha_\ell)_{\ell\in L}$ in the basis $(\alpha_k)_{k\in K}$, and $\mathbf{F}$ be the $J \times L$ matrix representing the L orthonormal eigenvectors $(\beta_\ell)_{\ell\in L}$ in the basis $(\beta_j)_{j\in J}$.
Then applying again the projection formula (p.434), we get

$$\begin{cases} \mathbf{M\,E} \;= \mathbf{F\,\Xi} \\ \mathbf{M^*\,F} = \mathbf{E\,\Xi} \end{cases} \quad \text{with } \mathbf{E'QE} = \mathbf{I}_L = \mathbf{F'RF}$$

or equivalently, since $\mathbf{M^*} = \mathbf{Q}^{-1}\mathbf{M'R}$ (cf 10.3.6), and letting $\mathbf{N} = \mathbf{MQ}^{-1}$, the spectral equations read:

$$\begin{cases} \mathbf{N\,Q\,E} = \mathbf{F\,\Xi} \\ \mathbf{N'R\,F} = \mathbf{E\,\Xi} \end{cases} \quad \text{with } \mathbf{E'Q\,E} = \mathbf{I}_L = \mathbf{F'R\,F}$$

The solution of spectral equations is the decomposition[9] of $\mathbf{N}$:

$$\mathbf{N} = \mathbf{F}\,\mathbf{\Xi}\,\mathbf{E}', \text{ with } \mathbf{E}'\mathbf{Q}\mathbf{E} = \mathbf{I}_L = \mathbf{F}'\mathbf{R}\mathbf{F}$$

Singular value decomposition (SVD). If the bases $(\alpha_\ell)_{\ell \in L}$ and $(\beta_\ell)_{\ell \in L}$ are both orthonormal, that is, if $\mathbf{Q}_K = \mathbf{I}_K$ and $\mathbf{R}_J = \mathbf{I}_J$, the spectral equations reduce to

$$\begin{cases} \mathbf{M}\,\mathbf{A} = \mathbf{B}\mathbf{\Xi} \\ \mathbf{M}'\,\mathbf{B} = \mathbf{A}\mathbf{\Xi} \end{cases} \text{ with } \mathbf{A}'\mathbf{A} = \mathbf{I}_L = \mathbf{B}'\mathbf{B}$$

The solution of spectral equations is the SVD of matrix $\mathbf{M}$:

$$\mathbf{M} = \sum_{\ell=1}^{L} \xi_\ell\, \mathbf{b}_\ell \mathbf{a}'_\ell \text{ or } \mathbf{M} = \mathbf{B}\mathbf{\Xi}\mathbf{A}' \text{ with } \mathbf{A}'\mathbf{A} = \mathbf{I}_L = \mathbf{B}'\mathbf{B}$$

The SVD of matrix $\mathbf{M}$ is equivalent to the diagonalization of either one of the symmetric matrices $\mathbf{M}'\mathbf{M}$ and $\mathbf{M}\,\mathbf{M}'$. One thus has: $\mathbf{M}'\mathbf{M}\mathbf{a}_\ell = \xi_\ell^2 \mathbf{a}_\ell$, therefore $\mathbf{a}_\ell$ (K– column) is eigenvector of the symmetric matrix $\mathbf{M}'\mathbf{M}$, associated with the eigenvalue $\lambda_\ell = \xi_\ell^2$. Similarly one has $\mathbf{M}\mathbf{M}'\mathbf{b}_\ell = \xi_\ell^2 \mathbf{b}_\ell$, therefore $\mathbf{b}_\ell$ (J–column) is an eigenvector of the symmetric matrix $\mathbf{M}\mathbf{M}'$ associated with the eigenvalue λ_ℓ. From the singular decomposition of $\mathbf{M}$, one deduces the *diagonalization* of the symmetric matrices $\mathbf{M}'\mathbf{M}$ and $\mathbf{M}\mathbf{M}'$:

$$\begin{cases} \mathbf{M}'\mathbf{M} = \sum_{\ell=1}^{L} \lambda_\ell \mathbf{a}_\ell \mathbf{a}'_\ell = \mathbf{A}\mathbf{\Lambda}\mathbf{A}' \text{ with } \mathbf{A}'\mathbf{A} = \mathbf{I}_L \\ \mathbf{M}\mathbf{M}' = \sum_{\ell=1}^{L} \lambda_\ell \mathbf{b}_\ell \mathbf{b}'_\ell = \mathbf{B}\mathbf{\Lambda}\mathbf{B}' \text{ with } \mathbf{B}'\mathbf{B} = \mathbf{I}_L \end{cases}$$

Conversely, from the diagonalization of $\mathbf{M}'\mathbf{M}$ (or of $\mathbf{M}\mathbf{M}'$), one deduces the singular decomposition of $\mathbf{M}$: If one has $\mathbf{A}$ and $\mathbf{\Lambda}$ such that $\mathbf{M}'\mathbf{M} = \mathbf{A}\mathbf{\Lambda}\mathbf{A}'$, by letting $\mathbf{\Xi} = \mathbf{\Lambda}^{1/2}$ and $\mathbf{B} = \mathbf{M}\mathbf{A}\mathbf{\Xi}^{-1}$, one gets the singular decomposition of $\mathbf{M}$, that is, $\mathbf{B}\mathbf{\Xi}\mathbf{A}'$.

For GDA procedures framed in terms of adjoint homomorphisms (see *CORRESPONDENCE*, *CLOUD* and *PRINCIPAL*), the SVD is a standard computational format (equivalent in practice to the diagonalization of positive symmetrical matrices).

Extremality properties in matrix form. Given a $(J \times K)$ matrix $\mathbf{M}$, a K–column $\mathbf{x}$ and a J–column $\mathbf{y}$:
— The ratio $(\mathbf{x}'\mathbf{M}'\mathbf{M}\mathbf{x})/(\mathbf{x}'\mathbf{x})$ is maximum for $\mathbf{x} = \mathbf{a}_1$, and the maximum is equal to λ_1 (largest eigenvalue of $\mathbf{M}'\mathbf{M}$);

[9] sometimes called "generalized SVD".

— the ratio $(\mathbf{y}'\mathbf{M}\mathbf{M}'\mathbf{y})/(\mathbf{y}'\mathbf{y})$ is maximum for $\mathbf{y} = \mathbf{b}_1$, and the maximum is equal to λ_1;
– the ratio $(\mathbf{y}'\mathbf{M}\mathbf{x})^2/(\mathbf{y}'\mathbf{y}\mathbf{x}'\mathbf{x})$ is maximum for $(\mathbf{a}_1, \mathbf{b}_1)$, and the maximum is equal to λ_1.

Postscript. The shortcomings of the Matrix Approach to Statistics

Using matrix notation as a shorthand is one thing; relying exclusively on matrix manipulations to characterize procedures and perform proofs is another thing, as currently done by the *matrix approach to statistics*. The matrix approach ignores *abstract* linear algebra. Vectors are confused with sets of numbers (rows or columns of matrices), and points are confused with vectors. This is a severe obstacle to a fair appreciation of GDA in the field of multivariate statistics. Let us take CA as a case in point[10].

• In the formalization of CA, marginal frequencies are firstly used to provide a reference measure, secondly to go from measures to variables and vice versa, and thirdly to define the chi–square metric. The matrix approach erases the rationale of the construction and the fact that the chi–square metric follows directly form the mathematical structures, without additional assumptions.

• Geometrically, the spectral decomposition is an orthogonal projection onto principal subspaces. Now in the matrix approach, the concept of "geometric orthogonal projection" does not exist, that is, it is confused with that of a "symmetric idempotent matrix". Whenever the reference basis is not orthonormal — as is often the case in multivariate statistics — the lack of invariance leads to meaningless statements and ruins the geometric interpretation.

In all cases, translating the geometric construction into matrix notation is an easy task, whereas the converse "translation" — i.e. starting from matrix formulas, deciphering the rationale of the procedure — is more of a headache. There is a haze of vagueness and arbitrariness around the choice of coordinates in graphical displays stemming from CA and related methods; this haze is entertained by the matrix approach to statistics, which is simply not powerful enough to cope with geometric structures.

[10]Cf. Benzécri & al (1973, Vol 2, p.58): "Critique du Calcul matriciel".

Bibliography

ADAMS J. (2001). *Party Competition and Responsible Party Government: A Theory of Spatial Competition Based upon Insights from Behavorial Voting Research.* Ann Arbor: The University of Michigan Press.

ALEVIZOS P. & MORINEAU A. (1992, 1993). Tests et valeurs–tests: application à l'étude des mastics utilisés dans la fabrication des vitraux, *Revue de Statistique Appliquée*, 40 (4), 27-43 & 41 (1), 8-22.

ANDERSON T.W. (1958). *An Introduction to Multivariate Statistical Analysis.* New York: Wiley.

ANDERSON T.W. (1963a). Asymptotic theory for principal component analysis, *Annals of Mathematical Statistics*, 34, 122-148.

ANDERSON T.W. & DASGUPTA S. (1963b). Some inequalities on characteristic roots of matrices, *Biometrika*, 50, 522-524.

ARIS E. (2001). *Statistical Causal Models for Categorical Data.* Tilburg: Dutch University Press.

BASTIN C., BENZÉCRI J-P., BOURGARIT C. & CAZES P. (1980). *Pratique de l'analyse des données, Vol. 2. Abrégé théorique: Etudes de cas modèle.* Paris: Dunod.

BÉNASSÉNI J. (1993). Perturbational aspects in correspondence analysis, *Computational Statistics & Data Analysis*, 15, 393-410.

BENZÉCRI J-P. (1964, 1965). Analyse factorielle des proximités, *Publications de l'Institut de Statistique de l'Université de Paris*, 13, 235-282; 14, 65-80.

BENZÉCRI J-P. (1966). Analyse statistique et modèle probabiliste en psychologie, *Revue de l'Institut International de Statistique*, Vol. 34 (2), 139-155.

BENZÉCRI J-P. (1969). Statistical analysis as a tool to make patterns emerge from data. In *Methodologies of Pattern Recognition* (Ed. Watanabe), 35-74. New York: Academic Press.

BENZÉCRI J-P. (1970). Sur l'analyse des matrices de confusion, *Revue de Statistique Appliquée*, Vol. 23, 5-62.

BENZÉCRI J-P. (1973). La place de l'a priori, *Encyclopedia Universalis*, 17, 11-24.

BENZÉCRI J-P. & Coll. (1973). *L'Analyse des Données. Vol. 1: Taxinomie. Vol. 2: Analyse des Correspondances.* Paris: Dunod.

BENZÉCRI J-P. (1977). Sur l'analyse des tableaux binaires associés à une correspondance multiple ["BinMult"], *Les Cahiers de l'Analyse des Données*, 2, 55-71 (from a mimeographed note of 1972).

BENZÉCRI J-P. (1981). *Pratique de l'Analyse des données. Vol. 3: Linguistique et lexicologie.* Paris: Dunod.

BENZÉCRI J-P. (1982a). *Histoire et Préhistoire de l'Analyse des Données.* Paris: Dunod (Chapters previously published in *Les Cahiers de l'Analyse des Données*, 1976 & 1977).

BENZÉCRI J-P. (1982b). Construction d'une classification ascendante hiérarchique par la recherche en chaîne des voisins réciproques, *Les Cahiers de l'Analyse des Données*, 7, 209-218.

BENZÉCRI J-P. (1986a). Elaboration statistique de données sur ordinateur; application à l'analyse des textes; contributions attendues à l'étude de la Bible, *Les Cahiers de l'Analyse des Données*, 11, 7-30.

BENZÉCRI J-P. (1986b). Mémoire reçu: Statistical patterns in Hebrew by M. Weitzman, *Les Cahiers de l'Analyse des Données*, 11, 497-500.

BENZÉCRI J-P. (1991). Comment on Goodman (1991), *Journal of the American Statistical Association*, 1112-1115

BENZÉCRI J-P. (1992). *Correspondence Analysis Handbook*, New York: Dekker (adapted from J.P & F. Benzécri, 1984).

BENZÉCRI J-P. & BENZÉCRI F. (1980, 1984). *Pratique de l'analyse des données, Vol. 1. Analyse des correspondances: Exposé élémentaire.* Paris: Dunod.

BENZÉCRI J-P., BENZÉCRI F. & MAÏTI G.D. (1992). *Pratique de l'Analyse des Données. Vol. 4: Médecine, Pharmacologie, Physiologie clinique.* Paris: Statmatic.

BERNARD J-M. (1994). Analyse des données planifiées, *Mathématiques Informatique et Sciences humaines* (special issue), 126, 7-102.

BERNARD J-M., BALDY R. & ROUANET H. (1988). The Language for Interrogating Data (LID). In *Data Analysis and Informatics* (Ed. Diday E.), 461-468.

BERNARD J-M., LE ROUX B., ROUANET H. & SCHILTZ M-A. (1989). L'analyse des données multidimensionnelles par le langage d'interrogation de données (LID): au delà de l'analyse des correspondances, *Bulletin de Méthodologie Sociologique*, 23, 3-46.

BERTHIER P. & BOUROCHE J-M. (1975). *Analyse des Données Multidimensionnelles.* Paris: PUF.

BLASIUS J. & GREENACRE M. (1998) (Eds.). *Visualization of Categorical Data.* San Diego: Academic Press.

BONNET P., LE ROUX B. & LEMAINE G. (1996). Analyse géométrique des données: une enquête sur le racisme, *Mathématiques et Sciences Humaines,* 136, 5-24.

BOURDIEU P. (1979). *La Distinction: Critique Sociale du Jugement.* Paris: Editions de Minuit (English translation: *Distinction* (1984). Boston (MA): Harvard University Press).

BOURDIEU P. (1984). *Homo Academicus.* Paris: Editions de Minuit.

BOURDIEU P. (1989). *La noblesse d'état: grands corps et esprit de corps.* Paris: Editions de Minuit (English translation. *The State Nobility: Elite Schools in the Field of Power.* Cambridge (U.K.): Polity Press).

BOURDIEU P. (1991). *Language and Symbolic Power.* Cambridge (U.K.): Polity Press.

BOURDIEU P. (1999). Une révolution conservatrice dans l'édition, *Actes de la Recherche en Sciences Sociales,* Vol. 126-127, 3-28.

BOURDIEU P. (2001). *Langage et pouvoir symbolique.* Paris: Fayard (adapted from Bourdieu, 1991).

BOURDIEU P. & SAINT–MARTIN M. (1976). Anatomie du goût, *Actes de la Recherche en Sciences Sociales,* Vol. 5, 1-110.

BOURDIEU P. & SAINT–MARTIN M. (1978). Le Patronat, *Actes de la Recherche en Sciences Sociales,* Vol. 20-21, 3-82.

BOX G.E.P. & TIAO G.C. (1973). *Bayesian Inference in Statistical Analysis.* Reading (MA). Addison–Wesley.

BREIMAN L., FRIEDMAN J.H., OHLSEN R.A. & STONE C.J. (1984). *Classification and Regression Trees.* Belmont (CA): Wadsworth.

BURT C. (1950). The factorial analysis of qualitative data, *British Journal of Psychology,* 3, 166-185.

CAILLIEZ F. & PAGÈS J-P. (1976). *Introduction à l'Analyse des Données.* Paris: SMASH.

CARROLL J.D. (1968). Generalization of canonical analysis to three or more sets of variables, *Proc. Amer. Psych. Assoc.,* 227-228.

CARROLL J.D. & GREEN P.E. (1988). An INDSCAL–based approach to Multiple Correspondence Analysis, *Journal of Marketing Research,* 193-203.

Caussinus H. (1986). Discussion of the paper by Goodman (1986), *International Statistical Review.*

Cazes P. (1982). Note sur les éléments supplémentaires en Analyse des Correspondances, *Les Cahiers de l'Analyse des Données*, 7, 9-23 & 133-154.

Cazes P. (1990). Codage d'une variable continue en vue de l'analyse des correspondances, *Revue de Statistique Appliquée*, 38 (3), 33-51.

Chiche J., Le Roux B., Perrineau P. & Rouanet H. (2000). L'espace politique des électeurs français à la fin des années 1990, *Revue française de sciences politiques,* 50, 463-487.

Cordier-Escofier B. (1964). L'Analyse des correspondances, Dissertation published in *Cahiers du BURO*, 13 (1969).

Cox D.R. (1977). The role of significance tests, *Scandinavian Journal of Statistics*, 4, 49-70.

Cox D.R. & Hinkley D.V. (1974). *Theoretical Statistics.* London: Chapman and Hall.

Coxeter H.S.M. (1961/1980). *Introduction to Geometry.* New York: Wiley.

Cramér H. (1946). *Mathematical Methods of Statistics.* Princeton: Princeton University Press.

Critchley F. (1985). Influence in Principal Component Analysis, *Biometrika*, 72, 627-636.

Daudin J-J., Duby C. & Trécourt P. (1988). Stability of principal components studied by bootstrap method, *Statistics*, 19, 241-258.

Davis C. & Kahan W. (1970). The rotation of eigenvectors by a perturbation, *SIAM Journal on Numerical Analysis* 7, 1-46.

de Finetti B. (1974). *Theory of Probability.* London: Wiley.

de Leeuw J. & van der Burg E. (1986). The Permutational Limit Distribution of Generalized Canonical Correlations, in *Data analysis and Informatics*, (Eds. Diday E. & *al.*). 4, 509-521.

Deheuvels R. (1981). *Formes quadratiques et groupes classiques.* Paris: PUF.

Dempster A.P. (1969). *Elements of Continuous Multivariate Analysis.* Reading: Addison–Wesley.

Deville J.C. & Malinvaud E. (1983). Data analysis in official socio–economic statistics, *Journal of the Royal Statistical Society*, A, 146, 335-361.

Diaconis P. & Efron B. (1983). Computer intensive methods in Statistics, *Scientific American*, 248, 116-130.

Diday E. (1971). La méthode des nuées dynamiques, *Revue de Statistique Appliquée*, 19 (2), 19-34.

DIXMIER J. (1948). Position relative de deux variétés linéaires fermées dans un espace de Hilbert, *Rev. Sci.*, 86, 387-399.

DONNAT O. (1998). *Les pratiques culturelles des Français: enquête 1997.* Paris: La Documentation française.

DURAND J-L. (1998). Taux de dispersion des valeurs propres en ACP, AC et ACM. *Mathématiques Informatique et Sciences humaines*, 144, 15-28.

EATON M.L. (1983). *Multivariate Statistics: A Vector Space Approach.* New York: Wiley.

ECKART C. & YOUNG G. (1936). The approximation of one matrix by another of lower rank, *Psychometrika*, 1, 211-218.

EDGINGTON E. (1987). *Randomization Tests.* New York: Dekker.

ESCOFIER B. & LE ROUX B., (1972). Etude de trois problèmes de stabilité en analyse factorielle, *Publications de l'Institut de Statistique de Paris*, Vol. 21, 1-48.

ESCOFIER B. & LE ROUX B. (1975a). Rotation du sous–espace invariant d'un endomorphisme symétrique de $\mathbb{R}^n$ par une pertubation symétrique, *R.A.I.R.O.*, 9 (R-1), 5-8.

ESCOFIER B. & LE ROUX B. (1975b). Etude des questionnaires par l'analyse des correspondances : modification du codage des questions ou de leur nombre et stabilité de l'analyse, *Mathématiques et Sciences humaines*, 49, 5-27.

ESCOFIER B. & LE ROUX B. (1976). Influence d'un élément sur les facteurs en analyse des correspondances, *Les Cahiers de l'Analyse des Données*, 1, 297-318.

ESCOFIER B. & LE ROUX B. (1977). Mesure de l'influence d'un descripteur en analyse en composantes principales, *Publications de l'Institut de Statistique de Paris*, Vol. 22, 25-44.

ESCOFIER B. & PAGÈS J. (1988), *Analyses factorielles simples et multiples.* Paris: Dunod.

FÉNELON J-P. (1981). *Qu'est-ce que l'Analyse des Données?* Paris: Lefonen.

FERRANDEZ A-M. & BLIN O. (1991). A comparison between the effect of intentional modulations and the action of L–Dopa on gait in Parkinson's disease, *Behav. Brain Research*, 45, 177-183

FISHER R.A. (1940). The precision of discriminant functions, *Ann. Eugen.*, 10, 422-429.

FREEDMAN D. & LANE D. (1983). A nonstochastic interpretation of reported significance levels. *Journal of Business and Economic Statistics*, 1, 292-298.

FREEDMAN D., PISANI R., PURVES R. & ADHIKARI A. (1991). *Statistics*, 2nd edition. New York: W.W.Norton & Co.

GABRIEL K.R. (1971). The biplot–graphic display of matrices with application to principal component analysis, *Biometrika*, 58, 453-467.

GIFI (1990). *Non Linear Multivariate Analysis.* Chichester: Wiley (adapted from Gifi, 1981).

GILULA Z. & HABERMAN S.J. (1986). Canonical analysis of contingency tables by maximum likelihood, *Journal of the American Statistical Association*, 83, 760-771.

GLASS D.V. (1954). *Social Mobility in Britain.* London: Routledge & Kegan Paul.

GODEMENT R. (1968). *Algebra.* Boston: Houghton Mifflin (translated from *Cours d'algèbre*, Paris: Hermann, 1966).

GOOD I.J. (1969). Some applications of the singular value decomposition of a matrix, *Technometrics*, 11 (4), 823-831 (with Erratum 12, p.722).

GOODMAN L.A. (1991). Measures, models and graphical displays in the analysis of cross–classification data, *Journal of the American Statistical Association*, 86, 416, 1085-1138.

GOWER J.C. (1966). Some distance properties of latent root and vector methods used in multivariate analysis, *Biometrika*, 53, 325-338.

GOWER J.C. & HAND D.J. (1996). *Biplots.* London: Chapman & Hall.

GRANGER C.W.J. (1969). Investigating causal relations by econometric models and cost–spectral methods, *Econometrica*, 37, 424-438.

GREENACRE M. (1984). *Theory and Applications of Correspondence Analysis.* London: Academic Press.

GREENACRE M. (2003). *International Conference on Correspondence Analysis and Related Methods.* Barcelona.

GUTTMAN L. (1941). The quantification of a class of attributes, in *The prediction of social adjustment* (Eds. Horst & *al.*) 319-348. New York: Social Science Research Council.

GUTTMAN L. (1959). Metricizing rank–ordered or ordered data for a linear factor analysis, *Sankhya*, 21, 257-268.

GUTTMAN L. (1977). What is not what in statistics, *The Statistician*, 26 (2), 81-107.

HALMOS P.R. (1958). *Finite Dimensional Vector Spaces.* New York: Van Nostrand Co. Inc.

HAMPEL F.R. (1974). The influence curve and its role in robust estimation. *Journal of the American Statistical Association*, 69, 383-393.

HAYASHI C. (1952). On the prediction of phenomena from qualitative data and the quantification of qualitative data from the mathematico–statistical point of view, *Annals of the Institute of Stat. Math.*, Tokyo, Vol.3.

HENLEY N. M. (1969). A psychological study of the semantics of animal terms, *Journal of Verbal Learning and Verbal Behavior*, 8, 176-184.

HÉRAN F. (1988). La sociabilité, une pratique culturelle, *Economie et Statistique*, 208, 33-44.

HILL M.O. (1974). Correspondence analysis: a neglected multivariate method, *Applied Statistics*, 23, 340-354.

HIRSCHFELD H.O. (1935). A connection between correlation and contingency, *Proc. Camb. Phil. Soc.*, 31, 520-524.

HJELLBREKKE J. & KORSNES O. (2003). "Le patronat norvégien": State vs. Market? Capital Structures, Oppositions and Political Position Taking in the Norwegian Field of Power, Economic Sociology, *European Electronic Newsletter*, Vol. 4, 2, `http://www.siswo.uva.nl/ES/newsletter4-2.pdf`.

HOC J-M (1983). *L'analyse des données planifiées*. Paris: PUF.

HOFFMAN D.L. & FRANKE G.R. (1986). Correspondence analysis: graphical representation of categorical data in marketing research, *Journal of Marketing Research*, 23, 213-27.

JAMBU M. (1991). *Exploratory and Multivariate analysis*. Boston: Academic Press.

JAMBU M. & LEBEAUX M-O. (1983). *Cluster Analysis and Data Analysis*. Amsterdam: North-Holland.

JEFFREYS H. (1961). *Theory of Probability*. Oxford University Press.

KAMINSKA A., ICKOWICZ A., PLOUIN P., BRU M-F., DELLATOLAS G. & DULAC O. (1999). Delineation of cryptogenic Lennox–Gastaut syndrome and myoclonic astatic epilepsy using Multiple Correspondence Analysis. *Epilepsy Res.*, 36, 15-29.

KENDALL M.G. & STUART A. (1973, 1984). *The Advanced Theory of Statistics*, Volumes 1, 2, 3. London: Griffin.

KRUSKAL J.B. & WISH M. (1978). *Multidimensional Scaling*. Beverly Hills: Sage.

KRUSKAL W.H. (1968). When are Gauss–Markov and least squares estimators identical? A coordinate–free approach, *Ann. Math. Statist.*, 39, 70-75.

KRZANOWSKI W.J. (1984). Sensitivity of Principal Components, *Journal of the Royal Statistical Society*, 46 (3), 558-563.

LAPLACE P.S. (1952). *A Philosophical Essay on Probability*. New York: Dover (translated from *Essai philosophique sur les probabilités*, 1825).

LE ROUX B. (1977). Effet du codage selon une partition (regroupement) sur les résultats d'une analyse factorielle. *Actes du colloque "Analyse des données et Informatique"*, INRIA, 581-587.

LE ROUX B. (1991). Sur la construction d'un protocole additif de référence, *Mathématiques Informatique et Sciences humaines*, 114, 57-62.

LE ROUX B. (1998). Inférence combinatoire en analyse géométrique des données, *Mathématiques Informatique et Sciences humaines*, 144, 5-14.

LE ROUX B. (1999). Analyse spécifique d'un nuage euclidien: application à l'étude des questionnaires, *Mathématiques Informatique et Sciences Humaines*, 146, 65-83.

LE ROUX B. & ROUANET H. (1983). L'analyse statistique des protocoles multidimensionnels : analyse des comparaisons (nuage pondéré sur le croisement de deux facteurs), *Publications de l'Institut de Statistique de Paris*, Vol.28, 7-70.

LE ROUX B. & ROUANET H. (1984). L'analyse multidimensionnelle des données structurées, *Mathématiques et Sciences Humaines*, 85, 5-18.

LE ROUX B. & ROUANET H. (1998). Interpreting axes in Multiple Correspondence Analysis: Method of the contributions of points and deviations, in *Visualization of Categorical Data*, (Eds. Blasius J. & Greenacre M.) 197-220, San Diego: Academic Press.

LE ROUX B. & ROUANET H. (2003). Geometric Analysis of Individual Differences in Mathematical Performance for EPGY Students in the Third Grade. `http://www-epgy.stanford.edu/research/`.

LEBARON F. (2000). *La croyance économique: les économistes entre science et politique.* Paris: Seuil.

LEBARON F. (2001). Economists and the economic order: The field of economists and the field of power in France, *European Societies*, 3 (1), 91-110.

LEBART L. (1975a). L'orientation du dépouillement de certaines enquêtes par l'analyse des correspondances multiples, *Consommation*, 2, 73-96.

LEBART L. (1975b). Validité des résultats en Analyse des Données. Rapport CREDOC–DGRST 4465.

LEBART L. (1976). The significance of eigenvalues issued from correspondence analysis, *Proceedings in Comp. Statist.*, *COMPSTAT*, 8-45. Vienna: Physica Verlag.

LEBART L. & FÉNELON J-P. (1971). *Statistique et Informatique Appliquées.* Paris: Dunod.

LEBART L. , MORINEAU A. & TABARD N. (1977). *Techniques de la description statistique: Méthodes et logiciels pour l'analyse des grands tableaux.* Paris: Dunod.

LEBART L., MORINEAU A. & WARWICK K.M. (1984). *Multivariate Descriptive Statistical Analysis: Correspondence Analysis and Related Techniques for Large Matrices.* New York: Wiley.

LEBART L., MORINEAU A. & PIRON M. (1995). *Statistique exploratoire multidimensionnelle.* Paris: Dunod.

LEBART L., PIRON M. & STEINER J-F. (2003). *La sémiométrie.* Paris: Dunod.

LEBART L. & SALEM A. (1988). *Analyse statistique des données textuelles.* Paris: Dunod.

LEBART L., SALEM A. & BERRY L. (1998). *Exploring Textual Data.* Dordrecht: Kluwer.

LECLERC A. (1975). L'analyse des correspondances sur juxtaposition de tableaux de contingence, *Revue de Statistique Appliquée*, 23 (3), 5-16.

LECOUTRE B. (1984). *L'Analyse Bayésienne des Comparaisons.* Presses Universitaires de Lille.

LECOUTRE B. (1999). Two useful distributions for Bayesian predictive procedures under normal modes, *Journal of Statistical Planning and Inference*, 77, 93-105.

LECOUTRE B., GUIGUES J-L. & POITEVINEAU J. (1992). Distributions of quadratic forms of multivariate Student variables, *Applied Statistics*, 41, 617-626.

LECOUTRE J-P. & TASSI P. (1987). *Statistique non paramétrique et robustesse.* Paris: Economica.

LEHMAN D. & BKOUCHE R. (1988). *Initiation à la Géométrie.* Paris: PUF.

LINDLEY D. (1965). *Introduction to Probability and Statistics from a Bayesian Viewpoint.* Cambridge University Press.

MACDONELL W.R. (1902). On criminal anthropometry and the identification of criminals, *Biometrika*, 1, 177-227.

MACLANE S. & BIRKHOFF G. (1967). *Algebra.* New York: Mac Millan.

MAHALANOBIS P.C. (1936). On the generalized distance in statistics, *Proc. Natl. Inst. Science, India*, 12, 49-55.

MAHALANOBIS P.C. (1948). Historical note on the D^2 statistic, *Sankhya*, 9, 238-240.

MAISONNEUVE J. (1978). *Recherches diachroniques sur la représentation sociale.* Paris: PUF.

MALINVAUD E. (1980). *Statistical Methods of Econometrics.* Chicago: Rand McNally.

MARKUS A.S. (1964). Eigenvalues and singular values of the sum and product of operators, ***Russian Mathematical Surveys***, 19, 91-119, MR29#6318.

MAUNG K. (1941). Measurement of association in a contingency table with special reference to the pigmentation of hair and eye colours of Scottish school children, *Ann. Eugen. Lond*, 11, 189-223.

MAXWELL A.E. (1977). ***Multivariate Analysis in Behavioural Research.*** London: Chapman & Hall.

MCQUITTY L.L. (1966). Single and Multiple Hierarchical Classification by Reciprocal Pairs and Rank Order Types, ***Educational and Psychological Measurement***, 26 (2), 253-265.

MORINEAU A. (1984). Note sur la caractérisation statistique d'une classe et les valeurs–tests, *Bull. techn. CISIA*, 2, 1-2 & 20-27.

NEVEU J. (1964). ***Bases mathématiques du calcul des probabilités***, Paris: Masson.

NISHISATO S. (1980). ***Analysis of Categorical Data: Dual Scaling and its Applications.*** Toronto: University of Toronto Press.

NISHISATO S. (1986). ***Quantification of categorical variables: a bibliography 1975-1986.*** Toronto: Microstats.

PACK P. & JOLIFFE I.T. (1992). Influence in Correspondence Analysis, ***Appl. Statist.***, 41, 2, 365-380.

PEARSON K. (1900). On the criterion that a given system of deviations from the probable in the case of a correlated system of variables is such that it can be reasonably supposed to have arisen from random sampling, ***Phil. Mag.***, 5, 50, 157.

PEARSON K. (1901). On lines and planes of closest fit to systems of points in space, ***Phil. Mag.***, 2, 6, 559-572.

PERRINEAU P., YSMAL C., & *al.* (1998). ***Le vote surprise. Les élections législatives des 25 Mai et 1er Juin 1997.*** Paris: Presses de Sciences Po.

PITMAN E.J.P. (1937). Significance tests which may be applied to samples from any populations, ***Journal of the Royal Statistical Society***, Suppl. 4, 119-130.

PRESS J. (1989). ***Bayesian Statistics: Principal Models and Applications.*** New York: Wiley.

QUEYSANNE M. (1964). ***Algèbre.*** Paris: Armand Colin.

RAO C.R. (1964). The use and interpretation of principal component analysis in applied research, ***Sankhya***, A, 26, 329-358.

RAO C.R. (1973). ***Linear Statistical Inference and its Applications*** (First edition 1965). New York: Wiley.

RAO C.R. (1995). A review of canonical coordinates and an alternative to correspondence analysis using Hellinger distance, *Qüestio*, 19, 23-63.

ROSENLUND L. (2000). Cultural change in Norway: Cultural and economic dimensions, *International Journal of Contemporary Sociology*.

ROUANET H. (1996). Bayesian methods for assessing importance of effects, *Psychological Bulletin*, 119 (1), 149-158.

ROUANET H., ACKERMANN W. & LE ROUX B. (2000). The geometric analysis of questionnaires: the Lesson of Bourdieu's La Distinction, *Bulletin de Méthodologie Sociologique*, 65, 5-18.

ROUANET H., BERNARD J-M., BERT M.C., LECOUTRE B., LECOUTRE M.P. & LE ROUX B. (1998). *New Ways in Statistical Inference: From Significance Tests to Bayesian Methods* (Foreword by P. Suppes). Bern: Peter Lang.

ROUANET H., BERNARD J-M. & LE ROUX B. (1990). *Analyse Inductive des Données*. Paris: Dunod.

ROUANET H., BERNARD J-M. & LECOUTRE B. (1986). Nonprobabilistic statistical inference: A set–theoretic approach, *The American Statistician*, 40, 60-65.

ROUANET H. & LE ROUX B. (1993). *Analyse des données multidimensionnelles*. Paris: Dunod.

ROUANET H., LE ROUX B., BERNARD J-M. & LECOUTRE B. (1998). Geometric Data: From Euclidean Clouds to Bayesian MANOVA, in *New Ways in Statistical Inference*, (Rouanet & *al.*). Bern: Peter Lang, 227-254.

ROUANET H., LEBARON F., LE HAY V., ACKERMANN W. & LE ROUX B. (2002). Régression et analyse géométrique des données: réflexions et suggestions, *Mathématiques & Sciences humaines*, 160, 13-45.

ROUANET H. & LECOUTRE B. (1983). Specific inference in ANOVA: from significance tests to Bayesian procedures, *Brit. J. Math. Stat. Psychol.*, 36, 252-268.

ROUANET H. & LÉPINE D. (1976). Structures linéaires et analyse des comparaisons, *Mathématiques & Sciences humaines*, 56, 5-46.

ROUANET H. & LÉPINE D. (1977). Introduction à l'analyse des comparaisons pour le traitement des données expérimentales, *Informatique & Sciences Humaines* (special issue), 33-34, 1-125.

ROUANET H., LÉPINE D. & HOLENDER D. (1978). Model acceptability and the use of Bayes–fiducial methods for validating models, in *Attention and performance VII* (Ed. Requin J.), 687-701. Hillsdale, NJ, Erlbaum.

ROUANET H., LÉPINE D. & PELNARD–CONSIDÈRE J. (1976). Bayes–fiducial procedures as practical substitutes for misplaced significance testing: An application to educational data, in *Advances in psychological and educational measurement* (Eds. de Gruijter D.N.M. & *al.*), 33-48. New York: Wiley.

SAPORTA G. (1990). *Probabilités, Analyse des Données et Statistique.* Paris: Technip.

SAPORTA G. & HATABIAN G. (1986). Régions de confiance en analyse factorielle, in *Data analysis and Informatics*, (Eds. Diday E. & *al.*) 499-508.

SHEPARD R.N. (1962). The Analysis of Proximities: Multidimensional Scaling with an unknown distance function, *Psychometrika*, 27, 125-139 & 219-246.

SHEPARD R.N. (1966). Metric structures in ordinal data, *Journal of Mathematical Psychology*, 3, 287-315.

SHEPARD R.N. (1980). Multidimensional Scaling, tree–fitting, and clustering, *Science*, 210, 390-398.

SHEPARD R.N., ROMNEY A.K. & NERLOVE S.B. (1972). *Multidimensional scaling: Theory and Applications in the Behavioral sciences. Vol. I: Theory, Vol. II: Applications.* New York: Seminar Press.

SPEARMAN C. (1904). 'General intelligence', objectively determined and measured, *Amer. J. Psychol.*, 15, 201-292.

STIGLER S.M. (1986). *The History of Statistics: The Measurement of Uncertainty before 1900.* Cambridge: MA, Harvard University Press.

STONE M. (1983). *Coordinate–Free Multivariate Statistics.* Oxford University Press.

STUDENT (1908). The probable error of a mean, *Biometrika*, 6, 1-25.

SUPPES P. (1993). *Models and Methods in the Philosophy of Science: Selected Essays.* Dordrecht: Kluwer.

SUPPES P. (2002). *Representation and Invariance of Scientific Structures.* Stanford: CSLI Press.

SUPPES P., KRANTZ D. H., LUCE R.D. & TVERSKY A. (1989). *Foundations of Measurement, Volume II: Geometrical, threshold and probabilistic representations.* San Diego: Academic Press.

TANAKA Y. (1988). Sensitivity Analysis in Principal Component Analysis: Influence on the Subspace Spanned by Principal Components, *Communs Statist. Theory Meth.*, 17, 3167-3175.

TENENHAUS M. & YOUNG F.W. (1985). An analysis and synthesis of multiple correspondence analysis, optimal scaling, dual scaling, homogeneity analysis and other methods for quantifying categorical data, *Psychometrika*, 50, 91-119.

TENENHAUS M., LEROUX Y., GUIMART C. & GONZALEZ P-L. (1993). Modèle linéaire généralisé et analyse des correspondances, *Revue de Statistique Appliquée*, 41 (2), 59-86.

TER BRAAK C.J.F. (1992). Permutation versus Bootstrap significance Tests, in *Multiple Regression and* ANOVA (Eds. Jöckel & *al.*). Berlin: Springer.

THOMSON R.C. & FREEDE L.J. (1971). On the eigenvalues of sums of Hermitian Matrix, *Linear Algebra and Applications*, 4, 369-376.

THURSTONE L.L. (1947). *Multiple Factor Analysis*. Chicago: University of Chicago Press.

TOCK K. & SUPPES P. (2002) The High Dimensionality of Students' Individual Differences in Performance in EPGY's K6 Computer-Based Mathematics Curriculum, `http://www-epgy.stanford.edu/research/`.

TORGERSON W.S. (1958). *Theory and Methods of Scaling*. New York: Wiley.

TUKEY J.W. (1960). Conclusions vs Decisions, *Technometrics*, 2, 423-433.

TUKEY J.W. (1991). The philosophy of multiple comparisons, *Statistical Science*, 6, 100-116.

VOLOCHINE C.A. (1995). Etude comparée des textes russes: le Don tranquille et autres œuvres de Cholokov, *Les Cahiers de l'Analyse des Données*, 20, 7-26.

WARD J.H. (1963). Hierarchical Grouping to optimize an Objective Function, *Journal of American Statistical Association*, 58, 236-244.

WIELANDT H. (1955). An extremum property of sums of eigenvalues, *Proc. Amer. Math. Soc.*, 6, 106-110.

WILKINSON J.H. (1965). *The Algebraic Eigenvalue Problem*. Oxford: Clarendon Press.

WELLER S.C. & ROMNEY A.K. (1990). *Metric Scaling: Correspondence Analysis*. Newbury Park (CA): Sage.

WOLFF M., ROUANET H. & GROSGEORGE B. (1998). Analyse d'une expertise professionnelle: l'évaluation des jeunes talents au basket–ball de haut niveau, *Le Travail Humain*, 61, 281-303.

Index

Name Index

Symbol Index

Matrices (rows×columns)

$\mathbf{a}_\ell$	$(K \times 1)$	coefficients a_ℓ^k of Axis ℓ
$\mathbf{A}$	$(K \times L)$	L columns $\mathbf{a}_\ell$
$\mathbf{C}$	$(K \times K)$	diagonal $(n_k)_{k\in K}$
$\mathbf{e}$	$(J \times 1)$	column of 1
$\mathbf{I}_J$	$(J \times J)$	identity matrix
$\mathbf{I}_K$	$(K \times K)$	identity matrix
$\mathbf{F}_J$	$(J \times J)$	diagonal $(f_j)_{j\in J}$
$\mathbf{\Lambda}$	$(L \times L)$	diagonal of L eigenvalues $(\lambda_\ell)_{\ell=1,\dots L}$
$\mathbf{M}$	$(J \times K)$	$= \mathbf{R}^{-1/2}\mathbf{N}(\mathbf{C}^{-1/2})$
$\mathbf{N}$	$(J \times K)$	general term n_{jk}
$\mathbf{\Omega}_K$	$(K \times K)$	diagonal $(\varpi_k)_{k\in K}$
$\mathbf{Q}$	$(K \times K)$	scalar products
$\mathbf{R}$	$(J \times J)$	diagonal $(n_j)_{j\in J}$
$\mathbf{T}$	$(J \times K)$	$= \mathbf{N}\mathbf{C}^{-1}$, transition f_J^K
$\mathbf{V}$	$(K \times K)$	matrix of covariances
$\mathbf{x}^j$	$(1 \times K)$	profile of j
$\mathbf{x}^k$	$(J \times 1)$	variable x^{Jk}
$\mathbf{X}$	$(J \times K)$	coordinates x^{JK}
$\mathbf{X}_0$	$(J \times K)$	$= (\mathbf{I}_J - \mathbf{e}\mathbf{e}'\mathbf{F}_J)\mathbf{X}$
$\mathbf{\Xi}$	$(L \times L)$	diagonal of L singular values ξ_ℓ
$\mathbf{z}_\ell$	$(J \times 1)$	ℓ–th standard principal variable z_ℓ^J
$\mathbf{Z}$	$(J \times L)$	L columns $\mathbf{z}_\ell$

Sets and Spaces

I	set of n individuals $i \in I$, 179
$I{<}k{>}$	subset of n_k individuals who have chosen k, 183
J	set of statistical units $j \in J$, 24
K	set of modalities k of categorized variable K, 31, *or* the overall set of modalities $k \in K$ of Q questions, 182
$K{<}i{>}$	subset of modalities chosen by individual i, 182
$K{<}q{>}$	set of response modalities of question q, 181
$\mathcal{M}$	affine support of cloud, 76
$\mathcal{L}$	L–dimensional vector subspace underlying $\mathcal{M}$
Q	set of questions $q \in Q$, 179
$\mathbb{R}_J$	vector space of measures over J, 27
$\mathbb{R}_J^0$	subspace of contrasts, 27
$\mathbb{R}_J^1$	set of normalized measures
$\mathbb{R}^J$	vector space of variables over J, 27
$\mathbb{R}_0^J$	subspace of centered variables, 95
$\mathcal{U}$	Euclidean, geometric space, 76, 439
$\mathcal{U}_I$, $\mathcal{U}_K$	I–dimensional, K–dimensional geometric spaces
$\mathcal{V}$, $\mathcal{V}_K$	K–dimensional vector spaces underlying $\mathcal{U}$, $\mathcal{U}_K$, 77, 183
$\mathcal{X}_0$	$\subset \mathbb{R}^J$ generated by the K centered initial variables over J

Abbrevations

AHC	Ascending Hierarchical Classification
CA	Correspondence Analysis
Corr	correlation coefficient
Cov	covariance
Cta	absolute contribution
Ctr	relative contribution
GDA	Geometric Data Analysis
IDA	Inductive Data Analysis
MCA	Multiple Correspondence Analysis
MDS	MultiDimensional Scaling
PCA	Principal Component Analysis
Qlt	quality of representation
SD	Standard Deviation
SVD	Singular Value Decomposition
Var	variance

Subject Index

Page numbers in slanted characters refer to places where the notion is applied.

MIX
Papier aus verantwortungsvollen Quellen
Paper from responsible sources
FSC® C105338

If you have any concerns about our products,
you can contact us on
ProductSafety@springernature.com

In case Publisher is established outside the EU,
the EU authorized representative is:
Springer Nature Customer Service Center GmbH
Europaplatz 3, 69115 Heidelberg, Germany

Printed by Libri Plureos GmbH
in Hamburg, Germany